COPULAS AND THEIR APPLICATIONS IN WATER RESOURCES ENGINEERING

Complex environmental and hydrological processes are characterized by more than one correlated random variable. These events are multivariate and their treatment requires multivariate frequency analysis. Traditional analysis methods are, however, too restrictive and do not apply in many cases. Recent years have therefore witnessed numerous applications of copulas to multivariate hydrologic frequency analyses. This book describes the basic concepts of copulas and outlines current trends and developments in copula methodology and applications. It includes an accessible discussion of the methods alongside simple step-by-step sample calculations. Detailed case studies with real-world data are included, and are organized based on applications, such as flood frequency analysis and water quality analysis. Illustrating how to apply the copula method to multivariate frequency analysis, engineering design, and risk and uncertainty analysis, this book is ideal for researchers, professionals, and graduate students in hydrology and water resources engineering.

DR. LAN ZHANG currently works as a postdoctorate research scholar in the Department of Agricultural and Biological Engineering at Texas A&M University. She received her BS in mechanical engineering, MS in water resources sciences, and PhD in civil and environmental engineering. She has written more than 40 publications in the areas of hydrology, copulas, water quality, entropy, and water resources. She has been working on copulas and their application in hydrology and water resource engineering for more than 10 years.

PROFESSOR V. P. SINGH is Distinguished Professor, Regents Professor, and Caroline and William N. Lehrer Distinguished Chair in Water Engineering at Texas A&M University. Professor Singh has published extensively in the areas of hydrology, groundwater, hydraulics, irrigation, pollutant transport, copulas, entropy, climate change, and water resources. He has received more than 90 national and international awards, including the Arid Lands Hydraulic Engineering Award, the Ven Te Chow Award, the Richard R. Torrens Award, the Norman Medal, and the Environmental and Water Resources Institute (EWRI) Lifetime Achievement Award, given by the American Society of Civil Engineers; the Ray K. Linsley Award and Founder's Award, given by the American Institute of Hydrology; the Crystal Drop Award; and the Ven Te Chow Memorial Award, given by the International Water Resources Association; the Merriam Improved Irrigation Award given by the US Committee on Irrigation and Drainage; the Hancor Soil and Water Engineering Award given by the American Society of Agricultural and Biological Engineers; and three honorary doctorates. He is a Distinguished Member of the American Society of Civil Engineers (ASCE) and a fellow of EWRI, the American Water Resources

Association (AWRA), the Indian Water Resources Society (IWRS), the Indian Society of Agricultural Engineers (ISAE), the Indian Association of Soil Water Conservationists (IASWC), and the Institution of Engineers (IE), as well as a member of 10 international science and engineering academies. He has also served as president of the American Institute of Hydrology (AIH).

COPULAS AND THEIR APPLICATIONS IN WATER RESOURCES ENGINEERING

LAN ZHANG
Texas A&M University

V. P. SINGH
Texas A&M University

CAMBRIDGE
UNIVERSITY PRESS

CAMBRIDGE
UNIVERSITY PRESS

University Printing House, Cambridge CB2 8BS, United Kingdom

One Liberty Plaza, 20th Floor, New York, NY 10006, USA

477 Williamstown Road, Port Melbourne, VIC 3207, Australia

314–321, 3rd Floor, Plot 3, Splendor Forum, Jasola District Centre, New Delhi – 110025, India

79 Anson Road, #06–04/06, Singapore 079906

Cambridge University Press is part of the University of Cambridge.

It furthers the University's mission by disseminating knowledge in the pursuit of education, learning, and research at the highest international levels of excellence.

www.cambridge.org
Information on this title: www.cambridge.org/9781108474252
DOI: 10.1017/9781108565103

First published 2019

Printed in the United Kingdom by TJ International Ltd. Padstow Cornwall

A catalogue record for this publication is available from the British Library.

Library of Congress Cataloging-in-Publication Data
Names: Zhang, Lan, 1972- author. | Singh, V. P. (Vijay P.), author.
Title: Copulas and their applications in water resources engineering / Lan Zhang and Vijay P. Singh (Texas A&M University).
Description: Cambridge ; New York, NY : Cambridge University Press, 2019. | Includes bibliographical references and index.
Identifiers: LCCN 2018026586 | ISBN 9781108474252 (hardback : alk. paper)
Subjects: LCSH: Copulas (Mathematical statistics) | Hydrology–Mathematics. | Water-supply engineering–Mathematical models. | Water resources development–Mathematical models.
Classification: LCC QA273.6 .Z53 2019 | DDC 519.2/40155148–dc23
LC record available at https://lccn.loc.gov/2018026586

ISBN 978-1-108-47425-2 Hardback

To

LZ: Mother Shuyuan, husband Bret, son Caelan

VPS: Wife Anita, son Vinay, daughter Arti, daughter-in-law Sonali, son-in-law Vamsi, and grandsons Ronin, Kayden, and Davin

Contents

Preface

Complex environmental and hydrological processes, such as floods, droughts, rainstorms, hurricanes, tornadoes, windstorms, weather extremes, and tides, are characterized by more than one correlated random variable. These events are multivariate and their treatment requires multivariate frequency analysis. Traditional multivariate frequency analysis methods are too restrictive and do not even apply in many cases. Recent years have therefore witnessed numerous applications of copulas to multivariate hydrologic frequency analyses.

Since the advent of Sklar theorem in 1959, several books have been written on copulas, but these books have been written by mathematicians and statisticians for students in mathematics and statistics. The book titled *Extremes in Nature: An Approach Using Copulas*, by Salvadori et al. (2007), is the only book discussing the copula theory and its application to natural events, but since its publication new types of copulas as well as new applications have been introduced. Therefore, there is a need for a book that describes basic concepts of copulas, illustrates them in an easy-to-understand manner, presents different types of copulas, and discusses their applications.

This book on copulas and their applications in water resources engineering covers current trends in copula applications in hydrological sciences and water engineering. Many copula-based approaches have been developed in econometrics that can be extended to hydrology and water resources engineering.

The book is organized into two parts. Part I introduces theoretical aspects of copulas, including copula properties and statistics, and different copula families. This part comprises nine chapters. Beginning with a short discussion of different methods of parameter estimation, Chapter 1 presents a short introduction to the history, development, and general applications of the copula theory. It also presents the theme of the entire book. Chapter 2 briefly discusses preliminaries for univariate and bivariate analyses. Chapter 3 deals with copulas and their properties. Starting with the definition of copula and its properties, it goes on to discussing bivariate copula, trivariate copula, methods of copula construction, copula families, dependence measures and properties, parameter estimation, copula simulation, goodness-of-fit tests, and return periods. Chapter 4 introduces the famous and well-accepted symmetric Archimedean copulas, including their properties and extension from two-dimensional to higher-dimensional analyses. Chapter 5 deals with asymmetric Archimedean copulas. Starting with nested Archimedean copula, this chapter discusses the properties, parameter estimation, copula random variable simulation, and goodness-of-fit statistics for both nested Archimedean copula and vine copulas. The Plackett copula family is presented in Chapter 6. This chapter also discusses the disadvantage of extending the two-dimensional Plackett copula to higher-dimensional analysis. Chapter 7 presents meta-elliptic copulas. The meta-elliptic copula (especially the famous meta-Gaussian and metastudent *t* copulas) are easy to construct and well accepted in spatial analysis with high dimensions. Defining univariate constraints based on the Shannon entropy theory, Chapter 8 discusses the constraints necessary to construct the most-entropic copula and presents the uniqueness of the most-entropic canonical copula with examples. Chapter 9 presents the theoretical aspects of applying the copula theory to study multivariate and univariate time series.

Part II, comprising eight chapters, covers applications of copulas with case studies. Chapter 10 focuses on rainfall analysis. The case studies in this chapter include the depth-duration-frequency analysis from partial durations series and spatial rainfall depth analysis. Chapter 11 deals with flood analysis for both at-site and spatial flood frequency analyses. Chapter 12 focuses on the copula application to water quality analysis, including multivariate and univariate water quality time series. Chapter 13 presents the application of copulas to drought analysis using at-site drought characteristics. Risk and compound extreme (i.e., temperature and precipitation) are presented in Chapter 14. Using rain gauges from Louisiana, Chapter 15 discusses network design using the copula approach. Chapter 16 introduces the application of copulas to sediment yield analysis through the construction of sediment discharge rating curve and at-site trivariate suspended sediment yield analysis. The last chapter of the book presents the application of copulas to interbasin water transfer analysis.

This book covers important theoretical and practical aspects of the copula theory and its applications. It is hoped that the book will be useful to graduate students and faculty members who are interested in stochastic hydrology and environmental research and risk analyses. In the long term, copula-based methodologies may help improve engineering design and risk analysis practice.

Acknowledgments

The authors wish to express their gratitude to researchers working on developing and applying the copula theory. The book would not be possible without following their expertise in statistics, econometrics, and hydrology and water resources engineering. The authors are especially thankful to: A. Sklar who developed the famous Sklar theorem; R. B. Nelson and H. Joe, whose copula books were the main source for better understanding the theoretical aspects of copulas; C. Genest and his research team, who made the formal goodness-of-fit statistics available and introduced the copula theory to the hydrologic community; and T. Bedford and R. M. Cooke, who first proposed the flexible vine copula model. They are also thankful to the Cambridge University Press Editorial Board for their patience and support.

Part One

Theory

1
Introduction

ABSTRACT

This chapter briefly reviews the development of the copula theory and its applications in the field of water resources engineering (flood, drought, rainfall, groundwater, etc.). It points out the need for applying the copula theory in hydrology and engineering. The chapter is concluded with an outline of the structure of the book.

1.1 Need for Copulas

Complex hydrological processes, such as floods, droughts, winds, rainstorms, and snowfall, are characterized by more than one correlated random variable. Hydrologic events emanating from these processes are multivariate and their treatment requires multivariate analysis. Yue (1999, 2000a, 2000b, 2000c), Yue et al. (2001), and Yue and Rasmussen (2002) reviewed some applications of multivariate hydrological analyses using traditional frequency analysis methods with multivariate distributions.

Multivariate frequency distributions have usually been derived using one of three fundamental assumptions (Zhang and Singh, 2006): (1) the random variables each have the same type of marginal probability distribution; (2) the variables are assumed to have a joint normal distribution or are transformed to have a joint normal distribution; or (3) the variables are assumed independent – a trivial case. In reality, the correlated random variables are generally dependent, do not follow the normal distribution, and/or do not have the same type of marginal distributions. In general, multivariate hydrological analyses are mathematically complicated, and the resulting joint distributions may be valid only in a limited solution space.

When deriving multivariate distributions, it has been demonstrated in the last two decades that the aforementioned difficulties can be overcome with the use of copulas because: (1) they separate the dependence function from the marginal distributions of random variables; (2) the dependence function represented by the copula function is the cumulative joint distribution of correlated random variables; and (3) the mutual information (bivariate/multivariate) may be expressed as the negative copula entropy that avoids the complexity of evaluating the uncertainty with the use of entropy theory (information theory). In what follows, we briefly summarize copulas and their applications.

1.2 Introduction of Copulas and Their Application

Copula was first introduced by Sklar (1959). Later on, Joe (1997) and Nelsen (2006) further discussed the dependence structure of multivariate random variables using the copula theory. The copula theory was first developed in the fields of statistics and finance (more specifically econometrics). In this section, we will first briefly introduce the history of development of copulas, followed by a brief introduction of copula properties, parameter estimation, and applications to the field of water resources engineering.

1.2.1 Development and Applications of Copulas in Statistics and Finance

Copula theory has been developed and applied in the fields of statistics and finance. Ali et al. (1978) proposed a bivariate distribution family, i.e., the bivariate logistic distribution by considering the survival odds ratio. They also studied the properties of the bivariate distribution. Now it is named the Ali–Mikhail–Haq (AMH) copula family. It is worth noting that this copula family may not be applicable, unless Kendall's tau rank correlation coefficient falls in the range of (–1/3 to 1/3).

Cook and Johnson (1981) proposed a simple bivariate distribution family to represent nonelliptical symmetric bivariate random variables. The proposed copula, however, may only be applied to the positively correlated random variables. They also proved that multivariate Pareto, Burr, and logistic distributions were special cases, and that copula is now named the Cook–Johnson (Clayton) Archimedean copula family.

Genest and McKay (1986) described bivariate distributions with uniform marginals on a unit interval. They discussed how bivariate distributions (copula) may be applied for singular components and the geometric interpretation of Kendall's tau. Genest (1987) studied the Frank family of bivariate distributions and concluded that it was appropriate to apply the Frank family to construct the bivariate distribution with any given marginals and cover all possible dependence structures. He then introduced three nonparametric estimators and one parametric estimator, i.e., the maximum likelihood estimation (MLE) method. Genest and Rivest (1993) studied the Archimedean one-parameter copula. They applied Kendall's tau for parameter estimation and found that Kendall's tau may also be applied for selecting the appropriate copula for certain multivariate random variables, and analyzed uranium exploration data to explain how to apply the estimation procedure.

Genest et al. (1995) investigated the properties of another semi-parametric estimation method to estimate copula parameters. This semi-parametric estimation method can be considered as a pseudo-likelihood method that is found to be consistent and asymptotically normal. The performance of the pseudo-likelihood method was investigated by analyzing the bivariate Clayton (Cook–Johnson) copula. Later, Caperaa et al. (1997) proposed a new nonparametric method and examined its asymptotic properties and small sample behavior compared to the estimation method through Kendall's tau statistic and maximum likelihood method. They found that the proposed method was strongly convergent and asymptotically unbiased.

Genest and Boies (2003) discussed the Kendall plot as a measure of dependence. Similar to chi-plot, the Kendall plot is invariant with respect to the monotone transformation of marginal distributions. They also found that the Kendall plot is easier to interpret than the chi-plot, which may also be extended to multivariate analysis (dimension ≥ 3). Genest et al. (2006, 2007a) investigated the formal goodness-of-fit statistical tests for copulas. Chakak and Koehler (1995) presented a procedure to construct families of multivariate distributions through specified univariate and bivariate margins. Their procedure constructs multivariate distributions through conditional distributions.

Zheng and Klein (1995) proposed a copula-graphic estimator, which is a maximum likelihood estimator. The copula-graphic estimator was applied for the estimation of marginal distributions from the given copula for survival analysis. Simulation was performed using the Monte Carlo method, and the robustness of the method showed that the assumption of completely specifying the copula allowed for estimating the complete joint survival function based only on the competing risk data.

Quesada-Molina and Rodriguez-Lallena (1995a, b) investigated bivariate copulas with quadratic and cubic sections, which were derived from simple univariate real-valued functions on the interval [0, 1]. They applied various positive dependence structures (i.e., quadrant dependence and total positivity), measures of association (i.e., Kendall's τ and Spearman's ρ), stochastic ordering, and various notions of symmetry, which were shown to be equivalent to certain simple properties of univariate functions used for constructing bivariate copulas. They applied several examples to illustrate how these copulas can be constructed.

Müller and Scarsini (2001) considered two random vectors X and Y with the component of X dominated in the convex order by the corresponding components of Y. They found that the positive linear combination of the components of X dominated in the convex order by the same positive linear combination of the components of Y had the properties as the two random vectors having the common copula and conditionally increasing.

Frees and Valdez (1997) applied copulas, i.e., the Archimedean copula in an actuarial study, and estimated their parameters by both nonparametric and parametric methods. It was concluded that the Archimedean copula could be used to represent the bivariate distribution in the actuarial study fairly well.

Sancetta and Satchell (2001) analyzed financial multivariate data whose marginals were not normally distributed. Based on the nice Bernstein properties, they applied the Bernstein polynomial approximation to copulas and then investigated the multivariate convergence properties. The portfolio data were applied to investigate statistical properties and applications of Bernstein copulas. Chen and Fan (2002) investigated the issue related to the density forecast by applying a copula. They proposed a parametric test for the correct density forecasts by nesting a series of independently identically distributed random variables from stationary Markov processes. By applying the copula, they found that this test exhibited a large variety of marginal properties. Coupling the same marginals with different copula functions, they found that the test again exhibited numerous dependence properties.

Fang et al. (2002) investigated the joint probability density function of continuous random variables with given marginals by analyzing elliptically contoured distributions, e.g., normal distribution. They named this joint density function as meta-elliptical distribution. The analytical formulation, conditional distribution, and dependence properties of this meta-elliptical density function were discussed. They found that meta-elliptical joint distribution held the same Kendall tau as did the meta-Gaussian joint distribution belonging to the meta-elliptical joint distribution. Brakekers and Veraverbeke (2005) extended the estimator proposed by Rivest and Wells (2001) to the fixed design regression application. In survival analysis, the variables were generally assumed independent, which may be invalid in certain practical applications.

1.2.2 Construction and Parameter Estimation of Copulas

With the development of copula theories in statistics, Nelsen (2006) summarized the four most efficient methods to construct the copulas: (1) inversion method, (2) geometric method, (3) algebraic method, and (4) with specified properties. A detailed discussion of the construction of copulas and their properties will be provided in Chapter 3.

For any given copulas, their parameters may be estimated non-parametrically, parametrically, or semi-parametrically. The nonparametric method estimates the parameters with the rank correlation coefficient, i.e., Kendall's τ or Spearman's ρ. This method yields the analytical solution if there is a closed-form solution between rank correlation coefficient and copula parameters (e.g., certain Archimedean copulas that will be discussed in Chapter 4).

The copula parameters may be estimated parametrically with the use of one of the following three methods:

- Full MLE, by which the parameters of marginal distributions and copulas are estimated simultaneously.
- Two-stage MLE, by which the parameters of marginal distributions and the parameters of copula function are estimated separately using MLE. In this case, the fitted parametric marginal distributions will be applied to estimate the copula parameters through MLE.
- The semi-parametric method (also called pseudo-MLE: PMLE), which applies the empirical distribution (computed using probability plotting-position formula or kernel density) to estimate the copula parameters using MLE. Unlike the parametric approach, the semi-parametric method is marginal free.

Details of the estimation methods will be discussed in Chapter 3 and the following chapters.

To assess the goodness-of-fit of the fitted or proposed copula functions, Genest and Boies (2003), Genest et al. (2006), and Genest et al. (2007a) proposed the graphical and numerical assessment tools. These goodness-of-fit measures will be further introduced and applied in the chapters that follow.

1.2.3 Application of Copulas in Water Resources Engineering

With the theoretical development of copula theory and its advancement in statistics and econometrics, copulas have been adopted and applied in the fields of hydrology, water resources, and environmental engineering. These applications are briefly reviewed in the following section.

Copula Applications in Flood Frequency Analysis

Salvadori and De Michele (2004) provided a general theoretical framework exploiting copulas to determine return periods of bivariate hydrological events. They concluded the following: (1) copula may greatly simply the calculations of return period and may even yield an analytical solution; (2) copula may be associated with the return period of specific events; (3) with the use of copula, one may define sub-, super-, and critical events as well as those of primary and secondary return periods; and (4) the copula approach may be easily generalized to multivariate cases. The proposed methodology was further illustrated using flood peak and flood volume in a river basin in southern Taiwan, the spillway design flood of an existing Italian dam, and the annual maximum peak flow at Chute-des-Passes. Using flood variables (i.e., peak discharge, flood volume, and flood duration) observed at Kanawa River as an example, Grimaldi and Serinaldi (2006a) showed that (1) the flood variables were correlated; and (2) the dependence may not be symmetric among the flood variables, depending on the threshold used to identify the flood event. Employing the asymmetric Frank copula, the symmetric Frank copula, and the logistic Gumbel distribution through case studies, they presented the following: (1) the possible improvement obtained using the asymmetric copula and (2) the advantages in using the asymmetric copula.

Zhang and Singh (2006) applied the copula method to derive bivariate distributions of flood peak and volume, and flood volume and duration, such that the mariginals may follow different probability distributions. The conditional return periods for hydrologic design were tested using flood data from Amite River at Denham Springs, Louisiana, and the Ashuapmushuan River at Saguenay, Quebec, Canada. Comparing the derived distributions with the Gumbel mixed distribution and the bivariate Box–Cox transformed normal distribution, the copula-based distributions were found to result in the best agreement with plotting position-based frequency estimates. Genest et al. (2007b) presented how meta-elliptical copulas could be used to model the dependence structure of random vectors when observed differences between their bivariate margins precluded the use of exchangeable copula families, e.g., the Archimedean copula family. A case of peak, volume, and duration of the annual spring flood for the Romaine River was employed to illustrate rank-based estimation and goodness-of-fit techniques for this broad extension of the multivariate normal distribution. Analysis of annual spring flood for the Romaine River suggested that in view of the short length of the series, any of the eight meta-elliptical copula models considered in their studies could be used for prediction purposes. Only with additional evidence could one hope to distinguish between these dependence structures.

Simonovic and Karmakar (2007) focused on the selection of marginal distribution functions for flood characteristics by parametric and nonparametric estimation procedures, and demonstrated how the concept of copula may be used for establishing a joint distribution function with mixed marginal distributions for 70 years of streamflow data of Red River at Grand Forks in North Dakota, United States. Zhang and Singh (2007b) employed the Gumbel–Hougaard copula to model trivariate distributions of flood peak, volume, and duration, and then obtained conditional return periods. The derived distributions were tested using flood data from the Amite River basin in Louisiana. A major advantage of the copula method is that marginal distributions of individual variables can be of any form and the variables can be correlated.

Grimaldi and Serinaldi (2006a) described the fully nested (asymmetric) Archimedean copula properties and the inference procedure, and applied the copulas to multivariate flood frequency analysis of the Kanawha River (Kanawha Falls, West Virginia, drainage area 21,681 km^2) recorded from 1877 to 2003, and multivariate sea wave frequency analysis of Rete Ondametrica Nazionale (RON) network off the La Spezia (Liguria region, Italy). They found the following: (1) the inference procedure via copulas was quite easy to perform; and (2) asymmetric Archimedean copulas were useful to describe trivariate structures of dependence of nonexchangeable variables with different mutual degrees of correlation fulfilling the conditions described in Section 5.2.1; and finally, (3) comparison between observed and synthetic samples generated by estimated trivariate distributions confirmed the satisfactory performance of the Chen–Fan–Patton (CFP) test in order to choose the best-fitting copula. But asymmetric Archimedean copulas were not able to describe all mutually different structures of dependence. In addition, since the CFP test is based on Rosenblatt's transformation, its application becomes difficult when the number of variables increases. Consequently, further studies are needed to find both families of copulas that are capable of describing more complex structures of dependence and goodness-of-fit tests suitable for application to every copula class and high dimensions.

Wang et al. (2009) used a copula-based flood frequency (COFF) approach to estimate the risk of floods at confluence points. The four often-used Archimedean copulas (Ali–Mikhail-Haq, Clayton, Frank, and Gumbel–Hougaard) were applied in a river basin for the joint probability estimation. The Frank copula and Gumbel–Hougaard copula performed the best for the discharge data collected at two United States Geological Survey (USGS) gauge stations located on the Des Moines River at Fort Dodge, Iowa (USGS 05480500; Station A) and the Boone River near Webster City, Iowa (USGS 05471000; Station B), upstream of Des Moines River basin near Stratford, Iowa. It was shown that the copula method for specifying the multivariate distribution function was powerful, because it avoided the requirement that the marginal distributions be of the same type, which is assumed in most studies of empirical multivariate distributions. They also explained that it avoided the complex formulas that arise for many multivariate distribution functions. Zhang and Singh (2014) studied the trivariate flood frequency analysis by allowing different lengths of the records for maximum daily discharge at different locations.

Copula Application to Precipitation and Storm Characteristics Analysis

Salvadori and De Michele (2006) presented a statistical procedure to estimate probability distributions of storm characteristics. They discussed a method to describe the temporal dynamics of rainfall via a reward alternating renewal process that describes wet and dry phases of storms. The dependence among the three variables of interest (I for average rainfall intensity, W for the wet phase, and D for the dry one) was given via a Frank 3-copula. Based on real data collected by the Italian Sea Wave Measurement Network, De Michele et al. (2007) focused on how copulas can be used for the multidimensional frequency analysis of sea storm significant wave height (H), storm duration (D), storm direction (A), and storm interarrival time (I) (i.e., the calm period separating two successive storms). These included the following analyses:

- The construction of a bivariate model for the pair (H, D). In turn, this yielded the statistics of the sea storm magnitude M.
- Calculation of the return period of multivariate events. This gives the possibility to calculate the probability of occurrence of supercritical events and yielded an estimate of the minimum energetic content of sea storms having an assigned (multivariate) return period.
- Construction of a trivariate model for a triplet (H, D, A). This provided useful indications about the relation between sea storm magnitude and direction.
- Extension to storm interarrival duration I. This yielded a trivariate model for the triple (D, I, A) that cast new light on the relation between sea storm timing and direction.
- The construction of a global model for the vector (H, D, I, A). The overall structure was that of a reward alternating renewal process, whose dynamics develops along a random direction. In turn, this gave the possibility to simulate a sequence of sea storm events, accounting for all the variables of interest and their mutual relations.

These statistical analyses are very important when dealing with coastal dynamics, marine structure reliability, or the planning of operations at sea.

Zhang and Singh (2007a) derived trivariate rainfall frequency distributions using the Gumbel–Hougaard copula, which does not assume the rainfall variables to be independent or normal or have the same type of marginal distributions. The trivariate distribution was then employed to determine joint conditional return periods and was tested using rainfall data from the Amite River basin in Louisiana. Zhang and Singh (2007c) derived bivariate rainfall frequency distributions using the copula method in which four Archimedean copulas (Gumbel–Hougaard, Ali–Mikhail–Haq, Frank, and Cook–Johnson) were examined and compared. Results indicated that the advantage of the copula method is that no assumption is needed for the rainfall variables to be independent or normal or have the same type of marginal distributions. They also used the aforementioned Archimedean copulas to determine joint and conditional return periods, and tested using rainfall data from the Amite River basin in Louisiana, United States. Salvadori and De Michele (2007) summarized a general theoretical framework for studying the return period of hydrological events and presented a trivariate Frank copula model for the temporal structure of the sequence of storms at the Scoffera station, located in the Bisagno River basin (Thyrrhenian Liguria, northwestern Italy). The model includes, simplifies, and generalizes many of the approaches already present in the literature. They also gave an explicit derivation of the storm volume statistics for any suitable copula and marginals and a copula-based procedure for estimating the probability law of antecedent moisture conditions. Results indicated that the copula may have important applications in many fields of water resources and hydrologic systems, as well as in several geophysical areas.

Using three different samples of extreme rainfall criteria, including annual maximum volume (AMV), annual maximum peak intensity (AMI), and annual maximum cumulative probability (AMP), Kao and Govindaraju (2007) characterized extreme rainfall events using hourly precipitation data from Indiana, United States. Results of their study have implications for current hydrologic design in that they provided better estimates of design rainfall. Gebremichael and Krajewski (2007) explored the use of copulas to construct the joint distribution between the sampling error and the corresponding rainfall rate. Taking 15-minute radar-rainfall data for the Mississippi River basin in the central United States as an example, the approach (1) estimated the marginal distribution functions in a parametric way; (2) used these with a number of copula functions in search of the one most appropriate; (3) used the maximum likelihood to estimate the parameters of copulas; and (4) selected the best-fitted parametric copula function as the one that gave the largest likelihood. Results showed that the approach had important implications for the interpretation and propagation of remote sensing precipitation uncertainties.

Based on a non-Archimedean Plackett copula family derived using the theory of constant cross-product ratio, Kao and Govindaraju (2008) showed that the Plackett family not only performed well at the bivariate level, but also allowed trivariate stochastic analysis where the lower-level dependencies between variables can be fully preserved while allowing for specificity at the trivariate level as well. The authors proposed a numerical method to estimate the feasible range of Plackett parameters. The trivariate Plackett family of copulas was then applied to study a total of 53 hourly rain gauges from the Hourly Precipitation Database (TD 3240) of the National Climate Data Center in Indiana. Results of this study suggested that while the constant cross-product ratio theory was conventionally applied to discrete type random variables, it was also applicable to continuous random variables, and that it provided further flexibility for multivariate stochastic analyses of rainfall.

Evin and Favre (2008) proposed a new stochastic point rainfall model (Neyman–Scott cluster process) considering the dependence between cell depth and duration using cubic copula, and explored the properties of this class of copulas and suggested several families of this kind attaining a large range of dependence. They derived first-, second-, and third-order moments of the modified Neyman–Scott rectangular pulses model. Hourly rainfall data from Belgium and America were employed to fit the model by these theoretical moments and obtained successful results for two rainfall series with different climates. Generating long series of synthetic rainfall and the observed rainfall data and under specific cubic families and exponential margins, the model fitting can be improved. Results also indicated that the independent Pareto distribution for cell intensity yielded interesting results, and both hourly and daily annual maxima were adequately reproduced by most of the models. Vandenbreghe et al. (2011) investigated the bivariate frequency of storms using the copula method.

Copula Application to Drought Characteristics Analysis

Shiau (2006) used the run theory to abstract the paired drought duration and severity data from observed drought events in Wushantou (Taiwan), which were defined as the Standardized Precipitation Index (SPI) continuously below 0. The exponential and gamma distributions were then used to model the drought duration and severity, respectively. Several two dimensional copulas, such as Ali–Mikhail–Haq, Clayton, Frank, Galambos, Gumbel–Hougaard, and Plackett copulas, were employed to construct the dependence structure for drought duration and severity, and the joint drought duration and severity distribution. A method of inference function for margins (IFM method), a two-step procedure, was employed to estimate the copula parameters. The Galambos copula (belonging to extreme value family) fitted the observed drought data best for the Wushantou case under consideration. The bivariate probabilistic properties of droughts, such as joint probabilities and bivariate return periods, were also investigated to demonstrate comprehensive drought assessments. Shiau (2006) showed that copulas were easily applied to construct the dependence structure of the bivariate correlated random variables that were often met in hydrology.

Dupuis (2007) discussed the bivariate modeling of extreme tails of correlated hydrological random variables and applied the copula approach to model the dependence structure independently of marginal distributions. Dupuis also applied results from the classical extreme value theory to choose marginal distributions for excesses of high thresholds. Using six copula families (Gumbel, Frank, Normal, Student t, Clayton, and associated Clayton), the author discussed pertinent copula properties and examined the effects of model misspecification and the impact of the chosen estimation method, targeting the estimated quantities frequently used in hydrology. Based on a simulation study, Dupuis showed not only the dangers of improper copula selection but also the possible benefits of using a bivariate approach to estimate univariate quantities. Finally, the author applied copulas to study low-flow events and analyzed two Canadian hydrometric datasets.

Using monthly medians of streamflow of the Yellow River in China as the truncation levels, Shiau et al. (2007) defined hydrological droughts to obtain drought duration and drought severity. Drought duration and drought severity were fitted by the mixture of exponential and gamma distributions. The observed drought duration was highly correlated with the observed drought severity. The Clayton copula was used to construct the bivariate drought distribution from the predetermined marginal distributions of drought duration and drought severity. Results showed that the most severe drought of the Yellow River occurring during the period 1919–2002 was the 1930–1933 drought with the drought duration of 36 months and drought severity of 5264.8 $m^3 s^{-1}$. The return period for this drought event was 105 years. The 1997–1998 drought had a return period of 4.4 years. It suggested that the dramatically reduced streamflow in the downstream Yellow River in 1997 deteriorated due to other factors, such as human activities.

Wong et al. (2007) employed the trivariate Gaussian copula and the Gumbel copula to fit drought data. Results showed that the drought data were best described by the Gumbel copula and three-parameter Weibull marginal distribution. Song and Singh (2009) modeled the joint probability distribution of periodic hydrologic data using meta-elliptical copulas. Monthly precipitation data from a gauging station (410120) in Texas, United States, were used to illustrate parameter estimation and goodness-of-fit for univariate drought distributions using the chi-square test, Kolmogorov–Smirnov test, Cramér–von Mises statistic, Anderson–Darling statistic, modified weighted Watson statistic, and Liao and Shimokawa statistic. Pearson's classical correlation coefficient r_n, Spearman's ρ_n, Kendall's τ, chi-plots, and K-plots were employed to assess the dependence of drought variables. The meta-elliptical copulas and Gumbel–Hougaard, Ali–Mikhail–Haq, Frank and Clayton copulas were tested to determine the best-fit copula. Based on the root mean square error and the Akaike information criterion, meta-Gaussian and t copulas yielded a better fit. A bootstrap version based on Rosenblatt's transformation was employed to test the goodness-of-fit for meta-Gaussian and t copulas. It was found that none of meta-Gaussian and t copulas considered could be rejected at the given significance level. The meta-Gaussian copula was then employed to model dependence due to its simplicity for parameter estimation, and results were found satisfactory. Mirabbasi et al. (2012) and Chen et al. (2013) investigated the copula applications for drought characteristics.

Copula Application in Other Fields Related to Water Resources Engineering

Using four copulas (independence/product, Farlie–Gumbel–Morgenstern, Frank, and Clayton), Favre et al. (2004) described the modeling of the combined risk in the framework of frequency analysis of peak flows from the watershed of Peribonka in Québec, Canada, and the joint modeling of peak flows and volumes of the watershed of Rimouski River in Québec, Canada, using three copulas (Independence, Frank, and Clayton). Results showed that the copula approach was promising, since it allowed the researchers to take into account a wide range of correlation that can happen in hydrology. De Michele et al. (2005) proposed a two-copula method to model a bivariate extreme value distribution with generalized extreme value marginals. The peak-volume pair can then be transformed to the corresponding flood hydrograph, representing the river basin response, through a simple linear model. The hydrological safety of dams was considered for checking the adequacy of dam spillway. The reservoir behavior was tested using a long synthetic series of flood hydrographs with application to an existing dam.

Bárdossy (2006) calculated empirical copulas for four water quality parameters, chloride, sulfate, pH, and nitrate, obtained from a large-scale groundwater quality measurement network in Baden-Württemberg (Germany). A Gaussian and a non-Gaussian copula were applied, and results indicated that the spatial dependence structure of the investigated parameters was not Gaussian. According to the bootstrap-based statistical tests using stochastic simulation of multivariate distributions, the Gaussian copula was rejected for most of the parameters, but the non-Gaussian alternative was not rejected in most cases. Grimaldi and Serinaldi (2006b) proposed a procedure to describe the trivariate cumulative distribution function (CDF) of critical depth, peak, and total depth. Seven three-copula functions were estimated with the canonical maximum likelihood (CML) method, and the best one was chosen for analyzing the CDF of copulas.

Bárdossy and Li (2008) used the Gaussian as well as non-Gaussian copulas to depict the dependence structure of the investigated parameters without the influence of marginal distributions. Division of observations into multipoint subsets and subsequent maximization of the corresponding likelihood function were employed to estimate copula parameters. Chloride, nitrate, pH, sulfate, and dissolved oxygen observations of a large-scale groundwater quality measurement network in Baden-Württemberg were used to demonstrate the methodology. Results showed that all five parameters showed non-Gaussian dependence, and the non-Gaussian copulas gave better results than the geostatistical interpolations. Meanwhile, validation of the confidence intervals showed that they were more realistic than the estimation variances obtained by ordinary kriging.

1.3 Theme of the Book

The goal of the book is to discuss for graduate level students and engineers how to appropriately apply the copula method. The book is divided into two parts. Part I introduces the copula theory, including copula properties, methods of construction, copula families, etc. Part II discusses applications of copulas in hydrology and water resources engineering with case studies.

More specifically, Part I includes the following chapters with regard to copula theory. Chapter 2 briefly reviews the preliminaries for univariate and multivariate frequency analysis. Chapter 3 discusses the important properties of copulas. Chapter 4 discusses the bivariate Archimedean copula families and multivariate symmetric Archimedean copula extensions. Chapter 5 discusses the nested (i.e., asymmetric) Archimedean copula and the vine copula through pair copula construction. Chapter 6 discusses the non-Archimedean Plackett copula family. Chapter 7 discusses meta-elliptical non-Archimedean copula families. Chapter 8 discusses the entropic copulas. Chapter 9 discusses the copula application in time series analysis.

Part II provides the following case studies. Chapter 10 discusses the copula application to rainfall analysis. Chapter 11 discusses the copula application to flood analysis. Chapter 12 discusses the copula application to water quality analysis. Chapter 13 discusses the copula application to drought analysis. Chapter 14 discusses the copula application to compound extremes. Chapter 15 discusses the copula application to network design. Chapter 16 discusses the river sediment transport. And Chapter 17 discusses the interbasin transfer.

References

Ali, M. M., Mikhail, N. N., and Haq, M. S. (1978). A class of bivariate distributions including the bivariate logistic. *Journal of Multivariate Analysis*, 8, 405–412.

Bárdossy, A. (2006). Copula-based geostatistical models for groundwater quality parameters. *Water Resources Research*, 42, W11416, doi:10.1029/2005WR004754.

Bárdossy, A. and Li, J. (2008). Geostatistical interpolation using copulas. *Water Resources Research*, 44, W07412, doi:10.1029/2007WR006115.

Braekers, R. and Veraverbeke, N. (2005). A copula-graphic estimator for the conditional survival function under dependent censoring. *Technical Report*, 0315. *Interuniversity Attraction Pole*.

Caperaa, P., Fougeres, A. L., and Genest, C. (1997). A nonparametric estimation procedure for bivariate extreme copulas. *Biometrika*, 84(3), 567–577.

Chakak, A. and Koehler, K. J. (1995). A strategy for constructing multivariate distributions. *Communicational Statistics (Simulation)*, 24(3), 537–550.

Chen, L., Singh, V. P., Guo, S., Mishra, A., and Guo, J. (2013) Drought analysis using copulas. *Journal of Hydrologic Engineering*, 18(7), 797–808. doi:10.1061/(ASCE) HE.1943–5584.0000697.

Chen, X. and Fan, Y. (2002). Evaluating density forecasts via the copula approach. www .vanderbilt.edu/Econ/wparchive/workpaper/vu02-w25R.pdf.

Cook, R. D. and Johnson, M. E. (1981). A family of distributions for modeling non-ellipitically symmetric multivariate data. *Journal of the Royal Statistical Society. Series B. (Methodological)*, 43(2), 210–218.

De Michele, C., Salvadori, G., Canossi, M., Petaccia, A., and Rosso, R. (2005). Bivariate statistical approach to check adequacy of dam spillway. *Journal of Hydrologic Engineering*, 10(1), 50–57.

De Michele, C., Salvadori, G., Passoni, G., and Vezzoli, R. (2007). A multivariate model of sea storms using copulas. *Coastal Engineering*, 54, 734–751.

Dupuis, D. J. (2007). Using copulas in hydrology: benefits, cautions, and issues. *Journal of Hydrologic Engineering*, 12(4), 381–393.

Evin, G. and Favre, A. C. (2008). A new rainfall model based on the Neyman–Scott process using cubic copulas. *Water Resources Research*, 44, W03433, doi:10.1029/ 2007WR006054.

Fang, H., Fang, K.T., and Kotz, S. (2002). The meta-elliptical distributions with given marginals. *Journal of Multivariate Analysis*, 82, 1–16.

Favre, A. C., Adlouni, S. E., Perreault, L., Thiémonge, N., and Bobeé, B. (2004). Multivariate hydrological frequency analysis using copulas. *Water Resources Research*, 40(1), W01101, doi:10.1029/2003WR002456.

Frees, E. W. and Valdez, E. A. (1997). Understanding relationships using copulas. *North American Acturial Journal*, 2(1), 1–37.

Gebremichael, M. and Krajewski, W. F. (2007). Application of copulas to modeling temporal sampling errors in satellite-derived rainfall estimates. *Journal of Hydrologic Engineering*, 12(4), 404–408.

Genest, C. (1987). Frank's family of bivariate distribution. *Biometrika*, 74(3), 549–555.

Genest, C. and Boies, J. C. (2003). Detecting dependence with Kendall plots. *American Statistician*, 57(4), 275–284.

Genest, C., Favre, A. C., Béliveau, J., and Jacques, C. (2007b). Meta-elliptical copulas and their use in frequency analysis of multivariate hydrological data. *Water Resources Research*, 43, W09401, doi:10.1029/2006WR005275.

Genest, C. and MacKay, J. (1986). The joy of copulas: bivariate distributions with uniform marginals. *American Statistician*, 40(4), 280–283.

Genest, C. and Rivest, L.-P. (1993). Statistical inference procedures for bivariate Archimedean copulas. *Journal of the American Statistical Association*, 88(423), 1034–1043.

Genest, C., Ghoudi, K., and Rivest, L.-P. (1995). A semiparametric estimation procedure of dependence parameters in multivariate families of distributions. *Biometrika*, 82(3), 543–552.

Genest, C., Quessy, J.-F., and Rémillard, B. (2006). Goodness-of-fit procedures for copula models based on the integral probability transformation. *Scandinavian Journal of Statistics*, 33, 337–366.

Genest, C., Rémillard, B., and Beaudoin, D. (2007a). Goodness-of-fit tests for copulas: a review and a power study. *Insurance: Mathematics and Economics*, doi:10.1016/j.insmatheco.2007.10.005.

Grimaldi, S. and Serinaldi, F. (2006a). Asymmetric copula in multi-variate flood frequency analysis. *Advances in Water Resources*, 29(8), 1155–1167.

Grimaldi, S. and Serinaldi, F. (2006b). Design hyetograph analysis with 3-copula function. *Hydrological Sciences Journal*, 51(2), 223–238.

Hosking, J. R. M. (1990). Fortran routines for use with the method of L-moments, Version 2. Research Report RC-17097, IBM Thomas J. Watson Research Center, Yorktown Heights.

Joe, H. (1997). *Multivariate Models and Dependence Concept*. Chapman & Hall, New York.

Kao, S. C. and Govindaraju, R. S. (2007). A bivariate rainfall frequency analysis of extreme rainfall with implications for design. *Journal of Geophysical Research*, 112, D13119, doi:10.1029/2007JD008522.

Kao, S. C. and Govindaraju, R. S. (2008). Trivariate statistical analysis of extreme rainfall events via the Plackett family of copulas. *Water Resources Research*, 44(2), W02415, doi:10.1029/2007WR006261.

Long, D. and Krzysztofowicz, R. (1995). A family of bivariate densities constructed from marginals. *Journal of the American Statistical Association*, 90(430), 739–746.

Mirabbasi, R., Fakheri-Fard, A., and Dinpashoh, Y. (2012). Bivaraite drought frequency analysis using the copula method. *Theoretical Applied Climatology*, 108(1–2), 191–206, doi:10.1007/s00704-011-0524-7.

Muller, A. and Scarsini, M. (2001). Stochastic comparison of random vectors with a common copula. *Mathematics of Operations Research*, 26(4), 723–740.

Nelsen, R. B. (2006). *An Introduction to Copulas*. Springer, New York.

Quesada-Molina, J. J. and Rodriguez-Lallena, J. A. (1995a). Bivariate copulas with quadratic sections. *Nonparametric Statistics*, 5, 323–337.

Quesada-Molina, J. J. and Rodriguez-Lallena, J. A. (1995b). Bivariate copulas with cubic sections. *Nonparametric Statistics*, 7, 205–220.

Rao, A. R. and Hamed, K. H. (2000). *Flood Frequency Analysis*. CRC Publications, Boca Raton, London, New York, Washington.

Rodriguez-Lallena, J. A. and Úbeda-Flores, M. (2004). A new class of bivariate copulas. *Statistics and Probability Letters*, 66, 315–325.

Salvadori, G. and Michele, C. D. (2003). A generalized Pareto intensity and duration model of storm rainfall exploiting 2-copulas. *Journal of Geophysical Research*, 108 (D2), doi:10,1029/2002JD002543.

Salvadori, G. and De Michele, C. (2004). Frequency analysis via copulas: theoretical aspects and applications to hydrological events. *Water Resources Research*, 40, W12511, doi:10.1029/2004WR003133.

Salvadori, G. and De Michele, C. (2007). On the use of copulas in hydrology: theory and practice. *Journal of Hydrologic Engineering*, 12(4), 369–380.

Sancetta, A. and Satchell, S. (2001). *Berstein Approximations to the Copula Function and Portfolio Optimization*. DAE Working Paper 0105, University of Cambridge. www.econ.cam.ac.uk/research-files/repec/cam/pdf/wp0105.pdf.

Shiau, J. T. (2006). Fitting drought duration and severity with two-dimensional copulas. *Water Resources Management*, 20, 795–815.

Shiau, J. T., Feng, S., and Nadarajah, S. (2007). Assessment of hydrological droughts for the Yellow River, China, using copulas. *Hydrological Processes*, 21(16), 2157–2163.

Simonovic, S. P. and Karmakar, S. (2007). *Flood Frequency Analysis Using Copula with Mixed Marginal Distribution*. Report No. 055. www.econ.cam.ac.uk/research-files/repec/cam/pdf/wp0105.pdf.

Singh, V. P. (1988). *Hydrologic Systems: Rainfall-Runoff Modeling*. Prentice Hall, Englewood Cliffs.

Singh, V. P. (1998). *Entropy-Based Parameter Estimation in Hydrology*. Kluwer Academic Publishers, Dordrecht, Boston, London.

Singh, V. P., Jain, S. K., and Tyagi, A. (2007). *Risk and Reliability Analysis*. ASCE Press, Reston.

Sklar, A. (1959). Fonctions de repartition à n dimensionls et leurs marges. *Publications de l'Institut de Statistique de l'Université de Paris, Paris*. 8, 229–231.

Song, S. B. and Singh, V. P. (2009). Meta-elliptical copulas for drought frequency analysis of periodic hydrologic data. *Stochastic Environmental Research and Risk Assessment*, doi:10.1007/s00477–009–0331–1.

Vandenberghe, S., Verhoest, N. E. C., Onof, C., and De Baets, B. (2011). A comparative Copula-based bivariate frequency analysis of observed and simulated storm events: a case study on Bartlett-Lewis modeled rainfall. *Water Resources Research*, 47. doi:10.1029/2009wr008388.

Wang, C., Chang, N. B., and Yeh, G. T. (2009). Copula-based flood frequency (COFF) analysis at the confluences of river systems. *Hydrological Processes*, 23, 1471–1486.

Wong, G., Lambert, M. F., and Metcalfe, A. V. (2007). Trivariate copulas for characterisation of droughts. *ANZIAM Journal*, 49, C306–C323.

Yue, S. (1999). Applying bivariate normal distribution to flood frequency analysis. *Water International*, 24(3), 248–254.

Yue, S. (2000a). Joint probability distribution of annual maximum storm peaks and amounts as represented by daily rainfalls. *Hydrologic Science Journal*, 45(2), 315–326.

Yue, S. (2000b). The Gumbel logistic model for representing a multivariate storm event. *Advances in Water Resources*, 24 (2), 179–185.

Yue, S. (2000c). The Gumbel mixed model applied to storm frequency analysis. *Water Resources Management*, 14(5), 377–389.

Yue, S., Ouarda, T. B. M. J., Bobée, B., Legendre, P., and Bruneau, P. (1999). The Gumbel mixed model for flood frequency analysis. *Journal of Hydrology*, 226, 88–100.

Yue, S., Ouarda, T. B. M. J., and Bobée B (2001). A review of bivariate gamma distributions for hydrological application. *Journal of Hydrology*, 246, 1–18.

Yue, S. and Rasmussen, P. (2002). Bivariate frequency analysis: discussion of some useful concepts for hydrological application. *Hydrological Processes*, 16(14), 811–819.

Zheng, M. and Klein, J. P. (1995). Estimates of marginal survival for dependent competing risk based on assumed copula. *Biometrika*, 82(1), 127–138.

Zhang, L. and Singh, V. P. (2006). Bivariate flood frequency analysis using the copula method. *Journal of Hydrologic Engineering*, 11(2), 150–164.
Zhang, L. and Singh, V. P. (2007a). Gumbel-Hougaard copula for trivariate rainfall frequency analysis. *Journal of Hydrologic Engineering*, 12(4), 409–419.
Zhang, L. and Singh, V. P. (2007b). Trivariate flood frequency analysis using the Gumbel–Hougaard copula. *Journal of Hydrologic Engineering*, 12(4), 431–439.
Zhang, L. and Singh, V. P. (2007c). Bivariate rainfall frequency distributions using Archimedean copulas. *Journal of Hydrology*, 332, 93–109.

Additional Reading

Adamson, P. T., Metcalfe, A. V., and Parmentier B. (1999). Bivariate extreme value distributions: an application of the Gibbs sampler to the analysis of floods. *Water Resources Research*, 35(9), 2825–2832.
Ashkar, F. (1980). Partial duration series models for flood analysis. PhD thesis. Ecole Polytechnique of Montreal, Montreal, Canada.
Ashkar, F., El Jabi, N., and Issa, M. (1998). A bivariate analysis of the volume and duration of low-flow events. *Stochastic Hydrology and Hydraulics*, 12, 97–116.
Bacchi, B., Becciu, G,. and Kottegoda, N. T. (1994). Bivariate exponential model applied to intensities and durations of extreme rainfall. *Journal of Hydrology*, 155, 225–236.
Choulakian, V., El Jabi, N., and Moussi, J. (1990). On the distribution of flood volume in partial duration series analysis of flood phenomena. *Stochastic Hydrology and Hydraulics*, 4, 217–226.
Correia, F. N. (1987). Multivariate partial duration series in flood risk analysis. In: Singh, V. P. (Ed) *Hydrologic Frequency Modeling*. Reidel, Dordrecht, 541–554.
Cunnane, C. (1987). Review of statistical models for flood frequency estimation. In: Singh, V. P. (Ed) *Hydrologic Frequency Modeling*, Reidel, Dordrecht, 49–95.
Durrans, S. R. (1998). Total probability methods for problems in flood frequency estimation. In: Parent, E., Hubert, P., Bobee, B., and Miquel, J. (Eds) *Statistical and Bayesian Methods in Hydrological Science*. International Hydrological Programme, Nairobi, Jakarta, Venice, Cairo, and Montevideo. Technical Documents in Hydrology, No. 20UNESCO, Paris, 299–326.
Futter, M. R., Mawdsley, J. A., and Metcalfe, A. V. (1991). Short-term flood risk prediction: a comparison of the Cox regression model and a conditional distribution model. *Water Resources Research*, 27(7), 1649–1656.
Goel, N. K., Seth, S. M., and Chandra, S. (1998). Multivariate modeling of flood flows. *Journal of Hydraulic Engineering*, 124(2), 146–155.
Goel, N. K., Kurothe, R. S., Mathur, B. S., and Vogel, R. M. (2000). A derived flood frequency distribution for correlated rainfall intensity and duration. *Journal of Hydrology*, 228, 56–67.
Grimaldi, S., Serinaldi, R., Napolitano, F., and Ubertini, L. (2005). A 3-copula function application or design hyetograph analysis. *Proceedings of Symposium S2*, Held during the Seventh IAHS Scientific Assembly at Foz do Iguacu, Brazil, April 2005. IAHS publ. 293. International Association of Hydrological Sciences (IAHS), London. https://iahs.info/uploads/dms/13113.33%20203-211%20s2-10%20Grimaldi%20et%20al%2066.pdf.

Haimes, Y. Y., Lambert, J. H., and Li, D. (1992). Risk of extreme events in a multi-objective framework. *Water Resources Bulletin*, 28(1), 201–209.

Hashino, M. (1985). Formulation of the joint return period of two hydrologic variates associated with a Poisson process. *Journal of Hydroscience and Hydraulic Engineering*, 3(2), 73–84.

Hosking, J. R. M. and Wallis, J. R. (1997). *Regional Frequency Analysis*. Cambridge University Press. Cambridge.

Kelly, K. S. and Krzysztofowicz, R. (1997). A bivariate meta-Gaussian density for use in hydrology. *Stochastic Hydrology and Hydraulics*, 11, 17–31.

Kite, G. W. (1978). *Frequency and Risk Analysis in Hydrology*. Water Resource Publications, Fort Collins.

Kurothe, R. S., Goel, N. K., and Mathur, B. S. (1997). Derived flood frequency distribution for negatively correlated rainfall intensity and duration. *Water Resources Research*, 33, 2103–2107.

Krstanovic, P. F. and Singh, V. P. (1987). A multivariate stochastic flood analysis using entropy. In: Singh, V. P. (Ed) *Hydrologic Frequency Modeling*. Reidel, Dordrecht, 515–539.

Lall, U. and Bosworth, K. (1994). Multivariate kernel estimation of functions of space and time. In: Hipel K. V., Mcleod, A. I., Panu, U. S., Singh, V. P. (Eds) *Time Series Analysis in Hydrology and Environmental Engineering*. Kluwer Academic Publications, Dordrecht, 301–315.

Loganathan, G. V., Kuo, C. Y., and Yannaccone, J. (1987). Joint probability distribution of streamflows and tides in estuaries. *Nordic Hydrology*, 18, 237–246.

Long, D. and Krzysztofowicz, R. (1996). Geometry of a correlation coefficient under a copula. *Communications in Statistics: Theory and Methods*, 25(6), 1397–1404.

Nachtnebel, H. P. and Konecny, F. (1987). Risk analysis and time-dependent flood models. *Journal of Hydrology*, 91, 295–318.

Renard, B. and Lang, M. (2007). Use of a Gaussian copula for multivariate extreme value analysis: some case studies in hydrology. *Advances in Water Resources*, 30, 897– 912.

Rényi, A. (1974). On measure of dependence. *Acta Mathematica Academiae Scientiarum Hungarica*, 10, 441–451.

Rivest, L.-P. and Wells, M. T. (2001). A martingale approach to the Copula-graphic estimator for the survival function under dependent censoring. *Journal of Multivariate Analysis*, 79, 138–155.

Sackl, B. and Bergmann, H. (1987). A bivariate flood model and its application. In: Singh, V. P. (Ed) *Hydrologic Frequency Modeling*. Reidel, Dordrecht, 571–582.

Salvadori, G. and De Michele, C. (2006). Statistical characterization of temporal structure of storms. *Advances in Water Resources*, 29(6), 827–842.

Schweizer, B. and Wolff, E. F. (1981). On nonparametric measures of dependence for random variables. *Annals of Statistics*, 9, 879–885.

Schweizer, B. (1991). Thirty years of copula. In: Dall'Aglio, G., Kotz, S., and Salinetti, G. (Eds) *Advances in Probability Distributions with Given Marginals: Beyond the Copulas*. Mathematics and Its Applications, 67, Kluwer Academic Publishers, Dordrecht, 13–50.

Serinaldi, F. and Grimaldi, S. (2007). Fully nested 3-copula: procedure and application on hydrological data. *Journal of Hydrologic Engineering*, 12(4), 420–430.

Singh, K. and Singh, V. P. (1991). Derivation of bivariate probability density functions with exponential marginals. *Journal of Stochastic Hydrology and Hydraulics*, 5, 55–68.

Wilks, D. S. (1998). Multisite generalization of a daily stochastic precipitation generation model. *Journal of Hydrology*, 210, 178–191.

Wolff, E. F. (1977). *Measures of Dependence Derived from Copulas*. PhD thesis, University of Massachusetts, Amherst.

Zhang, L. and Singh, V. P. (2014). Trivariate flood frequency analysis using discharge time series with possible different lengths: Cuyahoga River case study. *Journal of Hydrologic Engineering*. doi:10.1061/(ASCE)HR.1943-5584.0001003.

2

Preliminaries

ABSTRACT

Bivariate or multivariate frequency analysis entails univariate distributions that are determined by empirical fitting to data. The fitting, in turn, requires the determination of distribution parameters and the assessment of the goodness of fit. In practical applications, such as hydrologic design, risk analysis is also needed. The objective of this chapter, therefore, is to briefly discuss these basic elements, which are needed for frequency analysis and will be needed in subsequent chapters.

2.1 Univariate Probability Distributions

Among the univariate distributions, we will briefly discuss the most commonly applied continuous univariate distributions, especially in univariate hydrological frequency analyses (Kite, 1977; Singh, 1998; Rao and Hamed, 2000; Singh and Zhang, 2016). In what follows, we will use X as an independent identically distributed (IID) random variable with probability density function (PDF) $f(x)$ and cumulative distribution function (CDF) $F(x)$.

2.1.1 Normal Distribution

Normal distribution: The PDF and CDF of the normal distribution can be given as follows:

$$f(x) = \frac{1}{\sigma(2\pi)^{0.5}} \exp\left(-\frac{(x-\mu)^2}{2\sigma^2}\right);\ F(x) = \Phi\left(\frac{x-\mu}{\sigma}\right);\ \mu \in \mathbb{R},\ \sigma > 0 \tag{2.1}$$

In Equation (2.1), Φ represents the standard normal distribution, and μ, σ are the location and scale parameters having the connotation of mean and standard deviation of the random variable, respectively. Defining the standard normal variable $z = (x-\mu)/\sigma$, Equation (2.1) can be written as

$$f(z) = \frac{1}{\sqrt{2\pi}} \exp\left(-\frac{z^2}{2}\right); F(z) = \frac{1}{\sqrt{2\pi}} \int_{-\infty}^{z} \exp\left(-\frac{t^2}{2}\right) dt; F(-z) = 1 - F(z) \tag{2.1a}$$

Abramowitz and Stegun (1965) have numerically approximated F(z) with an error less than 7.5×10^{-5} as

$$F(z) = 1 - f(z)\left[a_1 z + a_2 z^2 + a_3 z^3 + a_4 z^4 + a_5 z^5\right] + \epsilon(z) \tag{2.1b}$$

where
$a_1 = -0.319381530, a_2 = -0.356563782,\ a_3 = 1.781477937,\ a_4 = -1.821255978, a_5 = 1.330274429$, and $\epsilon(z)$ is the error of approximation.

In hydrological frequency analysis, the normal distribution has been commonly applied in two scenarios:

1. Normal distribution with mean of zero is the classic assumption for time series analysis and regression analysis. As a simple example, let Y be the response or prediction variable and X be the predictor variable. Then, a simple linear regression can be expressed as

$$E(Y|X) = \hat{Y} = a + bx;\ \ e = Y - \hat{Y} \text{ and } \boldsymbol{e} \sim \boldsymbol{N}\left(0, \sigma_{\boldsymbol{e}}^2\right) \tag{2.2}$$

where e is the residual or error and $e \sim N\left(0, \sigma_e^2\right)$ denotes that e is distributed normally with mean 0 and variance σ_e^2. $E[Y|X]$ denotes the conditional expectation of Y given X. $\hat{Y}$ denotes the predicted response through simple linear regression with intercept of a and slope of b.

For example, a stationary time series $\{X_t, t = 1, 2, \ldots\}$ modeled by an Autoregressive and Moving Average (ARMA) model with (p, q) (Box et al., 2007) as follows:

$$x_t = c + \phi_1 x_{t-1} + \cdots + \phi_p x_{t-p} + e_t + \theta_1 e_{t-1} + \cdots + \theta_q e_{t-q}; \boldsymbol{e_t} \sim \boldsymbol{N}\left(0, \sigma_{e_t}^2\right) \tag{2.3}$$

In Equation (2.3), c is the long-term average of the time series, and $\phi_1, \ldots, \phi_p$; $\theta_1, \ldots, \theta_q$ are, respectively, the coefficients for autoregressive and moving average terms. More specifically, in Equations (2.2) and (2.3), the residual e, following normal distribution with mean of 0 is commonly called *white Gaussian noise*.

2. After certain monotone transformation (e.g., Box–Cox or probability integral transformation), the normal distribution (Equation (2.1)) may be applied to model the nonnormally distributed hydrologic variables (e.g., Hazen, 1914; Markovic, 1965).

2.1.2 Log-Normal Distribution

Let $Y = \ln(x)$. If X follows the log-normal distribution, then its logarithm follows the normal distribution, whose PDF can be written as follows:

$$f(x) = \frac{1}{x\sigma(2\pi)^{0.5}} \exp\left(-\frac{(\ln(x) - \mu)^2}{2\sigma^2}\right); x > 0 \tag{2.4}$$

The CDF of the log-normal distribution can be computed again through the standard normal distribution as follows:

$$F(x) = \Phi\left(\frac{\ln x - \mu}{\sigma}\right) \tag{2.5}$$

The logarithm of the random variable X is a special case of the Box–Cox transformation (Box and Cox, 1964) with $\lambda = 0$:

$$x^T = \begin{cases} \dfrac{x^\lambda - 1}{\lambda}, & \lambda \neq 0 \\ \ln x, & \lambda = 0 \end{cases} \tag{2.5a}$$

The log-normal distribution has been widely used in hydrological frequency analysis (e.g., Chow, 1954).

2.1.3 Student t Distribution

Similar to the normal distribution, the Student t distribution is also bell-shaped (Hogg and Craig, 1978). However, it possesses the heavy tail, i.e., excess kurtosis is greater than 0. The PDF of the standard Student t distribution is given as follows:

$$f(x, v) = \frac{\Gamma\left(\frac{v+1}{2}\right)}{(v\pi)^{0.5}\Gamma\left(\frac{v}{2}\right)}\left(1 + \frac{x^2}{v}\right)^{-\frac{v+1}{2}} \tag{2.6a}$$

And its CDF is given as follows:

$$F(x; v) = \frac{1}{2} + x\Gamma\left(\frac{v+1}{2}\right)\frac{{}_2F_1\left(\frac{1}{2}, \frac{v+1}{2}; \frac{3}{2}; -\frac{x^2}{v}\right)}{(v\pi)^{0.5}\Gamma\left(\frac{v}{2}\right)} \tag{2.6b}$$

In Equations (2.6a) and (2.6b), v represents the degree of freedom. It is worth to note that with the degree of freedom, the Student t distribution will converge to normal distribution, i.e., the excess kurtosis is approaching 0. It may be explained using the excess kurtosis of Student t distribution as follows: $\lim_{v\to\infty} exkurtosis = \lim_{v\to\infty} \frac{6}{v-4} = 0$. And ${}_2F_1$ represents the hypergeometric function as follows:

$${}_2F_1\left(\frac{1}{2}, \frac{v+1}{2}; \frac{3}{2}; -\frac{x^2}{v}\right) = \sum_{n=0}^{\infty} \frac{\left(\frac{1}{2}\right)_n \left(\frac{v+1}{2}\right)_n \left(-\frac{x^2}{v}\right)^n}{\left(\frac{3}{2}\right)_n n!} \tag{2.6c}$$

In Equation (2.6c), the Pochhammer symbol is defined as follows:

$$(a)_n = \begin{cases} 1 & n = 0 \\ a(a+1)\cdots(a+n-1) & n > 0 \end{cases} \tag{2.6d}$$

2.1.4 Exponential and Gamma Distributions

The exponential distribution is a special case of the gamma distribution (Hogg and Craig, 1978). These two distributions have been commonly applied in rainfall and flood frequency analyses. The gamma distribution can be given as follows:

$$f(x) = \frac{1}{\Gamma(\alpha)\beta^{\alpha}} x^{\alpha-1} \exp\left(-\frac{x}{\beta}\right); \ x \geq 0,\ \alpha,\ \beta : \text{shape and scale paramters},\ \alpha,\ \beta > 0 \quad (2.7)$$

When the shape parameter $\alpha = 1$, the gamma distribution is reduced to the exponential distribution as follows:

$$f(x) = \frac{1}{\beta} \exp\left(-\frac{x}{\beta}\right) \quad (2.7a)$$

whose CDF is simply

$$F(x) = 1 - \exp\left(-\frac{x}{\beta}\right) \quad (2.7b)$$

The CDF of the gamma distribution can be expressed as follows:

$$F(x) = \frac{\gamma\left(\alpha, \frac{x}{\beta}\right)}{\Gamma(\alpha)} \quad (2.8)$$

where

$$\gamma\left(\alpha, \frac{x}{\beta}\right) = \int_0^{\frac{x}{\beta}} t^{\alpha-1} e^{-t} dt - \text{Lower incomplete gamma function} \quad (2.8a)$$

The gamma function can be expressed as follows:

$$\Gamma(\alpha) = \int_0^{\infty} t^{\alpha-1} e^{-t} dt \quad (2.8b)$$

with the following properties:

$$\Gamma(\alpha+1) = \alpha\Gamma(\alpha),\ \alpha > 0; \Gamma(\alpha) = \frac{\Gamma(\alpha+1)}{\alpha},\ \alpha < 1 \text{ and}$$
$$\Gamma(n) = (n-1)!; \Gamma(2) = \Gamma(1) = 1; \Gamma\left(\frac{1}{2}\right) = \sqrt{\pi};$$

n is an integer. Abramowitz and Stegun (1965) have numerically approximated the gamma function for $0 < \alpha \leq 1$ with an absolute error less than 3×10^{-7} as $\Gamma(\alpha) = 1 + \sum_{i=1}^{8} a_i \alpha^i + \epsilon(\alpha)$,

where $a_1 = -0.57191652$, $a_2 = 0.988205891$, $a_3 = -0.897056937$, $a_4 = 0.918206857$
$a_5 = -0.756704078$, $a_6 = 0.482199394$, $\alpha\_7 = -0.193527818$, $a_8 = 0.035868343$

For other values of α, the gamma function properties can be used to compute the gamma function. For example,

$$\Gamma(4.25) = 3.25\Gamma(3.25) = 3.25(2.25)\Gamma(2.25) = 3.25(2.25)(1.25)\Gamma(1.25).$$

Besides the exponential distribution being a special case of Gamma distribution, the chi-square distribution is also a special case of gamma distribution by setting $\alpha = \frac{k}{2}$, where k denotes the degree of freedom and usually taking the integers, and $\beta = 2$.

2.1.5 Generalized Extreme Value (GEV) and Extreme Value (EV) Distributions

Introduced by Jenkinson (1955) and recommended by the Natural Environment Research Council (1975) of Great Britain, the GEV distribution has been widely applied for flood frequency analysis. The EV distributions may be directly obtained from the GEV distribution. The PDF and CDF of the GEV distribution can be written as follows:

$$f(x; a, b, c) = \frac{1}{a}\left(1 - \frac{b(x-c)}{a}\right)^{\frac{1-b}{b}} \exp\left(-\left(1 - \frac{b(x-c)}{a}\right)^{\frac{1}{b}}\right), \ 1 - \frac{b(x-c)}{a} > 0 \quad (2.9a)$$

$$F(x) = \exp\left(-\left(1 - \frac{b(x-c)}{a}\right)^{\frac{1}{b}}\right) \quad (2.9b)$$

In Equations (2.9a) and (2.9b), a, b, and c are the scale, shape, and location parameters, respectively, and the range of variable X depends on the sign of parameter b.

The EV distributions can be derived, depending on the shape parameter b.

EV I Distribution ($b = 0$)

The EV I distribution may also be called the Gumbel distribution (Gumbel, 1941). It is a popular distribution for flood, drought, and rainfall frequency analyses. The PDF and CDF of EV 1 distribution can be written as follows:

$$f(x; a, c) = \frac{1}{a}\exp\left[-\frac{x-c}{a} - \exp\left(-\frac{x-c}{a}\right)\right]; \ x \geq c \quad (2.10a)$$

$$F(x; a, c) = \exp\left(-\exp\left(-\frac{x-c}{a}\right)\right) \quad (2.10b)$$

The coefficient of skewness is 1.1396 and the X ranges as $x \in [c, \infty)$.

EV II Distribution ($b < 0$)

The EV II distribution is also called Fréchet distribution (Gumbel, 1958) that has also been applied to frequency analysis. The PDF and CDF of the EV II distribution can be written as follows:

$$f\left(x; a, c, \beta = -\frac{1}{b}\right) = \frac{\beta}{a}\left(\frac{x-c}{a}\right)^{-\beta-1} \exp\left(-\left(-\frac{x-c}{a}\right)^{-\beta}\right),\ a,\ \beta > 0 \tag{2.11a}$$

$$F(x; a, c, \beta) = \exp\left(-\left(\frac{x-c}{a}\right)^{-\beta}\right) \tag{2.11b}$$

The coefficient of skewness is greater than 1.1396 and *X* can take on values in the range $x \in \left(c + \frac{a}{k}, \infty\right)$, which makes it appropriate for flood frequency analysis.

*EV III Distribution (*b > 0*)*

Belonging to the Weibull family (i.e., inverse Weibull distribution), the EV III distribution is usually applied for low-flow frequency analysis (Singh, 1998). The PDF and CDF of the EV III distribution can be written as follows:

$$f\left(x; a, c, \beta = \frac{1}{b}\right) = \frac{\beta}{a}\left(-\frac{x-c}{a}\right)^{\beta-1} \exp\left(-\left(-\frac{x-c}{a}\right)^{\beta}\right); x \le c \tag{2.12a}$$

$$F(x; a, c, \beta) = \exp\left(-\left(-\frac{x-c}{a}\right)^{\beta}\right) \tag{2.12b}$$

The coefficient of skewness is less than 1.396 and variable X ranges as $x \in \left(-\infty, c + \frac{a}{\beta}\right)$, which does not render it suitable for flood frequency analysis.

2.1.6 Weibull Distribution

The Weibull distribution (Rosin and Rammler, 1933) is commonly applied for low-flow frequency analysis, hazard functional analysis, as well as risk and reliability analysis. The PDF and CDF of the Weibull distribution can be written as follows:

$$f(x; a, b) = \frac{a}{b}\left(\frac{x}{b}\right)^{a-1} \exp\left(-\left(\frac{x}{b}\right)^{a}\right); x > 0,\ a,\ b > 0 \tag{2.13a}$$

$$F(x; a, b) = 1 - \exp\left(-\left(\frac{x}{b}\right)^{a}\right) \tag{2.13b}$$

The Weibull distribution is a reverse GBV distribution.

Pearson and Log-Pearson Type III Distributions

These two distributions are commonly applied for flood frequency analysis (Singh, 1998). The log-Pearson type III distribution is the standard method for flood frequency analysis in the United States, whereas the Pearson type III distribution is the standard method in China.

Pearson Type III Distribution The PDF and CDF of Pearson type III distribution can be written as follows:

$$f(x;a,b,c) = \frac{1}{a\Gamma(b)}\left(\frac{x-c}{a}\right)^{b-1} \exp\left(-\frac{x-c}{a}\right); x \geq c,\ a > 0,\ b > 0 \tag{2.14a}$$

$$F(x) = \frac{1}{\Gamma(b)}\gamma\left(b, \frac{x-c}{a}\right) \tag{2.14b}$$

Using $y = (x-c)/a$ Equations (2.14a) and (2.14b) can be written as

$$f(y) = \frac{1}{a\Gamma(b)} y^{b-1} \exp(-y) \tag{2.14c}$$

$$F(y) = \frac{1}{\Gamma(b)}\int_0^y t^{b-1} \exp(-t)dt = \frac{\gamma(b,y)}{\Gamma(b)} \tag{2.14d}$$

The value of $F(y)$ can be determined in the same way as for the gamma distribution discussed earlier.

Log-Pearson Type III Distribution Similar to the log-normal distribution, if random variable X follows the log-Pearson type III distribution, then its logarithm $Y = \ln X$ follows the Pearson type III distribution. The PDF and CDF of log-Pearson type III distribution can be written as follows:

$$f(x;a,b,c) = \frac{1}{ax\Gamma(b)}\left(\frac{\ln x - c}{a}\right)^{b-1} \exp\left(-\frac{\ln x - c}{a}\right); x > \exp c,\ a > 0,\ b > 0 \tag{2.15a}$$

$$F(x;a,b,c) = \frac{1}{\Gamma(b)}\gamma\left(b, \frac{\ln x - c}{a}\right) \tag{2.15b}$$

2.1.7 *Burr XII Distribution*

The PDF and CDF of Burr XII distribution (Burr, 1942) can be written as follows:

$$f(x;a,b,c) = bc\left[1 + \left(\frac{x}{a}\right)^c\right]^{-k-1} \frac{x^{c-1}}{a^c},\ x \geq 0,\ a,\ b,\ c > 0 \tag{2.16a}$$

$$F(x;a,b,c) = 1 - \left[1 + \left(\frac{x}{a}\right)^c\right]^{-k} \tag{2.16b}$$

2.1.8 *Log-Logistic Distribution*

The log-logistic distribution is also known as Fisk distribution (Shoukri et al., 1988). Its PDF and CDF can be written as follows:

$$f(x;a,b) = \frac{\left(\frac{b}{a}\right)\left(\frac{x}{a}\right)^{b-1}}{\left(1+\left(\frac{x}{a}\right)^{b}\right)^{2}}; x > 0,\ a > 0,\ b > 0 \tag{2.17a}$$

$$F(x;a,b) = \frac{x^b}{a^b + x^b} \tag{2.17b}$$

Equation (2.17b) can be used to directly express a quantile. Equations (2.17) can also be generalized by including the location parameter.

2.1.9 Pareto Distribution

There are four distributions in the Pareto family (Arnold, 1983). The two- and three-parameter Pareto distributions have been used for modeling large floods. The PDF and CDF of the two-parameter Pareto distribution can be written as follows:

$$f(x) = \frac{a x_m^a}{x^{a+1}},\ x \geq x_m;\ f(x) = 0\ if\ x > x_m \tag{2.18a}$$

$$F(x) = 1 - \left(\frac{x_m}{x}\right)^a,\ x \geq x_m;\ \ F(x) = 0\ if\ x < x_m \tag{2.18b}$$

There are many other distributions that have been applied in frequency analysis (Singh and Zhang, 2016), besides the distributions illustrated in this section.

2.2 Bivariate Distributions

Here we discuss the commonly applied bivariate distributions in bivariate hydrologic analyses.

2.2.1 Bivariate Gamma Distribution

Several different bivariate gamma distributions have been applied in bivariate hydrological analyses. For all the bivariate gamma distributions introduced, their margins (or marginals) are univariate gamma distribution with the PDF and CDF given as Equations (2.7) and (2.8).

Izawa Bigamma Model

The joint PDF of Izawa bigamma model (Izawa, 1965) is given for random variables X and Y as follows:

$$f(x,y) = \frac{(xy)^{\frac{n-1}{2}} x^m \exp\left(-\frac{\beta_x x + \beta_y y}{1-\eta}\right)}{\Gamma(n)\Gamma(m)\left(\beta_x\beta_y\right)^{-\frac{n+1}{2}}\beta_x^{-m}(1-\eta)\eta^{\frac{n-1}{2}}} \times \int_0^1 (1-t)^{\frac{n-1}{2}} t^{m-1} \exp\left(\frac{\beta_x \eta x t}{1-\eta}\right) I_{n-1}\left(\frac{2\sqrt{\beta_x\beta_y\eta xy(1-t)}}{1-\eta}\right) dt \tag{2.19}$$

where

$$I_s(h) = \sum_{k=0}^{\infty} \frac{\left(\frac{h}{2}\right)^{s+2k}}{k!\Gamma(s+k+1)} \tag{2.19a}$$

$$\eta = \rho\sqrt{\frac{\alpha_x}{\alpha_y}};\quad 0 \leq \rho < 1; 0 \leq \eta < 1,\ \alpha_x \leq \alpha_y \tag{2.19b}$$

In the preceding expressions, $I_s(\cdot)$ is the modified Bessel function of the first kind; η is the association parameter between X and Y; ρ is Pearson's product-moment correlation coefficient of X and Y; $X \sim gamma(x; \alpha_x, \beta_x)$; and $Y \sim gamma(y; \alpha_y, \beta_y)$.

The limitations of the Izawa bigamma distribution are that (i) the shape parameter of X is less than that of Y; and (ii) it may only model the positively correlated random variables.

Moran Model

The PDF of the Moran model (Moran, 1969) of X and $Y$with the gamma marginals can be written as

$$f(x,y) = \frac{1}{\sqrt{1-\rho_N^2}} f_X(x; \alpha_x, \beta_x) f_Y(y; \alpha_y, \beta_y) \exp\left(-\frac{(\rho_N x')^2 - 2\rho_N x' y' + (\rho_N y')^2}{2(1-\rho_N^2)}\right) \tag{2.20}$$

where $x' = \Phi^{-1}(F_X(x; \alpha_x, \beta_x))$, $y' = \Phi^{-1}(F_Y(y; \alpha_y, \beta_y))$, ρ_N represents Pearson's product-moment correlation coefficient of the transformed variables x' and y'.

Smith–Adelfang–Tubbs (SAT) Model

Again with gamma marginals, Smith et al. (1982) developed the another bivariate model (i.e., the SAT model). Its PDF and CDF of the SAT model can be expressed as follows:

$$f(x,y) = \begin{cases} \frac{K_1}{K_2} \sum_{j=0}^{\infty} \sum_{k=0}^{\infty} c_{jk} (\beta_x x)^j (\eta \beta_y y)^{j+k},\ 0 < \eta < 1 \\ f_X(x; \alpha_x, \beta_x) f_Y(y; \alpha_y, \beta_y),\ \eta = 0 \end{cases} \tag{2.20a}$$

$$F(x,y) \begin{cases} J \sum_{j=0}^{\infty} \sum_{j=0}^{\infty} d_{jk} H\left(\frac{x}{1-\eta}, \alpha_x + j\right) H\left(\frac{y}{1-\eta}, \alpha_y + j + k\right); 0 < \eta < 1 \\ F_X(x; \alpha_x, \beta_x) F_Y(y; \alpha_y, \beta_y),\ \eta = 0 \end{cases} \tag{2.20b}$$

$$K_1 = (\beta_x x)^{\lambda_x - 1} (\beta_y y)^{\lambda_y - 1} \exp\left(-\frac{\beta_x + \beta_y}{1-\eta}\right) \tag{2.20c}$$

$$K_2 = (1-\eta)^{\alpha_x} \Gamma(\alpha_x) \Gamma(\alpha_y - \alpha_x) \tag{2.20d}$$

$$c_{jk} = \frac{\eta^{j+k}\Gamma(\alpha_y - \alpha_x + k)}{(1-\eta)^{2j+k}\Gamma(\alpha_y + j + k)j!k!} \tag{2.20e}$$

$$\eta = \rho\sqrt{\alpha_y/\alpha_x} \tag{2.20f}$$

$$J = \frac{(1-\eta)^{\alpha_y}}{\Gamma(\alpha_x)\Gamma(\alpha_y - \alpha_x)} \tag{2.20g}$$

$$d_{jk} = \frac{\eta^{j+k}\Gamma(\alpha_y - \alpha_x + k)}{\Gamma(\alpha_y + j + k)j!k!} \tag{2.20h}$$

$$H(z, a) = \int_0^z t^{a-1}e^{-t}dt \tag{2.20i}$$

Farlie–Gumbel–Morgenstern (FGM) Model

This bivariate model was first proposed by Morgenstern (1956). Its PDF and CDF of the FGM model for random variables X and Y can be expressed as follows:

$$f(x,y) = f_X(x)f_Y(y)(1 + \eta\ (2F_X(x) - 1)(2F_Y(y) - 1)) \tag{2.21a}$$

$$F(x,y) = F_X(x)F_Y(y)(1 + \eta(1 - F_X(x))(1 - F_Y(y))) \tag{2.21b}$$

where $\{f_X(x), f_Y(y)\}$ and $\{F_X(x), F_Y(y)\}$ are the marginal PDFs and CDFs of $X$and Y, respectively, and η is the correlation coefficient between X and Y.

Gumbel Mixed (GM) Model

The GM model has been applied to model the bivariate flood frequency analysis (Yue et al., 1999). The CDF of the GM model may be expressed as follows:

$$F(x,y) = F_X(x)F_Y(y)\exp\left(-\theta\left(\frac{1}{\ln F_X(x)} + \frac{1}{\ln F_Y(y)}\right)^{-1}\right); 0 \le \theta \le 1 \tag{2.22}$$

where θ is the association parameters of the GM model, which describes the dependence between random variables X and Y as follows:

$$\theta = 2\left(1 - \cos\left(\pi\sqrt{\frac{\rho}{6}}\right)\right),\ 0 \le \rho \le \frac{2}{3} \tag{2.22a}$$

where ρ is Pearson's product moment correlation coefficient.

It should be noted that the marginal CDFs of random variable X and Y are the Gumbel distribution (i.e., Equation (2.10b)) in the case of the conventional GM model.

Gumbel Logistic (GL) Model

The Gumbel logistic model was first proposed by Gumbel (Gumbel, 1960, 1961). With the Gumbel-distributed marginals (Equation (2.10)), the CDF of the GL model can be expressed as follows:

$$F(x,y) = \exp\left(-((-\ln F_X(x))^\eta + (-\ln F_Y(y))^\eta\,)^{\frac{1}{\eta}}\right); \eta \geq 1 \tag{2.23}$$

where

$$\eta = \frac{1}{\sqrt{1-\rho}}; 0 \leq \rho \leq 1 \tag{2.23a}$$

As the association parameter of the GL model, η describes the dependence between two random variables.

Bivariate Exponential Model

Marshall and Ingram (1967), Singh and Singh (1991), and Bacchi et al. (1994) proposed the bivariate exponential distribution that can be expressed as follows:

$$f_{XY}(x,y) = ab[(1+acx)(1+bcy) - \rho]\exp(-ax - by - abcxy) \tag{2.24a}$$

where X and Y are exponentially distributed as

$$f_X(x) = a\exp(-ax); f_Y(y) = b\,\exp(-by) \tag{2.24b}$$

and c represents the association between 0 and 1 between X and Y defined through the coefficient of correlation as

$$\rho = -1 + \int_0^\infty \frac{1}{1+cx}\exp(-x)dx \tag{2.24c}$$

This bivariate model is valid for ρ between 0 and –0.404.

Nagao–Kadoya Bivariate Exponential (BVE) Model

With the exponential distributed random variables X and Y (Equation (2.7a)), the PDF of the BEV model (Balakrisinan and Lai, 2009) can be expressed as follows:

$$f(x,y) = \frac{\alpha\beta}{(1-\rho)}\exp\left(-\frac{\alpha x + \beta y}{1-\rho}\right)I_0\left(\frac{2\sqrt{\rho\alpha\beta xy}}{1-\rho}\right), \quad x \geq 0, \quad y \geq 0, \quad 0 \leq \rho < 1 \tag{2.25}$$

where

$$I_0\left(\frac{2\sqrt{\rho\alpha\beta xy}}{1-\rho}\right) = \sum\nolimits_{k=0}^{\infty}\left(\frac{\rho\alpha\beta xy}{(1-\rho)^2}\right)^k \frac{1}{(k!)^2} \tag{2.25a}$$

In Equations (2.25) and (2.25a), ρ is the Pearson correlation coefficient between X and Y; α, β are the parameters of exponential variables X and Y, respectively, as

$X \sim \exp(\alpha)$, $Y \sim \exp(\beta)$from Equation (2.7a); and I_0 is the modified Bessel function of the first kind.

2.2.2 Bivariate Normal Distribution

The bivariate normal distribution is also applied in bivariate hydrological frequency analysis. Let X and Y follow normal distribution (Equation (2.1)). Then the bivariate normal distribution can be written as follows:

$$f(x,y) = \frac{1}{2\pi\sigma_X\sigma_Y\sqrt{1-\rho^2}}\exp\left(-\frac{q(x,y)}{2}\right) \tag{2.26}$$

$$q(x,y) = \frac{1}{1-\rho^2}\exp\left(\left(\frac{x-\mu_X}{\sigma_X}\right)^2 - 2\rho\left(\frac{x-\mu_X}{\sigma_X}\right)\left(\frac{y-\mu_Y}{\sigma_Y}\right) + \left(\frac{y-\mu_Y}{\sigma_Y}\right)^2\right) \tag{2.26a}$$

2.2.3 Bivariate Log-Normal Distribution

For the log-normally distributed random variables X and Y (Equation (2.4)), the joint distribution may be expressed with the bivariate log-normal distribution as follows:

$$f(x,y) = \frac{1}{2\pi\sigma_{X_1}\sigma_{Y_1}\sqrt{1-\rho^2}}\exp\left(\left(\frac{\ln x-\mu_{X_1}}{\sigma_{X_1}}\right)^2 - 2\rho\left(\frac{\ln x-\mu_{X_1}}{\sigma_{X_1}}\right)\left(\frac{\ln y-\mu_{Y_1}}{\sigma_{Y_1}}\right)\left(\frac{\ln y-\mu_{Y_1}}{\sigma_{Y_1}}\right)^2\right) \tag{2.27}$$

$$\mu_{X_1} = \ln\left(\frac{\mu_X}{\left(1+\frac{\sigma_X^2}{\mu_X^2}\right)^{0.5}}\right);\quad \mu_{Y_1} = \ln\left(\frac{\mu_Y}{\left(1+\frac{\sigma_Y^2}{\mu_Y^2}\right)^{0.5}}\right) \tag{2.27a}$$

$$\sigma_{X_1} = \left(\ln\left(1+\frac{\sigma_X^2}{\mu_X^2}\right)\right)^{\frac{1}{2}};\quad \sigma_{Y_1} = \left(\ln\left(1+\frac{\sigma_Y^2}{\mu_Y^2}\right)\right)^{\frac{1}{2}} \tag{2.27b}$$

where μ_X, σ_X; μ_Y, σ_Y are the mean and standard deviations of random variables X and Y; and ρ is the Pearson correlation coefficient of $(\ln X, \ln Y)$.

From the preceding commonly applied bivariate probability distribution models, it is seen that (1) the bivariate gamma and exponential family may only model the positive dependence; (2) the bivariate normal and log-normal distribution may model the dependence in the entire range; and (3) the marginal distributions of all the models belong to the same type of univariate distribution, i.e., gamma, exponential, normal, and log-normal distributions.

2.3 Estimation of Parameters of Probability Distributions

The dependence of the commonly applied conventional bivariate distributions are associated with the Pearson correlation coefficient of the bivariate random variables.

In this section, we will only briefly review the parameter estimation for univariate probability distributions.

There are a number of methods that may be applied to estimate the parameters of univariate distributions (Singh, 1998; Rao and Hamed, 2000). These methods are (1) method of moments (MOM), (2) method of maximum likelihood estimation (MLE), (3) method of probability weighted moments (PWM), (4) method of L-moments (LM), (5) method of least squares (LS), (6) method of maximum entropy (MAX_ENT), (7) method of mixed moments (MIX), (8) the generalized method of moments (GMM), and (9) incomplete means method (ICM). Let X be a random variable with density function $f(x; \alpha_1, \alpha_2, \ldots, \alpha_k)$ in which αs are the parameters and $\mathbf{X} = [x_1, x_2, \ldots, x_n]$ is the sample drawn from the population. In what follows, we will introduce the four most commonly applied methods in hydrology and water resources engineering, i.e., the MOM, MLE, PWM, and LM methods.

2.3.1 Method of Moments

The MOM is a natural and relatively easy parameter estimation method for univariate distributions. However, MOM is usually inferior in quality and not as efficient as the MLE, especially for distributions with a large number of parameters (three or more). This is partly because higher-order moments are more likely to be biased for relatively small samples (Rao and Hamed, 2000).

MOM assumes that the sample moments are equal to the population moments, that is, the sample is sufficiently large to be representative of the population. Given the probability distribution with k parameters $\alpha_1, \ldots, \alpha_k$, we can compute k sample moments from the sample $\mathbf{X} = \{x_1, \ldots, x_n\}$, such as sample mean $(\bar{X})$, sample standard deviation (S_X), sample skewness coefficient (g_1), and sample kurtosis. The relation between moments and parameters of the probability distribution is then established by simultaneously solving k equations for the unknown parameters: $\alpha_1, \ldots, \alpha_k$. It is worth noting that the first moment is computed about the origin, while the other sample moments are about the first moment (mean). We will illustrate parameter estimation by MOM for normal, gamma, Weibull, and Gumbel distributions as examples.

The rth-moment ratio, denoted as C_r, is defined as follows:

$$C_r(X) = \frac{\mu_r(X)}{[\mu_2(X)]^{\frac{r}{2}}} \tag{2.28}$$

From Equation (2.28), we can see the following:

$$\text{Coefficient of variation: } C_1 = \frac{\mu_1(X)}{[\mu_2(X)]^{0.5}} = \frac{\mu}{\sigma} \tag{2.28a}$$

$$\text{Coefficient of skewness: } C_3 = C_s = \frac{\mu_3(X)}{[\mu_2(X)]^{1.5}} \tag{2.28b}$$

$$\text{Coefficient of kurtosis: } C_4 = C'_\mathrm{k} = \frac{\mu_4(X)}{[\mu_2(X)]^2} \tag{2.28c}$$

In addition, the classical moment diagram is graphed using the possible pairs (β_1, β_2), which are related to C_3 and C_4 as follows:

$$\beta_1 = C_3^2 = C_s^2, \quad \beta_2 = C_4 = C'_k \tag{2.28d}$$

Example 2.1 **Estimate parameters of the normal distribution by MOM.**

Solution: With the PDF of normal distribution given in Equation (2.1) and letting $\alpha_1 = \mu; \alpha_2 = \sigma$, we can estimate parameters α_1 and α_2 by the solving the following two equations:

$$\mu_1 = \int_{-\infty}^{+\infty} x \frac{1}{\alpha_2 (2\pi)^{0.5}} \exp\left(-\frac{(x-\alpha_1)^2}{2\alpha_2^2} \right) dx : \text{population mean} \tag{2.29a}$$

$$\mu_2 = \int_{-\infty}^{+\infty} (x-\mu_1)^2 \frac{1}{\alpha_2 (2\pi)^{0.5}} \exp\left(-\frac{(x-\alpha_1)^2}{2\alpha_2^2} \right) dx : \text{population variance} \tag{2.29b}$$

The following equates the sample mean $\bar{X}$ to the population mean and the sample variance VAR (X) to the population variance:

$$\bar{X} = \frac{1}{N} \sum_{i=1}^{N} x_i = \mu_1 = \int_{-\infty}^{+\infty} x \frac{1}{\alpha_2 (2\pi)^{0.5}} \exp\left(-\frac{(x-\alpha_1)^2}{2\alpha_2^2} \right) dx \tag{2.29c}$$

$$\mathrm{VAR}(X) = \frac{1}{N} \sum_{i=1}^{N} (x_i - \bar{X})^2 = \mu_2 = \int_{-\infty}^{+\infty} (x-\mu_1)^2 \frac{1}{\alpha_2 (2\pi)^{0.5}} \exp\left(-\frac{(x-\alpha_1)^2}{2\alpha_2^2} \right) dx \tag{2.29d}$$

In Equation (2.29), N is replaced by (N−1) to correct for the bias due to sample size.
Solving Equations (2.29c) and (2.29d) simultaneously, we get the following:

$$\hat{\alpha}_1 = m_1 = \frac{1}{N} \sum_{i=1}^{N} x_i; \widehat{\alpha}_2^2 = m_2 = \frac{1}{N-1} \sum_{i=1}^{N} (x_i - m_1)^2 \tag{2.29e}$$

Example 2.2 **Estimate parameters of gamma distribution by MOM.**

Solution: The PDF of gamma distribution is given as Equation (2.7). Let $\alpha_1 = \alpha$ and $\alpha_2 = \beta$. The first moment of gamma distribution can be written as follows:

$$\mu_1 = \int_0^\infty x f(x) dx = \int_0^\infty \frac{\alpha_2^{\alpha_1} x^{\alpha_1}}{\Gamma(\alpha_1)} \exp(-\alpha_2 x) dx$$

$$= \frac{\alpha_2^{\alpha_1}}{\Gamma(\alpha_1)} \int_0^\infty x^{\alpha_1} \exp(-\alpha_2 x) dx = \frac{\alpha_2^{\alpha_1}}{\Gamma(\alpha_1)} \frac{\Gamma(\alpha_1 + 1)}{\alpha_2^{\alpha_1+1}} = \frac{\alpha_1}{\alpha_2} \tag{2.30a}$$

The variance of gamma distribution can be given as follows:

$$\mu_2 = \int_0^\infty (x-\mu_1)^2 f(x)dx = \int_0^\infty \left(x - \frac{\alpha_1}{\alpha_2}\right)^2 \frac{\alpha_2^{\alpha_1} x^{\alpha_1 - 1}}{\Gamma(\alpha_1)} \exp\left(-\alpha_2 x\right)dx = \frac{\alpha_1}{\alpha_2^2} \tag{2.30b}$$

Substituting the sample mean and variance as $m_1 = \mu_1; m_2 = \mu_2$, we can estimate the parameters by solving Equations (2.30a) and (2.30b) simultaneously as follows:

$$\alpha_2 = \frac{m_1}{m_2} = \frac{\sum_{i=1}^N x_i}{\sum_{i=1}^N (x_i - \bar{X})^2}; \alpha_1 = \frac{\sum_{i=1}^N x_i \bar{X}}{\sum_{i=1}^N (x_i - \bar{X})^2}; \bar{X} = \frac{1}{N}\sum_{i=1}^N x_i \tag{2.30c}$$

It is worth noting that the exponential distribution is a special case of gamma distribution with $\alpha_1 = 1$, and $\alpha_2 = 1/m_1$.

Example 2.3 **Estimate parameters of Weibull distribution using MOM.**

Solution: The PDF of Weibull distribution is given as Equation (2.13a). Let $\alpha_1 = a$, $\alpha_2 = b$. Then we can write the population mean as follows:

$$\mu_1 = \int_0^\infty x f(x)dx = \int_0^\infty x \frac{\alpha_1}{\alpha_2}\left(\frac{x}{\alpha_2}\right)^{\alpha_1 - 1} \exp\left(-\left(\frac{x}{\alpha_2}\right)^{\alpha_1}\right)dx \tag{2.31a}$$

After some simple algebra, Equation (2.27a) may be solved as follows:

$$\mu_1 = \alpha_2 \Gamma\left(1 + \frac{1}{\alpha_1}\right) \tag{2.32b}$$

Similarly, we can write the population variance as follows:

$$\mu_2 = \int_0^\infty (x-\mu_1)^2 f(x)dx$$
$$= \int_0^\infty \left(x - \alpha_2\Gamma\left(1+\frac{1}{\alpha_1}\right)\right)^2 \frac{\alpha_1}{\alpha_2}\left(\frac{x}{\alpha_2}\right)^{\alpha_1 - 1} \exp\left(-\left(\frac{x}{\alpha_2}\right)^{\alpha_1}\right)dx \tag{2.32c}$$

Again, with simple algebra, we obtain the following:

$$\mu_2 = \alpha_2^2 \Gamma\left(1 + \frac{2}{\alpha_2}\right) - \alpha_2\left(\Gamma\left(1 + \frac{1}{\alpha_1}\right)\right)^2 \tag{2.32d}$$

Replacing μ_1, μ_2 with their sample estimates m_1, m_2, the parameters of the Weibull distribution can be obtained by solving Equations (2.32b) and (2.32d) simultaneously numerically.

Example 2.4 **Estimate the parameters of Gumbel distribution with MOM.**

Solution: The Gumbel distribution is also called EV I distribution, which is given as Equation (2.10a). Let $\alpha_1 = a$, $\alpha_2 = c$. The first moment of the Gumbel distribution can be given as follows:

$$\mu_1 = \int xf(x)dx = \int \frac{x}{\alpha_1} \exp\left(-\left(\frac{x-\alpha_2}{\alpha_1}\right) + \exp\left(-\frac{x-\alpha_2}{\alpha_1}\right)\right)dx$$

$$= \alpha_1 + \gamma\alpha_2,\ \gamma \approx 0.5772,\ \gamma : \textit{Euler} - \textit{Mascheroni constant} \tag{2.33a}$$

The variance of Gumbel distribution can be given as follows:

$$\mu_2 = \int (x-\mu_1)^2 f(x)dx = \int \frac{(x-\mu_1)^2}{\alpha_2} \exp\left(-\left(\frac{x-\alpha_2}{\alpha_1}\right) + \exp\left(-\frac{x-\alpha_2}{\alpha_1}\right)\right)dx = \frac{\alpha_2\pi}{6} \tag{2.33b}$$

Replacing μ_1, μ_2 with their sample estimates m_1, m_2, the parameters of Gumbel distribution can be obtained by solving Equations (2.33a) and (2.33b) as follows:

$$\alpha_1 = \frac{0.5772\sqrt{6}m_2}{\pi};\ \alpha_1 = \frac{\sqrt{6}m_2}{\pi} \tag{2.33c}$$

2.3.2 Method of Maximum Likelihood Estimation

The MLE is considered as the most efficient parameter estimation method (Rao and Hamed, 2000). The reason is that MLE provides the smallest sampling variance of the estimated parameters and hence for the estimated quantiles. However, MLE has the following disadvantages: (1) in some particular cases, such as the Pearson type III distribution, the optimality of MLE is only asymptotic, and small sample estimates may lead to estimates of inferior quality; (2) MLE often results in the biased estimates, but these biases can be corrected; (3) in the case of small sample size, MLE may not yield parameter estimates, especially for probability distributions with multiple parameters; and (4) MLE often requires more computational effort with the increase in the number of parameters; however, this is no longer a problem with today's computing technology.

Parameters estimated with the use of MLE are obtained by maximizing the probability of occurrence of observations $\{x_1, \ldots, x_n\}$. Given a random variable X following a probability density function $f(x)$ with parameters $\alpha_1, \ldots, \alpha_k$, the likelihood function is defined as the joint PDF of the observations as follows:

$$L(\alpha_1, \ldots, \alpha_k) = \prod_{i=1}^{n} f(x_i; \alpha_1, \ldots, \alpha_k) \tag{2.34}$$

Applying the monotone transformation (i.e., natural logarithm transformation), Equation (2.34) can be rewritten as follows:

$$\ln L(\alpha_1, \ldots, \alpha_k) = \sum\nolimits_{i=1}^{n} \ln f(x_i; \alpha_1, \ldots, \alpha_k) \tag{2.35}$$

It is worth noting that Equation (2.35) may also be called log-likelihood (LL) and will not change the parameters that may be estimated using Equation (2.35). To this end, the parameters, i.e., $\alpha_1, \ldots, \alpha_k$, may be optimized by maximizing Equation (2.35), which may be computed by taking partial derivatives with respect to $\alpha_1, \ldots, \alpha_k$ and setting these partial derivatives equal to zero as follows:

$$\begin{cases} \dfrac{\partial \ln L(\alpha_1, \ldots, \alpha_k)}{\partial \alpha_1} = 0 \\ \cdots \\ \dfrac{\partial \ln L(\alpha_1, \ldots, \alpha_k)}{\partial \alpha_k} = 0 \end{cases} \tag{2.36}$$

The resulting set of equations is then solved simultaneously to obtain the estimated parameters: $\hat{\alpha}_1, \ldots, \hat{\alpha}_k$.

Example 2.5 **Estimate parameters of the normal distribution by MLE.**

Solution: The PDF of normal distribution is given in **Example 2.1.** The likelihood function of a sample of size n from a normal distribution is given by $L(\alpha_1, \alpha_2)$:

$$L(\alpha_1, \alpha_2) = \left(\frac{1}{\alpha_2\sqrt{2\pi}}\right)^n \exp\left(-\frac{1}{2\alpha_2^2}\sum\nolimits_{i=1}^{n}(x_i - \alpha_1)^2\right) \tag{2.37}$$

Taking the natural logarithm of Equation (2.37), we get the following:

$$\ln L(\alpha_1, \alpha_2) = -n \ln \alpha_2 - n \ln \sqrt{2\pi} - \frac{1}{2\alpha_2^2} \tag{2.38}$$

Taking the derivatives of $\ln L(\alpha_1, \alpha_2)$ with respect to α_1, α_2, and then setting these derivatives equal to zero, one gets the following:

$$\frac{\partial \ln L(\alpha_1, \alpha_2)}{\partial \alpha_1} = \frac{1}{2\alpha_2^2}\sum\nolimits_{i=1}^{n} 2(x_i - \alpha_1) = 0 \tag{2.38a}$$

$$\frac{\partial \ln L(\alpha_1, \alpha_2)}{\partial \alpha_2} = -\frac{n}{\alpha_2} + \frac{1}{\alpha_2^3}\sum\nolimits_{i=1}^{n}(x_i - \alpha_1)^2 = 0 \tag{2.38b}$$

Solving Equations (2.38a) and (2.38b) simultaneously, we get the following:

$$\begin{cases} \hat{\alpha}_1 = \dfrac{1}{n}\sum\nolimits_{i-1}^{n} x_i = m_1 \\ \hat{\alpha}_2 = \dfrac{1}{n}\sum\nolimits_{i=1}^{n}(x_i - \alpha_1)^2 = m_2 \end{cases} \tag{2.38c}$$

Equations (2.29e) and (2.38c) indicate that MOM and MLE yield the same parameter values for the normal distribution.

2.3.3 Probability Weighted Moments Method

Compared to MOM, the PWM is much less complicated with much simpler computation (Rao and Hamed, 2000). For small sample sizes, parameters estimated using PWM are sometimes more accurate than those estimated using MOM. Additionally, in some cases, e.g., the symmetric Lambda and Weibull distributions, explicit expressions of the parameters may be obtained using PWM, which may not be the case with MOM or MLE (Rao and Hamed, 2000).

For a random variable X with cumulative distribution function (CDF), $F(x)$, the probability weighted moment of the cumulative distribution function can be defined as follows:

$$M_{i,j,k} = E\left(x^i F^j (1-F)^k\right) = \int_0^1 [x(F)]^i F^j (1-F)^k dF \tag{2.39}$$

In Equation (2.39), $M_{i,j,k}$ is the probability weighted moment of order (i, j, k); E represents the expectation operator; and $i, j, k \in \mathbb{R}$. Based on Rao and Hamed (2000) and Singh et al. (2007), (1) $M_{i,0,0}$ represents the conventional ith moment of order i about the origin if i is a nonnegative integer; and (2) $M_{i,j,k}$ exists for all nonnegative real numbers j and k under the following two conditions: (a) $M_{i,0,0}$ exists and (b) X is a continuous function of F.

Considering the ordered sample, i.e., $x_{(1)} \leq x_{(2)} \leq \ldots \leq x_{(n)}$, the PWM for hydrologic applications (Singh et al., 2007) may be defined as follows:

$$M_{1,0,s} = a_s = \frac{\frac{1}{n}\sum_{i=1}^{n} \binom{n-i}{s} x_i}{\binom{n-1}{s}}; M_{1,r,s} = b_r = \frac{\frac{1}{n}\sum_{i=1}^{n} \binom{n-i}{r} x_i}{\binom{n-1}{r}} \tag{2.40}$$

The PWMs can also be expressed as follows:

$$a_s = M_{1,0,s} = \frac{1}{n}\sum_{i=1}^{n} (1-F_i)^s x_i; b_r = M_{1,r,s} = \frac{1}{n}\sum_{i=1}^{n} F_i^r x_i \tag{2.40a}$$

In Equation (2.40), $n > r$, s, r are nonnegative integers. Additionally, Equation (2.40) further indicates that a_s and b_r are functions of each other as follows:

$$a_s = \sum_{k=0}^{s} (-1)^k \binom{s}{k} b_k; b_r = \sum_{k=0}^{s} (-1)^k \binom{r}{k} a_k \tag{2.41}$$

Example 2.6 **Estimate the parameters of Weibull distribution using PWM.**

Solution: From Example 2.3, the CDF of the Weibull distribution is given as Equation (2.13b). Let $\alpha_1 = a$, $\alpha_2 = b$. Then, Equation (2.13b) can be rewritten as follows:

$$F(x;\alpha_1,\alpha_2) = 1 - \exp\left(-\left(\frac{x}{\alpha_2}\right)^{\alpha_1}\right) \tag{2.42}$$

Then, x can be expressed analytically through F as follows:

$$x = \alpha_2(-\ln(1-F))^{\frac{1}{\alpha_1}} \tag{2.42a}$$

$$a_0 = M_{1,0,0} = \int_0^1 x dF = \int_0^1 \alpha_2(-\ln(1-F))^{\frac{1}{\alpha_1}} dF \tag{2.42b}$$

With simple algebra, Equation (2.42b) may be integrated analytically as follows:

$$a_0 = M_{1,0,0} = \alpha_2 \Gamma\left(1 + \frac{1}{\alpha_1}\right) \tag{2.42c}$$

Similarly, we may solve for a_1 analytically as follows:

$$a_1 = M_{1,01} = \int_0^1 x(1-F)dF = \int_0^1 \alpha_2(-\ln(1-F))^{\frac{1}{\alpha_1}}(1-F)dF = \alpha_2 \Gamma\left(1+\frac{1}{\alpha_1}\right)\Big/2^{1+\frac{1}{\alpha_1}} \tag{2.42d}$$

Replacing a_0, a_1 with the sample estimates $\hat{a}_0$, $\hat{a}_1$, we can analytically solve Equations (2.42c) and (2.42d) simultaneously as follows:

$$\hat{\alpha}_1 = \frac{\ln 2}{\ln\left(\frac{\hat{\alpha}_0}{2\hat{\alpha}_1}\right)}; \quad \hat{\alpha}_2 = \frac{\hat{\alpha}_0}{\Gamma\left(\frac{\ln\left(\frac{\hat{\alpha}_0}{\hat{\alpha}_1}\right)}{\ln 2}\right)} \tag{2.43}$$

Compared with **Example 2.3**, it is seen that one may estimate the parameters analytically using PWM; however, this is not the case if MOM is applied to estimate the parameters for the Weibull distribution.

2.3.4 Method of L-Moments

Hosking (1990) developed the method of L-moments, which is simpler than the method of PWMs. He defined L-moments as linear combinations of probability-weight moments as follows:

$$\lambda_{r+1} = \sum\nolimits_{k=0}^{r} p_{r,k}^* \beta_k; a_k = \sum\nolimits_{k=0}^{r} p_{r,k}^* b_k \tag{2.44}$$

where $p_{r,k}^* = (-1)^{r-k}\binom{r}{k}\binom{r+k}{k}$; λ_1 is the mean of the distribution, a measure of location; λ_2 is a measure of scale; λ_3 is a measure of skewness; and λ_4 is a measure of kurtosis. In particular,

$$\begin{aligned}
\lambda_1 &= a_0 && = b_1 \\
\lambda_2 &= a_0 - 2a_1 && = 2b_1 - b_0 \\
\lambda_3 &= a_0 - 6a_1 + 6a_2 && = 6b_2 - 6b_1 + b_0 \\
\lambda_4 &= a_0 - 12a_1 + 30a_2 - 20a_3 && = 20b_3 - 30b_2 + 12b_1 - b_0
\end{aligned} \tag{2.44a}$$

The L-moment ratios are identified by $L - C_V; L - C_s; L - C_K$, respectively, and can be computed by the following:

$$\tau_2 = \frac{\lambda_2}{\lambda_1},\ \tau_3 = \frac{\lambda_3}{\lambda_2},\ \tau_4 = \frac{\lambda_4}{\lambda_2} \tag{2.45}$$

In practice, the L-moment ratios can be estimated for a given sample $x_1, \ \ldots, x_n$ of sample size n. Let $x_{(1)} \leq \cdots \leq x_{(n)}$ be arranged in ascending order. Define

$$l_{r+1} = \sum\nolimits_{k=0}^{r} p_{r,k}^{*} b_k \tag{2.46}$$

$$b_r = \frac{1}{n} \Sigma_{j=r+1}^{n} \frac{(j-1)(j-2)\cdots(j-r)}{(n-1)(n-2)\cdots(n-r)} x_j \tag{2.47}$$

where l_r is an unbiased estimator of λ_r,

$$b_0 = \frac{1}{n} \sum\nolimits_{j=1}^{n} x_j \tag{2.48}$$

$$b_1 = \frac{1}{n} \sum\nolimits_{j=2}^{n} \frac{j-1}{n-1} x_j \tag{2.49}$$

$$b_2 = \frac{1}{n} \sum\nolimits_{j=3}^{n} \frac{(j-1)(j-2)}{(n-1)(n-2)} x_j \tag{2.50}$$

$$b_3 = \frac{1}{n} \sum\nolimits_{j=4}^{n} \frac{(j-1)(j-2)(j-3)}{(n-1)(n-2)(n-3)} x_j \tag{2.51}$$

$$b_4 = \frac{1}{n} \sum\nolimits_{j=5}^{n} \frac{(j-1)(j-2)(j-3)(j-4)}{(n-1)(n-2)(n-3)(n-4)} x_j \tag{2.52}$$

$$l_1 = b_0 \tag{2.53}$$

$$l_2 = 2b_1 - b_0 \tag{2.54}$$

$$l_3 = 6b_2 - 6b_1 + b_0 \tag{2.55}$$

$$l_4 = 20b_3 - 30b_2 + 12b_1 - b_0 \tag{2.56}$$

Estimator t_r of τ_r is

$$t = \frac{l_2}{l_1} \tag{2.57}$$

$$t_r = \frac{l_r}{l_2},\ r = 3, 4, 5 \ldots \tag{2.58}$$

Example 2.7 **Estimate the parameters of the normal distribution by the L-moment method.**

Solution: The PDF of a normal distribution is given as Equation (2.1). Hosking (1990) gives the following properties of the normal distribution:

The first order L-moment equals the population mean of normal distribution as follows:

$$\lambda_1 = \beta_0 = \alpha_1 \tag{2.59}$$

The second-order L-moment relates to the standard deviation of normal distribution as follows:

$$\lambda_2 = 2\beta_1 - \beta_0 = \alpha_2/\sqrt{\pi} \tag{2.60}$$

$L\text{-}C_s$ equals the skewness of the normal distribution (i.e., skewness = 0), which leads to the third L-moment of normal distribution equal to 0 as follows:

$$\tau_3 = \frac{\lambda_3}{\lambda_2} = 0,\ or\ \lambda_3 = 0 \tag{2.61}$$

$L\text{-}C_K$ relates to the kurtosis of normal distribution, and $L\text{-}C_K$ of the normal distribution is a constant, as follows:

$$\tau_4 = \frac{\lambda_4}{\lambda_2} = \frac{30}{\pi} \tan^{-1}\sqrt{2} - 9 = 0.1226 \tag{2.62}$$

The parameter estimates by the method of L-moments can be given in terms of sample L-moments as follows:

$$\begin{cases} \hat{\alpha}_1 = l_1 \\ \hat{\alpha}_2 = \pi l_2^2 \end{cases} \tag{2.63}$$

2.4 Goodness-of-Fit Measures for Probability Distributions

To ensure the appropriateness of the selected univariate/bivariate (multivariate) distributions, it is usually recommended to apply formal goodness-of-fit statistical measures. Here we will briefly introduce the goodness-of-fit measures for both univariate and conventional bivariate probability distributions.

2.4.1 Goodness-of-Fit Measures for Univariate Probability Distributions

Let $X = \{x_1, \ldots, x_n\}$ be the IID random variable following the true probability distribution F. For a fitted distribution $\{\hat{F}(x; \hat{\boldsymbol{\alpha}}), \hat{\boldsymbol{\alpha}}\text{: fitted parameters}\}$ to random variable X, its goodness-of-fit may be expressed by testing the null hypothesis of $H_0 : F = \hat{F}$ versus the alternative $H_1 : F \neq \hat{F}$.

For testing, there are a number of formal goodness-of-fit statistics through measuring the distance between empirical CDF $[F_n(x)]$ and fitted parametric CDF $[\hat{F}(x;\hat{\boldsymbol{\alpha}})]$. These include Kolmogorov–Smirnov (KS) statistic D_N (Kolmogorov, 1933; Smirnov, 1948), Cramér–von Mises (CM) statistic W_N^2 (Cramér, 1928; von Mises, 1928), Anderson–Darling (AD) statistics A_N^2 (Anderson and Darling, 1952), modified weighted Watson statistic U_N^2 (Stock and Watson, 1989), and Liao and Shimokawa statistic L_N (Liao and Shimokawa, 1999). Also commonly applied is the chi-square goodness-of-fit test, which measures the difference between empirical frequency and the frequency computed from the fitted parametric distribution.

Kolmogorov–Smirnov (KS) Statistic D_N

The KS test statistic can be expressed theoretically as follows:

$$D_N = \sup_{x\in\mathbb{R}} |F_n(x) - F(x)| \tag{2.64}$$

where $F_n(x)$ is the fitted distribution estimated as n/N, and n is the cumulative number of sample events at class limit n. Applying the fitted distribution function $\hat{F}(x;\hat{\boldsymbol{\alpha}})$, Equation (2.64) can be rewritten as follows:

$$D_N = \max\left(\hat{\delta}_i\right),\ \hat{\delta}_i = \max\left[\frac{i}{N} - \hat{F}(x_i;\hat{\boldsymbol{\alpha}}), \hat{F}(x_i;\hat{\boldsymbol{\alpha}}) - \frac{i-1}{N}\right],\ i \in [1, N] \tag{2.64a}$$

Cramér–von Mises (CM) Statistic W_N^2

The CM test statistic can be expressed theoretically as follows:

$$W^2 = \int_{-\infty}^{\infty} [F_n(x) - F(x)]^2 dF(x) \tag{2.65}$$

Applying the fitted probability distribution $\hat{F}(x;\hat{\boldsymbol{\alpha}})$, Equation (2.65) can be rewritten as follows:

$$W_N^2 = \frac{1}{12N} + \sum_{i=1}^{N}\left[\hat{F}(x_i;\hat{\boldsymbol{\alpha}}) - \frac{2i-1}{2N}\right]^2 \tag{2.65a}$$

Anderson–Darling (AD) Statistic A_N^2

The AD test statistic can be expressed theoretically as follows:

$$A = n\int_{-\infty}^{\infty} \frac{(F_n(x) - F(x))^2}{F(x)(1 - F(x))} dF(x) \tag{2.66}$$

Applying the fitted probability distribution $\hat{F}(x;\hat{\boldsymbol{\alpha}})$, Equation (2.66) can be rewritten as follows:

$$A_N^2 = -N - \frac{1}{N}\sum_{i=1}^{N} (2i-1)\ln\left(\hat{F}(x_i;\hat{\boldsymbol{\alpha}})\left(1 - \hat{F}(x_{n+1-i};\hat{\boldsymbol{\alpha}})\right)\right) \tag{2.66a}$$

Modified Weighted Watson Statistic U_N^2

Applying the fitted probability distribution $\hat{F}(x;\hat{\boldsymbol{\alpha}})$, the U_N^2 test statistic can be expressed as follows:

$$U_N^2 = N^2\sum_{i=1}^{N} d_i^2 - N\left(\sum_{i=1}^{N} d_i\right)^2; \quad d_i = \frac{\hat{F}(x_i;\hat{\boldsymbol{\alpha}}) - \dfrac{i}{N+1}}{\sqrt{i(N-i+1)}} \tag{2.67}$$

Liao and Shimokawa Statistic L_N

Applying the fitted probability distribution $\hat{F}(x;\hat{\boldsymbol{\alpha}})$, the L_N test statistic can be expressed as follows:

$$L_N = \frac{1}{\sqrt{N}} + \sum_{i=1}^{N} \frac{\max\left[\dfrac{i}{N} - \hat{F}(x_i;\hat{\boldsymbol{\alpha}}), \hat{F}(x_i;\hat{\boldsymbol{\alpha}}) - \dfrac{i-1}{N}\right]}{\sqrt{\hat{F}(x_i;\hat{\boldsymbol{\alpha}})\left(1-\hat{F}(x_i;\hat{\boldsymbol{\alpha}})\right)}} \tag{2.68}$$

In Equations (2.67) and (2.68), N is the sample size.

Conventionally, the P-value of the preceding statistics is computed using the limiting probability distribution for each specific test statistic. To avoid the misidentification of the limiting probability distribution, the parametric bootstrap simulation method is widely applied to estimate the P-value with the following procedure:

1. Estimate the parameter vector $\hat{\boldsymbol{\alpha}}$ of the probability distribution $\hat{F}(x_i;\boldsymbol{\alpha})$.
2. Compute the test statistics of D_N, W_N^2, A_N^2, U_N^2, L_N.
3. With a larger number of M, for $k = 1 : M$ to proceed, follow these steps:
   a. Generate random variable $x^{(k)}$ with sample size N from the fitted probability distribution $\hat{F}(x_i;\hat{\boldsymbol{\alpha}})$.
   b. Reestimate the parameter vector $\hat{\boldsymbol{\alpha}}^*$ from the hypothesized distribution using the random sample generated from step a.
   c. Compute the test statistics of D_N^*, W_N^{2*}, A_N^{2*}, U_N^{2*}, L_N^* by following steps a and b.
   d. Repeat the steps a–c M times.
4. Compute the P-value using the following:

$$P_{value} = \frac{\sum_{i=1}^{M} 1\left(D_N^*(i) > D_N\right)}{M} \tag{2.69}$$

Replacing $\left(D_N^*, D_N\right)$ by $\left(W_N^{2*}, W_N^2\right)$, $\left(A_N^{2*}, A_N^2\right)$, $\left(U_N^{2*}, U_N^2\right)$, $\left(L_N^*, L_N\right)$ in Equation (2.69), we can simulate the P-values for other statistics.

From common practice, we may set $\alpha_{level} = 0.05$, which means the hypothesized parametric univariate distribution cannot be rejected if $P_{value} \geq 0.05 = \alpha_{level}$. Furthermore, the larger the M, the closer the simulated P-value to its true P-value.

Example 2.8 **Using the observed annual peak streamflow given in Table 2.1, compute the goodness-of-fit with the use of KS, CM, AD, modified weighted Watson, Liao, and Shimokawa tests, given the gamma distribution as the tested probability distribution.**

Table 2.1. *Observed annual peak streamflow.*

No.	Peak (cfs)	No.	Peak (cfs)
1	2,300	26	4,730
2	3,390	27	1,060
3	1,710	28	3,290
4	9,780	29	7,880
5	10,500	30	13,800
6	13,700	31	10,500
7	6,500	32	7,150
8	3,710	33	1,030
9	536	34	13,100
10	17,000	35	2,920
11	6,630	36	5,210
12	1,220	37	4,460
13	4,980	38	3,100
14	2,840	39	1,520
15	3,220	40	29,800
16	2,440	41	2,740
17	1,320	42	1,740
18	16,000	43	557
19	16,100	44	5,350
20	1,180	45	11,200
21	5,440	46	4,930
22	2,420	47	3,490
23	9,140	48	2,990
24	6,700	49	6,160
25	912	50	1,480
		51	496

Solution:

Gamma distribution is given as follows:

$$f(x;\alpha,\beta) = \frac{1}{\beta^{\alpha}\Gamma(\alpha)}x^{\alpha-1}\exp\left(-\frac{x}{\beta}\right)$$

Following the test procedures given previously, the following steps are needed for the goodness-of-fit test calculations.

Step 1: Order the streamflow values in increasing order and estimate the parameters for the probability distribution (as shown in Table 2.2). In Table 2.2, the parameters of gamma distribution are estimated using the MLE.

Step 2: Compute the corresponding test statistics:

1. Table 2.3 lists the CDF computed from increasingly ordered annual peak streamflow data for the fitted gamma distribution.
2. Compute the test statistics. The computation example is using $Q_{(1)} = 496$ cubic feet per second (cfs) for a sample size of $N = 51$. The full list of the computation is given in Table 2.3.

Table 2.2. *Ordered annual peak streamflow and parameter estimated with MLE.*

Order	Peak (cfs)	Order	Peak (cfs)
1	496	26	3,710
2	536	27	4,460
3	557	28	4,730
4	912	29	4,930
5	1,030	30	4,980
6	1,060	31	5,210
7	1,180	32	5,350
8	1,220	33	5,440
9	1,320	34	6,160
10	1,480	35	6,500
11	1,520	36	6,630
12	1,710	37	6,700
13	1,740	38	7,150
14	2,300	39	7,880
15	2,420	40	9,140
16	2,440	41	9,780
17	2,740	42	10,500
18	2,840	43	10,500
19	2,920	44	11,200
20	2,990	45	13,100
21	3,100	46	13,700
22	3,220	47	13,800
23	3,290	48	16,000
24	3,390	49	16,100
25	3,490	50	17,000
		51	29,800

Parameters: $\alpha = 1.3164$, $\beta = 4.4737 \times 10^3$.

Table 2.3. *CDF and corresponding statistics computed for the ordered annual peak streamflow.*

			Test statistics				
Order	Peak (cfs)	CDF	KS $\left(\hat{\delta}_i\right)$	CM (CMd_i)	AD (ADd_i)	U_N^2 (d_i)	L_N (Ld_i)
1	496	0.0441	0.0441	0.0012	–9.0297	0.0035	0.2147
2	536	0.0486	0.0290	0.0004	–18.6652	0.0010	0.1347
3	557	0.0510	0.0117	3.73E-06	–29.9324	–0.0006	0.0534
4	912	0.0933	0.0345	0.0006	–37.5253	0.0012	0.1185
5	1,030	0.1079	0.0295	0.0004	–42.8437	0.0008	0.0950
6	1,060	0.1117	0.0136	1.46E-05	–51.7633	–0.0002	0.0433
7	1,180	0.1267	0.0105	5.38E-07	–57.9304	–0.0004	0.0317
8	1,220	0.1318	0.0251	0.0002	–60.4520	–0.0012	0.0742
9	1,320	0.1444	0.0321	0.0005	–64.5542	–0.0015	0.0913
10	1,480	0.1646	0.0315	0.0005	–69.6576	–0.0014	0.0849
11	1,520	0.1697	0.0460	0.0013	–73.3189	–0.0020	0.1226
12	1,710	0.1936	0.0417	0.0010	–74.3267	–0.0017	0.1055
13	1,740	0.1974	0.0575	0.0023	–74.0839	–0.0023	0.1445
14	2,300	0.2665	0.0116	3.37E-06	–68.0529	–0.0001	0.0263
15	2,420	0.2810	0.0131	1.11E-05	–69.0385	–0.0003	0.0292
16	2,440	0.2834	0.0304	0.0004	–73.1189	–0.0010	0.0674
17	2,740	0.3187	0.0146	2.34E-05	–73.1350	–0.0003	0.0314
18	2,840	0.3302	0.0227	0.0002	–74.0442	–0.0006	0.0483
19	2,920	0.3393	0.0332	0.0005	–72.2151	–0.0010	0.0701
20	2,990	0.3473	0.0449	0.0012	–74.5592	–0.0015	0.0943
21	3,100	0.3596	0.0522	0.0018	–75.8782	–0.0017	0.1087
22	3,220	0.3728	0.0585	0.0024	–76.1786	–0.0020	0.1211
23	3,290	0.3805	0.0705	0.0037	–78.3911	–0.0024	0.1452
24	3,390	0.3913	0.0793	0.0048	–78.8187	–0.0027	0.1626
25	3,490	0.4019	0.0883	0.0062	–78.4213	–0.0030	0.1801
26	3,710	0.4248	0.0850	0.0056	–71.8666	–0.0029	0.1719
27	4,460	0.4979	0.0315	0.0005	–64.2055	–0.0008	0.0631
28	4,730	0.5222	0.0268	0.0003	–63.0316	–0.0006	0.0537
29	4,930	0.5396	0.0290	0.0004	–62.4556	–0.0007	0.0582
30	4,980	0.5439	0.0444	0.0012	–63.4603	–0.0013	0.0891
31	5,210	0.5630	0.0448	0.0012	–62.2242	–0.0013	0.0904
32	5,350	0.5743	0.0531	0.0019	–61.8107	–0.0016	0.1074
33	5,440	0.5815	0.0656	0.0031	–62.1859	–0.0021	0.1329
34	6,160	0.6349	0.0318	0.0005	–57.2926	–0.0008	0.0660
35	6,500	0.6579	0.0284	0.0003	–55.3675	–0.0006	0.0598
36	6,630	0.6664	0.0395	0.0009	–52.4782	–0.0011	0.0838
37	6,700	0.6708	0.0547	0.0020	–53.2252	–0.0017	0.1163

Table 2.3. (*cont.*)

Order	Peak (cfs)	CDF	Test statistics KS $(\hat{\delta}_i)$	CM (CMd_i)	AD (ADd_i)	U_N^2 (d_i)	L_N (Ld_i)
38	7,150	0.6983	0.0468	0.0014	–50.1841	–0.0014	0.1020
39	7,880	0.7383	0.0264	0.0003	–40.2878	–0.0005	0.0600
40	9,140	0.7960	0.0313	0.0005	–35.0231	0.0012	0.0777
41	9,780	0.8205	0.0362	0.0007	–31.0876	0.0015	0.0942
42	10,500	0.8446	0.0407	0.0010	–28.9422	0.0018	0.1124
43	10,500	0.8446	0.0211	0.0001	–27.6059	0.0009	0.0583
44	11,200	0.8651	0.0220	0.0001	–24.8973	0.0010	0.0643
45	13,100	0.9084	0.0457	0.0013	–20.6088	0.0024	0.1583
46	13,700	0.9190	0.0367	0.0007	–18.4602	0.0021	0.1344
47	13,800	0.9207	0.0187	7.92E-05	–18.3082	0.0011	0.0692
48	16,000	0.9497	0.0281	0.0003	–14.2122	0.0019	0.1284
49	16,100	0.9507	0.0101	8.56E-08	–9.9778	0.0007	0.0466
50	17,000	0.9591	0.0213	0.0001	–9.0618	–0.0002	0.1075
51	29,800	0.9973	0.0169	5.02E-05	–4.8272	0.0023	0.3244

- KS test (D_N) (Equation (2.64)), the distance for $i = 1$ is computed as follows:

$$\hat{\delta}_1 = \max\left[\frac{1}{N} - F\left(x_{(1)}\right), F\left(x_{(1)}\right) - \frac{1-1}{N}\right] = \max\left(\frac{1}{59} - 0.0441, 0.0441 - 0\right) = 0.0441$$

- CM test (W_N^2) (Equation (2.65)): the quantity inside of summation (i.e., CMd_i) for $I = 1$ is computed as follows:

$$CMd_1 = \left[F\left(x_{(1)}\right) - \frac{2(1)-1}{2N}\right]^2 = \left[0.0441 - \frac{1}{2(51)}\right]^2 = 0.0012$$

- AD test (A_N^2) (Equation (2.66)): the quantity inside of the summation (i.e., ADd_i) for $i = 1$ is computed as follows:

$$ADd_1 = (2(1)-1)\ln\left(F\left(x_{(1)}\left(1 - F\left(x_{(1)}\right)\right)\right) = \ln(0.0441(1-0.0441)) = -3.1670$$

- Modified weighted Watson test (U_N^2) (Equation (2.67)): the quantity inside of the summation for $i = 1$ is computed as follows:

$$d_1 = \frac{F\left(x_{(1)}\right) - \dfrac{1}{N+1}}{\sqrt{1(N-1+1)}} = \frac{0.0441 - \dfrac{1}{52}}{\sqrt{51}} = 0.0035$$

- Liao and Shimokawa test (L_N) (Equation (2.68)): the quantity inside of the summation (i.e., Ld_i) for $i = 1$ is computed as follows:

$$Ld_1 = \frac{\max\left(\frac{1}{N} - F(x_{(1)}), F(x_{(1)}) - \frac{1-1}{N}\right)}{\sqrt{F(x_{(1)})(1 - F(x_{(1)})}} = \max \frac{\left(\frac{1}{51} - 0.0441, 0.0441\right)}{\sqrt{0.0441(1 - 0.0441)}} = 0.2147$$

Now, substituting the quantities computed in Table 2.3 back into Equations (2.64)–(2.68), we can calculate the final test statistics for each goodness-of-fit test as follows:

KS test: $D_N = 0.0883$

CM test: $W_N^2 = 0.0558$

AD test: $A_N^2 = 0.3534$

Modified weighted Watson test: $U_N^2 = 0.2993$

Liao and Shimokawa test: $L_N = 5.1695$

3. Apply the parametric bootstrap method M times to approximate the P-value with given significance level α. Here we choose $M = 1{,}000$ and $\alpha = 0.05$. To illustrate the procedure, we will use one parametric bootstrap simulation as an example:
   a. Generate IID streamflow from the fitted gamma distribution (with parameters given in Table 2.2 of sample size $N = 51$), and sort the simulated streamflow values in increasing order (Table 2.4).
   b. Reestimate the parameters of gamma distribution and calculate the CDF and corresponding test statistics using the simulated streamflow. We have discussed how to compute the test statistics previously (steps 1 and 2), here we will only present the final results:
      i. Estimated parameters: $\alpha_1^* = 1.3241$, $\beta_1^* = 4.8206 \times 10^3$.
      ii. Test statistics computed from simulated streamflow with reestimated parameters:

$$D_{N1}^* = 0.1400; W_{N1}^{2*} = 0.1496; A_{N1}^{2*} = 0.8237; U_{N1}^{2*} = 0.6595; L_{N1}^* = 6.8445.$$

   c. Repeat the parametric bootstrap simulation 1,000 times. We can approximate the P-value and corresponding critical value using the KS test as an example:

$$P\text{-}value = \frac{\sum_{i=1}^{M} 1(D_{Ni}{}^* > D_N)}{M}$$

The critical value can be approximated by interpolation from computed D_{Ni}^*, $i = 1, \ldots, M$ and its empirical distribution.

KS test final result:

$$D_N = 0.0883,\ P = 0.222,\ Cri_{value} = 0.1156.$$

CM test final results:

$$W_N^2 = 0.0558,\ P = 0.456,\ Cri_{value} = 0.1327.$$

AD test final results:

$$A_N^2 = 0.3534,\ P = 0.489,\ Cri_{value} = 0.7549.$$

Table 2.4. *Generating gamma distributed streamflows and sorting in increasing order.*

No.	Generated	Order	Sorted
1	8,683.20	1	51.56
2	921.76	2	127.24
3	7,874.64	3	574.63
4	10,470.50	4	766.02
5	3,019.36	5	872.26
6	5,625.04	6	921.76
7	1,548.26	7	1,317.86
8	7,719.17	8	1,411.29
9	15,787.45	9	1,548.26
10	1,592.99	10	1,592.99
11	19,530.55	11	2,007.60
12	12,160.63	12	2,193.47
13	1,411.29	13	2,194.08
14	13,026.83	14	2,431.96
15	8,385.82	15	2,801.57
16	3,906.03	16	3,019.36
17	9,190.72	17	3,282.55
18	8,067.79	18	3,643.24
19	8,948.61	19	3,752.08
20	11,060.80	20	3,906.03
21	2,431.96	21	4,003.93
22	1,317.86	22	4,407.35
23	2,194.08	23	4,895.35
24	5,589.25	24	5,589.25
25	3,643.24	25	5,625.04
26	1,2416.01	26	6,351.30
27	872.26	27	6,756.52
28	4,003.93	28	7,025.33
29	3,752.08	29	7,581.81
30	6,756.52	30	7,719.17
31	12,419.87	31	7,789.85
32	9,953.94	32	7,874.64
33	10,547.60	33	8,067.79
34	4,895.35	34	8,329.23
35	13,512.85	35	8,385.82
36	2,193.47	36	8,683.20
37	51.56	37	8,872.19
38	7,025.33	38	8,948.61
39	574.63	39	9,190.72
40	8,329.23	40	9,953.94

Table 2.4. (*cont.*)

No.	Generated	Order	Sorted
41	4,407.35	41	10,131.30
42	7,581.81	42	10,470.50
43	127.24	43	10,547.60
44	2,801.57	44	11,060.80
45	7,789.85	45	12,160.63
46	2,007.60	46	12,416.01
47	766.02	47	12,419.87
48	10,131.30	48	13,026.83
49	6,351.30	49	13,512.85
50	8,872.19	50	15,787.45
51	3,282.55	51	19,530.55

Modified weighted Watson test final results:

$$U_N^2 = 0.2993,\ P = 0.532,\ Cri_{value} = 0.6821.$$

Liao and Shimokawa test final results:

$$L_N = 5.1695,\ P = 0.438,\ Cri_{value} = 6.9574.$$

Chi-Square Goodness-of-Fit Test

Rather than measuring the difference between the empirical CDF and the fitted parametric CDF, the chi-square goodness-of-fit test deals with the frequency directly. As its name indicates, the limiting distribution is the chi-square distribution with its statistic expressed as follows:

$$\chi^2_{K-m-1} = \sum\nolimits_{i=1}^{k} \frac{(o_i - e_i)^2}{e_i} \tag{2.70}$$

In Equation (2.70), o_i is the observed frequency count for the level-i of a variable; e_i is the corresponding expected frequency count from the fitted probability distribution; K is the number of levels of the random variable; m is the number of the parameters of the fitted probability distribution, and K-m-1 is the degree of freedom of the limiting chi-square distribution. In other words, Equation (2.70) is actually comparing the relative frequency computed from a histogram with K-bins to the fitted parametric distribution, i.e., (1) level-i is equivalent to the bin-i of the histogram and (2) number of level K is equivalent to the total number of bins (K) of the histogram.

The simplest rule of thumb to determine the number of bins for a histogram is given as follows:

$$K = \lceil 1 + \log_2 n \rceil \tag{2.71}$$

Example 2.9 **Rework Example 2.8 with the chi-square goodness-of-fit test.**

Solution:

Step 1: To apply the chi-square goodness-of-fit study, we will first study the frequency histogram.

Applying Equation (2.71), we obtain the number of bins for the frequency histogram as follows:

$k = \lceil 1 + \log_2 51 \rceil = 7$. The observed relative frequency is shown in Figure 2.1 and Table 2.5.

Table 2.5. *Relative frequency and corresponding data range.*

Relative frequency (observed)	Data interval	Estimated frequency computed from fitted gamma distribution
0.5294	[496, 4682.2857]	0.4739
0.2353	[4682.2857, 8868.5714]	0.2667
0.0980	[8868.5714, 13054.8571]	0.1228
0.1176	[13054.8571, 17241.1429]	0.0536
0	[17241.1429, 21427.4286]	0.0227
0	[21427.4286, 25613.7143]	0.0095
0.0196	[25613.7143, 29800]	0.0039

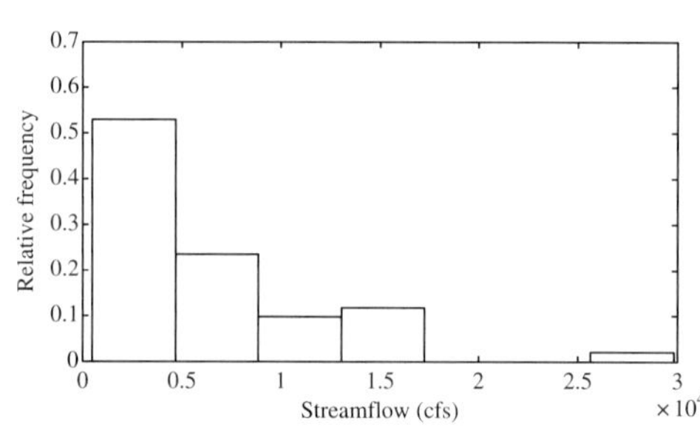

Figure 2.1 Relative frequency plot.

Step 2: Compute the estimated frequency with the fitted gamma distribution (parameters listed in Table 2.2) to compute the frequency of the corresponding data interval in Table 2.5). Using data interval of [496, 4682.2857], we have the following:

$$e_1 = F(4682.2857; 1.3164, 4.4737 \times 10^3) - F(496; 1.3164, 4.4737 \times 10^3) = 0.4739.$$

The rest of the results are listed in Table 2.5.

Step 3: Computing test statistics using Equation (2.70), we have the following:

$$Statistics = 0.1867.$$

From the chi-square goodness-of-fit, we know the test statistics should follow the chi-square distribution with the degree of freedom, i.e., $d.o.f. = K - m - 1 = 7 - 2 - 1 = 4$.

Choosing the significance level $\alpha = 0.05$, we can calculate the corresponding critical value as follows:

$$cri_{value} = \chi_4^{2(-1)}(0.95) = 9.4877.$$

2.4.2 Goodness-of-Fit Measures for Bivariate Probability Distributions

In this section, we briefly discuss two popular goodness-of-fit measures, both of which are based on the Rosenblatt transform (Rosenblatt, 1952). The Rosenblatt transform states that a bivariate random variable $\boldsymbol{Z} = [\boldsymbol{X}, \boldsymbol{Y}]$ may be modeled by the fitted joint distribution function of $\hat{F}_{X,Y}(x, y; \hat{\boldsymbol{\theta}})$. Let

$$T_1 = \hat{F}_X(x),\ T_2 = \hat{F}_{Y|X=x}(y|x). \tag{2.72}$$

Based on the Bayes theorem, the joint distribution function $\hat{F}_{X,Y}(x, y)$ may be expressed as follows:

$$\hat{F}_{X,Y}(x, y; \hat{\boldsymbol{\theta}}) = \hat{F}_X(x)\hat{F}_{Y|X=x}(y|x) = T_1 T_2 \tag{2.73}$$

In Equations (2.72) and (2.73), T_1, T_2 are independent and following a uniform distribution; $\hat{F}_X(x)$ is the fitted distribution of random variable X; and $\hat{F}_{Y|X=x}$ is the conditional distribution derived from the fitted joint distribution $\hat{F}_{X,Y}(x, y)$ and the fitted univariate distribution $\hat{F}_X(x)$.

Chi-Square Goodness-of-Fit Test

As stated in Rosenblatt (1952), the chi-square goodness-of-fit test for the univariate distribution can be extended to evaluate the goodness-of-fit for the multivariate distribution. The null hypothesis $\boldsymbol{T} = [T_1, T_2]$ is from the distribution on a unit square $[0, 1]^2$, if the hypothesized joint distribution is proper. Dividing the unite square into N^2 cells, the chi-square test may be generated as follows (Rosenblatt, 1952):

i. Define cell C_{j_1, j_2} as:

$$C_{j_1, j_2} = \left\{ T | \frac{j_i}{N} < t_i < \frac{j_i + 1}{N}; i = 1, 2 \right\} \tag{2.74}$$

where $j_1, j_2 \in [0, \ldots, N - 1]$, with each cell having the same probability mass as $1/N^2$.

ii. Let v_{j_1, j_2} be the number of T_i in cell C_{j_1, j_2}, the chi-square test statistics may be calculated to evaluate whether $\mathbf{Z} = [X, Y]$ may be drawn from the fitted distribution $\hat{F}_{X, Y}(x, y\hat{\boldsymbol{\theta}})$ using the following:

$$\chi^2 = \frac{\sum_{i=1}^{n} \left(v_{j_1, j_2} - \frac{n}{N^2} \right)^2}{\left(\frac{n}{N^2} \right)} \tag{2.75}$$

The test statistic computed using Equation (2.75) should follow the chi-square distribution with the degrees of freedom of $(N - 1)^2$.

Bivariate (Multivariate) KS Goodness-of-Fit Test

As with the univariate KS goodness-of-fit test, the bivariate KS goodness-of-fit test measures the distance of empirical joint distribution $F_n(x, y)$ from its true joint distribution $F_{X, Y}(x, y)$, expressed as follows:

$$D_N = \sup_{(x,y) \in \mathbb{R}^2} (|F_n(x, y) - F_{X, Y}(x, y)| \tag{2.76}$$

Applying the Rosenblatt transform to the fitted joint distribution (i.e., Equations (2.72) and (2.73)), the test of Equation (2.76) is equivalent to the following test:

$$D_N = \sup_{(x,y) \in \mathbb{R}^2} |G_n(T_1, T_2) - T_1 T_2| \tag{2.77}$$

where G_n is the empirical distribution of the transformed variables.

Given T_1, T_2 being independent random variables, the null hypothesis of $F_{X, Y} = \hat{F}_{X, Y}(x, y; \hat{\boldsymbol{\theta}})$ is equivalent to that of $F_{T_1, T_2} = T_1 \perp T_2 = \Pi(T_1, T_2)$.

To assess Equation (2.77), Justel et al. (1994) proposed the permutation method. One may also apply the same parametric bootstrap method as that for univariate analysis to approximate the P-value of the test statistic discussed for the univariate goodness-of-fit test.

Example 2.10 **Assess the goodness-of-fit for the bivariate data listed in Table 2.6, given that the data may be modeled with bivariate normal distribution (true population mean and population covariance matrix given as follows:** $\boldsymbol{\mu} = \begin{bmatrix} 100 \\ 1000 \end{bmatrix}$**; COV** $= \begin{bmatrix} 400 & 560 \\ 560 & 1600 \end{bmatrix}$**).**

Table 2.6. *Bivariate sample dataset.*

No.	X	Y	No.	X	Y
1	119	1,033	26	106	1,034
2	106	1,021	27	95	993
3	103	993	28	109	1,011
4	110	1,008	29	108	1,037
5	105	1,065	30	75	982
6	81	909	31	81	983
7	97	1,059	32	85	1,015
8	97	1,006	33	90	1,012
9	89	1,014	34	94	998
10	134	1,000	35	100	981
11	82	959	36	39	897
12	90	979	37	91	1,021
13	86	992	38	125	989
14	77	919	39	79	969
15	96	1,008	40	119	970
16	95	958	41	107	1,039
17	131	1,045	42	99	1,024
18	95	1,012	43	104	1,005
19	79	980	44	69	954
20	132	1,076	45	98	927
21	125	1,063	46	132	1,062
22	95	975	47	102	940
23	70	965	48	101	935
24	91	961	49	85	982
25	97	958	50	99	972

Solution: Applying the Rosenblatt transform (Equations (2.72) and (2.73)), we can compute T_1 and T_2 directly from the fitted bivariate normal distribution as follows:

$$\hat{T}_1 \sim N(x100, 400) \tag{2.78a}$$

$$\hat{T}_2 \sim N\left(y|X = x, 1,000 + \frac{560}{400}(x - 100), 1,600 - \frac{560^2}{400}\right) \tag{2.78b}$$

Table 2.7 lists the estimated $\hat{T}_1$, $\hat{T}_2$from Equations (2.78a) and (2.78b).

Table 2.7. *Estimated* $\hat{T}_1$, $\hat{T}_2$ from the bivariate normal distribution.

X	$\hat{T}_1$	Y	$\hat{T}_2$	X	$\hat{T}_1$	Y	$\hat{T}_2$
119	0.829	1,033	0.589	106	0.618	1,034	0.815
106	0.618	1,021	0.670	95	0.401	993	0.500

Table 2.7. (*cont.*)

X	$\hat{T}_1$	Y	$\hat{T}_2$	X	$\hat{T}_1$	Y	$\hat{T}_2$
103	0.560	993	0.348	109	0.674	1,011	0.478
110	0.691	1,008	0.417	108	0.655	1,037	0.817
105	0.599	1,065	0.979	75	0.106	982	0.724
81	0.171	909	0.012	81	0.171	983	0.632
97	0.440	1,059	0.987	85	0.227	1,015	0.896
97	0.440	1,006	0.639	90	0.309	1,012	0.819
89	0.291	1,014	0.848	94	0.382	998	0.589
134	0.955	1,000	0.048	100	0.500	981	0.253
82	0.184	959	0.290	39	0.001	897	0.269
90	0.309	979	0.403	91	0.326	1,021	0.880
86	0.242	992	0.658	125	0.894	989	0.054
77	0.125	919	0.044	79	0.147	969	0.478
96	0.421	1,008	0.683	119	0.829	970	0.024
95	0.401	958	0.110	107	0.637	1,039	0.847
131	0.939	1,045	0.522	99	0.480	1,024	0.813
95	0.401	1,012	0.747	104	0.579	1,005	0.492
79	0.147	980	0.629	69	0.061	954	0.464
132	0.945	1,076	0.863	98	0.460	927	0.007
125	0.894	1,063	0.837	132	0.945	1,062	0.726
95	0.401	975	0.264	102	0.540	940	0.014
70	0.067	965	0.597	101	0.520	935	0.010
91	0.326	961	0.178	85	0.227	982	0.542
97	0.440	958	0.093	99	0.480	972	0.176

Chi-Square Test

Applying Equation (2.74), Table 2.8 lists the numbers that fulfill the condition. Here $N = 6$ is chosen for the number of bins for both random variables X and Y. Applying Equation (2.75), we compute the chi-square test statistics as follows: $\chi^2_{test} = 26.64$.

With the chi-square distribution as the limiting distribution (d.f. = 25), we compute the critical value from the chi-square distribution with a significance level of $\alpha = 0.05$ as $\chi^2_{cri} = 37.65$. We obtain $\chi^2_{test} < \chi^2_{cri}$. Equivalently, we compute the P-value of the test statistics as follows:

$$P_{value} = 1 - \chi^2_{CDF}(26.64, 25) = 0.37 > \alpha = 0.05.$$

Thus, we reach the conclusion that the sample dataset listed in Table 2.6 may be modeled with the true population parameters.

Table 2.8. *Pairs of* $(\hat{T}_1, \hat{T}_2)$ within each interval.

	[0, 1/6]	[1/6, 1/3]	[1/3, 1/2]	[1/2, 2/3]	[2/3, 5/6]	[5/6, 1]
[0, 1/6]	1	1	2	2	1	0
[1/6, 1/3]	1	2	1	3	1	3
[1/3, 1/2]	3	3	1	3	3	1
[1/2, 2/3]	2	1	2	0	3	2
[2/3, 5/6]	1	0	2	1	0	0
[5/6, 1]	2	0	0	1	1	2

Bivariate KS goodness-of-fit test To apply the bivariate KS goodness-of-fit test, the hypothesis is that the variables (T_1, T_2) after Rosenblatt transformation are independent. This implies the joint distribution of T_1, T_2 may be simply expressed as $F = T_1T_2$. The KS statistic is computed by comparing the empirical joint distribution of T_1, T_2 with the hypothesized independence assumption.

Similar to the univariate goodness-of-fit test, the KS test statistics is evaluated as $D_n = 0.1684$. With parametric bootstrap simulation (N = 5,000), we obtain the corresponding *P*-value as *P*-value = 0.3140.

Both bivariate chi-square and KS goodness-of-fit tests suggest the data given in Table 2.6 may be sampled from the true population.

2.5 Quantile Estimation

In flood frequency analysis, the return period (T) of a given flood magnitude, called *quantile*, or the flood magnitude corresponding to a given return is needed. The return period is related to the probability of nonexceedance (F) as

$$F = 1 - \frac{1}{T} \; or \; T = \frac{1}{1 - F} \tag{2.78}$$

where $F = F(x_T)$, where x_T (quantile) corresponds to T, that is, the probability of a flood of magnitude smaller than or equal to x_T. If the CDF of a distribution can be expressed as explicitly in closed form, then x_T can be determined directly. Otherwise, it has to be computed numerically. Chow (1954) proposed a general formula for computing x_T as

$$x_T = \bar{x} + K_T\sigma \tag{2.78a}$$

where K_T is the frequency factor, which is a function of the return period and the distribution parameters, and $\bar{x}$ and σ are the mean and standard deviation of the distribution respectively. Chow (1964) has given K_T for different frequency distributions. For the normal distribution, it equals the standard normal variate.

2.6 Confidence Intervals

When estimating the quantile of a given return period, it is important to provide an estimate of the accuracy of the estimate. The accuracy of the estimate depends on the distribution parameters (method of parameter estimation), sample size, and dependence or independence of observed data. The variability of estimated value is measured by the standard error of estimate, which will depend on the distribution in use. There have been many studies that have computed the standard error of estimate of quantile for different distributions. It considers the error due to small sample size but not the error due to the use of an inappropriate distribution. Cunnane (1989) defined the standard error of estimate $s_T$of x_T as

$$S_T = \sqrt{E(x_T - E(x_T))^2} \tag{2.79}$$

where E is the expectation operator. Since s_T varies with the parameter estimation method, each method has its own standard error of estimate, so the method yielding the smallest error is considered the most efficient method. If the sample size n tends to infinity, then the distribution of x_T is asymptotically normal with mean $\overline{x_T}$ and variance s_T^2. Then, an approximate confidence interval $(1-\alpha)$ for $x_T$can be expressed as

$$CI = [\overline{x_T} - t_{\alpha 2} s_T; x_T + t_{\alpha 2} s_T] \tag{2.79a}$$

where t is the standard normal variate. Methods for computing confidence intervals for skewed distributions are available (USWRC, 1981).

2.7 Bias and Root Mean Square Error (RMSE) of Parameter Estimates

Let $\boldsymbol{\theta}$ and $\hat{\boldsymbol{\theta}}$ be the true and estimated parameter of a probability distribution respectively. The bias of the $\hat{\theta}$ with respect to $\boldsymbol{\theta}$ is defined as follows:

$$bias_{\theta}(\hat{\boldsymbol{\theta}}) = E_{x|\boldsymbol{\theta}}(\hat{\boldsymbol{\theta}}) - \boldsymbol{\theta} = E_{x|\boldsymbol{\theta}}(\hat{\boldsymbol{\theta}} - \boldsymbol{\theta}) \tag{2.80a}$$

In Equation (2.80a), the estimates are unbiased if the bias = 0.

In a similar vein, the RMSE of $\hat{\theta}$ with respect to $\boldsymbol{\theta}$ is defined as follows:

$$RMSE = \sqrt{MSE(\hat{\theta})} = \sqrt{E(\hat{\theta} - \boldsymbol{\theta})^2} \tag{2.80b}$$

Equation (2.80b) becomes the standard deviation of the estimator, if the estimator is unbiased.

2.8 Risk Analysis

In general, the probabilistic risk assessment and analysis are composed of two key components: (1) the severity of the possible consequence; and (2) the likelihood (probability) associated with the consequence. In other words, risk may be represented by the probability of loss ranging from [0, 1].

In water resources engineering, risk is one key component to the analysis of extreme events. Conveniently, the return period (i.e., univariate/multivariate) has been applied to represent risk. For example, the annual maximum discharge event with a 100-year return period (i.e., $P(Q > q) = 0.01$), representing the risk of the occurrence of peak discharge roughly about once a 100 year, is commonly used to design the designated infrastructure, such as a levee. The probable maximum precipitation (PMP) is required to analyze classified dams. For urban hydrology, storm events for a given return period are applied for highway drainage design (with different highway categories) and storm sewer (or combined sewer) design. In what follows, the concept of risk, through return period, is briefly reviewed for both univariate and multivariate cases.

2.8.1 Univariate Risk Analysis through Return Period

As discussed previously, the univariate risk may be expressed as the probability of the occurrence of the event of certain magnitude. With the assumption of continuous univariate variable, the risk may be represented as $P(X > x^*)$. For the univariate sequence (i.e., annual sequence or partial duration sequence) under the stationary assumption, the return period of $X > x^*$ is given as follows:

$$T = \frac{\mu}{P(X > x^*)} = \frac{\mu}{1 - F(x^*)} = \frac{n}{m(1 - F(x^*))} \tag{2.81}$$

where μ denotes the average interarrival time between two events (or realizations of the process); n denotes the length of time duration; m denotes the number of events (or realizations) of n length of time durations; x^* denotes the design value (or critical value); and F denotes the probability distribution function of X.

The probability that a value of X, x, will occur in n successive years can be given by $\left(1 - \frac{1}{T}\right)^n$. Hence, the probability that x will occur for the first time in n years can be expressed as $\frac{1}{T}\left(1 - \frac{1}{T}\right)^{n-1}$.

The probability that the value will occur at least once in n years can be given as

$$R = 1 - \left(1 - \frac{1}{T}\right)^n \tag{2.82}$$

Here R can be called risk. Equation (2.82) can be used to compute the probability that x will occur within its return period:

$$P_T = 1 - \left(1 - \frac{1}{T}\right)^T \tag{2.83}$$

If T is large, then

$$P_T = 1 - e^{-1} = 0.63 \tag{2.83a}$$

For practical applications, one can compute the values of T for different values of R and n.

2.8.2 Bivariate (Multivariate) Risk Analysis through Return Period

Unlike with the univariate risk analysis, one may select different scenarios for the bivariate (multivariate) risk analysis. Here, we will present the return period for the bivariate case only. Let random variables $\{X, Y\}$ with the marginal and joint distributions be denoted as $F_X(x)$, $F_Y(y)$, $F_{X,Y}(x, y)$, and we immediately have the univariate return period from Equation (2.81) as follows:

$$T_X = \frac{\mu}{1 - F_X(x)}, \quad T_Y = \frac{\mu}{1 - F_Y(y)} \tag{2.84}$$

Following Shiau (2003), the bivariate risk analysis may be evaluated through (1) an "OR" case $(X \geq x \cup Y \geq y)$; (2) an "AND" case $(X \geq x \cap Y \leq y)$; and (3) a "CONDITIONAL" case $(X \geq x|Y \geq y; or\ Y \geq y|X \geq x)$. In what follows, each case is further discussed.

"OR" Case $(X \geq x \cup Y \geq y)$

The risk of "OR" case can be expressed as the likelihood (probability) of either event X: $X \geq x$ or event Y: $Y \geq y$, i.e., $P(X \geq x \cup Y \geq y)$. This probability can be written as follows:

$$P(X \geq x, \cup Y \geq y) = 1 - F_{X,Y}(x, y) \tag{2.85}$$

The risk expressed through the return period of the "OR" case can then be given as follows:

$$T_{X,Y}^{OR} = \frac{\mu}{P(X \geq x \cup Y \geq y)} = \frac{\mu}{1 - F_{X,Y}(x, y)} \tag{2.86}$$

"AND" Case: $(X \geq x \cap Y \geq y)$

The risk for the "AND" case can be expressed as the likelihood (probability) of both events X and Y that exceed the given magnitude x, y, i.e., $P(X \geq x \cap Y \geq y)$. This probability can be written as follows:

$$P(X \geq x \cap Y \geq y) = 1 - F_X(x) - F_Y(y) + F_{X,Y}(x, y) \tag{2.87}$$

The risk expressed through the return period of the "AND" case can be given as follows:

$$T_{X,Y}^{AND} = \frac{\mu}{P(X \geq x \cap Y \geq y)} = \frac{\mu}{1 - F_X(x) - F_Y(y) + F_{X,Y}(x, y)} \tag{2.88}$$

"CONDITIONAL" Case

With the knowledge of event Y exceeding the magnitude of y, the risk of event X exceeding magnitude of x may be represented as the conditional likelihood (probability) of $P(X \geq x|Y \geq y)$. This probability can be given as follows:

$$P(X \geq x|Y \geq y) = \frac{P(X \geq x \cap Y \geq y)}{P(Y \geq y)} = \frac{1 - F_X(x) - F_Y(y) + F_{X,Y}(x,y)}{1 - F_y(y)} \tag{2.89}$$

Equation (2.84) may also be derived through the conditional probability distribution $F(x|Y \geq y)$ as follows:

$$F(x|Y \geq y) = \frac{F_X(x) - F_{X,Y}(x,y)}{1 - F_Y(y)} \tag{2.90}$$

Following Shiau (2003) and Salvadori (2004), the risk expressed through the conditional return period of $X \geq x \mid Y \geq y$ can be given as follows:

$$\begin{aligned} T_{X|Y \geq y} &= \frac{T_Y}{P(X \geq x \cap Y \geq y)} = \frac{T_y}{1 - F_X(x) - F_Y(y) + F_{X,Y}(x,y)} \\ &= \frac{\mu}{1 - F_Y(y)} \frac{1}{1 - F_X(x) - F_Y(y) + F_{X,Y}(x,y)} \end{aligned} \tag{2.91}$$

Similarly, the conditional return period of $Y \geq y \mid X \geq x$ can be given as follows:

$$\begin{aligned} T_{Y|X \geq x} &= \frac{T_X}{P(X \geq x \cap Y \geq y)} = \frac{T_X}{1 - F_X(x) - F_Y(y) + F_{X,Y}(x,y)} \\ &= \frac{\mu}{1 - F_X(x)} \frac{1}{1 - F_X(x) - F_Y(y) + F_{X,Y}(x,y)} \end{aligned} \tag{2.92}$$

In a similar vain, risk analysis may be extended to multivariate ($d \geq 3$) analysis.

References

Abramowitz, M. and Stegun, I. A. (1965). *Handbook of Mathematical Functions*. Dover Publications, New York.

Anderson, T. W. and Darling, D. A. (1952). Asymptotic theory of certain "goodness-of-fit" criteria based on stochastic processes. *Annals of Mathematical Statistics*, 23, 193–212.

Arnold, B. C. (1983). *Pareto Distributions*. International Co-operative Publishing House, Fairland.

Bacchi, B., Becciu, G., and Kottegoda, N. T. (1994). Bivariate exponential model applied to intensities and durations of extreme rainfall. *Journal of Hydrology*, 155, 225–236.

Balakrishnan, N. and Lai, C.-D. (2009). *Continuous Bivariate Distribution*, 2nd edition, Springer Science+Business Media, LLC, Berlin and Heidelberg.

Bobee, B., Perreault, L., and Ashkar, F. (1993). Two kinds of moment ratio diagrams and their applications in hydrology. *Stochastic Hydrology and Hydraulics*, 7, 41–65.

Box, G. E. P. and Cox, D. R. (1964). An analysis of transformations. *Journal of the Royal Statistical Society, Series B*, 26(2), 211–252.

Box, G. E. P., Jenkins, G. M., and Reinsel, G. C. (2007). *Time Series Analysis: Forecasting and Control*, 4th edition, John Wiley & Sons, Inc., Hoboken.

Burr, I. W. (1942). Cumulative frequency functions. *Annals of Mathematical Statistics*. 13(2), 215–232. doi:10.1214/aoms/1177731607.

Chow, V. T. (1954). The log-probability law and its engineering applications. *Proceedings of the ASCE*, 80(5), 1–25.

Chow, V. T. ed. (1964). *Handbook of Applied Hydrology*. McGraw-Hill, New York.
Chow, V. T., Maidment, D. R., and Mays, L. W. (1988). *Applied Hydrology*, McGraw-Hill, New York.
Cramér, H. (1928). On the composition of elementary errors. *Scandinavian Actuarial Journal*, 1, 13–74. doi:10.1080/03461238.1928.10416862.
Cunnane, C. (1989). Statistical distributions for flood frequency analysis. WMO Operational Hydrology Report No. 33, WMO-No.718, Geneva.
Gumbel, E. J. (1941). The return period of flood flows. *Annals of Mathematical Statistics*, 12, 163–190.
Gumbel, E. J. (1958). *Statistics of Extremes*. Columbia University Press, New York.
Gumbel, E. J. (1960). Distributions of del valeurs extremes an plusieurs dimensions. *Publications de L'Institute de Statistique, Paris*, 9, 171–173.
Gumbel, E. J. (1961). Bivariate logistic distributions. *Journal of the American Statistical Association*, 56, 335–349.
Hazen, A. (1914). Storage to be provided in impounding reservoirs for municipal water supply, *Transactions of the American Society of Civil Engineers*, 1308(77), 1547–1550.
Hosking, J. R. M. (1990). L-moments: analysis and estimation of distribution using linear combinations of order statistics. *Journal of the Royal Statistical Society, Series B (Methodological)*, 52(1), 105–124.
Hogg, R. V. and Craig, A. T. (1978). *Introduction to Mathematical Statistics*, 4th edition. Macmillan: New York.
Izawa, T. (1965). 2 or multi-dimensional gamma-type distribution and its application to rainfall data. *Meteorology and Geophysics*, 15, 167.
Jenkinson, A. F. (1955). The frequency distribution of the annual maximum (or minimum) values of meteorological elements. *Quarterly Journal of the Royal Meteorological Society*, 81(348), 158–171. doi:10.1002/qj.49708134804.
Justel, A., Peña, D., and Zamar, R. (1994). A multivariate Kolmogorov–Smirnov test of goodness of fit. Working paper 94–32, *Statistics and Econometrics Series* 13.
Kite, G. W. (1977) *Frequency and Risk Analysis in Hydrology*. Water Resources Publications, Fort Collins.
Kolmogorov, A. (1933). Sulla determinazione empirica di una legge di distribuzione. *Giornale dell'Istituto Italiano degli Attuari*, 4, 83–91.
Liao, M. and Shimokawa, T. (1999). A new goodness-of-fit for type I extreme value and 2-parameter Weibull distributions with estimated parameters. *Optimization*, 64(1), 23–48.
Markovic, R. D. (1965). Probability functions of the best fit to distributions of annual precipitation and runoff hydrology, Paper No. 8, Colorado State University.
Marshall, A. W. and Ingram, O. (1967). A multivariate exponential distribution. *Journal of American Statistical Association*, 62(317), 30–44.
Moran, P. A. P. (1969). Statistical inference with bivariate gamma distribution. *Biometrika*, 54, 385–394
Morgenstern, D. (1956). Einfache Beispiele zweidimensionaler Verteilungen. *Mitt. Math. Statistik*, 8, 234–235.
Natural Environmental Research Council (NERC) (1975). *Flood Studies Report 1*. NERC, London.
Rao, A. R. and Hamed, K. H. (2000). *Flood Frequency Analysis*. CRC Publications, New York.

Rosenblatt, M. (1952). Remarks on a multivariate transformation. *Annals of Mathematical Statistics*, 23(3), 470–472.

Rosin, P. and Rammler, E. (1933). The laws governing the fineness of powdered coal. *Journal of the Institute of Fuel*, 7, 29–36.

Salvadori, G. (2004). Bivariate return periods via 2-copulas. *Statistical Methodology*, 1, 129–144.

Serinaldi, F. (2015). Dismissing return periods! *Stochastic Environmental Research and Risk Assessment*, 29, 1179–1189. doi:10.1007/s00477–014-0916–1.

Shiau, J. T. (2003) Return period of bivariate distributed extreme hydrological events. *Stochastic Environmental Research and Risk Assessment*, 17(1–2), 42–57.

Shoukri, M. M., Mian, I. U. M., and Tracy, D. S. (1988). Sampling properties of estimators of the log-logistic distribution with application to Canadian precipitation data. *Canadian Journal of Statistics*, 16(3), 223–236. doi:10.2307/3314729.

Singh, V. P. (1998). *Entropy-Based Parameter Estimation in Hydrology*. Springer, Boston.

Singh, V. P., Jain, S. K., and Tyagi, A. (2007). *Risk and Reliability Analysis: A Handbook for Civil and Environmental Engineers*. ASCE Press, Reston.

Singh, K. and Singh, V. P. (1991). Derivation of bivariate probability density functions with exponential marginals. *Stochastic Hydrology and Hydraulics*, 5, pp. 55–68.

Singh, V. P. and Zhang, L. (2016). Frequency distributions. In: Singh, V. P. (Ed) *Handbook of Applied Hydrology*. McGraw Hill Education, New York.

Smith, O. E., Adelfang, S. I., and Tubbs, J. D. (1982). A bivariate gamma probability distribution with application to gust model. NASA technical memorandum, 82483, National Aeronautics and Space Administration, Houston.

Smirnov, N. (1948). Table for estimating the goodness-of-fit of empirical distributions. *Annals of Mathematical Statistics*, 19, 279–281. doi:10.1214/aoms/1177730256.

Stock, J. H. and Watson, M. W. (1989). Interpreting the evidence on money-income casualty. *Journal of Econometrics*, 40, 161–181.

USWRC (United States Water Resources Research Council) (1981). *Guidelines for Determining Flood Flow Frequency*. Bulletin 17B (revised), Hydrology Committee, Water Resources Research Council, Washington.

Von Mises, R. E. (1928). *Wahrscheinlichkeit*, Statistik und wahreit. Julius Springer, Berlin and Heidelberg.

Yue, S., Ouarda, T. B. M. J., Bobee, B., Legendre, P., and Bruneau, P. (1999). The Gumbel mixed model for flood frequency analysis. *Journal of Hydrology*, 226(1–2), 88–100, doi:10.1016/S0022–1694(99)00168–7.

3

Copulas and Their Properties

ABSTRACT

The term *copula* is derived from the Latin verb *copulare*, meaning "to join together." In the statistics literature, the idea of a copula can be dated back to the nineteenth century in modeling multivariate non-Gaussian distributions. By formulating a theorem, now called Sklar theorem, Sklar (1959) laid the theoretical foundation for the modern copula theory. In general, copulas couple multivariate distribution functions to their one-dimensional marginal distribution functions, which are uniformly distributed in [0, 1]. In other words, copula functions enable us to represent a multivariate distribution with the use of univariate probability distributions (sometimes simply called marginals, or margins), regardless of their forms or types. In this chapter, we will discuss the general concepts of copulas, including their definition, properties, composition and construction, dependence structure, and tail dependence.

3.1 Definition of Copulas

Based on the Sklar's theorem definition (Sklar, 1959), a copula has two or more dimensions. Let d be the dimension of a copula. Then, a d-dimensional copula can be defined as a mapping function of $[0,1]^d \rightarrow [0,1]$, i.e., a multivariate cumulative distribution function can be defined in $[0,1]^d$ with standard uniform univariate margins.

Copula has the following properties:

1. Let $\mathbf{u} = [u_1, \ldots, u_d]$, $u_i = F_i(x_i) \in [0,1]$, if $u_i = 0$ for any $i \leq d$ (at least one coordinate of u equals 0).

$$C(u_1, \ldots, u_d) = 0 \tag{3.1}$$

2. $C(\mathbf{u}) = u_i$, if all the coordinates are equal to 1 except u_i, i.e.,

$$C(1, 1, \ldots, u_i, \ldots, 1, 1) = u_i, \ \forall i \in \{1, 2, \ldots, d\}, \ u_i \in [0,1] \tag{3.2}$$

3. $C(u_1, \ldots, u_d)$ is bounded, i.e., $0 \leq C(u_1, \ldots, u_d) \leq 1$. This property represents the limit of the cumulative joint distribution, i.e., in the range of [0, 1].
4. $C(u_1, \ldots, u_d)$ is d-increasing. This means that the volume of any d-dimensional interval is nonnegative, $\forall\{(a_1, \ldots, a_d), (b_1, \ldots, b_d)\} \in [0,1]^d$, where $a_i \leq b_i$,

$$\sum_{i_1=1}^{2}\sum_{i_2=1}^{2}\cdots\sum_{i_1=1}^{2}(-1)^{(i_1+i_2+\cdots+i_d)}C(x_{1i_1},x_{2i_2},\ldots,x_{di_d})\geq 0 \tag{3.3}$$

This property indicates the monotone increasing property of the cumulative probability distribution.

5. For every copula $C(u_1,\ldots,u_d)$ and every $(u_1,\ldots,u_d)$ in $[0,1]^d$, the following version of the Fréchet–Hoeffding bounds hold:

$$W(u_1,\ldots,u_d)\leq C(u_1,\ldots,u_d)\leq M(u_1,\ldots,u_d);d\geq 2 \tag{3.4}$$

where $W(u_1,\ldots,u_d)=\max\left(1-d+\sum_{i=1}^{d}u_i,0\right)$ represents the perfectly negatively dependent random variables; $M(u_1,\ldots,u_d)=\min(u_1,\ldots,u_d)$ represents the perfectly positively dependent random variables.

Here, we will first explain the first two properties using the bivariate flood variables (i.e., peak discharge (Q) and flood volume (V)) as an example. Let $Q\sim F_Q(q)$, $V\sim F_V(v)$ in which $F_Q\equiv u_1$, $F_V\equiv u_2$ represent the probability distribution functions of $\{Q:Q\geq Q_{min}\}$, $\{V:V\geq V_{min}\}$, respectively.

To explain property (1), we set $u_1=F_Q(q)$, $q>Q_{min}$ and $u_2=F_V(v\leq V_{min})=0$. We have $C(u_1,0)=H(Q\leq q,V\leq V_{min})$. With the joint distribution being nondecreasing, we know the volume of the interval $[Q_{min},V_{min}]\times[q,V_{min}]=[0,0]\times[u_1,0]\equiv 0$ which means when the flood volume is lower than the minimum flood volume, the joint distribution of $H(Q\leq q,V\leq V_{min})=C(u_1,0)\equiv 0$. Similarly, we have the following:

$$H(Q\leq Q_{min},V<v)=C(0,u_2)\equiv 0.$$

To explain property (2), we will again use the bivariate flood variable (i.e., peak discharge and flood volume) as an example. Based on the probability theory, we have the following:

$$C(u_1,1)=H(Q\leq q,V<+\infty)=F_Q(q)\equiv u_1 \text{ and}$$

$$C(1,u_2)=H(Q<+\infty,V\leq v)=F_V(v)\equiv u_2$$

Example 3.1 **Explain and prove the first three copula properties.**

Solution: Proof of properties (1) and (2).

Properties (1) and (2) may be explained directly using the Fréchet–Hoeffding bounds.

a. $C(u_1,\ldots,0,\ldots,u_d)=0,\ if\ u_i=0.$

Since copula $C(u_1,\ldots,u_d)$ represents the joint cumulative probability distribution of random variables $\{X_1,\ldots,X_d\}$, from Equation (3.4), we have the following:

$$W(u_1,\ldots,0,\ldots,u_d)\leq C(u_1,\ldots,0,\ldots,u_d)\leq M(u_1,\ldots,0,\ldots u_d)$$

From

$$
\begin{aligned}
W(u_1, \ldots, 0, \ldots, u_d) &= \max\left(1 - d + \textstyle\sum_{i=1}^{d} u_i, 0\right) \\
&= \max\left(1 - d + u_1 + \cdots + u_{i-1} + u_{i+1} + \cdots + u_d, 0\right)
\end{aligned}
$$

$u_1 + \cdots + u_{i-1} + u_{i+1} + \cdots + u_d \leq d - 1; \exists u \in [0, 1]$; and we have

$1 - d + u_1 + \cdots + u_{i-1} + u_{i+1} + \cdots + u_d \leq 1 - d + d - 1 \leq 0$

$\Rightarrow W(u_1, \ldots, 0, \ldots u_d) = 0$

and

$$M(u_1, \ldots, 0, \ldots, u_d) = \min(u_1, \ldots, 0, \ldots, u_d) = 0$$

Now we have $C(u_1, \ldots, u_d) = 0$, $\exists\ u_i = 0$, $i \in [1, d]$. This proves property (1) with $u_i = 0$. Similarly, property (1) holds for more than one variable equal to zero.

b. $C(u_1, \ldots, u_d) = u_i,\ \exists u_j = 1\ j \in [1, d]\ and\ j \neq i$

Applying the Fréchet–Hoeffding bounds, we have the following:

$$
\begin{aligned}
W(u_1, \ldots, u_d) &= \max(1 - d + d - 1 + u_i, 0) = u_i \\
M(u_1, \ldots, u_d) &= \min(u_1, \ldots, u_d) = \min(1, \ldots, u_i, \ldots, 1) = u_i
\end{aligned}
$$

Thus, we have $C(1, \ldots, 1, u_i, 1, \ldots, 1) = u_i$. This proves property 2.

Proof of property (3): It can be shown that if the copula represents the joint cumulative probability distribution of d-dimensional variables, the limit of copula should be [0, 1]. Property (4), i.e., Fréchet–Hoeffding bounds, further ensures property (3).

Example 3.2 **Illustrate a case for d=2 in Equation (3.3) of property (4).**

Solution: For $d = 2$, we have (a_1, a_2), $(b_1, b_2) \in [0, 1]^2$and $a_1 \leq a_2$, $b_1 \leq b_2$ as shown in Figure 3.1(a):

$$\sum_{i_1=1}^{2} \sum_{i_2=1}^{2} (-1)^{i_1+i_2} C(x_{1i_1}, x_{2i_2}) \geq 0 \tag{3.5}$$

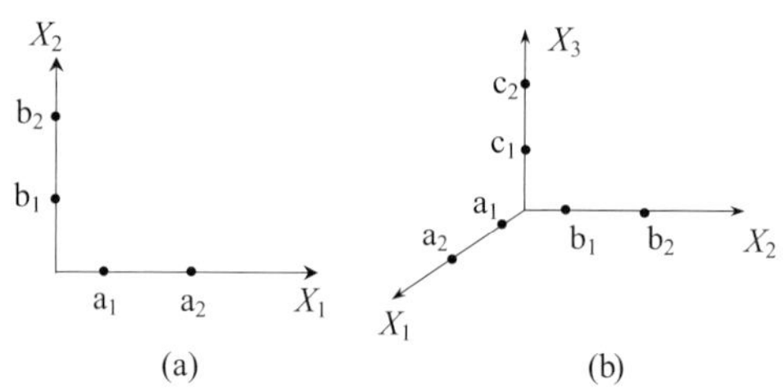

Figure 3.1 Schematic plots: (a) Example 3.2 and (b) Example 3.3.

$$\sum_{i_1=1}^{2}\sum_{i_2=1}^{2}(-1)^{i_1+i_2}C(x_{1i_1},x_{2i_2})$$
$$=\sum_{i_1=1}^{2}(-1)^{i_1+1}C(x_{1i_1},x_{21})+(-1)^{i_1+2}C(x_{1i_1},x_{22})$$
$$=(-1)^2C(x_{11},x_{21})+(-1)^3C(x_{11},x_{22})+(-1)^3C(x_{12},x_{21})+(-1)^4C(x_{12},x_{22})$$
$$=C(x_{11},x_{21})-C(x_{11},x_{22})-C(x_{12},x_{21})+C(x_{12},x_{22})$$

Therefore, Equation (3.5) follows:

$$C(a_1,a_2)-C(a_1,b_2)-C(b_1,a_2)+C(b_1,b_2)\geq 0 \tag{3.6}$$

Example 3.3 **Illustrate a case for d=3 in Equation (3.3) of property (4).**

Solution: For $d=3$ with $\left\{(x,y,z):(x_1,x_2),(y_1,y_2),(z_1,z_2)\in[0,1]^3\right\}$, where $x_1\leq x_2$, $y_1\leq y_2$, $z_1\leq z_2$ as shown in Figure 3.1(b),

$$\sum_{i_1=1}^{2}\sum_{i_2=1}^{2}\sum_{i_3=1}^{2}(-1)^{i_1+i_2+i_3}C(x_{1i_1},x_{2i_2},x_{3i_3})\geq 0 \tag{3.7}$$

and

$$\sum_{i_1=1}^{2}\sum_{i_2=1}^{2}\sum_{i_3=1}^{2}(-1)^{i_1+i_2+i_3}C(x_{1i_1},x_{2i_2},x_{3i_3})$$
$$=C(x_{12},x_{22},x_{32})-C(x_{12},x_{22},x_{31})-C(x_{12},x_{21},x_{32})-C(x_{11},x_{22},x_{32})$$
$$+C(x_{12},x_{21},x_{31})+C(x_{11},x_{22},x_{31})+C(x_{11},x_{21},x_{32})-C(x_{12},x_{21},x_{31}).$$

Using the notation in Figure 3.1(b) in Equation (3.7), we have the following:

$$C(a_2,b_2,c_2)-C(a_2,b_2,c_1)-C(a_1,b_2,c_2)+C(a_2,b_1,c_1)+C(a_1,b_2,c_1)$$
$$+C(a_1,b_1,c_2)-C(a_1,b_1,c_1)\geq 0;(a_1,a_2),(b_1,b_2),(c_1,c_2)\in[0,1]^2 \tag{3.8}$$

As introduced previously, copulas are multivariate distribution functions, and each copula induces a probability measure on $[0,1]^d$. In the bivariate case, $C(a_1,a_2)$ can be expressed as a joint probability in the rectangle $[0,a_1]\times[0,a_2]$. Thus, Equation (3.6) can be interpreted as follows:

$$C(a_1,a_2)-C(a_1,0)-C(0,a_2)+C(0,0)\geq 0 \tag{3.9}$$

Similarly in the trivaraite case, $C(a_1,a_2,a_3)$ can be expressed as a joint probability measure in the cube of $[0,a_1]\times[0,a_2]\times[0,a_3]$. Equation (3.8) can be interpreted as follows:

$$C(a_1,a_2,a_3)-C(a_1,a_2,0)-C(a_1,0,a_3)+C(a_2,0,0)+C(0,a_2,0)$$
$$+C(0,0,a_3)-C(0,0,0)=C(a_1,a_2,a_3)\geq 0 \tag{3.10}$$

6. Let $X_1,\ \ldots,\ X_d$ be random variables with margins $F_1,\ \ldots,\ F_d$ and joint distribution function $F(x_1,\ldots,x_d)$ and $u_i=F_i(x_i)$, $i=1,\ \ldots,\ d$. $X_1,\ \ldots,\ X_d$ are mutually independent if and only if $F(x_1,\ldots,x_d)=\prod_{i=1}^{d}F_i(x_i)$. Copula $C(u_1,\ldots,u_d)$ is called the independent or product copula and is defined as follows:

$$C(u_1,\ldots,u_d)=\prod_{i=1}^{d}u_i \tag{3.11}$$

According to Sklar's theorem, there exists a copula C such that for all $x \in R: R \in (-\infty, +\infty)$, the relation between cumulative joint distribution function $F(x_1, \ldots, x_d)$ and copula $C(u_1, \ldots, u_d)$ can be expressed as follows:

$$F(x_1, \ldots, x_d) = P(X_1 \leq x_1, \ldots, X_d \leq x_d) = C(F_1(x_1), \ldots, F_d(x_d)) = C(u_1, \ldots, u_d) \tag{3.12}$$

where $u_i = F(x_i) = P(X_i \leq x_i)$, $i = 1, \ldots, d; u_i \sim \mathrm{U}(0, 1)$, if F_i is continuous. Another way to think about the copula is as follows:

$$C(u_1, \ldots, u_d) = F\left(F_1^{-1}(u_1), \ldots, F_d^{-1}(u_d)\right); (u_1, \ldots, u_d) \in [0, 1]^d \tag{3.13}$$

where $x_i = F_i^{-1}(u_i)$ if $\boldsymbol{X}$ is continuous.

Example 3.4 **Illustrate Equation (3.12) using the Farlie–Gumbel–Morgenstern (FGM) model.**

The FGM model is as follows:

$$f(x, y) = f_X(x) f_Y(y)(1 + \eta(2F(x) - 1)(2F_Y(y) - 1)).$$

Solution: The joint CDF (JCDF) of the FGM model above can be expressed as follows:

$$F(x, y) = F_X(x) F_Y(y)\{1 + \eta[1 - F_X(x)][1 - F_Y(y)]\},\ |\eta| \leq 1$$

Let $u_1 = F_X(x)$, $u_2 = F_Y(y)$, and we have the following:

$$F(x, y) = C(u_1, u_2) = u_1 u_2[1 + \eta(1 - u_1)(1 - u_2)],\ |\eta| \leq 1.$$

The copula captures the essential features of the dependence of bivariate (multivariate) random variables. C is essentially a function that connects the multivariate probability distribution to its marginals. Then the problem of determining H (i.e., the joint cumulative distribution of correlated random variables) reduces to one of determining C.
Let $c(u_1, \ldots, u_d)$ denote the density function of copula $C(u_1, \ldots, u_d)$ as follows:

$$c(u_1, \ldots, u_d) = \frac{\partial C^d(u_1, \ldots, u_d)}{\partial u_1 \ldots \partial u_d} \tag{3.14a}$$

The mathematical relation between copula density function $c(u_1, \ldots, u_d)$ and joint density function $f(x_1, \ldots, x_d)$ can be expressed as follows:

$$\begin{aligned} f(x_1, \ldots, x_d) &= \frac{\partial F^d(x_1, \ldots, x_d)}{\partial x_1 \ldots \partial x_d} = \frac{\partial C^d(F_1(x_1), \ldots, F_d(x_d))}{\partial x_1 \ldots \partial x_d} \\ &= \frac{\partial C^d(F_1(x_1), \ldots, F_d(x_d))}{\partial F_1(x_1) \ldots \partial F_d(x_d)} \frac{\partial F_1(x_1)}{\partial x_1} \cdots \frac{\partial F_d(x_d)}{x_d} \\ &= \frac{\partial C^d(u_1, \ldots, u_d)}{\partial u_1 \ldots \partial u_d} \prod_{i=1}^{d} f_i(x_i) = c(u_1, \ldots, u_d) \prod_{i=1}^{d} f_i(x_i) \end{aligned} \tag{3.14b}$$

where f_i, F_i are, respectively, the probability density function and the probability distribution function for random variable X_i.

Equation (3.14b) may be rewritten as follows:

$$c(u_1, \ldots, u_d) = \frac{f(x_1, \ldots, x_d)}{\prod_{i=1}^{d} f_i(x_i)} \tag{3.14c}$$

Example 3.5 **Using the FGM model in Example 3.4, derive the copula density function and its relation to joint density function.**

Solution: From Example 3.4, the FGM model may be represented through the copula function as follows:

$C(u_1, u_2) = u_1 u_2(1 + \eta(1 - u_1)(1 - u_2))$. Then the copula density function can be derived using Equation (3.14a) as follows:

$$\begin{aligned} c(u_1, u_2) = \frac{\partial C^2(u_1, u_2)}{\partial u_1 \partial u_2} &= 1 + \eta(1 + 4u_1u_2 - 2u_1 - 2u_2) \\ &= 1 + \eta(2u_1 - 1)(2u_2 - 1),\ |\eta| \leq 1 \end{aligned} \tag{3.15a}$$

The relation between copula density function $c(u_1, u_2)$ and joint probability density function of the FGM model described in Example 3.4 can be expressed as follows:

$$f(x_1, x_2) = c(u_1, u_2) f_1(x_1) f_2(x_2) = f_1(x_1) f_2(x_2)[1 + \eta(2u_1 - 1)(2u_2 - 1)] \tag{3.15b}$$

where $u_i = F_i(x_i)$, $i = 1, 2$.

As an illustrative example, let $X_1 \sim \exp(\lambda)$ and $X_2 \sim gamma(\alpha, \beta)$, we may rewrite the probability density function of $f(x_1, x_2)$ as follows:

$$\begin{aligned} f(x_1, x_2) &= f_1(x_1) f_2(x_2)[1 + \eta(2u_1 - 1)(2u_2 - 1)] \\ &= \exp(-\lambda x_1) \frac{\beta^\alpha x^{\alpha - 1}}{\Gamma(\alpha)} \exp(-\beta x) \left[1 + \eta(1 - 2\exp(-\lambda x_1)) \left(\frac{2\gamma(\alpha, \beta x_2)}{\Gamma(\alpha)} - 1\right)\right] \end{aligned} \tag{3.15c}$$

3.1.1 Bivariate Copula

A bivariate copula $C(u_1, u_2)$ is a function from $[0, 1] \times [0, 1]$ into [0,1] to represent the joint cumulative probability distribution function of bivariate random variables with the following properties directly deduced from the discussions earlier as follows:

For every u_1, u_2 in [0, 1]:

$$C(u_1, 0) = C(0, u_2) = 0; C(u_1, 1) = u,\ C(1, u_2) = u_2 \tag{3.16}$$

1. For every $u_{11} \leq u_{12}$, $u_{21} \leq u_{22}$ in [0, 1]:

$$C(u_{12}, u_{22}) - C(u_{12}, u_{21}) - C(u_{11}, u_{22}) + C(u_{11}, u_{21}) \geq 0 \tag{3.17}$$

Equation (3.17) represents the volume:

$$V_C(B) = \Delta_{u_{11}}^{u_{12}} \Delta_{u_{21}}^{u_{22}} C(u_1, u_2) \geq 0 \tag{3.18}$$

Equation (3.18) represents the second-order derivative of function $C(u_1, u_2)$ (Nelsen, 2006). As the representation of the joint distribution of random variables X and Y (i.e., $C(u_1, u_2) \equiv C(F_X(x), F_Y(y)) \equiv H(x, y)$), the second-order derivative of $C(u_1, u_2)$ represents the copula density function of the bivariate random variable $c(u_1, u_2) = \frac{f(x,y)}{f_X(x) f_Y(y)} \geq 0$. This further explains Equations (3.17) and (3.18).

2. When random variables $X_1$and X_2 are independent, one obtains the so-called product copula:

$$C(u_1, u_2) = H(x_1, x_2) = u_1 u_2,\ u_i = F_i(x_i) \tag{3.19}$$

3. For every u_1, u_2 in [0, 1] with the corresponding copula $C(u_1, u_2)$, the following Fréchet–Hoeffding bounds hold:

$$\max\,(u_1 + u_2 - 1, 0) \leq C(u, v) \leq \min\,(u_1, u_2) \tag{3.20}$$

Example 3.6 **Express the bivariate Gaussian copula and its density function.**

Solution: The bivariate Gaussian copula is a distribution over the unit square $[0, 1]^2$, which is constructed from the bivariate normal distribution through the probability integral transform.

For a given correlation matrix, R, the bivariate Gaussian copula can be given as follows:

$$C_R^{GAU}(\mathbf{u}) = \Phi_R(\Phi^{-1}(u_1), \Phi^{-1}(u_2)),\ \mathbf{u} = [u_1, u_2] \tag{3.21}$$

where Φ^{-1}denotes the inverse cumulative distribution function of standard normal distribution; and Φ_R denotes the joint cumulative distribution function of bivariate normal distribution with mean vector of zero and covariance matrix of R.

The density function of bivariate Gaussian copula can be given as follows:

$$c_R^{GAU}(\mathbf{u}) = \frac{1}{2\pi\sqrt{1-\rho^2}} \exp\left(-\frac{(x^*)^2 - 2\rho x^* y^* + (y^*)^2}{2(1-\rho^2)}\right) \tag{3.22}$$

where x^*, y^* are the transformed variables as $x^* = \Phi^{-1}(u_1)$, $y^* = \Phi^{-1}(u_2)$; and ρ denotes the correlation coefficient of the bivariate random variable that may be expressed through the Kendall correlation coefficient as follows:

$$\rho = \sin\left(\frac{\pi\tau}{2}\right) \tag{3.22a}$$

It is worth noting that the Gaussian copula may also be called the meta-Gaussian distribution with no constraints on the type of marginal distributions. In what follows, we will further illustrate the bivariate Gaussian copula with two different marginal distributions:

$$X \sim N(\mu, \sigma^2),\ Y \sim \exp(\lambda).$$

Let $u_1 = F_X(x) = \mathrm{N}(x; \mu, \sigma^2)$ and $u_2 = F_Y(y) = 1 - \exp(-\lambda y)$. We have

$$x^* = \Phi^{-1}(u_1) = \Phi^{-1}\left(\mathrm{N}(x;\mu,\sigma^2)\right),\ y^* = \Phi^{-1}(u_2) = \Phi^{-1}(1 - \exp(-\lambda y)),\ \rho = \sin\left(\frac{\pi\tau_{XY}}{2}\right).$$

Finally, we obtain the bivariate Gaussian copula and its density function as follows:

$$C^{GAU}(\mathbf{u}) = \Phi\left(\Phi^{-1}(N(x;\mu,\sigma^2)), \Phi^{-1}(1-\exp(-\lambda y)); \sin\left(\frac{\pi\tau_{XY}}{2}\right)\right)$$

$$c^{GAU}(\mathbf{u}) = \frac{1}{2\pi\sqrt{1-\left(\sin\left(\frac{\pi\tau_{XY}}{2}\right)\right)^2}}$$

$$\exp\left(-\frac{\left[\Phi^{-1}(N(x;\mu,\sigma^2))\right]^2 - 2\Phi^{-1}(N(x;\mu,\sigma^2))\Phi^{-1}(1-\exp(-\lambda y)) + \left[\Phi^{-1}(1-\exp(-\lambda y))\right]^2}{2\left(1-\left(\sin\left(\frac{\pi\tau_{XY}}{2}\right)\right)^2\right)}\right)$$

Consider a simple numerical example with the random sample values $x = 2.5$ and $y = 4$ drawn from the probability distributions of $X \sim N(0, 2^2); Y \sim \exp(0.5)$. The rank based Kendall correlation coefficient of X, Y is $\tau_{XY} = 0.7$.

Applying Equation (3.22a), we may compute the Pearson correlation coefficient as follows:

$$\rho = \sin\left(\frac{\pi\tau}{2}\right) = \sin\left(\frac{0.7\pi}{2}\right) = 0.891.$$

From the parent normal and exponential distributions, we can compute the transformed variables:

$$\begin{aligned} X \sim N(0, 2^2) \quad & \Rightarrow F_X(2.5) = N(2.5, 0, 2^2) = 0.894 \\ & \Rightarrow x^* = \Phi^{-1}(F_X(2.5), 0, 1) = \Phi^{-1}(0.8944, 0, 1) = \mathbf{1.25} \\ Y \sim \exp(0.5) \quad & \Rightarrow F_Y(4) = 1 - \exp(-0.5(4)) = 0.8647 \\ & \Rightarrow y^* = \Phi^{-1}(F_Y(4), 0, 1) = \Phi^{-1}(0.8647, 0, 1) = \mathbf{1.1015} \end{aligned}$$

Substituting $x^* = 1.25$, $y^* = 1.1015$, $\rho = 0.891$ into the bivariate Gaussian copula and the corresponding density function, we have the following:

$$C^{GAU}(0.8944, 0.8647; 0.891) = 0.8406$$
$$c^{GAU}(0.8944, 0.8647; 0.891) = 4.0396$$

3.1.2 Trivariate Copula

A trivariate copula $C(u, v, w)$ is a function from $[0, 1]^3$ into $[0, 1]$. It should again satisfy all the properties discussed in the definition of copula such that the trivariate copula derived may represent the cumulative joint probability distribution of trivariate random variables.

1. For every u, v, w in [0, 1], use the following:

$$C(0, v, w) = C(u, 0, w) = C(u, v, 0) = 0 \tag{3.23}$$

$$C(u, 1, 1) = u,\ C(1, v, 1) = v,\ C(1, 1, w) = w \tag{3.24}$$

2. For every $u_1 \leq u_2$, $v_1 \leq v_2$, and $w_1 \leq w_2$ in [0, 1], use the following:

$$\begin{aligned} &C(u_2, v_2, w_2) - C(u_1, v_2, w_2) - C(u_2, v_1, w_2) - C(u_2, v_2, w_1) \\ &\quad + C(u_1, v_1, w_2) + C(u_1, v_2, w_2) + C(u_2, v_1, w_1) - C(u_1, v_1, w_1) \geq 0 \end{aligned} \tag{3.25}$$

Similar to the bivariate case, Equation (3.22) represents the volume:

$$V_C(B) = \Delta_{u_1}^{u_2} \Delta_{v_1}^{v_2} \Delta_{w_1}^{w_2} C(u, v, w) \geq 0 \tag{3.26}$$

For the function $C(u, v, w)$ to represent the trivariate joint distribution function, Equations (3.25) and (3.26) hold as a necessary condition, that is, the copula density is nonnegative.

3. When random variables $\{X_1, X_2, X_3\}$ are independent with $u = F_1(x_1)$, $v = F_2(x_2)$, $w = F_3(x_3)$, one obtains the so-called product copula

$$C(u, v, w) = uvw \tag{3.27}$$

4. For every u, v, w in [0, 1] with the copula function $C(u, v, w)$, the following Fréchet–Hoeffding bounds hold:

$$\max(u + v + w - 2, 0) \leq C(u, v, w) \leq \min(u, v, w) \tag{3.28}$$

The CDF and PDF of the trivariate copula can be written as follows:

$$C(u, v, w) = F(x_1, x_2, x_3) \tag{3.29}$$

$$c(u, v, w) = \frac{\partial C^3(u, v, w)}{\partial u \partial v \partial w} = \frac{f(x_1, x_2, x_3)}{f_1(x_1) f_2(x_2) f_3(x_3)} \tag{3.30}$$

Again, with the use of trivariate flood variables (i.e., peak discharge (Q), flood volume (V) and flood duration (D)), we may further illustrate these properties by setting the following:

$$u_1 = F_Q(q),\ u_2 = F_V(v),\ u_3 = F_D(d) \text{ and}$$
$$u_1 = 0 = F_Q(Q \leq q_{min}),\ u_2 = 0 = F_V(V \leq v_{min}),\ u_3 = 0 = F_D(D \leq d_{min}).$$

In case of property (1), we may evaluate $C(u_1, u_2, 0) = H(Q \leq q, V \leq v, D \leq d_{min})$ and $C(u_1, 1, 1) = H(Q \leq q, V < +\infty, D < +\infty)$ as an example.

$$H(Q \leq q, V \leq v, D \leq d_{min}) = P(D \leq d_{min} | Q \leq q, V \leq v) P(Q \leq q, V \leq v) \tag{3.31}$$

With the assumption of flood variables (i.e., $\{(Q, V, D) | Q \geq q_{min}, V \geq v_{min}, D \geq d_{min}\}$), we have $P(D \leq d_{min} | Q \leq q, V \leq v) = 0$, $0 \leq P(Q \leq q, V \leq v) < 1$ and $H(Q \leq q, V \leq v, D \leq d_{min}) = 0 = C(u_1, u_2, 0)$.

From the probability theory, it is obvious that $H(Q \leq q, V < +\infty, D < +\infty)$ reduces to the marginal probability distribution of peak discharge, i.e., $F_Q(q)$. Thus, we obtain the following:

$$C(u_1, 1, 1) = u_1.$$

In the same way as for the bivariate case, property (2) may be explained through the copula density function. Equation (3.26) may be rewritten as the third-order derivative of the copula function $C(u_1, u_2, u_3)$, i.e., $c(u_1, u_2, u_3)$. Related to the joint probability density function to Equations (3.14a)-(3.14c), it is clear that Equations (3.25) and (3.26) are nonnegative.

Example 3.7 **Express the trivariate Gaussian copula and its density function.**

Solution: The trivariate Gaussian copula is a distribution over the unit cube $[0, 1]^3$ which is constructed from the trivariate normal distribution through the probability integral transform.

For a given correlation matrix, R, the trivariate Gaussian copula can be given as follows:

$$C_R^{GAU}(\mathbf{u}) = \Phi_R\left(\Phi^{-1}(u_1), \Phi^{-1}(u_2), \Phi^{-1}(u_3)\right), \mathbf{u} = [u_1, u_2, u_3] \tag{3.32}$$

where Φ^{-1}denotes the inverse cumulative distribution function of the standard normal distribution; and Φ_R denotes the joint cumulative distribution function of trivariate normal distribution with a mean vector of zero and a covariance matrix of R.

The density function of trivariate Gaussian copula can be given as follows:

$$c_R^{GAU}(\mathbf{u}) = \frac{1}{\sqrt{|R|}} \exp\left(-\frac{1}{2}\begin{pmatrix}\Phi^{-1}(u_1)\\ \Phi^{-1}(u_2)\\ \Phi^{-1}(u_3)\end{pmatrix}^T \left(R^{-1} - \mathbf{I}\right)\begin{pmatrix}\Phi^{-1}(u_1)\\ \Phi^{-1}(u_2)\\ \Phi^{-1}(u_3)\end{pmatrix}\right) \tag{3.33}$$

where the mean vector is [0,0,0], R denotes the covariance matrix of the random variables, and $\mathbf{I}$ is the three-by-three identity matrix.

Similar to the bivariate Gaussian copula example (i.e., Example 3.6), there is no restriction in regard to the marginal distribution that the random variables may follow. More examples will be given in the chapter focused on meta-elliptical copulas.

3.2 Construction of Copulas

Copulas may be constructed using different methods, e.g., the inversion method, the geometric method, and the algebraic method. Nelsen (2006) discussed how to use these methods to construct copulas. In this section, these methods are briefly introduced.

3.2.1 Inversion Method

As the name of this method suggests, a copula is obtained through the joint distribution function F and its continuous maginals. Taking an example of a two-dimensional copula, the copula obtained by the inversion method can be expressed as follows:

$$C(u, v) = F\left(F_1^{-1}(u), F_2^{-1}(v)\right) \tag{3.34}$$

where $u = F_1(x_1)$, $v = F_2(x_2)$. The inversion method can be applied only if one knows the joint distribution of random variables X_1 and X_2.

Example 3.8 **Construct a copula using the Gumbel mixed distribution as joint distribution and the Gumbel distributions as marginals.**

Solution: Suppose that random variables X_1, X_2 each follow the Gumbel distribution as follows: X_1 ~ Gumbel (a_1, b_1), and X_2 ~ Gumbel (a_2, b_2). Their joint distribution follows the Gumbel mixed distribution. In this example, the univariate Gumbel distribution can be expressed as follows:

$$F(x) = \exp\left(-\exp\left(-\frac{x-b}{a}\right)\right) \tag{3.35}$$

and the bivariate Gumbel mixed distribution can be expressed as follows:

$$F(x_1, x_2) = F_1(x_1)F_2(x_2)\exp\left(-\alpha\left(\frac{1}{\ln F_1(x_1)} + \frac{1}{\ln F_2(x_2)}\right)^{-1}\right); \alpha \in [0, 1] \tag{3.36}$$

Again, let $u = F_1(x_1)$, $v = F_2(x_2)$ with $F_1(x_1)$ and $F_2(x_2)$ each following the Gumbel distribution given by Equation (3.35). Then, we have

$$C(u, v) = C(F_1(x_1), F_2(x_2)) = uv\exp\left(-\alpha\left((\ln u)^{-1} + (\ln v)^{-1}\right)^{-1}\right) \tag{3.37}$$

where α is the parameter of the copula.

The copula function derived as Equation (3.37) is actually the Gumbel-mixed model. Thus, it should be noted that Equations (3.37) can be successfully constructed to represent the joint distribution if and only if the random variables are positively correlated with the correlation coefficient not exceeding 2/3. This may be explained from the properties of the Gumbel-mixed model. Given by Oliveria (1982), the parameter of Gumbel-mixed model is related to the Pearson correlation coefficient as follows:

$$\alpha = 2\left[1 - \cos\left(\pi\sqrt{\frac{\rho}{6}}\right)\right]; \alpha = 0 \Rightarrow \rho = 0; \alpha = 1 \Rightarrow \rho = \frac{2}{3} \tag{3.38}$$

Example 3.9 **Construct a copula from bivariate exponential distribution with exponential marginals.**

Solution: Suppose that random variables X_1, X_2 with $X_1 \sim \exp(\theta_1)$, $X_2 \sim \exp(\theta_2)$, the joint distribution of X_1 and X_2, $F(x_1, x_2)$, follows the bivariate exponential distribution presented by Singh and Singh (1991) as follows:

$$F(x_1, x_2) = \left(1 - e^{-\frac{x_1}{\theta_1}}\right)\left(1 - e^{-\frac{x_2}{\theta_2}}\right)\left[1 + \delta e^{-\frac{x_1}{\theta_1} - \frac{x_2}{\theta_2}}\right] \tag{3.39}$$

Let $u = F_1(x_1) = 1 - e^{-\frac{x_1}{\theta_1}}$, $v = F_2(x_2) = 1 - e^{-\frac{x_2}{\theta_2}}$; then we have

$$C(u, v) = C(F_1(x_1), F_2(x_2)) = uv(1 + \delta(1 - u)(1 - v)) \tag{3.40}$$

where δ is the parameter of the copula in Equation (3.39).

In Equations (3.39) and (3.40), $|\delta| \leq 1$. In the case of the bivariate exponential distribution in this example, the correlation of bivariate random variables is in the range of [–0.25, 0.25] to guarantee that the bivariate distribution so derived is valid. In addition, the FGM copula is also expressed as Equation (3.40). In the case of the FGM copula, the correlation of bivariate random variables needs to be in the range of [–1/3, –1/3] (Schucany et al., 1978).

3.2.2 Geometric Method

Rather than deriving the copula functions by inverting the joint distribution functions based on the Sklar theorem, the geometric method derives the copula directly based on the definition of the copulas, e.g., the bivariate copula is 2-increasing and bounded. The geometric method does not require the knowledge of either distribution function or random variables. As the name of the method suggests, the geometric method requires the knowledge in regard to the geometric nature or support region of the random variables (Nelsen, 2006). In what follows, two bivariate copula examples borrowed from exercise problems (Nelsen, 2006) are used to illustrate the method.

Example 3.10 **Singular copula with prescribed support.**

Let (α, β) be a point in I2 such that $\alpha > 0$, $\beta > 0$, and $\alpha + \beta < 1$. Suppose that the probability mass α is uniformly distributed on the line segment joining (α,β) and (0, 1), the probability mass β is uniformly distributed on the line segment joining (α,β) and (1, 0), and the probability mass 1-α-β is uniformly distributed on the line segment joining (α,β) and (1, 1). Determine the copula function with these supports.

Solution: Based on the description of the problem statements, Figure 3.2(a) graphs the prescribed support (depicted by the solid line). It is seen from Figure 3.2(a) that (u, v) may be reside either in the upper triangle (i.e., Figure 3.2(b)) or in the lower triangle (i.e., Figure 3.2(c)).

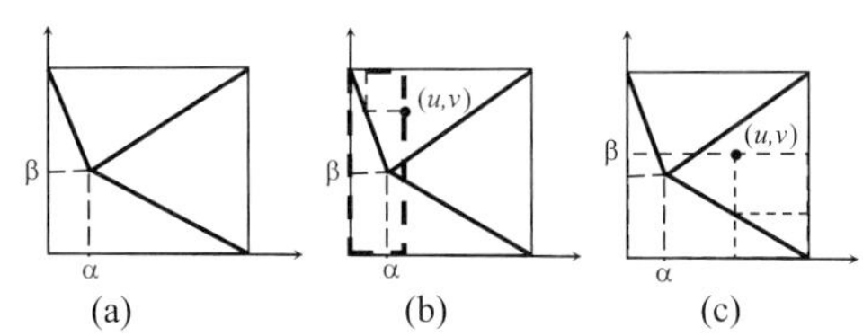

Figure 3.2 Schematic of singular copulas with prescribed support.

In addition, we will also check what may happen if (u, v) fall out of the prescribed support (i.e., beneath the two triangles). Now, to determine the copula function with the corresponding prescribed support, we will look at three different cases: (a) (u,v) is in the upper triangular support region (Figure 3.2(b)); (b) (u,v) is in the lower triangular support region (Figure 3.2(c)); and (c) (u,v) does not fall into either support region individually.

1. If (u,v) falls into the region bounded by the upper triangular region with vertices (α,β), (0, 1), and (1, 1), as shown in Figure 3.2(b), then according to the definition of the copula, Figure 3.2(b) clearly shows the following:

$$V_C([0,u]\times[v,1]) = V_C\left(\left[0,\frac{\alpha(1-v)}{1-\beta}\right]\times[v,1]\right) \tag{3.41}$$

$$V_C([0,u]\times[v,1]) = C(u,1) - C(u,v) - C(0,1) + C(0,v) = u - C(u,v) \tag{3.42}$$

$$\begin{aligned} V_C\left(\left[0,\frac{\alpha(1-v)}{1-\beta}\right]\times[v,1]\right) &= C\left(\frac{\alpha(1-v)}{1-\beta},1\right) - C\left(\frac{\alpha(1-v)}{1-\beta},v\right) - C(0,1) + C(0,v) \\ &= \frac{\alpha(1-v)}{1-\beta} - C\left(\frac{\alpha(1-v)}{1-\beta},v\right) \end{aligned} \tag{3.43}$$

Equating Equation (3.42) to Equation (3.43), we get the following:

$$C(u,v) = u - \frac{\alpha(1-v)}{1-\beta} \tag{3.44a}$$

In order to determine the copula function in this region, we can also look at the rectangle $\left[\frac{\alpha(1-v)}{1-\beta},u\right]\times[v,1]$. This rectangle is not intercepting any support line segment, thus we know the C-volume is zero, as follows:

$$\begin{aligned} V_C\left(\left[\frac{\alpha(1-v)}{1-\beta},u\right]\times[v,1]\right) &= C(u,1) - C(u,v) - C\left(\frac{\alpha(1-v)}{1-\beta},1\right) + C\left(\frac{\alpha(1-v)}{1-\beta},v\right) = 0 \\ &\Rightarrow C(u,v) = u - \frac{\alpha(1-v)}{1-\beta} \end{aligned} \tag{3.44b}$$

2. Similarly, If (u,v) falls into the region bounded by the lower triangular region with vertices (α,β), (1, 0), and (1, 1), as shown in Figure 3.2(c), then we can use the same approach to find the following:

$$C(u,v) = v - \frac{\beta}{1-\alpha}(1-u) \tag{3.45}$$

3. If (u,v) is not falling into the two triangles bounded by the support segment, then we immediately know that the C-volume is zero and $C(u,v)$ can be found as follows:

$$V_C([0,u]\times[0,v]) = C(u,v) - C(0,v) - C(u,0) - C(0,0) = 0 \Rightarrow C(u,v) = 0 \tag{3.46}$$

Note the following for the limiting cases:

1. If $\alpha = \beta = 0$, the support line segment is the main diagonal on $\mathbf{I}^2$. Nelsen (2006) proved that in this case, $C(u,v)$ is the Fréchet–Hoeffding upper bound, i.e., $C(u,v) = M(u,v) = \min(u,v)$.

2. If $\beta = 1 - \alpha$, Equation (3.44a) and Equation (3.45) reduce to the following:

$$C(u,v) = u - \frac{\beta}{1-\alpha}(1-v) = u + v - 1 \tag{3.47a}$$

and

$$C(u,v) = v - \frac{\alpha}{1-\beta}(1-u) = u + v - 1 \tag{3.47b}$$

Equation (3.47) represents the Fréchet–Hoeffding lower bound, i.e.,

$$C(u,v) = W(u,v) = \max(u+v-1,0). \tag{3.47}$$

Example 3.11 **Copulas with prescribed horizontal or vertical support.**

Show for each of the following choices of the Ψ function, the function C given as

$$C(u,v) = uv - \Psi(v)u(1-u) \tag{3.48}$$

is a copula:

a. $\Psi(v) = \left(\frac{\theta}{\pi}\right)\sin(\pi v); \theta \in [-1,1]$
b. $\Psi(v) = \theta[\zeta(v) + \zeta(1-v)]$, $\theta \in [-1,1]; \zeta$ is the piecewise linear function with the graph connecting [0, 0] to (1/4, 1/4) to (1/2, 0) to (1, 0).

Solution: According to Nelsen (2006), if Equation (3.48) is a copula, it is a copula with quadratic sections in u.

a. $\Psi(v) = \left(\frac{\theta}{\pi}\right)\sin(\pi v), \quad \theta \in [-1,1]$

Corollary 3.2.5 (from Nelsen, 2006) can be applied to prove that the C function with the Ψ function so defined is a copula. Corollary 3.2.5 states the necessary and sufficient conditions for the C function to be a copula:

1. $\Psi(v)$ is absolutely continuous on **I**.
2. $|\Psi'(v)| \le 1$ almost everywhere on **I**.
3. $|\Psi(v)| \le \min(v, 1-v)$.

Based on corollary 3.2.5, we conclude the following:

1. It is easy to see that $\Psi(v)$ is absolutely continuous on **I** with sine function being an absolutely continuous function.
2. It is seen that for $\theta \in [-1,1]$, $|\theta/\pi| < 1$, so we have the following: $|\Psi'(v)| = |\frac{\theta}{\pi}\cos(\pi v)| < 1$.

3. For $\Psi(v) = \left(\frac{\theta}{\pi}\right) \sin(\pi v)$, $v \in \mathbf{I}$, we have the following:

$$0 \le \pi v \le \pi,\ \sin(\pi v) \le \pi v \Rightarrow |\Psi(v)| = \left|\frac{\theta}{\pi} \sin(\pi v)\right| \le \left|\frac{\theta}{\pi}(\pi v)\right| = |\theta v| \le v \tag{3.49}$$

Similarly,

$$\begin{aligned} \sin(\pi v) &= \sin(\pi - \pi v) = \sin(\pi(1-v)) \le \pi(1-v) \\ &\Rightarrow |\Psi(v)| = \left|\frac{\theta}{\pi} \sin(\pi(1-v))\right| \le |\theta(1-v)| \le 1 - v \end{aligned} \tag{3.50}$$

From Equations (3.49) and (3.50), we have $|\Psi(v)| \le \min(v, 1-v)$ for $v \in \mathbf{I}$.

Now, all the conditions are satisfied and function C with $\Psi(v)$ defined in a. is a copula.

b. $\Psi(v) = \theta[\zeta(v) + \zeta(1-v)]$, $\theta \in [-1, 1]$; $\zeta$is the piecewise linear function with the graph connecting {[0, 0] to (1/4, 1/4)} to {(1/2, 0) to (1, 0)}.

Theorem 3.2.4 in Nelsen (2006) can be applied to prove that function C is a copula. Theorem 3.2.4 states the necessary and sufficient conditions for C to be a copula as follows:

1. $\Psi(0) = \Psi(1) = 0$
2. $\Psi(v)$ satisfies the Lipschitz condition: $|\Psi(v_2) - \Psi(v_1)| \le |v_2 - v_1|; v_1, v_2 \in \mathbf{I}$
3. C is absolutely continuous.

The schematic plot for the piecewise linear function is given in Figure 3.3(a).

The $\Psi(v)$ function can be written as follows:

$$\Psi(v) = \begin{cases} \theta v; & v \in \left[0, \frac{1}{4}\right] \\ \theta\left(\frac{1}{2} - v\right); & v \in \left[\frac{1}{4}, \frac{3}{4}\right] \\ \theta(v-1); & v \in \left[\frac{3}{4}, 1\right] \end{cases} \tag{3.51}$$

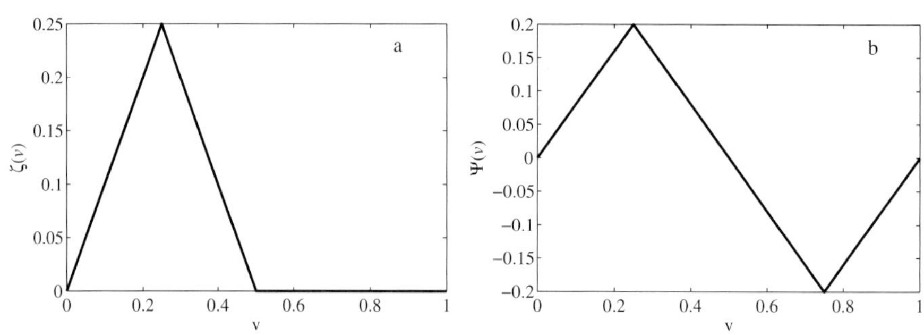

Figure 3.3 Plots of functions $\zeta(v)$ and $\Psi(v)$.

1. For $\theta \in [-1, 1]$, we have the following:

$$\Psi(0) = 0; \Psi(1) = \theta(1 - 1) = 0$$

2. Prove the Lipschitz condition with $v_1 \le v_2$ and $\theta \in [-1, 1]$.

   i. If $v_1 \in \left[0, \frac{1}{4}\right]$, we have the following:

$$|\Psi(v_2) - \Psi(v_2)|$$
$$= \begin{cases} |\theta(v_2 - v_1)| \le |v_2 - v_1|; v_2 \in \left[0, \frac{1}{4}\right] \\ \left|\theta\left(\frac{1}{2} - v_2\right) - \theta v_1\right| = \left|\theta\left(\frac{1}{2} - (v_1 + v_2)\right)\right| < |\theta(v_2 - v_1)| \le |v_2 - v_1|; v_2 \in \left[\frac{1}{4}, \frac{3}{4}\right] \\ |\theta(v_2 - 1) - \theta v_1| < |\theta(v_2 - v_1 - 1)| < |v_2 - v_1|; \quad v_2 \in \left[\frac{3}{4}, 1\right] \end{cases} \tag{3.52}$$

   ii. If $v_1 \in \left[\frac{1}{4}, \frac{3}{4}\right]$, we have the following:

$$|\Psi(v_2) - \Psi(v_1)| = \begin{cases} \left|\theta\left(\frac{1}{2} - v_2\right) - \theta\left(\frac{1}{2} - v_1\right)\right| = |\theta(v_2 - v_1)| \le |v_2 - v_1|; v_2 \in \left[\frac{1}{4}, \frac{3}{4}\right] \\ \left|\theta(v_2 - 1) - \theta\left(\frac{1}{2} - v_1\right)\right| = \left|\theta\left(v_2 + v_1 - \frac{3}{2}\right)\right| \le \left|\theta\left(v_2 - \frac{3}{4}\right)\right| \\ \le |\theta(v_2 - v_1)| \le |v_2 - v_1|; \quad v_2 \in \left[\frac{3}{4}, 1\right] \end{cases} \tag{3.53}$$

   iii. Similarly, it can be easily shown that the Lipschitz condition is also satisfied for $v_1 \in \left[\frac{3}{4}, 1\right]$.

3. Following Nelsen (2006), to prove the absolute continuity of C follows the absolute continuity of $\Psi(v)$ with the second condition. Figure 3.3(b) plots the $\Psi(v)$ function with $\theta = 0.8$; as an example, it is shown that there is no discontinuity in domain **I**:

$$\Psi'(v) = \begin{cases} \theta, & v \in \left[0, \frac{1}{4}\right] \\ -\theta, & v \in \left[\frac{1}{4}, \frac{3}{4}\right] \\ \theta, & v \in \left[\frac{3}{4}, 1\right] \end{cases} \tag{3.54}$$

   with $\theta \in [-1, 1]$, we have proved that $|\Psi'(v)| \le 1$ in domain **I**.

Now all the conditions are satisfied and the C function with the $\Psi(v)$ function defined in (b) is a copula.

It is worth noting that the copula defined as Equation (3.48) is a copula with quadratic sections in u. The reader can refer to Nelsen (2006) for more complete details of the geometric method and other types of geometric support to construct copulas.

3.2.3 Algebraic Method

Copulas may be constructed using the algebraic relationship between joint distribution and univariate distributions of random variables X_1 and X_2, which is called the algebraic method. Nelsen (2006) introduced this approach by constructing the Plackett and Ali–Mikhail–Haq copula through an "odd" ratio in which the Plackett copula is constructed by measuring the dependence of two-by-two contingency tables, and Ali–Mikhail–Haq copula is constructed by using the survival odds ratio. In order to discuss the method, the Ali–Mikhail–Haq copula construction example presented in Nelsen (2006) is used here.

The survival odds ratio for a univariate random variable X with $X \sim F(x)$ can be expressed as follows:

$$\frac{P(X > x)}{P(X \leq x)} = \frac{1 - F(x)}{F(x)} = \frac{\bar{F}(x)}{F(x)} \tag{3.55}$$

Similarly, the survival odds ratio for bivariate random variables X_1 and X_2 with joint distribution $F(x_1, x_2)$ and marginals $F_1(x_1)$, $F_2(x_2)$ can be expressed as follows:

$$\frac{P(X_1 > x_1 \text{ or } X_2 > x_2)}{P(X_1 \leq x_1 \text{ and } X_2 \leq x_2)} = \frac{1 - F(x_1, x_2)}{F(x_1, x_2)} = \frac{\bar{F}(x_1, x_2)}{F(x_1, x_2)} \tag{3.56}$$

Example 3.12 **The Ali–Mikhail–Haq copula.**

The Ali–Mikhail–Haq copula (Ali et al., 1978) can be expressed as follows:

$$C(u, v) = \frac{uv}{1 - \theta(1 - u)(1 - v)} \tag{3.57}$$

Construct the copula by using the algebraic method.

Solution: Ali et al. (1978) proposed that Ali–Mikhail–Haq copula belongs to the bivariate logistic distribution family with the standard bivariate logistic distribution and standard logistic marginals.

The standard bivariate logistic distribution can be given as follows:

$$F(x_1, x_2) = \left(1 + e^{-x_1} + e^{-x_2}\right)^{-1} \tag{3.58a}$$

The standard logistic marginal can be given as follows:

$$F(x) = \left(1 + e^{-x}\right)^{-1} \tag{3.58b}$$

The survival ratio of Equation (3.58a) is

$$\frac{1 - F(x_1, x_2)}{F(x_1, x_2)} = \frac{1 - \left(1 + e^{-x_1} + e^{-x_2}\right)^{-1}}{\left(1 + e^{-x_1} + e^{-x_2}\right)^{-1}} = e^{-x_1} + e^{-x_2} \tag{3.59}$$

From Equation (3.59), it is seen that the survival ratio of the standard bivariate logistic distribution can be rewritten as follows:

$$\frac{1-F(x_1,x_2)}{F(x_1,x_2)} = e^{-x_1} + e^{-x_2} = \frac{1-(1+e^{-x_1})^{-1}}{(1+e^{-x_1})^{-1}} + \frac{1-(1+e^{-x_2})^{-1}}{(1+e^{-x_2})^{-1}} \tag{3.60a}$$

Substituting Equation (3.58b) into Equation (3.60a), we have the following:

$$\frac{1-F(x_1,x_2)}{F(x_1,x_2)} = \frac{1-F_1(x_1)}{F_1(x_1)} + \frac{1-F_2(x_2)}{F_2(x_2)} \tag{3.60b}$$

In Ali et al. (1978), the Ali–Mikhail–Haq copula was considered a bivariate distribution satisfying the survival ratio as follows:

$$\frac{1-F(x_1,x_2)}{F(x_1,x_2)} = \frac{1-F_1(x_1)}{F_1(x_1)} + \frac{1-F_2(x_2)}{F_2(x_2)} + (1-\theta)\frac{1-F_1(x_1)}{F_1(x_1)}\frac{1-F_2(x_2)}{F_2(x_2)} \tag{3.61}$$

It is concluded from Equation (3.61) that $\theta = 1$ implies that the joint distribution $F(x, y)$ of random variables X_1 and X_2 follows the standard biviariate logistic distribution; and $\theta =$ 0 implies that X and Y are independent with the proof given in example 3.19 in Nelsen (2006).

Applying Sklar's theorem to Equation (3.59) and letting

$$C(u,v) = F(x_1,x_2),\ u = F_1(x_1),\ v = F_2(x_2)$$

Equation (3.61) can be rewritten as follows:

$$\frac{1-C(u,v)}{C(u,v)} = \frac{1-u}{u} + \frac{1-v}{v} + (1-\theta)\frac{1-u}{u}\frac{1-v}{v} \tag{3.62}$$

With simple algebra, we have

$$C(u,v) = \frac{uv}{1-\theta(1-u)(1-v)} \tag{3.63}$$

where θ is the parameter of the Ali–Mikhail–Haq copula.

3.3 Families of Copula

There are a multitude of copulas. Generally speaking, copulas may be grouped into the Archimedean copulas, meta-elliptical copulas, and copulas with prescribed geometric support (e.g., copulas with quadratic or cubic sections). According to their exchangeable properties, copulas may also be classified as symmetric copulas and asymmetric copulas. For example, one-parameter Archimedean copulas are symmetric copulas, and periodic copulas (Alfonsi and Brigo, 2005) and mixed copulas (Hu, 2006) are asymmetric copulas. Here we will only discuss the general concepts of each copula family.

The copula functions pertaining to a given copula family will be discussed in detail in subsequent chapters.

3.3.1 Archimedean Copulas

Archimedean copulas are widely applied in finance, water resources engineering, and hydrology due to their simple form, dependence structure, and other "nice" properties. Chapters 4 and 5 discuss the symmetric and asymmetric Archimedean copulas.

3.3.2 Plackette Copula

The Plackette copula has been applied in recent years. It will be discussed in Chapter 6.

3.3.3 Meta-elliptical Copulas

Meta-elliptical copulas are a flexible tool for modeling multivariate data in hydrology. They will be further discussed in Chapter 7.

3.3.4 Entropic Copula

Similar to the entropy-based univariate probability distributions, the entropy theory (e.g., Shannon entropy) may be applied to derive entropic copulas with the use of constraints in regard to the total probability theory, properties of marginals (i.e., $E(U^i) = \frac{1}{i+1}$), and the dependence measure (e.g., Spearman rank-based correlation coefficient). The entropic copula will be further discussed in Chapter 8.

3.3.5 Mixed Copulas

Parametric copulas place restrictions on the dependence parameter. When data are heterogeneous, it is desirable to have additional flexibility to model the dependence structure (Trivedi and Zimmer, 2007). A mixture model, proposed by Hu (2006), is able to measure dependence structures that do not belong to the aforementioned copula families. By choosing component copulas in the mixture, a model can be constructed that is simple and flexible enough to generate most dependence patterns and provide such a flexibility in practical data. This also facilitates the separation of the degree of dependence and the structure of dependence. These concepts are respectively embodied in two different groups of parameters: the association parameters and the weight parameters (Hu, 2006). For example, the given bivariate data may be modeled as a finite mixture with three bivariate copulas $C_I(u,v)$, $C_{II}(u_1,u_2)$, $C_{III}(u_1,u_2)$; the mixture model is defined as follows:

$$C_{mix}(u,v;\boldsymbol{\theta}_1,\boldsymbol{\theta}_2,\boldsymbol{\theta}_3,w_1,w_2,w_3) = w_1C_I(u,v;\boldsymbol{\theta}_1) + w_2C_{II}(u,v;\boldsymbol{\theta}_2) + w_3C_{III}(u,v;\boldsymbol{\theta}_3) \tag{3.64}$$

where $C_{mix}(u, v; \boldsymbol{\theta}_1, \boldsymbol{\theta}_2, \boldsymbol{\theta}_3, w_1, w_2, w_3)$ denotes the mixed copula; $C_I(u, v; \boldsymbol{\theta}_1)$, $C_{II}(u, v; \boldsymbol{\theta}_2)$, $C_{III}(u, v; \boldsymbol{\theta}_3)$ are the three bivariate copulas, each with $\boldsymbol{\theta}_1$, $\boldsymbol{\theta}_2$, $\boldsymbol{\theta}_3$ as the corresponding copula parameters; and w_1, w_2, w_3 may be interpreted as weights for each copula such that $0 < w_j < 1; j = 1, 2, 3, \sum_{j=1}^{3} w_j = 1, 0 < w_j < 1$.

3.3.6 Empirical Copula

Sometimes, we analyze data with an unknown underlying distribution. The empirical data distribution can be transformed into what is called an "empirical copula" by warping such that the marginal distributions become uniform. Let $\mathbf{x_1}$ and $\mathbf{x_2}$ be two samples each of size n. The empirical copula frequency function can often be computed for any pair (x_1, x_2) by

$$C_n\left(\frac{i}{n}, \frac{j}{n}\right) = \frac{\sum_{i=1}^{n} 1\left(x_1 \le x_{1j} \text{ and } x_2 \le x_{2j}\right)}{n} \tag{3.65}$$

where $\left\{(x_{1i}, x_{2j}) : 0 \le i, j \le n\right\}$ represent, respectively, the ith- and jth-order statistic of $\mathbf{x_1}$ and $\mathbf{x_2}$.

Example 3.13 **Using the peak discharge (Q: m3/s) and flood volume (V: m3) given in Table 3.1, calculate the empirical copula with the use of Equation (3.65).**

Table 3.1. *Peak discharge and flood volume data (from Yue, 2001).*

Pair	Year	V (m^3)	Q (cms)	Pair	Year	V (m^3)	Q (cms)
1	1942	8,704	371	28	1969	11,272	416
2	1943	6,907	245	29	1970	8,640	246
3	1944	4,189	189	30	1971	6,989	248
4	1945	8,637	229	31	1972	9,352	297
5	1946	8,409	240	32	1973	12,825	371
6	1947	13,602	331	33	1974	13,608	442
7	1948	8,788	206	34	1975	8,949	260
8	1949	5,002	157	35	1976	12,577	236
9	1950	5,167	184	36	1977	11,437	334
10	1951	10,128	275	37	1978	9,266	310
11	1952	12,035	286	38	1979	14,559	383
12	1953	10,828	230	39	1980	5,057	151
13	1954	8,923	233	40	1981	9,645	197
14	1955	11,401	351	41	1982	7,241	283
15	1956	6,620	156	42	1983	13,543	390
16	1957	3,826	168	43	1984	15,003	405
17	1958	8,192	343	44	1985	6,460	176

Table 3.1. (*cont.*)

Pair	Year	V (m^3)	Q (cms)	Pair	Year	V (m^3)	Q (cms)
18	1959	6,414	214	45	1986	7,502	181
19	1960	8,900	303	46	1987	5,650	233
20	1961	9,406	300	47	1988	7,350	187
21	1962	7,235	143	48	1989	9,506	216
22	1963	8,177	232	49	1990	6,728	196
23	1964	7,684	182	50	1991	13,315	424
24	1965	3,306	121	51	1992	8,041	255
25	1966	8,026	186	52	1993	10,174	257
26	1967	4,892	173	53	1994	14,769	232
27	1968	8,692	292	54	1995	8,711	286

Solution: To determine the empirical copula, we will first need to rank the flood volume and peak discharge variables in the increasing order. Then we can use Equation (3.65) to compute the empirical copula. Here we will use $C\left(\frac{1}{n}, \frac{1}{n}\right)$ as an illustration example. For the flood data in Table 3.1, Table 3.2 lists the order statistics of flood volume and peak discharge individually.

Table 3.2. *Order statistics of flood volume and peak discharge.*

Order	V (m^3 day/s)	Q (cms)	Order	V (m^3 day/s)	Q (cms)
1	3,306	121	28	8,704	245
2	3,826	143	29	8,711	246
3	4,189	151	30	8,788	248
4	4,892	156	31	8,900	255
5	5,002	157	32	8,923	257
6	5,057	168	33	8,949	260
7	5,167	173	34	9,266	275
8	5,650	176	35	9,352	283
9	6,414	181	36	9,406	286
10	6,460	182	37	9,506	286
11	6,620	184	38	9,645	292
12	6,728	186	39	10,128	297
13	6,907	187	40	10,174	300
14	6,989	189	41	10,828	303
15	7,235	196	42	11,272	310
16	7,241	197	43	11,401	331
17	7,350	206	44	11,437	334
18	7,502	214	45	12,035	343
19	7,684	216	46	12,577	351
20	8,026	229	47	12,825	371
21	8,041	230	48	13,315	371

Table 3.2. (*cont.*)

Order	V (m^3 day/s)	Q (cms)	Order	V (m^3 day/s)	Q (cms)
22	8,177	232	49	13,543	383
23	8,192	232	50	13,602	390
24	8,409	233	51	13,608	405
25	8,637	233	52	14,559	416
26	8,640	236	53	14,769	424
27	8,692	240	54	15,003	442

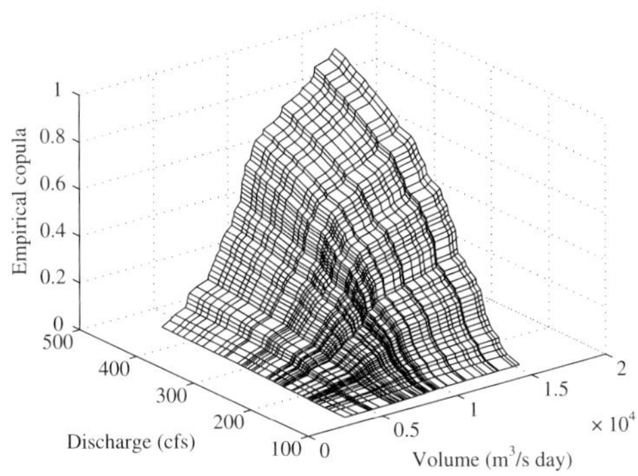

Figure 3.4 Empirical copula for peak discharge and flood volume.

To apply the empirical copula, using $x_{(1)} = 3360$, $y_{(1)} = 121$ as an example, we have $C\left(\frac{1}{n}, \frac{1}{n}\right)$ represent $(x_i \le x_{(1)} \ \& \ y_i \le y_{(1)} i = 12 \ldots 54)/54$. Looking up Table 3.1, we find that there is only one pair, i.e., pair 24 (3360, 121), that satisfies the condition $x_{(1)} = 3360$ and $y_{(1)} = 121$. Thus, we have $C\left(\frac{1}{54}, \frac{1}{54}\right) = \frac{1}{54}$. With this in mind, we can easily compute the empirical copula for the rest of the values, as shown in Figure 3.4.

3.4 Dependence Measure

There are several measures of dependence or association among variables. Five popular measures are Pearson's classical correlation coefficient r, Spearman's ρ, Kendall's τ, chi-plots, and K-plots. These dependence measures were originally developed in the field of nonparametric statistics. Pearson's classical correlation coefficient is also called the linear correlation coefficient or simply correlation coefficient (i.e., sensitive to linear dependence). Spearman's ρ and Kendall's τ are rank-based correlation coefficients based on the concordance and discordance of the dataset. Compared to the classic Pearson correlation coefficient, the rank-based correlation coefficients are more robust. Here, we first use the sample data to illustrate the dependence measurement, then we will show another example using the hydrological data.

3.4.1 Pearson's Classical Correlation Coefficient *r* and Spearman's ρ

Consider a continuous bivariate random variable (X_1, X_2) with marginal distributions $F_1(x_1)$ and $F_2(x_2)$. Spearman's ρ is given by

$$\rho = r(F_1(x_1), F_2(x_2)) \tag{3.66}$$

where r denotes Pearson's linear correlation coefficient.

In other words, rank-based Spearman's ρ represents Pearson's linear correlation coefficient between variables $u = F_1(x_1)$, $v = F_2(x_2)$. Because u and v are both uniform [0, 1] random variables with mean 1/2 and variance 1/12, Spearman's ρ in Equation (3.66) can be rewritten as

$$\begin{aligned}\rho &= \frac{E[F_1(x_1)F_2(x_2)] - E[F_1(x_1)]E[F_2(x_2)]}{\sqrt{Var[F_1(x_1)]Var[F_2(x_2)]}} = \frac{E[F_1(x_1)F_2(x_2)] - \frac{1}{2}\left(\frac{1}{2}\right)}{\sqrt{\frac{1}{12}\left(\frac{1}{12}\right)}} \\ &= 12E[F_1(x_1)F_2(x_2)] - 3\end{aligned} \tag{3.67}$$

In terms of copulas, substituting $u = F_1(x_1)$, $v = F_2(x_2)$ in Equation (3.67), Spearman's ρ is then

$$\rho = E(uv) - 3 = 12\int_{[0,1]^2} uv\, c(u,v)dudv - 3 = 12\int_{[0,1]^2} uv\frac{\partial C^2(u,v)}{\partial u \partial v}dudv - 3 \tag{3.68}$$

After some simple algebra, Equation (3.68) can be rewritten (Schweizer and Wolff, 1981) as follows:

$$\rho = 12\int_{[0,1]^2} C(u,v)dudv - 3 \tag{3.69}$$

The pairwise empirical Spearman ρ_n can be expressed as follows:

$$\rho_n = \frac{\sum_{i=1}^{n}(R_i - \bar{R})(S_i - \bar{S})}{\sqrt{\sum_{i=1}^{n}(R_i - \bar{R})^2\sum_{i=1}^{n}(S_i - \bar{S})^2}} = \frac{12}{n(n+1)(n-1)}\sum_{i=1}^{n} R_iS_i - 3\frac{n+1}{n-1} \tag{3.70}$$

where n is the sample size; R_i is the rank of x_i among $x_1, \ldots, x_n$; and S_i is the rank of y_i among $y_1, \ldots, y_n$; $\bar{R} = \bar{S} = \frac{1}{n}\sum_{i=1}^{n}$; $R_i = \frac{1}{n}\sum_{i=1}^{n}$; $S_i = \frac{n+1}{2}$.

Example 3.14 **Table 3.3 lists six learning datasets $\{(x_i, y_i): i = 1, \ldots, 6\}$. Calculate the rank-based correlation coefficient Spearman's ρ_n.**

Table 3.3. *Learning datasets.*

i	1	2	3	4	5	6
x_i	7.476	11.375	3.595	9.635	10.731	13.942
y_i	8.441	8.952	0.700	10.645	3.665	9.793

Solution: The rank of the dataset is computed as in Table 3.4 and Figure 3.5.

Table 3.4. *Rank of the learning datasets.*

i	1	2	3	4	5	6
x_i	7.476	11.375	3.595	9.635	10.731	13.942
R_i	2	5	1	3	4	6
y_i	8.441	8.952	0.700	10.645	3.665	9.793
S_i	3	4	1	6	2	5

Using Equation (3.70), we have the following:

$$\sum_{i=1}^{n} R_i S_i = [5(6) + 4(4) + 3(1) + 6(5) + 2(2) + 1(3)] = 83$$

$$\rho_n = \frac{12}{n(n+1)(n-1)} \sum_{i=1}^{n} R_i S_i - 3\frac{n+1}{n-1} = \frac{12 \times 86}{6(6+1)(6-1)} - 3\left(\frac{6+1}{6-1}\right) \approx 0.54$$

3.4.2 Kendall's τ

Consider two independent and identically distributed continuous bivariate random variables, (X_1, X_2) and (X_1^*, X_2^*), where $F_1(x_1)$ denotes the marginal distribution for X_1 and X_1^*, and the marginal distribution $F_2(x_2)$ for X_2 and X_2^*. Then, Kendall's τ is given by

$$\tau(X_1, X_2) = P\big((X_1 - X_1^*)(X_2 - X_2^*) > 0\big) - P\big((X_1 - X_1^*)(X_2 - X_2^*) < 0\big) \quad (3.71)$$

In Equation (3.71), the first term measures concordance, and the second term measures discordance. Therefore, Kendall's correlation coefficient τ can be rewritten as

$$\tau(X_1, X_2) = \mathrm{E}\big(\mathrm{sign}(X_1 - X_1^*)(X_2 - X_2^*)\big) \quad (3.72)$$

Now empirical Kendall's τ (τ_n) from bivariate observations can be written as

$$\tau_n = \frac{2}{n(n-1)} \sum_{i=1}^{n-1} \sum_{j=i+1}^{n} \mathrm{sign}\big((x_{1i} - x_{1j})(x_{2i} - x_{2j})\big) \quad (3.73)$$

where n is the number of observations. $\text{sign}(\cdot) = \begin{cases} 1; & \exists (x_{1i} - x_{1j})(x_{2i} - x_{2j}) > 0 \\ 0; & \exists (x_{1i} - x_{1j})(x_{2i} - x_{2j}) = 0 \\ -1; & \exists (x_{1i} - x_{1j})(x_{2i} - x_{2j}) \le 0 \end{cases}.$

In terms of the copula function, Kendall's τ can be expressed from Equation (3.71) as follows:

$$\begin{aligned} \tau(X_1, X_2) &= P\left((X_1 - X_1^*)(X_2 - X_2^*) > 0\right) - P\left((X_1 - X_1^*)(X_2 - X_2^*) < 0\right) \\ &= P\left((X_1 - X_1^*)(X_2 - X_2^*) > 0\right) - \left[1 - P\left((X_1 - X_1^*)(X_2 - X_2^*)\right) > 0\right] \\ &= 2P\left((X_1 - X_1^*)(X_2 - X_2^*) > 0\right) - 1 \end{aligned} \tag{3.74}$$

From Equation (3.74), we also know the following:

$$\begin{aligned} P\left((X_1 - X_1^*)(X_2 - X_2^*) > 0\right) &= P\left(X_1 > X_1^*, X_2 > X_2^*\right) + P\left(X_1 < X_1^*, X_2 < X_2^*\right) \\ &= 1 - P\left(X_1 \le X_1^*\right) - P\left(X_2 \le X_2^*\right) \\ &\quad + 2P\left(X_1 < X_1^*, X_2 < X_2^*\right) \end{aligned} \tag{3.75}$$

Let $u = F_1(x_1), v = F_2(x_2), C(u,v) = P(x_1,x_2)$ $P\left(X_1 < X_1^*, X_2 < X_2^*\right) = P\left(X_1 \le X_1^*, X_2 \le X_2^*\right)$ for continuous random variables. Substituting Equation (3.75) into Equation (3.74), we have the following:

$$\begin{aligned} \tau(X_1, X_2) &= 4E\left[P\left(X_1 \le X_1^*, X_2 \le X_2^*\right)\right] - 2E\left[P\left(X_1 \le X_1^*\right)\right] - 2E\left[P\left(X_2 \le X_2^*\right)\right] + 1 \\ &= 4E[C(u, v)] - 2E(u) - 2E(v) + 1 \\ &= 4\int_{[0,1]^2} C(u, v) dC(u, v) - 1 \end{aligned} \tag{3.76}$$

Example 3.15 **Calculate Kendall's τ_n for the data of Table 3.3.**

Solution: To calculate sample Kendall's τ_n, we will use Equation (3.73). To illustrate the calculation procedure, we will use the first pair ($x_1 = 15.237$, $y_1 = 19.2$) as an example shown in **Table 3.5**.

Table 3.5. *Sample results of computing Kendall's tau.*

Variable	$(x_1 - x_2)(y_1 - y_2)$	$(x_1 - x_3)(y_1 - y_3)$	$(x_1 - x_4)(y_1 - y_4)$	$(x_1 - x_5)(y_1 - y_5)$	$(x_1 - x_6)(y_1 - y_6)$
Result	> 0	> 0	< 0	> 0	> 0
Sign (•)	1	1	1	–1	1
Sum	3				

Similarly, we can compute the sum for the remaining pairs as follows:

Pair (x_2, y_2) compared to $\{(x_i, y_i): i = 3, \ldots, 6\}$, sum = 2;
Pair (x_3, y_3) compared to $\{(x_i, y_i): i = 4, \ldots, 6\}$, sum = 3;
Pair (x_4, y_4) compared to $\{(x_i, y_i): i = 5, 6\}$, sum = –2;
Pair (x_5, y_5) compared to (x_6, y_6), sum = –1;
Finally, using Equation (3.73), we have the following:

$$\tau_n = \frac{2}{6(6-1)} \sum_{i=1}^{5} \sum_{j=2}^{6} \operatorname{sign}\left((x_i - x_j)(y_i - y_j)\right) = \frac{2}{6(6-1)}(3 + 0 + 1 - 2 - 1) \approx 0.47$$

3.4.3 Chi-plot

The chi-plot is based on the chi-square statistic for independence in a two-way table. For bivariate random variables (X_1, X_2), let $H_i = \frac{\sum_{j=1}^{n} 1\left(x_{1j} \le x_{1i}, x_{2j} \le x_{2i}; j \ne i\right)}{n-1}$, $F_i = \frac{\sum_{j=1}^{n} 1\left(x_{1j} \le x_{1i}; j \ne i\right)}{n-1}$, and $G_i = \frac{\sum_{j=1}^{n} 1\left(x_{2j} \le x_{2i}; j \ne i\right)}{n-1}$; the chi-plot can be determined using pairs (λ_i, χ_i)

following Fisher and Switzer (2001) and Genest and Favre (2007) as follows:

$$\chi_i = \frac{H_i - F_i G_i}{\sqrt{F_i(1-F_i)G_i(1-G_i)}} \tag{3.77}$$

$$\lambda_i = 4\operatorname{sign}\left(\tilde{F}_i, \tilde{G}_i\right) \max\left(\tilde{F}_i^2, \tilde{G}_i^2\right) \tag{3.78}$$

where $\tilde{F}_i = F_i - \frac{1}{2}; \tilde{G}_i = G_i - \frac{1}{2}$.

To avoid outliers, Fisher and Switzer (2001) recommended that what should be plotted are only the pairs for which

$$|\lambda_i| \le 4\left(\frac{1}{n-1} - \frac{1}{2}\right)^2 \tag{3.79}$$

To detect how far apart the bivariate random variable is from independence, Fisher and Switzer (2001) also suggested the "control limit" estimated as follows:

$$CL = \pm \frac{c_p}{\sqrt{n}} \tag{3.80}$$

where CL stands for the "control limit" that may also be considered as the confidence bound for independence; n is the sample size; and c_p is the critical value to guarantee that the 100p% of the pair (λ_i, χ_i) falls into the control limit, i.e., for $p = 0.9$, 0.95, 0.99, $c_p = 1.54$, 1.78, 2.18, respectively.

Example 3.16 **Calculate the chi-plot for the data of Table 3.3.**

Solution: Using Equations (3.71) and (3.72), we have the pairs (λ_i, χ_i), as shown in Table 3.6 and Figure 3.5.

Table 3.6. *Coordinates of points, displayed on the chi-plot, for the data of Table 3.3.*

i	1	2	3	4	5	6
H_i	0.2	0.6	0	0.4	0.2	0.8
F_i	0.2	0.8	0	0.4	0.6	1
G_i	0.4	0.6	0	1	0.2	0.8
$\tilde{F}_i$	–0.3	0.3	–0.5	–0.1	0.1	0.5
$\tilde{G}_i$	–0.1	0.1	–0.5	0.5	–0.3	0.3
χ_i	0.61	0.61	—	—	0.41	—
λ_i	0.36	0.36	1	–1	–0.36	1

From Table 3.6, it is seen that only three pairs satisfy the condition, and the control limit for $p = 0.9, 0.95, 0.99$ will be $CL = \pm 0.63, \pm 0.73, \pm 0.89$, respectively.

3.4.4 K-plot

The K-plot was first proposed by Genest and Boies (2003). It is another rank-based graphical tool for detecting dependence. The K-plot consists in plotting pairs $\left(W_{i:n}, H_{(i)}\right)$, $i = 1, \ldots, n$, where $H_{(1)} < \ldots < H_{(n)}$ are the order statistics associated with quantities $H_1 < \ldots < H_n$, i.e.,

$$H_i = \frac{\sum_{j=1}^{n} 1\left(x_{1j} \le x_{1i}, x_{2j} \le x_{2i}; j \ne i\right)}{n-1}.$$

Based on the null hypothesis, i.e., H_0: U and V (or equivalently X and Y) are independent, Genest and Favre (2007) stated that $W_{i:n}$ is the expected value of the ith statistic from a random sample of size n from the random variable $W = C(U, V) = H(X, Y)$ as follows:

$$W_{i:n} = n\binom{n-1}{i-1}\int_0^1 wk_0(w)\{K_0(w)\}^{i-1}\{1 - K_0(w)\}^{n-1}dw \tag{3.81}$$

where

$$K_0(w) = P(UV < w) = \int_0^1 P\left(U \le \tfrac{w}{v}\right)dv = \int_0^w 1dv + \int_w^1 \tfrac{w}{v}dv = w - wln(w)$$

and

$$k_0(w) = \frac{dK_0(w)}{dw} = -\ln(w).$$

Similar to the chi-plot, the K-plot is also capable of detecting how far apart dependence is from independence. We already know the following relations in terms of copula function, i.e., $\Pi = uv; M = \min(u, v)$, $W = \max(u + v - 1, 0)$ for the independent, perfectly positively dependent, and perfectly negatively dependent bivariate random variables, respectively. Graphically, on the K-plot, (i) the $\left(W_{i:n}, H_{(i)}\right)$ pairs follow a straight line $x_2 = x_1$, i.e., $H_{(i)} = W_{i:n}$ if X and Y are independent; (ii) the $\left(W_{i:n}, H_{(i)}\right)$ pairs follow the $K_0(w)$ curve, if X and Y are perfectly positively dependent; and (iii) the $\left(W_{i:n}, H_{(i)}\right)$ pairs fall onto the x_1-axis, i.e., $W_{(i:n)}$, if X and Y are perfectly negatively dependent.

Example 3.17 **Calculate the K-plot for the data of Table 3.3.**

Solution: Let $f(x) = xk_0(w)\{K_0(x)\}^{i-1}\{1 - K_0(x)\}^{n-1}; n = 6$. We can obtain $W_{i:n}$ with the numerical integration. The results are given in Table 3.7 and Figure 3.5.

Table 3.7. *Coordinates of points displayed on the K-plot for the data of Table 3.3.*

i	1	2	3	4	5	6
$W_{i:n}$	0.04	0.09	0.16	0.26	0.38	0.57
$H_{(i)}$	0	0.2	0.2	0.4	0.6	0.8
K_0	0.16	0.31	0.46	0.60	0.75	0.89

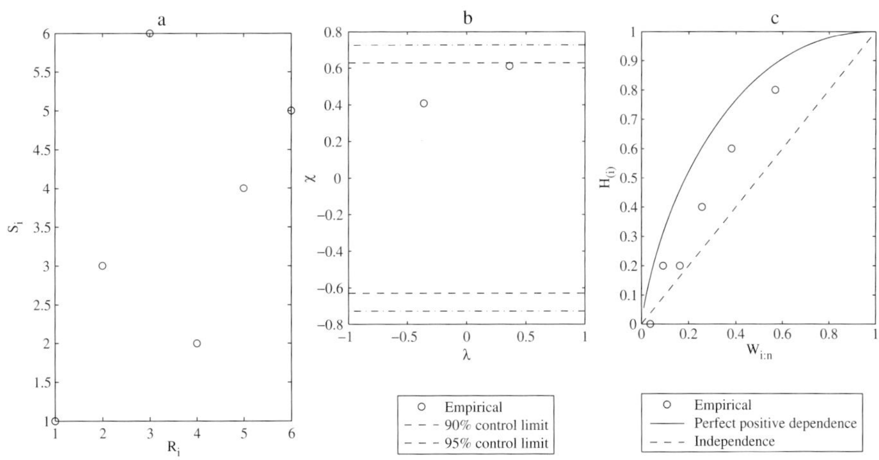

Figure 3.5 Scatter plot of ranked pairs (Ri, Si), chi-plot, and K-plot to detect the dependence for the dataset in Table 3.3. (a) Scatter plot of (Ri, Si); (b) chi-plot with control limit of $P = 0.9$, 0.95; and (c) K-plot with independent and perfectly positively dependent curves.

Example 3.18 **Using the peak discharge and flood volume data given in Table 3.1, (1) calculate sample Spearman's ρ_n and Kendall's τ_n; and (2) graph the chi-plot and K-plot.**

Solution: Table 3.8 lists the rank of flood volume (V) and peak discharge (Q). The computation procedure is exactly the same as that in Examples 3.14–3.17.

Table 3.8. *Rank* ($[R_V, R_Q]$) of the bivariate flood variables.

R_V	V(m^3)	R_Q	Q(cms)	R_V	V(m^3)	R_Q	Q(cms)
28	8,704	47.5	371	42	11,272	52	416
13	6,907	28	245	26	8,640	29	246
3	4,189	14	189	14	6,989	30	248
25	8,637	20	229	35	9,352	39	297
24	8,409	27	240	47	12,825	47.5	371
50	13,602	43	331	51	13,608	54	442
30	8,788	17	206	33	8,949	33	260
5	5,002	5	157	46	12,577	26	236
7	5,167	11	184	44	11,437	44	334
39	10,128	34	275	34	9,266	42	310
45	12,035	36.5	286	52	14,559	49	383
41	10,828	21	230	6	5,057	3	151
32	8,923	24.5	233	38	9,645	16	197
43	1,1401	46	351	16	7,241	35	283
11	6,620	4	156	49	13,543	50	390
2	3,826	6	168	54	15,003	51	405
23	8,192	45	343	10	6,460	8	176
9	6,414	18	214	18	7,502	9	181
31	8,900	41	303	8	5,650	24.5	233
36	9,406	40	300	17	7,350	13	187
15	7,235	2	143	37	9,506	19	216
22	8,177	22.5	232	12	6,728	15	196
19	7,684	10	182	48	13,315	53	424
1	3,306	1	121	21	8,041	31	255
20	8,026	12	186	40	10,174	32	257
4	4,892	7	173	53	14,769	22.5	232
27	8,692	38	292	29	8,711	36.5	286

1. Calculate the sample Spearman's ρ_n and Kendall's τ_n:

Using Equation (3.70) and the same procedure as in Example 3.14, we can easily compute sample $\rho_n = 0.7577$ using Table 3.8 with sample size of 54, as follows:

$$\rho_n = \frac{12}{54(54+1)(54-1)}[28(47.5) + 13(28) + 3(14) + \cdots + 29(36.5)] - 3\frac{54+1}{54-1}.$$

Using Equation (3.73) and the same procedure as in Example 3.15, we can compute sample Kendall's τ_n as $\tau_n = 0.5695$.

Given the double summation for the computation of sample Kendall's tau, here we will show the first intersummation (i.e., $i = 1, j = 2{:}54$), or in other words, comparing $(V, Q) = (8704, 371)$ to the rest of pairs:

$$j = 2,\ \text{sign}((V_1 - V_2)(Q_1 - Q_2)) = \text{sign}((8704 - 6907)(371 - 245) > 0) = 1\ldots$$
$$j = 6,\ \text{sign}((V_1 - V_6)(Q_1 - Q_6)) = \text{sign}((8704 - 13602)(371 - 331) < 0) = -1\ldots$$
$$j = 54,\ \text{sign}((V_1 - V_{54})(Q_1 - Q_{54})) = \text{sign}((8704 - 8711)(371 - 286) < 0) = -1$$

Taking the summation, we have $sum_{i=1} = 14$.

Proceeding with $i = 2$ till $i = 53$, we have the following:

$$sum_{i=2} = 22,\quad sum_{i=3} = 29,\quad \ldots sum_{i=53} = -1$$

and we have the following:

$$\tau_n = \frac{2}{54(54-1)}(14 + 22 + 29 + \ldots - 1) = 0.5695$$

2. Graph the chi-plot and K-plot:

Chi-plot: Using Equations (3.77)–(3.80) with the same procedure as given in Example 3.16, let $F_i = F_V$ and $G_i = F_Q$; F_i, G_i may be directly computed using $F_i = \frac{R_V}{53}$, $G_i = \frac{R_Q}{53}$ from the rank listed in Table 3.8. H_i is similar to the empirical copula, which is computed and listed in Table 3.8. Now we can compute and graph the chi-plot for correlated peak discharge and flood volume variables.

K-plot: Using Equation (3.81) with the same procedure as given in Example 3.17, we may compute and graph the K-plot for correlated peak discharge and flood volume variables. The K-plot involves integration; we can simply use the integral function in MATLAB to obtain results. Figure 3.6 graphs the scatter and chi- and K-plots for correlated peak discharge and flood volume variables.

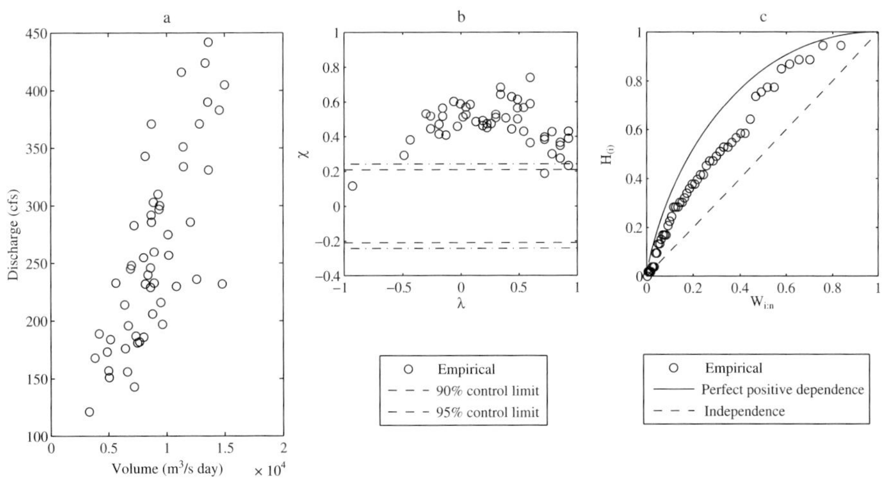

Figure 3.6 Scatter plot of observed data, chi-plot, and K-plot for the hydrological dataset. (a) Scatter plot of observed data; (b) chi-plot with $P = 0.9, 0.95$; and (c) K-plot with independent and perfectly positive dependent curves.

From this example, we see that from calculated sample Spearman's ρ_n and Kendall's τ_n, the peak discharge and flood volume are positively dependent. The chi-plot and K-plot also graphically indicate a positive dependence structure between peak discharge and flood volume.

3.5 Dependence Properties

The dependence between random variables is important for multivariate analysis. Joe (1997), Nelsen (2006), among others, studied the dependence properties of copulas in detail. Here we present the important dependence properties, including positive quadrant and orthant dependence, stochastically increasing positive dependence, right-tail increasing and left-tail decreasing dependence, positive function dependence, and tail dependence.

3.5.1 Positive Quadrant and Orthant Dependence

The positive quadrant dependence (PQD) may be expressed as follows:

$$P(X_1 > a, X_2 > b) \geq P(X_1 > a)P(X_2 > b) \ \ \forall a,\, b \in \mathfrak{R} \tag{3.82a}$$

or

$$P(X_1 \leq a, X_2 \leq b) \geq P(X_1 \leq a)P(X_2 \leq b) = F_1(a)F_2(b) \tag{3.82b}$$

in which X_1, X_2 are the random variables with margins $F_1(x_1)$ and $F_2(x_2)$, respectively.

Similarly, X_1, X_2 are negative quadrant dependent (NQD), if the following relationship is satisfied:

$$P(X_1 > a, X_2 > b) \leq P(X_1 > a)P(X_2 > b) \forall a,\, b \in \mathfrak{R} \tag{3.83a}$$

or

$$P(X_1 \leq a, X_2 \leq b) \leq P(X_1 \leq a)P(X_2 \leq b) = F_1(a)F_2(b) \tag{3.83b}$$

Considering multivariate variables (dimension ≥ 3), the positive upper/lower orthant dependent (PUOD/PLOD) may take place (Joe, 1997). Let $\boldsymbol{X}$ be a random vector with dimension n ($n \geq 3$) with multivariate distribution function H, then PUOD/PLOD states the following:

i. $\boldsymbol{X}$ or H is PUOD if for vector $\mathbf{a}$, $\mathbf{a} \in \mathfrak{R}^n$ such that

$$P(X_i > a_i, i = 1, \ldots, n) \geq \prod_{i=1}^{n} P(X_i > a_i) \tag{3.84}$$

ii. $\boldsymbol{X}$ or H is PLOD if for vector **a**, $\mathbf{a} \in \mathfrak{R}^n$ such that

$$P(X_i \leq a_i, i = 1, \ldots, n) \geq \prod\nolimits_{i=1}^{n} P(X_i \leq a_i) \tag{3.85}$$

Similarly, $\boldsymbol{X}$ or H is NUOD if for vector **a**, $\mathbf{a} \in \mathfrak{R}^n$ such that

$$P(X_i > a_i, i = 1, \ldots, n) \leq \prod\nolimits_{i=1}^{n} P(X_i > a_i) \tag{3.86}$$

and $\boldsymbol{X}$ or H is NLOD if for vector **a**, $\mathbf{a} \in \mathfrak{R}^n$ such that

$$P(X_i \leq a_i, i = 1, \ldots, n) \leq \prod\nolimits_{i=1}^{n} P(X_i \leq a_i) \tag{3.87}$$

It is seen from Equations (3.84) and (3.85) that multivariate random variables $X_1, \ldots, X_n$ are more likely having large values simultaneously, compared to the independence assumption. Similarly, Equations (3.86) and (3.87) show that multivariate random variables $X_1, \ldots, X_n$ are more likely having small values simultaneously, compared to the independence assumption.

Example 3.19 **Explain that the following Gumbel–Hougaard copula holds the positive quadrant dependence property.**

$$C(u, v) = \exp\left\{-\left[(-\ln u)^\theta + (-\ln v)^\theta\right]^{\frac{1}{\theta}}\right\}, \theta \geq 1; \tau = 1 - \frac{1}{\theta}; u = F_X(x), v = F_Y(y) \tag{3.88}$$

Solution: From Equation (3.88), with $\theta \geq 1$, we have the Kendall correlation coefficient $\tau \in [0, 1]$. With the robust Kendall correlation, it is guaranteed that the random variables are positively dependent. From the theorem of Fréchet–Hoeffding bounds, the product copula (i.e., $\Pi = uv$) represents independence (i.e., $\tau = 0$) and $M = \min(u, v)$ represents the perfectly correlated random variables (i.e., $\tau = 1$) with the relation of $\Pi \leq M$. Then we have the following: $\Pi < C(u, v) < M$ for the positively correlated random variables with $0 < \tau < 1$:

$$C(u, v) = F(x, y) = P(X \leq x, Y \leq y) > F_X(x) F_Y(y) = uv$$

The preceding relation aligns with Equation (3.82b) and holds the positive quadrant property. To illustrate this property graphically, we will use $\theta = 2.5$ as an example:

$$\theta = 2.5 \Rightarrow \tau = 1 - \frac{1}{2.5} = 0.6.$$

Figure 3.7 plots the comparison of Equation (3.88) and product copula with different pairs of (u, v). Figure 3.7 graphically shows that the JCDF computed using Equation (3.89) with $\theta = 2.5$ is greater than that computed from the product copula (i.e., fulfilling Equation (3.82b)).

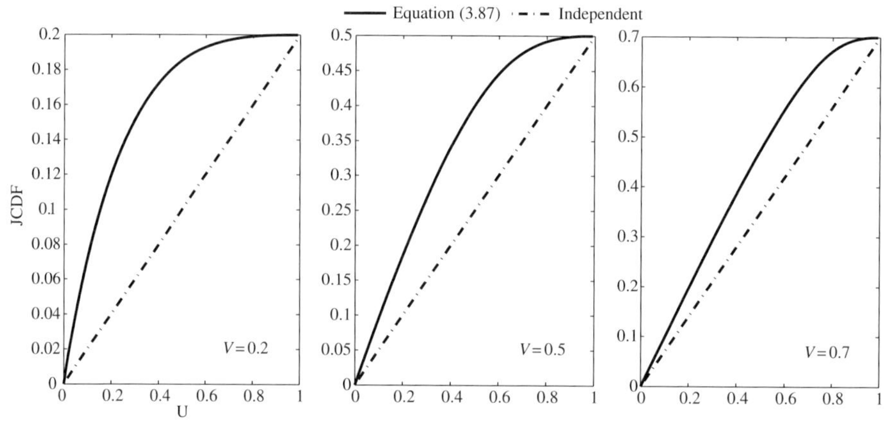

Figure 3.7 Comparison of Equation (3.88) with the product copula (independent).

3.5.2 Stochastic Increasing Positive Dependence

Bivariate Stochastic Positive Dependence

Let there be random variables X_1, X_2 with the joint probability distribution, $F(x_1, x_2)$, and marginal $F_1(x_1)$, $F_2(x_2)$. Then, X_1 is stochastically increasing (SI) in X_2, or, in other words, the conditional probability distribution $F(X_1|X_2)$ is stochastically increasing, if the following relationship exists:

$$P(X_1 > x_1|X_2 = x_2) = 1 - F(X_1 \leq x_1|X_2 = x_2)$$

is a nondecreasing function of x_1 for all x_2.

Similarly, we say X_1 is stochastically decreasing (SD) in X_2 if $P(X_1 > x|X_2 = x_2)$ is a nonincreasing function of x_1 for all x_2.

Multivariate Stochastic Positive Dependence

As introduced by Joe (1997), a random vector $\boldsymbol{X}$, $\boldsymbol{X} = (X_1, X_2, \ldots, X_n)$ is stochastically positive dependent if $\{X_i: i \neq j\}$ conditional on $X_j = x$ is increasing stochastically, as x increases for all $j = 1, \ldots, n$. The random vector $\boldsymbol{X}$ is conditional increasing in sequence in $X_1, \ldots, X_{i-1}$ for $i = 2, \ldots, n$, if $P\left(X_i > x_i|X_j = x_j, j = 1, 2, \ldots, i-1\right)$ is increasing in $x_1, \ldots, x_{i-1}$ for all x_i.

Example 3.20 **Rework Example 3.19 to evaluate the stochastic dependence property for the copula given in Equation (3.88) with $\theta = 2.5$.**

Solution: We first derive the conditional probability distribution as follows:

$$F(X \leq y|Y = y) = \left.\frac{\partial C(u, v)}{\partial v}\right|_{V=v} \tag{3.89}$$

Taking the partial derivative of Equation (3.89), we have the following:

$$F(X \leq x|Y = y) = C(U|V = v) = \frac{(-\ln v)^{1.5} \exp\left\{-\left[(-\ln u)^{2.5} + (-\ln v)^{2.5}\right]^{0.4}\right\}}{v\left[(-\ln v)^{2.5} + (-\ln u)^{2.5}\right]^{0.6}} \tag{3.90a}$$

$$F(X > x|Y = y) = 1 - C(U|V = v) = 1 - \frac{(-\ln v)^{1.5} \exp\left\{-\left[(-\ln u)^{2.5} + (-\ln v)^{2.5}\right]^{0.4}\right\}}{v\left[(-\ln v)^{2.5} + (-\ln u)^{2.5}\right]^{0.6}} \tag{3.90b}$$

Again let $v = 0.2, 0.5, 0.7$. Figure 3.8 plots Equation (3.90). Figure 3.8a plots the conditional copula (i.e., conditional cumulative distribution function) with different v. Figure 3.8b plots the exceedance conditional copula (i.e., the exceedance conditional distribution) with different v. Figure 3.8b clearly shows that the exceedance conditional copula is nondecreasing for any given u with increasing v, i.e., $C(u|V = 0.2) \leq C(u|V = 0.5) \leq C(u|V = 0.7)$. This indicates the stochastic increasing (SI) property of the copula function given in Equation (3.88).

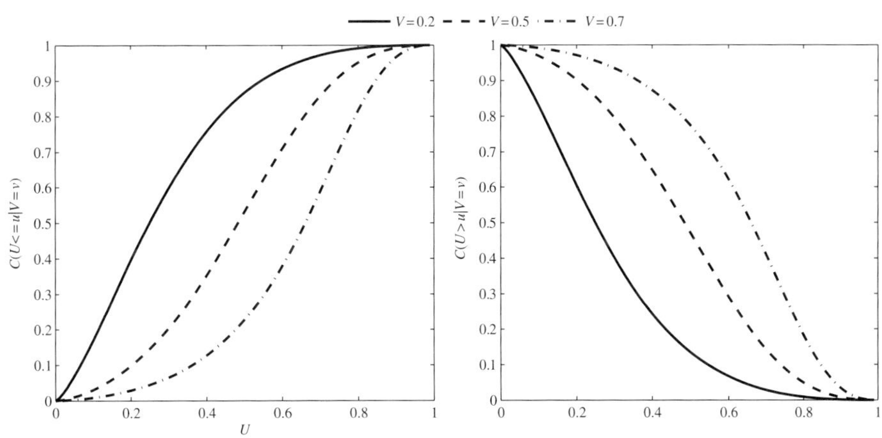

Figure 3.8 Comparison of the conditional copula (i.e., $\boldsymbol{P}(\boldsymbol{X} \leq \boldsymbol{x}|\boldsymbol{Y} = \boldsymbol{y})$) and the exceedance conditional copula (i.e., $\boldsymbol{P}(\boldsymbol{X} > \boldsymbol{x}|\boldsymbol{Y} = \boldsymbol{y})$).

3.5.3 Tail Dependence

Nelsen (2006) introduced the tail dependence as follows.

Population Version of Tail Dependence

Let X_1 and X_2 be two random variables. Then,

a. X_2 is left tail decreasing (LTD) in X_1, i.e., LTD $(X_2|X_1)$, if $P(X_2 \leq x_2|X_1 \leq x_1)$ is a nonincreasing function of x_1 for all x_2. Similarly, if $P(X_1 \leq x_1|X_2 \leq x_2)$ is a nonincreasing function of x_2 for all x_1, then there exists LTD $(X_1|X_2)$.
b. X_2 is left tail increasing (LTI) in X_1, i.e., LTI $(X_2|X_1)$, if $P(X_2 \leq x_2|X_1 \leq x_1)$ is a nondecreasing function of x_1 for all x_2. Similarly, if $P(X_1 \leq x_1|X_2 \leq x_2)$ is a nondecreasing function of x_2 for all x_1, then there exists LTI $(X_1|X_2)$.
c. X_2 is right tail increasing (RTI) in X_1, i.e., RTI $(X_2|X_l)$, if $P(X_2 > x_2|X_1 > x_1)$ is a nondecreasing function of x_1 for all x_2. Similarly, if $P(X_1 > x_1|X_2 > x_2)$ is a nondecreasing function of x_2 for all x_1, then there exists RTI $(X_1|X_2)$.
d. X_2 is right tail decreasing (RTD) in X_1, i.e., RTD $(X_2|X_1)$, if $P(X_2 > x_2|X_1 > x_1)$ is a nonincreasing function of x_1 for all x_2. Similarly, if $P(X_1 > x_1|X_2 > x_2)$ is a nonincreasing function of x_2 for all x_1, then there exists RTD $(X_1|X_2)$.

Copula Version of Tail Dependence

The copula version of tail dependence is given as Theorem 5.2.5 in Nelsen (2006).

Let X_l, X_2 be continuous random variables with margins $u = F_1(x_1); v = F_2(x_2)$ and the joint distribution represented by copula C. Then, the theorem says the following:

a. There exists LTD $(X_1|X_2)$ if and only if for any v in $\mathbf{I}$ = [0,1], such that $C(u, v)/v$ is nonincreasing in u. Similarly, LTD $(X_2|X_1)$ exists if and only if for any u in $\mathbf{I}$ = [0,1], such that $C(u, v)/u$ is nonincreasing in v.
b. There exists RTI $(X_1|X_2)$ if and only if for any v in $\mathbf{I}$ = [0,1], such that $[1 - u - v + C(u, v)]/(1 - v)$is nondecreasing in v, or equivalently, if $[u - C(u, v)]/(1 - v)$ is nonincreasing in v. Similarly, RTI $(X_2|X_1)$ exists if and only if for any u in $\mathbf{i}$ = [0,1], such that $[1 - u - v + C(u, v)]/(1 - u)$ is nondecreasing in u, or equivalently, if $[v - C(u, v)]/(1 - u)$ is nonincreasing in u.

Example 3.21 **Rework Example 3.20 to evaluate that the tail dependence of the copula function given in Equation (3.88) with parameter θ=2.5 holds the RTI property.**

Solution: To show the copula function (3.88) holds the RTI property, we need to show that $P(X_1 > x_1|X_2 > x_2)$ is a nondecreasing function of x_2 for all x_1 or equivalently to show that $[u - C(u, v)]/(1 - v)$ is nonincreasing in v. Similar to previous two examples, we will again use

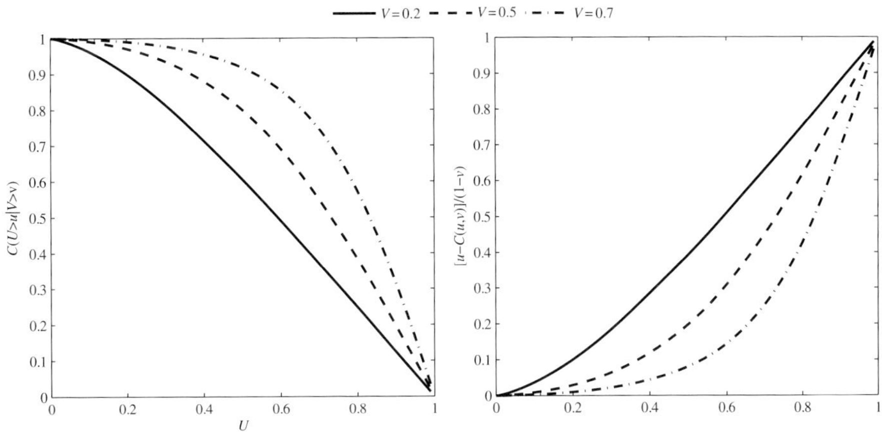

Figure 3.9 Graphical evaluation of tail dependence for copula function (Equation (3.90)).

$v = 0.2,\ 0.5,\ 0.7$ as an illustrative example. Figure 3.9 plots the exceedance joint distribution and corresponding $[u - C(u, v)]/(1 - v)$ for $u = 0{:}\ 0.01 : 0.99$. Figure 3.9(b) shows that given $V > v$, the conditional copula $C(U > u|V > v)$ is a nonincreasing function on v. In other words, $C(U > u|V > v)$ decreases for $V > v$ with the increase of v. Thus, the copula function given in Equation (3.88) holds the RTI property. Using three pairs of (u,v), (0.3,0.2), (0.3,0.5), and (0.3, 0.7), for an illustrative example, we have the following:

$$\left\{\frac{0.3 - C(0.3, 0.2; 2.5)}{1 - 0.2} = \frac{0.3 - 0.1519}{0.8} = 0.1852\right\}$$
$$> \left\{\frac{0.3 - C(0.3, 0.5; 2.5)}{1 - 0.5} = 0.0641\right\}$$
$$> \left\{\frac{0.3 - C(0.3, 0.7, 2.5)}{1 - 0.7} = 0.0224\right\}$$

Theoretically, we can prove the RTI property by taking the first-order derivative with respect to v and have the following:

$$\frac{dC(U > u|V > v)}{dv} = \frac{d\left[\dfrac{u - C(u, v)}{1 - v}\right]}{dv} = \frac{-(1 - v)\dfrac{dC}{dv} - (u - C(u, v))(-1)}{(1 - v)^2}$$
$$= \frac{(v - 1)\dfrac{dC}{dv} + u - C(u, v)}{(v - 1)^2} \tag{3.91}$$

To show the copula function (i.e., Equation (3.90)) holds the RTI property, we need to show that Equation (3.91) is equal to or less than 0 in what follows:

$$(v - 1)\frac{dC}{dv} + u - C(u, v) \leq 0 \tag{3.92}$$

Taking the first-order derivative of Equation (3.88) with respect to v, we have the following:

$$(v-1)\frac{dC}{dv}+u-C(u,v)=\frac{(v-1)C(u,v)(-\ln v)^{1.5}}{v\left[(-\ln u)^{2.5}+(-\ln v)^{2.5}\right]^{0.6}}+\frac{uv}{v}-C(u,v) \tag{3.93}$$

In Equation (3.93), we have the following inequalities:

$$\begin{cases} uv \le C(u,v) \\ \left[(-\ln u)^{2.5}+(-\ln v)^{2.5}\right]^{0.6} \ge (-\ln v)^{1.5},\ \forall 0 \le u < v \le 1 \\ \left[(-\ln u)^{2.5}+(-\ln v)^{2.5}\right]^{0.6} \ge (-\ln u)^{1.5},\ \forall 0 \le v < u \le 1 \end{cases} \tag{3.94}$$

Substituting Equation (3.94) back into Equation (3.93), Equation (3.92) may be rewritten as follows:

$$(v-1)\frac{dC}{dv}+u-C(u,v) \le \begin{cases} \frac{(v-1)C(u,v)}{v}\left[\left(\frac{-\ln v}{-\ln v}\right)^{1.5}-1\right]=0,\ \forall 0 \le u < v \le 1 \\ \frac{(v-1)C(u,v)}{v}\left[\left(\frac{-\ln v}{-\ln u}\right)^{1.5}-1\right] \le 0,\ \forall 0 \le v < u \le 1 \end{cases} \tag{3.95}$$

Equation (3.95) proves that the Equation (3.93) is equal to or less than 0, i.e., $[u-C(u,v)]/(1-v)$ is a nonincreasing function of v. Hence, the copula function in Equation (3.88) holds the RTI property.

3.5.4 Likelihood Ratio Dependence

The likelihood ratio dependence was discussed as Theorem 5.2.18 in Nelsen (2006). This theorem says that if there are two continuous random variables X_1 and X_2 whose joint density function is $f(x_1, x_2)$, then X_1 and X_2 are positively likelihood ratio dependent if the following inequality is satisfied:

$$f(x_1,x_2)f(x_1',x_2') \ge f(x_1,x_2')f(x_1',x_2) \tag{3.96}$$

for all x_1, x_2, x_1', x_2' in $\bar{\mathbf{R}}$ such that $x_1 \le x_1'$, $x_2 \le x_2'$. This is also called the total positivity of power 2 (TP2) (Joe, 1997).

The preceding discussion introduces the dependence structure and properties of the copulas that are most important for multivariate analysis in later chapters.

3.6 Copula Parameter Estimation

For a d-dimensional random sample $(X_1, X_2, \ldots, X_d)$ with marginal distribution $F_1(X_1)$, …, $F_d(X_d)$, let $f(X_1, \ldots, X_d)$ be the joint density function of the d-dimensional random variables and $F(X_1, \ldots, X_d)$ be the corresponding joint distribution. Here we discuss some general methods used to estimate copula parameters.

3.6.1 Exact Maximum Likelihood Estimation Method

The exact maximum likelihood estimation method is also called the one-stage method or full maximum likelihood (full ML). The full ML method estimates the parameters of marginal distributions and copula function simultaneously. Let $\boldsymbol{\Theta} = (\boldsymbol{\alpha}_1, \ldots, \boldsymbol{\alpha}_\mathbf{d}, \boldsymbol{\theta})$ be the parameters that need to be estimated in which $\boldsymbol{\alpha}_\mathbf{i}$, $i = 1, \ \ldots, \ d,$ is the parameter for marginal variable X_i and $\boldsymbol{\theta}$ is the copula parameter. Based on the relation between copula density and joint density function of d-dimensional variables, i.e., $f(X_1, \ldots, X_d; \boldsymbol{\Theta}) = c(U_1, \ldots, U_d; \boldsymbol{\theta})\prod_{i=1}^{d} f_i(X_i; \boldsymbol{\alpha}_i)$, the log-likelihood function is given as follows:

$$\begin{aligned}\mathrm{logL}(\boldsymbol{\Theta}) &= \sum\nolimits_{n=1}^{n} \ln\left[f(x_{1i}, \ldots, x_{di})\right] \\ &= \sum\nolimits_{i=1}^{n} \ln\left(c(F_1(x_{1i}; \boldsymbol{\alpha}_1), \ldots, F_d(x_{di}; \boldsymbol{\alpha}_\mathbf{d}); \boldsymbol{\theta})\right) + \sum\nolimits_{j=1}^{d}\sum\nolimits_{i=1}^{n} \ln f_i\left(x_{ji}; \boldsymbol{\alpha}_\mathbf{i}\right)\end{aligned} \tag{3.97}$$

The log-likelihood function can be maximized numerically by solving for $\boldsymbol{\Theta}$, i.e.,

$$\hat{\boldsymbol{\Theta}}_{FML} = \mathrm{argmax}\ (\log \mathrm{L}(\boldsymbol{\Theta}))$$

as follows:

$$\begin{cases} \dfrac{\partial logL(\boldsymbol{\Theta})}{\partial \boldsymbol{\alpha}_1} = 0 \\ \quad \ldots \\ \dfrac{\partial logL(\boldsymbol{\Theta})}{\partial \boldsymbol{\alpha}_\mathbf{d}} = 0 \\ \dfrac{\partial logL(\boldsymbol{\Theta})}{\partial \boldsymbol{\theta}} = 0 \end{cases} \tag{3.98}$$

Equation (3.98) shows that with increasing scale of the problem, the algorithm can be too burdensome computationally.

3.6.2 Inference Function for Marginal Method

In the inference function for margins (IFM) method, a two-stage method, i.e., the parameters of marginal distribution and copula function are estimated separately. First, parameter $\boldsymbol{\alpha}_i$ of the marginal distribution $\hat{F}_i(X_i; \boldsymbol{\alpha}_\mathbf{i})$ is estimated; and then the fitted marginal distribution is passed into the copula function to estimate its parameter θ.

1. The log-likelihood function of each of the marginal distributions is given as

$$\log \mathrm{L}(\boldsymbol{\alpha}_\mathbf{i}) = \sum_{j=1}^{n} \ln f\left(x_{ij}; \boldsymbol{\alpha}_\mathbf{i}\right); i = 1,\ 2,\ \ldots d \tag{3.99}$$

$\boldsymbol{\alpha}_{\mathbf{i}}$ is estimated by $\hat{\boldsymbol{\alpha}}_i = \text{argmax}(\log \mathrm{L}(\boldsymbol{\alpha}_{\mathbf{i}}))$, i.e.,

$$\frac{\partial \log \mathrm{L}(\boldsymbol{\alpha}_{\mathbf{i}})}{\partial \boldsymbol{\alpha}_{\mathbf{i}}} = 0 \tag{3.100}$$

2. With the fitted marginal distributions, the log-likelihood function for the copula can be given as

$$\log \mathrm{L}(\hat{\boldsymbol{\alpha}}_1, \ldots, \hat{\boldsymbol{\alpha}}_d; \boldsymbol{\theta}) = \sum_{i=1}^{n} \ln \left(F_1(x_{1i}; \hat{\boldsymbol{\alpha}}_1), \ldots, F_d(x_{di}; \hat{\boldsymbol{\alpha}}_d); \boldsymbol{\theta}\right) \tag{3.101}$$

Maximizing Equation (3.101) over θ,
$\hat{\boldsymbol{\theta}}_{IFM} = \boldsymbol{argmax}(\log\mathrm{L}(\boldsymbol{\theta}))$
which is estimated by setting the following:

$$\frac{\partial \log \mathrm{L}(\boldsymbol{\theta})}{\partial \boldsymbol{\theta}} = 0 \tag{3.102}$$

Comparing the IFM method with the full ML method, the IFM method is computationally more efficient than the full ML method. However, if the marginal distribution is misidentified, the accuracy of the copula (or joint distribution) estimated will be undermined.

The semiparametric approach provides the ability to avoid the misidentification of marginal distributions and is discussed in the following section.

3.6.3 Semiparametric Method

The semiparametric method is more flexible. In this method, copula parameters are estimated with the maximum likelihood estimation method, using nonparametric empirical distribution functions rather than the fitted parametric marginals. Using the commonly applied Weibull plotting position formula, the empirical probability is written as follows:

$$\hat{F}_i(x) = \frac{1}{n+1} \sum_{j=1}^{n} 1\left(X_{ij} \le x\right), \; i = 1, \; \ldots, \; d \tag{3.103}$$

Replacing the fitted marginal distribution in Equation (3.101) with the results obtained from Equation (3.103), the copula parameters can be estimated by maximizing the following pseudo-log-likelihood function:

$$\log \mathrm{L}(\boldsymbol{\theta}) = \sum_{j=1}^{n} \ln \left(c\left(\hat{F}_1\left(x_{1j}\right), \ldots, \hat{F}_d\left(x_{dj}\right); \boldsymbol{\theta}\right)\right) \tag{3.104}$$

For a set of copula candidates, the copula function reaching the largest log-likelihood is usually considered as the best-fitted copula to represent the multivariate distribution function for given multivariate continuous random variables.

In what follows, we will give one synthetic example to illustrate how to apply the preceding three methods to estimate the copula parameters.

Example 3.22 **Using the correlated random variables.**

This example uses the correlated random variables listed in Table 3.9 with the assumption of random variables X and Y following the gamma and Gumbel distributions, respectively, and the joint distribution following the Gumbel–Houggard copula as follows:

$$C(u, v; \theta) = \exp\left(-\left((-\ln u)^{\theta} + (-\ln v)^{\theta}\right)^{\frac{1}{\theta}}\right)$$

Estimate the parameters using the previously discussed full ML, IFM, and semiparametric methods.

Table 3.9. *Synthetic random variables.*

No.	X	Y
1	2.3284	16.2698
2	0.8867	8.6807
3	1.4106	11.2295
4	1.9654	12.1751
5	1.0221	7.5978
6	1.2089	8.8760
7	0.6915	9.0297
8	1.5375	10.2731
9	1.9472	13.4256
10	1.0080	8.9696
11	2.2308	10.2306
12	0.7600	7.4901
13	1.7782	11.1462
14	3.6810	15.2615
15	2.4564	13.1492
16	4.1957	19.5030
17	2.5038	12.4057
18	3.6670	16.4510
19	0.4646	5.9375
20	1.1004	10.1990
21	0.4608	10.1966
22	2.0799	11.5089
23	0.9049	9.2902
24	0.5785	7.4861
25	1.1199	9.1667
26	1.9836	13.0043
27	0.8940	8.6892
28	3.6308	17.6573
29	1.4556	10.5674
30	1.8813	9.4640

Solution: Before we proceed to estimate the parameters, we first give the density function for the Gumbel–Houggard copula as follows:

$$c(u,v;\theta)=\frac{\partial^2 C(u,v)}{\partial u\partial v}=e^{-S_1^{\frac{1}{\theta}}}(\ln u\ln v)^{\theta-1}\left(\frac{S_1^{\frac{2}{\theta}-2}-(1-\theta)S_1^{\frac{1}{\theta}-2}}{uv}\right),$$

$$S_1=(-\ln u)^{\theta}+(-\ln v)^{\theta}$$

Full ML method: The gamma and Gumbel density functions can be given as follows:

Gamma: $f_X(x;\alpha_x,\beta_x)=\frac{\beta_x^{\alpha_x}x^{\alpha_x-1}}{\Gamma(\alpha_x)}e^{-\beta_x x};$

Gumbel: $f_Y\left(y;\mu_y,\beta_y\right)=\frac{1}{\beta_y}\exp\left(-\left(\frac{y-\mu_y}{\beta_y}+\exp\left(-\frac{y-\mu_y}{\beta_y}\right)\right)\right)$

Now, the joint density function, i.e., $f(x,y)$ can be expressed as follows:

$$f(x,y)=f_X(x;\alpha_x,\beta_x)f_Y\left(y;\mu_y,\beta_y\right)c\left(F_X(x;\alpha_x,\beta_x),F_Y\left(y;\mu_y,\beta_y\right);\theta\right)$$

The log-likelihood function of $f(x,y)$ can be expressed as follows:

$$\log \mathrm{L}=\sum_{i=1}^{n}\ln\left(c\left(F_X(x_i;\alpha_x,\beta_x),F_Y\left(y_i;\mu_y,\beta_y\right);\theta\right)\right)$$
$$+\sum_{i=1}^{n}\ln f_X(x_i;\alpha_x,\beta_x)+\sum_{i=1}^{n}\ln f_Y\left(y_i;\mu_y,\beta_y\right)$$

Now, by maximizing the preceding log-likelihood function with the use of Equations (3.97) and (3.98), we can estimate all five parameters simultaneously. One may also use the optimization toolbox in MATLAB to estimate the parameters by minimizing the negative log-likelihood function. Here we use the optimization toolbox in MATLAB to estimate the parameters. The estimated parameters and corresponding log-likelihood value are listed in Table 3.10.

Table 3.10. *Estimated parameters using full ML, IFM, and semiparametric methods.*

	Univariate		Copula	
Method	$X\sim gamma(\alpha_x,\beta_x)$	$Y\sim gumbel\left(\mu_y,\beta_y\right)$	$GH\ (\theta)$	LL
Full ML	(3.0691, 0.5674)	(9.7235, 2.5083)	3.5236	–87.4934
IFM	(3.0782, 0.5613)	(9.7271, 2.4681)	3.4760	25.8129
Semiparametric	–	–	3.5570	23.6911

IFM method: The IFM method estimates the parameters of marginals and copula function separately.

First, we need to estimate the parameters for the marginal distributions using the ML method as follows:

$$X\sim Gamma(3.0782,0.5613);\ Y\sim Gumbel(9.7271,2.4681)$$

Second, we use the fitted probability distribution to compute the cumulative probability listed in Table 3.11.

Third, use the computed cumulative probability from the fitted probability distribution to estimate the copula parameter by maximizing the log-likelihood function of copula density function or minimizing its negative log-likelihood function. Again using the optimization

Table 3.11. *Estimated cumulative distributions using fitted and empirical probability distributions.*

X	Gamma	Empirical	Y	Gumbel	Empirical
2.3284	0.7695	0.7742	16.2698	0.9318	0.8710
0.8867	0.1967	0.1935	8.6807	0.2170	0.1613
1.4106	0.4404	0.4516	11.2295	0.5804	0.6129
1.9654	0.6628	0.6452	12.1751	0.6901	0.6774
1.0221	0.2583	0.3226	7.5978	0.0935	0.1290
1.2089	0.3464	0.4194	8.8760	0.2437	0.2258
0.6915	0.1166	0.1290	9.0297	0.2654	0.2903
1.5375	0.4969	0.5161	10.2731	0.4486	0.5161
1.9472	0.6567	0.6129	13.4256	0.7997	0.8065
1.0080	0.2517	0.2903	8.9696	0.2568	0.2581
2.2308	0.7439	0.7419	10.2306	0.4424	0.4839
0.7600	0.1431	0.1613	7.4901	0.0841	0.0968
1.7782	0.5954	0.5484	11.1462	0.5697	0.5806
3.6810	0.9550	0.9355	15.2615	0.8992	0.8387
2.4564	0.7999	0.8065	13.1492	0.7788	0.7742
4.1957	0.9773	0.9677	19.5030	0.9811	0.9677
2.5038	0.8103	0.8387	12.4057	0.7133	0.7097
3.6670	0.9542	0.9032	16.4510	0.9365	0.9032
0.4646	0.0458	0.0645	5.9375	0.0096	0.0323
1.1004	0.2950	0.3548	10.1990	0.4378	0.4516
0.4608	0.0448	0.0323	10.1966	0.4375	0.4194
2.0799	0.6999	0.7097	11.5089	0.6152	0.6452
0.9049	0.2047	0.2581	9.2902	0.3031	0.3548
0.5785	0.0777	0.0968	7.4861	0.0838	0.0645
1.1199	0.3042	0.3871	9.1667	0.2851	0.3226
1.9836	0.6690	0.6774	13.0043	0.7672	0.7419
0.8940	0.1999	0.2258	8.6892	0.2181	0.1935
3.6308	0.9520	0.8710	17.6573	0.9606	0.9355
1.4556	0.4607	0.4839	10.5674	0.4909	0.5484
1.8813	0.6336	0.5806	9.4640	0.3287	0.3871

toolbox in MATLAB, the fitted copula parameters and their log-likelihood value are listed in Table 3.10.

Semiparametric method: The semiparametric method estimates the parameter of copula function using the empirical marginal distributions, which is free of identification of marginal distributions.

First, we use the Weibull probability plotting-position formula to compute the empirical probabilities, which are listed in Table 3.11.

Second, we estimate the copula parameter using the computed empirical probabilities. Here we again use the optimization toolbox in MATLAB to estimate the parameters. The estimated parameter and the corresponding log-likelihood value are listed in Table 3.11.

Table 3.10 shows that the parameters of marginal distributions, estimated using the full ML method, are very close to those estimated separately by the IFM method. The copula parameter values estimated using all three methods are also very close to each other.

3.7 Copula Simulation

A common simulator for copula is the cumulative probability integral (CPI) Rosenblatt transformation. Let $\mathbf{X} = (X_1, X_2, \ldots, X_d)$ be a d-dimensional, absolutely continuous random variable, $H(x_1, x_2, \ldots, x_d)$ the joint distribution function, and $F_{X_i}(x_i) = P(X_i \le x_i)$, $i = 1, 2, \ldots, d$, the univariate marginals. In what follows, we introduce how to simulate copula samples using the CPI Rosenblatt transformation (Rosenblatt, 1952). The Rosenblatt transformation can be written as follows:

$$Z_1 = P(X_1 \le x_1) = F_{X_1}(x_1) \tag{3.105}$$

$$Z_2 = P(X_2 \le x_2 | X_1 = x_1) = F_{X_2|X_1}(x_2|x_1) = C_2(u_2|u_1) = \partial C_2(u_1, u_2)/\partial u_1 \tag{3.106}$$

$$\ldots$$

$$\begin{aligned} Z_d &= P(X_d \le x_d | X_1 = x_1, \ldots X_{d-1} = x_{d-1}) = C_d(u_d | u_1, \ldots, u_{d-1}) \\ &= \frac{\partial^{d-1} C_d(u_1, \ldots, u_d)}{\partial u_1 \ldots \partial u_{d-1}} \Bigg/ \frac{\partial^{d-1} C_{d-1}(u_1, \ldots, u_{d-1})}{\partial u_1 \ldots \partial u_{d-1}} \end{aligned} \tag{3.107}$$

Let U(0, 1) denote the uniform distribution on [0,1]. The following procedure generates a d-dimensional random variate $(u_1, \ldots, u_d)$ from copula $C(u_1, \ldots, u_d) = C_d(u_1, \ldots, u_d)$:

1. Simulate independent random variates $v_1, \ldots, v_d$ from $U(0, 1)$ and set $u_1 = v_1$.
2. Simulate random variate u_2 from $v_2 = C_2(u_2|u_1)$ by solving $u_2 = C_{2|1}^{-1}(v_2, u_1)$.

   ...

3. Simulate random variate u_d from $v_d = C_d(u_d|u_1, \ldots, u_{d-1})$ by solving

$$u_d = C_{d|1,\ldots,d-1}^{-1}(v_d, u_1, \ldots, u_{d-1}).$$

Example 3.23 **Simulate the bivariate random variable for the Clayton copula.**

The Clayton copula is as follows:

$$C(u_1, u_2; \theta) = \left(u_1^{-\theta} + u_2^{-\theta} - 1\right)^{-\frac{1}{\theta}}; \theta \geq -1$$

Solution: First, generate two independent random variates (v_1, v_2) from $U(0, 1)$, and set $u_1 = v_1$. Then,

$$C(u_2|u_1) = \frac{\partial C_2(u_1, u_2)}{\partial u_1} = \left(u_1^{-\theta} + u_2^{-\theta} - 1\right)^{-\frac{1}{\theta}-1} u_1^{-(\theta+1)} = \left(1 + u_1^{\theta}\left(u_2^{-\theta} - 1\right)\right)^{-\frac{1}{\theta}-1}$$

Solving the equation $v_2 = C(u_2|u_1)$ for u_2 yields

$$u_2 = C_{2|1}^{-1}(v_2, u_1) = \left(\left(v_2^{-\frac{\theta}{1+\theta}} - 1\right)u^{-\theta} + 1\right)^{-\frac{1}{\theta}}$$

Using a synthetic example with generated independently uniformly distributed random variate (0.6036, 0.4028) with the copula parameter $\theta = 0.5$ (Clayton copula), set the following:

$$u_1 = v_1 = 0.6036; v_2 = C_{2|1}(u_2|u_1) = 0.4028$$

Then we can compute the following:

$$u_2 = \left(\left(0.4028^{-\frac{0.5}{1.5}} - 1\right)\left(0.6036^{-0.5}\right) + 1\right)^{-\frac{1}{0.5}} = 0.4719$$

3.8 Goodness-of-Fit Tests for Copulas

Besides choosing the copula function that reaches the largest log-likelihood (or minimum negative log-likelihood, Akaike information criterion [AIC], Bayesian information criterion [BIC]) from possible copula functions tested, the goodness-of-fit test further ensures the appropriateness of the selected copula functions. Currently, there exist seven formal goodness-of-fit tests for copulas: (1) two tests based on the empirical copula with test statistics: S_n, T_n; (2) two tests based on Kendall's transform with test statistics: $S_n^{(K)}$, $T_n^{(K)}$; and (3) three tests based on Rosenblatt's transform with test statistics: A_n, $S_n^{(B)}$, $S_n^{(C)}$. S_n, $S_n^{(K)}$, $S_n^{(B)}$, and $S_n^{(C)}$, which are calculated based on Cramér–von Mises statistics; and T_n, $T_n^{(K)}$, which are calculated based on the Kolmogorov–Smirnov statistics. According to Genest et al. (2007), the preference ranking for these goodness-of-fit tests is $S_n^{(B)} \succ S_n \succ S_n^{(C)} \succ T_n \succ A_n \succ T_n^{(K)}$. In this section, we present procedures on how to calculate the goodness-of-fit statistics for bivariate random variables following Genest et al. (2007). All the preceding test statistics can be extended to higher dimensions using the same procedures. In what follows, we will discuss the goodness-of-fit procedures. The examples will be provided in the later chapters.

3.8.1 Goodness-of-Fit Test Based on Empirical Copula: S_n, T_n

The goodness-of-fit statistics $S_n$and T_n are based on the empirical copula. Similar to the univariate Cramér–von Mises and Kolmogorov–Smirnov goodness-of-fit tests, the test based on the empirical copula is to compare the distance between the empirical copula (C_n) and the parametric copula ($C_{\boldsymbol{\theta}}$) fitted to the pseudo-observations under the null hypothesis H_0 (the given parametric copula function cannot be rejected). The goodness-of-fit test statistics, i.e., Cramér–von Mises test statistic (S_n) and Kolmogorov–Smirnov test statistic (T_n) for an empirical copula can be written as follows:

$$S_n = \int_{[0,1]^2} \mathbb{C}_n(\mathbf{u})^2 dC_n(\mathbf{u}) \tag{3.108a}$$

$$T_n = \sup \mid \mathbb{C}_n(\mathbf{u}) \mid, \quad \mathbf{u} \in [0,1]^2 \tag{3.108b}$$

where

$$\mathbb{C}_n(\mathbf{u}) = \sqrt{n}\left(C_n - C_{\hat{\boldsymbol{\theta}}}\right) \tag{3.108c}$$

In Equations (3.108a)–(3.108c), $C_n(\mathbf{u})$is the empirical copula calculated from Equation (3.65) or using the following formula (Genest et al., 2007):

$$C_n(\mathbf{u}) = \frac{1}{n}\sum_{i=1}^{n} 1(U_{1i} \le u_1, U_{2i} \le u_2), \quad \mathbf{u} = (u_1, u_2) \in [0,1]^2 \tag{3.109}$$

This is the fitted copula function and n is the sample size.

If there is an analytical expression for $C_{\hat{\boldsymbol{\theta}}}$, S_n, T_n, then that may be calculated directly using Equations(3.108a)–(3.108c). Otherwise, Monte Carlo simulation is applied for $m > n$ as follows:

1. Generate a bivariate random sample $\mathbf{U}_1^*$, $\mathbf{U}_2^*$ from $C_{\hat{\boldsymbol{\theta}}}$.
2. Approximate $C_{\hat{\boldsymbol{\theta}}}$ by

$$B_m^* = \frac{1}{m}\sum_{i=1}^{m} 1\left(\mathbf{U}_i^* \le \mathbf{u}\right) \tag{3.110}$$

3. Approximate S_n by

$$S_n = \sum_{i=1}^{n} \left(C_n(\mathbf{U}_i) - B_m^*(\mathbf{U}_i)\right)^2 \tag{3.111a}$$

$$T_n = \sup_{\mathbf{u}\in[0,1]^2} \left|\sqrt{n}\left(C_n(\mathbf{U}_i) - B_m^*(\mathbf{U}_i)\right)\right| \tag{3.111b}$$

With the fitted copula function, the P-value of the test statistic is approximated using parametric bootstrap simulation repeated for some large integer N times as follows:

1. Generate a bivariate sample $\mathbf{X}_1^*$, $\mathbf{X}_2^*$ from the copula function $C_{\hat{\boldsymbol{\theta}}}$ and compute the associated rank vectors: $\mathbf{R}_1^*$, $\mathbf{R}_2^*$.

2. Compute $\mathbf{U}_i = \dfrac{\mathbf{R}_i}{n+1}$ and let

$$C_n^* = \frac{1}{n}\sum_{i=1}^{n} \left(\mathbf{U}_i^* \leq \mathbf{u}\right) \tag{3.111c}$$

3. Estimate the copula parameter from $\mathbf{U}_i^*$ for the tested copula function.
4. Calculate the test statistics either directly using Equations (3.108a)–(3.108c) or approximated using Equations (3.111a) and (3.111b).

Finally, the P-value of the test statistic is approximated as follows:

$$P_{value} = \frac{1}{N}\sum_{k=1}^{N} 1\left(S_{n,\,k}^* > S_n\right) \; or \; P_{value} = \frac{1}{N}\sum_{k=1}^{N} 1\left(T_{n,\,k}^* > T_n\right) \tag{3.112}$$

3.8.2 Goodness-of-Fit Test Based on Kendall's Transform: $S_n^{(K)}$, $T_n^{(K)}$

The goodness-of-fit test $S_n^{(K)}$ and $T_n^{(K)}$ are based on the probability integral transform, i.e., Kendall's transform with mapping as follows:

$$\mathbf{X} \mapsto V = H(\mathbf{X}) = C(\mathbf{u}); \forall \mathbf{X} = [X_1, X_2],\ \mathbf{u} = [\mathbf{u}_1, \mathbf{u}_2] \tag{3.113}$$

According to Genest et al. (2007), let K represent the univariate distribution function of V. K may be estimated nonparametrically using the empirical distribution function of the rescaled version of pseudo-observations $V_1 = C_n(\mathbf{u}_1),\ \ldots,\ V_n = C_n(\mathbf{u}_n)$ with the use of the following equation:

$$K_n(v) = \frac{1}{n}\sum_{i=1}^{n} (V_i \leq v), \quad v \in [0, 1] \tag{3.114}$$

The null hypothesis (H_0) is that $\mathbf{u} = [u_1, u_2]$ may be modeled by the copula function C_θ or equivalently, the Kendall transform of $C_\theta(\mathbf{u})$ follows the distribution K_θ. Measuring the distance between K_n(the empirical Kendall transform) and the parametric estimation $K_{\theta_n}$of K into the goodness-of-fit test may be performed through Cramér–von Mises ($S_n^{(K)}$) and Kolmogorov–Smirnov ($T_n^{(K)}$) statistics as follows:

$$S_n^{(K)} = \int_0^1 \mathbf{K}_n(v)^2 dK_{\theta_n}(v) \tag{3.115a}$$

$$T_n^{(K)} = \sup_{v\in[0,1]} \mid \mathbf{K}_n(v) \mid \tag{3.115b}$$

where

$$\mathbf{K}_n(v) = \sqrt{n}(K_n - K_{\theta_n}) \tag{3.115c}$$

In Equations (3.115a)–(3.115c), if there is an analytical expression for K_{θ_n}, the test statistics can be directly computed. Otherwise, Monte Carlo simulation with $m \geq n$ will be needed to approximate K_{θ_n}as follows:

1. Generate a random sample $\mathbf{u}_1^*,\ \mathbf{u}_2^*,\ \ldots,\ \mathbf{u}_\mathbf{m}^*$ from the fitted copula function C_{θ_n}.
2. Approximate K_{θ_n} using the following:

$$B_m^* = \frac{1}{m}\sum\nolimits_{i=1}^{m} 1\left(V_i^* \le t\right),\quad t \in [0, 1] \tag{3.116}$$

where
$V_i^* = \frac{1}{m}\sum_{j=1}^{m} 1\left(\mathbf{u}_j^* \le \mathbf{u}_i^*\right),\ i = 1,\ 2,\ \ldots,\ n$

3. Approximate $S_n^{(K)}$ and $T_n^{(K)}$ using the following equations:

$$S_n^{(K)} = \frac{n}{m}\sum\nolimits_{i=1}^{m} \left(K_n\left(V_i^*\right) - B_m^*\left(V_i^*\right)\right)^2 \tag{3.117a}$$

$$T_n^{(K)} = \sup_{V\in[0,1]}\left|\sqrt{n}\left(K_n(V^*) - B_m^*(V^*)\right)\right| \tag{3.117b}$$

For the fitted copula function, the P-value of the goodness-of-fit test is approximated using a similar parametric bootstrap simulation repeated for some large number N times as follows:

1. Generate random sample $\mathbf{X}_{1,k}^*,\ \ldots,\ \mathbf{X}_{n,k}^*$ from the fitted copula function C_{θ_n} and compute their associated rank $\mathbf{R}_{1,k}^*,\ \ldots,\ \mathbf{R}_{n,k}^*$.
2. Compute

$$V_{i,k}^* = \frac{1}{n}\sum\nolimits_{j=1}^{n} 1\left(\mathbf{X}_{j,k}^* \le \mathbf{X}_{i,k}^*\right),\ i = 1,\ \ldots,\ n \tag{3.118a}$$

$$K_{n,k}^*(t) = \frac{1}{n}\sum\nolimits_{i=1}^{n} 1\left(V_{i,k}^* \le t\right),\ t \in [0, 1] \tag{3.118b}$$

3. Assign $\mathbf{U}_{1,k}^* = \frac{\mathbf{R}_{1,k}^*}{n+1},\ \ldots,\ \mathbf{U}_{n,k}^* = \frac{\mathbf{R}_{n,k}^*}{n+1}$ and reestimate the parameters for the copula function.
4. If there is an analytical expression for K_θ, then calculate $S_{n,k}^{(K)*}$ and $T_{n,k}^{(K)*}$ using Equations (3.115a)–(3.115c). Otherwise, $K_{\theta_{n,k}}^*$ needs to be approximated using the procedure discussed earlier in this section to estimate $S_{n,k}^{(K)*}$ and $T_{n,k}^{(K)*}$.

Finally, the P-value of the test statistic can be written as follows:

$$P_{value} = \frac{1}{N}\sum\nolimits_{k=1}^{N} 1\left(S_{n,k}^{(K)*} > S_n^{(K)}\right),\ P_{value} = \frac{1}{N}\sum\nolimits_{k=1}^{N}\left(T_{n,k}^{(K)*} > T_n^{(K)}\right) \tag{3.119}$$

It is worth noting that this goodness-of-fit test is most sensitive to the copula functions with analytical Kendall's distribution, i.e., Archimedean copulas.

3.8.3 Goodness-of-Fit Test Based on Rosenblatt's Transform: A_n, $S_n^{(B)}$, $S_n^{(C)}$

Based on Rosenblatt's transform discussed through Equations (3.105)–(3.107), Rosenblatt's transform for the bivariate random variables with a joint distribution represented by a copula function $C_\mathbf{\theta}(u_1, u_2); u_1 = F_{X_1}(x_1); u_2 = F_{X_2}(x_2)$ can be expressed as follows:

$$Z_1 = u_1; Z_2 = \partial C_\theta(u_1, u_2)/\partial u_1 \tag{3.120}$$

The null hypothesis (H_0) of the goodness-of-fit test based on Rosenblatt's transform is that $\mathbf{u} = [u_1, u_2] \sim C_\theta$, i.e., Z_1, Z_2, is a bivariate independent copula, as follows:

$$C_\perp(Z_1, Z_2) = Z_1 \times Z_2 \tag{3.121}$$

In the preceding three test statistics, the A_n-test statistic is also called the Anderson–Darling test statistic such that the chi-square distribution is assumed as the limiting distribution. Compared to A_n, $S_n^{(B)}$ and $S_n^{(C)}$ do not assume the chi-square distribution as the limiting distribution; the latter two tests are also called the goodness-of-fit tests based on an improved Rosenblatt's transform (Genest et al., 2007). Cramér–von Mises statistic is considered for both $S_n^{(B)}$ and $S_n^{(C)}$. These two tests are further discussed in what follows.

Under the null hypothesis, let the empirical distribution be written as follows:

$$D_n(\mathbf{u}) = \frac{1}{n}\sum_{i=1}^{n} 1(\mathbf{Z}_i \le \mathbf{u}) \tag{3.122}$$

From Equation (3.121), it is known that $\mathbf{Z}_1$ and $\mathbf{Z}_2$ should be "close" to independently uniformly distributed random variables, i.e., $C_\perp$. Then the distance between $D_n(\mathbf{u})$ and $C_\perp$ is used to construct the goodness-of-fit test $S_n^{(B)}$ and $S_n^{(C)}$, as follows:

$$\begin{aligned} S_n^{(B)} &= n\int_{[0,1]^2} [D_n(\mathbf{u}) - C_\perp(\mathbf{u})]^2 d\mathbf{u} \\ &= \frac{n}{3^2} - \frac{1}{2}\sum_{i=1}^{n}\prod_{k=1}^{2}\left(1 - Z_{ik}^2\right) + \frac{1}{n}\sum_{i=1}^{n}\sum_{j=1}^{n}\prod_{k=1}^{2}\left(1 - Z_{ik} \vee Z_{jk}\right) \end{aligned} \tag{3.123}$$

where $Z_{ik} \vee Z_{jk} = \max\left(Z_{ik}, Z_{jk}\right)$.

$$S_n^{(C)} = n\int_{[0,1]^2} [D_n(\mathbf{u}) - C_\perp(\mathbf{u})]^2 dD_n(\mathbf{u}) = \sum_{i=1}^{n} \left(D_n(\mathbf{Z}_i) - C_\perp(\mathbf{Z}_i)\right)^2 \tag{3.124}$$

For the fitted copula function, the P-value of the statistic is also determined, based on the parametric bootstrap simulation repeated for some large integer N times as follows:

1. Generate a random sample $\{\mathbf{X}_1, \mathbf{X}_2\}$ with the same sample size as the original dataset, from the estimated copula function $C_{\hat{\boldsymbol{\theta}}}$ and compute the rank vectors: $\{\mathbf{R}_1^*, \mathbf{R}_2^*\}$.
2. Compute the intermediate variables as follows:

$$\mathbf{U}_1^* = \frac{\mathbf{R}_1^*}{n+1}, \ \mathbf{U}_2^* = \frac{\mathbf{R}_2^*}{n+1} \tag{3.125}$$

3. Reestimate the copula parameter $\hat{\boldsymbol{\theta}}^*$ using $\mathbf{U}_1^*$ and $\mathbf{U}_2^*$ with the same copula function, and compute $\mathbf{Z}_1^*$, $\mathbf{Z}_2^*$ using Equation (3.120).
4. Compute $S_{n,k}^{(B)*}$ and $S_{n,k}^{(C)*}$ using Equations (3.123) and (3.124), respectively.
5. After repeating steps 1 through 4 N times, the P-value can be given as follows:

$$P_{value} = \frac{1}{N}\sum_{k=1}^{N} 1\left(S_{n,k}^{(B)*} > S_n^{(B)}\right) \ or \ P_{vlaue} = \frac{1}{N}\sum_{k=1}^{N} \left(S_{n,k}^{(C)*} > S_n^{(C)}\right) \tag{3.126}$$

3.9 Procedure for Multivariate Frequency Analysis

The procedure for multivariate frequency analysis is sketched in Figure 3.10.

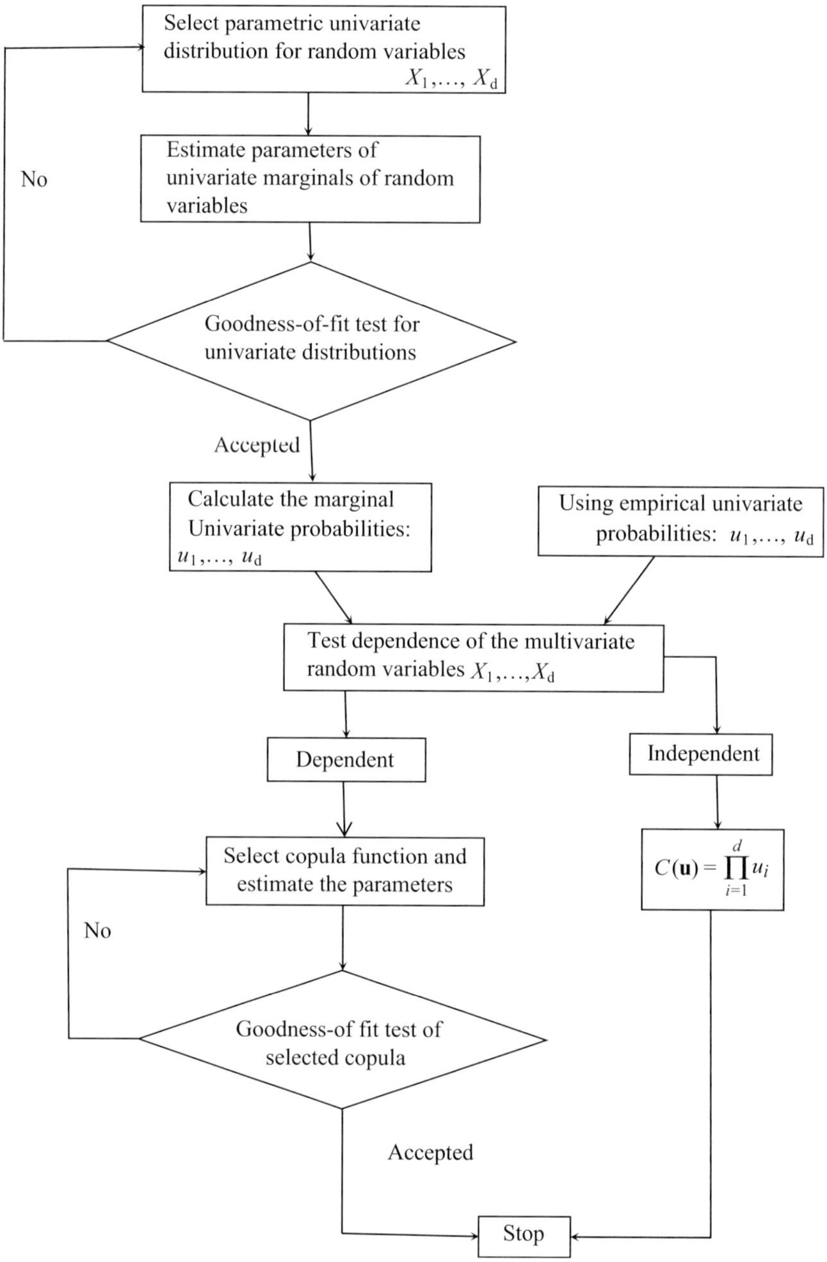

Figure 3.10 Procedure for multivariate frequency analysis.

3.10 Joint/Conditional Distributions and Corresponding Return Periods through Copulas

In multivariate frequency analysis, the following probability distributions are useful for hydrologic and environmental applications. In this section, applications are discussed for bivariate and trivariate cases using the following:

$$P(X_1 \le x_1) = F_1(x_1) = u; P(X_2 \le x_2) = F_2(x_2) = v; P(X_3 \le x_3) = F_3(x_3) = w;$$

$$C_{12}(u, v) = F_{12}(x_1, x_2) = P_{12}(X_1 \le x_1, X_2 \le x_2);$$

$$C_{13}(u, w) = F_{13}(x_1, x_3) = P_{13}(X_1 \le x_1, X_3 \le x_3);$$

$$C_{23}(v, w) = F_{23}(x_2, x_3) = P_{23}(X_2 \le x_2, X_3 \le x_3); \text{ and}$$

$$C(u, v, w) = F(x_1, x_2, x_3) = P(X_1 \le x_1, X_2 \le x_2, X_3 \le x_3)$$

3.10.1 Calculation of Joint Probability for Bivariate and Trivariate Cases

Joint Probability for Bivariate Events

Using bivariate variables X_1 and X_2 as an example, the joint probabilities can be expressed as follows:

$$\begin{aligned} P(X_1 > x_1, X_2 > x_2) &= 1 - P(X_1 \le x_1) - P(X_2 \le x_2) + P_{12}(X_1 \le x_1, X_2 \le x_2) \\ &= 1 - u - v + C_{12}(u, v) \end{aligned} \tag{3.127}$$

$$P(X_1 > x_1, X_2 \le x_2) = P(X_2 \le x_2) - P_{12}(X_1 \le x_1, X_2 \le x_2) = v - C_{12}(u, v) \tag{3.128}$$

$$P(X_1 \le x_1, X_2 \ge x_2) = P(X_1 \le x_1) - P_{12}(X_1 \le x_1, X_2 \le x_2) = u - C_{12}(u, v) \tag{3.129}$$

Joint Probability for Trivariate Events

For trivariate random variables X_1, X_2, and X_3, common formulas of trivariate probability distributions can be given as follows:

$$\begin{aligned} P(X_1 \le x_1, X_2 \le x_2, X_3 > x_3) &= P_{12}(X_1 \le x_1, X_2 \le x_2) - P(X_1 \le x_1, X_2 \le x_2, X_3 \le x_3) \\ &= C_{12}(u, v) - C(u, v, w) \end{aligned} \tag{3.130}$$

$$P(X_1 \le x_1, X_2 > x_2, X_3 \le x_3) = P_{12}(X_1 \le x_1, X_3 \le x_3) - P(X_1 \le x_1, X_2 \le x_2, X_3 \le x_3)$$
$$= C_{13}(u_1, u_3) - C(u_1, u_2, u_3) \tag{3.131}$$

$$P(X_1 > x_1, X_2 \le x_2, X_3 \le x_3) = P_{23}(X_2 \le x_2, X_3 \le x_3) - P(X_1 \le x_1, X_2 \le x_2, X_3 \le x_3)$$
$$= C_{23}(v, w) - C(u, v, w) \tag{3.132}$$

$$P(X_1 \le x_1, X_2 > x_2, X_3 > x_3)$$
$$= P(X_1 \le x_1) - P_{12}(X_1 \le x_1, X_2 \le x_2) - P_{13}(X_1 \le x_1, X_3 \le x_3) + P(X_1 \le x_1, X_2 \le x_2, X_3 \le x_3)$$
$$= u - C_{12}(u, v) - C_{13}(u, w) + C(u, v, w) \tag{3.133}$$

$$P(X_1 > x_1, X_2 \le x_2, X_3 > x_3)$$
$$= P(X_2 \le x_2) - P_{12}(X_1 \le x_1, X_2 \le x_2) - P_{23}(X_2 \le x_3, X_3 \le x_3) + P(X_1 \le x_1, X_2 \le x_2, X_3 \le x_3)$$
$$= v - C_{12}(u, v) - C_{23}(v, w) + C(u, v, w) \tag{3.134}$$

$$P(X_1 > x_1, X_2 > x_2, X_3 \le x_3)$$
$$= P(X_3 \le x_3) - P_{13}(X_1 \le x_1, X_3 \le x_3) - P_{23}(X_2 \le x_2, X_3 \le x_3) + P(X_1 \le x_1, X_2 \le x_2, X_3 \le x_3)$$
$$= w - C_{13}(u, w) - C_{23}(v, w) + C(u, v, w) \tag{3.135}$$

$$P(X_1 > x_1, X_2 > x_2, X_3 > x_3)$$
$$= 1 - P(X_1 \le x_1) - P(X_2 \le x_2) - P(X_3 \le x_3) + P_{12}(X_1 \le x_1, X_2 \le x_2)$$
$$+ P_{13}(X_1 \le x_1, X_3 \le x_3) + P_{23}(X_2 \le x_2, X_3 \le x_3) - P(X_1 \le x_1, X_2 \le x_2, X_3 \le x_3)$$
$$= 1 - u - v - w + C_{12}(u, v) + C_{13}(u, w) + C_{23}(v, w) - C_{123}(u, v, w) \tag{3.136}$$

3.10.2 Return Periods through Copulas

A return period, also known as a recurrence interval, is an estimate of the interarrival time between events, such as earthquakes, floods, or river discharge flows of a certain intensity or size. It is a statistical measurement denoting the average recurrence interval over an extended period of time and is usually required for risk analysis (i.e., whether a project should be allowed to go forward in a zone with a certain risk) and infrastructure engineering design purposes (i.e., designing structures so that they are capable of withstanding an event of a certain return period).

In hydrology and water resources engineering, the return period (also called recurrence interval) T is the average time interval between occurrences of the event $X \ge x$ for unviariate case. That is, T is defined as the average recurrence interval for events $X \ge x$ (i.e., Equation (2.81) in Chapter 2), as follows:

$$T = \frac{\mu}{P(X > x)} = \frac{\mu}{\bar{F}_X(x)} = \frac{\mu}{1 - F_X(x)} \tag{3.137}$$

Equation (3.137) also shows the relation among return period T, nonexceedance probability $F_X(x)$, and exceedance probability $\bar{F}_X(x)$. Hence, we also have

$$F_X(x) = \frac{T - \mu}{T} \tag{3.138}$$

Using the same concept for the univariate case, the return period can be estimated for multivariate cases. Here we present the bivariate and trivariate cases. Examples will be given in the later chapters.

Bivariate Case: Joint Return Period Using Copulas

- "AND" case: $X_1 > x_1$ *and* $X_2 > x_2$

The joint return period of the "AND" case can be expressed by substituting Equation (3.127) into Equation (3.137) as follows:

$$T_{AND}(x_1, x_2) = \frac{\mu}{P_{12}(X_1 > x_1, X_2 > x_2)} = \frac{\mu}{1 - u - v + C_{12}(u, v)} \tag{3.139}$$

- "OR" case: $X_1 > x_1$ *or* $X_2 > x_2$

The joint return period of the "OR" case is simply expressed as follows:

$$T_{OR} = \frac{\mu}{1 - P_{12}(X_1 \leq x_1, X_2 \leq x_2)} = \frac{\mu}{1 - C_{12}(u, v)} \tag{3.140}$$

Equation (3.140) indicates the combination of the following:

$$(X_1 > x_1, X_2 > x_2) \cup (X_1 > x_1, X_2 \leq x_2) \cup (X_1 \leq x_1, X_2 > x_2) \tag{3.141}$$

- Case: $X_1 > x_1$ *and* $X_2 \leq x_2$ (*or* $X_1 \leq x_1$ *and* $X_2 > x_2$):

For illustrative purposes, we use $X_1 > x_1$ and $X_2 \leq x_2$ as an example. From Equation (3.128),

$$T(X_1 > x_1, X_2 \leq x_2) = \frac{\mu}{v - C_{12}(u, v)} \tag{3.142}$$

Similarly, the return period of $X_1 > x_1$ and $X_2 \leq x_2$ is as follows:

$$T(X_1 \leq x_1, X_2 > x_2) = \frac{\mu}{u - C_{12}(u, v)} \tag{3.143}$$

Bivariate Case: Conditional Return Period Using Copulas

The copula can be used to determine the conditional distribution functions and conditional return periods under different conditions.

- Case: $X_2 > x_2 \mid X_1 = x_1$ (or $X_1 > x_1 \mid X_2 = x_2$):

Using $X_2 > x_2 \mid X_1 = x_1$ as an example, the conditional probability of $X_2 \leq x_2 \mid X_1 = x_1$ can be written as follows:

$$P(X_2 \leq x_2|X_1 = x_1) = C_{12}(V \leq v|U = u) = \left.\frac{\partial C_{12}(u, v)}{\partial u}\right|_{U=u} \tag{3.144}$$

Given that $P(X_2 \leq x_2|X_1 = x_1) + P(X_2 > x_2|X_1 = x_1) = 1$, we have the following:

$$P(X_2 > x_2|X_1 = x_1) = 1 - P(X_2 \leq x_2|X_1 = x_1) = 1 - \left.\frac{\partial C_{12}(u, v)}{\partial u}\right|_{U=u} \tag{3.145}$$

Then, the corresponding conditional return period is as follows:

$$T(X_2 > x_2|X_1 = x_1) = \frac{\mu}{1 - C_{12}(V \leq v|U = u)} = \frac{\mu}{1 - \left.\frac{\partial C_{12}(u, v)}{\partial u}\right|_{U=u}} \tag{3.146a}$$

Similarly, the conditional return period of $X_1 > x_1 \mid X_2 = x_2$ is as follows:

$$T(X_1 > x_1|X_2 = x_2) = \frac{\mu}{1 - \left.\frac{\partial C_{12}(u, v)}{\partial v}\right|_{V=v}} \tag{3.146b}$$

- Case: $X_2 > x_2 \mid X_1 \leq x_1$ (or $X_1 > x_1 \mid X_2 \leq x_2$)

Again, using $X_2 > x_2 \mid X_1 \leq x_1$, the conditional distribution of $X_2 \leq x_2 \mid X_1 \leq x_1$ is expressed using copula as follows:

$$P(X_2 \leq x_2|X_1 \leq x_1) = \frac{F(x_1, x_2)}{F_1(x_1)} = C_{12}(V \leq v|U \leq u) = \frac{C(u, v)}{u} \tag{3.147}$$

Then we have the following:

$$T(X_2 > x_2|X_1 \leq x_1) = \frac{\mu}{1 - P(X_2 \leq x_2|X_1 \leq x_1)} = \frac{\mu}{1 - \frac{C_{12}(u, v)}{u}} \tag{3.148a}$$

Likewise, we have the following:

$$T_1(X_1 > x_1|X_2 \leq x_2) = \frac{\mu}{1 - \frac{C_{12}(u, v)}{v}} \tag{3.148b}$$

Trivariate Case: Joint Return Period Using Copulas

Similar to the bivariate case, the joint return periods for the trivariate case are also discussed as the "AND" case:

$X_1 > x_1 \, \cap \, X_2 > x_2 \, \cap \, X_3 > x_3$

and the "OR" case:

$X_1 > x_1 \cup X_2 > x_2 \cup X_3 > x_3$.

- "AND" case: $X_1 > x_1 \cap X_2 > x_2 \cap X_3 > x_3$

In this case, all three values of X_1, X_2, and X_3 are exceeded. Applying Equation (3.136), the return period $T_{AND}(x_1, x_2, x_3)$ can be given as follows:

$$T_{AND}(x_1,x_2,x_3) = \frac{\mu}{P(X_1 > x_1, X_2 > x_2, X_3 > x_3)} = \frac{\mu}{1 - u - v - w + C_{12}(u,v) + C_{13}(u,w) + C_{23}(v,w) - C(u,v,w)} \quad (3.149)$$

- "OR" case: $X_1 > x_1 \cup X_2 > x_2 \cup X_3 > x_3$

In this case, at least one value of X_1, X_2, and X_3 is exceeded and the joint return period $T_{OR}(x_1, x_2, x_3)$ can be given as follows:

$$T_{OR}(x_1,x_2,x_3) = \frac{1}{1 - P(X_1 \le x_1, X_2 \le x_2, X_3 \le x_3)} = \frac{1}{1 - C(u,v,w)} \quad (3.150)$$

Trivariate Case: Conditional Return Periods through Copulas

- Case: $X_1 > x_1 \cup X_2 > x_2 \mid X_3 = x_3$ (or $X_1 > x_1 \cup X_3 > x_3 | X_2 = x_2, X_2 > x_2 \cup X_3 > x_3 | X_1 = x_1$)

In this case, under the condition $X_3 = x_3$, at least one value of X_1 and X_2 is exceeded. The conditional distribution function $F(X_1 \le x_1, X_2 \le x_2 | X_3 = x_3)$ can be written as follows:

$$F(X_1 \le x_1, X_2 \le x_2 | X_3 = x_3) = C(U \le u, V \le v | W = w) = \left.\frac{\partial C(u,v,w)}{\partial w}\right|_{W=w} \quad (3.151)$$

Then, the corresponding conditional return period can be expressed as follows:

$$T(X_1 > x_1 \cup X_2 > x_2 | X_3 = x_3) = \frac{\mu}{1 - C(u, v | W = w)} = \frac{\mu}{1 - \left.\frac{\partial C(u,v,w)}{\partial w}\right|_{W=w}} \quad (3.152)$$

Likewise, we have the following:

$$T(X_1 > x_1 \cup X_3 > x_3 | X_2 = x_2) = \frac{\mu}{1 - \left.\frac{\partial C(u,v,w)}{\partial v}\right|_{V=v}} \quad (3.152a)$$

$$T(X_2 > x_2 \cup X_3 > x_3 | X_1 = x_1) = \frac{\mu}{1 - \left.\frac{\partial C(u,v,w)}{\partial u}\right|_{U=u}} \quad (3.152b)$$

- Case: $X_1 > x_1 \cap X_2 > x_2 \mid X_3 = x_3$ (or $X_1 > x_1 \cap X_3 > x_3 | X_2 = x_2, X_2 > x_2 \cap X_3 > x_3 | X_1 = x_1$)

In this case, under the condition $X_3 = x_3$, both values of X_1 and X_2 are exceeded. Based on the probability theory, the conditional return period, i.e., $T(X_1 > x_1 \cap X_2 > x_2|X_3 = x_3)$, can be determined using the same approach as for bivariate analysis under the condition of $X_3 = x_3$ as follows:

$$P(X_1 \le x_1|X_3 = x_3) = \left.\frac{\partial C_{13}(u,w)}{\partial w}\right|_{W=w}, \quad P(X_2 \le x_2|X_3 = x_3) = \left.\frac{\partial C_{23}(v,w)}{\partial w}\right|_{W=w} \tag{3.153}$$

$$\begin{aligned} &P(X_1 > x_1 \cap X_2 > x_2|X_3 = x_3) \\ &\quad = 1 - P(X_1 \le x_1|X_3 = x_3) - P(X_2 \le x_2|X_3 = x_3) + P(X_1 \le x_1, X_2 \le x_2|X_3 = x_3) \end{aligned} \tag{3.154}$$

We have $T(X_1 > x_1 \cap X_2 > x_2|X_3 = x_3)$ as follows:

$$\begin{aligned} T(X_1 > x_1 \cap X_2 > x_2|X_3 = x_3) &= \frac{\mu}{P(X_1 > x_1 \cap X_2 > x_2|X_3 = x_3)} \\ &= \frac{\mu}{1 - \left.\frac{\partial C_{13}(u,w)}{\partial w}\right|_{W=w} - \left.\frac{\partial C_{23}(v,w)}{\partial w}\right|_{W=w} + \left.\frac{\partial C(u,v,w)}{\partial w}\right|_{W=w}} \end{aligned} \tag{3.155}$$

Likewise, we have the following:

$$\begin{aligned} T(X_1 > x_1 \cap X_3 > x_3|X_2 = x_2) &= \frac{\mu}{P(X_1 > x_1 \cap X_3 > x_3|X_2 = x_2)} \\ &= \frac{\mu}{1 - \left.\frac{\partial C_{12}(u,v)}{\partial v}\right|_{V=v} - \left.\frac{\partial C_{23}(v,w)}{\partial v}\right|_{V=v} + \left.\frac{\partial (u,v,w)}{\partial v}\right|_{V=v}} \end{aligned} \tag{3.155a}$$

$$\begin{aligned} T(X_2 > x_2 \cap X_3 > x_3|X_1 = x_1) &= \frac{\mu}{P(X_2 > x_2 \cap X_3 > x_3|X_1 = x_1)} \\ &= \frac{\mu}{1 - \left.\frac{\partial C_{12}(u,v)}{\partial v}\right|_{U=u} - \left.\frac{\partial C_{23}(u,w)}{\partial u}\right|_{U=u} + \left.\frac{\partial (u,v,w)}{\partial u}\right|_{U=u}} \end{aligned} \tag{3.155b}$$

- Case: $X_1 > x_1 \cup X_2 > x_2 \mid X_3 \le x_3$ (or $X_1 > x_1 \cup X_3 > x_3|X_2 \le x_2, X_2 > x_2 \cup X_3 > x_3|X_1 \le x_1$).

In this case, under the condition of $X_3 \le x_3$, at least one value of X_1 and X_2 is exceeded. Similar to the bivariate case, the conditional distribution function can be written as follows:

$$P(X_1 \le x_1, X_2 \le x_2|X_3 \le x_3) = \frac{P(X_1 \le x_1, X_2 \le x_2, X_3 \le x_3)}{F_3(x_3)} = \frac{C(u,v,w)}{w} \tag{3.156}$$

Then the return period $T(X_1 > x_1 \cup X_2 > x_2|X_3 \le x_3)$ can be given as follows:

$$T(X_1 > x_1 \cup X_2 > x_2|X_3 = x_3) = \frac{\mu}{1 - \dfrac{C(u, v, w)}{w}} \tag{3.157}$$

Likewise, we have the following:

$$T(X_1 > x_1 \cup X_3 > x_3|X_2 \le x_2) = \frac{\mu}{1 - \dfrac{C(u, v, w)}{v}} \tag{3.157a}$$

$$T(X_2 > x_2 \cup X_3 > x_3|X_1 \le x_1) = \frac{\mu}{1 - \dfrac{C(u, v, w)}{u}} \tag{3.157b}$$

- Case: $X_1 > x_1 \cap X_2 > x_2 \mid X_3 \le x_3$ (or $X_1 > x_1 \cap X_3 > x_3|X_2 \le x_2, X_2 > x_2 \cap X_3 > x_3|X_1 \le x_1$)

The return period for this case can be determined using an approach similar to that used in case $X_1 > x_1 \cap X_2 > x_2 \mid X_3 = x_3$, as follows.

The conditional probabilities of $X_1 \le x_1 \mid X_3 \le x_3$ and $X_2 \le x_2 \mid X_3 \le x_3$ can be written as follows:

$$P(X_1 \le x_1|X_3 \le x_3) = \frac{C_{13}(u, w)}{w}, \quad P(X_2 \le x_2|X_3 \le x_3) = \frac{C_{23}(v, w)}{w} \tag{3.158}$$

Then the return period of $T(X_1 > x_1 \cap X_2 > x_2|X_3 \le x_3)$ can be given as follows:

$$\begin{aligned} &T(X_1 > x_1 \cap X_2 > x_2|X_3 \le x_3) \\ &= \frac{\mu}{1 - P(X_1 \le x_1|X_3 \le x_3) - P(X_2 \le x_2|X_3 \le x_3) + P(X_1 \le x_1, X_2 \le x_2|X_3 \le x_3)} \\ &= \frac{\mu}{1 - \dfrac{C(u, w)}{w} - \dfrac{C(v, w)}{w} + \dfrac{C(u, v, w)}{w}} \end{aligned} \tag{3.159}$$

Likewise, we have the following:

$$T(X_1 > x_1 \cap X_3 > x_3|X_2 \le x_2) = \frac{\mu}{1 - \dfrac{C_{12}(u, v)}{v} - \dfrac{C_{23}(v, w)}{v} - \dfrac{C(u, v, w)}{v}} \tag{3.159a}$$

$$T(X_2 > x_2 \cap X_3 > x_3|X_1 \le x_1) = \frac{\mu}{1 - \dfrac{C_{12}(u, v)}{u} - \dfrac{C_{23}(u, w)}{u} - \dfrac{C(u, v, w)}{u}} \tag{3.159b}$$

- Case: $X_1 > x_1 \mid X_2 = x_2, X_3 = x_3$ (or $X_2 > x_2|X_1 = x_1, X_3 = x_3, X_3 > x_3|X_1 = x_1, X_2 = x_2$)

In the case of $X_1 > x_1 \mid X_2 = x_2,\ X_3 = x_3$, the conditional distribution of $X_1 \leq x_1 \mid X_2 = x_2,\ X_3 = x_3$ can be given as follows:

$$P(X_1 \leq x_1 | X_2 = x_2, X_3 = x_3) = \left. \frac{\dfrac{\partial^2 F(x_1, x_2, x_3)}{\partial x_2 \partial x_3}}{\dfrac{\partial^2 F_{23}(x_2, x_3)}{\partial x_2 \partial x_3}} \right|_{X_2 = x_2,\ X_3 = x_3}$$

$$= C(u | V = v, W = w) = \left. \frac{\dfrac{\partial^2 C(u, v, w)}{\partial v \partial w}}{\dfrac{\partial^2 C(v, w)}{\partial v \partial w}} \right|_{V = v,\ W = w} \tag{3.160}$$

Then the return period of $T(X_1 > x_1 | X_2 = x_2, X_3 = x_3)$ can be given as follows:

$$T(X_1 > x_1 | X_2 = x_2, X_3 = x_3) = \frac{\mu}{1 - \left. \dfrac{\dfrac{\partial^2 C(u, v, w)}{\partial v \partial w}}{\dfrac{\partial^2 C_{23}(v, w)}{\partial v \partial w}} \right|_{V = v,\ W = w}} \tag{3.161}$$

Likewise, we have the following:

$$T(X_2 > x_2 | X_1 = x_1, X_3 = x_3) = \frac{\mu}{1 - \left. \dfrac{\dfrac{\partial^2 C(u, v, w)}{\partial u \partial w}}{\dfrac{\partial^2 C_{13}(u, w)}{\partial u \partial w}} \right|_{U = u,\ W = w}} \tag{3.161a}$$

$$T(X_3 > x_3 | X_1 = x_1, X_2 = x_2) = \frac{\mu}{1 - \left. \dfrac{\dfrac{\partial^2 C(u, v, w)}{\partial u \partial v}}{\dfrac{\partial^2 C_{12}(u, v)}{\partial u \partial v}} \right|_{U = u,\ V = v}} \tag{3.161b}$$

- Case $X_1 > x_1 \mid X_2 \leq x_2,\ X_3 \leq x_3$ (or $X_2 > x_2 | X_1 \leq x_1, X_3 \leq x_3, X_3 > x_3 | X_1 \leq x_1,$ $X_2 \leq x_2$)

For $X_1 > x_1 \mid X_2 \leq x_2,\ X_3 \leq x_3$, the conditional probability of $P(X_1 \leq x_1 | X_2 \leq x_2, X_3 \geq x_3)$ can be written as follows:

$$P(X_1 \leq x_1 | X_2 \leq x_2, X_3 \leq x_3) = \frac{F(x_1, x_2, x_3)}{F_{23}(x_2, x_3)} = C(u | V \leq v, W \leq w) = \frac{C(u, v, w)}{C_{23}(v, w)} \tag{3.162}$$

Then the return period $T(X_1 > x_1 | X_2 \leq x_2, X_3 \leq x_3)$ can be expressed as follows:

$$T(X_1 > x_1 | X_2 \leq x_2, X_3 \leq x_3) = \frac{\mu}{1 - C(u | V \leq v, W \leq w)} = \frac{\mu}{1 - \dfrac{C(u, v, w)}{C_{23}(v, w)}} \tag{3.163}$$

Likewise, we have the following:

$$T(X_2 > x_2 | X_1 \le x_1, X_3 \le x_3) = \frac{\mu}{1 - C(v|U \le u, W \le w)} = \frac{\mu}{1 - \dfrac{C(u,v,w)}{C_{13}(u,w)}} \tag{3.163a}$$

$$T(X_3 > x_3 | X_1 \le x_1, X_2 \le x_2) = \frac{\mu}{1 - C(w|U \le u, V \le v)} = \frac{\mu}{1 - \dfrac{C(u,v,w)}{C_{12}(u,v)}} \tag{3.163b}$$

Relation between Univariate and Joint Return Periods

In what follows, we will discuss the relations between the univariate and joint return periods for the bivariate and trivariate cases.

Bivariate case: For bivariate random variables X_1 and X_2, with the joint distribution of $F(x_1, x_2)$, applying the Fréchet–Hoeffding bounds, we have the following:

$$\max(u + v - 1, 0) \le F(x_1, x_2) = C(u,v) \le \min(u,v) \tag{3.164}$$

Comparing Equation (3.140), i.e., the joint return period for the "OR" case, and Equation (3.137), i.e., the univariate return period, we have the following:

$$T_{OR}(x_1, x_2) \le \min(T_{X_1}, T_{X_2}) \tag{3.165}$$

Rearranging Equation (3.139) (i.e., the joint return period for the "AND" case), we have the following:

$$\begin{aligned} T_{AND}(x_1, x_2) &= \frac{\mu}{1 - u - v + C(u,v)} = \frac{\mu}{(1-u) + (1-v) - (1 - C(u,v))} \\ &= \frac{1}{\dfrac{1}{T_{X1}} + \dfrac{1}{T_{X_2}} - \dfrac{1}{T_{OR}(x_1, x_2)}} \end{aligned} \tag{3.166}$$

Substituting Equation (3.165) into Equation (3.166), we have the following inequality:

$$\max(T_{X_1}, T_{X_2}) \le T_{AND}(x_1, x_2) \tag{3.167}$$

Combining Equation (3.165) and Equation (3.167), we have the following:

$$T_{OR}(x_1, x_2) \le \min(T_{X_1}, T_{X_2}) \le \max(T_{X_1}, T_{X_2}) \le T_{AND}(x_1, x_2) \tag{3.168}$$

Trivariate case: For trivariate random variables X_1, X_2, and X_3, with a joint distribution of $F(x_1, x_2, x_3)$, we know the following:

$$F(x_1, x_2, x_3) = C(u,v,w) \le M = \min(u,v,w) \tag{3.169}$$

Comparing Equation (3.150), i.e., the joint return period for the "OR" case, and Equation (3.137), i.e., the univariate return period, we have the following:

$$T_{OR}(x_1, x_2, x_3) \le \min(T_{X_1}, T_{X_2}, T_{X_3}) \tag{3.170}$$

From Equation (3.168), we also have the following:

$$T_{OR}(x_1,x_2) \le \min(T_{x_1},T_{x_2}),\ T_{OR}(x_1,x_3) \le \min(T_{x_1},T_{x_3}),\ T_{OR}(x_2,x_3) \le \min(T_{x_2},T_{x_3}) \tag{3.171}$$

Rearranging Equation (3.150) for the "AND" case, we have the following:

$$\begin{aligned}T_{AND}(x_1,x_2,x_3) &= \frac{\mu}{(1-u-v-w+C_{12}(u,v)+C_{23}(v,w)+C_{13}(u,w)-C(u,v,w)}\\ &= \frac{\mu}{(1-u)+(1-v)+(1-w)-(1-C_{12}(u,v))-(1-C_{23}(v,w))-(1-C_{13}(u,w))+(1-C(u,v,w))}\\ &= \frac{1}{\frac{1}{T_{x_1}}+\frac{1}{T_{x_2}}+\frac{1}{T_{x_3}}-\frac{1}{T_{OR}(x_1,x_2)}-\frac{1}{T_{OR}(x_2,x_3)}-\frac{1}{T_{OR}(x_1,x_3)}+\frac{1}{T_{OR}(x_1,x_2,x_3)}}\end{aligned} \tag{3.172}$$

Substituting Equations (3.170) and (3.171) into Equation (3.172), the following inequality can be obtained:

$$\max(T_{x_1},T_{x_2},T_{x_3}) \le T_{AND}(x_1,x_2,x_3) \tag{3.173}$$

Thus, combining Equation (3.170) and Equation (3.173), we have the following:

$$T_{OR}(x_1,x_2,x_3) \le \min(T_{x_1},T_{x_2},T_{x_3}) \le \max(T_{x_1},T_{x_2},T_{x_3}) \le T_{AND}(x_1,x_2,x_3) \tag{3.174}$$

The inequalities, given as Equations (3.168) and (3.174), are valid if bivariate (trivariate) random variables are mutually independent random variables. In addition, the inequality is valid for the multivariate random variables for any dimension $d : d > 3$.

3.11 Summary

This chapter defines and summarizes the general concepts for copulas, including copula definition, copula properties, copula construction method and copula families, parameter estimation, simulation, goodness-of-fit study, and the risk measures using copulas. As the general discussion, this chapter does not provide detailed case study examples. Applications are provided in the later chapters, where the methodologies will be illustrated in detail.

References

Alfonsi, A. E. and Brigo, D. (2005). New families of copulas based on periodic functions. *Communications in Statistics: Theory and Methods*. 34(7), 1437–1447.

Ali, M. M., Mikhail, N. N., and Haq, M. S. (1978). A class of bivariate distributions including the bivariate logistic. *Journal of Multivariate Analysis*. 8, 405–412.

Genest, C. and Boies, J.-C. (2003). Detecting dependence with Kendall plots. *American Statistician*, 57(4), 275–284.

Genest, C. and Favre, A.-C. (2007). Everything you always wanted to know about copula modeling but were afraid to ask. *Journal of Hydrologic Engineering*. 12(4), 347–368.
Genest, C., Rémillard, B., and Beaudoin, D. (2007). Goodness-of-fit tests for copulas: A review and a power study. *Insurance: Mathematics and Economics*. doi:10.1016/j.insmatheco.2007.10.005.
Hu, L. (2006). Dependence patterns across financial markets: a mixed copula approach. *Applied Financial Economics*. 16, 717–729.
Joe, H. (1997). *Multivariate Models and Dependence Concepts*. Chapman & Hall/CRC, London.
Nelsen, R. B. (2006). *An Introduction to Copulas*, 2nd edition, Springer, New York.
Rosenblatt, M. (1952). Remarks on a Multivariate Transformation. *Annuals of Mathematical Statistics*. 23(3), 470–472.
Schucany, W., Parr, W., and Boyer, J. (1978). Correlation structure in Falie–Gumbel–Morgenstern Distributions. *Biometrika*. 65, 650–653.
Singh, K. and Singh, V. P. (1991). Derivation of bivariate probability density functions with exponential marginals. *Stochastic Hydrology and Hydraulics*. 5, 55–68.
Singh, K. and Singh, V. P. (1991). Derivation of bivariate exponential model applied to intensities and durations of extreme rainfall. *Journal of Hydrology*, 155, 225–236.
Trivedi, P. K. and Zimmer, D. M. (2007). Pitfalls in modeling dependence structures: explorations with copulas. www.economics.ox.ac.uk/hendryconference/Papers/Trivedi_DFHVol.pdf.
Wikipedia. Return period. http://en.wikipedia.org/wiki/Return_period.

Additional Reading

Bacchi, B., Becciu, G., and Kottegoda, N. T. (1994). Bivariate exponential model applied to intensities and durations of extreme rainfall. *Journal of Hydrology*, 155, 225–236.
Barbe, P., Genest, C., Ghoudi, K., and Rémillard, B. (1996). On Kendall's process. *Journal of Multivariate analysis*, 58, 197–229.
Breymann, W., Dias, A., and Embrechts, P. (2003). Dependence structures for multivariate high-frequency data in finance. *Quantitative Finance*, 3, 1–14.
Capéraà, P., Fougères, A.-L., and Genest, C. (1997). A nonparametric estimation procedure for bivariate extreme value copulas. *Biometrika*, 84(3), 567–577.
Coles, S., Heffernan, J., and Tawn, J. (1999). Dependence measures for extreme value analysis. *Extremes*, 2(4), 339–365.
Dobric, J. and Schmid, F. (2005). The goodness-of-fit for parametric families of copulas: application to financial data. *Communications in Statistics: Simulation and Computation*, 34, 1053–1068.
Dobric, J. and Schmid, F. (2007). A goodness of fit test for copulas based on Rosenblatt's transformation. *Computational Statistics & Data Analysis*, 51, 4633–4642.
Fermanian, J.-D. (2005). Goodness-of-fit test for copulas. *Journal of Multivariate Analysis*, 95, 119–152.
Fermanian, J.-D., Radulovic, D., and Wegkamp, M. H. (2004). Weak convergence of empirical copula processes. *Bernoulli*, 10, 847–860.
Fisher, N. I. and Switzer, P. (2001). Graphical assessment of dependence: is a picture worth 100 tests? *American Statistician*, 55(3), 233–239.
Frahm, G., Junker, M., and Schmidt, R. (2005). Estimating the tail-dependence coefficient: properties and pitfalls. *Insurance: Mathematics and Economics* 37, 80–100.

Francesco, S. and Salvatore, G. (2007). Fully nested 3-copula: procedure and application on hydrological data. *Journal of Hydrologic Engineering*, 12(4), 420–430.
Genest, C., Quessy, J.-F., and Rémillard, B. (2006). Goodness-of-fit procedures for copula models based on the integral probability transformation. *Scandinavian Journal of Statistics*, 33, 337–366.
Genest, C. and Rivest, L.-P. (1993). Statistical inference procedures for bivariate Archimedean copulas. *Journal of the American Statistical Association*, 88, 1034–1043.
Großmaß, T. (2007). Copulae and tail dependence. Diploma thesis. September 28, Berlin, Institute for Statistics and Econometrics School of Business and Economics, Humboldt-University, Berlin.
Marshall, A. W. and Ingram, O. (1967). A multivariate exponential distribution. *Journal of American Statistical Association*. 62(317), 30–44.
Oliveria, J. T. D. (1982). Bivariate extremes: extensions. *Bulletin of the International Statistical Institute*. 46(2), 241–251.
Schweizer, B. and Wolff, E. F. (1981). On nonparametric measures of dependence for random variables. *Annals of Statistics*, 9(4), 879– 885.
Sklar, A. (1959) Fonctions de repartition à n dimensions et leurs marges. *Publ. Inst. Statist. Univ. Paris*, 8, 229–231.
Wang, W. and Wells, M. T. (2000). Model selection and semiparametric inference for bivariate failure-time data. *Journal of the American Statistical Association*, 95, 62 72.
Yue, S. (2001). A bivariate gamma distribution for use in multivariate flood frequency analysis. *Hydrological Processes*. doi:10.1002/hyp.259.
Yue, S. and Rasmussen, P. (2002). Bivariate frequency analysis: discussion of some useful concept in hydrological application. *Hydrological Processes*. 16, 2881–2898.

4

Symmetric Archimedean Copulas

ABSTRACT

Symmetric Archimedean copulas are widely applied for hydrologic analyses for the following reasons: (1) they can be easily constructed with the given generating function; (2) a large variety of copulas belong to this class (Nelsen, 2006); and (3) the Archimedean copulas have nice properties, such as simple and elegant mathematical treatment. This chapter focuses on the symmetric Archimedean copulas.

4.1 Definition of Symmetric Archimedean Copulas

Formally, a d-dimensional Archimedean symmetric copula $C^d : [0,1]^d \to [0,1]$can be defined as follows (Nelsen, 2006; Salvadori et al, 2007; Savu and Trede, 2008). We first show it for a two-dimensional case and how it is constructed:

$$C(u_1, \ldots, u_d) = \phi^{[-1]}\left(\sum_{k=1}^{d} \phi(u_k)\right) = \phi^{[-1]}\left(\phi(u_1) + \cdots + \phi(u_d)\right); u_k \in [0,1], k = 1, 2, \ldots, d \tag{4.1}$$

In Equation (4.1), u_i, $i = 1, 2, \ldots, d$, the marginal cumulative distribution function (CDF) of the ith random variable; and $\phi(\cdot)$is the generating function of the Archimedean copula, which has the following properties:

- $\phi(\cdot)$ is a continuous strictly decreasing function from $[0,1] \to [0,\infty)$, we have $\phi(1) = 0$and $\phi(0) = \infty$, i.e., for $\phi(u_k), k = 1, \ldots, d; u_k \in [0,1], \phi(u_k) \in [0,\infty)$.
- $\phi^{[-1]}$ is the pseudo-inverse function of ϕ and nonincreasing on $[0,\infty)$. $\phi^{[-1]}$is strictly decreasing on $[0, \phi(0)]$ with Dom$\phi^{[-1]} \in [0,\infty)$and Ran$\phi^{[-1]} \in [0,1]$as follows:

$$\phi^{[-1]} = \begin{cases} \phi^{-1}(t); & 0 \le t \le \phi(0) \\ 0; & \phi(0) \le t < \infty \end{cases} \tag{4.2}$$

- $\phi^{[-1]}$ also has derivatives of all orders which alternate in sign, i.e., for all t to be in $[0,\infty)$.

With $k = 0, 1, \ldots,$ it satisfies the following:

$$(-1)^k \frac{d^k \phi^{[-1]}(t)}{dt^k} \ge 0 \tag{4.3}$$

Following Equation (4.1), the two- and three-dimensional symmetric Archimedean copulas can be written as follows:

$$C(u_1, u_2) = \phi^{[-1]}\Big(\phi(u_1) + \phi(u_2)\Big) \tag{4.4}$$

$$C(u_1, u_2, u_3) = \phi^{[-1]}\Big(\phi(u_1) + \phi(u_2) + \phi(u_3)\Big) \tag{4.5}$$

It should be noted that as the name of symmetric Archimedean copulas suggests, there is the same degree of dependence among all possible pairs for $d \geq 3$. This fact usually hinders the application of symmetric Archimedean copulas for multivariate analysis in higher dimensions, since the dependence among the possible pairs in reality is usually not the same. We will illustrate it in subsequent chapters.

Example 4.1 **Show that the function $\phi(t) = (-\ln t)^{\theta}, \theta \geq 1$is the generating function of Archimedean copula, and express the corresponding two- and three-dimensional copulas with this generating function.**

Solution: To show $\phi(t) = (-\ln t)^{\theta}, \theta \geq 1$is the generating function of Archimedean copulas, we need to show that it is a continuous strict decreasing function.

1. Let $f(t) = \ln t$. It is obvious that $f(t)$ is a strictly increasing function of t and thus $-f(t) = -\ln t$ is a strict decreasing function of t with $-\ln(0) \to \infty$, $-\ln(1) = 0$. Given $\theta \geq 1$, we have $(-\ln(0))^{\theta} = \infty^{\theta} = \infty$; $(-\ln(1))^{\theta} = 0^{\infty} = 0$. Now we show $\phi(t) = (-\ln t)^{\theta}$, $\theta \geq 1$ satisfies the generating function $\phi(t) = (-\ln t)^{\theta}, \theta \geq 1$ is a continuous strictly decreasing from $[0, 1] \to [0, \infty)$.
2. The inverse of function $\phi(t)$ can be given as follows:
   Let $u = (-\ln t)^{\theta}$. Then we have $\phi^{-1}(t) = \exp\left(-u^{\frac{1}{\theta}}\right)$. Now we need to show that $\phi^{-1}(t)$ is nonincreasing. And, it is obvious that the exponential function above is continuous and a nonincreasing function.

Applying Equation (4.1) for d = 2 or 3, we have the following:

$$\begin{aligned} &\phi(u_1) + \phi(u_2) = (-\ln u_1)^{\theta} + (-\ln u_2)^{\theta} \\ &C(u_1, u_2) = \phi^{[-1]}\Big(\phi(u_1) + \phi(u_2)\Big) = \exp\left(-\left((-\ln u_1)^{\theta} + (-\ln u_2)^{\theta}\right)^{\frac{1}{\theta}}\right) \end{aligned} \tag{4.6a}$$

and

$$\begin{aligned} \phi(u_1) + \phi(u_2) + \phi(u_3) &= (-\ln u_1)^{\theta} + (-\ln u_2)^{\theta} + (-\ln u_3)^{\theta} \\ C(u_1, u_2, u_3) &= \phi^{[-1]}\Big(\phi(u_1) + \phi(u_2) + \phi(u_3)\Big) \\ &= \exp\left(-\left((-\ln u_1)^{\theta} + (-\ln u_2)^{\theta} + (-\ln u_3)^{\theta}\right)^{\frac{1}{\theta}}\right) \end{aligned} \tag{4.6b}$$

To illustrate the copula with the random variables in a real domain, let random variables $\{X_1, X_2, X_3\}$ be positively dependent, and they may be modeled with the symmetric Archimedean copula in Equation (4.6b). In addition, X_1, X_2, X_3 follow the marginal distributions, respectively, of

$$X_1 \sim \exp(2), X_2 \sim \text{logistic}(4, 2), X_3 \sim Normal(3, 2^2)$$

We then have $u_1 = F_1(x_1) = 1 - \exp(-2x_1), u_2 = F_2(x_2) = \frac{1}{1 + \exp\left(-\frac{x_2 - 4}{2}\right)}$, $u_3 = \Phi\left(\frac{x_3 - 3}{2}\right)$. Now we have the following copula functions:

$$\begin{aligned} C(u_1, u_2; \theta) &= \exp\left(-\left((-\ln u_1)^\theta + (-\ln u_2)^\theta\right)^{\frac{1}{\theta}}\right) \\ &= \exp\left\{-\left[(-\ln(1 - \exp(-2x_1)))^\theta + \left(-\ln\left(\frac{1}{1 + \exp\left(\frac{x_2 - 4}{2}\right)}\right)\right)^\theta\right]^{\frac{1}{\theta}}\right\} \end{aligned} \tag{4.7a}$$

$$\begin{aligned} &C(u_1, u_2, u_3; \theta) = \exp\left(-\left((-\ln u_1)^\theta + (-\ln u_2)^\theta + (-\ln u_3)^\theta\right)^{\frac{1}{\theta}}\right) \\ &= \exp\left\{-\left[(-\ln(1 - \exp(-2x_1)))^\theta + \left(-\ln\left(\frac{1}{1 + \exp\left(\frac{x_2 - 4}{2}\right)}\right)\right)^\theta + \left(-\ln\left(\Phi\left(\frac{x_3 - 3}{2}\right)\right)\right)^\theta\right]^{\frac{1}{\theta}}\right\} \end{aligned} \tag{4.7b}$$

Equations (4.7a) and (4.7b) illustrate how to construct symmetric Archimedean copulas from the correlated random variables with following different marginal distributions. It may be worth noting here again that all cumulative marginal distribution $u_i \sim$uniform $(0, 1)$.

4.2 Properties of Symmetric Archimedean Copulas

Let C be an Archimedean copula with generating function ϕ. The Archimedean copula has the following properties (Nelsen, 2006; Salvadori et al., 2007; Savu and Trede, 2008):

- C is permutation-symmetric in its d arguments. This indicates that the Archimedean copula is the distribution function of d exchangeable uniform random variates.
- C is associative.
- If $\alpha > 0$ is any constant, then $\alpha\phi$ is also a generator of C.

Example 4.2 **Show for a given bivariate Archimedean copula function, one has $C(u_1, u_2) = C(u_2, u_1)$.**

Solution: Directly from Equation (4.1),

$$C(u_1, u_2) = \phi^{-1}(\phi(u_1) + \phi(u_2)) = \phi^{-1}(\phi(u_2) + \phi(u_1)) = C(u_2, u_1) \tag{4.8}$$

Example 4.3 **Show that the copula is associative.**

Suppose the symmetric Gumbel–Hougaard copula with parameter θ can be applied to study a given trivariate analysis. Show that the copula is associative, as follows:

$$C(u_1, u_2, u_3) = C(u_1, C(u_2, u_3)) = C(C(u_1, u_2), u_3) \tag{4.9}$$

Solution: The trivariate symmetric Gumbel–Hougaard copula can be expressed as follows:

$$C(u_1, u_2, u_3) = \phi^{-1}\left(\sum\nolimits_{k=1}^{3} \phi(u_k)\right) = \exp\left(-\left((-\ln u_1)^{\theta} + (-\ln u_2)^{\theta} + (-\ln u_3)^{\theta}\right)^{\frac{1}{\theta}}\right)$$

in which $\phi(u) = (-\ln u)^{\theta}$.

Now let's prove the associative property of the symmetric copula using $C(u_1, C(u_2, u_3))$as an example. The inner copula function $C(u_2, u_3)$is the bivariate Gumbel–Hougaard copula with the same parameter θ and can be written as follows:

$$C(u_2, u_3) = \exp\left(-\left((-\ln u_2)^{\theta} + (-\ln u_3)^{\theta}\right)^{\frac{1}{\theta}}\right)$$

Then $C(u_1, C(u_2, u_3))$ is also the bivariate Gumbel–Hougaard copula and can be written as follows:

$$C(u_1, C(u_2, u_3)) = \phi^{-1}\left(\phi(u_1) + \phi\left(C(u_2, u_3)\right)\right)$$

$$\phi(C(u_2, u_3)) = \left(-\ln\left(\exp\left(-\left((-\ln u_2)^{\theta} + (-\ln u_3)^{\theta}\right)^{\frac{1}{\theta}}\right)\right)\right)^{\theta} = (-\ln u_2)^{\theta} + (-\ln u_3)^{\theta}$$

Finally, we have the following:

$$\begin{aligned} C\left(u_1, C(u_2, u_3)\right) &= \phi^{-1}\left(\phi(u_1) + \phi\left(C(u_2, u_3)\right)\right) \\ &= \exp\left(-\left((-\ln u_1)^{\theta} + (-\ln u_2)^{\theta} + (-\ln u_3)^{\theta}\right)^{\frac{1}{\theta}}\right) = C(u_1, u_2, u_3) \end{aligned}$$

Similarly, we can prove that $C(u_1, u_2, u_3) = C(C(u_1, u_2), u_3)$.

Equation (4.9) implies that given three random variables u_1, u_2, u_3, the dependence between the first two random variables taken together and the third one alone is the same as the dependence between the first random variable taken alone and the two last ones taken together. This implies a strong symmetry between different variables in that they are exchangeable (Malevergne and Sornette, 2006). But the associative property of the Archimedean copula is not satisfied by other copula families in general (Embrechts et al., 2001).

Example 4.4 **Given the information $u = \frac{1}{2}, v = \frac{1}{4}, w = \frac{1}{6}$, and $\theta = \frac{1}{2}$, show that the associative property cannot be applied to the Farlie–Gumbel–Morgenstern copula.**

Solution: The bivariate Farlie–Gumbel–Morgenstern copula can be expressed as follows:

$$C(u,v) = uv + \theta uv(1-u)(1-v); \theta \in [-1,1]$$

With $u = \dfrac{1}{2}, v = \dfrac{1}{4}, w = \dfrac{1}{6},$ and $\theta = \dfrac{1}{2}$, we have

$$C(u,v) = \left(\frac{1}{2}\right)\left(\frac{1}{4}\right) + \left(\frac{1}{2}\right)\left(\frac{1}{2}\right)\left(\frac{1}{4}\right)\left(1-\frac{1}{2}\right)\left(1-\frac{1}{4}\right) = 0.1484$$

$$C(C(u_1,u_2),u_3) = 0.1484\left(\frac{1}{6}\right) + 0.1484\left(\frac{1}{2}\right)\left(\frac{1}{6}\right)(1-0.1484)\left(1-\frac{1}{6}\right) = 0.0335$$

and

$$C(v,w) = \left(\frac{1}{4}\right)\left(\frac{1}{6}\right) + \left(\frac{1}{2}\right)\left(\frac{1}{4}\right)\left(\frac{1}{6}\right)\left(1-\frac{1}{4}\right)\left(1-\frac{1}{6}\right) = 0.0547$$

$$C(u_1,C(u_2,u_3)) = 0.0547\left(\frac{1}{2}\right) + 0.0547\left(\frac{1}{2}\right)\left(\frac{1}{2}\right)\left(1-\frac{1}{2}\right)(1-0.0547) = 0.0338$$

Now we can reach the conclusion: $C(u_1, C(u_2, u_3)) \neq C(C(u_1, u_2), u_3)$.

Example 4.5 **Using the bivariate Gumbel–Hougaard copula, show $\alpha\phi$ is also a generator.**

Solution: The generating function and the corresponding Gumbel–Hougaard copula can be written as follows:

$$\phi(t) = (-\ln t)^{\theta}, \phi^{-1}(t) = \exp\left(-t^{\frac{1}{\theta}}\right) \text{ and}$$
$$C(u_1,u_2) = \exp\left(-\left((-\ln u_1)^{\theta} + (-\ln u_2)^{\theta}\right)^{\frac{1}{\theta}}\right)$$

in which θ is the copula parameter.

For any given $\alpha, \alpha > 0$and let $\psi(t) = \alpha\phi(t) = \alpha(-\ln t)^{\theta}$, we have the following:

$$\psi^{-1}(t) = \exp\left(-\left(\frac{t}{\alpha}\right)^{\frac{1}{\theta}}\right)$$

Rearranging the preceding Gumbel–Hougaard copula function, we have the following:

$$\begin{aligned} C(u_1,u_2) &= \exp\left(-\left((-\ln u_1)^{\theta} + (-\ln u_2)^{\theta}\right)^{\frac{1}{\theta}}\right) \\ &= \exp\left(-\left(\frac{a(-\ln u_1)^{\theta} + a(-\ln u_2)^{\theta}}{a}\right)^{\frac{1}{\theta}}\right) = \psi^{-1}\left(\psi(u_1) + \psi(u_2)\right) \end{aligned} \tag{4.10}$$

Now, we show that $\alpha\phi$ is also a generator of the Archimedean copula C, if $\alpha > 0$.

- Let $U_1, \ldots, U_d$ be d (dimensional) random variables with the joint distribution represented by the Archimedean copula C and generator ϕ. The distribution function of $C(U_1, \ldots, U_d)$, i.e., Kendall distribution, K_C, can be expressed as follows:

$$K_C(t) = P(C(U_1, \ldots, U_d) \leq t) = t + \sum_{i=1}^{d-1} (-1)^i \frac{\phi^i(t)}{i!} f_{i-1}(t) \tag{4.11}$$

where the auxiliary functions $f_0 = \frac{1}{\phi'(t)}$, and $f_i(t)$ for $i \geq 1$ are defined recursively as $f_i(t) = \frac{f'_{i-1}(t)}{\phi'(t)}$. For bivariate case, $K_C(t) = t - \frac{\phi(t)}{\phi'(t)}$. An Archimedean copula is determined by the function $K_C(t)$ defined on the unit interval [0,1]. This is a very useful result to determine which parametric copula family fits the data best (Savu and Trede, 2008).

From Section 3.4.2, we can derive the expression between Kendall's τ_n and parameter of symmetric Archimedean copulas using $K(t)$. Let bivariate random variables X and Y be modeled by the Archimedean copula, $u = F_X(x), v = F_Y(y)$. Then for the bivariate Archimedean copula $C(u, v)$, Equation (3.73) can be rewritten as follows:

$$\begin{aligned} \tau(X, Y) &= 4\int_{[0,1]^2} C(u, v) dC(u, v) - 1 = 4\int_{[0,1]} t dK_C(t) - 1 \\ &= 4\left(tK_C(t)|_0^1 - \int_0^1 K_C(t)dt\right) - 1 = 4\int_0^1 \frac{\phi(t)}{\phi'(t)} dt + 1 \end{aligned} \tag{4.12}$$

Example 4.6 **Consider the Gumbel–Hougaard copula with generator $\phi(t) = (-\ln t)^\theta, \theta \geq 1$. Derive Kendall's τ from the Gumbel–Hougaard copula.**

Solution: Taking the first derivative of $\phi(t)$, $\phi'(t) = -\frac{\theta(-\ln t)^{\theta-1}}{t}$, $\frac{\phi(t)}{\phi'(t)} = \frac{(-\ln t)^\theta}{-\frac{\theta(-\ln t)^{\theta-1}}{t}} = \frac{t \ln t}{\theta}$; Kendall's τ for the Gumbel–Hougaard copula is as follows:

$$\begin{aligned} \tau &= 1 + 4\int_0^1 \frac{t \ln t}{\theta} dt = 1 + \frac{4}{\theta}\int_0^1 t \ln t dt \\ &= 1 + \frac{4}{\theta}\left(\frac{t^2 \ln t}{2}\bigg|_0^1 - \int_0^1 \frac{t}{2} dt\right) = 1 + \frac{4}{\theta}\left(0 - \frac{1}{4}\right) = 1 - \frac{1}{\theta} \end{aligned} \tag{4.13}$$

Furthermore, in Equation (4.13) $\tau = 0$ if $\theta = 1$ (i.e., the bivariate random variable is independent), and the dependence increases with the increase of copula parameter θ.

Example 4.7 **Consider the Clayton copula with generator $\phi(t) = \frac{t^{-\theta} - 1}{\theta}$ and parameter $\theta : \theta \in [-1, \infty)$ 0. Derive Kendall's τ from the Clayton copula.**

Solution: Taking the first derivative of $\phi(t)$, we have the following:

$$\phi'(t) = -t^{-\theta-1} \text{ and } \frac{\phi(t)}{\phi'(t)} = \frac{\left(t^{-\theta}-1\right)/\theta}{-t^{-\theta-1}} = \frac{t^{\theta+1}-t}{\theta}$$

Kendall's τ for the Clayton copula can then be computed as follows:

$$\tau = 1 + 4\int_2^1 \frac{t^{\theta+1}-t}{\theta}dt = 1 + \frac{4}{\theta}\left(\frac{t^{\theta+2}}{\theta+2}\bigg|_0^1 - \frac{t^2}{2}\bigg|_0^1\right) = \frac{\theta}{\theta+2} \tag{4.14}$$

In Equation (4.14), $\tau = -1$ when $\theta = -1$ (i.e., perfectly negatively dependent). And $\tau \to 1$ when $\theta \to \infty$ (i.e., perfectly positively dependent). Similar to Example 4.6, the dependence of the bivariate random variable increases with the increase of parameter θ.

4.3 Archimedean Copula Families

4.3.1 Bivariate Archimedean Copula Families

There exists a large variety of symmetric Archimedean copula families that are used for constructing copulas to represent multivariate distributions. Table 4.1 lists the popularly applied one-parameter Archimedean copulas (Nelsen, 2006). Tables 4.2 and 4.3 list their first-order derivative of $\frac{\partial}{\partial u_1}C(u_1, u_2)$ and the copula density $c(u_1, u_2)$, respectively. One may refer to Nelsen (2006) for other one-parameter Archimedean copulas.

As discussed in Nelsen (2006), the Cook–Johnson (Clayton) family was derived by Clayton (1978), Oakes (1982, 1986), Cox and Oakes (1984), and Cook and Johnson (1981). This copula family can be used for modeling nonelliptically symmetric (nonnormal) multivariate data (Cook and Johnson, 1981). When $\theta = -1$, the copula represents the joint distribution of the perfectly negatively dependent bivariate random variables, i.e., the Fréchet–Hoeffding lower bound: $W : W = \max(u_1 + u_2 - 1, 0)$. When $\theta = 0$, the copula represents the joint distribution for independent bivariate random variables, i.e., product copula: $\Pi : \Pi = u_1u_2$. When $\theta \to \infty$, the copula represents the joint distribution for perfectly positively dependent bivariate random variables, i.e., the Fréchet–Hoeffding upper bound: $M : M = \min(u_1, u_2)$.

The Gumbel–Hougaard Archimedean copula was first introduced by Gumbel (1960). This copula family cannot be applied to model the negatively dependent bivariate random variables. Nelsen (2006) showed that the Gumbel–Hougaard copula belonged to the extreme value copula family. With this characteristic, the Gumbel–Hougaard Archimedean copula may be a suitable candidate for multivariate frequency analysis of extreme hydrological events, i.e., peak discharge and corresponding volume and duration.

The Ali–Mikhail–Haq Archimedean copula was developed by Ali et al. (1978). It was developed based on the concept of univariate logistic distribution that may be specified by considering a suitable form for the odds in favor of a failure against survival. The parameter of this copula is a measure of departure from independence or a measure of association between two random variables. In addition, the Ali–Mikhail–Haq copula can

Table 4.1. *Selected Archimedean copulas.*

Name	Copula function $C_\theta(u_1, u_2)$	Generating function $\phi(t)$	Parameter θ
Clayton	$\max\left(\left(u_1^{-\theta} + u_2^{-\theta} - 1\right)^{-\frac{1}{\theta}}, 0\right)$	$\frac{1}{\theta}\left(t^{-\theta} - 1\right)$	$[-1, \infty)\backslash\{0\}$
Ali–Mikhail–Haq	$\frac{u_1 u_2}{\left(1 + \left(1 - u_1^{\theta}\right)\left(1 - u_2^{\theta}\right)\right)^{\frac{1}{\theta}}}$	$\ln \frac{1 - \theta(1 - t)}{t}$	$[-1, 1]$
Gumbel–Hougaard	$\exp\left(-\left((-\ln u_1)^{\theta} + (-\ln u_2)^{\theta}\right)^{\frac{1}{\theta}}\right)$	$(-\ln t)^{\theta}$	$[1, \infty)$
Frank	$-\frac{1}{\theta}\ln\left(1 + \frac{\left(e^{-\theta u_1} - 1\right)\left(e^{-\theta u_2} - 1\right)}{e^{-\theta} - 1}\right)$	$-\ln \frac{e^{-\theta t} - 1}{e^{-\theta} - 1}$	$(-\infty, \infty)\backslash\{0\}$
Joe	$1 - \left((1 - u_1)^{\theta} + (1 - u_2)^{\theta} - (1 - u_1)^{\theta}(1 - u_2)^{\theta}\right)^{\frac{1}{\theta}}$	$-\ln\left(1 - (1 - t)^{\theta}\right)$	$[1, \infty)$
Survival	$u_1 u_2 e^{-\theta \ln u_1 \ln u_2}$	$\ln(1 - \theta \ln t)$	$(0, 1]$

Table 4.2. *First-order derivatives* $\frac{\partial}{\partial u_1} C_\theta(u_1, u_2)$ for the selected Archimedean copulas.

Name	$\frac{\partial}{\partial u_1} C_\theta(u_1, u_2)$
Clayton	$\frac{u_1^{-1-\theta}}{\left(-1+u_1^{-\theta}+u_2^{-\theta}\right)^{\frac{1+\theta}{\theta}}}, \quad \theta > 0$
Ali–Mikhail–Haq	$\frac{u_2+\theta u_2(-1+u_2)}{[-1+\theta(-1+u_1)(-1+u_2)]^2}$
Gumbel–Hougaard	$\frac{(-\ln u_1)^{-1+\theta}\left[(-\ln u_1)^\theta+(-\ln u_2)^\theta\right]^{-1+\frac{1}{\theta}}}{u_1 e^{\left[(-\ln u_1)^\theta+(-\ln u_2)^\theta\right]^{\frac{1}{\theta}}}}$
Frank	$\frac{e^{-\theta u_1}\left(e^{-\theta u_2}-1\right)}{e^{-\theta(u_1+u_2)}-e^{-\theta u_1}-e^{-\theta u_2}+e^{-\theta}}$
Joe	$-\left\{(1-u_1)^{-1+\theta}\left[-1+(1-u_2)^\theta\right]\left[(1-u_1)^\theta+(1-u_2)^\theta-(1-u_1)^\theta(1-u_2)^\theta\right]^{-1+\frac{1}{\theta}}\right\}$
Survival	$\frac{u_2-\theta u_2 \ln u_2}{e^{\theta \ln u_1 \ln u_2}}$

Table 4.3. *Copula density* $c_\theta(u_1, u_2)$ for the selected Archimedean copulas.

$\frac{\partial^2}{\partial u_1 \partial u_2} C_\theta(u_1, u_2)$
Clayton $\frac{(1+\theta)u_1^{-1-\theta}u_2^{-1-\theta}}{\left(-1+u_1^{-\theta}+u_2^{-\theta}\right)^{\frac{1+2\theta}{\theta}}}, \quad \theta > 0$
Ali–Mikhail–Haq $\frac{-1+\theta^2(-1+u_2+u_2-u_1u_2)-\theta(-2+u_1+u_2-u_1u_2)}{[-1+\theta(-1+u_1)(-1+u_2)]^3}$
Gumbel–Hougaard $\frac{(-\ln u_2)^{-1+\theta}(-\ln u_1)^{-1+\theta}\left[w^{\frac{2-2\theta}{\theta}}-(1-\theta)w^{\frac{1-2\theta}{\theta}}\right]}{u_1 u_2 e^{w^{\frac{1}{\theta}}}}, w = (-\ln u_1)^\theta + (-\ln u_2)^\theta$
Frank $\frac{\theta\left(e^\theta-1\right)e^{\theta(1+u_1+u_2)}}{\left(e^{\theta(u_1+u_2)}-e^{\theta(1+u_1)}-e^{\theta(1+u_2)}+e^\theta\right)^2}$
Joe $((1-u_1)(1-u_2))^{-1+\theta}(\theta-1+w)w^{\frac{1}{\theta}-2}$ $w = (1-u_1)^\theta-((1-u_1)(1-u_2))^\theta+(1-u_2)^\theta$
Survival $\frac{1-\theta-\theta \ln u_2+\theta \ln u_1(-1+\theta \ln u_2)}{e^{\theta \ln u_1 \ln u_2}}$

only capture the dependence within the range of $\tau \in [-0.182, 0.333]$, which limits the application of the Ali–Mikhail–Haq copula to bivariate frequency analysis.

The Frank Archimedean copula was developed by Frank (1979). The Frank copula satisfies all the conditions for the construction of bivariate distributions with fixed marginals except for independent variables ($\theta \neq 0$), for the Frank copula). However, if the bivariate random variables are independent, the copula function is the product copula. Thus, the Frank copula is also considered absolutely continuous with full support on the unit square as the Cook–Johnson (Clayton) copula family.

The Joe Archimedean copula was first introduced by Joe (1993). When $\theta = 1$, this copula represents the joint distribution for independent bivariate random variables. Similar to the Gumbel–Hougaard copula, the Joe copula cannot be applied to model negatively dependent bivariate random variables.

The survival copula is associated with Gumbel's bivariate exponential distribution. This family is the survival copula that is actually the survival probability distribution of the Gumbel bivariate exponential distribution.

Example 4.8 **Using the copulas given in Table 4.4, plot the density functions of bivariate Archimedean copulas. Can any conclusions be reached from these plots?**

Table 4.4. *Bivariate Archimedean copula parameters.*

Copula	Parameter θ	
Clayton	0.5	2
Gumbel–Houggard	2	5
Frank	–5	2
Joe	2	5

Solution: With the corresponding copula density function listed in Table 4.3, Figure 4.1 plots the copula density functions for the copulas listed in Table 4.4. In the case of the Clayton copula, when $\theta \in [-1, 0)$, its generating function is not strict. Thus, the Clayton copula is only sufficiently differentiable if $\theta > 0$. In addition, from Figure 4.1 and the discussion on tail dependence in Chapter 3, we can reach the following conclusions graphically: (1) the random variables are positively dependent and seem to have left (lower) tail dependence but no right (upper) tail dependence for the Clayton copula; (2) the random variables are positively dependent for the Gumbel–Hougaard and Joe copulas and exhibit the right (upper) tail dependence; and (3) the Frank copula does not seem to have either right (upper) or left (lower) tail dependence, and the random variables are negatively dependent when $\theta < 0$.

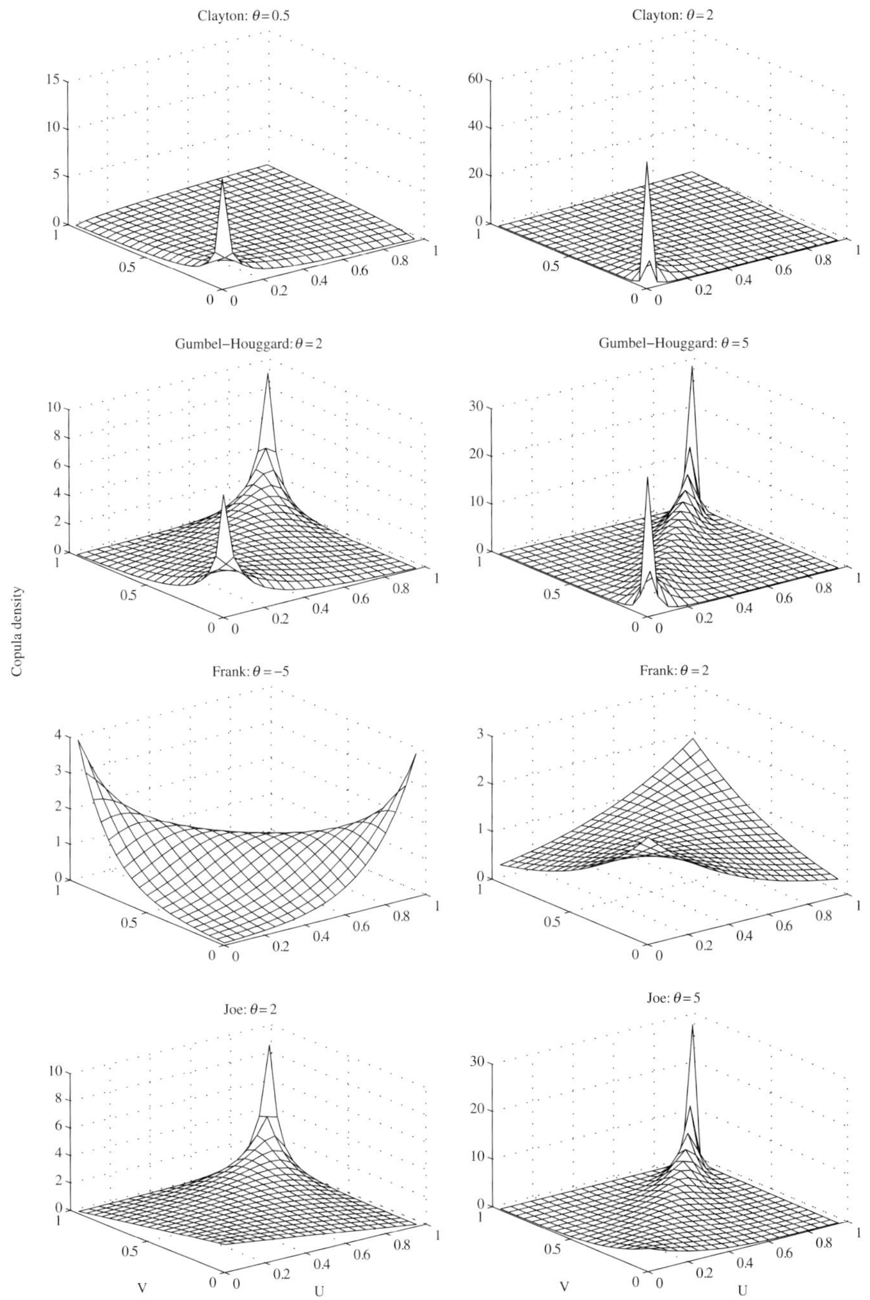

Figure 4.1 Copula density plots.

Table 4.5. *Relations between* τ and θ for selected symmetric Archimedean copulas.

Family	$\phi(t)$	$\phi'(t)$	$\tau = 1 + 4\int_0^1 \frac{\phi(t)}{\phi'(t)}dt$	Range of τ
Clayton	$\frac{1}{\theta}\left(t^{-\theta} - 1\right)$	$-t^{-1-\theta}$	$\frac{\theta}{\theta+2}$	$[-1, 1]\backslash 0$
Ali–Mikhail–Haq	$\ln\left(\frac{1-\theta(1-t)}{t}\right)$	$\frac{\theta - 1}{t - \theta t + \theta t^2}$	No analytical solution	$\left[-0.18, \frac{1}{3}\right]$
Gumbel–Hougaard	$(-\ln t)^{\theta}$	$-\frac{\theta}{t}(-\ln t)^{\theta-1}$	$1 - \frac{1}{\theta}$	$[0, 1]$
Frank	$-\ln\frac{e^{-\theta t} - 1}{e^{-\theta} - 1}$	$\frac{\theta}{1 - e^{\theta t}}$	$1 - \frac{4}{\theta}[D_1(-\theta) - 1]$, $D_1(\theta) = \frac{1}{\theta}\int_0^{\theta}\frac{t}{e^t - 1}dt$, $D_1(-\theta) = D_1(\theta) + \frac{1}{2}$	$[-1, 1]\backslash 0$
Joe	$-\ln\left[1 - (1-t)^{\theta}\right]$	$\frac{\theta(1-t)^{\theta-1}}{-1 + (1-t)^{\theta}}$	No analytical solution	[0, 1]
Survival	$\ln(1 - \theta\ln t)$	$\frac{\theta}{t[\ln t - 1]}$	No analytical solution	[−0.3613, 0]

4.3.2 Relation of Kendall's τ and Parameter θ for Bivariate Archimedean Copulas

Equation (4.9) presents the relation between Kendall's τ and the generating function of bivariate Archimedean copulas. In turn, the relation between Kendall's τ and parameter θ for a given bivariate Archimedean copula can be determined. Table 4.5 lists this relation for the selected Archimedean copulas.

4.4 Symmetric Multivariate Archimedean Copulas ($d \geq 3$)

Nelsen (2006) stated that the bivariate Archimedean copula may not be extended to a multivariate case, unless additional conditions are satisfied to construct the symmetric multivariate Archimedean copulas to represent the joint distribution of multivariate random variables (i.e., $d \geq 3$). Nelsen (2006) discussed the theorem and three additional useful results that are necessary to construct the appropriate multivariate Archimedean copula. These results are introduced in the following.

Theorem (Theorem 4.6.2, Nelsen, 2006)**:** Let ϕ be a continuous strictly decreasing function from **I** to $[0, \infty)$with $\phi(0) = \infty, \phi(1) = 0$, and ϕ^{-1}denote the inverse of ϕ. If C^{d} is the function from $\mathbf{I}^{\mathrm{d}}$ to **I**, given by Equation (4.1), then C^{d} is the copula function to

Table 4.6. *Multivariate ($d \geq 3$)* symmetric Archimedean copulas.

Family	$C(u_1, u_2, \ldots, u_d)$	$\phi(t)$	θ
Clayton	$\left(\sum_{i=1}^{d} u_i^{-\theta} - d + 1\right)^{-\frac{1}{\theta}}$	$\frac{1}{\theta}\left(t^{-\theta} - 1\right)$	$(0, +\infty)$
Ali–Mikhail–Haq	$\frac{\prod_{i=1}^{d} u_i}{1 - \theta \prod_{i=1}^{d}(1 - u_i)}$	$\ln \frac{1 - \theta(1 - t)}{t}$	$[-1, 1]$
Gumbel–Hougaard	$\exp\left(-\left(\sum_{i=1}^{d}(-\ln u_i)^{\theta}\right)^{\frac{1}{\theta}}\right)$	$(-\ln(t))^{\theta}$	$(1, +\infty)$
Frank	$-\frac{1}{\theta} \ln\left(1 + \frac{\prod_{i=1}^{d}(e^{-\theta u_i} - 1)}{(e^{-\theta} - 1)^{d-1}}\right)$	$-\ln \frac{e^{-\theta t} - 1}{e^{-\theta} - 1}$	$(0, +\infty)$
Joe	$1 - \left[1 - \prod_{i=1}^{d}\left(1 - (1 - u_i)^{\theta}\right)\right]^{\frac{1}{\theta}}$	$-\ln\left(1 - (1 - t)^{\theta}\right)$	$[1, +\infty)$
Survival	Not extendable		

represent the d-dimensional multivariate distribution if and only if ϕ is completely monotonic on $[0, \infty)$, as follows:

$$(-1)^k \frac{d^k \phi^{-1}(t)}{dt^k} \geq 0 \tag{4.15}$$

Result 1: If function f is absolutely monotonic, i.e.,

$$d^k f(x)/dx^k \geq 0, \quad k = 0, 1, 2, \ldots \tag{4.16}$$

and function g is completely monotonic, then the composite $f \circ g$is completely monotonic.
Result 2: If functions f and g are completely monotonic, then so is their product fg.
Result 3: If f is completely monotonic and g is a positive function with a completely monotone derivative, then the composite $f \circ g$is completely monotonic.

Table 4.6 lists the applicability to extend the selected bivariate Archimedean copula to higher dimension.

Example 4.9 **Show that the bivariate Clayton copula can be extended to higher dimension symmetric Clayton copulas for $\theta > 0$.**

Solution: It is known that if the Clayton copula can be extended to higher dimensions, i.e., $d \geq 3$, we need to satisfy the theorem (Theorem 4.6.2, Nelsen, 2006) discussed previously. The generating function for the Clayton copula can be written as follows:

$$\phi(t) = \frac{1}{\theta}\left(t^{-\theta} - 1\right) \Rightarrow \phi^{-1}(t) = (\theta t + 1)^{-\frac{1}{\theta}}$$

According to Nelsen (2006), we know that for $\theta \geq 0$, the generating function is strictly decreasing from **I** to $(0, \infty)$. Applying Equation (4.15), we have the following:

$$(-1)^1 \frac{d\phi^{-1}(t)}{dt} = (\theta t + 1)^{-\left(\frac{1+\theta}{\theta}\right)} \geq 0$$

$$(-1)^2 \frac{d^2\phi^{-1}(t)}{dt} = (1 + \theta)(\theta t + 1)^{\frac{1+2\theta}{\theta}} \geq 0$$

$$\ldots$$

$$(-1)^k \frac{d^k\phi^{-1}(t)}{dt} = (-1)^{2k}\left(\prod\nolimits_{j=2}^{k}(1 + (j-1)\theta)\right)(\theta t + 1)^{-\frac{1+k\theta}{\theta}}; k \geq 2$$

Now, we reach the conclusion that the bivariate Clayton copula can be extended to multivariate symmetric Clayton copula, as follows:

$$C_\theta^d(\mathbf{u}) = \left(\sum\nolimits_{i=1}^{d} u_i^{-\theta} - d + 1\right)^{-\frac{1}{\theta}}; \theta \geq 0$$

Note that the multivariate symmetric Clayton copula (i.e., $d \geq 3$) may only model the positive dependent/independent multivariate random variables. The reason is that if $\theta < 0$, Equation (4.15) cannot be guaranteed to be fully satisfied.

Example 4.10 **Show that the inverse of the generating function of the Ali–Mikhail–Haq copula is completely monotonic and thus the bivariate Ali–Mikhail–Haq copula can be extended to higher dimensions.**

Solution: Following Nelsen (2006), it is known that the generating function of the Ali–Mikhail–Haq copula is strictly decreasing from **I** to $(0, \infty)$. The generating function and its inverse function can be written as follows:

$$\phi(t) = \ln \frac{1 - \theta(1 - t)}{t}; \phi^{-1}(t) = \frac{\theta - 1}{\theta - \exp(t)}$$

Rather than directly applying the theorem as in Example 4.9, here we use the inequality proposed by Widder (1941) for function $\phi^{-1}$to be completely monotonic, as follows:

$$\phi^{-1}\left(\phi^{-1}\right)'' - \left(\left(\phi^{-1}\right)\right)^2 \geq 0 \tag{4.17}$$

The first and second derivative of the inverse function can be written as follows:

$$\left(\phi^{-1}\right)' = \frac{d\phi^{-1}}{dt} = \frac{(\theta - 1)\exp(t)}{(\theta - \exp(t))^2}; \left(\phi^{-1}\right)'' = \frac{d^2\phi(t)}{dt^2} = \frac{(\theta - 1)\exp(t)}{(\theta - \exp(t))^2} + \frac{2(\theta - 1)\exp(2t)}{(\theta - \exp(t))^3}$$

Substituting the first and second derivatives of the inverse function into Equation (4.17), we have the following:

$$\phi^{-1}(\phi^{-1})'' - \left((\phi^{-1})'\right)^2$$
$$= \frac{\theta-1}{\theta-\exp(t)}\left(\frac{(\theta-1)\exp(t)}{(\theta-\exp(t))^2} + \frac{2(\theta-1)\exp(2t)}{(\theta-\exp(t))^3}\right) - \left(\frac{(\theta-1)\exp(t)}{(\theta-\exp(t))^2}\right)^2 = \frac{(\theta-1)^2(\exp(t)+\theta)}{(\theta-\exp(t))^4}$$

Considering the Ali–Mikhail–Haq copula with $\theta \in [-1, 1)$, we have $\phi^{-1}(\phi^{-1})'' - \left((\phi^{-1})'\right)^2 \geq 0$ for the whole parameter range. Finally, we show that $\phi^{-1}$is completely monotonic in $t \in (0, \infty)$with $\theta \in [-1, 1)$. The bivariate Ali–Mikhail–Haq copula can be extended to higher dimensions as follows:

$$C_\theta^d(\mathbf{u}) = \frac{\prod_{i=1}^d u_i}{1 - \theta\prod_{i=1}^d (1 - u_i)}; \theta \in [-1, 1)$$

Example 4.11 **Show that the Joe copula can be extended to any dimension $d \geq 3$, for $\theta \in [1, \infty)$.**

Solution: We will solve this example using the result 1 introduced earlier, that is, for two given functions f and g, if f is absolutely monotonic and g is completely monotonic, then $f \circ g$ is completely monotonic.

The generating function and its inverse function of the Joe copula can be written as follows:

$$\phi(t) = -\ln\left(1 - (1-t)^\theta\right); \phi^{-1}(t) = 1 - (1 - \exp(-t))^{\frac{1}{\theta}}$$

To use two previously stated properties stated, we let $f(x) = 1 - (1 - x)^{\frac{1}{\theta}}, x \in (0, 1]$ and $g(t) = \exp(-t)$.

For function $f(x)$, applying Equation (4.11) we have the following:

$$f'(x) = \frac{df(x)}{dx} = \frac{1}{\theta}(1-x)^{\frac{1}{\theta}-1} \geq 0$$

$$f''(x) = \frac{d^2f(x)}{dx^2} = \frac{1}{\theta}\left(1 - \frac{1}{\theta}\right)(1-x)^{\frac{1}{\theta}-2}$$

$$\ldots$$

$$f^{(k)}(x) = \frac{d^kf(x)}{dx^k} = \frac{(1-x)^{\frac{1}{\theta}-k}}{\theta}\prod_{i=1}^{k-1}\left(i - \frac{1}{\theta}\right) \geq 0, k \geq 2$$

Thus, we know the function $f(x)$is absolutely monotonic.

For function $g(t)$, $g(t) = \exp(-t) \geq 0$, we also need to show that $g(t)$ is completely monotonic.

The first and second derivatives of function $g(t)$are as follows:

$$g'(t) = -\exp(-t); g''(t) = \exp(-t)$$

$$g(t)g''(t) - (g'(t))^2 = \exp(-2t) - \exp(-2t) = 0$$

We can also substitute function $g(t)$into Equation (4.15) and have the following:

$$\frac{(-1)dg(t)}{dt} = \exp(-t) > 0; (-1)^2 \frac{d^2 g(t)}{dt^2} = \exp(-t) > 0$$

$$\ldots, (-1)^k \frac{d^k g(t)}{dt^k} = \begin{cases} (-1)^{k+1} \exp(-t) > 0, & \text{if } k \textit{ is odd number} \\ (-1)^k \exp(-t) > 0, & \text{if } k \text{ is even number} \end{cases}$$

Now, we have $f \circ g$as completely monotonic. The bivariate Joe copula can be extended to higher dimensions as follows:

$$C_\theta^d(\mathbf{u}) = 1 - \left(1 - \prod_{i=1}^{d} \left(1 - (1 - u_i)^\theta\right)\right)^{\frac{1}{\theta}}$$

4.5 Identification of Symmetric Archimedean Copulas

The Archimedean copulas can be identified using nonparametric, semiparametric, and parametric estimation procedures.

4.5.1 Nonparametric Estimation Procedure for Bivariate Copulas

Genest and Rivest (1993) described a procedure to identify a copula function based on nonparametric estimation for bivariate Archimedean copulas. It is assumed that a random sample of bivariate observations $(x_{11}, x_{21}), (x_{12}, x_{22}), \ldots, (x_{1n}, x_{2n})$ is available and that its underlying distribution function $F(x_1, x_2)$has an associated Archimedean copula C, i.e., $C(F_{X1}(x_1), F_{X_2}(x_2)) = F(x_1, x_2)$. Then, the following steps can be followed to identify an appropriate copula:

1. Determine Kendall's τ (the dependence structure of the bivariate random variables) from observations using Equation (3.73).
2. Determine the copula parameter θ from the preceding value of τ according to the relation between Kendall's τ and the copula parameter θ (See Table 4.5), i.e., for the Gumbel–Hougaard copula family, the relation between Kendall's τ and the copula parameter θ is given as $\tau_n = 1 - 1/\theta$.
3. Obtain the generating function of the copula, ϕ, by inserting parameter θ obtained as in step 2.
4. Obtain the copula from its generating function ϕ.

Thus, copula functions based on different bivariate Archimedean copula families are obtained.

Now the identified copula needs to be tested, if it is adequate for given bivariate observations. This is accomplished using the following steps:

1. Define an intermediate random variable $Z = F(x_1, x_2)$, which has a distribution function $K(z) = P(Z \leq z)$. This distribution is related to the generator of the Archimedean copula through Equation (4.18).
2. Construct a nonparametric estimate of K_n as follows:
   a. Compute the following:

$$z_i = \frac{\sum_{j=1}^{n} 1\left(x_{1j} \leq x_{1i} \; and \; x_{2j} \leq x_{2i}\right)}{n-1}, \quad i = 1, \ldots, n \tag{4.18}$$

   b. Construct nonparametric Kendall distribution (K_n):

$$K_n(t) = \frac{\sum_{i=1}^{n} (z_i \leq t)}{n} \left(\text{i.e.,} \quad z_i's \leq z\right). \tag{4.19}$$

3. Construct a parametric estimate Kendall distribution (K) as follows:

$$K(t) = t - \frac{\phi(t)}{\phi'(t)} \tag{4.20}$$

Construct a plot of nonparametric $K_n(t)$ versus parametrically estimated K using Equation (4.20), which may also be called a *Q-Q* plot. If the plot is in agreement with a straight line passing through the origin at a 45 degree angle, then the generating function is satisfactory. The 45 degree angle indicates that the quantiles are equal. Otherwise, the copula function needs to be reidentified.

Example 4.12 **Using the bivariate sample data given in Table 4.7, (1) estimate the parameters if the Gumbel–Hougaard, Frank, and Clayton copulas are tested; (2) construct the Q-Q plot (i.e., nonparametric and parametric Kendall distribution), the K-plot and chi-square plot for each copula candidate; and (3) determine what can be concluded from the plots.**

Solution:

1. To determine the copula parameters nonparametrically using the relationship between Kendall's τ and copula parameter θ, we can proceed as follows:
   a. Calculate τ_n: using the flood data listed in Table 4.7, we can calculate τ_n from the sample data using exactly the same logic as in Example 3.15. τ_n is computed as 0.584. It indicates the positive dependence between random variables X and Y listed.
   b. Estimate copula parameter θ: Using Table 4.5 (i.e., the relation between Kendall's τ and copula parameter θ), we can estimate the copula parameter nonparametrically as follows:

Table 4.7. *Sample data:* X *and* Y *following gamma and normal distributions, respectively.*

No.	*X*	*Y*	No.	*X*	*Y*
1	11.68	7.67	51	12.82	6.28
2	18.01	15.54	52	7.79	7.49
3	9.15	3.03	53	16.02	13.94
4	16.56	12.49	54	14.03	12.99
5	7.80	7.41	55	8.53	8.46
6	13.11	6.36	56	10.45	9.44
7	9.81	8.03	57	18.71	13.97
8	11.76	11.63	58	13.60	10.89
9	20.59	16.60	59	9.02	15.45
10	21.60	16.00	60	12.05	2.15
11	7.05	8.00	61	12.65	11.01
12	16.44	14.32	62	10.79	7.24
13	16.91	16.68	63	11.24	9.48
14	13.94	13.28	64	12.66	10.39
15	12.74	10.81	65	10.56	9.77
16	10.75	10.43	66	15.56	13.47
17	8.63	8.37	67	15.74	13.64
18	26.09	20.42	68	16.45	16.24
19	8.47	6.47	69	7.64	6.57
20	18.33	14.25	70	6.37	6.31
21	7.28	5.30	71	10.36	11.21
22	18.81	13.14	72	8.59	6.05
23	8.63	9.70	73	16.45	12.99
24	18.29	16.51	74	5.72	1.37
25	17.24	10.94	75	14.38	11.53
26	20.95	13.47	76	9.71	3.11
27	8.65	7.91	77	12.75	9.22
28	6.84	7.13	78	10.29	8.66
29	8.40	9.02	79	13.01	10.17
30	11.32	10.79	80	8.09	8.65
31	11.69	9.16	81	7.06	7.38
32	12.80	11.71	82	13.63	10.13
33	7.07	1.34	83	12.76	11.56
34	7.96	11.08	84	7.86	5.65
35	4.76	2.34	85	20.31	15.68
36	9.18	7.98	86	7.14	10.80
37	14.80	12.87	87	12.06	11.11
38	8.95	7.27	88	15.57	11.14
39	18.26	15.09	89	13.75	10.87
40	5.92	3.40	90	7.36	3.41

Table 4.7. (*cont.*)

No.	X	Y	No.	X	Y
41	11.51	14.23	91	13.09	14.23
42	9.32	11.75	92	10.13	11.91
43	13.23	8.15	93	12.71	10.50
44	10.71	12.36	94	9.84	10.00
45	11.50	5.75	95	16.82	14.32
46	7.63	4.57	96	5.78	2.57
47	7.67	6.78	97	15.61	8.97
48	8.55	10.54	98	10.96	7.91
49	8.32	7.97	99	11.83	9.14
50	10.36	13.48	100	11.93	9.42

Gumbel–Hougaard copula: $\tau = 1 - \frac{1}{\theta_{GH}} \Rightarrow \theta_{GH} = \frac{1}{1-\tau} = \frac{1}{1-0.584} = 2.4038$

Clayton copula: $\tau = \frac{\theta_C}{\theta_C + 2} \Rightarrow \theta_C = \frac{2\tau}{1-\tau} = 2.8077$

Frank copula: $\tau = 1 - \frac{4}{\theta_F}[D_1(-\theta_F) - 1], D_1(-\theta_F) = D_1(\theta_F) + \frac{1}{2} \Rightarrow \theta_F = 7.5132$

where $D_1(\theta_F)$ is the first-order Debye function, i.e., $D_1(\theta_F) = \frac{1}{\theta_F}\int_0^{\theta_F} \frac{t}{e^t - 1} dt$. Unlike the Gumbel–Houggard and Clayton copulas, the parameters of the Frank copula need to be estimated numerically:

1. Construct the Q-Q plot of nonparametric and parametric Kendall distributions.

   Applying Equation (4.20), the parametric Kendall distribution for the Gumbel–Hougaard, Clayton, and Frank copulas may be written as follows:

   Gumbel–Hougaard copula:

$$\phi(t) = (-\ln t)^{\theta}; \phi'(t) = -\frac{\theta(-\ln t)^{\theta-1}}{t}; \quad K(t) = t - \frac{\phi(t)}{\phi'(t)} = \frac{t(\theta - \ln(t))}{\theta} \tag{4.21}$$

   Clayton copula:

$$\phi(t) = \frac{1}{\theta}\left(t^{-\theta} - 1\right); \phi'(t) = -t^{-\theta-1}; \quad K(t) = t - \frac{\phi(t)}{\phi'(t)} = t - \frac{t^{\theta+1} - t}{\theta} \tag{4.22}$$

   Frank copula:

$$\phi(t) = -\ln\left(\frac{e^{-\theta t} - 1}{e^{-\theta} - 1}\right); \quad \phi'(t) = \frac{\theta e^{-\theta t}}{e^{-\theta t} - 1}; \quad K(t) = t + \frac{e^{\theta t}\ln\left(\frac{e^{-\theta t} - 1}{e^{-\theta} - 1}\right)\left(e^{-\theta t} - 1\right)}{\theta} \tag{4.23}$$

Table 4.8 lists the nonparametric and parametric Kendall distributions computed using the sample data. Figure 4.2 plots the nonparametric and parametric Kendall distributions.

Table 4.8. *Nonparametric and parametric estimates of the Kendall distribution.*

No.	X	Y	V_i	Kn	Gumbel–Hougaard	Clayton	Frank
1	11.68	7.67	0.24	0.39	0.38	0.32	0.36
2	18.01	15.54	0.88	0.95	0.93	0.97	0.96
3	9.15	3.03	0.05	0.08	0.11	0.07	0.12
4	16.56	12.49	0.73	0.81	0.83	0.88	0.85
5	7.80	7.41	0.14	0.24	0.25	0.19	0.25
6	13.11	6.36	0.17	0.28	0.30	0.23	0.28
7	9.81	8.03	0.26	0.41	0.41	0.35	0.38
8	11.76	11.63	0.48	0.64	0.63	0.63	0.61
9	20.59	16.60	0.96	0.99	0.98	1.00	0.99
10	21.60	16.00	0.95	0.98	0.97	1.00	0.99
11	7.05	8.00	0.07	0.15	0.15	0.09	0.15
12	16.44	14.32	0.82	0.88	0.89	0.94	0.92
13	16.91	16.68	0.88	0.95	0.93	0.97	0.96
14	13.94	13.28	0.70	0.79	0.80	0.86	0.82
15	12.74	10.81	0.51	0.69	0.65	0.66	0.64
16	10.75	10.43	0.36	0.52	0.51	0.48	0.49
17	8.63	8.37	0.21	0.35	0.35	0.28	0.33
18	26.09	20.42	1.00	1.00	1.00	1.00	1.00
19	8.47	6.47	0.11	0.21	0.21	0.15	0.21
20	18.33	14.25	0.85	0.92	0.91	0.96	0.94
21	7.28	5.30	0.06	0.12	0.13	0.08	0.14
22	18.81	13.14	0.78	0.85	0.86	0.92	0.89
23	8.63	9.70	0.25	0.40	0.39	0.34	0.37
24	18.29	16.51	0.91	0.96	0.95	0.99	0.98
25	17.24	10.94	0.61	0.74	0.74	0.77	0.74
26	20.95	13.47	0.80	0.87	0.87	0.93	0.90
27	8.65	7.91	0.19	0.32	0.32	0.26	0.31
28	6.84	7.13	0.06	0.12	0.13	0.08	0.14
29	8.40	9.02	0.20	0.34	0.33	0.27	0.32
30	11.32	10.79	0.41	0.59	0.56	0.54	0.54
31	11.69	9.16	0.36	0.52	0.51	0.48	0.49
32	12.80	11.71	0.59	0.73	0.72	0.75	0.72
33	7.07	1.34	0.01	0.03	0.03	0.01	0.04
34	7.96	11.08	0.19	0.32	0.32	0.26	0.31
35	4.76	2.34	0.01	0.03	0.03	0.01	0.04
36	9.18	7.98	0.23	0.37	0.37	0.31	0.35
37	14.80	12.87	0.71	0.80	0.81	0.87	0.83
38	8.95	7.27	0.16	0.26	0.28	0.22	0.27
39	18.26	15.09	0.87	0.93	0.92	0.97	0.95
40	5.92	3.40	0.04	0.06	0.09	0.05	0.10
41	11.51	14.23	0.50	0.68	0.64	0.65	0.63

Table 4.8. (*cont.*)

No.	*X*	*Y*	V_i	*Kn*	Gumbel–Hougaard	Clayton	Frank
42	9.32	11.75	0.33	0.46	0.48	0.44	0.46
43	13.23	8.15	0.34	0.48	0.49	0.46	0.47
44	10.71	12.36	0.42	0.60	0.57	0.56	0.55
45	11.50	5.75	0.12	0.22	0.23	0.16	0.22
46	7.63	4.57	0.07	0.15	0.15	0.09	0.15
47	7.67	6.78	0.11	0.21	0.21	0.15	0.21
48	8.55	10.54	0.23	0.37	0.37	0.31	0.35
49	8.32	7.97	0.17	0.28	0.30	0.23	0.28
50	10.36	13.48	0.40	0.58	0.55	0.53	0.53
51	12.82	6.28	0.15	0.25	0.27	0.20	0.26
52	7.79	7.49	0.14	0.24	0.25	0.19	0.25
53	16.02	13.94	0.79	0.86	0.87	0.93	0.90
54	14.03	12.99	0.70	0.79	0.80	0.86	0.82
55	8.53	8.46	0.20	0.34	0.33	0.27	0.32
56	10.45	9.44	0.32	0.45	0.47	0.43	0.45
57	18.71	13.97	0.83	0.89	0.89	0.95	0.93
58	13.60	10.89	0.57	0.71	0.70	0.73	0.70
59	9.02	15.45	0.31	0.43	0.46	0.42	0.44
60	12.05	2.15	0.03	0.05	0.07	0.04	0.08
61	12.65	11.01	0.49	0.66	0.64	0.64	0.62
62	10.79	7.24	0.18	0.30	0.31	0.24	0.29
63	11.24	9.48	0.35	0.49	0.50	0.47	0.48
64	12.66	10.39	0.45	0.61	0.60	0.59	0.58
65	10.56	9.77	0.34	0.48	0.49	0.46	0.47
66	15.56	13.47	0.74	0.83	0.83	0.89	0.85
67	15.74	13.64	0.78	0.85	0.86	0.92	0.89
68	16.45	16.24	0.84	0.91	0.90	0.96	0.93
69	7.64	6.57	0.10	0.19	0.20	0.14	0.20
70	6.37	6.31	0.05	0.08	0.11	0.07	0.12
71	10.36	11.21	0.37	0.54	0.52	0.49	0.50
72	8.59	6.05	0.10	0.19	0.20	0.14	0.20
73	16.45	12.99	0.74	0.83	0.83	0.89	0.85
74	5.72	1.37	0.01	0.03	0.03	0.01	0.04
75	14.38	11.53	0.64	0.76	0.76	0.80	0.76
76	9.71	3.11	0.06	0.12	0.13	0.08	0.14
77	12.75	9.22	0.39	0.57	0.54	0.52	0.52
78	10.29	8.66	0.30	0.42	0.45	0.40	0.43
79	13.01	10.17	0.47	0.63	0.62	0.62	0.60
80	8.09	8.65	0.18	0.30	0.31	0.24	0.29
81	7.06	7.38	0.07	0.15	0.15	0.09	0.15
82	13.63	10.13	0.49	0.66	0.64	0.64	0.62

Table 4.8. (*cont.*)

No.	X	Y	V_i	Kn	Gumbel–Hougaard	Clayton	Frank
83	12.76	11.56	0.57	0.71	0.70	0.73	0.70
84	7.86	5.65	0.09	0.16	0.18	0.12	0.18
85	20.31	15.68	0.93	0.97	0.96	0.99	0.98
86	7.14	10.80	0.10	0.19	0.20	0.14	0.20
87	12.06	11.11	0.50	0.68	0.64	0.65	0.63
88	15.57	11.14	0.63	0.75	0.75	0.79	0.75
89	13.75	10.87	0.58	0.72	0.71	0.74	0.71
90	7.36	3.41	0.06	0.12	0.13	0.08	0.14
91	13.09	14.23	0.66	0.77	0.77	0.82	0.78
92	10.13	11.91	0.37	0.54	0.52	0.49	0.50
93	12.71	10.50	0.47	0.63	0.62	0.62	0.60
94	9.84	10.00	0.32	0.45	0.47	0.43	0.45
95	16.82	14.32	0.84	0.91	0.90	0.96	0.93
96	5.78	2.57	0.03	0.05	0.07	0.04	0.08
97	15.61	8.97	0.39	0.57	0.54	0.52	0.52
98	10.96	7.91	0.24	0.39	0.38	0.32	0.36
99	11.83	9.14	0.36	0.52	0.51	0.48	0.49
100	11.93	9.42	0.38	0.55	0.53	0.51	0.51

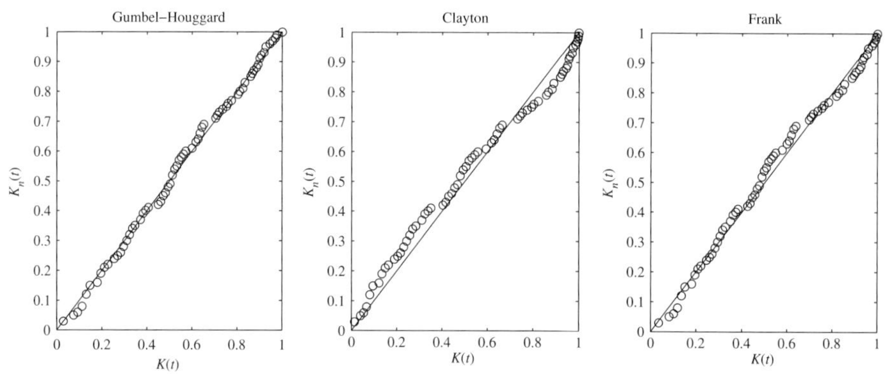

Figure 4.2 Nonparametric and parametric Kendall distribution plots for bivariate random variables X and Y.

To illustrate how to obtain the results listed in Table 4.8, we will use $\{(x_1, y_1) : (11.68, 7.67)\}$ as an example.

Compare $\{(x_1, y_1) : (11.68, 7.67)\}$ with all other bivariate pairs. We have $(x_3, y_3) < (x_1, y_1), (x_5, y_5) < (x_1, y_1), \ldots, (x_{96}, y_{96}) < (x_1, y_1)$with the total number of 23. Applying Equation (4.19) we have $z_1 = 23/(100 - 1) \approx 0.23$. Following the same procedure, we can compute $z_2, \ldots, z_{100}$. Applying Equation (4.19), $K_n(t = 0.23) = \dfrac{39}{100} = 0.39$.

Now applying the Kendall distribution equations just derived for the Gumbel–Hougaard, Clayton, and Frank copulas using $z_1 = 23/(100-1) \approx 0.23$, we have the following:

Gumbel–Houggard: $K_{GH}(t=0.23) = \dfrac{0.23(2.4029 - \ln(0.23))}{2.4029} \approx 0.37$

Clayton: $K_C(t = 0.23) = t - \dfrac{t^{\theta+1} - t}{\theta} = 0.23 - \dfrac{0.23^{2.8058+1} - 0.23}{2.8058} \approx 0.31$

Frank:

$$K_F(t = 0.23) = t + \frac{e^{\theta t} \ln\left(\dfrac{e^{-\theta t} - 1}{e^{-\theta} - 1}\right)\left(e^{\theta t} - 1\right)}{\theta} = 0.23$$

$$+ \frac{e^{7.5132(0.23)} \ln\left(\dfrac{e^{-7.5132(0.23)} - 1}{e^{-7.5132} - 1}\right)\left(e^{-7.5132(0.23)} - 1\right)}{7.5132} \approx 0.35$$

2. Construct the K-plot for bivariate sample data. Following Example 3.17 and using Equations (3.81) introduced in Section 3.4.4, the K-plot of the bivariate sample data is shown in Figure 4.3.

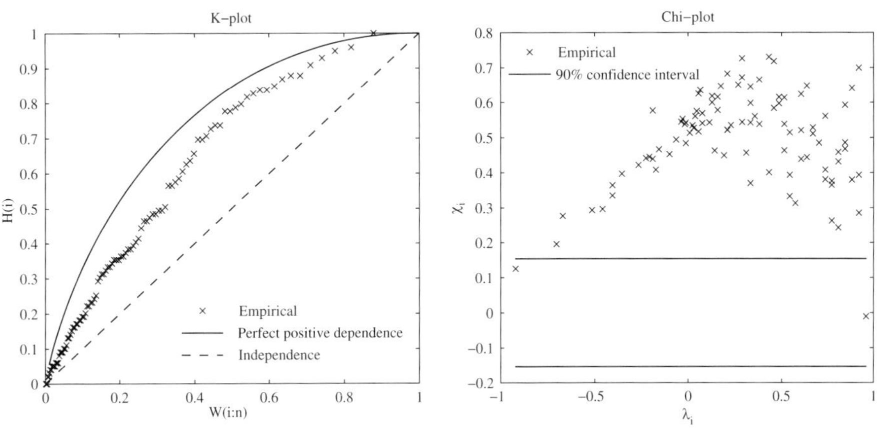

Figure 4.3 K-plot and chi-plot for the bivariate sample data.

3. Construct the chi-plot for the bivariate sample data. Following Example 3.16 and using Equations (3.77)–(3.80) introduced in Section 3.4.3, the chi-plot is shown in Figure 4.3.

   Now from this example, we can reach the following conclusions:

- The empirical Kendall correlation coefficient calculated, K-plot, and chi-plot in Figure 4.3 graphically indicate the positive dependence of the bivariate sample data.
- From the Q-Q plots (Figure 4.2), graphically the Gumbel–Hougaard and Frank copulas seem to have a better fit than does the Clayton copula in the case of modeling the bivariate sample data.

4.5.2 *MLE for Two- or d*-Dimensional Symmetric Archimedean Copulas

In Section 3.6, we introduced three procedures to estimate the copula parameter θ using maximum likelihood estimation (MLE): (i) find the exact MLE in which the parameters of marginal distribution and copula function are estimated simultaneously using MLE; (ii) estimate the parameters for marginal distributions first and then estimate the copula parameter using the fitted marginal distributions using MLE, i.e., two-stage ML; and (iii) estimate the copula parameter directly from empirical marginal distributions using MLE. For the first and second marginal-dependent procedures, the copula function is more likely to be misidentified if the marginal distributions are misidentified. For the third marginal free procedure, the copula function is less likely to be misidentified. Table 4.3 lists the copula density functions needed for the parameter estimation using the maximum likelihood method. We present the parameter estimation using MLE with two examples.

Example 4.13 **Using the same dataset as those in Example 4.12, estimate the copula parameters for the Gumbel–Hougaard, Clayton, Frank, and Joe copulas.**

Solution: We will use all three procedures to estimate the copula parameters with the detailed derivation given for the Gumbel–Hougaard copula as an example.

Exact ML: From Table 4.3, we have the copula density function of the Gumbel–Hougaard copula as follows:

$$c_{GH}(u_1, u_2) = \frac{(\ln u_1 \ln u_2)^{-1+\theta}\left(w^{\frac{2-2\theta}{\theta}} - (1-\theta)w^{\frac{2-2\theta}{\theta}}\right)}{u_1 u_2 \exp\left(w^{\frac{1}{\theta}}\right)}, \quad w = (-\ln u_1)^{\theta} + (-\ln u_2)^{\theta}; \theta \geq 1 \tag{4.24}$$

Its logarithm can be written as follows:

$$\ln(c_{GH}(u_1, u_2)) = (\theta - 1)\ln(\ln u_1 \ln u_2) + \ln\left(w^{\frac{2-2\theta}{\theta}} + (\theta-1)w^{\frac{1-2\theta}{\theta}}\right) - \ln(u_1 u_2) - w^{\frac{1}{\theta}} \tag{4.25}$$

As shown in Table 4.8, X and Y follow the gamma and normal distributions, respectively, as follows:

$$f_X(x) = \frac{x^{\alpha-1}\beta^{\alpha}}{\Gamma(\alpha)}\exp\left(-\frac{x}{\beta}\right); \; f_Y(y) = \frac{1}{\sigma\sqrt{2\pi}}\exp\left(-\frac{(y-\mu)^2}{2\sigma^2}\right)$$

Using Equations (3.97) and (3.98), we can rewrite the joint density function and its log-likelihood function as follows:

$$f\left(x, y; \alpha, \beta, \mu, \sigma^2, \theta\right) = c_{GH}\left(F_X(x; \alpha, \beta), F_Y\left(y; \mu, \sigma^2\right); \theta\right) f_X(x; \alpha, \beta) f_Y\left(y; \mu, \sigma^2\right)$$

Let $\Theta = [\alpha, \beta, \mu, \sigma^2, \theta]$.We have the following:

$$logL(\Theta) = \sum_{i=1}^{n} \ln\left(c_{GH}\left(\hat{F}_X(x;\alpha,\beta),\ \hat{F}_Y(y;\mu,\sigma^2);\theta\right)\right) + \sum_{i=1}^{n}\left(\ln\left(\hat{f}_X(x;\alpha,\beta)\right) + \ln\left(\hat{f}_Y(y;\mu,\sigma^2)\right)\right) \tag{4.26}$$

Taking the partial derivative of $logL(\Theta)$with respect to parameter $\Theta = [\alpha, \beta, \mu, \sigma^2, \theta]$and setting the derivative as zero, we can optimize the parameter as $\hat{\Theta} = [\hat{\alpha}, \hat{\beta}, \hat{\mu}, \hat{\sigma}^2, \hat{\theta}]$.

Two-stage ML: To apply this method, first we estimate the parameters of marginal distributions using MLE. Second, let $u_1 = \hat{F}_X\left(x\hat{\alpha}, \hat{\beta}\right), u_2 = \hat{F}_Y\left(y\hat{\mu}, \hat{\sigma}^2\right)$, and substitute u_1, u_2 into Equation (4.24). Third, optimize the log-likelihood function to estimate the copula parameter in which the log-likelihood function can be written as follows:

$$logL(\theta) = \sum_{i=1}^{n} \ln\left(c_{GH}\left(\hat{F}_X\left(x;\hat{\alpha},\hat{\beta}\right), \hat{F}_Y\left(y;\hat{\mu},\hat{\sigma}^2\right);\theta\right)\right) \tag{4.27}$$

Semiparametric ML: To apply the semiparametric ML method, first we need to calculate the empirical probability distribution. For example, the commonly applied Weibull plotting-position formula can be given as follows:

$$F_n(x_i) = \frac{1}{n+1}\sum_{j=1}^{n} 1\left(x_j \le x_i\right),\ j \neq i \tag{4.28}$$

Second, let $u_1 = F_n(x_1), u_2 = F_n(x_2)$and substitute u_1, u_2into Equation (4.24). Third, optimize the likelihood function as in the two-stage ML solution to estimate the copula parameter.

Table 4.9 lists the parameters estimated using all three procedures for the bivariate random variables.

Table 4.9. *Parameters estimated using MLE.*

Methods	Copulas	Marginal distributions		Copula parameter: θ	Log-likelihood
		$X : (\alpha, \beta)$	$Y : (\mu, \sigma^2)$		
Exact ML	Gumbel–Houggard	(8.408, 1.424)	$(9.939, 3.757^2)$	2.401	–501.85
Two-stage ML		(8.655, 1.379)	$(9.944, 3.853^2)$	2.037	52.837
Pseudo-ML		—	—	2.390	48.578
Exact ML	Clayton	(8.381, 1.437)	$(10.181, 4.114^2)$	1.773	–519.472
Two-stage ML		(8.655, 1.379)	$(9.944, 3.853^2)$	2.439	47.931
Pseudo-ML		—	—	1.712	33.844
Exact ML	Frank	(7.927, 1.534)	$(10.195, 3.83^2)$	7.569	–509.77
Two-stage ML		(8.655, 1.379)	$(9.944, 3.853^2)$	10.155	72.455
Pseudo-ML				7.474	43.775
Exact ML	Joe	(7.945, 1.5)	$(9.836, 3.855^3)$	3.077	–506.285
Two-stage ML		(8.655, 1.379)	$(9.944, 3.853^2)$	2.068	41.44
Pseudo-ML		—	—	2.952	43.776

Example 4.14 **Using the sample data given in Table 4.10: (1) estimate the trivariate copula parameters for the Clayton, Gumbel–Houggard, Frank, and Joe trivariate copula candidates using two-stage and semiparametric ML methods; (2) plot the empirical and parametric Kendall distributions.**

Table 4.10. X, Y, Z *sampled from gamma, exponential, and extreme value populations.*

No.	*X*	*Y*	*Z*	No.	*X*	*Y*	*Z*
1	10.32	1.88	18.84	26	8.26	0.49	18.80
2	17.61	2.16	19.49	27	16.08	3.03	20.15
3	16.03	2.03	19.89	28	11.21	0.80	19.91
4	13.49	1.32	18.55	29	15.15	1.93	18.68
5	19.23	3.13	19.45	30	20.54	3.38	19.17
6	16.79	2.95	18.78	31	10.82	0.87	19.01
7	17.06	2.04	20.11	32	16.84	1.29	18.12
8	12.31	2.99	19.27	33	15.54	2.14	19.36
9	39.47	20.13	21.17	34	34.68	14.16	20.93
10	8.25	0.08	18.03	35	17.40	2.86	20.03
11	10.06	0.23	18.24	36	28.96	10.19	20.60
12	16.91	2.96	19.71	37	16.37	2.37	19.27
13	35.41	15.32	20.99	38	12.60	0.40	18.81
14	21.27	2.79	19.91	39	17.26	2.61	19.81
15	17.30	1.94	19.40	40	6.34	0.29	17.74
16	19.04	1.48	20.45	41	32.84	13.70	20.78
17	14.18	1.04	18.36	42	29.37	11.89	20.39
18	32.88	14.65	20.58	43	19.45	4.40	19.93
19	6.68	0.18	16.10	44	18.03	1.75	18.94
20	20.22	6.23	19.15	45	14.01	2.29	17.83
21	17.28	1.59	19.12	46	9.07	0.11	18.48
22	12.29	1.79	18.88	47	15.66	2.89	19.01
23	14.38	1.98	19.00	48	20.64	8.27	19.99
24	17.56	3.08	19.91	49	10.79	0.33	18.72
25	10.95	0.61	18.79	50	28.63	10.59	20.52

Solution: As discussed earlier, the Clayton copula can be extended to multivariate dimensions when $\theta > 0$ with strict generating function. The Gumbel–Hougaard and Joe bivariate copulas can be fully extended to multivariate dimensions with strict generating function in full parameter range. Even though the Frank copula also has strict generating function in full parameter range, the condition is only satisfied if $\theta > 0$. These multivariate copula functions are listed in Table 4.6.

1. Estimate the copula parameters using the two-stage and semiparametric ML methods.

   The copula density function for each copula candidate can be written as follows:

- Trivariate Clayton copula:

$$c_\theta(u_1, u_2, u_3) = \frac{(2\theta + 1)(\theta + 1)}{(u_1 u_2 u_3)^{\theta+1}\left(u_1^{-\theta} + u_2^{-\theta} + u_3^{-\theta}\right)^{\frac{1}{\theta}+3}} \tag{4.29}$$

- Trivariate Gumbel–Houggard copula:

$$c_\theta(u_1, u_2, u_3) = \frac{we^{-w_1^{\frac{1}{\theta}}} w_1^{\frac{1}{\theta}}\left(3\theta w_1^{\frac{1}{\theta}} - 3\theta - 3w_1^{\frac{1}{\theta}} + w_1^{\frac{2}{\theta}} + 1\right)}{u_1 u_2 u_3 (\ln u_1)(\ln u_2)(\ln u_3) w_1^3} \tag{4.30}$$

where: $w = (-(\ln u_1)(\ln u_2)(\ln u_3))^\theta$; $w_1 = (-\ln u_1)^\theta + (-\ln u_2)^\theta + (-\ln u_3)^\theta$

- Trivariate Frank copula

$$c_\theta(u_1, u_2, u_3) = \frac{\theta^2 e^{-\theta(u_1+u_2+u_3)}}{(e^{-\theta} - 1)^2 w_1} - \frac{3\theta^2 we^{-\theta(u_1+u_2+u_3)}}{(e^{-\theta} - 1)^4 w_1^2} + \frac{2\theta^2 w^2 we^{-\theta(u_1+u_2+u_3)}}{(e^{-\theta} - 1)^6 w_1^3} \tag{4.31}$$

where: $w = \left(e^{-\theta u_1} - 1\right)\left(e^{-\theta u_2} - 1\right)\left(e^{-\theta u_3} - 1\right)$; $w_1 = \frac{w}{(e^{-\theta}-1)^2} + 1$

- Trivariate Joe copula:

$$\begin{aligned} c_\theta(u_1, u_2, u_3) = {} & \theta^2 w(w_1 + 1)^{\frac{1}{\theta}-1} + 3\theta^2\left(\theta^{-1} - 1\right) ww_1 (w_1 + 1)^{\frac{1}{\theta}-2} \\ & + \theta^2\left(\theta^{-1} - 1\right)\left(\theta^{-1} - 2\right) ww_1^2 (w_1 + 1)^{\frac{1}{\theta}-3} \end{aligned} \tag{4.32}$$

where: $w = (1 - u_1)^{\theta-1}(1 - u_2)^{\theta-1}(1 - u_3)^{\theta-1}$;

$$w_1 = \left((1 - u_1)^\theta - 1\right)\left((1 - u_2)^\theta - 1\right)\left((1 - u_3)^\theta - 1\right)$$

Now, to apply the two-stage ML method, the marginal distributions need to be estimated first. From Table 4.10, we know that random variables *X*, *Y*, and *Z* are sampled from the gamma, exponential, and extreme value populations. We have shown the gamma density function in Example 4.13. The exponential distribution is a special case of gamma distribution with parameter $\alpha = 1$. Thus, we only show the extreme value probability density function with location (μ) and scale (σ) parameters as follows:

$$f(x; \mu, \sigma) = \frac{1}{\sigma} \exp\left(\frac{x - \mu}{\sigma}\right) \exp\left(-\exp\left(\frac{x - \mu}{\sigma}\right)\right) \tag{4.33}$$

Applying the MLE for univariate probability distribution, the parameters are estimated and listed in Table 4.11.

Table 4.11. *Parameters estimated for random variables* X, Y, *and* Z.

X~Gamma (α, β)	Y~Exponential (β)	Z~Extreme value (μ, σ)
(5.9251, 2.9824)	3.9519	(19.8077, 0.8634)

Again, in the case of semiparametric ML method, the marginal probability is estimated nonparametrically using the Weibull plotting-position formula (i.e., Equation (4.28)). Table 4.12 lists the marginals computed parametrically and non-parametrically.

Finally, maximizing the log-likelihood function of copula density functions, we are able to estimate the parameters for each copula candidate given in Table 4.13.

(2) Graphical comparison of nonparametric and parametric Kendall distributions:

From Equation (4.11), the parametric Kendall distribution for trivariate Archimedean copula may be simplified as follows:

$$K_C(t) = P(C(U_1, U_2, U_3) \leq t) = t - \frac{\phi(t)}{\phi'(t)} - \frac{\phi^2(t)\phi''(t)}{2\left(\phi'(t)\right)^3} \tag{4.34}$$

Now, substituting the generating functions for Clayton, Gumbel–Houggard, Frank, and Joe copulas into Equation (4.34), we obtain the Kendall distribution function as follows:

- Trivariate Clayton copula:

$$K_C(t) = t + \frac{t\left(t^{2\theta} - 4t^{\theta} + 3\right)}{2\theta} + \frac{t\left(t^{\theta} - 1\right)^2}{2\theta^2} \tag{4.35}$$

- Trivariate Gumbel–Houggard copula:

$$K_C(t) = \frac{t\left(2\theta^2 - 3\theta \ln t + \ln^2 t + \ln t\right)}{2\theta^2} \tag{4.36}$$

- Trivariate Frank copula:

$$K_C(t) = t + \frac{1 - e^{\theta t}}{\theta} \ln\left(\frac{e^{-\theta t} - 1}{e^{-\theta} - 1}\right) - \frac{e^{\theta t} - e^{2\theta t}}{2\theta}\left(\ln\left(\frac{e^{-\theta t} - 1}{e^{-\theta} - 1}\right)\right)^2 \tag{4.37}$$

- Trivariate Joe copula:

$$K_C(t) = t + \frac{s_1 s_2 (1-t)^{1-\theta}}{\theta} + \frac{s_2^2 s_1^3 (1-t)^{3-3\theta}\left(\frac{\theta^2 (1-t)^{2\theta-2}}{s_1^2} - \frac{\theta(\theta-1)(1-t)^{\theta-2}}{s_1}\right)}{2\theta^3} \tag{4.38}$$

where: $s_1 = (1-t)^{\theta} - 1, s_2 = \ln\left(1 - (1-t)^{\theta}\right)$

Table 4.12. *Marginal distribution estimated parametrically and nonparametrically.*

No.	Random variables			Parametric			Nonparametric		
	X	Y	Z	$F(x)$	$F(y)$	$F(z)$	$F_n(x)$	$F_n(y)$	$F_n(z)$
1	10.32	1.88	18.84	0.15	0.38	0.28	0.14	0.37	0.31
2	17.61	2.16	19.49	0.55	0.42	0.50	0.67	0.51	0.59
3	16.03	2.03	19.89	0.46	0.40	0.67	0.43	0.45	0.65
4	13.49	1.32	18.55	0.31	0.28	0.21	0.29	0.27	0.18
5	19.23	3.13	19.45	0.63	0.55	0.48	0.73	0.75	0.57
6	16.79	2.95	18.78	0.51	0.53	0.26	0.49	0.65	0.24
7	17.06	2.04	20.11	0.52	0.40	0.76	0.55	0.47	0.78
8	12.31	2.99	19.27	0.25	0.53	0.41	0.25	0.69	0.49
9	39.47	20.13	21.17	0.99	0.99	0.99	0.98	0.98	0.98
10	8.25	0.08	18.03	0.07	0.02	0.12	0.06	0.02	0.08
11	10.06	0.23	18.24	0.13	0.06	0.15	0.12	0.08	0.12
12	16.91	2.96	19.71	0.51	0.53	0.59	0.53	0.67	0.61
13	35.41	15.32	20.99	0.98	0.98	0.98	0.96	0.96	0.96
14	21.27	2.79	19.91	0.73	0.51	0.68	0.82	0.59	0.71
15	17.30	1.94	19.40	0.53	0.39	0.46	0.61	0.41	0.55
16	19.04	1.48	20.45	0.63	0.31	0.88	0.71	0.29	0.84
17	14.18	1.04	18.36	0.35	0.23	0.17	0.33	0.24	0.14
18	32.88	14.65	20.58	0.97	0.98	0.91	0.92	0.94	0.88
19	6.68	0.18	16.10	0.03	0.04	0.01	0.04	0.06	0.02
20	20.22	6.23	19.15	0.68	0.79	0.37	0.76	0.80	0.45
21	17.28	1.59	19.12	0.53	0.33	0.36	0.59	0.31	0.43
22	12.29	1.79	18.88	0.24	0.36	0.29	0.24	0.35	0.33
23	14.38	1.98	19.00	0.37	0.39	0.32	0.35	0.43	0.37
24	17.56	3.08	19.91	0.55	0.54	0.67	0.65	0.73	0.69
25	10.95	0.61	18.79	0.17	0.14	0.26	0.20	0.18	0.25
26	8.26	0.49	18.80	0.07	0.12	0.27	0.08	0.16	0.27
27	16.08	3.03	20.15	0.47	0.54	0.77	0.45	0.71	0.80
28	11.21	0.80	19.91	0.19	0.18	0.67	0.22	0.20	0.67
29	15.15	1.93	18.68	0.41	0.39	0.24	0.37	0.39	0.20
30	20.54	3.38	19.17	0.70	0.57	0.38	0.78	0.76	0.47
31	10.82	0.87	19.01	0.17	0.20	0.33	0.18	0.22	0.39
32	16.84	1.29	18.12	0.51	0.28	0.13	0.51	0.25	0.10
33	15.54	2.14	19.36	0.43	0.42	0.45	0.39	0.49	0.53
34	34.68	14.16	20.93	0.98	0.97	0.97	0.94	0.92	0.94
35	17.40	2.86	20.03	0.54	0.52	0.73	0.63	0.61	0.76
36	28.96	10.19	20.60	0.93	0.92	0.92	0.86	0.84	0.90
37	16.37	2.37	19.27	0.48	0.45	0.41	0.47	0.55	0.51
38	12.60	0.40	18.81	0.26	0.10	0.27	0.27	0.14	0.29
39	17.26	2.61	19.81	0.53	0.48	0.63	0.57	0.57	0.63
40	6.34	0.29	17.74	0.02	0.07	0.09	0.02	0.10	0.04

Table 4.12. (*cont.*)

No.	Random variables			Parametric			Nonparametric		
	X	Y	Z	$F(x)$	$F(y)$	$F(z)$	$F_n(x)$	$F_n(y)$	$F_n(z)$
41	32.84	13.70	20.78	0.97	0.97	0.95	0.90	0.90	0.92
42	29.37	11.89	20.39	0.93	0.95	0.86	0.88	0.88	0.82
43	19.45	4.40	19.93	0.65	0.67	0.68	0.75	0.78	0.73
44	18.03	1.75	18.94	0.57	0.36	0.31	0.69	0.33	0.35
45	14.01	2.29	17.83	0.34	0.44	0.10	0.31	0.53	0.06
46	9.07	0.11	18.48	0.09	0.03	0.19	0.10	0.04	0.16
47	15.66	2.89	19.01	0.44	0.52	0.33	0.41	0.63	0.41
48	20.64	8.27	19.99	0.70	0.88	0.71	0.80	0.82	0.75
49	10.79	0.33	18.72	0.17	0.08	0.25	0.16	0.12	0.22
50	28.63	10.59	20.52	0.92	0.93	0.90	0.84	0.86	0.86

Table 4.13. *Copula parameters estimated for trivariate analysis.*

	Clayton θ, log$L$	Gumbel–Houggard θ, log$L$	Frank θ, log$L$	Joe θ, log$L$
Two-stage	2.112, 52.034	3.132, 86.533	9.796, 71.995	4.252, 80.813
Semiparametric	2.042, 49.606	3.034, 76.136	8.673, 60.927	4.213, 73.519

According to Equation (4.19) in Section 4.5.1, the nonparametric estimation of Kendall distribution can be given as follows:

i. Obtain $z_i = \dfrac{\sum_{j=1}^{n} 1\left(x_{1j} \leq x_{1i}, x_{2j} \leq x_{2i}, x_{3j} \leq x_{3i}\right)}{n-1}; i = 1, \ldots, n, \;\; j \neq i$

ii. Construct $K_n(z) = \dfrac{\sum_{i=1}^{n} 1(z_i \leq z)}{n}$

Using the fitted copula parameter given in Table 4.13, Figures 4.4 and 4.5 plot the nonparametric and parametric Kendall distributions using the parameters estimated with two-stage and pseudo-MLE, respectively. From Figures 4.4 and 4.5, we see that nonparametric and parametric Kendall distributions have the best match for the Gumbel–Hougaard copula.

4.6 Simulation of Symmetric Archimedean Copulas

In Section 3.7, we discussed the general procedure to simulate the random variables from any given copula function. For the symmetric Archimedean copulas, the simulation procedure can be revised based on the general simulation technique as follows.

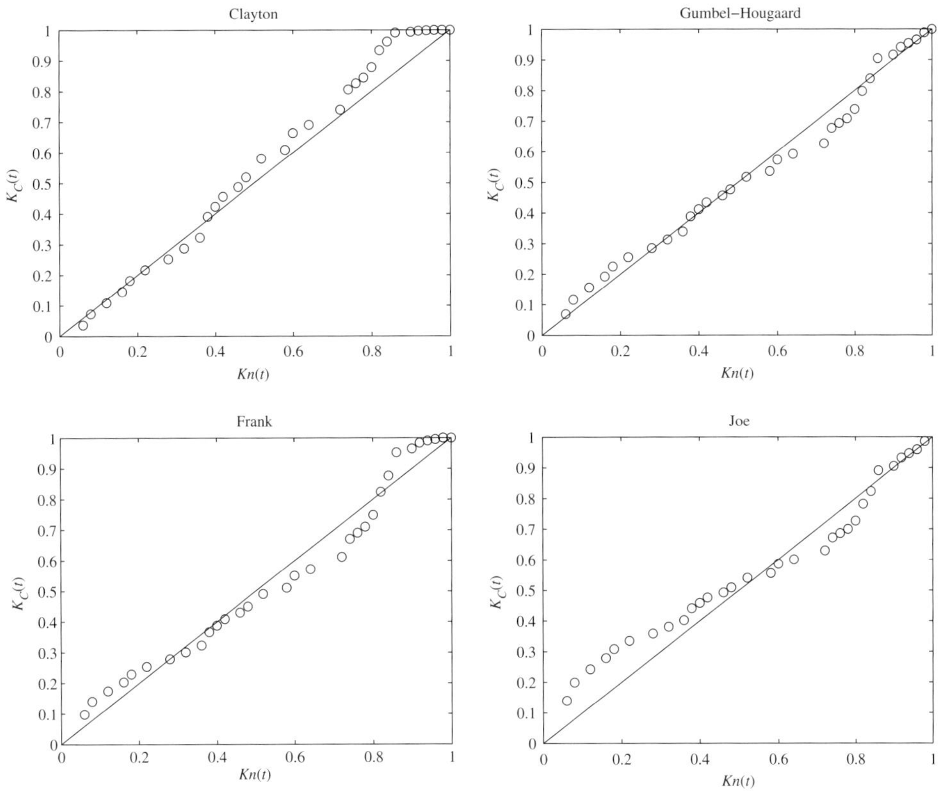

Figure 4.4 Comparison of nonparametric and parametric Kendall distributions with parameters estimated using two-stage MLE.

Let the joint distribution multivariate random variables $(\mathbf{x}_1, \mathbf{x}_2, \ldots, \mathbf{x}_d)$ be modeled by a symmetric Archimedean copula with generating function ϕ. Then we have the following:

$$F(x_1, \ldots, x_d) = C_\theta(F_{X_1}(x_1), \ldots, F_{X_d}(x_d)) \tag{4.39}$$

Let $u_1 = F_{X_1}(x_1), \ldots, u_d = F_{X_d}(x_d)$, and the copula function can be written using the generating function ϕ as follows:

$$C_\theta(u_1, \ldots, u_d) = \phi^{-1}(\phi(u_1) + \cdots + \phi(u_d))$$

From the definition of the copula discussed in Section 3.1, we also have the following:

$$\begin{aligned} C_1(u_1, 1, \ldots, 1) = u_1; C_i(u_1, \ldots, u_i) = C_i(u_1, \ldots, u_i, 1, \ldots, 1); \ldots; C_d(u_1, \ldots, u_d) \\ = C(u_1, , , , .u_d) \end{aligned} \tag{4.40}$$

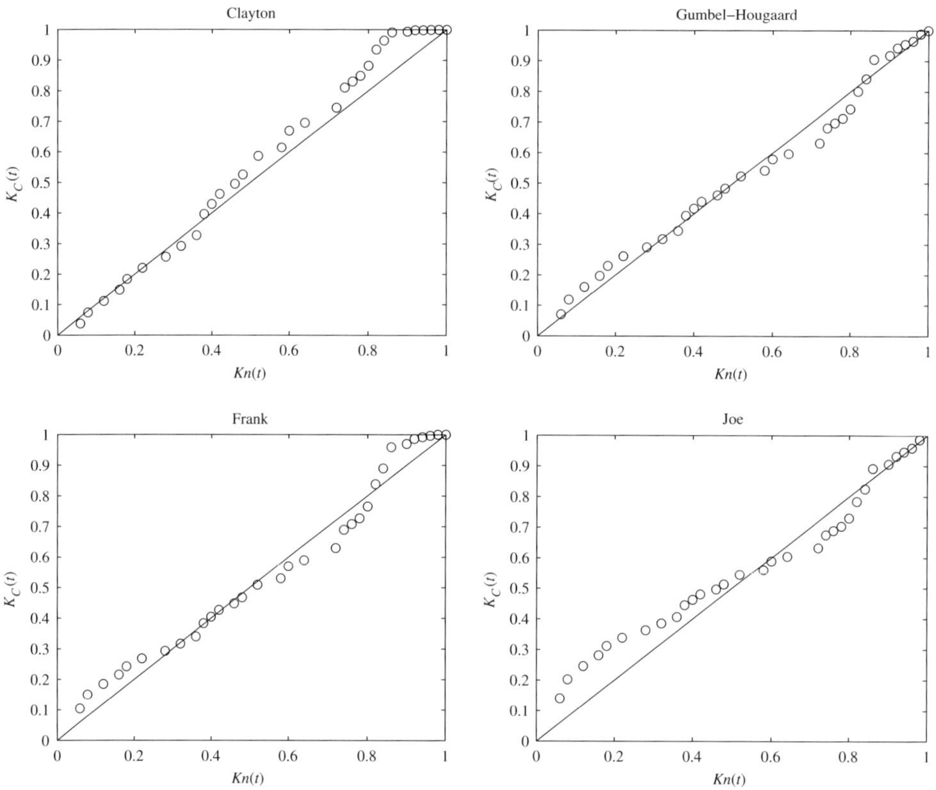

Figure 4.5 Comparison of nonparametric and parametric Kendall distributions with parameters estimated using pseudo-MLE.

Let the conditional distribution of U_i, given the values of $U_1, \ldots, U_{i-1}$, be

$$C_i(u_i|U_1 = u_1, \ldots, U_{i-1} = u_{i-1}) = \frac{\dfrac{\partial C^{i-1} C_i(u_1, \ldots, u_i)}{\partial u_1 \ldots \partial u_{i-1}}}{\dfrac{\partial^{i-1} C_{i-1}(u_1, \ldots, u_{i-1})}{\partial u_1 \ldots \partial u_{i-1}}}; \quad i = 2, 3, \ldots, d \tag{4.41}$$

Substituting Equation (4.40) into Equation (4.41) and applying the associative property of the symmetric Archimedean copulas, we have the following:

$$C_i(u_i|U_1 = u_1, \ldots, U_{i-1} = u_{i-1}) = \frac{\phi^{-1(i-1)}(\phi(u_1) + \cdots + \phi(u_i))}{\phi^{-1(i-1)}(\phi(u_1) + \cdots + \phi(u_{i-1}))} = \frac{\phi^{-1(i-1)}(t_i)}{\phi^{-1(i-1)}(t_{i-1})} \tag{4.42}$$

where $t_i = \phi(u_1) + \phi(u_2) + \cdots + \phi(u_i) = \sum_{k=1}^{i} \phi(u_k); \phi^{-1(i-1)}(t_i) = \dfrac{\partial^{i-1}\phi^{-1}(t_i)}{\partial t_i^{i-1}}, i = 2,$ $\ldots, d$ Obviously, in Equations (4.41) and (4.42), the (partial) derivative exists for both the numerator and the denominator. More specifically, the (partial) derivative of the denominator is not zero.

Following the preceding derivation, the general simulation algorithm can be written as follows:

1. Simulate a d-independent random variable $(v_1, v_2, \ldots, v_d)$from the uniform distribution $U(0, 1)$.
2. Set $u_1 = v_1$.
3. Set $v_2 = C_2(u_2|U_1 = u_1) = \frac{\phi^{-1(1)}(t_2)}{\phi^{-1(1)}(t_1)}$; $t_1 = \phi(u_1), t_2 = \phi(u_1) + \phi(u_2)$. Solve for u_2 using the equation $v_2 = \frac{\phi^{-1(1)}(t_2)}{\phi^{-1(1)}(t_1)}$.
4. Set $v_3 = C_3(u_3|U_1 = u_1, U_2 = u_2) = \frac{\phi^{-1(2)}(t_3)}{\phi^{-1(2)}(t_2)}$; $t_3 = \phi(u_1) + \phi(u_2) + \phi(u_3)$ and $t_2 = \phi(u_1) + \phi(u_2)$. Solve for $u_3$using the equation $v_3 = \frac{\phi^{-1(2)}(t_3)}{\phi^{-1(2)}(t_2)}$.

   ……
5. Set $v_d = C_d(u_d|U_1 = u_1, \ldots, U_d = u_d) = \frac{\phi^{-1(d-1)}(t_d)}{\phi^{-1(d-1)}(t_{d-1})}$; $t_{d-1} = \phi(u_1) + \phi(u_2) + \ldots + \phi(u_{d-1}), t_d = \phi(u_1) + \phi(u_2) + \ldots + \phi(u_d)$. Solve for u_d using $v_d = \frac{\phi^{-1(d-1)}(t_d)}{\phi^{-1(d-1)}(t_{d-1})}$.

Here we summarize $\phi^{-1(2)}(t)$ of the Gumbel–Hougaard, Frank, Clayton, and Ali–Mikail–Haq copulas:

Gumbel-Hougaard copula:
The generating function of the Gumbel–Houggard copula is given by $\phi(t) = (-\ln(t))^{\theta}$. Hence,

$$\phi^{-1}(t) = e^{-t^{\frac{1}{\theta}}} \tag{4.43a}$$

$$\phi^{-1(1)} = \frac{-t^{\frac{1}{\theta}-1}e^{-t^{\frac{1}{\theta}}}}{\theta} \tag{4.43b}$$

$$\phi^{-1(2)} = \frac{t^{\frac{2}{\theta}-2}e^{-t^{\frac{1}{\theta}}} - (1-\theta)t^{\frac{1}{\theta}-2}e^{-t^{\frac{1}{\theta}}}}{\theta^2} \tag{4.43c}$$

Frank copula
The generating function of the Frank copula is given by $\phi(t) = -\ln\left(\frac{e^{-\theta u} - 1}{e^{-\theta} - 1}\right)$. Hence,

$$\phi^{-1}(t) = -\frac{1}{\theta}\ln\left(1 + e^{-t}\left(e^{-\theta} - 1\right)\right) \tag{4.44a}$$

$$\phi^{-1(1)}(t) = \frac{e^{-t}\left(e^{-\theta} - 1\right)}{\theta(e^{-t}(e^{-\theta} - 1) + 1)} \tag{4.44b}$$

$$\phi^{-1(2)}(t) = \frac{e^{-2t}\left(e^{-\theta}-1\right)^2}{\theta(e^{-t}(e^{-\theta}-1)+1)^2} - \frac{e^{-t}\left(e^{-\theta}-1\right)}{\theta(e^{-t}(e^{-\theta}-1)+1)} \tag{4.44c}$$

Clayton copula

The generating function of the Clayton copula is given by $\phi(t) = \frac{1}{\theta}\left(t^{-\theta}-1\right)$. Hence,

$$\phi^{-1}(t) = (\theta t+1)^{-\frac{1}{\theta}} \tag{4.45a}$$

$$\phi^{-1(1)}(t) = -(\theta t+1)^{-\frac{1}{\theta}-1} \tag{4.45b}$$

$$\phi^{-1(2)}(t) = (\theta+1)(\theta t+1)^{\left(-\frac{1}{\theta}-2\right)} \tag{4.45c}$$

Ali–Mikhail–Haq copula

The generating function of the Ali–Mikail–Haq copula is given by $\phi(t) = \ln\left(\frac{1-\theta(1-t)}{t}\right)$. Hence, we have the following:

$$\phi^{-1}(t) = \frac{e^t(\theta-1)}{(\theta-e^t)^2} \tag{4.46a}$$

$$\phi^{-1(1)}(t) = \frac{e^t(\theta-1)}{(\theta-e^t)^2} \tag{4.46b}$$

$$\phi^{-1(2)}(t) = \frac{e^t(\theta-1)(\theta+e^t)}{(\theta-e^t)^3} \tag{4.46c}$$

Example 4.15 **Show how to generate the random variable for the bivariate (trivariate) Joe copula using the simulation procedure discussed previously.**

Solution: The generating function of Joe copula is written as follows: $\phi(t) = -\ln\left(1-(1-t)^\theta\right)$. Hence, the inverse of ϕ can be written as follows: $\phi^{-1}(t) = 1-(1-\exp(t))^{\frac{1}{\theta}}$.

Bivariate case:

1. Generate two independent random variables $[v_1, v_2]$ from $U(0,1)$.
2. Set $u_1 = v_1$.
3. Set $v_2 = \frac{\phi^{-1(1)}(t_2)}{\phi^{-1(1)}(t_1)}$, and we have the following:

$$\phi^{-1(1)}(t) = \frac{\partial^{-1}\phi(t)}{\partial t} = \frac{\exp(-t)}{\theta}(1-\exp(-t))^{\frac{1}{\theta}-1} \tag{4.47}$$

4. Let $t_1 = -\ln\left(1-(1-u_1)^\theta\right)$, $t_2 = -\ln\left(1-(1-u_1)^\theta\right) - \ln\left(1-(1-u_2)^\theta\right)$. Then we have the following:

$$\phi^{-1(1)}(t_1) = \frac{\left((1-u_1)^\theta - 1\right)\left((1-u_1)^\theta\right)^{\frac{1}{\theta}-1}}{\theta} \tag{4.48a}$$

$$\phi^{-1(1)}(t_2) = \frac{\left((1-u_1)^\theta - 1\right)\left((1-u_2)^\theta - 1\right)\left((1-u_1)^\theta(1-u_1)^\theta(1-u_2)^\theta + (1-u_2)^\theta\right)^{\frac{1}{\theta}}}{\theta\left((1-u_1)^\theta - (1-u_1)^\theta(1-u_2)^\theta + (1-u_2)^\theta\right)} \tag{4.48b}$$

$$v_2 = \frac{\left((1-u_1)^\theta\right)^{1-\frac{1}{\theta}}\left((1-u_2)^\theta - 1\right)\left((1-u_1)^\theta - (1-u_1)^\theta(1-u_2)^\theta + (1-u_2)^\theta\right)^{\frac{1}{\theta}}}{(1-u_1)^\theta - (1-u_1)^\theta(1-u_2)^\theta + (1-u_2)^\theta} \tag{4.48c}$$

Now u_2 can be calculated numerically.

Trivariate case:

1. Generate independent random variables $[v_1, v_2, v_3]$ from $U(0,1)$.
2. Use Equation (4.48c) to numerically calculate u_2.
3. For the trivariate case, we need to determine $\phi^{-1(2)}(t)$, which is given as follows:

$$\phi^{-1(2)}(t) = \frac{\partial^2 \phi^{-1}(t)}{\partial t^2} = \frac{(1-\exp(-t))^{\frac{1}{\theta}}(\theta\exp(t) - 1)}{\theta^2(\exp(t)-1)^2} \tag{4.49}$$

4. Let $t_3 = -\ln\left(1-(1-u_1)^\theta\right) - \ln\left(1-(1-u_2)^\theta\right) - \ln\left(1-(1-u_3)^\theta\right)$, and we have the following:

$$\phi^{-1(2)}(t_2) = \frac{\left(1-\bar{u}_1^\theta\right)\left(1-\bar{u}_2^\theta\right)\left(\theta - \left(1-\bar{u}_1^\theta\right)\left(1-\bar{u}_2^\theta\right)\left(\bar{u}_1^\theta - \bar{u}_1^\theta\bar{u}_2^\theta + \bar{u}_2^\theta\right)^{(1/\theta)}\right.}{\theta^2\left(\bar{u}_1^\theta - \bar{u}_1^\theta\bar{u}_2^\theta + \bar{u}_2^\theta\right)^2} \tag{4.49a}$$

$$\phi^{-1(2)}(t_3) = \frac{\left(1-\bar{u}_1^\theta\right)\left(1-\bar{u}_2^\theta\right)\left(1-\bar{u}_3^\theta\right)\left(\theta - \left(1-\bar{u}_1^\theta\right)\left(1-\bar{u}_2^\theta\right)\left(1-\bar{u}_3^\theta\right)\right)\left(1-\left(1-\bar{u}_1^\theta\right)\left(1-\bar{u}_2^\theta\right)\left(1-\bar{u}_3^\theta\right)\right)^{\frac{1}{\theta}}}{\theta^2\left(1-\left(1-\bar{u}_1^\theta\right)\left(1-\bar{u}_2^\theta\right)\left(1-\bar{u}_3^\theta\right)\right)^2} \tag{4.49b}$$

$$v_3 = \frac{\left(1-\bar{u}_3^\theta\right)\left(\theta - \left(1-\bar{u}_1^\theta\right)\left(1-\bar{u}_2^\theta\right)\left(1-\bar{u}_3^\theta\right)\right)}{\theta\left(1-\bar{u}_1^\theta\right)\left(1-\bar{u}_2^\theta\right)} \times \left(\frac{1-\left(1-\bar{u}_1^\theta\right)\left(1-\bar{u}_2^\theta\right)\left(1-\bar{u}_3^\theta\right)}{\bar{u}_1^\theta - \bar{u}_1^\theta\bar{u}_2^\theta + \bar{u}_2^\theta}\right)^{\frac{1}{\theta}-2} \tag{4.49c}$$

where, in Equations (4.49a)–(4.49c): $\bar{u}_1 = 1-u_1, \bar{u}_2 = 1-u_2, \bar{u}_3 = 1-u_3$.

Now u_3 can be calculated numerically.

Example 4.16 Simulate bivariate random variables (sample size of 200) with the parameters estimated in Example 4.13 based on the semiparametric ML method for the Gumbel–Hougaard and Frank copulas, and compare the simulated random variables with the empirical marginal variables.

Solution: According to the previous discussion of the Gumbel–Hougaard (Equations (4.43a) and (4.43b)) and Frank (Equations (4.44a) and (4.44b)) copulas, we can generate the bivariate random variables with the fitted parameter using the simulation procedure for symmetric Archimedean copulas. Here we will illustrate the simulation procedure using the Gumbel–Hougaard copula as an example:

1. Generate two independent, uniformly distributed variables

   One can generate the independent, uniformly distributed random variables using the rand function in MATLAB:

   $[v_1, v_2] = \text{rand}(2, 1)$, and we have: $[v_1, v_2] = [0.1270, 0.9134]$.

   Notice that the random variables generated are subjected to change for each generation.
2. Set $u_1 = v_1 = 0.1270, C(u_2|u_1) = v_2 = 0.9134$.

   Substituting Equations (4.43a)–(4.43b) into Equation (4.42), we have the following:

$$C(u_2|u_1) = \frac{\phi^{-1(1)}(t_2)}{\phi^{-1(1)}(t_1)} = \left(t_1^{1-\frac{1}{\theta}} e^{t_1^{(1/\theta)}}\right) t_2^{\frac{1}{\theta}-1} e^{-t_2^{\frac{1}{\theta}}}$$

Applying $\theta = 2.39$, we have the following:

$$t_1 = \phi(u_1) = [-\ln(0.1270)]^{2.39} = 5.6486; t_1^{1-\frac{1}{\theta}} e^{t_1^{\frac{1}{\theta}}} = 21.5534; \text{ and}$$

$$C(u_2|u_1) = 21.5534 t_2^{\frac{1}{2.39}-1} e^{-\left(t_2^{\frac{1}{2.39}}\right)} = 0.9134$$

Now we need to solve for t_2. It is seen that the preceding equation does not have a closed-form inverse, and we will need to solve the equation numerically. In MATLAB, we can use the fsolve function to solve the general function of $f(x) = 0$. Thus, here we are solving $f(t_2) = C(u_2|u_1) - 0.9134 = 0$ as follows:

t_2 = fsolve(@(t2)21.5534*t2^(1/2.39–1).*exp(-t2^(1/2.39))-0.9134,10), where @ is the function handle and 10 is the initial value. We obtain $t_2 = 6.0111$.

Applying $t_2 = \phi(u_1) + \phi(u_2)$we have the following:

$\phi(u_2) = t_2 - \phi(u_1) = 6.0111 - 5.6486 = 0.3625$

Finally, we have $u_2 = e^{-0.3625^{\frac{1}{2.39}}} = 0.5199$.

With the same procedure, we will be able to simulate the rest of the bivariate random variables.

Figure 4.6 compares simulated copula random variables with their corresponding empirical distributions. Figure 4.7 compares simulated X and Y from the fitted gamma and normal distributions (Table 4.9) with the sample random variables.

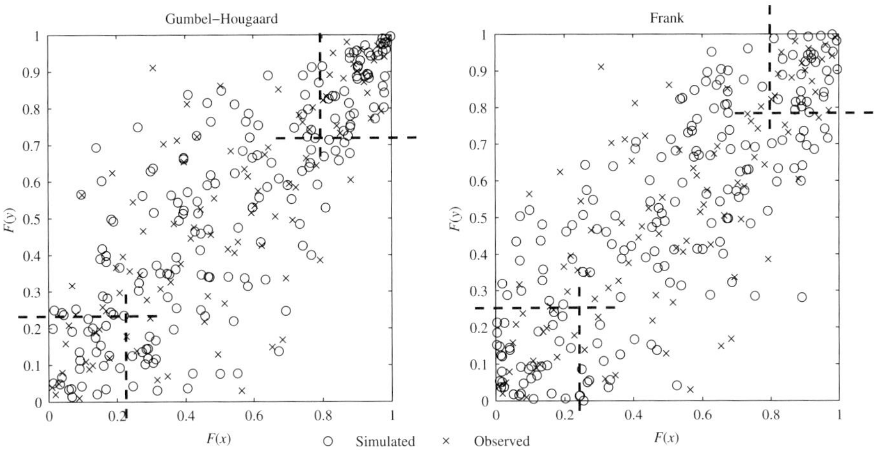

Figure 4.6 Comparison of simulated random variables with empirical marginal variables.

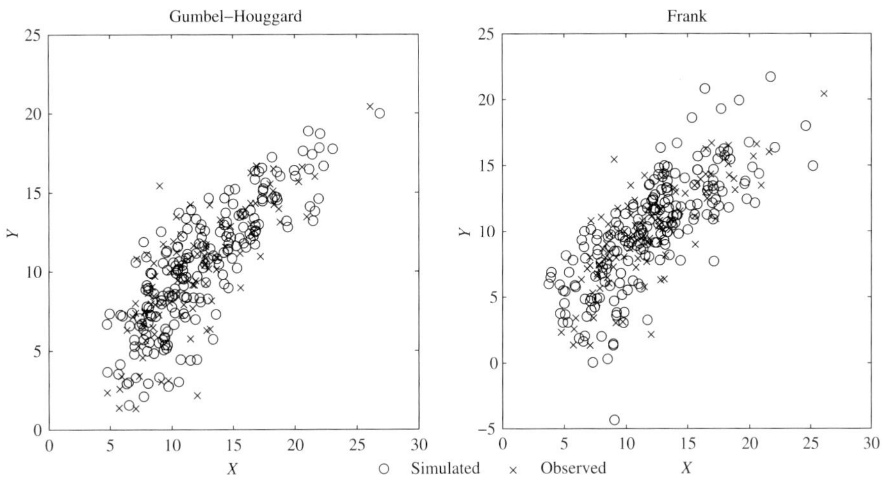

Figure 4.7 Comparison of simulated peak discharge and flood volume with observations.

From the simulation with Gumbel–Hougaard copula shown in Figures 4.6 and 4.7, we see that there exists an upper-tail dependence for the Gumbel–Hougaard copula and no visual effects of lower-tail dependence. From the simulation with Frank copula, we see that there does not exist significant dependence for either an upper- (upper-right corner) or a lower- (lower-left corner) tail dependence for the Frank copula.

Example 4.17 **Simulate trivariate random variables (sample size of 200) with the parameters estimated in Example 4.14 based on the semiparametric ML for the Gumbel–Houggard and Clayton copulas, and compare the simulated random variables with the empirical marginal variables.**

Solution: According to the general copula simulation procedure discussed in Equations (4.39)–(4.42), (4.43a)–(4.43c), (4.45a)–(4.45c) are derived from and can be applied to simulate the random variables from Gumbel–Houggard and Clayton copulas as follows: (1) generate independent trivariate random variables $[v_1, v_2, v_3]$; (2) set $u_1 = v_1$; (3) solve for $u_2$using u_1, v_2, and Equation (4.43b) or Equation (4.45b); and (4) solve for $u_3$using u_1, u_2, v_3, and Equation (4.43c) or Equation (4.45c). Here we will illustrate how to simulate the random variables with an example using the Clayton copula:

1. Generate three independent, uniformly distributed random variables:
   Using the rand function to generate uniformly distributed random variables as follows:
   V=rand(3,1)
   Then we have the following:
   V=rand(3,1)
   $[v_1, v_2, v_3] = [0.8147, 0.9058, 0.1270]$.
2. Set $u_1 = v_1 = 0.8147$and solve $u_2$from $v_2 = C(u_2|u_1) = 0.9508$. From the procedure described for the simulation for the Archimedean copulas, applying Equation (4.45a) with $\theta = 0.532$, estimated using semiparametric MLE (or pseudo-MLE), we have the following:

$$t_1 = \phi(u_1; 0.532) = \frac{1}{0.532}\left(0.8147^{-0.532} - 1\right) = 0.2165$$

$$t_2 = \phi(u_1) + \phi(u_2) = 0.2165 + \phi(u_2)$$

$$v_2 = \frac{\phi^{-1(1)}(t_2)}{\phi^{-1(1)}(t_1)} = \frac{(0.532t_2 + 1)^{-2.8797}}{(0.532(0.2165) + 1)^{-2.8797}} = 0.9058$$

$$\Rightarrow t_2 = \phi(u_2) + 0.2165 = 0.2898 \Rightarrow \phi(u_2) = 0.2898 - 0.2165 = 0.0733$$

   Finally, we can solve for $u_2$as follows: $u_2 = (0.532(0.0733) + 1)^{-\left(\frac{1}{0.532}\right)} = 0.9306$
3. Now set $u_1 = 0.8147, u_2 = 0.9306$to solve for u_3 from
   $v_3 = C(u_3|U_1 = u_1, U_2 = u_2) = 0.1270$. Applying Equation (4.45b), we have the following:
   $t_2 = \phi(u_1) + \phi(u_2) = 0.2898$, which is computed from the previous step, and

$$t_3 = \phi(u_3) + 0.2898$$

$$v_3 = \frac{\phi^{-1(2)}(t_3)}{\phi^{-1(2)}(t_2)} = \frac{(0.532t_3 + 1)^{-\frac{1}{0.532}-2}}{(0.532(0.298) + 1)^{-\frac{1}{0.532}-2}} = 0.1270$$

$$\Rightarrow t_3 = 1.8131, u_3 = 0.2810.$$

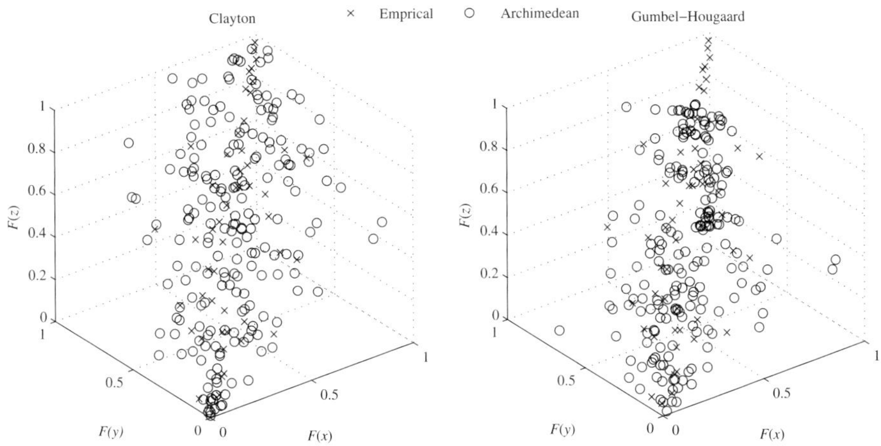

Figure 4.8 Comparison of simulated trivariate random variables and empirical maginals.

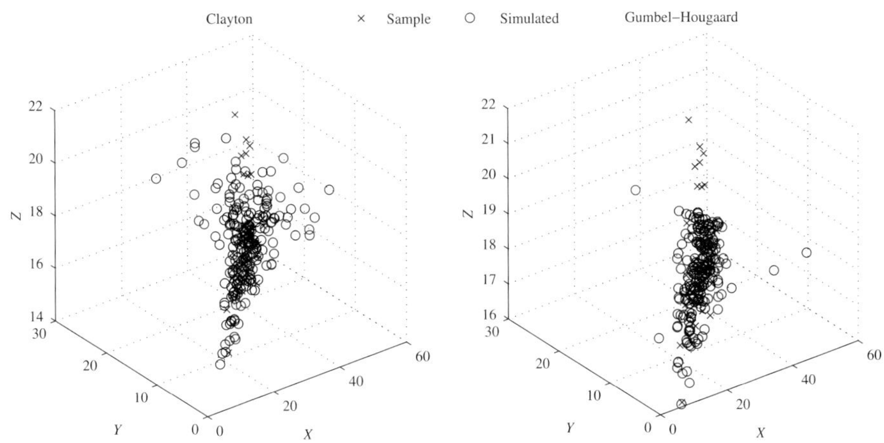

Figure 4.9 Comparison of simulated trivariate random variables and samples.

Similarly, one can perform the simulation using the Gumbel–Hougaard copula. Figure 4.8 plots the marginal random variables simulated from the copula function and empirical marginal random variables. Figure 4.9 plots the simulated random variables from the fitted marginal distributions and the samples given in Table 4.10.

We can see from Figure 4.8 that visually, the Clayton and Gumbel–Hougaard copulas have similar performance.

4.7 Goodness-of-Fit Statistics Test for Archimedean Copulas

Usually, the best-fitted copula is considered as the copula function with the largest log-likelihood. However, it is needed to further ensure the appropriateness of the chosen copula function with the use of the formal goodness-of-fit (GoF) test statistics besides the visual comparison. In Section 3.8, we introduced two of the most powerful GoF test statistics: $S_n^{(B)}$ based on Rosenblatt's transform and S_n based on the empirical copula for bivariate random variables. Here, we will discuss the procedure to construct the goodness-of-fit test statistics $S_n^{(B)}$ and S_n for multivariate symmetric Archimedean copulas (i.e., $d \geq 3$).

4.7.1 Goodness-of-Fit Statistics $S_n^{(B)}$ for Multivariate Symmetric Archimedean Copulas

Let multivariate random variable $X_1, X_2, \ldots, X_d$ be modeled by the symmetric Archimedean copulas function $C_\theta(u_1, \ldots, u_d); u_1 = F_1(x_1), \ldots, u_d = F_d(x_d)$; then, based on Rosenblatt's transform, i.e., Equations (4.41) and (4.42), we have the following:

$$Z_1 = u_1$$
$$Z_i = C_\theta(u_i | U_1 = u_1, \ldots, U_{i-1} = u_{i-1}) = \frac{\phi^{-1(i-1)}\left(\sum_{j=1}^{i} \phi(u_j)\right)}{\phi^{-1(i-1)}\left(\sum_{j=1}^{i-1} \phi(u_j)\right)}; i = 2, \ldots, d \tag{4.50}$$

Then Equation (3.122) is rewritten as follows:

$$D_n(\mathbf{u}) = \frac{1}{n} \sum_{i=1}^{n} 1(\mathbf{Z}_i < \mathbf{u}), \mathbf{u} \in [0, 1]^d \tag{4.51}$$

In the same way as the goodness-of-fit statistics test for bivariate case, $\mathbf{Z}_1, \ldots, \mathbf{Z}_d$ should be "close" to independently uniformly distributed as $C_\perp$. Then, according to Genest et al. (2007), Equation (3.123) for the construction of goodness-of-fit statistics can be rewritten as follows:

$$\begin{aligned} S_n^{(B)} &= n \int_{[0,1]^d} \{D_n(\mathbf{u}) - C_\perp(\mathbf{u})\}^d d\mathbf{u} \\ &= \frac{1}{3^d} - \frac{1}{2^{d-1}} \sum_{i=1}^{n} \prod_{k=1}^{d} (1 - Z_{ik}^2) + \frac{1}{n} \sum_{i=1}^{n} \sum_{j=1}^{n} \prod_{k=1}^{d} (1 - Z_{ik} \vee Z_{jk}) \end{aligned} \tag{4.52}$$

The P-value of the statistics is again determined, based on the parametric bootstrap simulation, by simply extending the bivariate case to a multivariate case with the same simulation procedure, except that this case is in d dimension.

4.7.2 Goodness-of-Fit Statistic S_n for Multivariate Symmetric Archimedean Copulas

Following Genest et al. (2007) and Genest and Rémillard (2008), again let multivariate random variables $X_1, \ldots, X_d$ be modeled by the symmetric Archimedean copula function: $C_\theta(u_1, u_2, \ldots, u_d); u_1 = F_1(x_1), \ldots, u_d = F_d(x_d)$.

The empirical d-dimensional copula function can be given as follows:

$$C_n(\mathbf{u}) = \frac{1}{n}\sum_{i=1}^{n} (\mathbf{U}_i \leq \mathbf{u}) \tag{4.53}$$

Now the goodness-of-fit test statistic and the P-value can be estimated using the same procedure as that discussed in Section 3.8.1.

4.7.3 Goodness-of-Fit Test Statistic $S_n^{(K)}$ Based on the Kendall Probability Transform

Besides the GoF statistics $S_n^{(B)}, S_n$; $S_n^{(K)}$ are the test statistics based on the Kendall probability transform. It is a powerful and convenient test for the symmetric Archimedean copulas. In what follows, we discuss the procedure to construct $S_n^{(K)}$ and the corresponding P-value based on the discussion in Genest et al. (2007).

Similar to the test statistics $S_n^{(B)}$ and S_n, its null hypothesis is that the fitted copula function (i.e., here fitted symmetric Archimedean copula function) can appropriately represent the multivariate distribution function of the multivariate random variable. In Section 4.1, we have introduced the Kendall distribution function $K_C(t)$(i.e., Equation (4.11)). Based on the Kendall distribution for bivariate and trivariate random variables introduced in Sections 4.5.1 and 4.5.2, the nonparametric Kendall distribution for multivariate random variable of d-dimension can be given as

$$z_i = \frac{\sum_{j=1}^{n} 1(\mathbf{X}_j \leq \mathbf{x}_i)}{n-1}, i = 1, \ldots, n\, j \neq i, \mathbf{x}_i = [x_{1i}, x_{2i}, \ldots, x_{di}] \tag{4.54a}$$

$$K_n(z) = \frac{\sum_{i=1}^{n} (z_i \leq z)}{n} \tag{4.54b}$$

Now the test statistic can be written as follows:

$$S_n^{(K)} = \int_0^1 \mathbb{K}_n(v)^2 dK_{\hat{\theta}}(v) \tag{4.55}$$

where

$$\mathbb{K}_n(v) = \sqrt{n}\left(K_n(v) - K_{\hat{\theta}}(v)\right) \tag{4.55a}$$

Genest et al. (2007) showed that Equation (4.55) can be calculated as follows:

$$\begin{aligned} S_n^{(K)} &= \frac{n}{3} + n\sum_{i=1}^{n-1} K_n^2\left(\frac{i}{n}\right)\left(K_{\hat{\theta}}\left(\frac{i+1}{n}\right) - K_{\hat{\theta}}\left(\frac{i}{n}\right)\right) \\ &- n\sum_{i=1}^{n-1} K_n\left(\frac{i}{n}\right)\left(K_{\hat{\theta}}^2\left(\frac{i+1}{n}\right) - K_{\hat{\theta}}^2\left(\frac{i}{n}\right)\right) \end{aligned} \tag{4.56}$$

Finally, with the fitted symmetric Achimedean copula, the P-value of the test statistic is again approximated using parametric bootstrap simulation as follows:

1. Generate a multivariate sample $\mathbf{X}_1^*, \ldots, \mathbf{X}_\text{d}^*$ (with the same sample of the tested dataset) from the fitted Archimedean copula $C_{\hat{\theta}}$ and compute their associated rank $\mathbf{R}_1^*, \ldots, \mathbf{R}_\text{d}^*$.
2. Estimate the copula parameter $\hat{\theta}^*$ using $\left\{\frac{\mathbf{R}_1^*}{n+1}, \ldots, \frac{\mathbf{R}_\text{d}^*}{n+1}\right\}$.
3. Compute K_n^* using Equation (4.54) from the generated multivariate sample $\mathbf{X}_1^*, \ldots, \mathbf{X}_\text{d}^*$ and $S_n^{(K)*}$ using Equation (4.56), replacing $\hat{\theta}$ with $\hat{\theta}^*$.

Repeating steps 1 through 3 for a larger integer number N, we can approximate the P-value as follows:

$$P_{value} = \frac{\sum_{i=1}^{N} 1\left(S_{n,i}{}^{(K)} > S_n^{(K)}\right)}{N} \tag{4.57}$$

Example 4.18 **Goodness-of-fit statistics.**

In this example, we generate GoF statistics for both bivariate and trivariate cases:

- Bivariate case: Compute the goodness-of-fit statistics $S_n^{(B)}$ and S_n, and the corresponding P-value using parametric bootstrap simulation for the parameters (the Gumbel–Hougaard and Frank copulas) estimated with semiparametric ML in Example 4.13.
- Trivariate case: Compute the goodness-of-fit statistics $S_n^{(K)}$ and the corresponding P-value using parametric bootstrap simulation for the parameters of the Gumbel–Houggard and Clayton copulas with semiparametric ML in Example 4.14.

Solution:

- Bivariate case:

  For bivariate random variables given in Example 4.13, we have estimated the parameters using the semiparametric ML as the Gumbel–Hougaard copula ($\theta = 2.390$) and the Frank copula ($\theta = 7.474$). Let $u_1 = F_X(x), u_2 = F_Y(y)$; we can construct test statistics for bivariate frequency analysis.

  i. Goodness-of-fit statistics $S_n^{(B)}$ for the Gumbel–Hougaard and Frank copulas:
     From Equation (4.41), we have the following:
     Gumbel–Hougaard copula:

$$Z_1 = u_1$$

$$Z_2 = \frac{\phi^{-1(1)} \sum_{i=1}^{2} (\phi(u_1) + \phi(u_2))}{\phi^{-1(1)}(\phi(u_1))}$$

$$= \frac{e^{\left(-\left((-\ln u_1)^\theta + (-\ln u_2)^\theta\right)^{\frac{1}{\theta}}\right)} (-\ln u_1)^{\theta-1}\left((-\ln u_1)^\theta + (-\ln u_2)^\theta\right)^{\frac{1}{\theta}-1}}{u_1} \tag{4.58}$$

Frank copula:

$$Z_1 = u_1$$
$$Z_2 = \frac{\phi^{-1(1)}\sum_{i=1}^{2}(\phi(u_1)+\phi(u_2))}{\phi^{-1(1)}(\phi(u_1))} = \frac{e^{-\theta u_1}\left(e^{-\theta u_2}-1\right)}{(e^{-\theta u_1}-1)(e^{-\theta u_2}-1)(e^{-\theta}-1)} \quad (4.59)$$

Now, we can compute $\{Z_1, Z_2\}$ using Equations (4.58) and (4.59) as shown in Table 4.14.

Inserting the computed $\{Z_1, Z_2\}$ into Equation (4.52), we can compute the test statistic $S_n^{(B)}$ as follows:

Gumbel–Hougaard: $S_n^{(B)} = 0.0483$
and Frank: $S_n^{(B)} = 0.0414$
With 5,000 bootstrap parametric simulations as an example, the P-values can be approximated using the procedure discussed in Section 3.8.1, as follows:

$$P_{Gumbel-Hougaard\ (\theta=2.390)} = 0.202; P_{Frank(\theta=7.474)} = 0.28$$

ii. Goodness-of-fit statistics S_n for Gumbel–Hougaard and Frank copulas:
The empirical copula function is estimated using Equation (4.53) and the copula function, with the estimated parameter calculated using the Gumbel–Hougaard (Frank) copula function.
Now, the test statistics S_n can be estimated as follows:
Gumbel–Hougaard: $S_n = 0.0141$
and Frank: $S_n = 0.0153$
With 5,000 bootstrap parametric simulations, the P-values can be approximated using the procedure discussed in Section 3.8.2, as follows:

$$P_{Gumbel-Hougaard(\theta=1.889)} = 0.714; P_{Frank(\theta=5.606)} = 0.567$$

From Example 4.12, we have shown that the log-likelihood estimated from the Frank copula is slightly higher than that estimated from the Gumbel–Hougaard copula. However, the goodness-of-fit tests indicate that the Gumbel–Hougaard copula reached a higher P-value than did Frank copula for both $S_n^{(B)}$ (Rosenblatt transform) and S_n (empirical copula). This is because the Frank copula cannot capture the upper-tail dependence embedded in the flood peak and flood volume (i.e., Figures 4.6 and 4.7).

- Trivariate case:

From Example 4.14, we have estimated the copula parameters for trivariate flood frequency analysis using semiparametric ML as the Gumbel–Hougaard copula ($\theta = 1.368$) and the

Table 4.14. $\{Z_1, Z_2\}$ computed from Equations (4.58) and (4.59).

	Marginals		Gumbel–Hougaard		Frank			Marginals		Gumbel–Hougaard		Frank	
No.	$F_n(x)$	$F_n(y)$	Z_1	Z_2	Z_1	Z_2	No	$F_n(x)$	$F_n(y)$	Z_1	Z_2	Z_1	Z_2
1	0.515	0.267	0.515	0.163	0.515	0.092	51	0.653	0.149	0.653	0.027	0.653	0.031
2	0.891	0.921	0.891	0.790	0.891	0.972	52	0.158	0.257	0.158	0.573	0.158	0.085
3	0.317	0.059	0.317	0.044	0.317	0.009	53	0.812	0.832	0.812	0.685	0.812	0.917
4	0.851	0.733	0.851	0.300	0.851	0.814	54	0.743	0.762	0.743	0.653	0.743	0.850
5	0.168	0.248	0.168	0.537	0.168	0.079	55	0.238	0.356	0.238	0.631	0.238	0.176
6	0.683	0.168	0.683	0.028	0.683	0.038	56	0.416	0.446	0.416	0.549	0.416	0.303
7	0.356	0.327	0.356	0.419	0.356	0.144	57	0.931	0.842	0.931	0.248	0.931	0.925
8	0.535	0.683	0.535	0.796	0.535	0.742	58	0.703	0.594	0.703	0.375	0.703	0.583
9	0.960	0.970	0.960	0.785	0.960	0.991	59	0.307	0.911	0.307	0.998	0.307	0.967
10	0.980	0.941	0.980	0.194	0.980	0.980	60	0.564	0.030	0.564	0.004	0.564	0.004
11	0.069	0.317	0.069	0.805	0.069	0.134	61	0.584	0.614	0.584	0.614	0.584	0.620
12	0.822	0.891	0.822	0.848	0.822	0.957	62	0.455	0.218	0.455	0.151	0.455	0.061
13	0.871	0.980	0.871	0.994	0.871	0.994	63	0.475	0.455	0.475	0.486	0.475	0.319
14	0.733	0.782	0.733	0.721	0.733	0.872	64	0.594	0.515	0.594	0.417	0.594	0.428
15	0.614	0.574	0.614	0.492	0.614	0.544	65	0.426	0.475	0.426	0.587	0.426	0.354
16	0.446	0.525	0.446	0.645	0.446	0.447	66	0.772	0.802	0.772	0.693	0.772	0.891
17	0.267	0.347	0.267	0.575	0.267	0.165	67	0.802	0.822	0.802	0.680	0.802	0.909
18	0.990	0.990	0.990	0.666	0.990	0.997	68	0.832	0.950	0.832	0.971	0.832	0.984
19	0.228	0.178	0.228	0.304	0.228	0.042	69	0.139	0.188	0.139	0.463	0.139	0.047
20	0.921	0.871	0.921	0.393	0.921	0.945	70	0.050	0.158	0.050	0.596	0.050	0.035
21	0.109	0.109	0.109	0.317	0.109	0.020	71	0.396	0.653	0.396	0.868	0.396	0.692
22	0.941	0.772	0.941	0.108	0.941	0.861	72	0.257	0.139	0.257	0.193	0.257	0.028
23	0.277	0.465	0.277	0.746	0.277	0.337	73	0.842	0.752	0.842	0.370	0.842	0.839
24	0.911	0.960	0.911	0.924	0.911	0.988	74	0.020	0.020	0.020	0.179	0.020	0.003
25	0.881	0.604	0.881	0.097	0.881	0.602	75	0.752	0.663	0.752	0.405	0.752	0.709

26	0.970	0.792	0.970	0.047	0.970	0.882	76	0.347	0.069	0.347	0.046	0.347	0.011
27	0.287	0.277	0.287	0.421	0.287	0.100	77	0.624	0.426	0.624	0.243	0.624	0.271
28	0.059	0.208	0.059	0.670	0.059	0.056	78	0.386	0.376	0.386	0.468	0.386	0.200
29	0.218	0.396	0.218	0.716	0.218	0.227	79	0.663	0.505	0.663	0.298	0.663	0.409
30	0.485	0.554	0.485	0.645	0.485	0.505	80	0.198	0.366	0.198	0.698	0.198	0.188
31	0.525	0.416	0.525	0.352	0.525	0.256	81	0.079	0.238	0.079	0.678	0.079	0.072
32	0.644	0.693	0.644	0.676	0.644	0.757	82	0.713	0.495	0.713	0.218	0.713	0.390
33	0.089	0.010	0.089	0.027	0.089	0.001	83	0.634	0.673	0.634	0.652	0.634	0.726
34	0.188	0.624	0.188	0.941	0.188	0.639	84	0.178	0.119	0.178	0.237	0.178	0.022
35	0.010	0.040	0.010	0.389	0.010	0.005	85	0.950	0.931	0.950	0.484	0.950	0.976
36	0.327	0.307	0.327	0.423	0.327	0.124	86	0.099	0.564	0.099	0.947	0.099	0.525
37	0.762	0.743	0.762	0.560	0.762	0.826	87	0.574	0.634	0.574	0.665	0.574	0.657
38	0.297	0.228	0.297	0.315	0.297	0.067	88	0.782	0.644	0.782	0.308	0.782	0.675
39	0.901	0.901	0.901	0.645	0.901	0.962	89	0.723	0.584	0.723	0.323	0.723	0.563
40	0.040	0.079	0.040	0.398	0.040	0.013	90	0.119	0.089	0.119	0.243	0.119	0.015
41	0.505	0.861	0.505	0.978	0.505	0.939	91	0.673	0.851	0.673	0.921	0.673	0.932
42	0.337	0.703	0.337	0.934	0.337	0.772	92	0.376	0.713	0.376	0.927	0.376	0.786
43	0.693	0.337	0.693	0.099	0.693	0.154	93	0.604	0.535	0.604	0.436	0.604	0.466
44	0.436	0.723	0.436	0.910	0.436	0.800	94	0.366	0.485	0.366	0.677	0.366	0.372
45	0.495	0.129	0.495	0.053	0.495	0.025	95	0.861	0.881	0.861	0.715	0.861	0.951
46	0.129	0.099	0.129	0.255	0.129	0.017	96	0.030	0.050	0.030	0.312	0.030	0.007
47	0.149	0.198	0.149	0.467	0.149	0.051	97	0.792	0.386	0.792	0.067	0.792	0.213
48	0.248	0.545	0.248	0.859	0.248	0.486	98	0.465	0.287	0.465	0.230	0.465	0.107
49	0.208	0.297	0.208	0.570	0.208	0.116	99	0.545	0.406	0.545	0.311	0.545	0.241
50	0.406	0.812	0.406	0.972	0.406	0.901	100	0.554	0.436	0.554	0.344	0.554	0.286

Clayton copula ($\theta = 0.721$). The corresponding Kendall distribution is given as Equation (4.32) for the Gumbel–Hougaard copula and Equation (4.31) for the Clayton copula. The test statistics are determined using Equations (4.51a)–(4.53). Table 4.15 lists the computed nonparametric and parametric Kendall distribution.

Table 4.15. *Nonparametric and parametric Kendall distribution estimation for trivariate random variables.*

X	*Y*	*Z*	*V*	*Kn*	*K*(gumbel)	*K*(clayton)
10.32	1.88	18.84	0.14	0.28	0.29	0.26
17.61	2.16	19.49	0.42	0.64	0.60	0.70
16.03	2.03	19.89	0.34	0.52	0.52	0.59
13.49	1.32	18.55	0.12	0.22	0.26	0.22
19.23	3.13	19.45	0.54	0.76	0.70	0.83
16.79	2.95	18.78	0.22	0.38	0.39	0.40
17.06	2.04	20.11	0.4	0.6	0.58	0.67
12.31	2.99	19.27	0.24	0.4	0.42	0.43
39.47	20.13	21.17	1	1	1.00	1.00
8.25	0.08	18.03	0.02	0.06	0.07	0.04
10.06	0.23	18.24	0.06	0.12	0.16	0.11
16.91	2.96	19.71	0.46	0.72	0.63	0.75
35.41	15.32	20.99	0.98	0.98	0.99	1.00
21.27	2.79	19.91	0.56	0.78	0.71	0.85
17.3	1.94	19.4	0.36	0.58	0.54	0.62
19.04	1.48	20.45	0.3	0.48	0.48	0.53
14.18	1.04	18.36	0.1	0.18	0.23	0.18
32.88	14.65	20.58	0.9	0.92	0.94	1.00
6.68	0.18	16.1	0.02	0.06	0.07	0.04
20.22	6.23	19.15	0.46	0.72	0.63	0.75
17.28	1.59	19.12	0.28	0.46	0.46	0.49
12.29	1.79	18.88	0.18	0.36	0.34	0.33
14.38	1.98	19	0.28	0.46	0.46	0.49
17.56	3.08	19.91	0.6	0.8	0.74	0.88
10.95	0.61	18.79	0.14	0.28	0.29	0.26
8.26	0.49	18.8	0.08	0.16	0.20	0.15
16.08	3.03	20.15	0.46	0.72	0.63	0.75
11.21	0.8	19.91	0.18	0.36	0.34	0.33
15.15	1.93	18.68	0.16	0.32	0.32	0.29
20.54	3.38	19.17	0.46	0.72	0.63	0.75
10.82	0.87	19.01	0.16	0.32	0.32	0.29
16.84	1.29	18.12	0.08	0.16	0.20	0.15
15.54	2.14	19.36	0.34	0.52	0.52	0.59
34.68	14.16	20.93	0.94	0.96	0.97	1.00

Table 4.15. (*cont.*)

X	Y	Z	V	Kn	K(gumbel)	K(clayton)
17.4	2.86	20.03	0.52	0.74	0.68	0.81
28.96	10.19	20.6	0.86	0.9	0.92	0.99
16.37	2.37	19.27	0.36	0.58	0.54	0.62
12.6	0.4	18.81	0.14	0.28	0.29	0.26
17.26	2.61	19.81	0.42	0.64	0.60	0.70
6.34	0.29	17.74	0.02	0.06	0.07	0.04
32.84	13.7	20.78	0.92	0.94	0.95	1.00
29.37	11.89	20.39	0.84	0.86	0.91	0.99
19.45	4.4	19.93	0.68	0.82	0.80	0.94
18.03	1.75	18.94	0.26	0.42	0.44	0.46
14.01	2.29	17.83	0.06	0.12	0.16	0.11
9.07	0.11	18.48	0.04	0.08	0.12	0.07
15.66	2.89	19.01	0.36	0.58	0.54	0.62
20.64	8.27	19.99	0.74	0.84	0.84	0.96
10.79	0.33	18.72	0.12	0.22	0.26	0.22
28.63	10.59	20.52	0.86	0.9	0.92	0.99

The test statistics are computed using Equation (4.56). The corresponding P-values are approximated with 5,000 parametric bootstrap simulations using the procedure discussed in Section 4.7.3.

Gumbel–Hougaard: $S_n^K = 0.0796; P_{value} = 0.664$

Clayton: $S_n^K = 0.209, P_{value} = 0.827$

4.8 Summary

This chapter focuses on the symmetric Archimedean copulas. As its name, the symmetric copulas are exchangeable. We discuss generating functions of Archimedean copulas and their properties, parameter estimation, simulation, and goodness-of-fit statistical tests. Regarding the applicability, the Archimedean copula may be easily constructed with the generating function. In addition, the Archimedean copula may cover the entire range of the independence. The Archimedean copula can be properly applied to model the bivariate random variables. While only certain bivariate Archimedean copulas (i.e., fulfilling the conditions: strictly decreasing generating function, positive dependence structure) may be extended to the symmetric Archimedean copula in a higher dimension. Moreover, the symmetric Archimedean copula in a higher dimension (i.e., $d \geq 3$) assumes that variables share the same degree of dependence. For example, (X_1, X_2), (X_1, X_3), and (X_2, X_3) have the same Kendall's tau ($\tau_{12} = \tau_{13} = \tau_{23}$) for the trivariate random variables (X_1, X_2, X_3).

Forcing all the variables to share the same degree of dependences limits the application of symmetric Archimedean copulas into a higher dimension. In the later chapters, we will discuss the alternative approaches for the analysis in higher dimensions.

References

Ali, M. M., Mikhail, N. N., and Haq, M. S. (1978). A class of bivariate distributions including the bivariate logistic. *Journal of Multivariate Analysis*, 8, 405–412.

Antonio, J., Manuel, R. L., and Úbeda-Flores, M. (2004). A new class of bivariate copulas. *Statistics and Probability Letters*, 66, 315–325.

Caperaa, P., Fougeres, A. L., and Genest, C. (1993). A nonparametric estimation procedure for bivariate extreme value copulas. *Biometrika*, 84(3), 567–577.

Clayton, D. G. (1978). A model for association in bivariate life tables and its application in epidemiological studies of familial tendency in chronic disease incidence. *Biometrika*, 65(1), 141–151.

Cook, R. D. and Johnson, M. W. (1981). A family of distribus for modeling nonelliptically symmetric multivariate data. *Journal of the Royal Statistical Society. Series B (Methodological)*, 43(2), 210–218.

Cox, D. R. and Oaks, D. (1984). *Analysis of Survival Data*. Chapman and Hall, London.

De Matteis, R. (2001). *Fitting Copulas to Data*. Diploma Thesis, Institute of Mathematics of the University of Zurich, http://89.179.245.94/svn/study/copulas/copulas-fitting .pdf.

Embrechts, P., Lindskog, F., and McNeil, A. (2001). Modelling dependence with copulas and applications to risk management. www.risklab.ch/ftp/papers/Dependence WithCopulas.pdf.

Favre, A.-C., Adlouni, S. E., Perreault, L., Thiémonge, N., and Bobée, B. (2004). Multivariate hydrological frequency analysis using copulas. *Water Resources Research*, 40. W01101. doi:10.1029/2003WR002456.

Francesco, S. and Salvatore, G. (2007). Fully nested 3-copula: procedure and application on hydrological data. *Journal of Hydrologic Engineering*, 12(4), 420–430.

Frank, M. J. (1979). On the simultaneous associativity of F(x, y) and x + y - F(x, y). *Aequationes Mathematics*, 19, 617–627.

Frees, E. W. and Valdez, E. A. (1997). Understanding relationships using copulas. *North American Actuarial Journal*, 2(1), 1–25.

Genest, C., Ghoudi, K., and Rivest, L.-P. (1995). A semipraramentric estimation procedure of dependence parameters in multivariate families of distributions. *Biometrika*, 82(3), 543–552.

Genest, C. and MacKay, J. (1986). The joy of copulas: bivariate distributions with uniform marginals. *American Statistician*, 40(4), 280–283.

Genest, C. and Rémillard, B. (2008). Validity of the parametric bootstrap for goodness-of-fit testing in semiparametric models. *Annales de l'Institue Henri Poincaré–Probabilités et Statistiques*, 44(6), 1096–1127.

Genest, C., Rémillard, B., and Beaudoin, D. (2007). Goodness-of-fit tests for copulas: a review and a power study. *Insurance: Mathematics and Economics*. doi:10.1016/j. insmatheco.2007.10.1005.

Genest, C. and Rivest, L.-P. (1993). Statistical inference procedures for bivariate Archimedean copulas. *Journal of the American Statistical Association*, 88, 1034–1043.

Gumbel, E. J. (1960a). Bivariate exponential distributions. *Journal of the American Statistical Association*, 55(292), 698–707.

Gumbel, E. J. (1960b). Distributions del valeurs extremes en plusieurs dimensions. *Publ. l'Inst. de Statistique, Paris*, 9, 171–173.

Joe, H. (1993). Parametric families of multivariate distributions with given margins. *Journal of Multivariate Analysis*, 46(2), 262–282.

Malevergne, Y. and Sornette, D. (2006). *Extreme Financial Risks from Dependence to Risk Management*. Springer, Netherlands.

Nelsen, R. B. (2006). *An Introduction to Copulas*, 2nd edition. Springer, New York, NY.

Oakes, D. (1982). A model for association in bivariate survival data. *Journal of the Royal Statistical Society. Series B (Methodological)*, 44(3), 414–422.

Oakes, D. (1986). Semiparametric inference in a model for association in bivariate survival data. *Biometrika*, 73, 353–361.

Rodriguez-Lallena, J. A. and Úbeda-Flores, M. (2004). A new class of bivariate copulas. *Statistics and Probability Letters*, 66, 315–325.

Rosenblatt, M. (1952). Remarks on a multivariate transformation. *Annals of Mathematical Statistics*, 23(3), 470–472.

Salvadori, G., De Michele, C., Kottegoda, N. T., and Rosso, R. (2007). *Extremes in Nature: an Approach Using Copulas*. Springer, Dordrecht.

Savu, C. and Trede, M. (2008). Goodness-of-fit tests for parametric families of Archimedean copulas. *Quantitative Finance*, 8 (2), 109–116

Sklar, A. (1959). Fonctions de repartition à n dimensions et leurs marges. *Publ. l'Inst. de Statistique Univ., Paris* 8, 229–231.

Widder, D. V. (1941). *The Laplace Transform*. Princeton University Press, Princeton.

5

Asymmetric Copulas

High Dimension

ABSTRACT

Much of the literature on copulas, discussed in the previous chapters, is limited to the bivariate cases. The Gaussian and student copulas have been commonly applied to model the dependence in higher dimensions (Genest and Favre, 2007; Genest et al., 2007a). In Chapter 4, we discussed the extension of symmetric bivariate Archimedean copulas as well as their major restrictions to model high-dimensional dependence (i.e., $d \geq 3$). Through the extension of the bivariate Archimedean copula, the multivariate Archimedean copula is symmetric and denoted as exchangeable Archimedean copula (EAC). EAC allows for the specification of only one generating function and only one set of parameters $\boldsymbol{\theta}$. In other words, random variates by pair share the same degree of dependence. Using the trivariate random variable $\{X_1, X_2, X_3\}$ as an example, $\{X_1, X_2\}$, $\{X_2, X_3\}$, and $\{X_1, X_3\}$ should have the same degree of dependence. However, this assumption is rarely valid. This chapter discusses the following two approaches of constructing asymmetric multivariate copulas: nested Archimedean copula construction (NAC) and the vine copulas through pair-copula construction (PCC).

5.1 Construction of Higher-Dimensional Copulas

In general, there are $\frac{d(d-1)}{2}$ pairs of variables for a given d-dimensional multivariate problem. The NAC approach constitutes a significant improvement over EAC; however, it is still not rich enough to model all possible mutual dependencies among the d dimensional random variables (Berg and Aas, 2007). Based on the multivariate probability density function decomposition (Joe, 1997), the PCC approach allows for the free specification of $\frac{d(d-1)}{2}$ copulas that are hierarchical in nature. Further, it allows for selecting copulas from different families to model the dependence structure (Berg and Aas, 2007; Aas et al., 2009). Hence, the NAC approach is introduced first, followed by the PCC approach.

5.2 Nested Archimedean Copulas (NAC)

Representing one type of multivariate extension, NAC constitutes a significant improvement over EAC. We first review the fully nested Archimedean construction (FNAC) and the partially nested Archimedean construction (PNAC), and then turn to the general nested Archimedean copula.

5.2.1 Fully Nested Archimedean Copulas (FNAC)

For d-dimensional random variables modeled with FNAC, there are $d - 1$ bivariate copula functions, which result in dependence structure with partial exchangeability (Joe, 1997; Embrechts et al., 2003; Whelan, 2004; McNeil, 2007; Savu and Trede, 2010; among others). Figure 5.1 presents an example of a four-dimensional FNAC structure. The bivariate copula is the building block for FNAC. The FNAC structure is constructed, based on the degree of dependence between the pair variables, with the following procedures:

i. Choose the variables with the highest degree of dependence (rank-based) as the first two variables (1 and 2).
ii. Compute the empirical copula using variables 1 and 2.
iii. Evaluate the degree of dependence (rank-based) between empirical copula from step ii with the remaining variables.
iv. Choose variable 3, i.e., yielding the highest degree of dependence (rank-based) with the empirical copula built with variables 1 and 2.
v. Continue the process until the last variable is considered.

From Figure 5.1, it is seen that three bivariate copulas are needed to represent the dependence for the four-dimensional random variables through FNAC as follows. First, random variables u_1 and u_2 are coupled through copula C_3. Second, random variable u_3 is coupled with $C_3(u_1, u_2)$ through copula C_2. Third, random variable u_4 is coupled with $C_2(u_3, C_3(u_1, u_2))$ through copula C_1. Hence, a four-dimensional copula requires three bivariate copulas C_1, C_2, and C_3, with generators ϕ_1, ϕ_2, and ϕ_3 and may be written as follows:

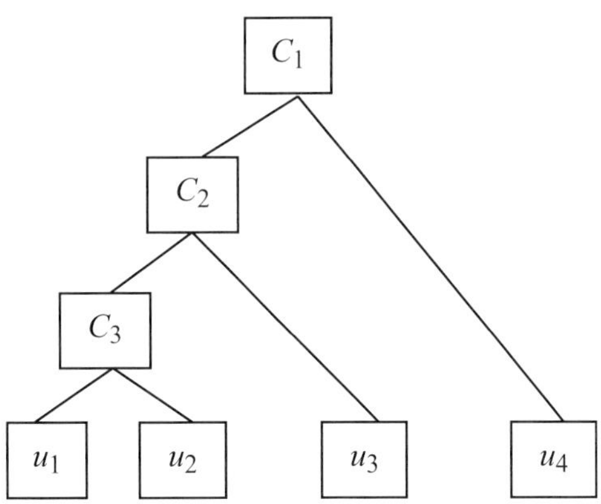

Figure 5.1 Four-dimensional FNAC structure.

$$
\begin{aligned}
&C(u_1,u_2,u_3,u_4)\\
&= C_1\Big(u_4, C_2\big(u_3, C_3(u_1,u_2)\big)\Big)\\
&= \phi_1^{-1}\left(\phi_1(u_4)+\phi_1\left(\phi_2^{-1}\left(\phi_2(u_3)+\phi_2\left(\phi_3^{-1}(\phi_3(u_1)+\phi_3(u_2))\right)\right)\right)\right)\\
&= \phi_1^{-1}\left(\phi_1(u_4)+\phi_1\circ\phi_2^{-1}\left(\phi_2(u_3)+\phi_2\circ\phi_3^{-1}(\phi_3(u_1)+\phi_3(u_2))\right)\right)
\end{aligned}
\tag{5.1}
$$

where $\circ$ represents the composition of functions.

Similarly, the FNAC for d-dimensional random variables (e.g., Joe, 1997; Embrechts et al., 2003; Whelan, 2004; Nelsen, 2006) may be generated as follows:

$$
\begin{aligned}
&C(u_1,\ldots,u_d)\\
&= \phi_1^{-1}\left(\phi_1(u_d)+\phi_1\circ\phi_2^{-1}\left(\phi_2(u_{d-1})+\phi_2\circ\ \ldots\circ\phi_{d-1}^{-1}\left(\phi_{d-1}(u_1)+\phi_{d-1}(u_2)\right)\right)\right)
\end{aligned}
\tag{5.2}
$$

It is worth noting that Equation (4.1) in Chapter 4, i.e., the exchangeable symmetric Archimedean copula, is a special case of Equation (5.2) if $\phi(\theta_1)=\phi_2(\theta_2)=\cdots=\phi_{d-1}(\theta_{d-1})=\phi(\theta), \theta_1=\theta_2=\cdots=\theta_{d-1}$. For the d-dimensional FNAC, the bivariate margins themselves are also Archimedean copulas that allow for free specification of $d-1$ copulas with the remaining identified implicitly through FNAC (Whelan, 2004; Berg and Aas, 2007). Using Equation (5.1) (Figure 5.1) as an example, this statement may be expressed as follows: (i) there are three Archimedean copulas of free specification, i.e., C_3 with parameter $\boldsymbol{\theta}_3$ for variables u_1,u_2; C_2 with parameter $\boldsymbol{\theta}_2$ for variables $\{u_3, C_3(u_1,u_2;\boldsymbol{\theta}_3\}$; and C_1 with parameter $\boldsymbol{\theta}_1$ for variables $\{u_4, C_2(u_3, C_3(u_1,u_2;\boldsymbol{\theta}_3);\boldsymbol{\theta}_2\}$; (ii) pairs (u_1,u_3), (u_2,u_3) have copula C_2 with parameter $\boldsymbol{\theta}_2$; and (iii) pairs $(u_1,u_4),(u_2,u_4),(u_3,u_4)$ have copula C_3 with parameter $\boldsymbol{\theta}_1$. The decreasing degree of dependence for the increasing levels of nesting (i.e., $\boldsymbol{\theta}_1\le\boldsymbol{\theta}_2\le\cdots\le\boldsymbol{\theta}_{\mathrm{d}-1}$ with $\boldsymbol{\theta}_1$ and $\boldsymbol{\theta}_{\mathrm{d}-1}$ representing the parameters for the highest and lowest levels, respectively) is another technical condition for proper construction of the d-dimensional fully nested asymmetric Archimedean copula.

It should also be pointed out that the following conditions need to be satisfied for the nested generating functions:

- $\phi_1^{-1},\phi_2^{-1},\ldots,\phi_{d-1}^{-1}$ must satisfy the necessary conditions for being completely monotonic.
- According to Embrechts et al. (2003), the coupling of functions $w_k=\phi_k\circ\phi_{k+1}^{-1}$ belongs to a class of functions $\mathcal{L}_\infty^*$ defined as follows:

$$
\mathcal{L}_\infty^*=\left\{\omega\colon[0,\infty)\to[0,\infty)|\omega(0)=0,\omega(\infty)=\infty,(-1)^{k-1}\frac{d^k\omega(t)}{dt}\ge 0; k=1,2,\ldots,\infty\right\}
\tag{5.3}
$$

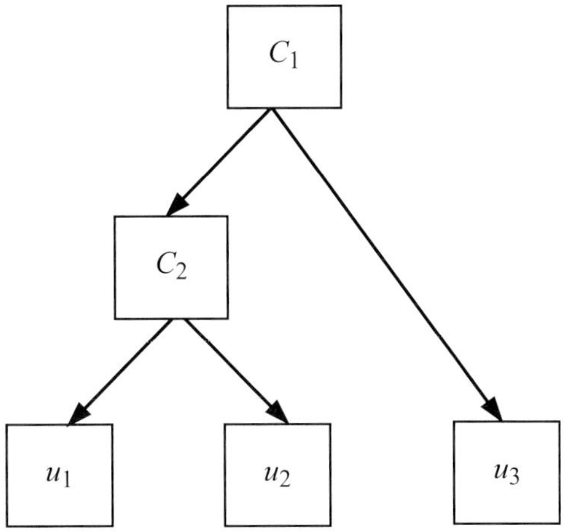

Figure 5.2 Three-dimensional FNAC structure.

Based on Equation (5.2), the simplest three-dimensional FNAC (shown in Figure 5.2) can be written as follows:

$$C(u_1, u_2, u_3) = \phi_1^{-1}(\phi_1(u_3) + \phi_1 \circ \phi_2^{-1}(\phi_2(u_1) + \phi_2(u_2)) \quad (5.4)$$

In accordance with Equation (5.4), we outline here the derivation of five three-dimensional asymmetric Archimedean copulas that are commonly applied.

M3 (Joe, 1997):

$$C_2(u_1, u_2) = -\frac{1}{\theta_2} \ln\left(1 - \frac{\left(1 - e^{-\theta_2 u_1}\right)\left(1 - e^{-\theta_2 u_2}\right)}{1 - e^{-\theta_2}}\right)$$

Let $t = C_2(u_1, u_2)$. Then we have $C_1(u_3, t) = -\frac{1}{\theta_1} \ln\left(1 - \frac{\left(1 - e^{-\theta_1 u_3}\right)\left(1 - e^{-\theta_1 t}\right)}{1 - e^{-\theta_1}}\right)$

$$\begin{aligned} C(u_1, u_2, u_3) &= C_1(u_3, C_2(u_1, u_2)) = C_1(u_3, t) \\ &= -\frac{1}{\theta_1} \ln\left(1 - \frac{\left(1 - e^{-\theta_1 u_3}\right)\left(1 - \frac{\left(1 - e^{-\theta_2 u_1}\right)\left(1 - e^{-\theta_2 u_2}\right)}{1 - e^{-\theta_2}}\right)}{1 - e^{-\theta_1}}\right) \end{aligned} \quad (5.5)$$

$\theta_2 \geq \theta_1 \in [0, \infty), \tau_{12}, \tau_{13}, \tau_{23} \in [0, 1]$ for positive dependent trivariate variables.

The M3 copula may be also called the asymmetric trivariate Frank copula.

We now use the following specific examples to illustrate these marginal distributions.

Example 5.1 **Derive the M3 copula for $\theta_1 = 2.0$ and $\theta_2 = 3.0$ by setting $u_3 = 0.6$. Assuming $u_1 \sim F_1(x_1)$: $X_1 \sim gamma(2, 4)$; $u_2 \sim F_2(x_2)$: $X_2 \sim normal(1, 3^2)$; $u_3 \sim F_3(x_3)$: $X_3 \sim EV1(10, 7)$, and $\{X_1, X_2\}$ has a higher pairwise dependence.**

Solution: With $\{X_1, X_2\}$ having higher pairwise dependence, we first couple X_1 and X_2 and build the copula function from the marginals as follows:

$$u_1 = F_1(x_1) = \frac{1}{\Gamma(2)}\gamma(4x_1) \quad \gamma(): \text{incomplete gamma function}$$

$$u_2 = F_2(x_2) = \Phi\left(\frac{x_2 - 1}{3}\right), \Phi(\,): \text{Standard normal distribution}$$

$$u_3 = F_3(x_3) = \exp\left(-\exp\left(-\frac{x_3 - 10}{7}\right)\right)$$

Since we already set $u_3 = 0.6$, then we have $x_3 \approx 9.388$ from the EV1 population.

Finally, we can write the fully nested copula using the M3 copula as follows:

$$C_2(u_1, u_2; 3) = -\frac{1}{3.0}\ln\left(1 - \frac{(1 - e^{-3.0u_1})(1 - e^{-3.0u_2})}{1 - e^{-3.0}}\right)$$

$$= -\frac{1}{3.0}\ln\left(1 - \frac{\left(1 - e^{-3.0\left(\frac{1}{\Gamma(2)}\gamma(4x_1)\right)}\right)\left(1 - e^{-3.0\left(\Phi\left(\frac{x_2-1}{3}\right)\right)}\right)}{1 - e^{-3.0}}\right)$$

$$C(u_1, u_2, 0.6; 3, 2) = C(0.6, C_2(u_1, u_2; 3); 2)$$

$$= -\frac{1}{2.0}\ln\left(1 - \frac{\left(1 - e^{-2.0(0.6)}\right)\left(1 - \frac{(1 - e^{-3.0u_1})(1 - e^{-3.0u_2})}{1 - e^{-3.0}}\right)}{1 - e^{-2.0}}\right)$$

$$= -\frac{1}{2.0}\ln\left(1 - \frac{\left(1 - e^{-2.0(0.6)}\right)\left(1 - \frac{\left(1 - e^{-3.0\left(\frac{1}{\Gamma(2)}\gamma(4x_1)\right)}\right)\left(1 - e^{-3.0\left(\Phi\left(\frac{x_2-1}{3}\right)\right)}\right)}{1 - e^{-3.0}}\right)}{1 - e^{-2.0}}\right)$$

Figure 5.3(a) plots the corresponding joint CDF for the derived M3 copula with $u_3 = 0.6$.

M4 (Joe, 1997):

$$C_2(u_1, u_2) = \left(u_1^{-\theta_2} + u_2^{-\theta_2} - 1\right)^{-\frac{1}{\theta_2}}$$

Let $t = C_2(u_1, u_2)$. Then we have $C_1(u_3, t) = \left(u_3^{-\theta_1} + t^{-\theta_1} - 1\right)^{-\frac{1}{\theta_1}}$

$$C(u_1, u_2, u_3) = C_1(u_3, C_2(u_1, u_2))$$
$$= \left(\left(u_1^{-\theta_2} + u_2^{-\theta_2} - 1\right)^{-\frac{\theta_1}{\theta_2}} + u_3^{-\theta_1} - 1\right)^{-\frac{1}{\theta_1}} \tag{5.6}$$

$\theta_2 \geq \theta_1 \in [0, \infty), \tau_{12}, \tau_{13}, \tau_{23} \in [0, 1]$ for positive dependent trivariate variables. The M4 copula may also be called the trivariate asymmetric Clayton copula.

Example 5.2 **Derive the M4 copula using information given in Example 5.1.**

Solution: In Example 5.1, we have $\theta_1 = 2.0, \theta_2 = 3.0$ by setting $u_3 = 0.6$. Thus, we have the following:

$$C_2(u_1, u_2; 3) = \left(u_1^{-3.0} + u_2^{-3.0} - 1\right)^{-\frac{1}{3.0}} = \left\{\left[\frac{1}{\Gamma(2)}\gamma(4x_1)\right]^{-3.0} + \left[\Phi\left(\frac{x_2 - 1}{3}\right)\right]^{-3.0} - 1\right\}^{-\frac{1}{3.0}}$$

$$C(u_1, u_2, 0.6; 3, 2) = C_1(C_2(u_1, u_2; 3), 0.6) = \left(\left(u_1^{-3.0} + u_2^{-3.0} - 1\right)^{\frac{2}{3}} + 0.6^{-2.0} - 1\right)^{-\frac{1}{2.0}}$$
$$= \left(\left\{\left[\frac{1}{\Gamma(2)}\gamma(4x_1)\right]^{-3.0} + \left[\Phi\left(\frac{x_2 - 1}{3}\right)\right]^{-3.0} - 1\right\}^{\frac{2}{3}} + 0.6^{-2.0} - 1\right)^{-\frac{1}{2.0}}$$

Figure 5.3(b) plots the corresponding joint CDF for the derived M4 copula with $u_3 = 0.6$.

M5 (Joe, 1997):

$$C_2(u_1, u_2) = 1 - \left((1 - u_1)^{\theta_2} + (1 - u_2)^{\theta_2} - (1 - u_1)^{\theta_2}(1 - u_2)^{\theta_2}\right)^{\frac{1}{\theta_2}}$$

Let $t = C_2(u_1, u_2), 1 - t = \left((1 - u_1)^{\theta_2} + (1 - u_2)^{\theta_2} - (1 - u_1)^{\theta_2}(1 - u_2)^{\theta_2}\right)^{\frac{1}{\theta_2}}$. Then we have the following:

$$C(u_1, u_2, u_3) = C_1(u_3, C_2(u_1, u_2))$$
$$= 1 - \left(\left((1 - u_1)^{\theta_2}\left(1 - (1 - u_2)^{\theta_2}\right) + (1 - u_2)^{\theta_2}\right)^{\frac{\theta_1}{\theta_2}}\left(1 - (1 - u_3)^{\theta_1}\right) + (1 - u_3)^{\theta_1}\right)^{\frac{1}{\theta_1}} \tag{5.7}$$

$\theta_2 \geq \theta_1 \in [1, \infty), \tau_{12}, \tau_{13}, \tau_{23} \in [0, 1]$. The M5 copula may also be called the trivariate asymmetric Joe copula.

Example 5.3 **Derive M5 copula using the information given in Example 5.1.**

Solution: In Example 5.1, we have $\theta_1 = 2.0, \theta_2 = 3.0$ by setting $u_3 = 0.6$. Thus we have the following:

$$C_2(u_1,u_2;3.0) = 1 - \left((1-u_1)^{3.0} + (1-u_2)^{3.0} - (1-u_1)^{3.0}(1-u_2)^{3.0}\right)^{\frac{1}{3.0}}$$
$$= 1 - \left(\left(1 - \frac{1}{\Gamma(2)}\gamma(4x_1)\right)^{3.0} + \left(1 - \Phi\left(\frac{x_2-1}{3}\right)\right)^{3.0} - \left[\left(1 - \frac{1}{\Gamma(2)}\gamma(4x_1)\right)\left(1 - \Phi\left(\frac{x_2-1}{3}\right)\right)\right]^{3.0}\right)^{\frac{1}{3.0}}$$

$$C(u_1,u_2,0.6;3,2) = 1 - \left(\left((1-u_1)^{3.0}\left(1-(1-u_2)^{3.0}\right) + (1-u_2)^{3.0}\right)^{\frac{2.0}{3.0}}\left(1-0.4^{2.0}\right) + 0.4^{2.0}\right)^{\frac{1}{2.0}}$$

Figure 5.3(c) plots the corresponding joint CDF for the derived M5 copula with $u_3 = 0.6$.
M6 (Joe, 1997; Embrechts, 2003):

Let $C_2(u_1,u_2) = e^{-\left((-\ln u_1)^{\theta_2}+(-\ln u_2)^{\theta_2}\right)^{\frac{1}{\theta_2}}}$, and
$t = C_2(u_1,u_2), \quad -\ln t = \left((-\ln u_1)^{\theta_2} + (-\ln u_2)^{\theta_2}\right)^{\frac{1}{\theta_2}}$. Then we have

$$C(u_1,u_2,u_3) = C_1(u_3, C_2(u_1,u_2)) = e^{-\left(\left((-\ln u_1)^{\theta_2}+(-\ln u_2)^{\theta_2}\right)^{\frac{\theta_1}{\theta_2}}+(-\ln u_3)^{\theta_1}\right)^{\frac{1}{\theta_1}}} \tag{5.8}$$

$\theta_2 \geq \theta_1 \in [1,\infty), \tau_{12}, \tau_{13}, \tau_{23} \in [0,1]$ for positive dependent trivariate variables. The M6 copula may also be called the trivariate asymmetric Gumbel–Hougaard copula.

Example 5.4 **Derive the M6 copula using the information given in Example 5.1.**

Solution: In Example 5.1, we have $\theta_1 = 2.0, \theta_2 = 3.0$ by setting $u_3 = 0.6$. Thus we have the following:

$$C_2(u_1,u_2;3) = e^{-\left((-\ln u_1)^{3.0}+(-\ln u_2)^{3.0}\right)^{\frac{1}{3.0}}}$$
$$C(u_1,u_2,0.6;3,2) = e^{-\left(\left((-\ln u_1)^{3.0}+(-\ln u_2)^{3.0}\right)^{\frac{2}{3}}+(-\ln 0.6)^{2.0}\right)^{\frac{1}{2.0}}}$$

Figure 5.3(d) plots the corresponding joint CDF for the derived M6 copula with $u_3 = 0.6$.
M12 (Embrechts, 2003):

$$C_2(u_1,u_2) = \frac{1}{1 + \left(\left(\frac{1}{u_1}-1\right)^{\theta_2} + \left(\frac{1}{u_2}-1\right)^{\theta_2}\right)^{\frac{1}{\theta_2}}}$$

Let $t = C_2(u_1,u_2), \ \frac{1}{t} - 1 = \left(\left(\frac{1}{u_1}-1\right)^{\theta_2} + \left(\frac{1}{u_2}-1\right)^{\theta_2}\right)^{\frac{1}{\theta_2}}$. Then we have

$$C(u_1,u_2,u_3) = \frac{1}{1+\left(\left(\left(\frac{1}{u_1}-1\right)^{\theta_2}+\left(\frac{1}{u_2}-1\right)^{\theta_2}\right)^{\frac{\theta_1}{\theta_2}}+\left(\frac{1}{u_3}-1\right)^{\theta_1}\right)^{1/\theta_1}} \tag{5.9}$$

$$\theta_2 \geq \theta_1 \in [1,\infty), \tau_{12}, \tau_{13}, \tau_{23} \in \left[\frac{1}{3}, 1\right].$$

Example 5.5 **Derive the M12 copula using the information given in Example 5.1.**

Solution:

$$C_2(u_1,u_2;3) = \frac{1}{1+\left(\left(\frac{1}{u_1}-1\right)^{3.0}+\left(\frac{1}{u_2}-1\right)^{3.0}\right)^{\frac{1}{3.0}}}$$

$$C(u_1,u_2,0.6;3,2) = \frac{1}{1+\left(\left(\left(\frac{1}{u_1}-1\right)^{3.0}+\left(\frac{1}{u_2}-1\right)^{3.0}\right)^{\frac{2}{3}}+\left(\frac{1}{0.6}-1\right)^{2.0}\right)^{1/2.0}}$$

Figure 5.3(e) plots the joint CDF for the derived M12 copula with $u_3 = 0.6$.

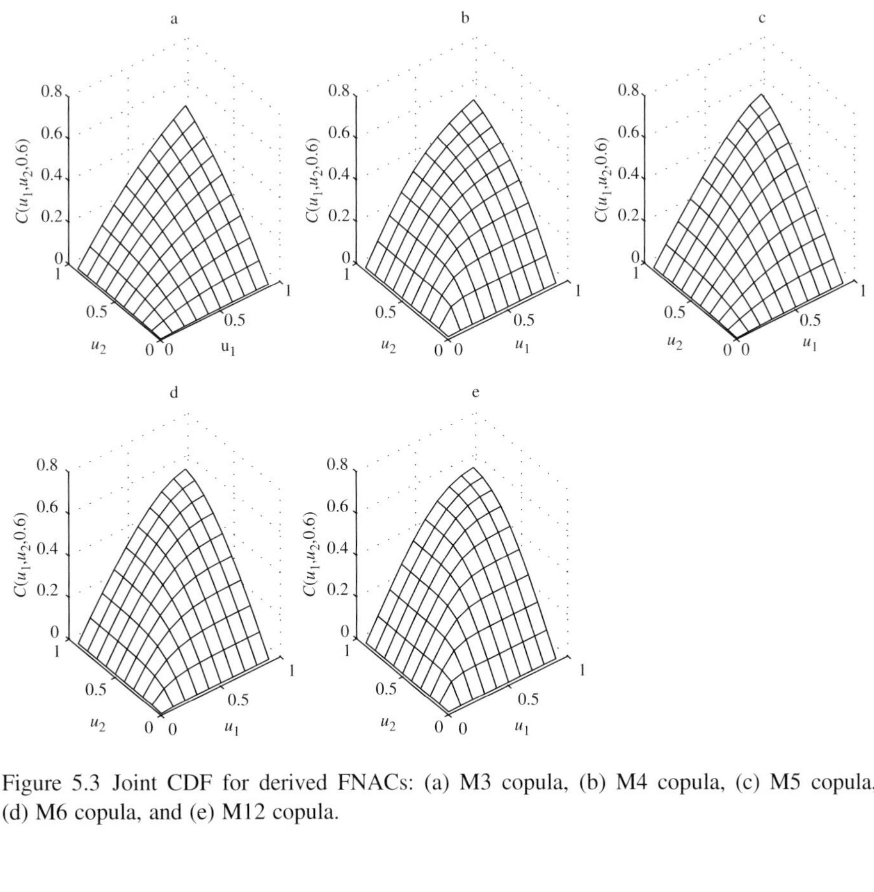

Figure 5.3 Joint CDF for derived FNACs: (a) M3 copula, (b) M4 copula, (c) M5 copula, (d) M6 copula, and (e) M12 copula.

Example 5.6 **Derive a four-dimensional FNAC copula function based on the bivariate Frank copula.**

Solution: From Figure 5.1, we have the following:

$$C_3(u_1, u_2; \theta_3) = -\frac{1}{\theta_3} \ln\left(1 - \frac{\left(1 - e^{-\theta_3 u_1}\right)\left(1 - e^{-\theta_3 u_2}\right)}{1 - e^{-\theta_3}}\right)$$

C_3 and u_3 are coupled as copula $C_2(C_3, u_3)$ with parameter θ_2, which can be written as follows:

$$C_2(u_1, u_2, u_3) = C_2(C_3, u_3; \theta_2)$$
$$= -\frac{1}{\theta_2} \ln\left(1 - \frac{\left(1 - \left(1 - \frac{\left(1 - e^{-\theta_3 u_1}\right)\left(1 - e^{-\theta_3 u_2}\right)}{1 - e^{-\theta_3}}\right)^{\frac{\theta_2}{\theta_3}}\right)\left(1 - e^{-\theta_2 u_3}\right)}{1 - e^{-\theta_2}}\right)$$

Finally, C_2 and u_4 are defined as copula $C_1(C_2, u_4)$ with parameter θ_1, which results in $C_1(C_2, u_4; \theta_1) - C(u_1, u_2, u_3, u_4; \theta_1, \theta_2, \theta_3)$ as follows:

$$C(u_1, u_2, u_3, u_4; \theta_1, \theta_2, \theta_3)$$
$$= -\frac{1}{\theta_1} \ln\left(1 - \frac{\left(1 - e^{-\theta_1 u_4}\right)}{1 - e^{-\theta_1}}\left(1 - \left(1 - \frac{\left(1 - \left(1 - \frac{\left(1 - e^{-\theta_3 u_1}\right)\left(1 - e^{-\theta_3 u_2}\right)}{1 - e^{-\theta_3}}\right)^{\frac{\theta_2}{\theta_3}}\right)\left(1 - e^{-\theta_2 u_3}\right)}{1 - e^{-\theta_2}}\right)^{\frac{\theta_1}{\theta_2}}\right)\right)$$

In the same way as for the previous examples, for the four-dimensional random variables $\{X_i, i = 1, \ldots, 4\}$, the random variable X_i may follow different marginal distributions as follows:

$u_1 = F_1(x_1); u_2 = F_2(x_2); u_3 = F_3(x_3); u_4 = F_4(x_4)$. As an illustration , we can say,

$$X_1 \sim \exp(\lambda_1) \Rightarrow u_1 = F_1(x_1) = 1 - \exp(-\lambda_1 x_1);$$

$$X_2 \sim gamma(\alpha, \beta) \Rightarrow u_2 = F_2(x_2) = \frac{1}{\Gamma(\alpha)} \gamma(\beta x_2);$$

$$X_3 \sim logistic(a, b) \Rightarrow u_3 = F_3(x_3) = \frac{1}{1 + \exp\left(\frac{x - a}{b}\right)};$$

$$X_4 \sim Pearson\ III\ (c, \alpha, \beta) \Rightarrow u_4 = F_4(x_4) = \frac{1}{\Gamma(\alpha)} \gamma(\beta(x - c)).$$

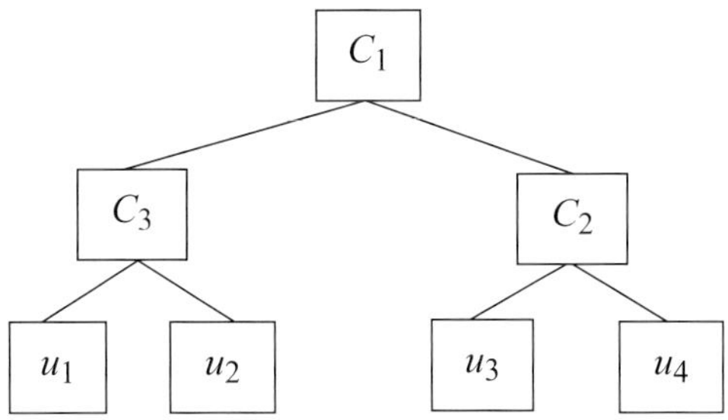

Figure 5.4 Partially nested Archimedean construction.

5.2.2 Partially Nested Archimedean Copulas (PNAC)

Originally, Joe (1997) proposed the structure of PNAC as an alternative approach for FNAC. PNAC may be considered a composite of EAC and FNAC (Berg and Aas, 2007).Similar to FNAC, PNAC also has $d - 1$ bivariate copulas that are partially exchangeable. As a simple example, Figure 5.4 illustrates the PNAC structure for four-dimensional random variables: (1) couple the two pairs (u_1, u_2) and (u_3, u_4) with copula C_3 with parameter θ_3 and C_2 with parameter θ_2, respectively, at the first level; and (2) the third copula C_1 with parameter θ_1 will be applied to couple C_2 and C_3 at the second level (Berg and Aas, 2007). Figure 5.4 also shows (1) exchangeability between u_1 and u_2, as well as between u_3 and u_4; and (2) four pairs $(u_1, u_3), (u_1, u_4), (u_2, u_3)$, and (u_2, u_4) all have copula C_1. Furthermore, the same constraints on parameters for FNAC are required to be satisfied for PNAC (Berg and Aas, 2007), i.e., (i) PNAC may be used to model the positively dependent variables, and (ii) the dependence decreases with the increase of nesting levels (i.e., the parameters of a higher level are smaller than those of a lower level).

Example 5.7 **Using the bivariate Frank copula as the building block to derive a four-dimensional PNAC function for the structure given in Figure 5.4.**

Solution: As shown in Figure 5.4, (u_1, u_2) and (u_3, u_4) can be represented through the Frank copula as follows:

$$C_3(u_1, u_2; \theta_3) = -\frac{1}{\theta_3} \ln\left(1 - \frac{\left(1 - e^{-\theta_3 u_1}\right)\left(1 - e^{-\theta_3 u_2}\right)}{e^{-\theta_3}}\right)$$

$$C_4(u_3, u_4; \theta_2) = -\frac{1}{\theta_2} \ln\left(1 - \frac{\left(1 - e^{-\theta_2 u_3}\right)\left(1 - e^{-\theta_2 u_4}\right)}{e^{-\theta_2}}\right)$$

Then C_1 can be represented through C_3, C_2 as follows:

$$C(u_1,u_2,u_3,u_4;\theta_1,\theta_2,\theta_3)$$
$$=C_1(C_3,C_2;\theta_1)$$
$$=-\frac{1}{\theta_1}\ln\left(1-\frac{\left(1-e^{-\theta_1 C_3}\right)\left(1-e^{-\theta_1 C_2}\right)}{1-e^{-\theta_1}}\right)$$
$$=-\frac{1}{\theta_1}\ln\left(1-\frac{\left(1-\left(1-\frac{\left(1-e^{-\theta_3 u_1}\right)\left(1-e^{-\theta_3 u_2}\right)}{1-e^{-\theta_3}}\right)^{\frac{\theta_1}{\theta_3}}\right)\left(1-\left(1-\frac{\left(1-e^{-\theta_2 u_3}\right)\left(1-e^{-\theta_2 u_4}\right)}{1-e^{-\theta_2}}\right)^{\frac{\theta_1}{\theta_2}}\right)}{1-e^{-\theta_1}}\right)$$

with the parameters: $0 \leq \theta_1 \leq \theta_2, \theta_3$.

In the same manner for FNAC, random variables $\{X_1 : i = 1,2,3,4\}$ may follow different marginal distributions as $u_i = F_i(x_i)$.

5.2.3 General Case

Originating in Joe (1997), the general nested Archimedean copula (GNAC) construction was further developed by Whelan (2004) and Savu and Trede (2006). Savu and Trede (2006) first introduced the notation for arbitrary nesting and the procedure for calculating the d-dimensional probability density function in general. To build a hierarchy of Archimedean copulas, they also applied the notation for the hierarchical Archimedean copula for GNAC. The main idea of the generally nested Archimedean construction is presented in this section (Berg and Aas, 2007).

For the GNAC with L levels, there are n_l distinct objects (an object is either a copula or a variable) at each level l. At level $l = 1$, variables $u_1, \ldots, u_d$ are grouped into n_1 exchangeable multivariate Archimedean copulas. These copulas are, in turn, coupled with n_2 copula at level $l = 2$, and so on. Berg and Aas (2007) presented an example of a nine-dimensional copula to explain this structure (Figure 5.5).

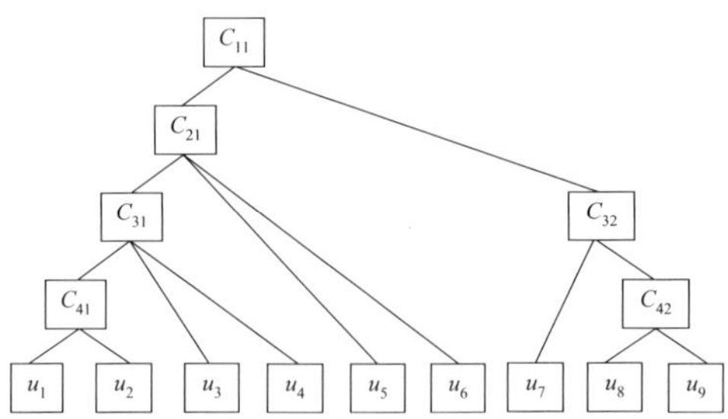

Figure 5.5 Hierarchically nested Archimedean copula construction.

Following Figure 5.5, the nine-dimensional copula can be written as

$$C(u_1, \ldots, u_9) = C_{11}\Big(C_{21}(C_{31}(C_{41}(u_1, u_2), u_3, u_4), u_5, u_6), C_{32}\big(u_7, C_{42}(u_8, u_9)\big)\Big). \tag{5.10}$$

At the first level, there are two two-dimensional EACs, i.e., $C_{41}(u_1, u_2)$ with parameter θ_{41} and $C_{42}(u_8, u_9)$ with parameter θ_{42}. There are one three-dimensional and one two-dimensional EACs at the second level, i.e., $C_{31}(C_{41}, u_3, u_4)$ with parameter θ_{31} and $C_{32}(u_7, C_{42})$ with parameter θ_{32}. At the third level, there is only one copula, $C_{21}(C_{31}, u_5, u_6)$ with parameter θ_{21}. At the top (fourth) level, the copula C_{11}, with parameter θ_{11}, is applied to model the dependence between C_{21} and C_{32}.

To ensure that GNAC is a valid Archimedean copula, there are a number of conditions that need to be satisfied (Savu and Trede, 2006; Berg and Aas, 2007):

a. The number of copulas must decrease with the increasing level of nesting. The top level may contain only one copula, and the inverse of the generating functions (ϕ^{-1}) must be completely monotonic.
b. The dependence of GNAC must decrease with the increasing level of nesting. For example, in Figure 5.5, parameters must be stratified following the condition $\theta_{41} \geq \theta_{32} \geq \theta_{21} \geq \theta_{11}$ and $\theta_{42} \geq \theta_{32} \geq \theta_{11}$. However, when mixing copula generators that belong to different Archimedean copula families, this requirement might not be sufficient. Two Archimedean copulas from different families (i.e., Fam1 and Fam2) can only be nested if the derivative of the product $\phi_1 \circ \phi_2^{-1}$ is completely monotonic. Joe (1997) presented details about copula families that can be mixed and explored structures where all the generators are from the same family are explored, and the other structures are still not fully explored.

5.2.4 Parameter Estimation for Nested Copulas

For NAC with an explicit density expression, the maximum likelihood estimation method is commonly applied to estimate the copula parameters; however, the NAC density function may not be straightforwardly derived. Savu and Trede (2006) proposed a recursive approach to derive the density function for general NAC. With this approach, the number of computational steps for evaluating the density increases rapidly with the copula complexity, and parameter estimation becomes very time consuming in higher dimensions (Savu and Trede, 2006; Berg and Aas, 2007).

The density function of NAC can be derived using the chain rule as discussed by Savu and Trede (2006). We will use the following examples to illustrate the general procedure on how to apply the chain rule. Furthermore, we derive the density functions for the M3, M4, M5, M6, and M12 copulas (Joe, 1997) in the appendix as specific examples.

Example 5.8 **Derive the density function for three-dimensional FNAC (Equation (5.4) corresponding to Figure 5.2).**

Solution: Equation (5.4) may be rewritten as follows:
$C(u_1,u_2,u_3) = C_1(C_2(u_1,u_2),u_3)$ and its density, i.e., $c(u_1,u_2,u_3)$, may be derived as follows:

$$\frac{\partial C(u_1,u_2,u_3)}{\partial u_1} = \frac{\partial C_1(C_2(u_1,u_2),u_3)}{\partial u_1} = \frac{\partial C_1}{\partial C_2}\frac{\partial C_2}{\partial u_1};\ \frac{\partial C^2(u_1,u_2.u_3)}{\partial u_1 \partial u_2} = \frac{\partial^2 C_1}{\partial C_2^2}\frac{\partial C_2}{\partial u_2}\frac{\partial C_2}{\partial u_1} + \frac{\partial^2 C_1}{\partial C_2}\frac{\partial^2 C_2}{\partial u_1 \partial u_2}$$

Finally, we have the following:

$$c(u_1,u_2,u_3) = \frac{\partial^3 C(u_1,u_2,u_3)}{\partial u_1 \partial u_2 \partial u_3} = \frac{\partial^3 C_1}{\partial C_2^2 \partial u_3}\frac{\partial C_2}{\partial u_2}\frac{\partial C_2}{\partial u_1} + \frac{\partial^2 C_1}{\partial C_2 \partial u_3}\frac{\partial^2 C_2}{\partial u_1 \partial u_2}$$

Example 5.9 **Derive the density function for four-dimensional FNAC (i.e., Equation (5.1) corresponding to Figure 5.1).**

Solution: Following Equation (5.1) and Figure 5.1, we have the following:
$C(u_1,u_2,u_3,u_4) = C_1(u_4,C_2) = C_1(u_4,C_2(u_3,C_3(u_1,u_2)))$ and its density $c(u_1,u_2,u_3,u_4)$ may be derived as follows:

$$\frac{\partial C(u_1,u_2,u_3,u_4)}{\partial u_1} = \frac{\partial C_1\left(u_4, C_2\left(u_3, C_3(u_1,u_2)\right)\right)}{\partial u_1} = \frac{\partial C_1}{\partial C_2}\frac{\partial C_2}{\partial C_3}\frac{\partial C_3}{\partial u_1}$$

$$\frac{\partial^2 C(u_1,u_2,u_3,u_4)}{\partial u_1 \partial u_2} = \frac{\partial^2 C_1}{\partial C_2^2}\left(\frac{\partial C_2}{\partial C_3}\right)^2\left(\frac{\partial C_3}{\partial u_1}\right)\left(\frac{\partial C_3}{\partial u_2}\right) + \frac{\partial C_1}{\partial C_2}\frac{\partial^2 C_2}{\partial C_3^2}\frac{\partial C_3}{\partial u_1}\frac{\partial C_3}{\partial u_2} + \frac{\partial C_1}{\partial C_2}\frac{\partial C_2}{\partial C_3}\frac{\partial^2 C_3}{\partial u_1 \partial u_2}$$

$$\begin{aligned}
\frac{\partial^3 C(u_1,u_2,u_3,u_4)}{\partial u_1 \partial u_2 \partial u_3} &= \frac{\partial^3 C_1}{\partial C_2^3}\left(\frac{\partial C_2}{\partial C_3}\right)^2\frac{\partial C_2}{\partial u_3}\frac{\partial C_3}{\partial u_1}\frac{\partial C_3}{\partial u_2} + 2\frac{\partial^2 C_1}{\partial C_2^2}\frac{\partial C_2}{\partial C_3}\frac{\partial^2 C_2}{\partial C_3 \partial u_3}\frac{\partial C_3}{\partial u_1}\frac{\partial C_3}{\partial u_2} \\
&= \frac{\partial^2 C_1}{\partial C_2^2}\frac{\partial C_2}{\partial u_3}\frac{\partial^2 C_2}{\partial C_3^2}\frac{\partial C_3}{\partial u_1}\frac{\partial C_3}{\partial u_2} \\
&= \frac{\partial C_1}{\partial C_2}\frac{\partial^3 C_2}{\partial C_3^2 \partial u_3}\frac{\partial C_3}{\partial u_1}\frac{\partial C_3}{\partial u_2} + \frac{\partial^2 C_1}{\partial C_2^2}\frac{\partial C_2}{\partial u_3}\frac{\partial C_2}{\partial C_3}\frac{\partial^2 C_3}{\partial u_1 \partial u_2} + \frac{\partial C_1}{\partial C_2}\frac{\partial^2 C_2}{\partial C_3 \partial u_3}\frac{\partial^2 C_3}{\partial u_1 \partial u_2}
\end{aligned}$$

Finally, we have the following:

$$\begin{aligned}
c(u_1,u_2,u_3,u_4) &= \frac{\partial^4 C(u_1,u_2,u_3,u_4)}{\partial u_1 \partial u_2 \partial u_3 \partial u_4} \\
&= \frac{\partial^4 C_1}{\partial C_2^3 \partial u_4}\left(\frac{\partial C_2}{\partial C_3}\right)^2\frac{\partial C_2}{\partial u_3}\frac{\partial C_3}{\partial u_1}\frac{\partial C_3}{\partial u_2} + 2\frac{\partial^3 C_1}{\partial C_2^2 \partial u_4}\frac{\partial C_2}{\partial C_3}\frac{\partial^2 C_2}{\partial C_3 \partial u_3}\frac{\partial C_3}{\partial u_1}\frac{\partial C_3}{\partial u_2} \\
&\quad + \frac{\partial^3 C_1}{\partial C_2^2 \partial u_4}\frac{\partial C_2}{\partial u_3}\frac{\partial^2 C_2}{\partial C_3^2}\frac{\partial C_3}{\partial u_1}\frac{\partial C_3}{\partial u_2} + \frac{\partial^2 C_1}{\partial C_2 \partial u_4}\frac{\partial^3 C_2}{\partial C_3^2 \partial u_3}\frac{\partial C_3}{\partial u_1}\frac{\partial C_3}{\partial u_2} \\
&\quad + \frac{\partial^3 C_1}{\partial C_2^2 \partial u_4}\frac{\partial C_2}{\partial u_3}\frac{\partial C_2}{\partial C_3}\frac{\partial^2 C_3}{\partial u_1 \partial u_2} + \frac{\partial^2 C_1}{\partial C_2 \partial u_4}\frac{\partial^2 C_2}{\partial C_3 \partial u_3}\frac{\partial^2 C_3}{\partial u_1 \partial u_2}
\end{aligned}$$

Example 5.10 **Derive the density function for the copula function represented by Figure 5.4.**

Solution: According to Figure 5.4, we have the following:
$C(u_1,u_2,u_3,u_4) = C_1(C_3(u_1,u_2),C_2(u_3,u_4))$. Then its density function $c(u_1,u_2,u_3,u_4)$ may be expressed as follows:

$$\partial C(u_1,u_2,u_3,u_4) = \frac{\partial C_1}{\partial C_3}\frac{\partial C_3}{\partial u_1}$$

$$\frac{\partial^2 C(u_1,u_2,u_3,u_4)}{\partial u_1 \partial u_2} = \frac{\partial^2 C_1}{\partial C_3^2}\frac{\partial C_3}{\partial u_2}\frac{\partial C_3}{\partial u_1} + \frac{\partial C_1}{\partial C_3}\frac{\partial^2 C_3}{\partial u_1 \partial u_2}$$

$$\frac{\partial^3 C(u_1,u_2,u_3,u_4)}{\partial u_1 \partial u_2 \partial u_3} = \frac{\partial^3 C_1}{\partial C_3^2 \partial C_2}\frac{\partial C_2}{\partial u_3}\frac{\partial C_3}{\partial u_2}\frac{\partial C_3}{\partial u_1} + \frac{\partial^2 C_1}{\partial C_3 \partial C_2}\frac{\partial C_2}{\partial u_3}\frac{\partial^2 C_3}{\partial u_1 \partial u_2}$$

Finally, we have the following:

$$\begin{aligned} c(u_1,u_2,u_3,u_4) &= \frac{\partial^4 C(u_1,u_2,u_3,u_4)}{\partial u_1 \partial u_2 \partial u_3 \partial u_4} \\ &= \frac{\partial^4 C_1}{\partial C_3^2 \partial C_2^2}\frac{\partial C_2}{\partial u_4}\frac{\partial C_2}{\partial u_3}\frac{\partial C_3}{\partial u_2}\frac{\partial C_3}{\partial u_1} + \frac{\partial^3 C_1}{\partial C_3^2 \partial C_2}\frac{\partial^2 C_2}{\partial u_3 \partial u_4}\frac{\partial C_3}{\partial u_2}\frac{\partial C_3}{\partial u_1} \\ &\quad + \frac{\partial^3 C_1}{\partial C_3 \partial C_2^2}\frac{\partial C_2}{\partial u_4}\frac{\partial C_2}{\partial u_3}\frac{\partial^2 C_3}{\partial u_1 \partial u_2} + \frac{\partial^2 C_1}{\partial C_3 \partial C_2}\frac{\partial^2 C_2}{\partial u_3 \partial u_4}\frac{\partial^2 C_3}{\partial u_1 \partial u_2} \end{aligned}$$

With the copula density function derived, we can then apply MLE to estimate parameters simultaneously with the constraints of parameters at a lower level being larger than those at a higher level. However, the copula parameters may also be estimated sequentially with the use of MLE as follows:

i. Estimate the copula parameter at the lowest level.
ii. Estimate the copula parameter for the second-lowest level by fixing the parameters estimated for the lowest level.
iii. Repeat the preceding steps until we reach the top level of the NAC structure.

5.2.5 Simulation for Nested Copulas

In the previous chapters, we have shown that EAC may be simulated with several methods, such as Laplace transform (LT) and CPI Rosenblatt's transform, and through its unique generating function ϕ with a simple algorithm. Frees and Valdez (1998) showed how to use the LT method to simulate NACs for the generators taken from either the Gumbel–Hougaard or the Clayton copula family. However, Berg and Aas (2007) have pointed out that the LT method is limited to the copulas such that we can find a distribution that equals the LT of the inverse generating function and from which we can easily sample. In most cases, the LT method needs to obtain the $d-1$ first derivatives of the copula function, which usually yield extremely complex expressions under higher-order derivatives. The limitation of LT method may cause the simulation to become inefficient for high dimensions (Berg and Aas, 2007).

Compared to the LT method, the CPI Rosenblatt transform method is more universal and will be introduced to simulate from NAC. Let $X = \{X_1, X_2, \ldots, X_d\}$ be a d-dimensional random vector with marginal distributions $F(x_i)$ and conditional distributions $F(x_i|x_1, \ldots, x_{i-1}), i = 1, \ldots, d$. The CPI Rosenblatt's transform of X is defined as $T(X) = \{T(X_1), \ldots, T(X_d)\}$:

$$T(X_1) = F_1(x_1), T(X_2) = F_{2|1}(x_2|x_1), \ldots, T(X_d) = F_{d|1,2,\ldots,d-1}(x_d|x_1, x_2, \ldots, x_{d-1}). \tag{5.11}$$

With the use of CPI method, random variables are simulated with the following procedure:

i. Generate $W = \{w_1, w_2, \ldots, w_d\}$ independent random variables following the uniform distribution [0, 1].
ii. Set $x_1 = w_1$.
iii. Set $w_2 = T(X_2) = F_{2|1}(x_2|x_1)$ to obtain $x_2 = F_{2|1}^{-1}(w_2|x_1)$.
iv. Set $w_3 = T(X_3) = F_{3|1,2}(w_3|x_1, x_2)$ to obtain $x_3 = F_{3|1,2}^{-1}(w_3|x_1, x_2)$.

...

Set $w_d = T(X_d) = F_{d|1,2,\ldots d-1}(w_d|x_1, x_2, \ldots, x_d)$.

Example 5.11 **Assuming the pseudo-observations given in Table 5.1 may be modeled with the M6 copula, (1) estimate the copula parameters both simultaneously and sequentially using MLE; and (2) simulate the random variables with a sample size of 50.**

Table 5.1. *Trivariate pseudo-observations.*

	u_1	u_2	u_3
1	0.241	0.138	0.103
2	0.241	0.172	0.172
3	0.241	0.241	0.276
4	0.241	0.586	0.655
5	0.793	0.828	0.897
6	0.483	0.345	0.379
7	0.931	0.914	0.621
8	0.724	0.759	0.724
9	0.414	0.621	0.586
10	0.759	0.414	0.310
11	0.862	0.793	0.793
12	0.655	0.517	0.448

Table 5.1. (*cont.*)

	u_1	u_2	u_3
13	0.414	0.379	0.552
14	0.569	0.448	0.414
15	0.569	0.690	0.690
16	0.414	0.310	0.241
17	0.241	0.552	0.862
18	0.069	0.035	0.035
19	0.241	0.276	0.345
20	0.069	0.069	0.069
21	0.897	0.914	0.931
22	0.655	0.655	0.483
23	0.069	0.103	0.138
24	0.241	0.207	0.207
25	0.655	0.724	0.759
26	0.517	0.483	0.517
27	0.828	0.862	0.828
28	0.966	0.966	0.966

Solution: Estimate the copula parameters.

To estimate the parameters for the fitted M6 copula, we use Figure 5.2 as the FNAC scheme.

- Estimate the copula parameters simultaneously.

  To estimate the copula parameters simultaneously, the copula density function (i.e., Equation (M6–3) in the appendix) is applied to write the log-likelihood function as follows:

$$
\begin{aligned}
\log L = \sum \frac{1}{u_1 u_2 u_3} (-\ln u_1)^{\theta_2 - 1} (-\ln u_2)^{\theta_2 - 1} (-\ln u_3)^{\theta_3 - 1} e^{-w^{\frac{1}{\theta_1}}} \Big(G^{\frac{2\theta_1}{\theta_2} - 2} w^{\frac{3}{\theta_1} - 3} \\
+ (2\theta_1 - 2) w^{\frac{2}{\theta_1} - 2} + (\theta_2 - \theta_1) G^{\frac{\theta_1}{\theta_2} - 2} w^{\frac{2}{\theta_1} - 2} + (\theta_1 - 1)(2\theta_1 - 1) G^{\frac{2\theta_1}{\theta_2} - 2} w^{\frac{1}{\theta_1} - 3} \\
+ (\theta_1 - 1) G^{\frac{2\theta_1}{\theta_2} - 2} w^{\frac{2}{\theta_1} - 3} + (\theta_1 - 1)(\theta_2 - \theta_1) G^{\frac{\theta_1}{\theta_2} - 2} w^{\frac{1}{\theta_1} - 2} \Big)
\end{aligned}
$$

  where $G = (-\ln u_1)^{\theta_2} + (-\ln u_2)^{\theta_2}; w = (-\ln u_3)^{\theta_1} + \left[(-\ln u_1)^{\theta_2} + (-\ln u_2)^{\theta_1}\right]^{\frac{\theta_1}{\theta_2}}$.

  The parameter constraint is given as $1 \le \theta_1 \le \theta_2$, where θ_2 corresponds to the parameters for the first level.

  Maximizing the log-likelihood function numerically (e.g., using genetic algorithm ga function in MATLAB), the parameters are estimated as follows:

$$\theta_2 = 4.4158; \theta_1 = 3.3532.$$

  It is worth noting that to properly estimate the parameters simultaneously, the linear constraint needs to be applied with vector A = [–1,1] B = 0, which represents $-\theta_2 + \theta_1 \le 0$.

- Estimate the copula parameters sequentially.

  To estimate the copula parameters sequentially, the density function for the bivariate Gumbel–Hougaard copula is applied (Chapter 4).

  Step 1: Maximizing the log-likelihood function for (u_1, u_2), we have $\theta_2 = 4.4682$.

  Step 2: Compute $C(u_1, u_2; \theta_2 = 4.4682)$ and estimate the parameter for $(u_3, C(u_1, u_2; \theta_2 = 4.4682))$. Again using MLE, we have $\theta_1 = 3.2088$. It is worth noting that to estimate the parameter (i.e., the Gumbel–Hougaard copula) for the top level, the lower and upper bounds are $[1, \theta_2]$.

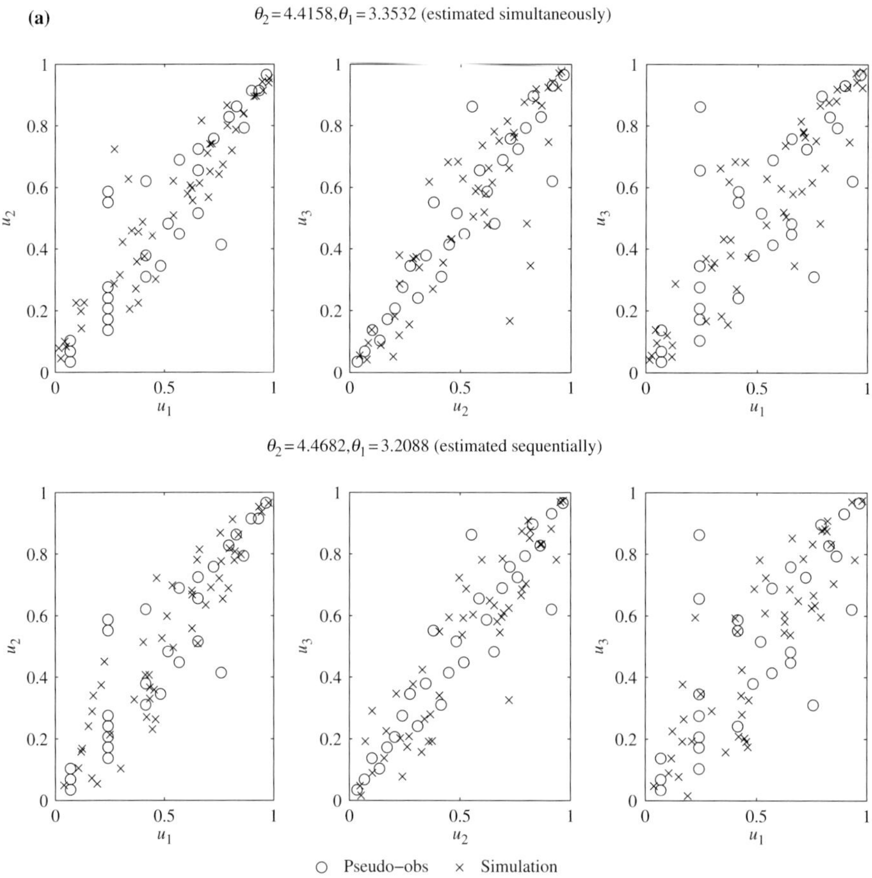

Figure 5.6 (a) Comparison of pseudo-observations with those simulated from M6 copula;

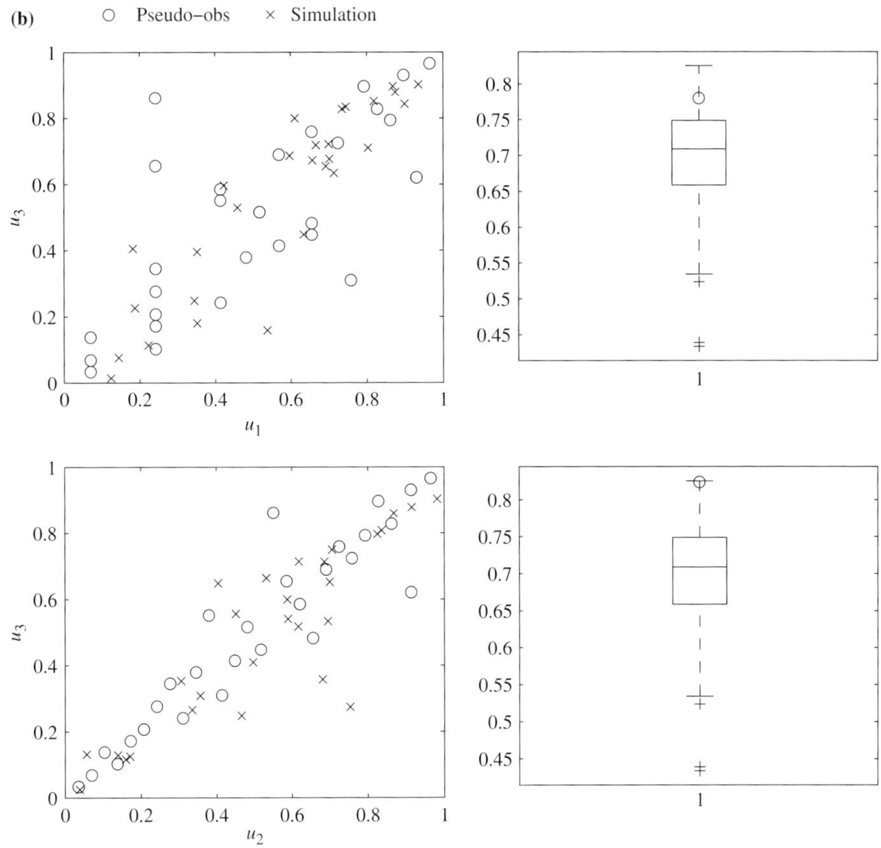

Figure 5.6 (*cont.*) (b) simulation comparison from the Gumbel–Hougaard copula with parameter $\boldsymbol{\theta}_1$ for $(\boldsymbol{u}_1, \boldsymbol{u}_3), (\boldsymbol{u}_2, \boldsymbol{u}_3)$ directly; (c) comparison of sample Kendall's tau with simulated Kendall's tau from Gumbel–Hougaard copula with parameter $\boldsymbol{\theta} = 2.8816$.

Finally, for both simultaneous and sequential estimation, the parameters estimated are coded as follows:

$param = [param(1), param(2)] = [\theta_2, \theta_1]$; param(1) and param(2) represents bottom and top levels, respectively.

- Simulation from the fitted M6 copula.

  As discussed previously, the random variates are simulated using the CPI Rosenblatt transform, as shown in Figure 5.6(a).

In addition, we have discussed previously that $[u_1, u_3]$ and $[u_2, u_3]$ may be modeled with the Gumbel–Hougaard copula with parameter θ_1. Figure 5.6(b) compares the simulation as well as the box plot of simulated and sample Kendall's tau (100 simulations with a sample size of 28).

Example 5.12 **Assuming the Gumbel–Hougaard copula may be applied as a biviarate building block, and using the scheme shown in Figure 5.4 and the pseudo-observations listed in Table 5.2, (1) estimate the copula parameters; and (2) simulate random variates with fitted copula for a sample size of 100.**

Table 5.2. *Pseudo-observations for Example 5.12.*

	u_1	u_2	u_3	u_4
1	0.194	0.338	0.421	0.545
2	0.819	0.901	0.743	0.705
3	0.614	0.639	0.615	0.662
4	0.235	0.208	0.298	0.292
5	0.792	0.755	0.865	0.894
6	0.433	0.517	0.559	0.480
7	0.130	0.197	0.095	0.087
8	0.570	0.583	0.802	0.680
9	0.128	0.274	0.256	0.137
10	0.218	0.116	0.262	0.481
11	0.468	0.367	0.367	0.439
12	0.490	0.434	0.391	0.515
13	0.194	0.083	0.019	0.042
14	0.120	0.227	0.178	0.289
15	0.676	0.601	0.759	0.673
16	0.990	0.990	0.991	0.993
17	0.657	0.777	0.942	0.950
18	0.226	0.174	0.284	0.134
19	0.828	0.857	0.836	0.916
20	0.373	0.367	0.151	0.249
21	0.698	0.656	0.727	0.584
22	0.645	0.738	0.641	0.787
23	0.025	0.051	0.034	0.199
24	0.298	0.300	0.470	0.394
25	0.906	0.936	0.955	0.950
26	0.658	0.476	0.556	0.647
27	0.302	0.158	0.224	0.105
28	0.581	0.393	0.733	0.779
29	0.371	0.433	0.179	0.145
30	0.169	0.537	0.213	0.344
31	0.041	0.083	0.009	0.059
32	0.982	0.978	0.928	0.935
33	0.585	0.162	0.326	0.312
34	0.618	0.753	0.661	0.633
35	0.280	0.622	0.400	0.574

Table 5.2. (*cont.*)

	u_1	u_2	u_3	u_4
36	0.902	0.969	0.879	0.904
37	0.440	0.648	0.587	0.811
38	0.243	0.147	0.281	0.524
39	0.044	0.081	0.177	0.052
40	0.122	0.149	0.229	0.180
41	0.497	0.645	0.528	0.545
42	0.701	0.644	0.745	0.599
43	0.323	0.538	0.806	0.796
44	0.013	0.044	0.063	0.041
45	0.651	0.721	0.774	0.646
46	0.190	0.298	0.773	0.841
47	0.520	0.772	0.636	0.542
48	0.926	0.943	0.900	0.812
49	0.468	0.447	0.518	0.633
50	0.868	0.894	0.893	0.905
51	0.422	0.710	0.727	0.560
52	0.888	0.835	0.868	0.823
53	0.372	0.590	0.734	0.792
54	0.132	0.116	0.095	0.041
55	0.429	0.288	0.219	0.125
56	0.390	0.366	0.375	0.172
57	0.983	0.986	0.991	0.990
58	0.980	0.988	0.976	0.974
59	0.308	0.318	0.147	0.193
60	0.932	0.913	0.943	0.933

Solution:

1. Estimate the copula parameters.

   According to Figure 5.4, let us use θ_{12}, θ_{34} to represent the copula parameters of $[u_1, u_2], [u_3, u_4]$ at the bottom level and θ to represent the copula parameter at the top level.

   - Estimate the parameters simultaneously.

     Given the Gumbel–Hougaard copula as a bivariate building block, the copula density function for the four-dimensional PNAC Gumbel–Hougaard copula may be derived based on the chain rule following the procedure given in Example 5.10. With the parameter constraints $1 \le \theta \le \theta_{12}, \theta_{34}$, i.e.,

     $\begin{cases} \theta - \theta_{12} \le 0 \\ \theta - \theta_{34} \le 0 \end{cases}$, the inequality vector is then given as $A = \begin{bmatrix} -1, 0, 1 \\ 0, -1, 1 \end{bmatrix}, B = \begin{bmatrix} 0 \\ 0 \end{bmatrix}$, with the parameter set as param $= [\theta_{12}, \theta_{34}, \theta]$.

The parameters can be estimated numerically by maximizing the log-likelihood function with the preceding linear constraint as follows:

$$\theta_{12} = 3.6949,\ \theta_{34} = 4.5035,\ \theta = 2.8816.$$

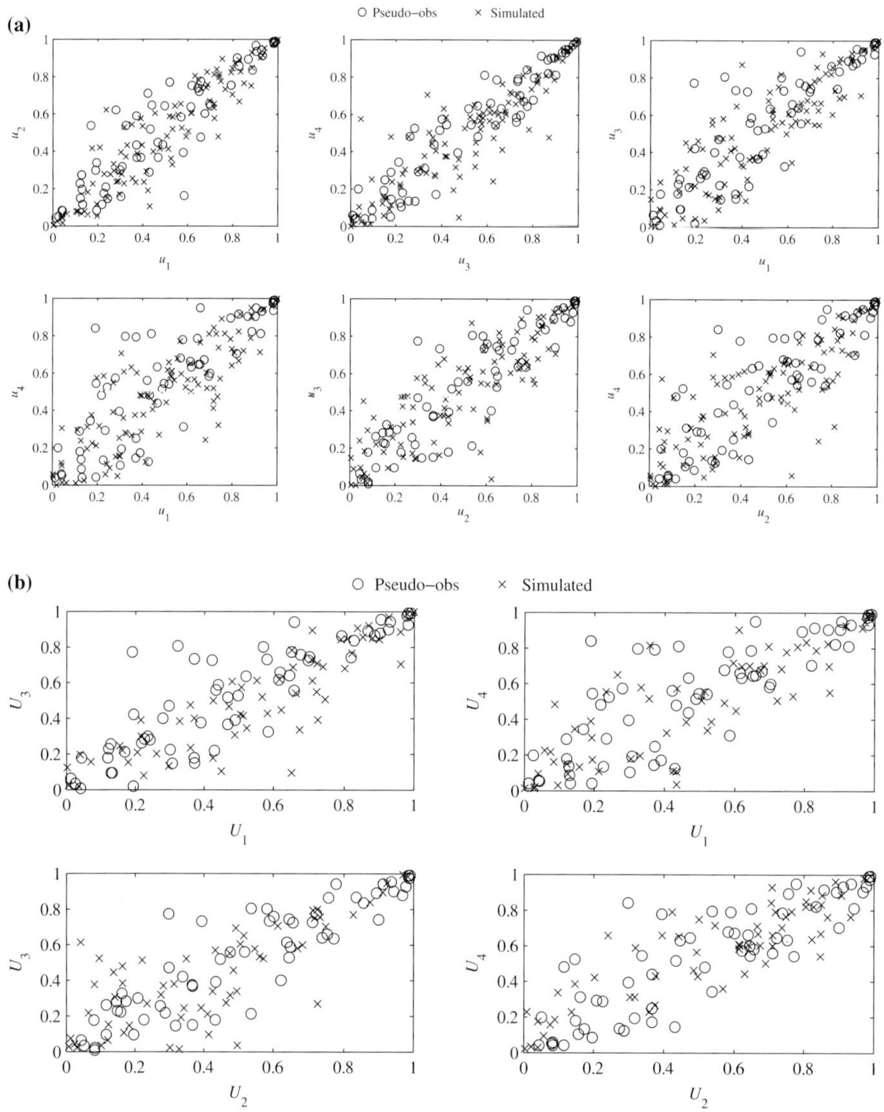

Figure 5.7 (a) Comparison of pseudo-observations with those simulated with the parameters estimated simultaneously ($\boldsymbol{\theta}_{12} = 3.6949, \boldsymbol{\theta}_{34} = 4.5035, \boldsymbol{\theta} = 2.8816$); (b) comparison of observed variables with simulated variables with $\boldsymbol{\theta} = 2.8816$; (c) comparison of sample Kendall's tau with the simulated Kendall's taus.

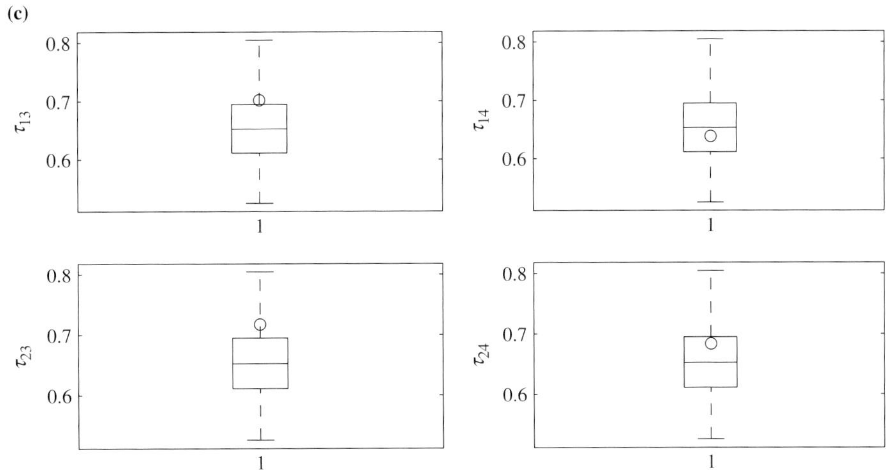

Figure 5.7 (*cont.*)

- Estimate the parameters sequentially.

  With the same estimation procedures shown in Example 5.11:

  The parameter for (u_1, u_2) is estimated as $\theta_{12} = 3.8545$.

  The parameter for (u_3, u_4) is estimated as $\theta_{34} = 4.3949$.

  The parameter for $\{C_3(u_1, u_2; \theta_{12}), C_2(u_3, u_4; \theta_{34})\}$ is estimated by fixing θ_{12}, θ_{34} as $\theta = 3.3297$.

2. Simulate random variates.

   Using the CPI Rosenblatt transform, Figure 5.7(a) compares the pseudo-observations in Table 5.2 with those simulated from the fitted PNAC Gumbel–Hougaard copula function.

As discussed previously for the PNAC structure, we know $(u_1, u_3), (u_1, u_4), (u_2, u_3)$, (u_2, u_4) should have the same joint distribution that may be modeled using the Gumbel–Hougaard copula with parameter at the top level, i.e., $\theta = 2.8816$ with the comparison of simulated random variable and Kendall's tau as shown in Figure 5.7(b) and 5.7(c). Figure 5.7(b) and 5.7(c) indicates that the preceding four pairs may be modeled using the same Gumbel–Hougaard copula.

5.3 Pair-Copula Construction (PCC)

PCCs are also hierarchical in nature. Compared to EAC and NAC, a large improvement is made in PCCs that allows for the free specification of $\frac{d(d-1)}{2}$ copulas. The modeling scheme of PCCs is based on the decomposition of a multivariate density function. The d-dimensional probability density function may be decomposed to $\frac{d(d-1)}{2}$ bivariate density functions, where the first $d - 1$ density functions are unconditional and the rest are conditional (Berg and Aas, 2007). First proposed by Joe (1997), there are two main types of PCCs, canonical (C)-vines and D-vines, in the literature (e.g., Bedford and Cooke, 2001, 2002; Kurowicka and Cooke, 2004, 2006; Aas et al., 2009).

5.3.1 Principle of Pair-Copula Decomposition of General Multivariate Distribution

Following Aas et al. (2009), we introduce the pair-copula decomposition of general multivariate distributions.

Let $\mathbf{X} = (X_1, X_2, \ldots, X_d)$ be a vector of random variables with a joint density function $f(x_1, \ldots, x_d)$. According to the conditional probability theory, the joint density function can be defined as follows:

$$f(x_1, x_2, \ldots, x_d) = f(x_1) f(x_2|x_1) \cdots f(x_d|x_1, \ldots, x_{d-1}) \tag{5.12}$$

In Chapters 3 and 4, the multivariate distribution F with marginals $F_1(x_1), \ldots, F_d(x_d)$ is defined using Sklar's theorem as follows:

$$F(x_1, \ldots, x_d) = C(F_1(x_1), \ldots, F_d(x_d)) \text{ or } C(u_1, \ldots, u_d) = F\left(F_1^{-1}(x_1), \ldots, F_d^{-1}(x_d)\right) \tag{5.13}$$

where $u_i = F_i(x_i)$; $F_i^{-1}(u_i)$ is the inverse distribution of marginal $F_i(x_i)$.

Then, for an absolutely continuous F with strictly increasing, continuous marginal probability densities $f_1(x_1), \ldots, f_d(x_d)$, applying $\frac{\partial^d}{\partial x_1 \ldots \partial x_d}$ to Equation (5.13), we have

$$f(x_1, \ldots, x_d) = \frac{\partial^d}{\partial F_1(x_1) \cdots \partial F_d(x_d)} C(F_1(x_1), \ldots, F_d(x_d)) \frac{\partial F_1(x_1)}{\partial x_1} \cdots \frac{\partial F_d(x_d)}{\partial x_d} \tag{5.14a}$$

$$f(x_1, \ldots, x_d) = c_{1,2,\ldots,d}(F_1(x_1), \ldots, F_d(x_d)) \prod_{i=1}^{d} f_i(x_i) \tag{5.14b}$$

where $c_{1,2,\ldots,d}(\cdot)$ stands for the d-dimensional copula density function.

In the bivariate case, Equation (5.14b) can be simplified to

$$f(x_1, x_2) = c_{12}(F_1(x_1), F_2(x_2)) f_1(x_1) f_2(x_2) \tag{5.15}$$

where $c_{12}(\cdot)$ is the appropriate pair-copula density.

Using the conditional probability in Equation (5.12), the conditional probability density function can be easily written as

$$f(x_1|x_2) = \frac{f(x_1, x_2)}{f_2(x_2)} = \frac{c_{12}(F_1(x_1), F_2(x_2)) f_1(x_1) f_2(x_2)}{f_2(x_2)} = c_{12}(F_1(x_1), F_2(x_2)) f_1(x_1) \tag{5.16}$$

Likewise, we have

$$f(x_{d-1}|x_d) = c_{d-1,d}(F_{d-1}(x_{d-1}), F_d(x_d)) f_{d-1}(x_{d-1}). \tag{5.17}$$

Similarly, in the trivariate case, we can obtain the conditional probability density function:

$$f(x_1|x_2,x_3)=\frac{f(x_1,x_2,x_3)}{f(x_2,x_3)}=\frac{f(x_3)f(x_1,x_2|x_3)}{f(x_3)f(x_2|x_3)}=\frac{f(x_1,x_2|x_3)}{f(x_2|x_3)} \tag{5.18}$$

According to the definition of conditional copula, we have

$$\begin{aligned} f(x_1,x_2|x_3) &= \frac{\partial^2 F(x_1,x_2|x_3)}{\partial x_1 \partial x_2} = \frac{\partial^2}{\partial x_1 \partial x_2} C_{12|3}\left(F_{1|3}(x_1|x_3), F_{2|3}(x_2|x_3)\right) \\ &= \frac{\partial^2 C_{12|3}\left(F_{1|3}(x_1|x_3), F_{2|3}(x_2|x_3)\right)}{\partial F_{1|3}(x_1|x_3)\partial F_{2|3}(x_2|x_3)} \frac{\partial F_{1|3}(x_1|x_3)}{\partial x_1} \frac{\partial F_{2|3}(x_2|x_3)}{\partial x_2} \\ &= c_{12|3}\left(F_{1|3}(x_1|x_3), F_{2|3}(x_2|x_3)\right) f_{1|3}(x_1|x_3) f_{2|3}(x_2|x_3) \end{aligned} \tag{5.19}$$

Thus,

$$\begin{aligned} f(x_1|x_2,x_3) &= \frac{f(x_1,x_2|x_3)}{f(x_2|x_3)} = \frac{c_{12|3}\left(F_{1|3}(x_1|x_3), F_{2|3}(x_2|x_3)\right) f(x_1|x_3) f(x_2|x_3)}{f(x_2|x_3)} \\ &= c_{12|3}\left(F_{1|3}, F_{2|3}\right) f_{1|3} \end{aligned} \tag{5.20}$$

Alternatively, $f(x_1|x_2,x_3)$ may be also written as follows:

$$f(x_1|x_2,x_3) = c_{13|2}\left(F_{1|2}, F_{3|2}\right) f_{1|2} \tag{5.21}$$

Equations (5.20) and (5.21) can be further decomposed as follows:

$$f(x_1|x_2,x_3) = c_{13|2}\left(F_{1|3}, F_{2|3}\right) c_{13}(F_1, F_3) f(x_1) \tag{5.22a}$$

$$f(x_1|x_2,x_3) = c_{13|2}\left(F_{1|2}, F_{3|2}\right) c_{12}(F_1, F_2) f(x_1) \tag{5.22b}$$

From the expression of the appropriate pair-copula, a conditional marginal density function can be expressed in a general form as follows:

$$f(x|\mathbf{v}) = c_{xv_j|\mathbf{v}_{-j}}\left(F\left(x|\mathbf{v}_{-j}\right), F\left(v_j|\mathbf{v}_{-j}\right)\right) f\left(x|\mathbf{v}_{-j}\right) \tag{5.23}$$

where $\mathbf{v}$ is a d-dimensional vector; v_j is one arbitrarily chosen component of $\mathbf{v}$; and $\boldsymbol{v}_{-j}$ denotes the $\mathbf{v}$ vector except v_j, i.e., $\boldsymbol{v}_{-j} = \boldsymbol{v} \backslash v_j$.

Under appropriate conditions, a multivariate probability density function may be expressed through the product of pair-copulas, acting on several different conditional probability distributions (Aas et al., 2009).

Joe (1997) showed a conditional marginal distribution for the appropriate pair-copula for every j as

$$F(x|\mathbf{v}) = \frac{\partial C_{x,v_j|\mathbf{v}_{-j}}\left(F\left(x|\mathbf{v}_{-j}\right), F\left(v_j|\mathbf{v}_{-j}\right)\right)}{\partial F\left(v_j|\mathbf{v}_{-j}\right)} \tag{5.24}$$

where $C_{x,v_j|\mathbf{v}_{-j}}$ is a bivariate copula function with the conditional marginals. For the special case where $\mathbf{v}$ is univariate, Equation (5.24) can be rewritten as follows:

$$F(x|v) = \frac{\partial C_{x,v}(F_X(x), F_V(v))}{\partial F_V(v)} \tag{5.25}$$

In Equation (5.25), when x and v are copula random variables (i.e., the margins following the uniform [0,1] as $f(x) = f(v) = 1, F_X(x) = x, F_V(v) = v$), Equation (5.25) can be rewritten as follows:

$$h(x, v, \Theta) = F(x|v) = \frac{\partial C_{x,v}(x, v; \Theta)}{\partial v} \tag{5.26}$$

where the second variable of $h(\cdot)$ function represents the conditional variable, and Θ denotes the set of copula parameters to model the joint distribution function of x and v. Letting $u = x$, Equation (5.26) is essentially the conditional copula function of $C(u|V = v; \Theta)$.

Example 5.13 **Derive the *h* function for the bivariate Gumbel–Hougaard copula.**

Solution: As seen in the previous chapters, the bivariate Gumbel–Hougaard copula can be written as follows:

$$C(u_1, u_2; \theta) = e^{-\left((-\ln u_1)^\theta + (-\ln u_2)^\theta\right)^{\frac{1}{\theta}}}$$

Then the h function, i.e., $h(u_1, u_2, \theta)$, can be expressed as follows:

$$\begin{aligned} h(u_1, u_2, \theta) = F(u_1|U_2 = u_2; \theta) &= \frac{\partial C(u_1, u_2; \theta)}{\partial u_2} \\ &= \frac{C(u_1, u_2)}{u_2} (-\ln u_2)^{\theta-1} \left((-\ln u_1)^\theta + (-\ln u_2)^\theta\right)^\wedge \left(\frac{1}{\theta} - 1\right) \\ &= \frac{e^{-\left((-\ln u_1)^\theta + (-\ln u_2)^\theta\right)^{\frac{1}{\theta}}} (-\ln u_2)^{\theta-1} \left((-\ln u_1)^\theta + (-\ln u_2)^\theta\right)^{\frac{1}{\theta}-1}}{u_2} \end{aligned}$$

5.3.2 Vines

High-dimensional distributions have a significant number of possible pair-copula constructions. The regular vine, introduced by Bedford and Cooke (2001, 2002), is used to organize the general structure and embrace a large number of possible pair-copula decompositions. Two special types of regular vines, the C-vine and the D-vine (Kurowicka and Cooke, 2004), are given in the form of a nested set of trees and are used to decompose the multivariate density function. Figure 5.8 shows one sample specification corresponding to a five-dimensional D-vine that can be explained with Table 5.3.

Table 5.3. *Five-dimensional D-vine.*

Tree T_j	Nodes	Edges
T_1	1, 2, 3, 4, 5	12, 23, 34, 45
T_2	12, 23, 34, 45	13|2, 24|3, 35|4
T_3	13|2, 24|3, 35|4	14|23, 25|34
T_4	14|23, 25|34	15|234

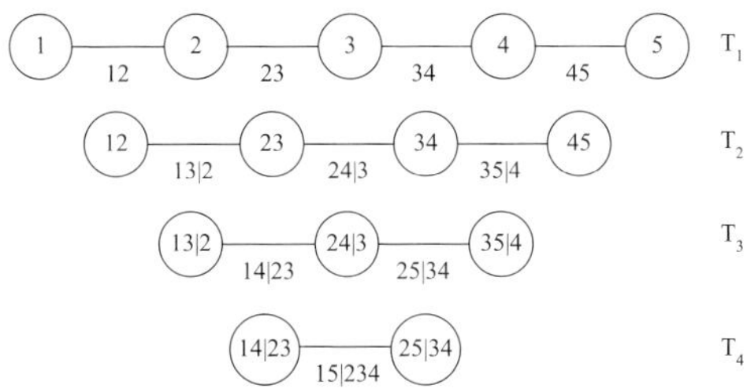

Figure 5.8 A D-vine with five variables, four trees, and 10 edges.

In Figure 5.8 and Table 5.3, each edge represents a pair-copula density, and the edge label corresponds to the subscript of the pair-copula density. For example, 14|23 corresponds to the copula density $c_{14|23}\left(C_{13|2}, C_{24|3}\right)$. The entire decomposition is defined by $\frac{d(d-1)}{2} = \frac{5(5-1)}{2} = 10$ edges as well as the density functions of random variables.

The density function of random variable $\boldsymbol{X} = \{X_1, X_2, \ldots, X_d\}$ with a D-vine copula can be written as

$$f(x_1, \ldots, x_d) = \prod_{k=1}^{d} f(x_k) \prod_{j=1}^{d-1} \prod_{i=1}^{d-j} c_{i,i+j|j+1,\ldots,i+j-1}\left(F\left(x_i|x_{i+1}, \ldots, x_{i+j-1}\right), F\left(x_{i+j}|x_{i+1}, \ldots, x_{i+j-1}\right)\right) \tag{5.27}$$

where index j identifies the trees, and i identifies the edges in each tree.

A sample of C-vine with five variables is given in Figure 5.9. The meanings of symbols are the same as in Figure 5.8. We can see that each tree T_j has a unique node connecting to $d - j$ edges in tree T_j. For example, node 1 of tree T_1 is connected to nodes 2, 3, 4, and 5 and forms the edges 12, 13, 14, and 15. Similarly, node 12 of T_2 is connected to nodes 13, 14, and 15 and forms the edges 23|1, 24|1 and 25|1.

In general, the d-dimensional density function corresponding to a C-vine is defined as

$$f(x_1, \ldots, x_d) = \prod_{k=1}^{d} f(x_k) \prod_{j=1}^{d-1} \prod_{i=1}^{d-j} c_{j,i+j|1,\ldots,j-1}\left(F\left(x_j|x_1, \ldots, x_{j-1}\right), F\left(x_{i+j}|x_1, \ldots, x_{j-1}\right)\right) \tag{5.28}$$

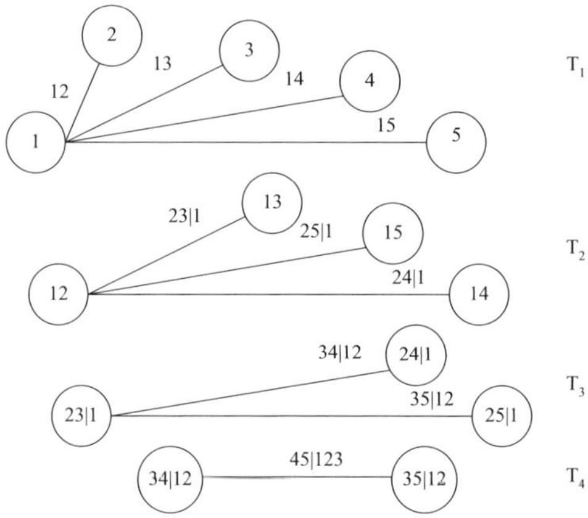

Figure 5.9 A C-vine with five variables, four trees, and 10 edges.

Looking at Figures 5.8 and 5.9, it is seen that the D-vine is more flexible than the C-vine. However, the C-vine might be advantageous if a particular variable is known to be the key variable governing interactions among the variables. In such a situation, one may decide to locate this variable at the root of the C-vine.

Following Aas et al. (2009), we present several typical pair-copulas.

Three Variables

For three-dimensional variables, there should be a total of six different pair-copula decompositions, including three D-vines and three C-vines. However, for three-dimensional variables, the D-Vine and C-vine are exactly the same, i.e., there are three different decompositions whose structures are both canonical vine and D-vine, as shown in Figure 5.10.

According to the decomposition schemes in Figure 5.10 and using Figure 5.10(a) as an example, the probability density function for both C-vine and D-vine structures can be written for three-dimensional random variables as

$$f(x_1, x_2, x_3) = \prod_{i=1}^{3} f_i(x_i) c_{12}(F_1(x_1), F_2(x_2)) c_{23}(F_2(x_2), F_3(x_3)) C_{13|2}\left(F_{1|2}(x_1|x_2), F_{3|2}(x_3|x_2)\right) \tag{5.29}$$

where f_1, f_2, f_3 and F_1, F_2, F_3 represent the univariate PDF and CDF for variables x_1, x_2, x_3, respectively.

Four Variables

For four-dimensional variables, we can construct a total of 24 different pair-copula decompositions, including 12 D-vines and 12 C-vines, as shown in Figure 5.11 (examples for one

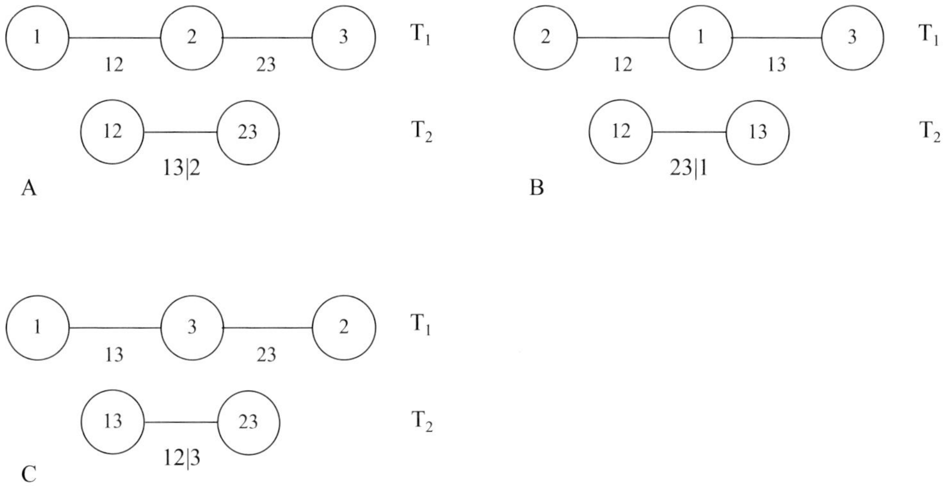

Figure 5.10 Decomposition schemes for three-dimensional variables using vines.

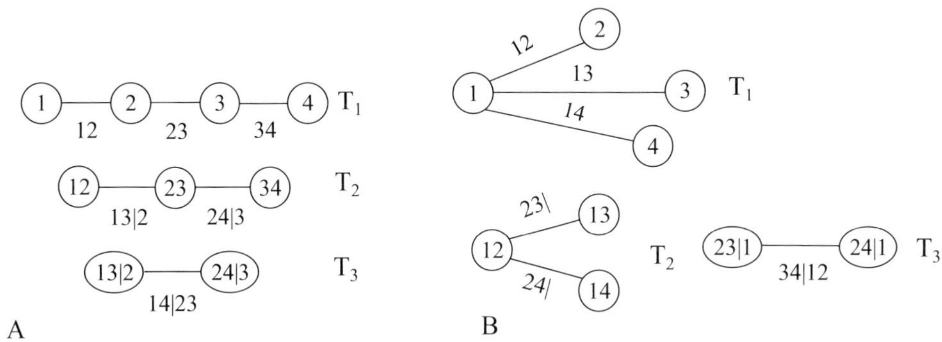

Figure 5.11 Vines for four-dimensional variables: (a) D-vine; (b) C-vine).

D-vine and one C-vine construction). Following the scheme, one may easily construct the rest D-vine and C-vine structures for four-dimensional variables.

According to Figure 5.11(a), the four-dimensional D-vine structure can be expressed as

$$
\begin{aligned}
f(x_1,x_2,x_3,x_4) = \prod\nolimits_{i=1}^{4} f_i(x_i)\cdot c_{12}(F_1(x_1),F_2(x_2))\cdot c_{23}(F_2(x_2),F_3(x_3))\cdot c_{34}(F_3(x_3),F_4(x_4)) \\
\cdot c_{13|2}\left(F_{1|2}(x_1|x_2),F_{3|2}(x_3|x_2)\right)\cdot c_{24|3}\left(F_{2|3}(x_2|x_3),F_{4|3}(x_4|x_3)\right) \\
\cdot c_{14|23}\left(F_{1|23}(x_1|x_2,x_3),F_{4|23}(x_4|x_2,x_3)\right)
\end{aligned}
\tag{5.30}
$$

and according to Figure 5.11(b), the four-dimensional C-vine structure can be expressed as follows:

$$f(x_1,x_2,x_3,x_4) = \prod_{i=1}^{4} f_i(x_i)\cdot c_{12}(F_1(x_1),F_2(x_2))\cdot c_{13}(F_1(x_1),F_3(x_3))\cdot c_{14}(F_1(x_1),F_4(x_4))$$
$$\cdot c_{23|1}\left(F_{2|1}(x_2|x_1),F_{3|1}(x_3|x_1)\right)\cdot c_{24|1}\left(F_{2|1}(x_2|x_1),F_{4|1}(x_4|x_1)\right)$$
$$\cdot c_{34|12}\left(F_{3|12}(x_3|x_1,x_2),F_{4|12}(x_4|x_1,x_2)\right) \quad (5.31)$$

Five Variables

For five-dimensional variables, there are 240 different possible pair-copula decompositions, including 60 C-vines (Figure 5.8, for example), 60 D-vines (Figure 5.9 is an example), and 120 other regular vine decompositions (Aas et al., 2009; shown in Figure 5.12 with two examples)

According to Figure 5.8, the general expression for the five-dimensional D-vine structure can be given as follows:

$$f(x_1,x_2,x_3,x_4,x_5) = f_1(x_1)f_2(x_2)f_3(x_3)f_4(f_4)f_5(x_5)c_{12}(F_1(x_1),F_2(x_2))$$
$$\cdot c_{23}(F_2(x_2),F_3(x_3))c_{34}(F_3(x_3),F_4(x_4))c_{45}(F_4(x_4),F_5(x_5))$$
$$\cdot c_{13|2}\left(F_{1|2}(x_1|x_2),F_{3|2}(x_3|x_2)\right)c_{24|3}\left(F_{2|3}(x_2|x_3),F_{4|3}(x_4|x_3)\right)$$
$$\cdot c_{35|4}\left(F_{3|4}(x_3|x_4),F_{5|4}(x_5|x_4)\right)c_{14|23}\left(F_{1|23}(x_1|x_2,x_3),F_{4|23}(x_4|x_2,x_3)\right)$$
$$\cdot c_{25|34}\left(F_{2|34}(x_2|x_3,x_4),F_{5|34}(x_5|x_3,x_4)\right)$$
$$\cdot c_{15|234}\left(F_{1|234}(x_1|x_2,x_3,x_4),F_{5|234}(x_5|x_2,x_3,x_4)\right) \quad (5.32)$$

According to Figure 5.9, the general expression for the five-dimensional C-vine structure can be given as

$$f(x_1,x_2,x_3,x_4,x_5) = f_1(x_1)f_2(x_2)f_3(x_3)f_4(x_4)f_5(x_5)c_{12}(F_1(x_1),F_2(x_2))c_{13}(F_1(x_1),F_3(x_3))$$
$$\cdot c_{14}(F_1(x_1),F_4(x_4))c_{15}(F_1(x_1),F_5(x_5))c_{23|1}\left(F_{2|1}(x_2|x_1),F_{3|1}(x_3|x_1)\right)$$
$$\cdot c_{24|1}\left(F_{2|1}(x_2|x_1),F_{4|1}(x_4|x_1)\right)c_{25|1}\left(F_{2|1}(x_2|x_1),F_{5|1}(x_5|x_1)\right)$$
$$\cdot c_{34|12}\left(F_{3|12}(x_3|x_1,x_2),F_{4|12}(x_4|x_1,x_2)\right)c_{35|12}\left(F_{3|12}(x_3|x_1,x_2),F_{5|12}(x_5|x_1,x_2)\right)$$
$$\cdot c_{45|123}\left(F_{4|123}(x_4|x_1,x_2,x_3),F_{5|123}(x_5|x_1,x_2,x_3)\right) \quad (5.33)$$

According to Figure 5.12(a), the density function for a five-dimensional regular vine structure can be expressed as follows:

$$f(x_1,x_2,x_3,x_4,x_5) = f_1(x_1)f_2(x_2)f_3(x_3)f_4(x_4)f_5(x_5)c_{12}(F_1(x_1),F_2(x_2))c_{25}(F_2(x_2),F_5(x_5))$$
$$\cdot c_{23}(F_2(x_2),F_3(x_3))c_{34}(F_3(x_3),F_4(x_4)),c_{15|2}\left(F_{1|2}(x_1|x_2),F_{5|2}(x_5|x_2)\right)$$
$$\cdot c_{13|2}\left(F_{1|2}(x_1|x_2),F_{3|2}(x_3|x_2)\right)c_{24|3}\left(F_{2|3}(x_2|x_3),F_{4|3}(x_4|x_3)\right)$$
$$\cdot c_{35|12}\left(F_{3|12}(x_3|x_1,x_2),F_{5|12}(x_5|x_1,x_2)\right)c_{14|23}\left(F_{1|23}(x_1|x_2,x_3),F_{4|23}(x_4|x_2,x_3)\right)$$
$$\cdot c_{45|123}\left(F_{4|123}(x_4|x_1,x_2,x_3),F_{5|123}(x_5|x_1,x_2,x_3)\right) \quad (5.34a)$$

According to Figure 5.12(b), the density function for the five-dimensional regular vine can be expressed as follows:

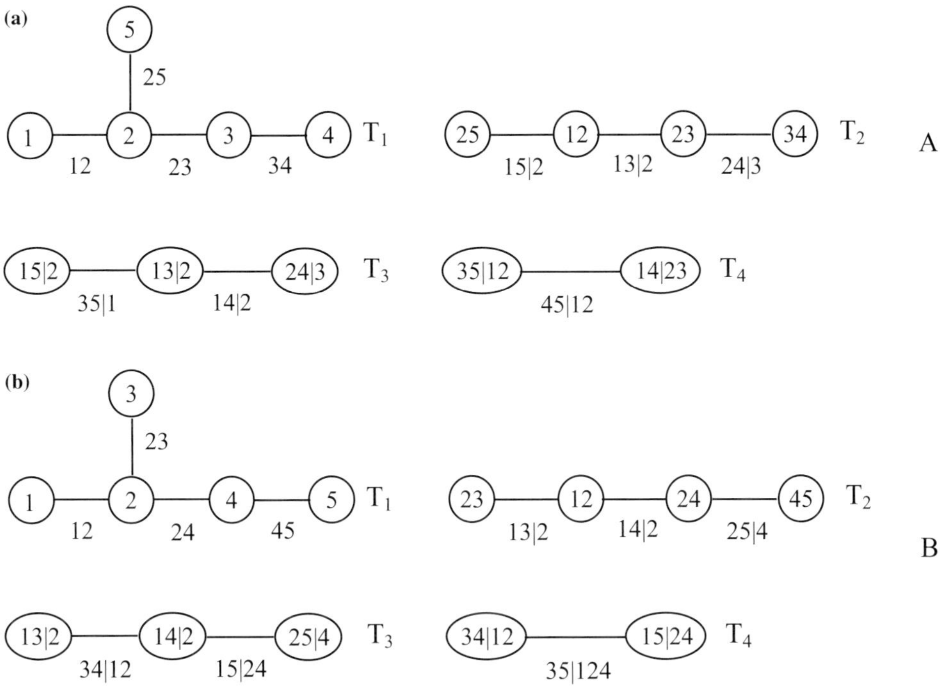

Figure 5.12 Two regular-vine examples for five-dimensional variables.

$$
\begin{aligned}
f(x_1,x_2,x_3,x_4,x_5) = & f_1(x_1)f_2(x_2)f_3(x_3)f_4(x_4)f_5(x_5)c_{12}(F_1(x_1),F_2(x_2))c_{23}(F_2(x_2),F_3(x_3)) \\
& \cdot c_{24}(F_2(x_2),F_4(x_3))c_{45}(F_4(x_4),F_5(x_5)),c_{13|2}\left(F_{1|2}(x_1|x_2),F_{3|2}(x_3|x_2)\right) \\
& \cdot c_{14|2}\left(F_{1|2}(x_3|x_2),F_{4|2}(x_4|x_2)\right)c_{25|4}\left(F_{2|4}(x_2|x_4),F_{5|4}(x_5|x_4)\right) \\
& \cdot c_{34|12}\left(F_{3|12}(x_3|x_1,x_2),F_{4|12}(x_5|x_1,x_2)\right)c_{15|24}\left(F_{1|24}(x_1|x_2,x_4),F_{5|24}(x_5|x_2,x_4)\right) \\
& \cdot c_{35|124}\left(F_{3|124}(x_3|x_1,x_2,x_4),F_{5|124}(x_5|x_1,x_2,x_4)\right)
\end{aligned}
\tag{5.34b}
$$

d-*Dimensional Variables*

For a d-dimensional D-vine, Aas et al. (2009) concluded that there are $d!$ possible ways of ordering the variables in tree T_1. But only $d!/2$ are different trees on the first level. Given such a tree T_1, trees $T_1, T_2, \ldots, T_{d-1}$ are completely determined. This implies that the number of distinct D-vines on d nodes is given by $d!/2$. For a d-dimensional C-vine, there are also $d!/2$ distinctive vine structures.

5.3.3 Conditional Independence and the Pair-Copula Decomposition

First, let us consider the three-dimensional case in Equation (5.29). If X_1 and X_3 are independent, conditioned on random variable X_2, i.e., $c_{13|2}\left(F_{1|2}(x_1|x_2),F_{3|2}(x_3|x_2)\right) = 1$, the density function in Equation (5.29) can be simplified as

$$f(x_1,x_2,x_3)=f_1(x_1)f_2(x_2)f_3(x_3)c_{12}(F_1(x_1),F_2(x_2))c_{23}(F_2(x_2),F_3(x_3)) \tag{5.35}$$

Equation (5.35) indicates that the number of levels reduces to one with the assumption of conditional independence imposed for the three-dimensional variable.

Similarly, if X and Y are independent conditioned on any vector $\mathbf{v}$, we have the following:

$$c_{xy|\mathbf{v}}(F(x|\mathbf{v}),F(y|\mathbf{v}))=1 \tag{5.36}$$

5.3.4 Simulation from Vine Copulas

As discussed previously in Section 5.2.5, the CPI Rosenblatt transformation is commonly applied for the simulation (or sampling) from vine copulas. The conditional probability of the jth variable conditioned on the previous j–1 variables, i.e., $F(x_j|x_1,\ldots,x_{j-1})$, can be written using Equations (5.37) and (5.38) for C-vine and D-vine copulas, respectively, as follows.

For the C-vine copula, the conditional probability is

$$F(x_j|x_1,\ldots,x_{j-1})=\frac{\partial C_{j,j-1|1,\ldots,j-2}(F(x_j|x_1,\ldots,x_{j-2}),F(x_{j-1}|x_1,\ldots,x_{j-2}))}{\partial F(x_{j-1}|x_1,\ldots,x_{j-2})} \tag{5.37}$$

For the D-vine copula structure, we use

$$F(x_j|x_1,\ldots,x_{j-1})=\frac{\partial C_{j,1|2,\ldots,j-1}(F(x_j|x_2,\ldots,x_{j-1}),F(x_1|x_2,\ldots,x_{j-1}))}{\partial F(x_1|x_2,\ldots,x_{j-1})} \tag{5.38}$$

Here, we give the simulation procedure of the C-vine and D-vine copulas (Aas et al., 2009). In these algorithms, we first define that $\boldsymbol{x}=\{x_1\ldots,x_d\}$ are pseudo-observations (i.e., the maringal CDF: copula variables); we also define the parameters as T_1: $\boldsymbol{\theta}_{11},\ldots,\boldsymbol{\theta}_{1(d-1)}$, T_2: $\boldsymbol{\theta}_{21},\ldots,\boldsymbol{\theta}_{2(d-2)}$,…, T_{d-1}: $\boldsymbol{\theta}_{(d-1)1}$.

Simulation from a C-Vine Copula

The procedure for sampling from a C-vine copula can be described as algorithm 1 in Aas et al. (2009). This algorithm applies the margins (i.e., marginal CDF) as variable x and variable 1 as the center variable. In other words, the algorithm simulates the pseudorandom variables rather than the random variables in a real domain. Algorithm 1 involves the following steps:

i. Generate d independent random numbers $\boldsymbol{W}=\{w_1,\ldots,w_d\}$ from uniform $[0,1]$ distribution. And we have $x_1=u_1=w_1$ and $w_i=F(x_i|x_1,\ldots,x_{i-1}), i=2,\ldots,d$.
ii. Simulate $x_2=u_2$ from u_1 and w_2 as $x_2=u_2=h^{-1}(w_2,u_1;\boldsymbol{\theta}_{11})$.
iii. Simulate $x_3=u_3$ from u_1,u_2 and w_3, where $w_3=C(u_3|u_1,u_2)$ as follows:

- Simulating $C(u_3|u_1)$:

$$w_3 = C(u_3|u_1, u_2) = \frac{\partial C_{2,3|1}(C(u_3|u_1; \boldsymbol{\theta}_{12}), C(u_2|u_1, \boldsymbol{\theta}_{11}); \boldsymbol{\theta}_{21})}{\partial C(u_2|u_1)} = h\left(C_{3|1}, C_{2|1}; \boldsymbol{\theta}_{21}\right)$$

$$C_{3|1}(u_3|u_1; \boldsymbol{\theta}_{12}) = h^{-1}\left(w_3, C_{2|1}; \boldsymbol{\theta}_{21}\right) = h^{-1}(w_3, w_2; \boldsymbol{\theta}_{21})$$

- Simulating u_3 using $C_{3|1}$, which we just simulated, as follows:

$$u_3 = h^{-1}\left(C_{3|1}, u_1; \boldsymbol{\theta}_{12}\right)$$

iv. Simulate $x_4 = u_4$ from $u_1, u_2, u_3,$ and w_4 with the following procedures:

- Simulating $C(u_4|u_1, u_2)$:

$$w_4 = C(u_4|u_1, u_2, u_3; \boldsymbol{\theta}_{31}) = \frac{\partial C_{34|12}(C(u_4|u_1, u_2; \boldsymbol{\theta}_{22}), C(u_3|u_1, u_2; \boldsymbol{\theta}_{21}); \boldsymbol{\theta}_{31})}{\partial C(u_3|u_1, u_2; \boldsymbol{\theta}_{21})}$$
$$= \frac{\partial C_{34|12}(C(u_4|u_1, u_2; \boldsymbol{\theta}_{22}), w_3; \boldsymbol{\theta}_{31})}{\partial w_3}$$

$$C_{4|12}(u_4|u_1, u_2) = h^{-1}(w_4, w_3; \boldsymbol{\theta}_{31})$$

- Simulating u_4 using u_1 and $C_{2|1} = w_2$ as follows:

$$C_{4|12} = \frac{\partial C_{2,4|1}\left(C_{4|1}, C_{2|1}; \boldsymbol{\theta}_{22}\right)}{\partial C_{2|1}} = \frac{\partial C_{2,4|1}\left(C_{4|1}, w_2; \boldsymbol{\theta}_{22}\right)}{\partial w_2}$$

$$C_{4|1} = h^{-1}\left(h^{-1}(w_4, w_3; \boldsymbol{\theta}_{31}), w_2; \boldsymbol{\theta}_{22}\right) \Rightarrow u_4 = h^{-1}\left(C_{4|1}, u_1; \boldsymbol{\theta}_{13}\right)$$

...

Carry on the logic for simulation until we reach the dimension d. And one may refer to Aas et al. (2009) for the exact algorithm.

Simulating the Random Variables for a D-Vine Copula

Algorithm 2 in Aas et al. (2009) provided the simulation procedure for the D-vine copula. As stated in Aas et al. (2009), algorithm 2 is less efficient than that for the C-vine copula. To simulate a d-dimensional D-vine copula, we will need to compute $(d-2)^2$ conditional copulas, while we only need to compute$(d-2)(d-1)/2$ for a C-vine. Again, as with algorithm 1, algorithm 2 simulates the pseudorandom variables and includes the following steps:

i. Generate d-independent random numbers $\boldsymbol{W} = \{w_1, \ldots, w_d\}$ from uniform [0, 1] distribution. And we have $x_1 = u_1 = w_1$ and $w_i = F(x_i|x_1, \ldots, x_{i-1}), i = 2, \ldots, d$;

ii. Simulate $x_2 = u_2$ from u_1 and w_2 as $x_2 = u_2 = h^{-1}(w_2, u_1; \boldsymbol{\theta}_{11})$.

iii. Simulate $x_3 = u_3$ from u_1, u_2 and w_3 where $w_3 = C(u_3|u_1, u_2)$ as follows:

- Compute the conditional copula $C_{1|2} = \dfrac{\partial C_{12}(u_1|u_2; \boldsymbol{\theta}_{11})}{\partial u_2}$

- Simulate $C(u_3|u_2)$:

$$w_3 = C(u_3|u_1, u_2) = \frac{\partial C_{1,3|2}(C(u_3|u_2;\boldsymbol{\theta}_{12}), C(u_1|u_2, \boldsymbol{\theta}_{11}); \boldsymbol{\theta}_{21})}{\partial C(u_1|u_2)} = h(C_{3|2}, C_{1|2}; \boldsymbol{\theta}_{21})$$

$$C_{3|2}(u_3|u_2; \boldsymbol{\theta}_{12}) = h^{-1}(w_3, C_{1|2}; \boldsymbol{\theta}_{21}) = h^{-1}(w_3, C_{1|2}; \boldsymbol{\theta}_{21})$$

- Simulate u_3 using $C_{3|2}$, which we just simulated, as follows:

$$u_3 = h^{-1}(C_{3|2}, u_2; \boldsymbol{\theta}_{12})$$

iv. Simulate $x_4 = u_4$ from $u_1, u_2, u_3,$ and w_4 with the following procedures:

- Compute the conditional copula $C_{1|23}$:

$$C_{1|23} = \frac{\partial C_{13|2}(C_{1|2}, C_{3|2}; \boldsymbol{\theta}_{21})}{\partial C_{3|2}}$$

- Simulate $C(u_4|u_2, u_3)$:

$$\begin{aligned} w_4 = C(u_4|u_1, u_2, u_3; \boldsymbol{\theta}_{31}) &= \frac{\partial C_{14|23}(C(u_4|u_2, u_3; \boldsymbol{\theta}_{22}), C(u_1|u_2, u_3; \boldsymbol{\theta}_{21}); \boldsymbol{\theta}_{31})}{\partial C(u_1|u_2, u_3; \boldsymbol{\theta}_{21})} \\ &= \frac{\partial C_{14|23}(C(u_4|u_2, u_3; \boldsymbol{\theta}_{22}), C_{1|23}; \boldsymbol{\theta}_{31})}{\partial C_{1|23}} \end{aligned}$$

$$C_{4|23}(u_4|u_2, u_3) = h^{-1}(w_4, C_{1|23}; \boldsymbol{\theta}_{31})$$

- Compute $C_{2|3}$:

$$C_{2|3} = \frac{\partial C_{23}(u_2, u_3; \boldsymbol{\theta}_{12})}{\partial u_3}$$

- Simulate u_4 using u_3 and $C_{2|3}$ as follows:

$$\begin{aligned} C_{4|23} &= \frac{\partial C_{2,4|3}(C_{4|3}, C_{2|3}; \boldsymbol{\theta}_{22})}{\partial C_{2|3}} \Rightarrow C_{4|3} = h^{-1}(C_{4|23}, C_{2|3}, \boldsymbol{\theta}_{22}) \Rightarrow u_4 \\ &= h^{-1}(C_{4|3}, u_3; \boldsymbol{\theta}_{13}) \end{aligned}$$

...

Carry on the computation until we reach the d-dimension using Equation (5.38). Refer to Aas et al. (2009) for the exact algorithm.

Example 5.14 **Simulate the random variables for the Clayton–Clayton C-vine copula with the following information: $\Theta = (\theta_{11}, \theta_{12}, \theta_{21}) = (2.0, 5.0, 2.0)$ and the independent variables of $(x_1, F(x_2|x_1), F(x_3|x_1, x_2)) = (w_1, w_2\ w_3) = (0.1858, 0.1930, 0.3416)$, where $\{x_1, x_2, x_3\} \in$ uniform [0, 1].**

Solution: According to the sampling procedure discussed, we can simulate the random variables from the vine copula using Figure 5.8(b) in what follows.

As shown in Chapter 4, the bivariate Clayton copula is given as follows:
$C(u,v;\theta) = \left(u^{-\theta} + v^{-\theta} - 1\right)^{-\frac{1}{\theta}}$

a. Set $x_1 = w_1 = 0.1858$
b. From $w_2 = F(x_2|x_1) = h(x_2, x_1, \theta_{11}) = \dfrac{\partial C(x_1, x_2; \theta_{11})}{\partial x_1}$, we have the following:

$$w_2 = \frac{\partial C\left(x_1^{-\theta_{11}} + x_2^{-\theta_{11}} - 1\right)^{-\frac{1}{\theta_{11}}}}{\partial x_1} = x_1^{-\theta_{11}-1}\left(x_1^{-\theta_{11}} + x_2^{-\theta_{11}} - 1\right)^{-1-\frac{1}{\theta_{11}}}$$

$$\Rightarrow x_2 = h^{-1}(w_2, x_1, \theta_{11}) = \left[1 + x_1^{-\theta_{11}} w_2^{-\frac{\theta_{11}}{1+\theta_{11}}} - x_1^{-\theta_{11}}\right]^{-\frac{1}{\theta_{11}}}$$

Substituting $x_1 = 0.1858, w_2 = 0.1930, \theta_{11} = 2.0$ into the preceding equation, we have the following:

$$x_2 = 0.1304.$$

c. Set $w_3 = F(x_3|x_1, x_2) = h\{h(x_3, x_1; \theta_{12}), h(x_2, x_1; \theta_{11}); \theta_{21}\}$, where

$$h(x_3, x_1; \theta_{12}) = t_2 = x_1^{-\theta_{12}-1}\left(x_1^{-\theta_{12}} + x_3^{-\theta_{12}} - 1\right)^{-1-\frac{1}{\theta_{12}}};$$

$$h(x_2, x_1; \theta_{11}) = t_1 = x_1^{-\theta_{11}-1}\left(x_1^{-\theta_{11}} + x_2^{-\theta_{11}} - 1\right)^{-1-\frac{1}{\theta_{11}}};$$

$$h\{h(x_3, x_1; \theta_{12}), h(x_2, x_1; \theta_{11}); \theta_{21}\} = t_1^{-\theta_{21}-1}\left(t_1^{-\theta_{21}} + t_2^{-\theta_{21}} - 1\right)^{-1-\frac{1}{\theta_{21}}}$$

Substitute $x_1 = 0.1858, x_2 = 0.1304, w_3 = 0.3416, \theta_{11} = 2.0, \theta_{12} = 5.0, \theta_{21} = 2.0$ to solve the nonlinear equation
$x_3 = h^{-1}\{h^{-1}(0.3416, h(0.1304, 0.1858; 2.0); 2.0), 0.1858; 5.0\}$, and we have the following:

$$x_3 = 0.1484.$$

Finally, we get the following:

$$(x_1, x_2, x_3) = (0.1858, 0.1304, 0.1484).$$

5.3.5 Parameter Estimation for a Specified Pair-Copula Decomposition

Parameter estimation for specified pair-copula decomposition can be obtained using the log-likelihood method for the C-vine copula using the density function given by Equation (5.28) or D-vine copula with the density function given by Equation (5.27).

Parameter Estimation for a C-Vine Copula

From Equation (5.28), the log-likelihood expression of the C-vine copula is given as

$$LogL(x, v; \mathbf{\Theta}) = \sum_{j=1}^{d-1} \sum_{i=1}^{d-j} \sum_{t=1}^{T} \ln \left(c_{j,j+i|1,\ldots,j-1}\left(F\left(x_{j,t}|x_{1,t}, \ldots, x_{j-1,t}\right), F\left(x_{j+i,t}|x_{1,t}, \ldots, x_{j-1,t}\right)\right)\right) \quad (5.39)$$

The log-likelihood in Equation (5.39) must be numerically maximized over all parameters using the algorithm 3 (Aas et al., 2009). As discussed earlier, for the d-dimensional Vine copula, we have $T = \{T_i : i = 1, \ldots d-1\}$ levels. Within each level T_i, we have $Edge_{T_i} = \{E_j : j = 1, \ldots, d-i\}$. In other words, we have $d-i$ bivariate unconditional/conditional copulas for each level T_i. There are two loops in algorithm 3. The outer loop identifies the tree level, while the inner loop identifies the edges (i.e., the bivariate copulas) of each level. Using variable 1 as the center variable, the algorithm can be explained as follows:

Setting
$\boldsymbol{x}_0 = [x_1, \ldots, x_d] = [u_1, \ldots, u_d], \boldsymbol{\theta} = \left[\boldsymbol{\theta}_{11}, \boldsymbol{\theta}_{12}, \ldots \boldsymbol{\theta}_{(d-1)1}\right]$ and LL=0

```
Outer Loop: i = 1 to d − 1 (for level T)
   Inner Loop: j = 1 to d − i (edges for each level)
      c = copulapdf(x_{i−1,1}, x_{i−1,j+1}, θ_ij);
      LL = LL + ∑ ln(c);
      x_ij = h(x_{i−1,j+1}, x_{i−1,1}; θ_ij)
   End Inner Loop
End Outer Loop
```

Parameter Estimation for a D-Vine Copula

For the D-vine copula, the log-likelihood function is given by

$$LogL(x, v; \mathbf{\Theta}) = \sum_{j=1}^{d-1} \sum_{i=1}^{d-1} \sum_{t=1}^{T} \ln \left(c_{i,i+j|1,\ldots,j-1}\left(F\left(x_{i,t}|x_{i+1,t}, \ldots, x_{i+j-1,t}\right), F\left(x_{i+j,t}|x_{i+1,t}, \ldots, x_{i+j-1,t}\right)\right)\right) \quad (5.40)$$

Let $\Theta_{j,i}$ be the set of parameters of the copula density $C_{i,i+j|i+1,\ldots,i+j-1}(\cdot,\cdot)$. Algorithm 4 (Aas et al., 2009) evaluates the likelihood, which can be explained as follows:

Setting
$s_0 = [s_{01}, s_{02}, \ldots, s_{0d}] = [x_1, \ldots, x_d] = [u_1, \ldots, u_d], \boldsymbol{\theta} = \left[\boldsymbol{\theta}_{11}, \boldsymbol{\theta}_{12}, \ldots \boldsymbol{\theta}_{(d-1)1}\right]$ and $\mathrm{LL} = 0$

Compute the log-likelihood (LL) for T_1 and start the computation of conditional copulas:

for $i = 1$ *to* $d - 1$

$c = c(x_i, x_{i+1}; \theta_{1i}), LL = LL + \sum (\ln c)$

end

$s_{11} = h(s_{01}, s_{02}; \boldsymbol{\theta}_{11})$

Prepare the conditional probability for a higher level:

$$for\ i = 1\ to\ d - 3$$

$$s_{1(2i)} = h\left(s_{0(i+2)}, s_{0(i+1)}; \boldsymbol{\theta}_{1(i+1)}\right), s_{1(2i+1)} = h\left(s_{0(i+1)}, s_{0(k+2)}; \boldsymbol{\theta}_{1(i+1)}\right)$$

$$end$$

$$s_{1(2d-4)} = h\left(s_{0d}, s_{0(d-1)}; \boldsymbol{\theta}_{1(d-1)}\right)$$

Update the log-likelihood as well as the conditional probability for a higher level:

$$for\ i = 2\ to\ d - 1$$

$$for\ j = 1\ to\ d - i$$

$$c = copulapdf\left(s_{(i-1)(2j-1)}, s_{(i-1)(2j)}; \boldsymbol{\theta}_{ij}\right)$$

$$LL = LL + \sum (\ln c)$$

$$end$$

stop the loop if $i = d - 1$; otherwise, we will continue the loop

$$s_{i1} = h\left(s_{(i-1)1}, s_{(i-1)2}; \boldsymbol{\theta}_{i1}\right)$$

again stop the loop if $d \leq 4$; otherwise we will continue on

$$for\ j = 1\ to\ d - i - 2$$

$$s_{i,2j} = h\left(s_{(i-1)(2j+2)}, s_{(i-1)(2j+1)}; \boldsymbol{\theta}_{i(j+1)}\right),$$

$$s_{i(2j+1)} = h\left(s_{(i-1)(2j+1)}, s_{(i-1)(2j+2)}; \boldsymbol{\theta}_{i(j+1)}\right)$$

$$end$$

$$s_{i(2d-2i-2)} = h\left(s_{(i-1)(2d-2i)}, s_{(i-1)(2d-2i-1)}; \boldsymbol{\theta}_{i(n-i)}\right)$$

$$end$$

To apply algorithms 3 and 4 to optimize the parameters, the initial values of the parameters are needed, which may be determined as follows (Aas et al., 2009):

a. Estimate parameters of the copulas in T_1 from the original data.
b. Compute observations (i.e., conditional distribution functions) for T_2 using the copula parameters from T_1 and the corresponding h-function.
c. Estimate parameters of the copulas in T_2 using the results computed from step b.
d. Compute observations for T_3 using the copula parameters at T_2 and the corresponding h-function.
e. Estimate the parameters of copulas in T_3 using the results computed from step d.

. . .

f. Repeat the previous steps sequentially until we teach the top level of the vine tree, i.e., T_{d-1}.

Parameter Estimation for Basic Three-Variable Model

For a three-dimensional special case (i.e., Figure 5.10(a)), the log-likelihood in Equation (5.39) and Equation (5.40) can be simply written as

$$\mathrm{Log}L(x,v;\boldsymbol{\Theta}) = \sum\nolimits_{i=1}^{n} \left(\ln c_{12}(x_{1,i},x_{2,i};\boldsymbol{\Theta}_{11}) + \ln c_{23}(x_{2,i},x_{3,i};\boldsymbol{\Theta}_{12}) + \ln c_{13|2}(v_{1,i},v_{2,i},\boldsymbol{\Theta}_{21})\right) \tag{5.41}$$

where $v_{1,i} = F(x_{1,i}|x_{2,i}) = h(x_{1,i},x_{2,i};\boldsymbol{\Theta}_{11})$ and $v_{2,i} = F(x_{3,i}|x_{2,i}) = h(x_{3,i},x_{2,i};\boldsymbol{\Theta}_{12})$; $\boldsymbol{\Theta}_{ji}$ are the set of parameters of the corresponding copula density $c_{j,j+i|1,\ldots,j-1}(\cdot|\cdot)$. Here we give some common h-functions.

For the Gumbel–Hougaard copula, the h-function can be given as

$$h(u_1,u_2;\theta) = \frac{\partial C(u_1,u_2;\theta)}{\partial u_1} = \frac{C(u_1,u_2;\theta)}{u_2}(-\ln u_2)^{\theta-1}\left((-\ln u_1)^{\theta} + (-\ln u_2)^{\theta}\right)^{\frac{1}{\theta}-1} \tag{5.42}$$

where $C(u_1,u_2;\theta) = e^{-\left((-\ln u_1)^{\theta}+(-\ln u_2)^{\theta}\right)^{\frac{1}{\theta}}}$.

For the Clayton copula, the h-function can be expressed as

$$h(u_1,u_2;\theta) = \frac{\partial C(u_1,u_2;\theta)}{\partial u_2} = u_2^{-\theta-1}\left(u_1^{-\theta} + u_2^{-\theta} - 1\right)^{-1-\frac{1}{\theta}} \tag{5.43}$$

For the Frank copula, the h-function can be written as

$$h(u_1,u_2;\theta) = \frac{\partial C(u_1,u_2;\theta)}{\partial u_2} = \frac{e^{-\theta u_2}}{\dfrac{1-e^{-\theta}}{e^{-\theta u_1}} + e^{-\theta u_2} - 1} \tag{5.44}$$

For the Ali–Mikhail–Haq copula, the h-function can be cast as

$$h(u_1,u_2;\theta) = \frac{\partial C(u_1,u_2;\theta)}{\partial u_2} = \frac{u_2 + \theta u_2(-1+u_2)}{(-1+\theta(-1+u_1)(-1+u_2))^2} \tag{5.45}$$

For the Gaussian copula, the h-function can be written as

$$h(u_1,u_2;\rho_{12}) = \frac{\partial C(u_1,u_2;\rho_{12})}{\partial u_2} = \Phi\left(\frac{\Phi^{-1}(u_1) - \rho_{12}\Phi^{-1}(u_2)}{\sqrt{1-\rho_{12}^2}}\right) \tag{5.46}$$

In Equation (5.46), ρ_{12} is the parameter of copula, i.e., the correlation coefficient for the bivariate random variables after meta-Gaussian transformation, and $\Phi^{-1}()$ is the inverse of the standard univariate Gaussian distribution function.

For the Student t copula, the h-function can be given as

$$h(u_1,u_2;\rho_{12},v_{12}) = \frac{\partial C(u_1,u_2;\rho_{12},v_{12})}{\partial u_2} = T_{v_{12}+1}\left(\frac{T_{v_{12}}^{-1}(u_1) - \rho_{12}T_{v_{12}}^{-1}(u_2)}{\sqrt{\dfrac{\left(v_{12}+T_{v_{12}}^{-1}(u_2)\right)^2(1-\rho_{12})^2}{v_{12}+1}}}\right) \tag{5.47}$$

In Equation (5.47), ρ_{12} and v_{12} are the parameters of Student t copula, i.e., the correlation coefficient and degree of freedom for the transformed variables using Student distribution with degree of freedom (d.f.) of v_{12}; and $T_{v_{12}}^{-1}(\cdot)$ is the inverse of Student T distribution with d.f. of v_{12}, expectation 0, and variance $\frac{v_{12}}{v_{12}-2}$.

Example 5.15 **Assuming that the trivariate random variable given in Table 5.4 may be modeled by the Clayton–Clayton–Frank vine copula with the vine scheme shown in Figure 5.10(a), (1) estimate the parameters using the sequential MLE; and (2) simulate 50 samples from the fitted vine-copula function.**

Table 5.4. *Data and results for Example 5.14.*

u_1	u_2	u_3	$h(u_1, u_2; \theta_{11})$	$h(u_3, u_2; \theta_{12})$
0.241	0.138	0.103	0.892	0.061
0.241	0.172	0.172	0.762	0.460
0.241	0.241	0.276	0.424	0.729
0.241	0.586	0.655	0.010	0.696
0.793	0.828	0.897	0.503	0.741
0.483	0.345	0.379	0.771	0.660
0.931	0.914	0.621	0.767	0.026
0.724	0.759	0.724	0.452	0.379
0.414	0.621	0.586	0.102	0.344
0.759	0.414	0.310	0.936	0.061
0.862	0.793	0.793	0.705	0.500
0.655	0.517	0.448	0.716	0.195
0.414	0.379	0.552	0.526	0.954
0.569	0.448	0.414	0.699	0.297
0.569	0.690	0.690	0.254	0.472
0.414	0.310	0.241	0.727	0.083
0.241	0.552	0.862	0.013	0.981
0.069	0.035	0.035	0.935	0.460
0.241	0.276	0.345	0.287	0.852
0.069	0.069	0.069	0.424	0.460
0.897	0.914	0.931	0.661	0.694
0.655	0.655	0.483	0.473	0.053
0.069	0.103	0.138	0.100	0.908
0.241	0.207	0.207	0.593	0.460
0.655	0.724	0.759	0.364	0.587
0.517	0.483	0.517	0.517	0.609
0.828	0.862	0.828	0.539	0.431
0.966	0.966	0.966	0.854	0.776

Solution:

1. Estimate the parameters. For the bivariate Clayton copula $C(u, v; \theta)$, its copula density function can be given as follows:

$$c(u, v; \theta) = \frac{1+\theta}{u^{\theta+1}v^{\theta+1}\left(u^{-\theta}+v^{-\theta}-1\right)^{2+\frac{1}{\theta}}} \tag{5.48}$$

For the bivariate Frank copula, its copula density function can be given as follows:

$$c(u,v;\theta) = \frac{\theta e^{-\theta(u+v)}\left(e^{-\theta u}-1\right)\left(e^{-\theta v}-1\right)}{\left(e^{-\theta}-1\right)^2 s_1^2} - \frac{\theta e^{-\theta(u+v)}}{(e^{-\theta}-1)s_1}; \; s_1 = \frac{\left(e^{-\theta u}-1\right)\left(e^{-\theta v}-1\right)}{e^{-\theta}-1}+1 \tag{5.49}$$

   a. Estimate the parameters for T_1.
   Using the maximum likelihood estimation for the Clayton copula, the copula parameters estimated for T_1 can be estimated as follows:
   $\theta_{11} = 4.1728; \theta_{12} = 8.3834$ for (u_1, u_2) and (u_2, u_3), respectively.
   b. Compute the conditional distribution functions for T_2 using the copula parameters estimated from T_1. Using the h-function for the Clayton copula (Equation (5.43)) and parameters estimated for T_1, we have the following:

$$h(u_1, u_2; \theta_{11}) = u_2^{-5.1728}\left(u_1^{-4.1728}+u_2^{-4.1728}-1\right)^{-1-\frac{1}{4.1728}}$$

$$h(u_3, u_2; \theta_{12}) = u_2^{-9.3834}\left(u_2^{-8.3834}+u_3^{-8.3834}-1\right)^{-1-\frac{1}{8.3834}}$$

   Table 5.4 lists the original datasets with the fourth and fifth columns as the computed conditional probabilities.
   c. Estimate the parameter for T_2 using the computed conditional probabilities from step b.

   Similar to step a, using the maximum likelihood estimation for the Frank copula, the parameter estimated for T_2 is estimated as $\theta_{21} = -3.8431$.

2. Simulate 50 samples from the fitted vine-copula function:
   Based on the algorithm 2 for sampling from the D-vine copula, we can simulate the samples from the fitted vine-copula as follows:
   a. Generate independently uniform random variables $\{w_1, w_2, w_3\}$.
   b. Set $u_1 = w_1$.
   c. Use $w_2 = C(u_2|u_1) = h_{12}(u_2, u_1; 4.1728)$ to compute $u_2 = h_{12}^{-1}(w_2, u_1; 4.1728)$ using the h-function of the Clayton copula (Equation (5.43)).
   d. Compute u_3 with the following procedure:

$$C(u_3|u_1, u_2) = \frac{\partial C_{13|2}\left(C_{1|2}(u_1|u_2), C_{3|2}(u_3|u_2); \theta_{21}\right)}{\partial C_{1|2}(u_1|u_2)}$$
$$= h_{13|2}(h_{23}(u_3, u_2; 8.3834), h_{12}(u_1, u_2; 4.1728); -3.8431)$$

$$u_3 = h_{23}^{-1}\left\{h_{13|2}^{-1}(w_3, h_{12}(u_1, u_2; 4.1728); -3.8431), u_2; 8.3834\right\}$$

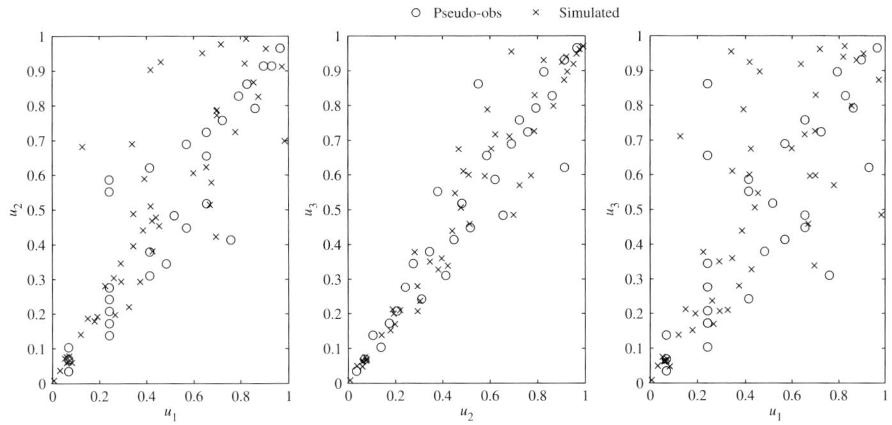

Figure 5.13 Comparison of observed variables with those simulated from vine copula.

where h_{12}, h_{23} are h-functions for the Clayton copula at T_1; $h_{13|2}$ is the h-function for the Frank copula (Equation (5.44)) at T_2.

Using the simulated samples and pseudo-observations, Figure 5.13 evaluated the performance of the fitted vine copula. it is seen that the pair-wise dependence is well preserved

Example 5.16 **Using the four-dimensional pseudo-observations in Example 5.12 to (1) estimate the copula parameters using sequential MLE if D-vine copula (Figure 5.11(a)) with the specified copula (i.e., the Gumbel– Hougaard copula for T1 and the Frank copula for T2 and T3) and C-vine copula (Figure 5.11(b)) with specified copula (i.e., the Gumbel– Hougaard copula for T1, T2, and T3); and (2) simulate the random variates for the sample size of 100 from the fitted copulas.**

Solution:

I. D-Vine Copula

1. Estimate the copula parameters:

The density function of the biviariate Gumbel–Hougaard and Frank copulas are given in Chapter 4 as follows:

Gumbel–Hougaard copula:

$$c(u,v;\theta) = \frac{(\ln u \ln v)^{\theta-1} e^{-S_1^{\frac{1}{\theta}}} \left(S_1^{\frac{2}{\theta}-2} - (1-\theta) S_1^{\frac{1}{\theta}-2} \right)}{uv}; S_1 = (-\ln u)^{\theta} + (-\ln v)^{\theta} \quad (5.50)$$

Frank copula: The same as the previous example, its copula density is given as Equation (5.49).

a. Estimate the parameters for the D-vine copula.

Estimation of copula parameters (the Gumbel–Hougaard copula) for T_1:

For T_1, applying the MLE, we have: $\theta_{11} = 3.8545, L_{11} = 59.783$ for (u_1, u_2); $\theta_{12} = 3.0942, L_{12} = 49.653$ for (u_2, u_3); $\theta_{13} = 4.3949, L_{13} = 71.727$ for (u_3, u_4).

Estimation of copula parameters (Frank copula) for T_2:

i. Compute the conditional distribution $C_{1|2}(u_1|U_2 = u_2; \theta_{11} = 3.8545)$, $C_{3|2}(u_3|U_2 = u_2; \theta_{12} = 3.0942)$; $C_{2|3}(u_2|U_3 = u_3; \theta_{12} = 3.0942)$; and $C_{4|3}(u_4|U_3 = u_3; \theta_{13} = 4.3949)$.

ii. Apply the MLE to estimate the parameters for T_2 as follows:
$\theta_{21} = 1.9708$, $L_{21} = 3.032$ for $\left(C_{1|2}, C_{3|2}\right)$;
$\theta_{22} = 0.7916$, $L_{22} = 0.565$ for $\left(C_{2|3}, C_{4|3}\right)$.

Estimation of copula parameters (the Frank copula) for T_3:

According to Figure 5.11(a), the copula function for T_3 is given as follows: $C_{14|23}(F(u_1|u_2, u_3), F(u_4|u_2, u_3))$

From Equation (5.24), we have the following:

$$F(u_1|u_2, u_3) = \frac{\partial C_{13|2}(F(u_1|u_2), F(u_3|u_2))}{\partial F(u_3|u_2)};$$

$$F(u_4|u_2, u_3) = \frac{\partial C_{24|3}(F(u_2|u_3), F(u_4|u_3))}{\partial F(u_2|u_3)}$$

Using the parameters estimated for T_1 and T_2, we can easily calculate the conditional probability distribution needed for parameter estimation in T_3. Maximizing the log-likelihood for the specified Frank copula, we have $\theta_{31} = -0.4281, L_{31} = 0.173$.

Finally, we have the following:

T_1: $\theta_{11} = 3.8545; \theta_{12} = 3.0942; \theta_{13} = 4.3949$
T_2: $\theta_{21} = 1.9708; \theta_{22} = 0.7916$
T_3: $\theta_{31} = -0.4281$

The overall log-likelihood is computed as the sum of all L s: $L = 184.933$. Table 5.5 lists the conditional probability distributions computed for T_2 and T_3 using the fitted copula of the previous level.

II: C-Vine Copula

a. Estimation of copula parameters (the Gumbel–Hougaard copula) for T_1:

According to Figure 5.11(b), we have the parameters estimated for T_1 as follows: $\theta_{11} = 3.8545, L_{11} = 59.783$ for (u_1, u_2); $\theta_{12} = 3.0834, L_{12} = 47.245$ for (u_1, u_3); $\theta_{13} = 2.5704, L_{13} = 38.08$ for (u_1, u_4).

b. Estimation of copula parameters (the Gumbel–Hougaard copula) for T_2:

From Figure 5.11(b), we need to compute the conditional distribution using the parameter estimated from T_1 first, and then we will be able to estimate the copula parameters for T_2 as follows:

i. Compute the conditional distribution $C_{2|1}(u_2 \mid U_1 = u_1; \theta_{11} = 3.8545)$, $C_{3|1}(u_3|U_1 = u_1; \theta_{12} = 3.0834)$ and $C_{4|1}(u_4|U_1 = u_1; \theta_{13} = 2.5704)$.

ii. Apply the MLE to estimate the parameters for T2 as follows:
$\theta_{21} = 1.2618, L_{21} = 4.265$ for $\left(C_{2|1}, C_{3|1}\right)$;
$\theta_{22} = 1.267, L_{22} = 4.356$ for $\left(C_{2|1}, C_{4|1}\right)$.

c. Estimation of copula parameters (the Gumbel–Hougaard copula) for T_3:

According to Figure 5.11(b), the copula function for T_3 is given as $C_{34|12}(F(u_3|u_1, u_2), F(u_4|u_1, u_2))$.

From Equation (5.24), we have the following:

$$F(u_3|u_1, u_2) = \frac{\partial C_{23|1}(F(u_3|u_1), F(u_2|u_1))}{\partial F(u_2|u_1)}; F(u_4|u_1, u_2) = \frac{\partial C_{24|1}(F(u_4|u_1), F(u_2|u_1))}{\partial F(u_2|u_1)}$$

Using the parameters estimated for T_1 and T_2, we will first compute the conditional probability needed for parameter estimation in T_3. Maximizing the log-likelihood for the specified Frank copula, we have $\theta_{31} = 1.959, L_{31} = 27.687$.

Finally, we have the following:

T1: $\theta_{11} = 3.8545; \theta_{12} = 3.0834; \theta_{13} = 2.5704$
T2: $\theta_{21} = 1.2618; \theta_{22} = 1.2672$
T3: $\theta_{31} = 1.959$

The overall log-likelihood is computed as $L = 181.416$. Table 5.6 lists the conditional probability distributions computed for T_2 and T_3.

Table 5.5. *Conditional probability distributions computed for T_2 and T_3 for fitted* D-*Vine copula*

T_2				T_3	
$C_{u_2\|u_1}$	$C_{u_2\|u_3}$	$C_{u_3\|u_2}$	$C_{u_4\|u_3}$	$C_{u_1\|u_2,u_3}$	$C_{u_4\|u_2,u_3}$
0.143	0.327	0.654	0.830	0.099	0.851
0.134	0.971	0.089	0.387	0.237	0.302
0.470	0.613	0.499	0.703	0.469	0.687
0.524	0.258	0.638	0.456	0.458	0.504
0.722	0.200	0.910	0.799	0.554	0.837
0.307	0.445	0.625	0.291	0.246	0.298
0.220	0.665	0.149	0.346	0.340	0.316
0.500	0.106	0.949	0.118	0.292	0.152
0.102	0.487	0.409	0.119	0.107	0.118
0.736	0.122	0.749	0.929	0.648	0.948
0.742	0.486	0.487	0.701	0.761	0.705
0.654	0.588	0.414	0.821	0.702	0.814
0.773	0.651	0.058	0.575	0.899	0.546
0.143	0.535	0.307	0.780	0.179	0.777
0.760	0.196	0.888	0.220	0.615	0.261
0.601	0.529	0.713	0.712	0.505	0.710

Table 5.5. (*cont.*)

T_2				T_3	
$C_{u_2\|u_1}$	$C_{u_2\|u_3}$	$C_{u_3\|u_2}$	$C_{u_4\|u_3}$	$C_{u_1\|u_2,u_3}$	$C_{u_4\|u_2,u_3}$
0.177	0.040	0.991	0.725	0.068	0.794
0.593	0.209	0.674	0.084	0.516	0.101
0.401	0.697	0.497	0.965	0.395	0.960
0.507	0.843	0.092	0.756	0.698	0.706
0.690	0.376	0.749	0.125	0.592	0.134
0.254	0.802	0.313	0.945	0.313	0.933
0.143	0.386	0.193	0.943	0.214	0.949
0.467	0.191	0.796	0.296	0.326	0.347
0.265	0.376	0.821	0.482	0.151	0.506
0.908	0.361	0.693	0.806	0.884	0.825
0.809	0.261	0.575	0.096	0.801	0.112
0.891	0.052	0.969	0.761	0.786	0.821
0.345	0.880	0.081	0.305	0.534	0.243
0.015	0.934	0.055	0.817	0.031	0.763
0.145	0.754	0.020	0.833	0.281	0.806
0.762	0.985	0.071	0.683	0.890	0.597
0.989	0.140	0.771	0.440	0.984	0.510
0.164	0.801	0.319	0.443	0.202	0.384
0.026	0.887	0.144	0.902	0.045	0.873
0.032	0.991	0.048	0.792	0.065	0.724
0.101	0.693	0.409	0.985	0.106	0.983
0.713	0.159	0.722	0.948	0.632	0.961
0.163	0.137	0.659	0.042	0.113	0.054
0.309	0.231	0.609	0.286	0.254	0.329
0.177	0.784	0.293	0.579	0.226	0.524
0.731	0.301	0.812	0.113	0.613	0.128
0.099	0.071	0.968	0.508	0.037	0.592
0.072	0.211	0.399	0.209	0.077	0.246
0.320	0.413	0.737	0.126	0.218	0.132
0.190	0.015	0.991	0.874	0.075	0.912
0.047	0.873	0.224	0.245	0.069	0.193
0.363	0.899	0.250	0.086	0.472	0.063
0.566	0.375	0.664	0.843	0.492	0.857
0.383	0.614	0.602	0.692	0.328	0.675
0.041	0.525	0.625	0.095	0.028	0.091
0.863	0.437	0.750	0.267	0.813	0.274
0.096	0.225	0.862	0.805	0.044	0.840
0.461	0.424	0.290	0.108	0.560	0.112
0.805	0.595	0.297	0.149	0.871	0.137
0.555	0.470	0.505	0.058	0.557	0.058

Table 5.5. (*cont.*)

T_2				T_3	
$C_{u_2\|u_1}$	$C_{u_2\|u_3}$	$C_{u_3\|u_2}$	$C_{u_4\|u_3}$	$C_{u_1\|u_2,u_3}$	$C_{u_4\|u_2,u_3}$
0.414	0.312	0.873	0.486	0.248	0.523
0.191	0.929	0.210	0.510	0.274	0.425
0.449	0.783	0.124	0.602	0.628	0.549
0.785	0.327	0.852	0.403	0.666	0.435

Note: $C_{u_1|u_2,u_3} = \partial C_{13|2}(C_{u_1|u_2}, C_{u_3|u_2})/\partial C_{u_3|u_2}$; $C_{u_4|u_2,u_3} = \partial C_{24|3}(C_{u_4|u_3}, C_{u_2|u_3})/\partial C_{u_2|u_3}$.

Table 5.6. *Conditional probability distributions computed for T_2 and T_3 of a fitted C-Vine copula.*

T_2			T_3	
$C_{u_2\|u_1}$	$C_{u_3\|u_1}$	$C_{u_4\|u_1}$	$C_{u_3\|u_1,u_2}$	$C_{u_4\|u_1,u_2}$
0.804	0.853	0.910	0.805	0.887
0.939	0.323	0.303	0.157	0.143
0.620	0.556	0.663	0.539	0.658
0.368	0.585	0.536	0.654	0.603
0.406	0.855	0.905	0.905	0.946
0.722	0.762	0.591	0.731	0.530
0.637	0.258	0.225	0.224	0.191
0.574	0.954	0.762	0.972	0.784
0.819	0.712	0.370	0.609	0.260
0.145	0.535	0.841	0.667	0.922
0.263	0.307	0.464	0.378	0.557
0.372	0.316	0.572	0.357	0.640
0.101	0.014	0.056	0.025	0.094
0.744	0.548	0.717	0.472	0.664
0.322	0.790	0.573	0.862	0.656
0.594	0.710	0.781	0.721	0.800
0.894	0.998	0.997	0.998	0.996
0.287	0.569	0.213	0.660	0.260
0.743	0.634	0.913	0.568	0.911
0.477	0.089	0.259	0.086	0.264
0.412	0.656	0.346	0.716	0.379
0.830	0.550	0.859	0.420	0.797
0.530	0.319	0.789	0.314	0.825
0.477	0.797	0.631	0.843	0.672
0.863	0.938	0.898	0.904	0.834
0.127	0.321	0.551	0.441	0.691
0.122	0.286	0.098	0.400	0.152

Table 5.6. (*cont.*)

	T_2		T_3	
$C_{u_2\|u_1}$	$C_{u_3\|u_1}$	$C_{u_4\|u_1}$	$C_{u_3\|u_1,u_2}$	$C_{u_4\|u_1,u_2}$
0.129	0.869	0.900	0.940	0.959
0.655	0.125	0.109	0.099	0.085
0.979	0.520	0.727	0.213	0.352
0.594	0.050	0.330	0.040	0.307
0.413	0.046	0.115	0.048	0.121
0.008	0.108	0.137	0.232	0.283
0.898	0.661	0.596	0.471	0.402
0.977	0.709	0.882	0.355	0.558
0.992	0.455	0.646	0.138	0.224
0.925	0.801	0.973	0.599	0.936
0.180	0.529	0.863	0.650	0.933
0.567	0.775	0.293	0.801	0.276
0.503	0.671	0.495	0.707	0.516
0.867	0.594	0.617	0.435	0.456
0.365	0.702	0.367	0.773	0.417
0.906	0.994	0.983	0.991	0.970
0.622	0.625	0.406	0.615	0.376
0.774	0.859	0.563	0.830	0.470
0.736	0.996	0.996	0.998	0.998
0.975	0.778	0.576	0.429	0.256
0.793	0.397	0.173	0.297	0.112
0.459	0.627	0.801	0.672	0.851
0.766	0.745	0.794	0.686	0.748
0.973	0.958	0.743	0.779	0.393
0.251	0.488	0.360	0.586	0.444
0.918	0.973	0.976	0.942	0.950
0.335	0.252	0.090	0.293	0.104
0.178	0.125	0.062	0.175	0.089
0.434	0.461	0.130	0.500	0.134
0.762	0.928	0.888	0.925	0.873
0.918	0.518	0.527	0.313	0.317
0.503	0.133	0.235	0.128	0.233
0.370	0.731	0.645	0.800	0.717

(2) Simulate random variates from fitted copulas:
- Simulation from a fitted D-vine copula

According to algorithm 2, we can simulate the random variates from the fitted D-vine copula as follows:

Step 1: Generate independent uniformly distributed random variables: $\{w_1, w_2, w_3, w_4\}$.
Step 2: Simulate u_1 by setting $u_1 = v_{11} = w_1$.
Step 3: Simulate u_2 by setting $u_2 = v_{21} = h^{-1}(w_2, u_1; 3.8545)$, where h is the conditional probability distribution for the Gumbel–Hougaard copula.
Step 4: Simulate u_3:

- Calculate $v_{22} = h(v_{11}, v_{21}; 3.8545) = h(u_1, u_2; 3.8545)$.
- Simulate u_3 in the same way as in Example 5.14:

$$\begin{aligned} u_3 = v_{31} &= h^{-1}\left(h^{-1}(w_3, v_{22}; \theta_{21}), v_{21}; \theta_{12}\right) \\ &= h^{-1}\left\{h^{-1}[w_3, h(u_1, u_2; 3.8545); 1.9708], u_2; 3.0942\right\} \end{aligned}$$

- Simulate u_4 using the following procedure:
  ✓ Calculate v_{32}, v_{33}, and v_{34} using

$$\begin{aligned} v_{32} &= h(v_{21}, v_{31}; \theta_{12}) = h(u_2, u_3; 3.0942) \\ v_{33} &= h(v_{31}, v_{21}; \theta_{12}) = h(u_3, u_2; 3.0942) \\ v_{34} &= h(v_{22}, v_{33}; \theta_{21}) = h\{h(u_1, u_2; 3.8545), h(u_3, u_2; 3.0942); 1.9708\} \end{aligned}$$

  ✓ Finally simulate u_4 using:

$$\begin{aligned} temp1 &= h^{-1}(w_4, v_{34}; \theta_{31}) = h^{-1}(w_4, v_{34}; -0.4281) \\ temp2 &= h^{-1}(temp1, v_{32}; \theta_{22}) = h^{-1}(temp1, v_{32}; 0.7916) \\ u_4 = v_{41} &= h^{-1}(temp2, u_3; \theta_{13}) = h^{-1}(temp2, u_3; 4.3949) \end{aligned}$$

  To this end, we simulate random variates from the fitted D-vine copula. As discussed earlier, for every h function (i.e., the conditional copula function of the corresponding bivariate copula functions: the Gumbel–Hougaard copula for T1 and T2, and the Frank copula for T3), the second variable is the conditioning variable. Figure 5.14(a) compares the pseudo-observations with those simulated from the D-vine copula.

- Simulation from a fitted C-vine copula

  To simulate random variates from the fitted C-vine copula, algorithm 1 is applied. By generating independent uniformly distributed random variables $\{w_1, w_2, w_3, w_4\}$, we can simulate $u_1 = v_{11}$ and $u_2 = v_{21}$ using the exact same procedure as that for simulation from the fitted D-vine copula. In what follows, we will discuss how to generate u_3 and u_4 using algorithm 1 in detail:

i. Simulate u_3:
  ✓ Calculate v_{22}, i.e., $C_{2|1}$:

$$v_{22} = h(v_{21}, v_{11}; \theta_{11}) = h(u_2, u_1; 3.8545)$$

  ✓ Simulate u_3 by computing $temp = C_{3|1}$ first:

  From $w_3 = C_{3|1,2} = \frac{\partial C_{23|1}\left(C_{3|1}, C_{2|1}\right)}{\partial C_{2|1}} = h\left(C_{3|1}, C_{2|1}, \theta_{21}\right) = h\left(C_{3|1}, v_{22}, \theta_{21}\right)$, we have the following:$C_{3|1} = temp = h^{-1}(w_3, v_{22}; \theta_{21}) = h^{-1}(w_3, v_{22}; 1.2618)$, and

$$u_3 = v_{31} = h^{-1}(temp, v_{11}; \theta_{12}) = h^{-1}(temp, u_1; 3.0834)$$

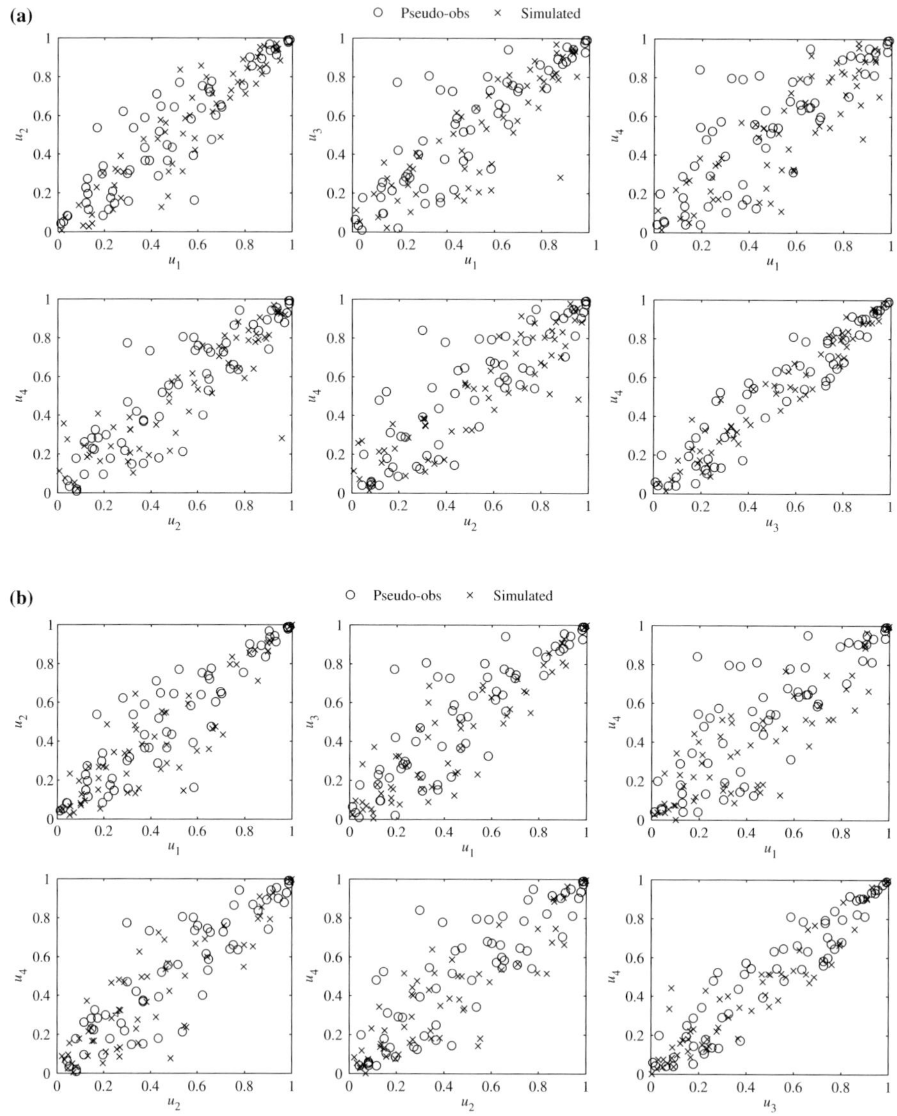

Figure 5.14 (a) Comparison of pseudo-observations with those simulated from the fitted D-vine copula; (b) comparison of pseudo-observations with those simulated from the fitted C-vine copula.

ii. Similarly, we can simulate u_4 as follows:

✓ Calculate v_{32} and v_{33}:

$$v_{32} = h(v_{31}, v_{11}; \theta_{12}) = h(u_3, u_1; 3.0834)$$
$$v_{33} = h(v_{32}, v_{22}; \theta_{21}) = h(v_{32}, v_{22}; 1.2618)$$

✓ Simulate u_4:

$$temp1 = h^{-1}(w_4, v_{33}; \theta_{31}) = h^{-1}(w_4, v_{33}; 1.9590)$$
$$temp2 = h^{-1}(temp1, v_{22}; \theta_{22}) = h^{-1}(temp1, v_{22}; 1.2672)$$
$$u_4 = v_{41} = h^{-1}(temp2, v_{11}; \theta_{13}) = h^{-1}(temp2, u_1; 2.5704)$$

Figure 5.14(b) compares the pseudo-observations with those simulated from the fitted C-vine copula.

For the simulation of random variates, the inverse of the h function is evaluated numerically for both D-vine and C-vine copulas.

Based on the overall log-likelihood computed in this example, we see that the log-likelihood value for the D-vine copula is slightly higher than that for the C-vine copula. Simulation plots show similar results between the fitted D-vine and C-vine copulas.

5.3.6 Selection of Vine Copula Structure

Previously, we have discussed how to estimate the parameters for the specified vine copula structure. Following Aas et al. (2009), for the estimation of pair-copula decomposition, we should consider (i) the selection of pair-copula decompositions; (ii) the selection of pair-copula types; and (iii) the estimation of copula parameters. In principle, we may use all the possible decompositions to estimate the copula parameters and to choose the best-fitted vine copula structure for a given d-dimensional variable. However, in reality with higher dimensions (i.e., $d \geq 3$), the number of possible decompositions increases significantly as $d!/2$ (i.e., 3 C-Vine (D-Vine) copulas for three-dimensional variables, 12 D-vine and 12 C-vine copulas for four-dimensional variables, 60 D-vine and 60 C-vine copulas for five-dimensional variables, etc.). To avoid the evaluations for all possible decompositions, we may first look at the rank-based correlation structure, starting from $T_{1,}$ to achieve the proper vine decomposition.

Similar to the discussion in Section 5.3.5, with the proper study of rank-based correlation structure, we can modify the model selection using sequential MLE (Aas et al., 2009) for decomposition with the tree levels $\{T_1, T_2, \ldots, T_{d-1}\}$ in what follows:

1. Select the copula family and estimate the parameters for T_1 using the original data: (a) the parameters may be estimated using MLE; (b) the best-fitted copula can be selected by minimizing AIC or BIC and assessed with the goodness-of-fit study that will be discussed in Section 5.3.7.
2. Transform observations required in T_2 with the use of the copula fitted in T_1 and its corresponding $h(\cdot)$ function.
3. Select the copula family and estimate the parameters for T_2. The best-fitted copula in T_2 is selected in the same way as in T_1.
4. Repeat steps 2 and 3 until we reach T_{d-1}.

Based on the previously discussed model selection, we know the copulas selected do not need to belong to the same copula families (D-vine copula in Example 5.15, as an example). In addition, we should note that the sequential MLE may not result in a globally optimal solution. To avoid this problem, we may estimate all the parameters simultaneously using algorithm 3 for C-vine (algorithm 4 for D-vine) copulas for the selected vine structure with the parameters estimated using the sequential MLE as the initial estimates. Here, we will show how to estimate the parameters simultaneously.

Example 5.17 **Re-work Example 5.16: (1) estimate the copula parameters simultaneously using the same decomposition and copula families as Example 5.16; and (2) simulate the random variates for the sample size of 100 from the fitted copula functions.**

Solution:

- Estimate the copula parameters simultaneously.

Estimate the parameters for D-vine copula.

In Example 5.16, we have estimated the copula parameters sequentially for the D-vine copula as follows:
T_1: $\theta_{11} = 3.8545; \theta_{12} = 3.0942; \theta_{13} = 4.3949$ (the Gumbel–Hougaard copula family)
T_2: $\theta_{21} = 1.9708; \theta_{22} = 0.7916$ (the Frank copula family)
T_3: $\theta_{31} = -0.4281$ (the Frank copula family)

To estimate the parameters simultaneously, we apply algorithm 4 (Equation (5.41)) to write the log-likelihood function for the D-vine copula as follows:

$$L_1 = \sum_{i=1}^{n} [\ln(c_{12}(u_{1i}, u_{2i}; \theta_{11})) + \ln(c_{23}(u_{2i}, u_{3i}; \theta_{12})) + \ln(c_{34}(u_{3i}, u_{4i}; \theta_{13}))]$$

$$v_{11} = h(u_1, u_2; \theta_{11}); v_{12} = h(u_3, u_2; \theta_{12}); v_{13} = h(u_2, u_3; \theta_{12}); v_{14} = h(u_4, u_3; \theta_{13})$$

$$L_2 = \sum_{i=1}^{n} \left[\ln\left(c_{13|2}(v_{11i}, v_{12i}; \theta_{21})\right) + \ln\left(c_{34|2}(v_{13i}, v_{14i}; \theta_{22})\right)\right]$$

$$v_{21} = h(v_{11}, v_{12}; \theta_{21}); v_{22} = h(v_{14}, v_{13}; \theta_{22})$$

$$L_3 = \sum_{i=1}^{n} \left[\ln\left(c_{14|23}(v_{21i}, v_{22i}; \theta_{31})\right)\right]$$

Finally, we have the overall log-likelihood as $L = L_1 + L_2 + L_3$, where n is the sample size.

Using the parameters estimated sequentially as initial estimates, we obtain the parameters simultaneously by maximizing the final L (or equivalently minimizing $-L$):

$$\theta_{11} = 3.7723, \theta_{12} = 3.1705, \theta_{13} = 4.3913, \theta_{21} = 1.9931, \theta_{22} = 0.7811, \theta_{31} = -0.4325$$

Overall log-likelihood is $L = L_1 + L_2 + L_3 = 184.988$

$$AIC = -2L + 2\text{length}(\boldsymbol{\Theta}) = -2(184.988) + 2(6) = -357.976$$

$$BIC = -2L + \ln(n)\text{length}(\boldsymbol{\Theta}) = -2(184.988) + \ln(60)(6) = -345.409$$

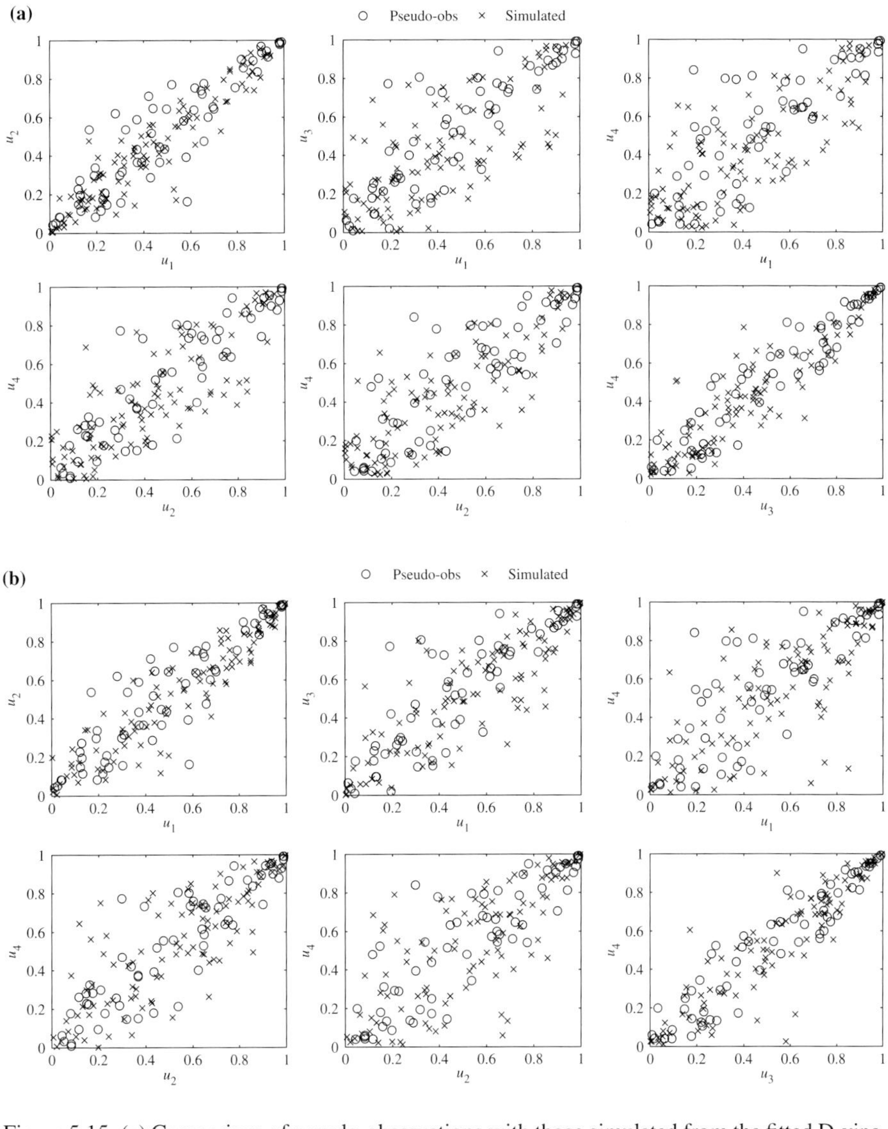

Figure 5.15 (a) Comparison of pseudo-observations with those simulated from the fitted D-vine copula; (b) comparison of pseudo-observations with those simulated from the fitted C-vine copula.

- Estimate the parameters for C-vine copula.

In Example 5.16, we have estimated the copula parameters sequentially for the C-vine copula as follows:

T_1: $\theta_{11} = 3.8545; \theta_{12} = 3.0834; \theta_{13} = 2.5704$ (the Gumbel–Hougaard copula family)

T_2: $\theta_{21} = 1.2618; \theta_{22} = 1.2672$ (the Gumbel–Hougaard copula family)
T_3: $\theta_{31} = 1.9590$ (the Gumbel–Hougaard copula family)

To estimate parameters simultaneously using the C-vine copula, we apply algorithm 3 (Equation (5.40)) to write the likelihood function for the C-vine copula as follows:

$$L_1 = \sum_{i=1}^{n} [\ln(c_{12}(u_{1i}, u_{2i}; \theta_{11})) + \ln(c_{13}(u_{1i}, u_{3i}; \theta_{12})) + \ln(c_{14}(u_{1i}, u_{4i}; \theta_{13}))]$$
$$v_{11} = h(u_2, u_1; \theta_{11}); v_{12} = h(u_3, u_1, \theta_{12}); v_{13} = h(u_4, u_1; \theta_{13})$$
$$L_2 = \sum_{i=1}^{n} \left[\ln\left(c_{23|1}(v_{11i}, v_{12i}; \theta_{21})\right) + \ln\left(c_{24|1}(v_{11}, v_{13}; \theta_{22})\right)\right]$$
$$v_{21} = h(v_{12}, v_{11}; \theta_{21}); v_{22} = h(v_{13}, v_{11}; \theta_{22})$$
$$L_3 = \ln\left(c_{34|12}(v_{21}, v_{22}; \theta_{31})\right)$$

Finally, we have the overall log-likelihood as $L = L_1 + L_2 + L_3$.

Again, using the parameters estimated sequentially as initial estimates from Example 5.16, we can estimate the parameters simultaneously by maximizing L (or minimizing $-L$) as follows:

$$\theta_{11} = 3.9280, \theta_{12} = 2.9592, \theta_{13} = 2.5509, \theta_{21} = 1.2463, \theta_{22} = 1.2285, \theta_{31} = 2.0333$$

The log-likelihood is evaluated as follows:

$$L = 181.673, AIC = -351.346, BIC = -338.780.$$

From the log-likelihood value, we see that the log-likelihood value obtained from the D-vine copula is slightly higher than that obtained from the C-vine copula. The AIC and BIC values (D-vine) are slightly smaller than those for the C-vine copula.

- Simulate random variates

  Using the same procedure as in Example 5.16, Figures 5.15(a) and 5.15(b), compare pseudo-observations with those simulated from the D-vine and C-vine copulas, respectively. The simulation plots show a similar comparison between the fitted D-vine and C-vine copulas.

Comparing with Example 5.16, there are minimal differences for the log-likelihood value, AIC and BIC obtained for D-vine and C-vine copulas. In addition, the sequential estimation method is more direct and easier to apply than is the simultaneous estimation method.

5.3.7 Goodness-of-Fit Test

Aas et al. (2009) proposed to use the probability integral transform (PIT, i.e., Rosenblatt's transform) to test the goodness-of-fit for the pair-copula decomposition. Previously, in Section 5.2.5 we have discussed Rosenblatt's transform. In what follows, we will illustrate the PIT algorithm for the C-vine and D-vine copulas (Aas et al., 2009). For a d-dimensional random variable $\mathbf{x} = \{\mathbf{x}_1, \mathbf{x}_2, \ldots \mathbf{x}_i \ldots, \mathbf{x}_n\}, \mathbf{x}_i = \{x_{i,1}, \ldots, x_{i,d}\}$, the PIT is defined as follows:

$$\begin{aligned} Z_1 &= F(X_1 \le x_1), \\ Z_2 &= F(X_2 \le x_2 | X_1 = x_1) \\ &\ldots \\ Z_i &= F(X_i \le x_i | X_1 = x_1, \ldots, X_{i-1} = x_{i-1}) \end{aligned} \tag{5.51}$$

For a C-vine copula, the conditional distribution is computed using Equation (5.37) as given in algorithm 5 (Aas et al., 2009) that may be explained with d-dimensional copula variable of sample size n, as follows:

1. Set $z_{1,1} = x_{1,1} = u_{1,1}$. Here the first subscript represents the dimension, and second represents the sample considered.
2. Use loops to compute $z_i, i = 2, \ldots, d$.

   for i $= 2$ *to d*

   $z_{i,1} = x_{i,1}$

       for j $= 1$ *to d* $- 1$

       $z_{i,1} = h\left(temp, z_{j,1}; \boldsymbol{\theta}_{j,i-j}\right)$

       end

   end

3. Repeat steps 1 and 2 n times.

The D-vine copula applies Equation (5.38) to compute the conditional distributions for PIT, which is given as algorithm 6 in Aas et al. (2009). It again may be explained for a d-dimensional D-vine copula variables of sample size n using $\boldsymbol{x}_1 = [x_{11}, x_{21}, \ldots, x_{d1}]$ as follows:

1. Set $z_{11} = x_{11} = u_{11}$. The subscripts are defined exactly same as those in algorithm 5.
2. Compute the conditional distribution of $z_{2,1} = C_{2|1}$ and $C_{1|2}$:

   $z_{21} = h(x_{21}, x_{11}; \boldsymbol{\theta}_{11})$;

   setting $s_{21} = x_{21}$;

   computing $s_{22} = h(x_{11}, x_{21}; \boldsymbol{\theta}_{11})$

3. Compute the conditional distribution for $x_{31}|x_{11}, x_{21}, \ldots x_{d1}|x_{11}, \ldots x_{(d-1)1}$:

   for i $= 3$ *to d*

     $z_{i1} = h\left(x_{i1}, x_{(i-1)1}; \boldsymbol{\theta}_{1(i-1)}\right)$ % temporary: representing $C_{i|i-1}$.

     for j $= 2$ *to i* $- 1$

       $z_{i1} = h\left(z_{i1}, s_{i-1,2(j-1)}; \boldsymbol{\theta}_{j,i-j}\right)$

     end

       stop if $i = d$. Otherwise we need to continue the loop

       set $s_{i1} = x_{i1}; s_{i2} = h\left(s_{(i-1)1}, s_{i1}; \boldsymbol{\theta}_{1(i-1)}\right); s_{i3} = h\left(s_{i1}, s_{(i-1)1}; \boldsymbol{\theta}_{1(i-1)}\right)$

     for j $= 1$ *to i* $- 3$

       $s_{i(2j+2)} = h\left(s_{(i-1)2j}, s_{i(2j+1)}; \boldsymbol{\theta}_{(j+1)(i-j-1)}\right)$;

       $s_{i(2j+3)} = h\left(s_{i(2j+1)}, s_{(i-1)2j}; \boldsymbol{\theta}_{(j+1)(i-j-1)}\right)$

     end

       $s_{i(2i-2)} = h\left(s_{(i-1)(2i-4)}, s_{i(2i-3)}; \boldsymbol{\theta}_{(i-1)1}\right)$

   end

4. Repeat steps 1–3 n times.

With the use of the PIT, the goodness-of-fit test may be performed in two ways: by applying the Anderson–Darling test and by applying the new procedure based on PIT proposed by Genest et al. (2007b).

Applying the Anderson–Darling Test

Compared to the new procedure proposed by Genest et al. (2007b), the Anderson–Darling test has inferior performance. However, we are still going to introduce this formal test here. Using the variables after PIT, we define the following:

$$\chi^2 = \left\{\chi_i^2 = \sum\nolimits_{j=1}^{d} \left[\Phi^{-1}(Z_{ij})\right]^2, i = 1, 2, \ldots, n\right\}, \tag{5.52a}$$

where χ^2 follows the chi-square distribution with the degree of freedom (d.f.= d; i.e., the dimension of the multivariate random variable). The nonparametric CDF of $\chi 2$ computed from Equation (5.52a) may then be estimated as follows:

$$G_n(t) = \frac{1}{n+1}\sum\nolimits_{i=1}^{n} 1\left(\chi^2 \le t\right), t > 0 \tag{5.52b}$$

Under the null hypothesis of **Z**s being independent and uniformly distributed, the Anderson–Darling test statistic is given as (Genest et al., 2007a):

$$A_k = -n - \frac{1}{n}\sum\nolimits_{i=1}^{n} (2i-1)\left[\ln\left(G\left(\chi_{(i)}^2\right)\right) + \ln\left(1 - G\left(\chi_{(n+1-i)}^2\right)\right)\right], \tag{5.53}$$

where $\chi_{(1)}^2 \le \cdots \le \chi_{(n)}^2$ are the order statistics corresponding to $\chi_1^2, \ldots \chi_n^2$.

To avoid the misidentification of the limiting probability distribution, the P-value is estimated using the parametric bootstrap method for large integer N. Repeat the following steps for every $k \in \{1, \ldots, N\}$:

a. Generate a random sample $\mathbf{X}_{1,k}^*, \ldots, \mathbf{X}_{n,k}^*$ from the vine copula $C_{\boldsymbol{\theta}_n}$ and compute their associated rank vectors: $\mathbf{R}_{1,k}^*, \ldots, \mathbf{R}_{n,k}^*$.
b. Compute $\mathbf{U}_{i,k}^* = \mathbf{R}_{i,k}^*/(n+1)$ for $i \in \{1, \ldots, n\}$.
c. Reestimate parameters (i.e., $\boldsymbol{\theta}_{n,k}^*$) for the vine copula using $\left\{\mathbf{U}_{1,k}^*, \ldots, \mathbf{U}_{n,k}^*\right\}$ and compute $\boldsymbol{\chi}_k^* = \left\{\chi_{1,k}^{2*}, \ldots, \chi_{n,k}^{2*}\right\}$, where $\chi_{i,k}^{2*} = \sum\nolimits_{j=1}^{d} \left[\Phi^{-1}\left(Z_{ij,k}^*\right)\right]^2$ with $Z_{ij,k}^*$ is determined from algorithm 5 or 6 (or simply using Equation (5.11)).
d. Compute the Anderson–Darling test statistics A_k^* using $\boldsymbol{\chi}_k^*$ from Equation (5.53).

The approximate P-value for the test is then given by $\sum\nolimits_{k=1}^{N} 1\left(A_k^* > A\right)/N$.

Applying the New Procedure Based on PIT Proposed by Genest et al. (2007b)

As discussed in Section 4.7.1, the null hypothesis is **Z** (after Rosenblatt's transform), being close to $C_\perp$, where $\mathbf{Z} = \{\mathbf{Z}_1, \ldots \mathbf{Z}_\mathbf{i}, \ldots, \mathbf{Z}_\mathbf{n}\}, \mathbf{Z}_\mathbf{i} = \{Z_1, Z_2, \ldots, Z_d\}$ as follows:

1. Compute D_n and test statistics $S_n^{(B)}$ using the fitted copula model as follows:

$$D_n(\mathbf{u}) = \frac{1}{n}\sum\nolimits_{i=1}^{n} (\mathbf{Z}_i \leq \boldsymbol{u}), \quad \mathbf{u} \in [0, 1]^d \tag{5.54}$$

$$\begin{aligned} S_n^{(B)} &= n\int_{[0,1]^d} [D_n(\mathbf{u}) - C_\perp(\mathbf{u})]^2 d\mathbf{u} \\ &= \frac{n}{3^d} - \frac{1}{2^{d-1}}\sum\nolimits_{i=1}^{n}\prod\nolimits_{k=1}^{d}\left(1 - Z_{ik}^2\right) + \frac{1}{n}\sum\nolimits_{i=1}^{n}\sum\nolimits_{j=1}^{n}\prod\nolimits_{k=1}^{d}\left(1 - Z_{ik}\vee Z_{jk}\right) \end{aligned} \tag{5.55}$$

where $Z_{ik}\vee Z_{jk} = \max\left(Z_{ik}, Z_{jk}\right)$.

2. For some large integer N, repeat the following steps for $k = \{1, 2, \ldots, N\}$:
   a. Generate a random sample $\mathbf{X}_{1,k}^*, \ldots, \mathbf{X}_{n,k}^*$ from the vine copula $C_{\boldsymbol{\theta}_n}$ and compute their associated rank vectors: $\mathbf{R}_{1,k}^*, \ldots, \mathbf{R}_{n,k}^*$.
   b. Compute $\mathbf{U}_{i,k}^* = \mathbf{R}_{i,k}^*/(n+1)$ for $i \in \{1, \ldots, n\}$.
   c. Reestimate parameters (i.e., $\boldsymbol{\theta}_{n,k}^*$) for the vine copula using $\left\{\mathbf{U}_{1,k}^*, \ldots, \mathbf{U}_{n,k}^*\right\}$ and compute $\mathbf{Z}_{1,k}^*, \ldots, \mathbf{Z}_{n,k}^*$ using an appropriate algorithm (algorithm 5 or 6) or simply using Equation (5.11).
   d. Compute $D_{n,k}^*$ and $S_{n,k}^{(B)*}$ using Equations (5.54) and (5.55) with reestimated parameter $\boldsymbol{\theta}_{n,k}^*$.

The appropriate P-value for the test is then given as follows: $\sum_{k=1}^{N} 1\left(S_{n,k}^{(B)*} > S_n^{(B)}\right)/N$.

Example 5.18 **Assess the GoF for the C- or D-vine copula constructed in Example 5.15 for trivariate analysis with both the Anderson–Darling test and the new procedure based on PIT proposed by Genest et al. (2007b) discussed in the preceding section.**

Solution: Previously, we have shown that in the case of trivariate random variables, it is indifferent between C- and D-vine copulas. From Example 5.15, we have estimated the parameters for the Clayton–Clayton–Frank copula sequentially as follows:

T_1: Clayton copula— $\theta_{11} = 4.1728\ (u_1, u_2); \theta_{12} = 8.3834\ (u_2, u_3)$
T_2: Frank copula— $\theta_{21} = -3.8431\ \left(C_{1|2}, C_{3|2}\right)$

Based on the Rosenblatt transform, Equation (5.53) may be rewritten for the three-dimensional C- or D-vine copulas as follows:

$$\begin{cases} Z_1 = u_1 \\ Z_2 = C(u_2|U_1 = u_1) \\ Z_3 = C(u_3|U_1 = u_1, U_2 = u_2) = \dfrac{\partial C_{13|2}\left(C_{3|2}, C_{1|2}\right)}{\partial C_{1|2}} \end{cases} \tag{5.56}$$

Table 5.7. *Computed Zs and corresponding test statistics for three-dimensional C- or D-vine copulas.*

Observed					
u_1	u_2	u_3	Z_1	Z_2	Z_3
0.241	0.138	0.103	0.241	0.049	0.152
0.241	0.172	0.172	0.241	0.134	0.690
0.241	0.241	0.276	0.241	0.424	0.723
0.241	0.586	0.655	0.241	0.973	0.304
0.793	0.828	0.897	0.793	0.627	0.793
0.483	0.345	0.379	0.483	0.135	0.869
0.931	0.914	0.621	0.931	0.696	0.042
0.724	0.759	0.724	0.724	0.575	0.306
0.414	0.621	0.586	0.414	0.833	0.087
0.759	0.414	0.310	0.759	0.041	0.176
0.862	0.793	0.793	0.862	0.458	0.687
0.655	0.517	0.448	0.655	0.211	0.281
0.414	0.379	0.552	0.414	0.336	0.974
0.569	0.448	0.414	0.569	0.204	0.419
0.569	0.690	0.690	0.569	0.687	0.252
0.414	0.310	0.241	0.414	0.164	0.120
0.241	0.552	0.862	0.241	0.965	0.933
0.069	0.034	0.034	0.069	0.026	0.813
0.241	0.276	0.345	0.241	0.572	0.790
0.069	0.069	0.069	0.069	0.423	0.378
0.897	0.914	0.931	0.897	0.730	0.841
0.655	0.655	0.483	0.655	0.473	0.029
0.069	0.103	0.138	0.069	0.811	0.771
0.241	0.207	0.207	0.241	0.267	0.538
0.655	0.724	0.759	0.655	0.611	0.482
0.517	0.483	0.517	0.517	0.362	0.653
0.828	0.862	0.828	0.828	0.665	0.448
0.966	0.966	0.966	0.966	0.854	0.949

Notes:
Anderson–Darling test statistic: An = 0.3572, $P = 0.878$ (with $N = 1{,}000$).
Rosenblatt (SnB) test statistic: SnB = 0.0417, $P = 0.532$ (with $N = 1{,}000$).

With the estimated parameters using the sequential MLE and Equation (5.56), Table 5.7 lists *Zs* along with test statistics.

The formal GoF results using the Anderson–Darling and SnB tests show that with 1,000 parametric bootstrap simulations, the fitted Clayton–Clayton–Frank copula may properly model the dependence of the studied trivariate random variables.

Example 5.19 **Assess the GoF for the D- and C-vine copulas constructed in Example 5.16 with both of the two GoF approaches previously discussed.**

Solution:

1. D-vine copula

For the four-dimensional random variable, the parameters were estimated sequentially for the D-vine copula in Example 5.16 as follows:

T_1: Gumbel–Hougaard copula
$\theta_{11} = 3.8545\ (u_1, u_2), \theta_{12} = 3.0942\ (u_2, u_3), \theta_{13} = 4.3949\ (u_3, u_4)$;

T_2: Frank copula
$\theta_{21} = 1.9708\ \left(C_{1|2}, C_{3|2}\right), \theta_{22} = 0.7916\ \left(C_{2|3}, C_{4|3}\right)$;

T_3: Frank copula
$\theta_{31} = -0.4281\ \left(C_{1|23}, C_{4|23}\right)$.

Now based on the PIT, Equation (5.53) can be rewritten for the four-dimensional D-vine copula as follows:

$$\begin{cases} Z_1 = u_1 \\ Z_2 = C(u_2|U_1 = u_1) = \dfrac{\partial C(u_1, u_2; \theta_{11})}{\partial u_1} \\ Z_3 = C(u_3|U_1 = u_1, U_2 = u_2) = \dfrac{\partial C_{13|2}\left(C_{3|2}, C_{1|2}\right)}{\partial C_{1|2}} \\ Z_4 = C(u_4|U_1 = u_1, U_2 = u_2, U_3 = u_3) = \dfrac{\partial C_{14|23}\left(C_{4|23}, C_{1|23}\right)}{\partial C_{1|23}} \end{cases} \tag{5.57}$$

Table 5.8 lists the computed values from PIT using Equation (5.57) with the fitted four-dimensional D-vine copula.

To approximate the *P*-value using the parametric bootstrap simulation method, we will use $N = 1{,}000$ as an example. It is known that the larger the N value, the closer to the true *P*-value for the GoF study.

2. C-vine copula

For the four-dimensional random variable, parameters were estimated sequentially for the C-vine copula in Example 5.15 as follows:

T_1 (Gumbel): $\theta_{11} = 3.8545, (u_1, u_2)$; $\theta_{12} = 3.0834, (u_1, u_3)$; $\theta_{13} = 2.5704, (u_1, u_4)$.
T2 (Gumbel): $\theta_{21} = 1.2618\ \left(C_{2|1}, C_{3|1}\right); \theta_{22} = 1.2672\ \left(C_{2|1}, C_{4|1}\right)$
T3 (Gumbel): $\theta_{31} = 1.9590\ \left(C_{3|12}, C_{4|12}\right)$

Table 5.8. *Computed Zs and corresponding test statistics for the D-vine copula.*

Z_1	Z_2	Z_3	Z_4
0.194	0.804	0.801	0.828
0.819	0.939	0.150	0.278
0.614	0.620	0.513	0.685
0.235	0.368	0.638	0.499
0.792	0.406	0.882	0.841
0.433	0.722	0.717	0.275
0.130	0.638	0.214	0.301
0.570	0.574	0.956	0.141
0.128	0.819	0.596	0.101
0.218	0.145	0.670	0.951
0.468	0.263	0.370	0.729
0.490	0.373	0.337	0.827
0.194	0.102	0.030	0.588
0.120	0.744	0.458	0.754
0.676	0.322	0.843	0.271
0.990	0.589	0.683	0.713
0.657	0.894	0.996	0.762
0.226	0.287	0.646	0.101
0.828	0.743	0.546	0.959
0.373	0.477	0.081	0.724
0.698	0.412	0.690	0.138
0.645	0.830	0.410	0.928
0.025	0.531	0.304	0.943
0.298	0.477	0.820	0.330
0.906	0.863	0.890	0.470
0.658	0.127	0.519	0.848
0.302	0.122	0.430	0.125
0.581	0.129	0.943	0.839
0.371	0.655	0.095	0.245
0.169	0.980	0.115	0.726
0.041	0.594	0.034	0.791
0.982	0.410	0.039	0.637
0.585	0.008	0.584	0.562
0.618	0.898	0.462	0.354
0.280	0.977	0.276	0.851
0.902	0.992	0.098	0.686
0.440	0.925	0.596	0.981
0.243	0.180	0.647	0.963
0.044	0.567	0.798	0.046
0.122	0.503	0.701	0.306
0.497	0.867	0.423	0.495

Table 5.8. (*cont.*)

Z_1	Z_2	Z_3	Z_4
0.701	0.365	0.752	0.133
0.323	0.906	0.987	0.544
0.013	0.622	0.599	0.214
0.651	0.774	0.812	0.118
0.190	0.736	0.996	0.897
0.520	0.975	0.393	0.165
0.926	0.793	0.287	0.062
0.468	0.459	0.646	0.858
0.868	0.766	0.663	0.660
0.422	0.973	0.811	0.076
0.888	0.251	0.613	0.301
0.372	0.918	0.940	0.812
0.132	0.335	0.293	0.114
0.429	0.178	0.179	0.156
0.390	0.434	0.478	0.059
0.983	0.765	0.902	0.495
0.980	0.917	0.310	0.401
0.308	0.503	0.122	0.562
0.932	0.370	0.786	0.452

Notes:
An (Equation 4.55): An = 0.7411, *P*-value = 0.261.
SnB (Equation 4.56): SnB = 0.0362, *P*-value = 0.08.

Table 5.9. *Computed Zs and the corresponding test statistics for the fitted C-vine copula.*

Z_1	Z_2	Z_3	Z_4
0.194	0.804	0.805	0.846
0.819	0.939	0.157	0.296
0.614	0.620	0.539	0.727
0.235	0.368	0.654	0.511
0.792	0.406	0.905	0.859
0.433	0.722	0.731	0.306
0.130	0.638	0.223	0.319
0.570	0.574	0.972	0.102
0.128	0.819	0.609	0.140
0.218	0.145	0.667	0.971
0.468	0.263	0.378	0.724
0.490	0.373	0.357	0.832

Table 5.9. (*cont.*)

Z_1	Z_2	Z_3	Z_4
0.194	0.102	0.025	0.409
0.120	0.744	0.472	0.786
0.676	0.322	0.862	0.257
0.990	0.589	0.719	0.771
0.657	0.894	0.998	0.461
0.226	0.287	0.660	0.114
0.828	0.743	0.568	0.978
0.373	0.477	0.086	0.617
0.698	0.412	0.717	0.169
0.645	0.830	0.420	0.936
0.025	0.531	0.314	0.969
0.298	0.477	0.843	0.314
0.906	0.863	0.904	0.448
0.658	0.127	0.441	0.838
0.302	0.122	0.401	0.137
0.581	0.129	0.940	0.816
0.371	0.655	0.098	0.231
0.169	0.980	0.213	0.596
0.041	0.594	0.041	0.749
0.982	0.410	0.048	0.416
0.585	0.008	0.232	0.469
0.618	0.898	0.471	0.416
0.280	0.977	0.354	0.743
0.902	0.992	0.138	0.476
0.440	0.925	0.599	0.987
0.243	0.180	0.650	0.981
0.044	0.567	0.801	0.062
0.122	0.503	0.707	0.318
0.497	0.867	0.435	0.531
0.701	0.365	0.774	0.154
0.323	0.906	0.991	0.289
0.013	0.622	0.614	0.249
0.651	0.774	0.830	0.140
0.190	0.736	0.998	0.728
0.520	0.975	0.429	0.250
0.926	0.793	0.297	0.135
0.468	0.459	0.673	0.895
0.868	0.766	0.686	0.717
0.422	0.973	0.779	0.133
0.888	0.251	0.586	0.360
0.372	0.918	0.942	0.742
0.132	0.335	0.293	0.125

Table 5.9. (*cont.*)

Z_1	Z_2	Z_3	Z_4
0.429	0.178	0.175	0.168
0.390	0.434	0.500	0.081
0.983	0.765	0.925	0.468
0.980	0.917	0.314	0.442
0.308	0.503	0.128	0.506
0.932	0.370	0.800	0.474

Notes:
An (Equation 4.53): An = 0.7365, P-value = 0.276 (with N = 1,000).
SnB (Equation 4.54): SnB = 0.03, P-value = 0.415 (with N = 1,000).

According to the C-vine structure, the PIT of Equation (5.57) is rewritten as follows:

$$\begin{cases} Z_1 = u_1 \\ Z_2 = C(u_2|U_1 = u_1) = \dfrac{\partial C(u_1, u_2)}{\partial u_1} \\ Z_3 = C(u_3|U_1 = u_1, U_2 = u_2) = \dfrac{\partial C\left(C_{3|1}, C_{2|1}\right)}{\partial C_{2|1}} \\ Z_4 = C(u_4|U_1 = u_1, U_2 = u_2, U_3 = u_3) = \dfrac{\partial C\left(C_{4|21}, C_{3|21}\right)}{\partial C_{3|21}} \end{cases} \tag{5.58}$$

Table 5.9 lists the computed Zs and corresponding test statistics for the fitted C-vine copula.

5.3.8 *JCDF for d*-Dimensional Vine Copulas

Let $\boldsymbol{X} = \{X_1, \ldots, X_d\}$ be a random vector with marginal distributions $F_i(x_i) = u_i = P(X_i \leq x_i)$ and conditional distributions $F(x_i|x_1, \ldots, x_{i-1}) = P(X_i \leq x_i|X_1 \leq x_1, \ldots, X_{i-1} \leq x_{i-1})$. From the probability theory, the joint probability distribution $F(x_1, x_2, \ldots, x_d)$ can be expressed as follows:

$$F(x_1, \ldots, x_d) = P(X_1 \leq x_1, \ldots, X_d \leq x_d) = C(U_1 \leq u_1, \ldots, U_d \leq u_d)$$

Then, with the given vine-copula structure, the joint probability distribution may be evaluated starting from the top-level T_{d-1} of the given pair-copula decomposition. In what follows, we illustrate how to derive the JCDF for C-vine and D-vine copulas using three-dimensional, four-dimensional, and five-dimensional random variables as examples.

JCDF for Three-Dimensional Variables

Using Figure 5.10(a) as an example and applying the total probability theory, we have the following:

$$F(x_1, x_2, x_3) = P(X_1 \leq x_1, X_3 \leq x_3|X_2 \leq x_2)P_2(x_2) \tag{5.59a}$$

Let $u_1 = F_1(x_1), u_2 = F_2(x_2), u_3 = F_3(x_3)$ and $\theta_{11}, \theta_{12}, \theta_{21}$ represent the copula parameters for $(u_1, u_2); (u_2, u_3); and\ (u_1|u_2, u_3|u_2)$, respectively. Then, we have the following:

$$P(X_1 \le x_1, X_3 \le x_3 | X_2 \le x_2) = C_{1,3|2}\left(C_{1|2}(U_1 \le u_1|U_2 \le u_2), C_{3|2}(U_3 \le u_3|U_2 \le u_2); \theta_{21}\right) \tag{5.59b}$$

$$C_{1|2}(U_1 \le u_1|U_2 \le u_2) = \frac{C(u_1, u_2; \theta_{11})}{u_2}; C_{3|2}(U_3 \le u_3|U_2 \le u_2) = \frac{C(u_2, u_3; \theta_{12})}{u_2} \tag{5.59c}$$

JCDF for Four-Dimensional D-Vine Variables

Using Figure 5.11(a) as an example, we have the following:

$$F(x_1, x_2, x_3, x_4) = P(X_1 \le x_1, X_4 \le x_4 | X_2 \le x_2, X_3 \le x_3) C(u_2, u_3) \tag{5.60a}$$

Let $\theta_{11}, \theta_{12}, \theta_{13}, \theta_{21}, \theta_{22}, \theta_{31}$ represent the copula parameters for T_1, T_2, and T_3, respectively. Then we have the following:

$$P(X_1 \le x_1, X_4 \le x_4 | X_2 \le x_2, X_3 \le x_3)$$
$$= C_{14|23}\left(C_{1|23}(u_1|U_2 \le u_2, U_3 \le u_3), C_{4|23}(u_4|U_2 \le u_2, U_3 \le u_3); \theta_{31}\right) \tag{5.60b}$$

$$C_{1|23}(u_1|U_2 \le u_2, U_3 \le u_3) = C_{13|2}\left(\frac{C(u_1, u_2; \theta_{11})}{u_2}, \frac{C(u_2, u_3; \theta_{12})}{u_2}; \theta_{21}\right) \Big/ \frac{C(u_2, u_3; \theta_{12})}{u_2} \tag{5.60c}$$

$$C_{4|23}(u_4|U_2 \le u_2, U_3 \le u_3) = C_{24|3}\left(\frac{C(u_3, u_4; \theta_{13})}{u_3}, \frac{C(u_2, u_3; \theta_{12})}{u_3}; \theta_{22}\right) \Big/ \frac{C(u_2, u_3; \theta_{12})}{u_3} \tag{5.60d}$$

JCDF for Four-Dimensional C-Vine Variables

Using Figure 5.11(b) as an example, Equation (5.60a) can be rearranged as follows:

$$F(x_1, x_2, x_3, x_4) = P(X_3 \le x_3, X_4 \le x_4 | X_1 \le x_1, X_2 \le x_2) C(u_1, u_2) \tag{5.61a}$$

Let $\theta_{11}, \theta_{12}, \theta_{13}, \theta_{21}, \theta_{22}, \theta_{31}$ represent the copula parameters for T_1, T_2, and T_3, respectively. Then we have the following:

$$P(X_3 \le x_3, X_4 \le x_4 | X_1 \le x_1, X_2 \le x_2)$$
$$= C_{34|12}\left(C_{3|12}(u_3|U_1 \le u_1, U_2 \le u_2), C_{4|12}(u_4|U_1 \le u_1, U_2 \le u_2); \theta_{31}\right) \tag{5.61b}$$

$$C_{3|12}(u_3|U_1 \le u_1, U_2 \le u_2) = C_{23|1}\left(\frac{C(u_1,u_2;\theta_{11})}{u_1}, \frac{C(u_1,u_3;\theta_{12})}{u_1}; \theta_{21}\right) \Big/ \frac{C(u_1,u_2;\theta_{11})}{u_1} \tag{5.61c}$$

$$C_{4|12}(u_4|U_1 \le u_1, U_2 \le u_2) = C_{24|1}\left(\frac{C(u_1,u_4;\theta_{13})}{u_1}, \frac{C(u_1,u_2;\theta_{12})}{u_1}; \theta_{22}\right) \Big/ \frac{C(u_1,u_2;\theta_{11})}{u_1} \tag{5.61d}$$

JCDF for Five-Dimensional D-Vine Variables

Using Figure 5.8 as an example, we have the following:

$$F(x_1,x_2,x_3,x_4,x_5) = P(X_1 \le x_1, X_5 \le x_5|X_2 \le x_2, X_3 \le x_3, X_4 \le x_4)P(x_2,x_3,x_4) \tag{5.62a}$$

Let $\{\theta_{11},\theta_{12},\theta_{13},\theta_{14}\}, \{\theta_{21},\theta_{22},\theta_{23}\}, \{\theta_{31},\theta_{32}\}, \theta_{41}$ represent the copula parameters for T_1, T_2, T_3, and T_4 respectively. Then we have the following:

$$\begin{aligned} P(x_2,x_3,x_4) &= C(u_2,u_3,u_4) \\ &= C_{24|3}(u_2,u_4|U_3 \le u_3)u_3 = C_{24|3}\left(\frac{C(u_2,u_3;\theta_{12})}{u_3}, \frac{C(u_3,u_4;\theta_{13})}{u_3}; \theta_{22}\right) \end{aligned} \tag{5.62b}$$

$$\begin{aligned} &P(X_1 \le x_1, X_5 \le x_5|X_2 \le x_2, X_3 \le x_3, X_4 \le x_4) \\ &\quad = C_{15|234}(P(x_1|X_2 \le x_2, X_3 \le x_3, X_4 \le x_4), P(x_5|X_2 \le x_2, X_3 \le x_3, X_4 \le x_4), \theta_{41}) \end{aligned} \tag{5.62c}$$

$$\begin{aligned} &P(X_1 \le x_1|X_2 \le x_2, X_3 \le x_3, X_4 \le x_4) \\ &\quad = C_{14|23}(P(x_1|X_2 \le x_2, X_3 \le x_3), P(x_4|X_2 \le x_2, X_3 \le x_3), \theta_{31}) \end{aligned} \tag{5.62d}$$

$$\begin{aligned} &P(X_5 \le x_5|X_2 \le x_2, X_3 \le x_3, X_4 \le x_4) \\ &\quad = C_{25|34}(P(x_5|X_3 \le x_3, X_4 \le x_4), P(x_2|X_3 \le x_3, X_4 \le x_4), \theta_{32}) \end{aligned} \tag{5.62e}$$

$$P(X_1 \le x_1|X_2 \le x_2, X_3 \le x_3) = C_{13|2}\left(\frac{C(u_1,u_2;\theta_{11})}{u_2}, \frac{C(u_2,u_3;\theta_{12})}{u_2}; \theta_{21}\right) \Big/ \left[\frac{C(u_2,u_3;\theta_{12})}{u_2}\right] \tag{5.62f}$$

$$P(X_4 \le x_4|X_2 \le x_2, X_3 \le x_3) = C_{24|3}\left(\frac{C(u_3,u_4;\theta_{13})}{u_3}, \frac{C(u_2,u_3;\theta_{12})}{u_3}; \theta_{22}\right) \Big/ \left[\frac{C(u_2,u_3;\theta_{12})}{u_3}\right] \tag{5.62g}$$

$$P(X_2 \le x_2|X_3 \le x_3, X_4 \le x_4) = C_{24|3}\left(\frac{C(u_2,u_3;\theta_{12})}{u_3}, \frac{C(u_3,u_4;\theta_{13})}{u_3}; \theta_{22}\right) \Big/ \left[\frac{C(u_3,u_4;\theta_{13})}{u_3}\right] \tag{5.62h}$$

$$P(X_5 \le x_5 | X_3 \le x_3, X_4 \le x_4) = C_{35|4}\left(\frac{C(u_3,u_4;\theta_{13})}{u_4}, \frac{C(u_4,u_5;\theta_{14})}{u_4}; \theta_{23}\right) \Big/ \left[\frac{C(u_4,u_5;\theta_{14})}{u_4}\right] \tag{5.62i}$$

JCDF for Five-Dimensional C-Vine Variables

Using Figure 5.9 as an example, we have the following:

$$\begin{aligned} F(x_1,x_2,x_3,x_4,x_5) &= P(X_1 \le x_1, \ldots, X_5 \le x_5) \\ &= P(X_4 \le x_4, X_5 \le x_5 | X_1 \le x_1, X_2 \le x_2, X_3 \le x_3) P(x_1,x_2,x_3) \end{aligned} \tag{5.63a}$$

$$F(x_1,x_2,x_3) = C(u_1,u_2,u_3) = C_{23|1}\left(\frac{C(u_1,u_2;\theta_{11})}{u_1}, \frac{C(u_1,u_3;\theta_{12})}{u_1}; \theta_{21}\right) u_1 \tag{5.63b}$$

$$\begin{aligned} &P(X_4 \le x_4, X_5 \le x_5 | X_1 \le x_1, X_2 \le x_2, X_3 \le x_3) \\ &\quad = C_{45|123}(P(X_4 \le x_4 | X_1 \le x_1, X_2 \le x_2, X_3 \le x_3), \\ &\qquad P(X_5 \le x_5 | X_1 \le x_1, X_2 \le x_2, X_3 \le x_3); \theta_{41}) \end{aligned} \tag{5.63c}$$

$$\begin{aligned} &P(X_4 \le x_4 | X_1 \le x_1, X_2 \le x_2, X_3 \le x_3) \\ &\quad = C_{34|12}(P(X_4 \le x_4 | X_1 \le x_1, X_2 \le x_2), P(X_3 \le x_3 | X_1 \le x_1, X_2 \le x_2), \theta_{31}) \end{aligned} \tag{5.63d}$$

$$\begin{aligned} &P(X_5 \le x_5 | X_1 \le x_1, X_2 \le x_2, X_3 \le x_3) \\ &\quad = C_{35|12}(P(X_5 \le x_5 | X_1 \le x_1, X_2 \le x_2), P(X_3 \le x_3 | X_1 \le x_1, X_2 \le x_2), \theta_{32}) \end{aligned} \tag{5.63e}$$

$$P(X_3 \le x_3 | X_1 \le x_1, X_2 \le x_2) = \frac{C_{23|1}\left(\frac{C(u_1,u_3;\theta_{12})}{u_1}, \frac{C(u_1,u_2;\theta_{11})}{u_1}, \theta_{21}\right)}{\frac{C(u_1,u_2;\theta_{11})}{u_1}} \tag{5.63f}$$

$$P(X_4 \le x_4 | X_1 \le x_1, X_2 \le x_2) = \frac{C_{24|1}\left(\frac{C(u_1,u_4;\theta_{13})}{u_1}, \frac{C(u_1,u_2;\theta_{11})}{u_1}, \theta_{22}\right)}{\frac{C(u_1,u_2;\theta_{11})}{u_1}} \tag{5.63g}$$

$$P(X_5 \le x_5 | X_1 \le x_1, X_2 \le x_2) = \frac{C_{25|1}\left(\frac{C(u_1,u_5;\theta_{14})}{u_1}, \frac{C(u_1,u_2;\theta_{11})}{u_1}, \theta_{23}\right)}{\frac{C(u_1,u_2;\theta_{11})}{u_1}} \tag{5.63h}$$

Example 5.20 **Compute the JCDF and compare it with the empirical JCDF, using the data and vine copula constructed in Example 5.15.**

Solution: The empirical copula can be computed using the following:

$$C_n(\boldsymbol{u}) = \frac{1}{n}\sum\nolimits_{i=1}^{n} (u_{i1} \le u_1, u_{i2} \le u_2, u_{i3} \le u_3); \boldsymbol{u} = [u_1, u_2, u_3] \tag{5.64}$$

Applying the parameters estimated for the vine structure in Example 5.14, we have the joint distribution function for the given Clayton–Clayton–Frank vine copula as follows:

$$JCDF = \frac{u_2}{3.8431} \ln\left(1 + \frac{(e^{3.8431A} - 1)(e^{3.8431B} - 1)}{e^{3.8431} - 1}\right)$$

where

$$A = C(u_1|U_2 \le u_2) = \frac{\left(u_1^{-4.1728} + u_2^{-4.1728} - 1\right)^{-\frac{1}{4.1728}}}{u_2}$$

$$B = C(u_3|U_2 \le u_2) = \frac{\left(u_2^{-8.3834} + u_3^{-8.3834} - 1\right)^{-\frac{1}{8.3834}}}{u_2}$$

The quantile-quantile (QQ) plot shown in Figure 5.16 shows that the JCDF estimated from the vine copula underestimates the joint distribution.

It should be noted that we have only shown how to compute the joint CDF from vine copula in this chapter. In the application chapters that follow, we will further discuss joint and conditional return periods obtained from copula using real-world examples.

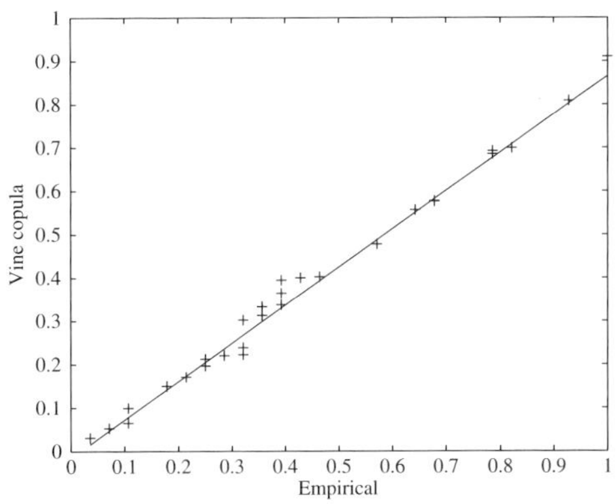

Figure 5.16 Comparison of empirical JCDF versus JCDF computed from the vine copula.

5.4 Summary

This chapter focuses on the theoretical aspects of the asymmetric Archimedean copula for the analysis in higher dimensions. Two types of asymmetric Archimedean copulas are discussed: (1) nested Archimedean copulas; and (2) vine copulas.

The nested Archimedean copulas include fully nested, partially nested, and general nested Archimedean copulas. Nested Archimedean copulas (NAC) requires the following: (i) the nested generating function must be completely monotonic; and (ii) with the increasing levels in the NAC structure, the dependence of the upper level needs to be weaker than the lower level. Compared to the symmetric Archimedean copulas (i.e., EAC forcing all the variables to share the same degree of pair dependence), the NAC is more flexible and may better model the dependence structure.

Vine copula includes D-vine, C-vine, and R-vine copulas. A vine copula is constructed based on the multivariate probability density decomposition. With the bivariate copula as the building block for the vine copula, the vine copula allows the free identification of the bivariate copula for each pair of variables for each level in the vine structure. Compared to EAC and NAC, the vine copula is most flexible, with D-vine copulas being more flexible than C-vine copulas. With the flexibility offered by the vine copula, the copula modeling in higher dimensions may also be computationally time consuming.

References

Aas, K., Czado, C., Frigessi, A., and Bakken, H. (2009). Pair-copula constructions of multiple dependence. *Insurance: Mathematics and Economics*, 44, 182–198, doi:10.1016/j.insmatheco.2007.02.001.

Bedford, T. and Cooke, R. M. (2001). Probability density decomposition for conditionally dependent random variables modeled by vines. *Annals of Mathematics and Artificial Intelligence*, (32), 245–268.

Bedford, T. and Cooke, R. M. (2002). Vines – a new graphical model for dependent random variables. *Annals of Statistics*, (30), 1031–1068.

Berg, D. and Aas, K. (2007), Models for construction of multivariate dependence, Technical report, Norwegian Computing Center.

Embrechts, P., Lindskog, F., and McNeil, A. (2003). Modelling dependence with copulas and applications to risk management. In Rachev, S. T. ed. *Handbook of Heavy Tailed Distributions in Finance*. North-Holland: Elsevier.

Frees, E. W. and Valdez, E. A. (1998). Understanding relationships using copulas. *North American Actuarial Journal*, 2(1), 1–25

Genest, C. and Favre, A.-C. (2007). Everything you always wanted to know about copula modeling but were afraid to ask. *Journal of Hydrologic Engineering*, 12(4), 347–368.

Genest, C., Favre, A.-C., Beliveau, J., and Jacques, C. (2007a). Metaelliptical copulas and their uses in frequency analysis of multivariate hydrological data. *Water Resources Research*, 43, W09401, doi:10.1029/2006WR005275.

Genest, C., Rémillard, B., and Beaudoin, D. (2007b). Goodness-of-fit tests for copulas: A review and a power study. *Insurance: Mathematics and Economics*. doi:10.1016/j.insmatheco.2007.10.005.

Joe, H. (1996). Families of m-variate distributions with given margins and m(m-1)/2 bivariate dependence parameters. In R¨uschendorf, L., Schweizer B., and Taylor, M. D., ed. *Distributions with Fixed Marginals and Related Topics*. Institute of Mathematical Statistics, Hayward, CA, 120–141.

Joe, H. (1997). *Multivariate Models and Dependence Concept*. Chapman & Hall, New York.

Kurowicka, D. and Cooke, R. M. (2004). Distribution – free continuous Bayesian belief nets. In *Fourth International Conference on Mathematical Methods in Reliability Methodology and Practice*. Wiley, Santa Fe, 309–322.

Kurowicka, D. and Cooke, R. M. (2006). *Uncertainty Analysis with High Dimensional Dependence Modelling*. Wiley, New York.

McNeil, A. J. (2007). Sampling nested Archimedean copulas. http://citeseerx.ist.psu.edu/viewdoc/download?doi=10.1.1.395.5498&rep=rep1&type=pdf Nelsen, R. B. (2006). *An Introduction to Copulas*. Springer-Verlag, New York.

Rosenblatt, M. (1952). Remarks on a multivariate transformation. *Annals of Mathematical Statistics,* 23(3), 470–472.

Savu, C. and Trede, M. (2010). Hierarchies of Archimedean copulas. *Quantative Finance*, 10, 295–304.

Whelan, N. (2004). Sampling from Archimedean copulas. *Quantitat Finance*, 4(3), 339–52.

Additional Reading

Francesco, S. and Salvatore G. (2007). Fully nested 3-copula: procedure and application on hydrological data. *Journal of Hydrologic Engineering,* 12(4), 420–430.

Salvatori, G. and Francesco, S. (2006). Asymmetric copula in multivariate flood frequency analysis. *Advanced in Water Resources,* 29, 1155–1167.

Salvadori, G., De Michele, C., Kottegoda, N., and Rosso, R. (2007). *Extremes in Nature: An Approach Using Copulas.* Water Science and Technology Library, Vol. 56, Springer, Dordrecht.

Salvadori, G. and De Michele, C. (2007), On the use of copulas in hydrology: theory and practice. *Journal of Hydrologic Engineering*, 12(4), 369– 380.

Appendix

With the use of Example 5.8, the density functions for M3, M4, M5, M6, and M12 copulas are derived.

M3 Copula

$$\frac{\partial C}{\partial u_1} = \frac{e^{-\theta_2 u_1} S_2 \left(e^{-\theta_1 u_3} - 1\right)\left(e^{-\theta_2 u_2} - 1\right)}{(e^{-\theta_1} - 1)(e^{-\theta_2} - 1) S_1 \left(\frac{(S_2 - 1)\left(e^{-\theta_1 u_3} - 1\right)}{e^{-\theta_1} - 1} + 1\right)} \tag{M3–1}$$

$$\begin{aligned}\frac{\partial^2 C}{\partial u_1 \partial u_2} = & \frac{\theta_1 e^{-\theta_2 (u_1 + u_2)} S_2^2 \left(e^{-\theta_2 u_1} - 1\right)\left(e^{-\theta_1 u_3} - 1\right)^2 \left(e^{-\theta_2 u_2} - 1\right)}{\left(e^{-\theta_2} - 1\right)^2 S_1^2 ((s_2 - 1)(e^{-\theta_3 u_3} - 1) + (e^{-\theta_1} - 1))^2} \\ & - \frac{\theta_2 e^{-\theta_2 (u_1 + u_2)} S_2 \left(e^{-\theta_1 u_3} - 1\right)}{(e^{-\theta_2} - 1) S_1 ((s_2 - 1)(e^{-\theta_1 u_3} - 1) + (e^{-\theta_1} - 1))} \\ & + \frac{(\theta_2 - \theta_1) e^{-\theta_2 (u_1 + u_2)} S_2 \left(e^{-\theta_2 u_1} - 1\right)\left(e^{-\theta_2 u_2} - 1\right)\left(e^{-\theta_1 u_3} - 1\right)}{\left(e^{-\theta_2} - 1\right)^2 S_1^2 ((S_2 - 1)(e^{-\theta_1 u_3} - 1) + (e^{-\theta_1} - 1))}\end{aligned} \tag{M3–2}$$

$$\begin{aligned}\frac{\partial^3 C}{\partial u_1 \partial u_2 \partial u_3} = & \frac{\theta_1 \theta_2 e^{-\theta_2 (u_1 + u_2) - \theta_1 u_3} S_2}{(e^{-\theta_2} - 1) S_1 S_3} + \frac{\left(\theta_1^2 - \theta_1 \theta_2\right) e^{-\theta_2 (u_1 + u_2) - \theta_1 u_3} S_2 \left(e^{-\theta_2 u_1} - 1\right)\left(e^{-\theta_2 u_2} - 1\right)}{\left(e^{-\theta_2} - 1\right)^2 S_1^2 S_3} \\ & - \frac{\theta_1 \theta_2 e^{-\theta_2 (u_1 + u_2) - \theta_1 u_3} S_2 (S_2 - 1)\left(e^{-\theta_1 u_3} - 1\right)}{e^{(-\theta_2 - 1)} S_1 S_3} \\ & + \frac{2\theta_1^2 e^{-\theta_2 (u_1 + u_2) - \theta_1 u_3} S_2 (S_2 - 1)\left(e^{-\theta_2 u_1} - 1\right)\left(e^{-\theta_2 u_2} - 1\right)\left(e^{-\theta_1 u_3} - 1\right)}{\left(e^{-\theta_1} - 1\right)\left(e^{-\theta_2} - 1\right)^2 S_1^2 S_3^3} \\ & + \frac{\theta_1 S_2 (\theta_2 S_2 - 3\theta_1 S_2 + \theta_1 - \theta_2) e^{-\theta_2 (u_1 + u_2) - \theta_1 u_3} \left(e^{-\theta_2 u_1} - 1\right)\left(e^{-\theta_2 u_2} - 1\right)\left(e^{-\theta_1 u_3} - 1\right)}{\left(e^{-\theta_2} - 1\right)^2 S_1^2 S_3^2}\end{aligned} \tag{M3–3}$$

where

$$S_1 = \frac{\left(e^{-\theta_2 u_1} - 1\right)\left(e^{-\theta_2 u_2} - 1\right)}{e^{-\theta_2} - 1} + 1; S_2 = S_1^{\frac{\theta_1}{\theta_2}}; S_3 = (S_2 - 1)\left(e^{-\theta_1 u_3} - 1\right) + \left(e^{-\theta_1} - 1\right)$$

M4 Copula

$$\frac{\partial C}{\partial u_1} = u_1^{-\theta_2 - 1}\left(u_1^{-\theta_2} + u_2^{-\theta_2} - 1\right)^{\frac{\theta_1}{\theta_2} - 1}\left(\left(u_1^{-\theta_2} + u_2^{-\theta_2} - 1\right)^{\frac{\theta_1}{\theta_2}} + u_3^{-\theta_1} - 1\right)^{-\frac{1}{\theta_1} - 1} \quad \text{(M4–1)}$$

$$\frac{\partial^2 C}{\partial u_1 \partial u_2} = u_1^{-\theta_2 - 1} u_2^{-\theta_2 - 1}\left(u_1^{-\theta_2} + u_2^{-\theta_2} - 1\right)^{\frac{\theta_1}{\theta_2} - 2}\left(\left(u_1^{-\theta_2} + u_2^{-\theta_2} - 1\right)^{\frac{\theta_1}{\theta_2}} + u_3^{-\theta_1} - 1\right)^{-\frac{1}{\theta_1} - 1}$$
$$\left(-(\theta_1 - \theta_2) + (1 + \theta_1)\left(\left(u_1^{-\theta_2} + u_2^{-\theta_2} - 1\right)^{\frac{\theta_1}{\theta_2}}\left(\left(u_1^{-\theta_2} + u_2^{-\theta_2} - 1\right)^{\frac{\theta_1}{\theta_2}} + u_3^{-\theta_1} - 1\right)^{-1}\right)\right)$$
$$\text{(M4–2)}$$

$$\frac{\partial^3 C}{\partial u_1 \partial u_2 \partial u_3} =$$
$$(1 + \theta_1)(u_1 u_2)^{-\theta_2 - 1} u_3^{-\theta_1 - 1}\left(u_1^{-\theta_2} + u_2^{-\theta_2} - 1\right)^{\frac{\theta_1}{\theta_2} - 2}\left(\left(u_1^{-\theta_2} + u_2^{-\theta_2} - 1\right)^{\frac{\theta_1}{\theta_2}} + u_3^{-\theta_1} - 1\right)^{-\frac{1}{\theta_1} - 2}$$

$$\left(-(\theta_1 - \theta_2) + (1 + 2\theta_1)\left(u_1^{-\theta_2} + u_2^{-\theta_2} - 1\right)^{\frac{\theta_1}{\theta_2}}\left(\left(u_1^{-\theta_2} + u_2^{-\theta_2} - 1\right)^{\frac{\theta_1}{\theta_2}} + u_3^{-\theta_1} - 1\right)^{-1}\right) \quad \text{(M4–3)}$$

M5 Copula

$$\frac{\partial C}{\partial u_1} = \left((1 - u_1)^{\theta_2 - 1} - (1 - u_1)^{\theta_2 - 1}(1 - u_2)^{\theta_2}\right)$$
$$\left((1 - u_1)^{\theta_2} + (1 - u_2)^{\theta_2} - (1 - u_1)^{\theta_2}(1 - u_2)_2^{\theta}\right)^{\frac{\theta_1}{\theta_2} - 1}$$
$$\left\{\left((1 - u_1)^{\theta_2} + (1 - u_2)^{\theta_2} - (1 - u_1)^{\theta_2}(1 - u_2)^{\theta_2}\right)^{\frac{\theta_1}{\theta_2}}\left(1 - (1 - u_3)^{\theta_1}\right)\right.$$
$$\left. + (1 - u_3)^{\theta_1}\right\}^{\frac{1}{\theta_1} - 1}\left(1 - (1 - u_3)^{\theta_1}\right) \quad \text{(M5–1)}$$

$$\frac{\partial^2 C}{\partial u_1 \partial u_2} = G_1(G_2 + G_3) w^{\frac{1}{\theta_1} - 1}\left(-1 + (1 - u_3)^{\theta_1}\right) + G_4 G_5 w^{\frac{1}{\theta_1} - 2}\left(-1 + (1 - u_3)^{\theta_1}\right)^2$$
$$\text{(M5–2)}$$

$$\frac{\partial^3 C}{\partial u_1 \partial u_2 \partial u_3} = G_1(G_2+G_3)\left(-\theta_1(1-u_3)^{\theta_1-1}w^{\frac{1}{\theta_1}-1} + \left(-1+(1-u_3)^{\theta_1}\right)\left(\frac{1}{\theta_1}-1\right)w^{\frac{1}{\theta_1}-2}\frac{\partial w}{\partial u_3}\right)$$
$$+ G_4G_5\left(2\left(-1+(1-u_3)^{\theta_1}\right)\left(-\theta_1(1-u_3)^{\theta_1-1}\right)w^{\frac{1}{\theta_1}-2}\right.$$
$$\left. + \left(-1+(1-u_3)^{\theta_1}\right)^2\left(\frac{1}{\theta_1}-2\right)w^{\frac{1}{\theta_1}-2}w^{\frac{1}{\theta_1}-3}\frac{\partial w}{\partial u_3}\right) \quad \text{(M5–3)}$$

where

$$w = \left((1-u_1)^{\theta_2} + (1-u_2)^{\theta_2} - (1-u_1)^{\theta_2}(1-u_2)^{\theta_2}\right)^{\frac{\theta_1}{\theta_2}}\left(-1+(1-u_3)^{\theta_1}\right) + (1-u_3)^{\theta_1}$$

$$G_1 = (1-u_1)^{\theta_2-1}(1-u_2)^{\theta_2-1}\left((1-u_1)^{\theta_2} + (1-u_2)^{\theta_2} - (1-u_1)^{\theta_2}(1-u_2)^{\theta_2}\right)^{\frac{\theta_1}{\theta_2}-2}$$

$$G_2 = (\theta_1 - 1)\left(1-(1-u_1)^{\theta_2} - (1-u_2)^{\theta_2} + (1-u_1)^{\theta_2}(1-u_2)^{\theta_2}\right)$$

$$G_3 = -\theta_2 + 1 - (1-u_1)^{\theta_2} - (1-u_2)^{\theta_2} + ((1-u_1)(1-u_2))^{\theta_2}$$

$$G_4 = (\theta_1 - 1)(1-u_1)^{\theta_2-1}(1-u_2)^{\theta_2-1}\left(-1+(1-u_1)^{\theta_2}\right)\left(-1+(1-u_2)^{\theta_2}\right)$$

$$G_5 = \left((1-u_1)^{\theta_2} + (1-u_2)^{\theta_2} - (1-u_1)^{\theta_2}(1-u_2)^{\theta_2}\right)^{\frac{2\theta_1}{\theta_2}-2}$$

$$\frac{\partial w}{\partial u_3} = \theta_1(1-u_3)^{\theta_1-1}\left(\left((1-u_1)^{\theta_2} + (1-u_2)^{\theta_2} - (1-u_1)^{\theta_2}(1-u_2)^{\theta_2}\right)^{\frac{\theta_1}{\theta_2}-1} - 1\right)$$

M6 Copula

$$\frac{\partial C}{\partial u_1} = \frac{1}{u_1}(-\ln u_1)^{\theta_2-1}G^{\frac{\theta_1}{\theta_2}-1}w^{\frac{1}{\theta_1}-1}e^{-w^{-\frac{1}{\theta_1}}} \quad \text{(M6–1)}$$

$$\frac{\partial^2 C}{\partial u_1 \partial u_2} = \frac{1}{u_1u_2}(-\ln u_1)^{\theta_2-1}(-\ln u_2)^{\theta_2-1}e^{-w^{-\frac{1}{\theta_1}}}$$
$$\left(G^{\frac{2\theta_1}{\theta_2}-2}w^{\frac{2}{\theta_1}-2} + (\theta_2-\theta_1)G^{\frac{\theta_1}{\theta_2}-2}w^{\frac{1}{\theta_1}-1} + (\theta_1-1)G^{\frac{2\theta_1}{\theta_2}-2}w^{\frac{1}{\theta_1}-2}\right) \quad \text{(M6–2)}$$

$$\frac{\partial^3 C}{\partial u_1 \partial u_2 \partial u_3} = \frac{1}{u_1u_2u_3}(-\ln u_1)^{\theta_2-1}(-\ln u_2)^{\theta_2-1}(-\ln u_3)^{\theta_1-1}e^{-w^{\frac{1}{\theta_1}}}\left(G^{\frac{2\theta_1}{\theta_2}-2}w^{\frac{3}{\theta_1}-3}\right.$$
$$+ (2\theta_1-2)w^{\frac{2}{\theta_1}-3} + (\theta_2-\theta_1)G^{\frac{\theta_1}{\theta_2}-2}w^{\frac{2}{\theta_1}-2} + (\theta_1-1)(2\theta_1-1)G^{\frac{2\theta_1}{\theta_2}-2}w^{\frac{1}{\theta_1}-3}$$
$$\left. + (\theta_1-1)G^{\frac{2\theta_1}{\theta_2}-2}w^{\frac{2}{\theta_1}-3} + (\theta_1-1)(\theta_2-\theta_1)G^{\left(\frac{\theta_1}{\theta_2}-2\right)}w^{\frac{1}{\theta_1}-2}\right) \quad \text{(M6-3)}$$

where

$$G = (-\ln u_1)^{\theta_2} + (-\ln u_2)^{\theta_2}; w = (-\ln u_3)^{\theta_1} + \left((-\ln u_1)^{\theta_2} + (-\ln u_2)^{\theta_2}\right)^{\frac{\theta_1}{\theta_2}}$$

M12 Copula

$$\frac{\partial C}{\partial u_1} = \frac{\left(u_1^{-1}-1\right)^{\theta_2-1}\left(\left(u_1^{-1}-1\right)^{\theta_2}+\left(u_2^{-1}-1\right)^{\theta_2}\right)^{\frac{\theta_1}{\theta_2}-1} w^{\frac{1}{\theta_1}-1}}{u_1^2\left(1+w^{\frac{1}{\theta_1}}\right)^2} \tag{M12–1}$$

$$\frac{\partial^2 C}{\partial u_1 \partial u_2} = \frac{\left(u_1^{-1}-1\right)^{\theta_2-1}\left(u_2^{-1}-1\right)^{\theta_2-1}}{u_1^2 u_2^2}\left(\left(u_1^{-1}-1\right)^{\theta_2}+\left(u_2^{-1}-1\right)^{\theta_2}\right)^{\frac{\theta_1}{\theta_2}-2}$$
$$\left(\frac{(\theta_2-\theta_1)w^{\frac{1}{\theta_1}-1}}{\left(1+w^{\frac{1}{\theta_1}}\right)^2}+\frac{(\theta_1-1)\left(\left(u_1^{-1}-1\right)^{\theta_2}+\left(u_2^{-1}-1\right)^{\theta_2}\right)^{\frac{\theta_1}{\theta_2}}w^{\frac{1}{\theta_1}-2}}{\left(1+w^{\frac{1}{\theta_1}}\right)^2}\right.$$
$$\left.+\frac{2\left(\left(u_1^{-1}-1\right)^{\theta_2}+\left(u_2^{-1}-1\right)^{\theta_2}\right)^{\frac{\theta_1}{\theta_2}}w^{\frac{2}{\theta_1}-2}}{\left(1+w^{\frac{1}{\theta_1}}\right)^3}\right) \tag{M12–2}$$

$$\frac{\partial^3 C}{\partial u_1 \partial u_2 \partial u_3} = \frac{\left(u_1^{-1}-1\right)^{\theta_2-1}\left(u_2^{-1}-1\right)^{\theta_2-1}\left(u_3^{-1}-1\right)^{\theta_1-1}}{u_1^2 u_2^2 u_3^2}\left(\left(u_1^{-1}-1\right)^{\theta_2}+\left(u_2^{-1}-1\right)^{\theta_2}\right)^{\frac{\theta_1}{\theta_2}-2}$$
$$\left((\theta_2-\theta_1)\frac{(\theta_1-1)\left(1+w^{\frac{1}{\theta_1}}\right)^2 w^{\frac{1}{\theta_1}-2}+2\left(1+w^{\frac{1}{\theta_1}}\right)w^{\frac{2}{\theta_1}-2}}{\left(1+w^{\frac{1}{\theta_1}}\right)^4}+(\theta_1-1)\left(\left(u_1^{-1}-1\right)^{\theta_2}+\left(u_2^{-1}-1\right)^{\theta_2}\right)^{\frac{\theta_1}{\theta_2}}\right.$$
$$\frac{(2\theta_1-1)\left(1+w^{\frac{1}{\theta_1}}\right)^2 w^{\frac{1}{\theta_1}-3}+2\left(1+w^{\frac{1}{\theta_1}}\right)w^{\frac{2}{\theta_1}-3}}{\left(1+w^{\frac{1}{\theta_1}}\right)^4}+2\left(\left(u_1^{-1}-1\right)^{\theta_2}+\left(u_{2_1}^{-1}-1\right)^{\theta_2}\right)^{\frac{\theta_1}{\theta_2}}$$
$$\left.\frac{(2\theta_1-2)\left(1+w^{\frac{1}{\theta_1}}\right)^3 w^{\frac{2}{\theta_1}-3}+3\left(1+w^{\frac{1}{\theta_1}}\right)^2 w^{\frac{3}{\theta_1}-3}}{\left(1+w^{\frac{1}{\theta_1}}\right)^6}\right)$$

where: $w = \left(\left(u_1^{-1}-1\right)^{\theta_2}+\left(u_2^{-1}-1\right)^{\theta_2}\right)^{\frac{\theta_1}{\theta_2}}+\left(u_3^{-1}-1\right)^{\theta_1}$

6

Plackett Copula

ABSTRACT

Similar to the Archimedean copulas, the non-Archimedean copulas can be classified as one-parameter non-Archimedean bivariate copulas, two-parameter non-Archimedean bivariate copulas, and multivariate ($d \geq 3$) non-Archimedean copulas. In recent years, successful applications of non-Archimedean copulas, such as meta-elliptical copulas and Plackett copulas, have been reported in hydrology and water resources management. In this chapter, we will focus on Plackett copulas and more specifically bivariate and trivariate Plackett copula.

6.1 Bivariate Plackett Copula

In this section, we will introduce the definition, parameter estimation, as well as the random variate simulation with the use of bivariate Plackett copulas.

6.1.1 Definition of Bivariate Plackett Copula

As discussed in Chapter 3, the Plackett copula is constructed using the algebraic method. The cross-product ratio θ, or odds ratio, is a measure of "association" or "dependence" in 2×2 contingency tables. Here, we label the categories for each variable as "low" and "high" and give four categories in Table 6.1, where a, b, c, and d represent the observed counts in the four categories, respectively. From Table 6.1, the cross-product ratio $(\theta : \theta > 0)$ is defined as $\theta = \frac{ad}{bc}$. Following Palaro and Hotta (2006), the dependence may be explained through θ as follows:

1. $0 < \theta < 1$ corresponds to negative dependence, i.e., observations are more concentrated in the "low-high" and "high-low" cells.
2. $\theta = 1$ corresponds to independence, each "observed" entry; for example, a is equal to its "expected value" under independence $\left[\text{i.e., } \frac{(a+b)(a+c)}{a+b+c+d}\right]$.
3. $\theta > 1$ corresponds to positive dependence, i.e., observations are more concentrated in the "low-low" and "high-high" cells.

Table 6.1. *Two-by-two contingency table.*

		Column variable		
Row variable		**Low** $(X \leq x)$	**High** $(X > x)$	
	Low $(Y \leq y)$	a	b	$a+b$
	High $(Y > y)$	c	d	$c+d$
		$a+c$	$b+d$	$a+b+c+d$

With the use of the 2×2 contingency table, Plackett (1965) developed what is now called the Plackett copula for bivariate continuous random variables. Assuming the continuous random variables X and Y with marginals F_X and F_Y and the joint distribution function $H(x, y) = P(X \leq x, Y \leq y)$, then the "low" and "high" categories for the column and row variables are replaced by events $X \leq x, X > x$ and $Y \leq y, Y > y$, respectively. According to the definition of cross-product ratio $\theta = \frac{ad}{bc}$, it is clear that a, b, c, and d denote the probabilities of $P(X \leq x, Y \leq y), P(X > x, Y \leq y), P(X \leq x, Y > y)$, and $P(X > x, Y > y)$, respectively.

Now, based on the bivariate probability relation discussed in Chapter 3, we have the following:

$$a = P(X \leq x, Y \leq y) \tag{6.1a}$$

$$b = F_Y(y) - H(x, y) \tag{6.1b}$$

$$c = F_X(x) - H(x, y) \tag{6.1c}$$

$$d = 1 - F_X(x) - F_Y(y) + H(x, y) \tag{6.1d}$$

Replacing the values of a, b, c, and d, we obtain the expression of parameter θ as follows:

$$\theta = \frac{H(x, y)[1 - F_X(x) - F_Y(y) + H(x, y)]}{[F_X(x) - H(x, y)][F_Y(y) - H(x, y)]} \tag{6.1e}$$

Let $u = F_X(x)$ and $v = F_Y(y)$. Equation (6.1e) may be written in the copula form by applying Sklar's theorem as follows:

$$\theta = \frac{C(u, v)[1 - u - v + C(u, v)]}{[u - C(u, v)][v - C(u, v)]} \tag{6.2}$$

Solving for C in Equation (6.2), we obtain the Plackett copula:

$$C(u, v; \theta) = \frac{[1 + (\theta - 1)(u + v)] - \sqrt{[1 + (\theta - 1)(u + v)]^2 - 4\theta(\theta - 1)uv}}{2(\theta - 1)}; \theta > 0 \,\&\, \theta \neq 1 \tag{6.3a}$$

$$C(u, v; \theta) = uv; \theta = 1 \tag{6.3b}$$

Taking the partial derivatives with respect to u and v, its copula density function can be written as follows:

$$c(u, v; \theta) = \frac{\partial^2 C(u, v; \theta)}{\partial u \partial v} = \frac{\theta[1 + (\theta - 1)(u + v - 2uv)]}{\left\{[1 + (\theta - 1)(u + v)]^2 - 4\theta(\theta - 1)uv\right\}^{1.5}} \tag{6.4}$$

Taking the partial derivative of equation (6.3a) with respect to u or v, the conditional probability distributions can be obtained as follows:

$$C(V \le v | U = u) = P(Y \le y | X = x) = \frac{\partial C(u, v; \theta)}{\partial u}$$

$$= \frac{1}{2} + \frac{-1 + u + v - u\theta + v\theta}{2\sqrt{[1 + (\theta - 1)(u + v)]^2 - 4\theta(\theta - 1)uv}} \tag{6.5}$$

$$C(U \le u | V = v) = P(X \le x | Y = y) = \frac{\partial C(u, v; \theta)}{\partial v}$$

$$= \frac{1}{2} + \frac{-1 + u + v + u\theta - v\theta}{2\sqrt{[1 + (\theta - 1)(u + v)]^2 - 4\theta(\theta - 1)uv}} \tag{6.6}$$

Example 6.1 **Graph the Plackett copula function and its density function with $\theta = 20, \theta = 1$, and $\theta = 0.5$.**

Solution: Using Equations (6.3) and (6.4), we can graph the Plackett copula function and its density function in Figure 6.1 using $u, v \in [0, 1]$. From the copula density function plots with different parameters in Figure 6.1, it is seen that (i) the density is higher if both u and v take on smaller or bigger values at the same time for $\theta = 20$, i.e., high follows high and low follows low as the representation of positive dependence; (ii) the density is constant, i.e., 1, if $\theta = 1$ for the independent random variables; and (iii) the negative dependence is observed from the density function plot for $\theta = 0.5$, in this case, smaller u and bigger v reach higher density and vice versa.

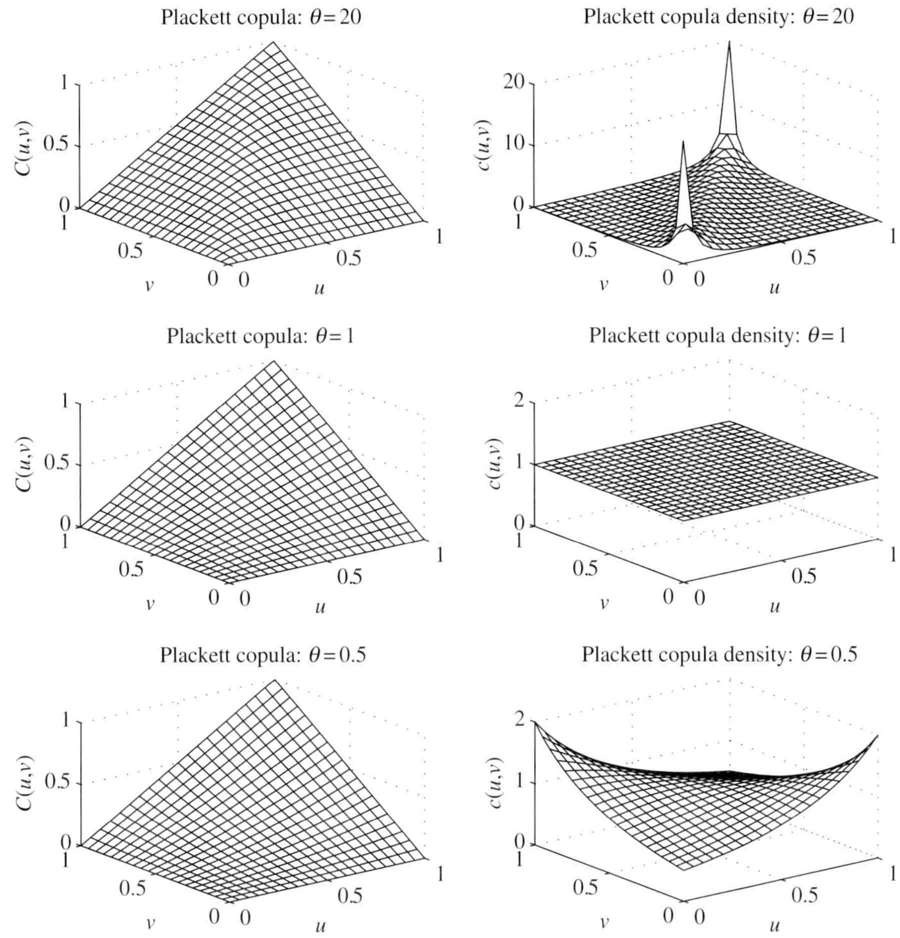

Figure 6.1 Plackett copula function and its density function plot for $\boldsymbol{\theta} = 20, \boldsymbol{\theta} = 1$ ***and*** $\boldsymbol{\theta} = 0.5$.

6.1.2 Simulation of Bivariate Plackett Copula

Following the Rosenblatt transform (Rosenblatt, 1952), the random variable can be simulated as follows:

1. Simulate two independent random variables (w_1, w_2) from the uniform distribution $U(0, 1)$.
2. Set $u = w_1$.
3. Using Equation (6.5a) and set $w_2 = C(v|u)$, i.e.,

$$w_2 = \frac{\partial C(u, v; \theta)}{\partial u} = \frac{1}{2} + \frac{-1 + u + v - u\theta + v\theta}{2\sqrt{[1 + (\theta - 1)(u + v)]^2 - 4\theta(\theta - 1)uv}} \tag{6.7}$$

After some algebraic manipulation of Equation (6.7), v can be solved as follows:

$$v = \frac{c - (1 - 2w_2)d}{2b} \tag{6.8}$$

where

$$b = \theta + S(\theta - 1)^2; c = 2S\left(u\theta^2 + 1 - u\right) + \theta(1 - 2S);$$
$$d = \theta^{0.5}\left(\theta + 4Su(1 - u)(1 - \theta)^2\right)^{0.5} S = w_2(1 - w_2).$$

Example 6.2 **Generate the random variables from the Plackett copula function.**

To generate the variables, use the following information:

1. Simulate Plackett random variables from the uniformly distributed independent random variables $w_1 = 0.1645$, $w_2 = 0.9629$, and $\theta = 50$.
2. Given $\theta = 50$, $\theta = 2.5$, and $\theta = 0.1$, graph the the random variables generated from the Plackett copula with a sample size of 100.

Solution: We can use the procedure discussed in Section 6.1.2 to generate the random variables from Plackett copula:

1. $w_1 = 0.1645$, $w_2 = 0.9629$, and $\theta = 50$.
   Set $u = w_1 = 0.1645$. We may then compute the random variate v using $w_2 = C(v|U = u; \theta)$.
   Solving Equation (6.8), we have the following:
   $$S = 0.0357; b = 135.7723; c = 75.8700; d = 69.6972.$$
   Then we have the following:
   $$v = \frac{c - (1 - 2w_2)d}{2b} = \frac{75.8700 - [1 - 2(0.9629)](69.6972)}{2(135.7723)} = 0.5170$$
   Thus, the generated random variables are $(u, v) = (0.1645, 0.5170)$.
2. Set $\theta = 50$, $\theta = 2.5$ and $\theta = 0.1$ with a sample size of 100.
   Using the same procedure as in step 1, we graph the simulated random variables with a sample size of 100 in Figure 6.2. Again, Figure 6.2 clearly shows that (i) the random variables generated are positively dependent with $\theta = 50$; (ii) the random variables generated are negatively dependent with $\theta = 0.1$; and (iii) the random variables generated are more scattered within $[0, 1]^2$ that are near independent when $\theta = 2.5$.

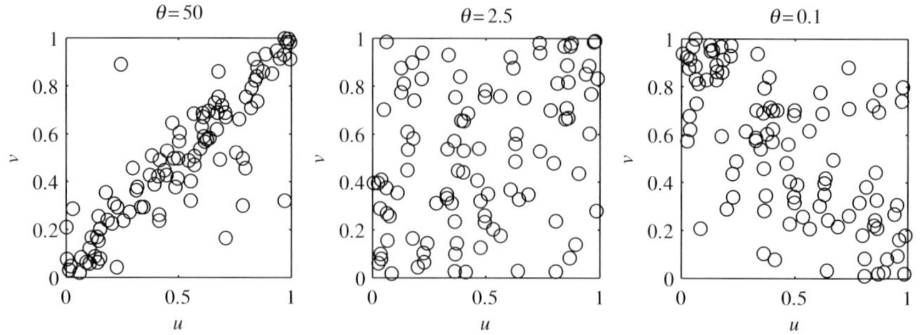

Figure 6.2 Scatter plot of simulated random variables from the Plackett copula.

6.1.3 Parameter Estimation for Bivariate Plackett Copulas

As discussed in Section 3.6, the full ML, IFM, and semiparametric (pseudo-ML) methods may be applied to estimate the parameter numerically for the Plackett copula function. Here, without further discussion, we will give one example to illustrate the procedure of parameter estimation.

Example 6.3 **Using the random variables (Table 6.2) and assuming (a) random variables X and Y are sampled from the normal distribution and gamma distribution, respectively, and (b) the joint distribution may be modeled using Plackett copula, estimate the parameters using full ML, IFM, and semiparametric methods.**

Table 6.2. *Sample data for Example 6.3.*

No.	*X*	*Y*	No.	*X*	*Y*
1	11.276	5.049	26	12.793	12.942
2	19.570	12.015	27	16.772	4.140
3	10.864	3.691	28	12.215	4.522
4	14.517	9.233	29	24.909	7.689
5	17.512	6.862	30	17.580	12.331
6	14.312	5.343	31	17.200	7.060
7	17.785	12.689	32	10.621	5.583
8	9.457	8.182	33	10.310	19.026
9	13.290	8.531	34	8.957	3.648
10	15.470	31.129	35	18.735	7.534
11	18.392	20.848	36	11.536	7.519
12	9.411	8.567	37	16.264	10.727
13	18.883	15.874	38	21.382	21.947
14	11.749	12.142	39	19.153	11.813
15	14.173	10.224	40	17.355	7.988
16	14.044	6.223	41	17.877	12.159
17	13.032	7.594	42	14.799	9.622
18	18.374	14.827	43	11.457	11.147
19	17.979	14.283	44	18.601	14.626
20	7.656	4.639	45	11.636	4.732
21	14.642	10.039	46	11.427	6.263
22	19.871	16.856	47	15.067	11.378
23	7.769	17.575	48	16.328	14.778
24	12.870	7.763	49	21.471	29.678
25	14.119	6.964	50	15.327	9.639

Solution: With the assumption of X following the Gumbel distribution (Equation (2.10)) and Y following the gamma distribution (Equation (2.8)), applying MLE, we can initially estimate the parameters of random variables X and Y as follows:

Random variable X: $\mu_X = 14.9358; \sigma_X = 3.8484$.
Random variable Y: $\alpha_Y = 4.0031; \beta_Y = 0.3668$.

In addition, using Equation (3.72), we can compute the sample Kendall correlation coefficient as $\tau_n = 0.3690$.

1. **Full ML Method:**
   As discussed in Section 3.6.1, we will need to estimate the parameters of marginal distributions and copula function simultaneously with the full log-likelihood function given as follows:

$$LL = \sum_i \ln\left(c^{plackett}\left(F_X^{Normal}(x_i; \mu_X, \sigma_X), F_Y^{Gamma}(y_i; \alpha_Y, \beta_Y); \theta\right)\right. \\ + \sum_i \ln\left(f_X^{Normal}(x_i; \mu_X, \sigma_X)\right) + \sum_i \ln\left(f_Y^{Gamma}(y_i; \alpha_Y, \beta_Y)\right)$$

   Using the parameters initially estimated for marginal distributions and assuming the initial estimate of the Plackett copula parameter $\theta = 10$, we can use optimization toolbox in MATLAB to estimate the full set of parameters. The fitted marginal distribution is listed in Table 6.3 with the estimated parameters listed in Table 6.4.

Table 6.3. *Cumulative probability computed using the fitted normal and gamma distributions and Weibull probability plotting-position formula.*

X	FMLE	IFM	Empirical	Y	FMLE	IFM	Empirical
19.570	0.871	0.882	0.900	12.015	0.638	0.633	0.640
10.864	0.129	0.141	0.160	3.691	0.047	0.045	0.040
14.517	0.427	0.449	0.460	9.233	0.434	0.428	0.460
17.512	0.724	0.742	0.680	6.862	0.242	0.237	0.220
14.312	0.406	0.428	0.440	5.343	0.132	0.129	0.140
17.785	0.747	0.764	0.720	12.689	0.680	0.675	0.720
9.457	0.067	0.075	0.100	8.182	0.348	0.343	0.400
13.290	0.308	0.328	0.360	8.531	0.377	0.371	0.420
15.470	0.526	0.548	0.560	31.129	0.996	0.996	0.980
18.392	0.795	0.810	0.800	20.848	0.946	0.944	0.920
9.411	0.065	0.073	0.080	8.567	0.380	0.374	0.440
18.883	0.829	0.843	0.860	15.874	0.830	0.827	0.840
11.749	0.183	0.199	0.260	12.142	0.646	0.641	0.660
14.173	0.392	0.414	0.420	10.224	0.512	0.506	0.540
14.044	0.380	0.401	0.380	6.223	0.193	0.189	0.180
13.032	0.284	0.304	0.340	7.594	0.300	0.295	0.320
18.374	0.794	0.809	0.780	14.827	0.789	0.785	0.820
17.979	0.763	0.780	0.760	14.283	0.764	0.760	0.760

Table 6.3. (*cont.*)

X	FMLE	IFM	Empirical	Y	FMLE	IFM	Empirical
7.656	0.025	0.028	0.020	4.639	0.091	0.088	0.100
14.642	0.440	0.462	0.480	10.039	0.498	0.492	0.520
19.871	0.887	0.897	0.920	16.856	0.863	0.860	0.860
7.769	0.026	0.030	0.040	17.575	0.883	0.881	0.880
12.870	0.270	0.289	0.320	7.763	0.314	0.309	0.360
14.119	0.387	0.409	0.400	6.964	0.250	0.245	0.240
12.793	0.264	0.282	0.300	12.942	0.694	0.690	0.740
16.772	0.656	0.676	0.620	4.140	0.066	0.064	0.060
12.215	0.217	0.234	0.280	4.522	0.085	0.082	0.080
24.909	0.994	0.995	0.980	7.689	0.308	0.303	0.340
17.580	0.730	0.748	0.700	12.331	0.658	0.653	0.700
17.200	0.696	0.715	0.640	7.060	0.257	0.253	0.260
10.621	0.116	0.127	0.140	5.583	0.148	0.145	0.160
10.310	0.101	0.111	0.120	19.026	0.916	0.914	0.900
8.957	0.052	0.058	0.060	3.648	0.045	0.044	0.020
18.735	0.819	0.833	0.840	7.534	0.295	0.290	0.300
11.536	0.169	0.184	0.220	7.519	0.294	0.289	0.280
16.264	0.607	0.628	0.580	10.727	0.549	0.544	0.560
21.382	0.945	0.951	0.940	21.947	0.959	0.958	0.940
19.153	0.847	0.859	0.880	11.813	0.625	0.620	0.620
17.355	0.710	0.729	0.660	7.988	0.332	0.327	0.380
17.877	0.755	0.772	0.740	12.159	0.647	0.642	0.680
14.799	0.456	0.478	0.500	9.622	0.465	0.459	0.480
11.457	0.164	0.178	0.200	11.147	0.579	0.574	0.580
18.601	0.810	0.824	0.820	14.626	0.780	0.776	0.780
11.636	0.175	0.191	0.240	4.732	0.096	0.093	0.120
11.427	0.162	0.176	0.180	6.263	0.196	0.192	0.200
15.067	0.484	0.506	0.520	11.378	0.596	0.590	0.600
16.328	0.613	0.634	0.600	14.778	0.787	0.783	0.800
21.471	0.948	0.953	0.960	29.678	0.995	0.994	0.960
15.327	0.511	0.533	0.540	9.639	0.466	0.461	0.500

Table 6.4. *Estimated parameters using the preceding three methods.*

Method	Univariate		Copula	
	X~normal	*Y*~Gamma	θ	LL
Full ML	(15.224, 3.846)	(4.039, 0.369)	7.500	–275.327
IFM	(15.011, 3.851)	(4.069, 0.369)	7.167	8.106
Semiparametric	–	–	7.759	8.464

2. **IFM Method:**

   As discussed in Section 3.6.2, the parameters of marginal distributions and copulas are estimated separately with the use of IFM method. We will first compute the cumulative probability using the parameters initially estimated for the marginal distributions listed in Table 6.3. Then we will estimate the parameter of the Plackett copula using the ML method (the optimization toolbox in MATLAB) and the computed cumulative probabilities as random variates as follows:

$$LL = \sum_i ln\left(c^{plackett}\left(\hat{F}_X(x_i;\hat{\mu}_X,\hat{\sigma}_X),\hat{F}_Y\left(y_i;\hat{\alpha}_Y,\hat{\beta}_Y\right);\theta\right)\right)$$

   The estimated copula parameter is listed in Table 6.4.

3. **Semiparametric Method:**

   As discussed in Section 3.6.3, the semiparametric method is also called the pseudo-ML method. The marginal distributions are estimated nonparametrically using the Weibull plotting-position formula (Equation (3.92)) as listed in Table 6.3. Now with the use of the probability estimated nonparametrically, the pseudo-log-likelihood function can be written as follows:

$$LL = \sum_i \ln\left(c^{plackett}\left(\hat{F}_n(x_i),\hat{F}_n(y_i);\theta\right)\right)$$

   The estimated parameter is again estimated using the optimization toolbox in MATLAB and listed in Table 6.4. From Table 6.4, it is seen that there is minimal difference in regard to the parameters of the marginal distributions estimated separately from the copula using the IFM method and those estimated simultaneously using the full ML method. Figure 6.3

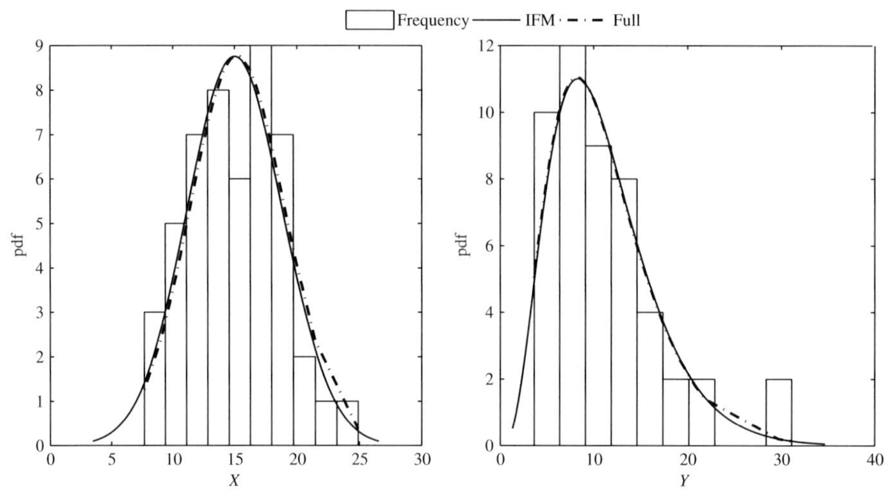

Figure 6.3 Comparison of frequency and the fitted probability distributions using IFM and Full MLE.

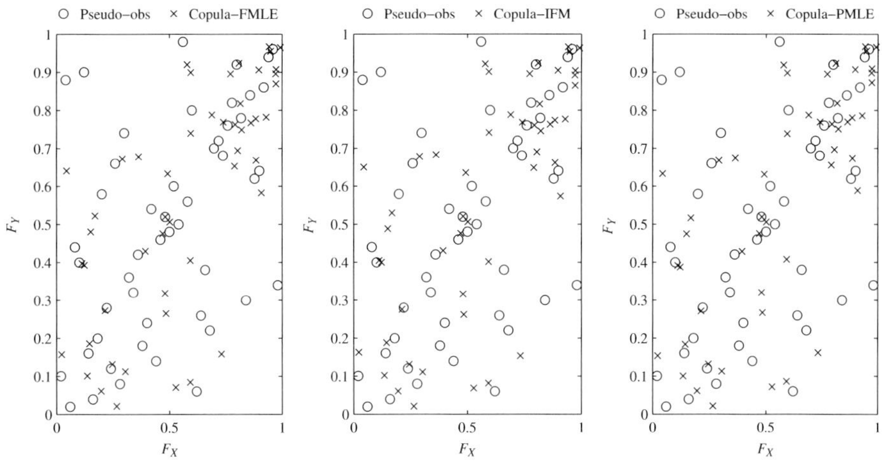

Figure 6.4 Comparison of observations with simulated random variables with three estimation methods.

further indicates this similarity through the univariate probability density comparison. Figure 6.4 compares the observed variates with the simulated variates from the fitted copula function. Figure 6.4 shows that the performances are very similar for the copulas with parameters estimated using three different techniques.

Example 6.4 **Using the sample data and the parameters estimated with the IFM method in Example 6.3, compute the joint return period and conditional return period of**

$$T(X > 19 \cap Y > 21), T(X > 19 | Y > 21), T(X > 19 | Y = 21).$$

Solution: Applying the parameters estimated for the marginal distributions listed in Table 6.4 for the IFM method, we have

$$F_X(X \leq 19; \mu = 15.011, \sigma = 3.852) \approx 0.850 \text{ (Normal distribution)}$$
$$F_Y(Y \leq 21; \alpha = 4.069, \beta = 2.712) \approx 0.946 \text{ (Gamma distribution)}$$

i. $T(X > 19 \cap Y > 21)$

In this case, we are evaluating the recurrence interval if both X and Y exceed the value given in the preceding. Applying Equation (3.127) for the "and" case, we have the following:

$$F(X>19\bigcap Y>21)=1-F_X(X\le 19)-F_Y(Y\le 21)+F(X\le 19,Y\le 21)$$
$$=1-F_X(X\le 19)-F_Y(Y\le 21)+C^{plackett}\Big(F_X(X\le 19),F_Y(Y\le 21);\theta\Big)$$
$$=1-0.850-0.946+C^{plackett}(0.850,0.946;7.167)$$
$$=1-0.850-0.946+0.824=0.028$$
$$T(X>19\cap Y>21)=\frac{1}{F(X>19\cap Y>21)}=\frac{1}{0.028}=36.10(\text{time units})$$

ii. $T(X>19|Y>21)$

In this case, we are evaluating the recurrence interval of $X>19$ under the condition of $Y>21$:

$$F(X>19|Y>21)=\frac{F(X>19\bigcap Y>21)}{F(Y>21)}$$
$$=\frac{1-F_X(X\le 19)-F_Y(Y\le 21)+C^{plackett}\Big(F_X(X\le 19),F_Y(Y\le 21);\theta\Big)}{1-F(Y\le 21)}$$
$$=\frac{1-0.850-0.946+0.824}{1-0.946}=\frac{0.028}{0.054}=0.516$$

Applying Equation (2.91), we have the following:

$$T(X>19|Y>21)=\frac{1}{1-F_Y(y\le 21)}\frac{1}{1-F_X(x\le 19)-F_Y(Y\le 21)+C^{plackett}(F_X,F_Y;\theta)}$$
$$=\frac{1}{0.0537}\frac{1}{0.0277}\approx 672(\text{time units})$$

iii. $T(X>19|Y=21)$

In this case, we are evaluating the recurrence interval of $X>19$ under the condition that Y is exactly equal to 21:

$$F(X>19|Y=21)=1-F(X\le 19|Y=21)$$
$$=1-\frac{\partial C\Big(F_X(X\le 19),F_Y(Y\le 21);\theta\Big)}{\partial F_Y(Y\le 21)}=1-0.528=0.472$$
$$T(X>19|Y=21)=\frac{1}{F(X>19|Y=21)}=\frac{1}{0.472}\approx 2.12(\text{time units})$$

Comparing the joint return period with the two conditional return periods we calculated, it is seen that the recurrence interval is longest for the conditional return period of $(X>19|Y>21)$.

6.2 Trivariate Plackett Copula

In this section, we will focus on the trivariate Plackett copula, including its definition, the derivation of trivariate Plackett copula density, and a brief introduction of the parameter estimation method. Given the complexity of parameter estimation for the trivariate Plackett copula and the simplicity of other multivariate copula approaches, we will not further discuss the simulation as well as the formal goodness-of-fit measure in detail.

6.2.1 Definition of Cross-Product Ratio for the Trivariate Plackett Copula

For the given (u, v, w), there are three compatible bivariate copulas: C_{UV}, C_{VW}, and C_{UW}. Analogous to the bivariate case, the trivariate constant cross-product ratio θ_{UVW} can be defined following Kao and Govindaraju (2008), Song and Singh (2010) as

$$\theta_{UVW} = \frac{P_{000}P_{011}P_{101}P_{110}}{P_{111}P_{100}P_{010}P_{001}} \tag{6.9}$$

where

$$\begin{cases} P_{000} & = C_{UVW}(u, v, w) \\ P_{100} & = C_{VW}(v, w) - C_{UVW}(u, v, w) \\ P_{010} & = C_{UW}(u, w) - C_{UVW}(u, v, w) \\ P_{001} & = C_{UV}(u, v) - C_{UVW}(u, v, w) \\ P_{110} & = w - C_{UV}(u, w) - C_{VW}(v, w) + C_{UVW}(u, v, w) \\ P_{101} & = v - C_{UV}(u, v) - C_{VW}(v, w) + C_{UVW}(u, v, w) \\ P_{011} & = u - C_{UV}(u, v) - C_{VW}(u, w) + C_{UVW}(u, v, w) \\ P_{111} & = 1 - u - v - w + C_{UV}(u, v) + C_{VW}(v, w) + C_{UW}(u, w) - C_{UVW}(u, v, w) \end{cases} \tag{6.10}$$

Here C_{UV}, C_{VW}, and C_{UW} are bivariate Plackett copulas with dependence parameters θ_{UV}, θ_{VW}, and θ_{UW}, for the given θ_{UVW}. Denoting $z = C_{UVW}(u, v, w)$, one can compute $C_{UVW}(u, v, w)$ as follows:

$$\theta_{UVW}(a_1 - z)(a_2 - z)(a_3 - z)(a_4 - z) - z(z - b_1)(z - b_2)(z - b_3) = 0 \tag{6.11}$$

where

$$\begin{cases} a_1 & = C_{VW}(v, w), a_2 = C_{UW}(u, w), a_3 = C_{UV}(u, v) \\ a_4 & = 1 - u - v - w + C_{UV}(u, v) + C_{VW}(v, w) + C_{UW}(u, w) \\ b_1 & = C_{UW}(u, w) + C_{VW}(v, w) - w \\ b_2 & = C_{UV}(u, v) + C_{VW}(v, w) - v \\ b_3 & = C_{UW}(u, w) + C_{UV}(u, v) - u \end{cases} \tag{6.12}$$

For the given $\theta_{UV}, \theta_{VW}, \theta_{UW}$, and θ_{UVW}, the corresponding trivariate Plackett copula may be obtained from Equations (6.11) and (6.12). For $C_{UVW}(u, v, w)$ to be a valid three-copula, the following conditions needs to be satisfied:

1. Since each component in Equation (6.11) is a probability measure, we have the following:

$$C_{UVW}(u, v, w) \in [b, a], b = \max(0, b_1, b_2, b_3); a = \min(a_1, a_2, a_3, a_4) \tag{6.13}$$

2. Equation (6.13) is the Fréchet–Hoeffding bounds for trivairate joint distributions with the known bivariate joint distributions (Joe, 1997).
3. As discussed in Section 3.1.2 of Chapter 3, Equations (3.23)–(3.26) need to be satisfied.
4. The copula density is $C_{UVW}(u,v,w) = \dfrac{\partial^3 C_{UVW}(u,v,w)}{\partial u \partial v \partial w} \geq 0$. Following Kao and Govidaraju (2008), the derivation of the density function will be discussed in Section 6.2.2.

With the fulfillment of the preceding four conditions, for the given cross-product ratio parameters $\theta_{UV}, \theta_{VW}, \theta_{UW}$, and θ_{UVW}, $z = C_{UVW}(u,v,w)$ may be computed numerically with the following steps:

1. Compute C_{UV}, C_{VW}, and C_{UW} using Equation (6.3).
2. To compute C_{UVW}, Equation (5.11) can be rewritten as follows:

$$\begin{aligned}(\theta_{UVW}-1)z^4 &+ [-\theta_{UVW}(a_1+a_2+a_3+a_4)+(b_1+b_2+b_3)]^2 \\ &+ \{\theta_{UVW}[a_1a_2+(a_1+a_2)(a_3+a_4)+a_3a_4]-[b_1b_2+b_3(b_1+b_2)]\}z^2 \\ &+ \{-\theta_{UVW}[a_1a_2(a_3+a_4)+a_3a_4(a_1+a_2)]+b_1b_2b_3\}z+\theta_{UVW}a_1a_2a_3a_4=0\end{aligned} \tag{6.14}$$

Let $f(z)$ represent the left side of Equation (6.14). We may use Newton's iterative method to compute z numerically as follows:

$$z_{n+1} = z_n - \frac{f(z_n)}{f'(z_n)} \tag{6.15}$$

where $f'(z)$ is the first derivative of $f(z)$ with respect to z; z_n and z_{n+1} are the nth and $(n+1)$th iteratively computed values of z.

6.2.2 Derivation of Density Function of the Trivariate Plackett Copula

Following Kao and Govindaraju (2008), the density function of trivariate Plackett copula may be derived in the following manner:

1. Solve C_{UVW} using given parameter θ_{UVW} and known bivariate copulas from Equation (6.11) or equivalently Equation (6.14).
2. Compute first-order derivatives of $\dfrac{\partial C_{UV}}{\partial u}, \dfrac{\partial C_{UV}}{\partial v}, \dfrac{\partial C_{UW}}{\partial u}, \dfrac{\partial C_{UW}}{\partial w}, \dfrac{\partial C_{VW}}{\partial v}$, and $\dfrac{\partial C_{VW}}{\partial w}$ from the corresponding known bivariate copulas. Similar to the vine copula discussed in Chapter 4, these bivariate copulas are not required to belong to the Plackett copula family, and each may belong to a different copula family.
3. Compute the first-order derivatives of $P_{000}, P_{010}, P_{100}, P_{011}, P_{110}, P_{101}, P_{011}$, and P_{111} with respect to u, v, w, respectively, as follows:

$$\begin{cases} \dfrac{\partial P_{000}}{\partial u} = -\dfrac{\partial P_{100}}{\partial u} = \dfrac{\partial C_{UVW}}{\partial u} \\ \dfrac{\partial P_{010}}{\partial u} = -\dfrac{\partial P_{110}}{\partial u} = -\dfrac{\partial C_{UVW}}{\partial u} + \dfrac{\partial C_{UW}}{\partial u} \\ \dfrac{\partial P_{001}}{\partial u} = -\dfrac{\partial P_{101}}{\partial u} = -\dfrac{\partial C_{UVW}}{\partial u} + \dfrac{\partial C_{UV}}{\partial u} \\ \dfrac{\partial P_{011}}{\partial u} = -\dfrac{\partial P_{111}}{\partial u} = \dfrac{\partial C_{UVW}}{\partial u} - \dfrac{\partial C_{UV}}{\partial u} - \dfrac{\partial C_{UW}}{\partial u} + 1 \end{cases} \tag{6.16}$$

$$\begin{cases} \dfrac{\partial P_{000}}{\partial v} = -\dfrac{\partial P_{010}}{\partial v} = \dfrac{\partial C_{UVW}}{\partial v} \\ \dfrac{\partial P_{100}}{\partial v} = -\dfrac{\partial P_{110}}{\partial v} = -\dfrac{\partial C_{UVW}}{\partial v} + \dfrac{\partial C_{VW}}{\partial v} \\ \dfrac{\partial P_{001}}{\partial v} = -\dfrac{\partial P_{011}}{\partial v} = -\dfrac{\partial C_{UVW}}{\partial v} + \dfrac{\partial C_{UV}}{\partial v} \\ \dfrac{\partial P_{101}}{\partial v} = -\dfrac{\partial P_{111}}{\partial v} = \dfrac{\partial C_{UVW}}{\partial v} - \dfrac{\partial C_{UV}}{\partial v} - \dfrac{\partial C_{VW}}{\partial v} + 1 \end{cases} \tag{6.17}$$

$$\begin{cases} \dfrac{\partial P_{000}}{\partial w} = -\dfrac{\partial P_{001}}{\partial w} = \dfrac{\partial C_{UVW}}{\partial w} \\ \dfrac{\partial P_{100}}{\partial w} = -\dfrac{\partial P_{101}}{\partial w} = -\dfrac{\partial C_{UVW}}{\partial w} + \dfrac{\partial C_{VW}}{\partial w} \\ \dfrac{\partial P_{010}}{\partial w} = -\dfrac{\partial P_{011}}{\partial w} = -\dfrac{\partial C_{UVW}}{\partial w} + \dfrac{\partial C_{UW}}{\partial w} \\ \dfrac{\partial P_{110}}{\partial w} = -\dfrac{\partial P_{111}}{\partial w} = \dfrac{\partial C_{UVW}}{\partial w} - \dfrac{\partial C_{UW}}{\partial w} - \dfrac{\partial C_{VW}}{\partial w} + 1 \end{cases} \tag{6.18}$$

4. Compute $\dfrac{\partial C_{UVW}}{\partial u}, \dfrac{\partial C_{UVW}}{\partial v}, \dfrac{\partial C_{UVW}}{\partial w}$ as follows:

$$\begin{aligned} &\frac{\partial P_{000}}{\partial u} P_{011} P_{101} P_{110} + P_{000} \frac{\partial P_{011}}{\partial u} P_{101} P_{110} + P_{000} P_{011} \frac{\partial P_{101}}{\partial u} P_{110} + P_{000} P_{011} P_{101} \frac{\partial P_{110}}{\partial u} \\ &\quad - \theta_{UVW} \left(\frac{\partial P_{111}}{\partial u} P_{100} P_{010} P_{001} + P_{111} \frac{\partial P_{100}}{\partial u} P_{010} P_{001} + P_{111} P_{100} \frac{\partial P_{010}}{\partial u} P_{001} \right. \\ &\quad \left. + P_{111} P_{110} P_{101} \frac{\partial P_{110}}{\partial u} \right) = 0 \end{aligned} \tag{6.19}$$

$$\frac{\partial P_{000}}{\partial v}P_{011}P_{101}P_{110}+P_{000}\frac{\partial P_{011}}{\partial v}P_{101}P_{110}+P_{000}P_{011}\frac{\partial P_{101}}{\partial v}P_{110}+P_{000}P_{011}P_{101}\frac{\partial P_{110}}{\partial v}$$
$$-\theta_{UVW}\left(\frac{\partial P_{111}}{\partial v}P_{100}P_{010}P_{001}+P_{111}\frac{\partial P_{100}}{\partial v}P_{010}P_{001}+P_{111}P_{100}\frac{\partial P_{010}}{\partial v}P_{001}\right.$$
$$\left.+P_{111}P_{110}P_{101}\frac{\partial P_{110}}{\partial v}\right)=0 \tag{6.20}$$

$$\frac{\partial P_{000}}{\partial w}P_{011}P_{101}P_{110}+P_{000}\frac{\partial P_{011}}{\partial w}P_{101}P_{110}+P_{000}P_{011}\frac{\partial P_{101}}{\partial w}P_{110}+P_{000}P_{011}P_{101}\frac{\partial P_{110}}{\partial w}$$
$$-\theta_{UVW}\left(\frac{\partial P_{111}}{\partial w}P_{100}P_{010}P_{001}+P_{111}\frac{\partial P_{100}}{\partial w}P_{010}P_{001}+P_{111}P_{100}\frac{\partial P_{010}}{\partial w}P_{001}\right.$$
$$\left.+P_{111}P_{110}P_{101}\frac{\partial P_{110}}{\partial w}\right)=0 \tag{6.21}$$

5. Compute the bivariate density function of c_{UV}, c_{VW}, c_{UW}.
6. Compute the second-order derivative of $P_{000}, P_{010}, P_{100}, P_{011}, P_{110}, P_{101}, P_{011}$, and P_{111} with respect to *u, v, w*, respectively, as follows:

$$\begin{cases}\dfrac{\partial^2 P_{000}}{\partial u\partial v}=-\dfrac{\partial^2 P_{100}}{\partial u\partial v}=-\dfrac{\partial^2 P_{010}}{\partial u\partial v}=\dfrac{\partial^2 P_{110}}{\partial u\partial v}=\dfrac{\partial^2 C_{UVW}}{\partial u\partial v}\\[2ex] \dfrac{\partial^2 P_{001}}{\partial u\partial v}=-\dfrac{\partial^2 P_{101}}{\partial u\partial v}=-\dfrac{\partial^2 P_{011}}{\partial u\partial v}=\dfrac{\partial^2 P_{111}}{\partial u\partial v}=-\dfrac{\partial^2 C_{UVW}}{\partial u\partial v}+\dfrac{\partial^2 C_{UV}}{\partial u\partial v}\end{cases} \tag{6.22}$$

$$\begin{cases}\dfrac{\partial^2 P_{000}}{\partial u\partial w}=-\dfrac{\partial^2 P_{100}}{\partial u\partial w}=-\dfrac{\partial^2 P_{001}}{\partial u\partial w}=\dfrac{\partial^2 P_{101}}{\partial u\partial w}=\dfrac{\partial^2 C_{UVW}}{\partial u\partial w}\\[2ex] \dfrac{\partial^2 P_{010}}{\partial u\partial w}=-\dfrac{\partial^2 P_{110}}{\partial u\partial w}=-\dfrac{\partial^2 P_{011}}{\partial u\partial w}=\dfrac{\partial^2 P_{111}}{\partial u\partial v}=-\dfrac{\partial^2 C_{UVW}}{\partial u\partial w}+\dfrac{\partial^2 C_{UW}}{\partial u\partial w}\end{cases} \tag{6.23}$$

$$\begin{cases}\dfrac{\partial^2 P_{000}}{\partial v\partial w}=-\dfrac{\partial^2 P_{010}}{\partial v\partial w}=-\dfrac{\partial^2 P_{001}}{\partial v\partial w}=\dfrac{\partial^2 P_{011}}{\partial v\partial w}=\dfrac{\partial^2 C_{UVW}}{\partial v\partial w}\\[2ex] \dfrac{\partial^2 P_{100}}{\partial v\partial w}=-\dfrac{\partial^2 P_{110}}{\partial v\partial w}=-\dfrac{\partial^2 P_{101}}{\partial v\partial w}=\dfrac{\partial^2 P_{111}}{\partial v\partial w}=-\dfrac{\partial^2 C_{UVW}}{\partial v\partial w}+\dfrac{\partial^2 C_{VW}}{\partial v\partial w}\end{cases} \tag{6.24}$$

7. Compute $\dfrac{\partial^2 C_{UVW}}{\partial u\partial v}, \dfrac{\partial^2 C_{UVW}}{\partial v\partial w}, \dfrac{\partial^2 C_{UVW}}{\partial u\partial w}$. As an example, $\dfrac{\partial^2 C_{UVW}}{\partial u\partial v}$ may be computed by applying $\dfrac{\partial}{\partial v}$ to Equation (6.19) as follows:

$$\frac{\partial^2 P_{000}}{\partial u\partial v}P_{011}P_{101}P_{110}+\frac{\partial P_{000}}{\partial u}\frac{\partial P_{011}}{\partial v}P_{101}P_{110}+\frac{\partial P_{000}}{\partial u}P_{011}\frac{\partial P_{101}}{\partial v}P_{110}+\frac{\partial P_{000}}{\partial v}P_{011}P_{101}\frac{\partial P_{110}}{\partial v}$$
$$+\frac{\partial P_{000}}{\partial v}\frac{\partial P_{011}}{\partial u}P_{101}P_{110}+P_{000}\frac{\partial^2 P_{011}}{\partial u\partial v}P_{101}P_{110}+P_{000}\frac{\partial P_{011}}{\partial u}\frac{\partial P_{101}}{\partial v}P_{110}+P_{000}\frac{\partial P_{011}}{\partial u}P_{101}\frac{\partial P_{110}}{\partial v}$$
$$+\frac{\partial P_{000}}{\partial v}P_{011}\frac{\partial P_{101}}{\partial u}P_{110}+P_{000}\frac{\partial P_{011}}{\partial v}\frac{\partial P_{101}}{\partial u}P_{110}+P_{000}P_{011}\frac{\partial^2 P_{101}}{\partial u\partial v}P_{110}+P_{000}P_{011}\frac{P_{101}}{\partial u}\frac{\partial P_{110}}{\partial v}$$
$$+\frac{\partial P_{000}}{\partial v}P_{011}P_{101}\frac{\partial P_{110}}{\partial u}+P_{000}\frac{\partial P_{011}}{\partial v}P_{101}\frac{\partial P_{110}}{\partial u}+P_{000}P_{011}\frac{\partial P_{101}}{\partial v}\frac{\partial P_{110}}{\partial u}+P_{000}P_{011}P_{101}\frac{\partial^2 P_{110}}{\partial u\partial v}$$
$$-\theta_{UVW}\left(\frac{\partial^2 P_{111}}{\partial u\partial v}P_{100}P_{010}P_{001}+\frac{\partial P_{111}}{\partial u}\frac{\partial P_{100}}{\partial v}P_{010}P_{001}+\frac{\partial P_{111}}{\partial u}P_{100}\frac{\partial P_{010}}{\partial v}P_{001}\right.$$
$$+\frac{\partial P_{111}}{\partial u}P_{100}P_{010}\frac{\partial P_{001}}{\partial v}+\frac{\partial P_{111}}{\partial v}\frac{\partial P_{100}}{\partial u}P_{010}P_{001}+P_{111}\frac{\partial^2 P_{100}}{\partial u\partial v}P_{010}P_{001}$$
$$+P_{111}\frac{\partial P_{100}}{\partial u}\frac{\partial P_{010}}{\partial v}P_{001}+P_{111}\frac{\partial P_{100}}{\partial u}P_{010}\frac{\partial P_{001}}{\partial v}+\frac{\partial P_{111}}{\partial v}P_{100}\frac{\partial P_{010}}{\partial u}P_{001}$$
$$+P_{111}\frac{\partial P_{100}}{\partial v}\frac{\partial P_{010}}{\partial u}P_{001}+P_{111}P_{100}\frac{\partial^2 P_{010}}{\partial u\partial v}P_{001}+P_{111}P_{100}\frac{\partial P_{010}}{\partial u}\frac{\partial P_{001}}{\partial v}$$
$$+\frac{\partial P_{111}}{\partial v}P_{100}P_{010}\frac{\partial P_{001}}{\partial u}+P_{111}\frac{\partial P_{100}}{\partial v}P_{010}\frac{\partial P_{001}}{\partial u}+P_{111}P_{100}\frac{\partial P_{010}}{\partial v}\frac{\partial P_{001}}{\partial u}$$
$$\left.+P_{111}P_{100}P_{010}\frac{\partial^2 P_{001}}{\partial u\partial v}\right) \tag{6.25}$$

Similarly, applying $\frac{\partial}{\partial w}, \frac{\partial}{\partial u}$, we can obtain $\frac{\partial^2 C_{UVW}}{\partial v\partial w}, \frac{\partial^2 C_{UVW}}{\partial u\partial w}$ from Equations (6.20) and (6.21), respectively.

8. Compute the probability density function $\frac{\partial^3 C_{UVW}}{\partial u\partial v\partial w}$ for the trivariate Plackett copula.

   Applying $\frac{\partial}{\partial w}$ to Equation (6.22), we have the following:

$$\begin{cases} \frac{\partial^3 P_{000}}{\partial u\partial v\partial w}=-\frac{\partial^3 P_{100}}{\partial u\partial v\partial w}=-\frac{\partial^3 P_{010}}{\partial u\partial v\partial w}=\frac{\partial^3 P_{110}}{\partial u\partial v\partial w}=\frac{\partial^3 C_{UVW}}{\partial u\partial v\partial w} \\ \frac{\partial^3 P_{001}}{\partial u\partial v\partial w}=-\frac{\partial^3 P_{101}}{\partial u\partial v\partial w}=-\frac{\partial^3 P_{011}}{\partial u\partial v\partial w}=\frac{\partial^3 P_{111}}{\partial u\partial v\partial w}=-\frac{\partial^3 C_{UVW}}{\partial u\partial v\partial w} \end{cases} \tag{6.26}$$

   Applying $\frac{\partial}{\partial w}$ to Equation (6.25), we obtain a new third-order derivative equation (we omit the derivative here). Substituting Equation (6.26) into the new equation derived for the third-order derivative, we have the density function as a function of $P_{000}, P_{011}, P_{101}, P_{110}, P_{111}, P_{010}, P_{010}, P_{001}$.

Example 6.5 **Express the PDF of trivariate Plackette copula with the following information: $\theta_{UVW} = 20; \theta_{UV} = 15; \theta_{UW} = 1.3;$ $\theta_{VW} = 1.4; u = 0.5; v = 0.975; w = 0.975$**

Solution: Applying the equations derived for the trivariate Plackett copula, we can compute the trivariate Plackett copula density function by following these procedure and steps:

1. Compute the bivariate Plackett copula for the paired variables with Equation (6.3); using bivariate variable (u, v) as an example, we have the following:

$$\begin{aligned} a_3 &= C_{UV} = C_{UV}(0.5, 0.975; 15) \\ &= \frac{[1+(15-1)(0.5+0.975)] - \sqrt{[1+(15-1)(0.5+0.975)]^2 - 4(15)(15-1)(0.5)(0.975)}}{2(15-1)} \\ &= 0.498 \end{aligned}$$

Similarly, we have the following:

$$a_2 = C_{UW} = C_{UW}(0.5, 0.975; 1.3) = 0.489$$
$$a_1 = C_{VW} = C_{VW}(0.975, 0.975; 1.4) = 0.951$$

2. Compute the trivariate Plackett copula value using Equation (6.14), and solve it numerically as follows: $C_{UVW}(0.5, 0.975, 0.975; [15, 1.4, 1.3, 20]) = 0.488$
   where the remaining $a's$ and $b's$ needed in Equation (6.12) are computed as follows:

$$a_4 = 1 - 0.5 - 0.975 - 0.975 + 0.498 + 0.589 + 0.951 = 0.488$$
$$b_1 = 0.489 + 0.951 - 0.975 = 0.465$$
$$b_2 = 0.498 + 0.951 - 0.975 = 0.474$$
$$b_3 = 0.498 + 0.489 - 0.5 = 0.487$$

3. Compute the derivatives needed to compute the trivariate Plackett density:

$$P_{000} = C_{UVW} = 0.488, P_{100} = C_{VW} - C_{UVW} = 0.463,$$
$$P_{010} = C_{UW} - C_{UVW} = 7.903 \times 10^{-4}, \quad P_{001} = C_{UV} - C_{UVW} = 0.0074,$$
$$P_{110} = w - C_{UW} - C_{VW} + C_{UVW} = -0.0234, \quad P_{101} = v - C_{UV} - C_{VW} + C_{UVW} = 0.017$$
$$P_{011} = u - C_{UV} - C_{UW} + C_{UVW} = 0.004$$
$$P_{111} = 1 - u - v - w + C_{UV} + C_{VW} + C_{UW} - C_{UVW} = -0.003$$

The rest computation will need to apply the numerical method (i.e., Newton's method). Here we will only lists the final results:

$$\frac{\partial^2 C_{UV}}{\partial u \partial v} = 0.594; \frac{\partial^2 C_{UW}}{\partial u \partial w} = 0.985; \frac{\partial^2 C_{VW}}{\partial v \partial w} = 1.348$$
$$\frac{\partial P_{000}}{\partial u} = 0.015; \frac{\partial P_{111}}{\partial v} = 0.697; \frac{\partial P_{111}}{\partial w} = 1.9461$$

$$\frac{\partial C_{UVW}}{\partial u} = 0.946; \frac{\partial C_{UVW}}{\partial v} = -0.555; \frac{\partial C_{UVW}}{\partial w} = 1.542$$

$$\frac{\partial^2 C_{UVW}}{\partial u \partial v} = 9.101; \frac{\partial^2 C_{UVW}}{\partial u \partial w} = 29.094; \frac{\partial^2 C_{UVW}}{\partial v \partial w} = 458.057$$

Finally, we have the trivariate Plackett copula density as follows:

$$c_{UVW} = \frac{\partial^3 C_{UVW}}{\partial u \partial v \partial w} = 8.412.$$

6.2.3 *Estimation of Cross-Product Ratio (Copula Parameter) for the Trivariate Plackett Copula*

Following the same procedure for the bivariate Plackett copula, the parameter for the trivariate Plackett copula may be estimated. For a trivariate sample of $\boldsymbol{X} = \{x_{i1}, x_{i2}, x_{i3}, i = 1, \dots n\}$ with $u_{i1} = F_1(x_{i1}), u_{i2} = F_2(x_{i2}), u_{i3} = F_3(x_{i3})$, we can then write the pseudo-MLE as follows:

$$L(\theta_{UVW}) = \sum\nolimits_{i=1}^{n} \log\left(c_{UVW}(u_{i1}, u_{i2}, u_{i3}; \theta_{UVW})\right) \tag{6.27}$$

Taking the derivative with respect to θ_{UVW} and setting the derivative equal to 0, we have the following:

$$\frac{1}{n}\frac{\partial L}{\partial \theta_{UVW}} = \frac{1}{n}\sum\nolimits_{i=1}^{n} \frac{\partial \log\left(c_{UVW}(u_{i1}, u_{i2}, u_{i3}; \theta_{UVW})\right)}{\partial \theta_{UVW}} = 0 \tag{6.27a}$$

As shown in the previous section, the trivariate Plackett copula does not have an analytical form of the trivariate Plackett copula density function, and the parameter may be optimized by the numerical scheme (e.g., central differencing). Compared to the bivariate case, the parameter estimation of the trivariate Plackett copula is more tedious. It holds true, compared to the asymmetric Archimedean, vine, and meta-elliptical copulas.

6.3 Summary

In this chapter, we introduce the bivariate and trivariate Plackett copulas with the focus on the bivariate Plackett copula. The parameter estimation for the trivariate Plackett copulas is rather complex, compared to the trivariate asymmetric Archimedean, vine, and meta-elliptical copulas. Additionally, there does not exist the analytical form for the trivariate Plackett copula density. In general, it is recommended to apply asymmetric Archimedean, vine, and meta-elliptical copulas to model the multivariate dimensional dependence.

References

Joe, H. (1997). *Multivariate Models and Dependence Concept*. Chapman & Hall, New York.

Kao, S. C. and Govindaraju, R. S. (2008). Trivarariate statistical analysis of extreme rainfall events via the Plackett family of copulas. *Water Resources Research*, 44(2), W02415, doi:10.1029/2007WR006261.

Palaro, H. P. and Hotta L. K. (2006). Using conditional copula to estimate Value at Risk. http://papers.ssrn.com/sol3/papers.cfm?abstract_id=818884.

Plackett, R. L. (1965). A class of bivariate distributions. *Journal of the American Statistical Association*, 60, 516–522.

Rosenblatt, M. (1952). Remarks on a multivariate transformation. *Annals of Mathematical Statistics*, 470–472.

Song, S. and Singh, V. P. (2010). Frequency analysis of droughts using the Plackett copula and parameter estimation by genetic algorithm. *Stochastic Environmental Research and Risk Assessment*, 24, 783–805, doi:10.1007/s00477–010–0364–5.

7

Non-Archimedean Copulas

Meta-Elliptical Copulas

ABSTRACT

Meta-elliptical copulas are derived from elliptical distributions. Kotz and Nadarajah (2001) and Nadarajah (2006) made solutions of meta-elliptical copulas available. In this chapter, we will review the definition and probability distributions as well as other properties of meta-elliptical copulas.

7.1 Meta-Elliptical Copulas

7.1.1 d-Dimensional Symmetric Elliptical Type Distribution

In previous chapters, we have discussed symmetric and asymmetric (i.e., nested) Archimedean copulas for multivariate modeling (i.e., $d \geq 3$). We have shown that (i) symmetric multivariate Archimedean copulas require that all the correlated variables share the same dependence structure, and (ii) the nested asymmetric multivariate copulas still require that some variables share the same dependence structure. Compared to the symmetric and nested asymmetric Archimedean copulas, the meta-elliptical copulas are more flexible than the symmetric (or nested) Archimedean copulas for modeling multivariate hydrological variables.

Following Genest et al. (2007), a d-dimensional random vector $\mathbf{z}$ ($\mathbf{z} = [z_1, \ldots, z_d]^T$) is said to have an elliptical joint distribution, i.e., $\mathcal{E}_d(\boldsymbol{\mu}, \Sigma, g)$ with mean vector $\boldsymbol{\mu}(d \times 1)$, covariance matrix $\Sigma(d \times d)$, and generator $g : [0, \infty) \rightarrow [0, \infty)$, if there exists a stochastic representation, as follows:

$$\mathbf{z} = \boldsymbol{\mu} + r\mathbf{A}\mathbf{u} \tag{7.1}$$

where $r \geq 0$ is a random variable with the probability density function as

$$f_g(r) = \frac{2\pi^{\frac{d}{2}}}{\Gamma\left(\frac{d}{2}\right)} r^{d-1} g\left(r^2\right) \tag{7.1a}$$

$\mathbf{u}$ (independent of r) is uniformly distributed on the sphere as follows:

$$\mathbb{S}_d = \left\{(u_1, \ldots, u_d) \in \mathbb{R}^d : u_1^2 + u_2^2 + \ldots + u_d^2 = 1\right\} \tag{7.1b}$$

Table 7.1. *Common probability density function generators [$g(t)$] for elliptical copulas.*

Copula	$g(t)$
Normal	$(2\pi)^{-\frac{d}{2}}\exp\left(-\frac{t}{2}\right)$
Student	$\frac{(\pi v)^{-\frac{d}{2}}\Gamma\left(\frac{d+v}{2}\right)}{\Gamma\left(\frac{d}{2}\right)}\left(1+\frac{t}{v}\right)^{-\frac{d+v}{2}}$
Cauchy	$\frac{\pi^{-\frac{d}{2}}\Gamma\left(\frac{d+1}{2}\right)}{\Gamma\left(\frac{1}{2}\right)}(1+t)^{-\frac{d+1}{2}}$
Kotz[a]	$\frac{s\Gamma\left(\frac{d}{2}\right)r^{\frac{2N+d-2}{2s}}t^{N-1}\exp\left(-rt^{s}\right)}{\pi^{\frac{d}{2}}\Gamma\left(\frac{2N+d-2}{2s}\right)};r,\ s>0,\ 2N+d>2$
Pearson type II	$\frac{\Gamma\left(\frac{d}{2}+m+1\right)}{\pi^{\frac{d}{2}}\Gamma(m+1)}(1-t)^{m};t\in[-1,1],\ m>-1$
Pearson type VII[b]	$\frac{\Gamma(N)}{\Gamma\left(N-\frac{d}{2}\right)(\pi m)^{\frac{d}{2}}}\left(1+\frac{t}{m}\right)^{-N};N>\frac{d}{2},\ m>0$

Notes: [a] Kotz type copula reduces to normal copula if $N=s=1,\ r=1/2$.
[b] Pearson type VII copula reduces to Cauchy copula if $m=1,\ N=3/2$ and reduces to Student copula if $N=\frac{m}{2}+1$.

A is Cholesky decomposition of Σ, i.e., $\mathbf{AA}^{T}=\Sigma$
and the joint probability density function of **z** can be written as follows:

$$|\Sigma|^{-\frac{1}{2}}g\left[(\mathbf{z}-\boldsymbol{\mu})^{T}\Sigma^{-1}(\mathbf{z}-\boldsymbol{\mu})\right] \tag{7.2}$$

In Equations (7.1a) and (7.2), $g(\cdot)$ is a scale function uniquely determined by the distribution of r and referred to as the probability density function generator. Common d-dimensional symmetric elliptical type distribution generators are given in Table 7.1.

To build the meta-elliptical copula using the $g(\cdot)$ function (listed in Table 7.1) and Equation (7.2), we should note that there is one limitation of these elliptical distributions, that is, the scaled variables $\frac{z_1}{\sqrt{\sigma_{11}}},\ \frac{z_2}{\sqrt{\sigma_{22}}},\ \ldots,\ \frac{z_d}{\sqrt{\sigma_{dd}}}$ are identically distributed with the density function as follows:

$$q_g\left(\frac{z_k}{\sqrt{\sigma_{kk}}}=x\right)=\frac{\pi}{\Gamma\left(\dfrac{d-1}{2}\right)}\int_{u^2}^{\infty}\left(y-x^2\right)^{\frac{d-1}{2}-1}g(y)dy;k=1,\ldots,d \tag{7.3}$$

and the CDF of the scaled variables given as follows:

$$Q_g\left(\frac{z_k}{\sqrt{\sigma_{kk}}}\le x\right)=\frac{1}{2}+\frac{\pi^{\frac{d-1}{2}}}{\Gamma\left(\frac{d-1}{2}\right)}\int_0^x\int_{u^2}^{\infty}\left(y-u^2\right)^{\frac{d-1}{2}-1}g(y)dydu \tag{7.4}$$

From Equations (7.3) and (7.4), it is known that $q_g(x)=q_g(-x)$ and $Q_g(x)=1-Q_g(x)$ for $x>0$.

Example 7.1 **Derive the d-dimensional multivariate normal density function: $\mathbf{z}=[z_1,\ldots,z_d]$.**

Solution: As listed in Table 7.1, the probability density function generator for the multivariate normal distribution is $g(t)=(2\pi)^{-\frac{d}{2}}\exp\left(-\frac{t}{2}\right)$. Applying Equation (7.2), we have the following:

$$f(\mathbf{z})=|\Sigma|^{-\frac{1}{2}}(2\pi)^{-\frac{d}{2}}\exp\left(-\frac{(\mathbf{z}-\boldsymbol{\mu})^T\Sigma^{-1}(\mathbf{z}-\boldsymbol{\mu})}{2}\right),\ \mathbf{z}\sim\mathcal{E}_d(\boldsymbol{\mu},\Sigma,g) \tag{7.5}$$

If $\boldsymbol{\mu}=0$ in Equation (7.5), we have

$$f(\mathbf{z})=|\Sigma|^{-\frac{1}{2}}(2\pi)^{-\frac{d}{2}}\exp\left(-\frac{\mathbf{z}^T\Sigma^{-1}\mathbf{z}}{2}\right),\ \mathbf{z}\sim\mathcal{E}_d(\mathbf{0},\Sigma,g) \tag{7.6a}$$

By applying Equation (7.1), we have $\mathbf{z}^T\Sigma^{-1}\mathbf{z}=r^2(\mathbf{Au})^T\Sigma^{-1}(\mathbf{Au})=r^2$ and Equation (7.6a) may be rewritten as follows:

$$f(\mathbf{z})=|\Sigma|^{-\frac{1}{2}}(2\pi)^{-\frac{d}{2}}\exp\left(-\frac{r^2}{2}\right),\ \mathbf{z}\sim\mathcal{E}_d(\mathbf{0},\Sigma,g) \tag{7.6b}$$

where

$$\Sigma=\begin{pmatrix}\rho_{11} & \cdots & \rho_{1d}\\ \vdots & \ddots & \vdots\\ \rho_{d1} & \cdots & \rho_{dd}\end{pmatrix},\ \rho_{ii}=1;\ |\rho_{ij}|<1,\ i\ne j; i,j=1,..,d,\ \text{correlation matrix.}$$

Example 7.2 **Derive the d-dimensional multivariate Cauchy density function for $\mathbf{z}=[z_1,\ldots,z_d]$.**

Solution: Using the probability density function generator for the multivariate Cauchy distribution listed in Table 7.1:

$$g(t)=\frac{\pi^{-\frac{d}{2}}\Gamma\left(\frac{d+1}{2}\right)}{\Gamma\left(\frac{1}{2}\right)}(1+t)^{-\frac{d+1}{2}}$$

Applying Equation (7.2), we have the following:

$$f(\mathbf{z}) = |\Sigma|^{-\frac{1}{2}} 2^{-\frac{d}{2}} \pi^{-d} \frac{\Gamma\left(\frac{d+1}{2}\right)}{\Gamma\left(\frac{1}{2}\right)} \left(1 + \frac{(\mathbf{z}-\boldsymbol{\mu})^T \Sigma^{-1} (\mathbf{z}-\boldsymbol{\mu})}{2}\right)^{-\frac{d+1}{2}}, \ \mathbf{z} \sim \mathcal{E}_d(\boldsymbol{\mu}, \Sigma, g) \tag{7.7}$$

Similarly, if $\boldsymbol{\mu} = \mathbf{0}$, we have from Equation (7.7) the following:

$$f(\mathbf{z}) = |\Sigma|^{-\frac{1}{2}} 2^{-\frac{d}{2}} \pi^{-d} \frac{\Gamma\left(\frac{d+1}{2}\right)}{\Gamma\left(\frac{1}{2}\right)} \left(1 + \frac{\mathbf{z}^T \Sigma^{-1} \mathbf{z}}{2}\right)^{-\frac{d+1}{2}}, \ \mathbf{z} \sim \mathcal{E}_d(\mathbf{0}, \Sigma, g) \tag{7.7a}$$

Or equivalently

$$f(\mathbf{z}) = |\Sigma|^{-\frac{1}{2}} 2^{-\frac{d}{2}} \pi^{-d} \frac{\Gamma\left(\frac{d+1}{2}\right)}{\Gamma\left(\frac{1}{2}\right)} \left(1 + \frac{r^2}{2}\right)^{-\frac{d+1}{2}}, \ \mathbf{z} \sim \mathcal{E}_d(\mathbf{0}, \Sigma, g) \tag{7.7b}$$

Without loss of generality, we will only investigate the case $\mathcal{E}_d(\mathbf{0}, \Sigma, g)$. Let $\mathbf{z} = [z_1, z_2, \ldots, z_d]^T$ be a random vector with each component z_i with given continuous PDF $f_i(z_i)$ and CDF $F_i(z_i)$. Suppose

$$x_i = Q_g^{-1}(F_i(z_i)), \ i = 1, \ 2, \ \ldots d \tag{7.8}$$

where Q_g^{-1} is the inverse of Q_g.

Then, the probability density function of $\mathbf{z}$ is given by

$$f(z_1, \ldots, z_d) = f(x_1, \ldots, x_d) \ |J| \tag{7.9}$$

where the Jacobian matrix J is given as follows:

$$J = \begin{pmatrix} \frac{\partial x_1}{\partial z_1} & \cdots & \frac{\partial x_d}{\partial z_d} \\ \vdots & \ddots & \vdots \\ \frac{\partial x_1}{\partial z_d} & \cdots & \frac{\partial x_d}{\partial z_d} \end{pmatrix}$$

Since $x_i = Q_g^{-1}(F_i(z_i))$, we have $\frac{\partial x_i}{\partial z_j} = \begin{cases} \frac{dx_i}{dz_i}, \ i = j \\ 0, \ i \neq j \end{cases}$. Rewriting matrix J, we have the following:

$$J = \begin{pmatrix} \frac{dx_1}{dz_1} & \cdots & 0 \\ \vdots & \ddots & \vdots \\ 0 & \cdots & \frac{dx_d}{dz_d} \end{pmatrix}; \ \ |J| = \frac{dx_1}{dz_1} \cdot \frac{dx_2}{dz_2} \cdots \frac{dx_d}{dz_d} = \prod_{i=1}^{d} \frac{dx_i}{dz_i}$$

From $x_i = Q_g^{-1}(F_i(z_i))$, we have $F_i(z_i) = Q_g(x_i)$. Differentiation on both sides leads to $f_i(z_i)dz_i = q_g(x_i)dx_i; \frac{dx_i}{dz_i} = \frac{f_i(z_i)}{q_g(x_i)} = \frac{f_i(z_i)}{q_g(Q_g^{-1}(F_i(z_i))} \Rightarrow |J| = \prod_{i=1}^{d} \frac{f_i(z_i)}{q_g\left[Q_g^{-1}(F_i(z_i))\right]}$

Then, we have the following:

$$f(z_1, z_2, \ldots, z_d) = f(x_1, x_2, \ldots, x_d) \prod_{i=1}^{d} \frac{f_i(z_i)}{q_g[Q_g^{-1}(F_i(z_i))]} = \frac{f(x_1, \ldots x_d)}{\prod_{i=1}^{d} q_g(x_i)} \prod_{i=1}^{d} f_i(z_i) \tag{7.10}$$

For $\mathbf{x} = (x_1, \ldots, x_d)^T \sim \mathcal{E}_d(\mathbf{0}, \Sigma, g)$, we have the following:

$$f(x_1, \ldots, x_d) = |\Sigma|^{-\frac{1}{2}} g\left(\mathbf{x}^T \Sigma^{-1} \mathbf{x}\right) \tag{7.11}$$

Inserting Equation (7.11) in Equation (7.10), we have the following:

$$f(z_1, \ldots, z_d) = \frac{|\Sigma|^{-\frac{1}{2}} g\left(\mathbf{x}^T \Sigma^{-1} \mathbf{x}\right)}{\prod_{i=1}^{d} q_g(x_i)} \prod_{i=1}^{d} f_i(z_i) \tag{7.12}$$

Using $\mathcal{H}$ to represent the d-variant probability density function as

$$\mathcal{H}\left(Q_g^{-1}(F_1(z_1)), \ldots, Q_g^{-1}(F_d(z_d))\right) = \frac{|\Sigma|^{-\frac{1}{2}} g\left(\mathbf{x}^T \Sigma^{-1} \mathbf{x}\right)}{\prod_{i=1}^{d} q_g(x_i)}$$

Equation (7.12) may be written as follows:

$$f(z_1, \ldots, z_d) = \mathcal{H}\left(Q_g^{-1}(F_1(z_1)), \ldots, Q_g^{-1}(F_d(z_d))\right) \prod_{i=1}^{d} f_i(z_i) \tag{7.12a}$$

To this end, the d-dimensional random vector $\mathbf{z}$ is said to have a meta-elliptical distribution, if its probability density function is given by Equation (7.12). Denote $\mathbf{x} \sim \mathcal{ME}_d(0, \Sigma, g; F_1, \ldots, F_d)$. The function $\mathcal{H}\left(Q_g^{-1}(F_1(z_1)), \ldots, Q_g^{-1}(F_d(z_d))\right)$ is referred to as the probability density function weighting function. The class of meta-elliptical distributions includes various distributions, such as elliptically contoured distributions, the meta-Gaussian distributions, and various asymmetric distributions. The marginal distributions $F_i(.)$ can be arbitrarily chosen (Fang et al., 2002). The meta-elliptical distributions allow for the possibility of capturing tail dependence (Joe, 1997), which will be discussed later.

7.1.2 Bivariate Symmetric Elliptical Type Distribution

Suppose $\mathbf{x} \sim \mathcal{ME}_2(0, \Sigma, g)$, we have the following:

$$\Sigma = \begin{pmatrix} 1 & \rho \\ \rho & 1 \end{pmatrix}, \Sigma^{-1} = \begin{pmatrix} \frac{1}{1-\rho^2} & -\frac{\rho}{1-\rho^2} \\ -\frac{\rho}{1-\rho^2} & \frac{1}{1-\rho^2} \end{pmatrix}$$

$$[x_1, x_2]\Sigma^{-1}\begin{bmatrix} x_1 \\ x_2 \end{bmatrix} = \frac{x_1^2 + x_2^2 - 2\rho x_1 x_2}{1-\rho^2}$$

Vector **x** has the following probability density function:

$$f(x_1, x_2; \Sigma, g) = \frac{1}{\sqrt{1-\rho^2}} g\left(\frac{x_1^2 + x_2^2 - 2\rho x_1 x_2}{1-\rho^2}\right) \tag{7.13}$$

The marginal PDF and CDF of **x** are

$$q_g(x) = \int_{x^2}^{\infty} \left(y - x^2\right)^{-\frac{1}{2}} g(y) dy \tag{7.14}$$

$$Q_g(x) = \frac{1}{2} + \int_{x^2}^{\infty} \arcsin\left(\frac{x}{\sqrt{y}}\right) dy \tag{7.15}$$

A two-dimensional random vector $(\mathbf{x_1}, \mathbf{x_2})$ follows an elliptically contoured distribution, if its joint PDF takes on the form of Equation (7.13). Its copula function can be written as follows:

$$C_X(u, v) = F(x_1, x_2) = \frac{1}{\sqrt{1-\rho^2}} \int_{-\infty}^{Q_g^{-1}(u)} \int_{-\infty}^{Q_g^{-1}(v)} g\left(\frac{s^2 + t^2 - 2\rho st}{1-\rho^2}\right) dsdt \tag{7.16}$$

where $u = F_1(x_1)$, $v = F_2(x_2)$, $s = Q_g^{-1}(u)$, $t = Q_g^{-1}(v)$.

The copula density function can be given as follows:

$$c_X(u, v) = \mathcal{H}(s, t; \rho) \tag{7.17}$$

where

$$\mathcal{H}(s, t; \rho) = \frac{f(s, t; \rho)}{q_g(s) q_g(t)} \tag{7.18}$$

Now, let $\mathbf{z} = (z_1, z_2)^T \sim \mathcal{ME}_2(0, \Sigma, g; F_1, F_2)$. Its probability density function may then be expressed as follows:

$$f(x_1, x_2) = \mathcal{H}\left(Q_g^{-1}(F_1(x_1)), Q_g^{-1}(F_2(x_2))\right) f_1(x_1) f_2(x_2) \tag{7.19}$$

Take simple examples to illustrate the preceding.

Symmetric Kotz Type Distribution

Let **x** be distributed according to a bivariate symmetric Kotz type distribution. Inserting the density generator for Kotz type distribution (listed in Table 7.1) in Equation (7.2), we obtain the joint probability density function as follows:

$$f(x_1, x_2) = \frac{s r^{\frac{N}{s}} \left(x_1^2 + x_2^2 - 2\rho x_1 x_2\right)^{N-1}}{\pi \Gamma\left(\frac{N}{s}\right) (1-\rho^2)^{N-\frac{1}{2}}} \exp\left(-r\left(\frac{x_1^2 + x_2^2 - 2\rho x_1 x_2}{1-\rho^2}\right)^s\right) \tag{7.20}$$

where $r > 0$, $s > 0$, $N > 0$ are the parameters.

The marginal PDF (i.e., $q_1(x)$) can be written as follows:

$$q_1(x) = \frac{2sr^{\frac{N}{s}}}{\pi\Gamma\left(\frac{N}{s}\right)} \int_0^{\infty} \left(t^2 + x^2\right)^{N-1} \exp\left(-r\left(t^2 + x^2\right)^s\right) dt \tag{7.21}$$

The corresponding CDF (i.e., $Q_1(x)$) can be written as follows:

$$Q_1(x) = \frac{1}{2} + \frac{2sr^{\frac{N}{s}}}{\pi\Gamma\left(\frac{N}{s}\right)} \int_0^{x} \int_0^{\infty} \left(t^2 + x^2\right)^{N-1} \exp\left(-r\left(t^2 + x^2\right)^s\right) dt dx \tag{7.22}$$

Then, the copula probability density function can be given as follows:

$$c(u, v) = \frac{f\left(Q_1^{-1}(u), Q_1^{-1}(v)\right)}{q_1\left(Q_1^{-1}(u)\right) q_2\left(Q_2^{-1}(v)\right)} \tag{7.23}$$

Example 7.3 **Show that the bivariate Kotz type distribution converges to the bivariate Gaussian distribution as noted in Table 7.1, i.e., $N = s = 1$, $r = 1/2$.**

Solution: Substituting $N = s = 1$, $r = \frac{1}{2}$ into the probability density function generator of symmetric Kotz type distribution, we have

$$g_2(t) = \frac{s\Gamma\left(\frac{d}{2}\right) r^{\frac{2N+d-2}{2s}} t^{N-1} \exp\left(-rt^s\right)}{\pi^{\frac{d}{2}} \Gamma\left(\frac{2N+d-2}{2s}\right)} = \frac{t \exp\left(-t/2\right)}{2\pi}, \ d = 2$$

Comparing with the probability density function generator for the normal copula in the bivariate case, we have the following:

$$g_2^N = (2\pi)^{-1} \exp\left(-\frac{t}{2}\right)$$

Now we show that the bivariate Kotz type distribution reduces to the bivariate normal distribution if $N = s = 1$, and $r = \frac{1}{2}$. The same conclusion is reached for higher dimensional cases.

Example 7.4 **Compute the copula density function for symmetric Kotz type distribution with the information given as $N = 2.0$, $s = 1.0$, $r = 0.5$, $\rho = 0.1$, $u = 0.4$, $v = 0.3$.**

Solution: Using Equation (7.22), we can calculate $Q^{-1}(u)$, $Q^{-1}(v)$ numerically as follows:

$$Q^{-1}(0.4) = -0.4843; Q^{-1}(0.3) = -0.9158$$

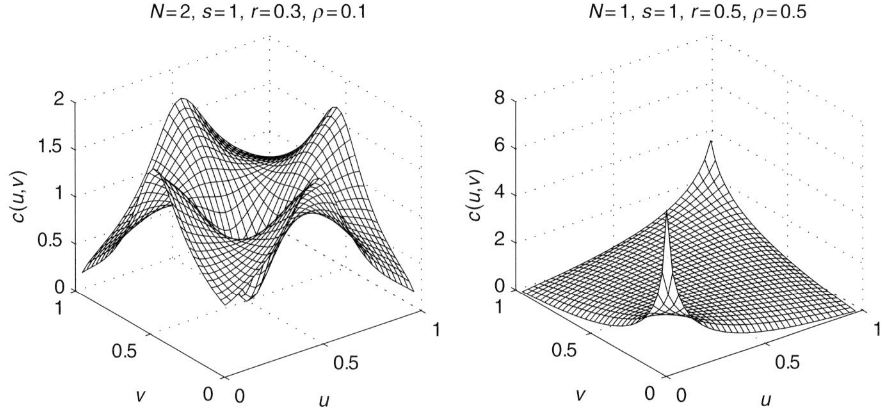

Figure 7.1 Copula density plots for Kotz type bivariate distribution.

Using Equation (7.20), we can compute the joint density function as follows:

$$f\left(Q^{-1}(0.4), Q^{-1}(0.3)\right) = f(-0.4843, -0.9158) = 0.0484$$

Using Equation (7.21), we can compute the univariate density as follows:

$$q\left(Q^{-1}(0.4)\right) = q(-0.4843) = 0.2190; q(Q(v)) = q(-0.9158) = 0.2411$$

Finally, substituting the computed quantities above into Equation (7.23), we have the following:

$$c(u, v) = c(0.4, 0.3) = \frac{0.0484}{0.2190 * 0.2411} = 0.9160$$

To further illustrate the shape of the bivariate symmetric Kotz type density function, Figure 7.1 graphs the bivariate density function for the following:

1. $N = 2.0$, $s = 1.0$, $r = 0.5$, $\rho = 0.1$.
2. $N = s = 1$, $r = 0.5$, $\rho = 0.5$: bivariate normal distribution.

Symmetric Bivariate Pearson Type VII Distribution

The PDF of symmetric bivariate Pearson type VII distribution can be given as follows:

$$f(x_1, x_2) = \frac{N - 1}{\pi m \sqrt{1 - \rho^2}} \left(1 + \frac{1}{m(1 - \rho^2)} \left(x_1^2 + x_2^2 - 2\rho x_1 x_2\right)\right)^{-N} \tag{7.24}$$

where $N > 1$, and $m > 0$ are parameters.

When $N = \frac{m}{2} + 1$, Equation (7.24) is the bivariate t-distribution with m degrees of freedom. When $m = 1$, $n = \frac{3}{2}$, Equation (7.24) is the bivariate Cauchy distribution.

The marginal PDF of symmetric bivariate Pearson type VII distribution is as follows:

$$q(x) = \frac{\Gamma\left(n - \frac{1}{2}\right)}{\sqrt{\pi m}\Gamma(N-1)} \left(1 + \frac{x^2}{m}\right)^{-\left(N-\frac{1}{2}\right)} \tag{7.25}$$

The corresponding CDF of symmetric bivariate Pearson type VII distribution can be written as follows:

$$Q(x) = \frac{\Gamma\left(n - \frac{1}{2}\right)}{\sqrt{\pi m}\Gamma(N-1)} \int_{-\infty}^{x} \left(1 + \frac{t^2}{m}\right)^{-\left(N-\frac{1}{2}\right)} dt \tag{7.26}$$

where $x \in (-\infty, \infty)$, $m > 0$, $N > 1$.

Then, the copula density function $c(u, v)$ can be given as follows:

$$c(u,v) = \frac{\Gamma(N-1)\Gamma(N)}{\left[\Gamma\left(N-\frac{1}{2}\right)\right]^2} \frac{\left[\left(1+\frac{Q_1^{-1}(u)}{m}\right)\left(1+\frac{Q_1^{-1}(v)}{m}\right)\right]^{N-\frac{1}{2}}}{\sqrt{1-\rho^2}\left[1+\frac{1}{m(1-\rho^2)}\left(\left(Q_1^{-1}(u)\right)^2+\left(Q_1^{-1}(v)\right)^2-2\rho\left(Q_1^{-1}(u)\right)\left(Q_1^{-1}(v)\right)\right)\right]^N} \tag{7.27}$$

Example 7.5 **Show the following bivariate Pearson type VII distribution cases are true.**

Show the following cases are true:

1. $N = \frac{m}{2} + 1$, the bivariate Pearson type VII distribution is the bivariate Student t-distribution with m degrees of freedom.
2. $m = 1$, $N = \frac{3}{2}$, the bivariate Pearson type VII distribution is the bivariate Cauchy distribution.

Solution:

1. $N = \frac{m}{2} + 1$

When $N = \frac{m}{2} + 1$, the probability density function generator of the Pearson type VII distribution may be rewritten as follows:

$$g_2^{PVII} = \frac{\Gamma\left(\frac{m}{2}+1\right)}{\Gamma\left(\frac{m}{2}-1\right)\pi m} \left(1 + \frac{t}{m}\right)^{-\left(\frac{m}{2}+1\right)}; N > 1,\, m > 0,\, N = \frac{m}{2} + 1$$

Comparing with the probability density function generator for the bivariate Student t-distribution $g_2^t = \frac{(\pi v)^{-1}\Gamma\left(\frac{v}{2}+1\right)}{\Gamma\left(\frac{v}{2}\right)} \left(1 + \frac{t}{v}\right)^{-\left(\frac{v}{2}+1\right)}$, we show that when $N = \frac{m}{2} + 1$, the bivariate Pearson type VII distribution reduces to the bivariate Student t-distribution. The same conclusion is reached for higher-dimensional cases.

2. $m = 1,\ N = \frac{3}{2}$

When $m = 1,\ N = \frac{3}{2}$, the probability density function generator of the Pearson type VII distribution may be rewritten as follows:

$$g_2^{PVII}(t) = \frac{\Gamma\left(\frac{3}{2}\right)}{\Gamma\left(\frac{1}{2}\right)\pi}(1+t)^{-\frac{3}{2}},\ m = 1,\ N = \frac{3}{2}$$

Comparing with the probability density function generation for the Cauchy distribution $g_2^{Cauchy} = \frac{\pi^{-1}\Gamma\left(\frac{3}{2}\right)}{\Gamma\left(\frac{1}{2}\right)}(1+t)^{-\frac{3}{2}}$, we show that when $m = 1,\ N = \frac{3}{2}$, the bivariate Pearson type VII distribution reduces to the bivariate Cauchy distribution. The same conclusion is reached for higher-dimensional cases.

Example 7.6 **Compute the Pearson type VII copula density with the information given as follows: $m = 0.5$, $N = 2.0$, $\rho = 0.1$, $u = 0.4$, $v = 0.3$.**

Solution: Applying Equation (7.26), we can compute $Q^{-1}(u)$, $Q^{-1}(v)$ numerically as follows:

$$Q^{-1}(u) = -0.1443; Q^{-1}(v) = -0.3086$$

Substituting $Q^{-1}(u) = -0.1443; Q^{-1}(v) = -0.3086$ into Equation (7.24), we can compute the copula density function $c(0.4, 0.3)$ as $c(0.4, 0.3) = 1.1941$.

To illustrate the shape of Pearson type VII distribution, we graph the Pearson type VII copula density function for the following parameters in Figure 7.2:

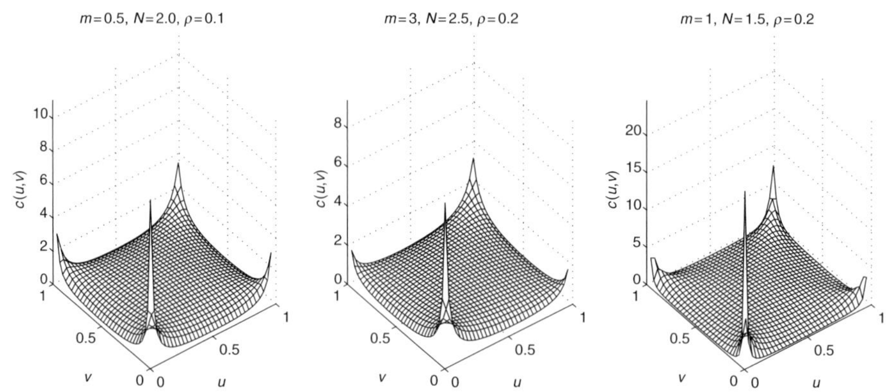

Figure 7.2 Pearson type VII copula density plots.

1. $m = 0.5, \quad N = 2.0, \quad \rho = 0.1.$
2. $m = 3, \quad N = 2.5, \quad \rho = 0.2$: bivariate Student t-distribution with degrees of freedom as 3.
3. $m = 1$, $N = \frac{3}{2}$, $\rho = 0.2$: bivariate Cauchy distribution.

Symmetric Bivariate Pearson Type II Distribution

The PDF of symmetric bivariate Pearson type II distribution can be expressed as

$$f(x_1, x_2) = \begin{cases} \frac{m+1}{\pi\sqrt{1-\rho^2}}\left(1 - \frac{x_1^2 + x_2^2 - 2\rho x_1 x_2}{1-\rho^2}\right)^m, & \forall (x_1, x_2)\Sigma^{-1}(x_1, x_2)^T \le 1 \\ 0 & otherewise \end{cases} \tag{7.28}$$

where $m > -1$.

The marginal PDF can be given as follows:

$$q(x) = \frac{\Gamma(m+2)}{\sqrt{\pi}\Gamma\left(m + \frac{3}{2}\right)}\left(1 - x^2\right)^{m+\frac{1}{2}}; \; x \in [-1, 1] \tag{7.29}$$

The corresponding CDF can be given as follows:

$$Q(x) = \frac{\Gamma(m+2)}{\sqrt{\pi}\Gamma\left(m + \frac{3}{2}\right)}\int_{-1}^{x}\left(1 - t^2\right)^{m+\frac{1}{2}}dt; \; x \in [-1, 1] \tag{7.30}$$

The copula probability density function can then be given as follows:

$$c(u,v) = \frac{(m+1)\Gamma^2\left(m+\frac{3}{2}\right)}{\Gamma^2(m+2)} \frac{\left(1 - \left(\frac{1}{1-\rho^2}\right)\left(\left(Q^{-1}(u)\right)^2 + \left(Q^{-1}(v)\right)^2 - 2\rho\left(Q^{-1}(u)\right)\left(Q^{-1}(v)\right)\right)\right)^m}{\sqrt{1-\rho^2}\left(1 - \left(Q^{-1}(u)\right)^2\right)^{m+\frac{1}{2}}\left(1 - \left(Q^{-1}(v)\right)^2\right)^{m+\frac{1}{2}}} \tag{7.31}$$

Example 7.7 **Compute the bivariate Pearson type II copula density function with information given as follows: $m = -0.5$, $\rho = 0.1$, $u = 0.4$, $v = 0.3$.**

Solution: Applying Equation (7.30), we can compute $Q^{-1}(0.4)$, $Q^{-1}(0.3)$ numerically as follows:

$$Q^{-1}(0.4) = -0.2; Q^{-1}(0.3) = -0.4$$

Applying Equation (7.31), we can compute the bivariate Pearson type II copula density function as follows:

$$c(0.4, 0.3) = 0.7091.$$

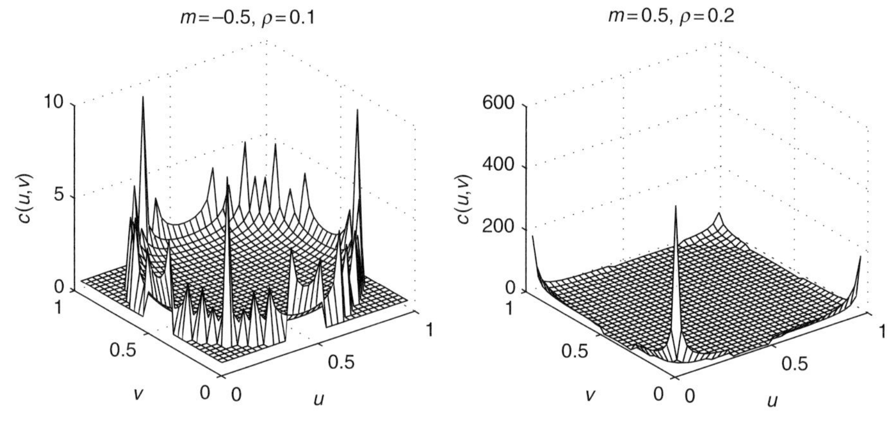

Figure 7.3 Pearson type II copula density function plots.

To further illustrate the shape of the bivariate Pearson type II copula density function, we graph the Pearson type II copula density function for the following parameters in Figure 7.3:

$m = -0.5$, $\rho = 0.1$; (2) $m = 0.5$, $\rho = 0.2$.

7.2 Two Most Commonly Applied Meta-Elliptical Copulas

In Section 7.1, we have stated that (1) the symmetric meta-Kotz type distribution reduces to the meta-Gaussian distribution if $N = s = 1,\ r = 0.5$; (2) the symmetric meta-Pearson distribution reduces to the meta-Student t-distribution if $N = \frac{m}{2} + 1$. In this section, we will start to focus on the discussion of two most commonly applied meta-elliptical copulas, and these are meta-Gaussian and meta-Student t copulas.

7.2.1 Meta-Gaussian Copula

A d-dimensional meta-Gaussian copula can be expressed as follows:

$$
\begin{aligned}
C(u_1, \ldots, u_d; \Sigma) &= \Phi_\Sigma\left(\Phi^{-1}(u_1), \ldots, \Phi^{-1}(u_d)\right) \\
&= \int_{-\infty}^{\Phi^{-1}(u_1)} \cdots \int_{-\infty}^{\Phi^{-1}(u_d)} \frac{1}{(2\pi)^{\frac{d}{2}}|\Sigma|^{\frac{1}{2}}} \exp\left(-\frac{1}{2}\mathbf{w}^T\Sigma^{-1}\mathbf{w}\right) d\mathbf{w}
\end{aligned}
\tag{7.32}
$$

where $\Phi^{-1}(\cdot)$ represents the inverse function of standard normal distribution; $\Phi_\Sigma\left(\Phi^{-1}(u_1), \ldots, \Phi^{-1}(u_d)\right)$ represents multivariate standard normal distribution function;

$\boldsymbol{\Sigma}$ represents the correlation matrix; $\Sigma = \begin{pmatrix} 1 & \cdots & \rho_{1d} \\ \vdots & \ddots & \vdots \\ \rho_{d1} & \cdots & 1 \end{pmatrix}$,

$\rho_{ij} = \begin{cases} 1 & i = j \\ \rho_{ji} & i \neq j \end{cases}$, $\rho_{ij} = \sin\left(\frac{\pi\tau_{ij}}{2}\right)$, $\tau_{i,j}$ the rank correlation coefficient; d the dimension of continuous multivariate random variables; and $\mathbf{w}$ the integral matrix: $\mathbf{w} = [w_1, \ldots, w_d]^T$.

Let $g(w_1, \ldots, w_d) = \frac{1}{(2\pi)^{\frac{d}{2}}|\Sigma|^{\frac{1}{2}}} \exp\left(-\frac{1}{2}\mathbf{w}^T\Sigma^{-1}\mathbf{w}\right)$, $x_1 = \Phi^{-1}(u_1), \ldots, x_d = \Phi^{-1}(u_d)$, Equation (7.32) may be rewritten as follows:

$$C(u_1, \ldots, u_d) = \int_{-\infty}^{x_1} \cdots \int_{-\infty}^{x_d} g(w_1, \ldots, w_d) dw_1 \ldots dw_d \tag{7.32a}$$

and its copula density function can be given as follows:

$$\begin{aligned} c(u_1, \ldots, u_d; \Sigma) &= \frac{\partial^d}{\partial u_1 \ldots \partial u_d} C(u_1, \ldots, u_d; \Sigma) \\ &= \frac{\partial^d}{\partial u_1 \ldots \partial u_d} \int_{-\infty}^{\Phi^{-1}(u_1)} \cdots \int_{-\infty}^{\Phi^{-1}(u_d)} \frac{1}{(2\pi)^{\frac{d}{2}}|\Sigma|^{\frac{1}{2}}} \exp\left(-\frac{1}{2}\mathbf{w}^T\Sigma^{-1}\mathbf{w}\right) d\mathbf{w} \end{aligned} \tag{7.33}$$

or equivalently

$$c(u_1, \ldots, u_d; \Sigma) = \frac{\partial^d}{\partial u_1 \ldots \partial u_d} \int_{-\infty}^{x_1} \cdots \int_{-\infty}^{x_d} g(w_1, \ldots, w_d) dw_1 \ldots dw_d \tag{7.33a}$$

Applying the partial derivative rule of inverse function,

$$\begin{cases} u_1 = \Phi(x_1) \\ \cdots \\ u_d = \Phi(x_d) \end{cases} \Rightarrow \begin{cases} \frac{du_1}{dx_1} = \phi(x_1) \\ \cdots \\ \frac{du_d}{dx_d} = \phi(x_d) \end{cases} \Rightarrow \begin{cases} \frac{\partial x_1}{\partial u_1} = \frac{dx_1}{du_1} = \frac{1}{\frac{du_1}{dx_1}} = \frac{1}{\phi(x_1)} \\ \cdots \\ \frac{\partial x_d}{\partial u_d} = \frac{dx_d}{du_d} = \frac{1}{\frac{du_d}{dx_d}} = \frac{1}{\phi(x_d)} \end{cases} \tag{7.34}$$

In Equation (7.34), $\Phi(\cdot)$ is the CDF of standard normal distribution: $\Phi(x) = \int_{-\infty}^{x} \frac{1}{\sqrt{2\pi}} e^{-\frac{t^2}{2}} dt$; and $\phi(\cdot)$ is the PDF of standard normal distribution: $\phi(x) = \frac{1}{\sqrt{2\pi}} e^{-\frac{x^2}{2}}$.

Now substituting Equation (7.34) back into Equation (7.32) or (7.32a), we can calculate the partial derivatives for the d-dimensional meta-Gaussian copula in what follows.

First-Order Partial Derivative

Using $\frac{\partial C}{\partial u_1}$ as an example, the first-order partial derivative of the meta-Gaussian copula may be derived as follows:

$$\begin{aligned}\frac{\partial C}{\partial u_1} &= \frac{\partial}{\partial u_1}\int_{-\infty}^{x_1}\cdots\int_{-\infty}^{x_d} g(w_1,\ldots,w_d)dw_1\ldots dw_d \\ &= \left(\frac{\partial}{\partial x_1}\int_{-\infty}^{x_1}\cdots\int_{-\infty}^{xd} g(w_1,\ldots,w_d)dw_1\ldots dw_d\right)\frac{\partial x_1}{\partial u_1} \\ &= \frac{1}{\phi(x_1)}\int_{-\infty}^{x_2}\cdots\int_{-\infty}^{x_d} g(x_1,\ldots,w_d)dw_2\ldots dw_d\end{aligned} \tag{7.35}$$

Second-Order Partial Derivative

Using $\frac{\partial^2 C}{\partial u_1 \partial u_2}$ as an example, the second-order partial derivative of the meta-Gaussian copula may be derived as follows:

$$\begin{aligned}\frac{\partial^2 C}{\partial u_1\partial u_2} &= \frac{\partial}{\partial u_2}\left(\frac{\partial C}{\partial u_1}\right) = \frac{\partial}{\partial u_2}\left[\frac{1}{\phi(x_1)}\int_{-\infty}^{x_2}\cdots\int_{-\infty}^{x_d} g(x_1,\ldots,w_d)dw_2\ldots dw_d\right] \\ &= \frac{\partial}{\partial x_2}\left[\frac{1}{\phi(x_1)}\int_{-\infty}^{x_2}\cdots\int_{-\infty}^{x_d} g(x_1,\ldots,w_d)dw_2\ldots dw_d\right]\left(\frac{\partial x_2}{\partial u_2}\right) \\ &= \frac{1}{\phi(x_1)\phi(x_2)}\int_{-\infty}^{x_3}\cdots\int_{-\infty}^{x_d} g(x_1,x_2\ldots,w_d)dw_3\ldots dw_d\end{aligned} \tag{7.36}$$

d*th-Order Partial Derivative*

Repeating the derivative *d*-times, we obtain the meta-Gaussian copula density function as follows:

$$c(u_1,\ldots,u_d;\Sigma) = \frac{\partial^d(u_1,\ldots u_d;\Sigma)}{\partial u_1\ldots\partial u_d} = \frac{1}{\phi(x_1)\ldots\phi(x_d)}g(x_1,\ldots x_d) \tag{7.37}$$

Let $\varsigma = [x_1,\ldots,x_d]^T = \left[\Phi^{-1}(u_1),\ldots,\Phi^{-1}(u_d)\right]^T$. Equation (7.37) may be rewritten as follows:

$$\begin{aligned}c(u_1,\ldots,u_d;\Sigma) &= \frac{1}{\left(\frac{1}{\sqrt{2\pi}}\right)e^{-\frac{x_1^2}{2}}\ldots\left(\frac{1}{\sqrt{2\pi}}\right)e^{-\frac{x_d^2}{2}}}\cdot\frac{1}{(2\pi)^{\frac{d}{2}}|\Sigma|^{\frac{1}{2}}}\exp\left(-\frac{1}{2}\varsigma^T\Sigma^{-1}\varsigma\right) \\ &= \frac{1}{\left(\frac{1}{\sqrt{2\pi}}\right)e^{-\frac{[\Phi^{-1}(u_1)]^2}{2}}\ldots\left(\frac{1}{\sqrt{2\pi}}\right)e^{-\frac{[\Phi^{-1}(u_d)]^2}{2}}}\cdot\frac{1}{(2\pi)^{\frac{d}{2}}|\Sigma|^{\frac{1}{2}}}\exp\left(-\frac{1}{2}\varsigma^T\Sigma^{-1}\varsigma\right) \\ &= \frac{1}{\prod_{i=1}^d\left(\frac{1}{\sqrt{2\pi}}\right)e^{-\frac{[\Phi^{-1}(u_i)]^2}{2}}}\cdot\frac{1}{(2\pi)^{\frac{d}{2}}|\Sigma|^{\frac{1}{2}}}\exp\left(-\frac{1}{2}\varsigma^T\Sigma^{-1}\varsigma\right) \\ &= |\Sigma|^{-\frac{1}{2}}\cdot\frac{1}{\prod_{i=1}^d e^{-\frac{[\Phi^{-1}(u_i)]^2}{2}}}\exp\left(-\frac{1}{2}\varsigma^T\Sigma^{-1}\varsigma\right)\end{aligned} \tag{7.38}$$

Note that in Equation (7.38), $\prod_{i=1}^{d} e^{-\frac{[\Phi^{-1}(u_i)]^2}{2}}$ may be rewritten as follows:

$$\prod_{i=1}^{d} e^{-\frac{[\Phi^{-1}(u_i)]^2}{2}} = \exp\left\{-\left[\left(\Phi^{-1}(u_1)\right)^2 + \ldots + \left(\Phi^{-1}(u_d)\right)^2\right]\right\} \tag{7.38a}$$

$$\left(\Phi^{-1}(u_1)\right)^2 + \ldots + \left(\Phi^{-1}(u_d)\right)^2 = \left[\Phi^{-1}(u_1) \; \ldots \; \Phi^{-1}(u_d)\right] \begin{bmatrix} \Phi^{-1}(u_1) \\ \ldots \\ \Phi^{-1}(u_d) \end{bmatrix} = \varsigma^T \varsigma \tag{7.38b}$$

Substituting Equations (7.38a) and (7.38b) into Equation (7.38), Equation (7.38) may be simplified as follows:

$$c(u_1, \ldots, u_d; \Sigma) = |\Sigma|^{-\frac{1}{2}} \frac{1}{e^{-\frac{\varsigma^T \varsigma}{2}}} \exp\left(-\frac{1}{2}\varsigma^T \Sigma^{-1} \varsigma\right) = |\Sigma|^{-\frac{1}{2}} \exp\left(-\frac{1}{2}\varsigma^T \Sigma^{-1} \varsigma + \frac{\varsigma^T \varsigma}{2}\right) \tag{7.39}$$

Recall that $\varsigma^T \varsigma = \varsigma^T \mathbf{I} \varsigma$, where $\mathbf{I}$ is d by d identity matrix. Equation (7.39) may also be rewritten as follows:

$$c(u_1, \ldots, u_d; \Sigma) = |\Sigma|^{-\frac{1}{2}} \exp\left(-\frac{1}{2}\varsigma^T \left(\Sigma^{-1} - \mathbf{I}\right)\varsigma\right) \tag{7.39a}$$

Example 7.8 **(Bivariate meta-Gaussian copula): Compute the bivariate meta-Gaussian copula and its copula density function with the given information: $\Sigma = \begin{bmatrix} 1 & 0.2 \\ 0.2 & 1 \end{bmatrix}$, $u_1 = 0.4$, $u_2 = 0.3$, and show the first-order derivative of the bivariate meta-Gaussian copula.**

Solution: Applying Equation (7.32) for $d = 2$, we have the bivariate meta-Gaussian copula as follows:

$$\begin{aligned} C(u_1, u_2; \Sigma) &= \Phi_\Sigma\left(\Phi^{-1}(u_1), \Phi^{-1}(u_2)\right) \\ &= \int_{-\infty}^{\Phi^{-1}(u_1)} \int_{-\infty}^{\Phi^{-1}(u_2)} \frac{1}{(2\pi)|\Sigma|^{\frac{1}{2}}} \exp\left(-\frac{1}{2}[x_1, x_2]\Sigma^{-1}[x_1, x_2]^T\right) dx_1 dx_2 \end{aligned} \tag{7.40}$$

From standard normal distribution, we have the following:

$$\Phi^{-1}(0.4) = -0.2533; \Phi^{-1}(0.3) = -0.5244.$$

$$\Sigma^{-1} = \begin{pmatrix} 1.0417 & -0.2083 \\ -0.2083 & 1.0417 \end{pmatrix}; |\Sigma| = 0.96.$$

Substituting $\Phi^{-1}(0.4)$, $\Phi^{-1}(0.3)$, Σ^{-1} and $|\Sigma|$ into Equation (7.40), we have the following:

$$C(0.4, 0.3; \Sigma) = 0.1474.$$

Applying Equation (7.39a) for $d = 2$, we have the meta-Gaussian copula density function as follows:

$$c(u_1,u_2;\Sigma)=|\Sigma|^{-\frac{1}{2}}\exp\left(-\frac{1}{2}\left[\Phi^{-1}(u_1)\quad \Phi^{-1}(u_2)\right]\left(\Sigma^{-1}-\mathbf{I}\right)\begin{bmatrix}\Phi^{-1}(u_1)\\ \Phi^{-1}(u_2)\end{bmatrix}\right);\mathbf{I}=\begin{pmatrix}1&0\\0&1\end{pmatrix}$$
$$=|\Sigma|^{-\frac{1}{2}}\exp\left(-\frac{1}{2}[x_1\quad x_2](\Sigma^{-1}-\mathbf{I})\begin{bmatrix}x_1\\ x_2\end{bmatrix}\right) \tag{7.40a}$$

Substituting $\Phi^{-1}(0.4)$, $\Phi^{-1}(0.3)$, Σ^{-1} and $|\Sigma|$ into Equation (7.40a), we have the following:

$$c(0.4,0.3;\Sigma)=1.0419.$$

Applying Equation (7.35) for $d=2$, we have the first-order derivative of the bivariate meta-Gaussian copula function as follows:

$$\frac{\partial C}{\partial u_1}=\frac{1}{\phi(x_1)}\int_{-\infty}^{x_2}g(x_1,w_2)dw_2=\frac{1}{\phi(x_1)}\int_{-\infty}^{x_2}\frac{1}{2\pi|\Sigma|^{\frac{1}{2}}}\exp\left(-\frac{1}{2}[x_1\quad w_2]\Sigma^{-1}\begin{bmatrix}x_1\\ w_2\end{bmatrix}\right)dw_2 \tag{7.41}$$

Substituting $|\Sigma|=1-\rho^2$, $\Sigma^{-1}=\dfrac{\begin{pmatrix}1&-\rho\\-\rho&1\end{pmatrix}}{1-\rho^2}$ into Equation (7.41), we have the following:

$$\frac{\partial C}{\partial u_1}=\frac{1}{\phi(x_1)}\int_{-\infty}^{x_2}\frac{1}{2\pi\sqrt{1-\rho^2}}\exp\left(-\frac{1}{2(1-\rho^2)}(x_1^2-2\rho x_1w_2+w_2^2)\right)dw_2$$
$$=\frac{1}{\phi(x_1)}\frac{1}{\sqrt{2\pi}\sqrt{1-\rho^2}}\exp\left(-\frac{x_1^2}{2(1-\rho^2)}\right)\left\{\frac{1}{\sqrt{2\pi}}\int_{-\infty}^{x_2}\exp\left(-\frac{1}{2(1-\rho^2)}(w_2^2-2\rho x_1w_2)\right)dw_2\right\} \tag{7.41a}$$

In Equation (7.41a), $\frac{1}{\sqrt{2\pi}}\int_{-\infty}^{x_2}\exp\left(-\frac{1}{2(1-\rho^2)}(w_2^2-2\rho x_1w_2)\right)dw_2$ may be further simplified as follows:

$$\frac{1}{\sqrt{2\pi}}\int_{-\infty}^{x_2}\exp\left(-\frac{1}{2(1-\rho^2)}(w_2^2-2\rho x_1w_2)\right)dw_2$$
$$=\frac{1}{\sqrt{2\pi}}\int_{-\infty}^{x_2}\exp\left(-\frac{1}{2(1-\rho^2)}\left[(w_2-\rho x_1)^2-\rho^2x_1^2\right]\right)dw_2$$
$$=\frac{1}{\sqrt{2\pi}}\exp\left(\frac{\rho^2x_1^2}{2(1-\rho^2)}\right)\int_{-\infty}^{x_2}\exp\left(-\frac{1}{2}\left(\frac{w_2-\rho x_1}{\sqrt{1-\rho^2}}\right)^2\right)dw_2$$

Let $y=\dfrac{w_2-\rho x_1}{\sqrt{1-\rho^2}}$. We have the following:

$$\frac{1}{\sqrt{2\pi}}\int_{-\infty}^{x_2}\exp\left(-\frac{1}{2}\left(\frac{w_2-\rho x_1}{\sqrt{1-\rho^2}}\right)^2\right)dw_2=\frac{\sqrt{1-\rho^2}}{\sqrt{2\pi}}\int_{-\infty}^{\frac{x_2-\rho x_1}{\sqrt{1-\rho^2}}}\exp\left(-\frac{y^2}{2}\right)dy$$
$$=\sqrt{1-\rho^2}\,\Phi\left(\frac{x_2-\rho x_1}{\sqrt{1-\rho^2}}\right)$$

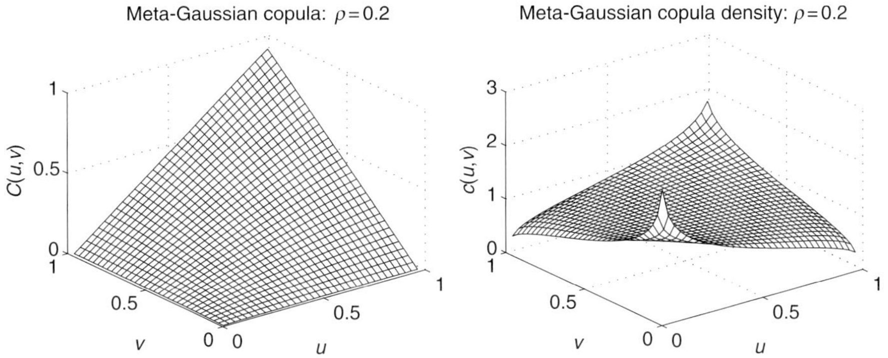

Figure 7.4 Meta-Gaussian copula and its copula density plots.

Finally, Equation (7.41a) is rewritten as follows:

$$\begin{aligned}\frac{\partial C}{\partial u_1} &= \frac{1}{\phi(x_1)}\frac{1}{\sqrt{2\pi}\sqrt{1-\rho^2}}\exp\left(-\frac{x_1^2}{2(1-\rho^2)}\right)\exp\left(\frac{\rho^2 x_1^2}{2(1-\rho^2)}\right)\sqrt{1-\rho^2}\Phi\left(\frac{x_2-\rho x_1}{\sqrt{1-\rho^2}}\right)\\ &= \Phi\left(\frac{x_2-\rho x_1}{\sqrt{1-\rho^2}}\right)\end{aligned} \tag{7.42}$$

or equivalently

$$\frac{\partial C}{\partial u_1} = \Phi\left(\frac{\left(\Phi^{-1}(u_2) - \rho\Phi^{-1}(u_1)\right)}{\sqrt{1-\rho^2}}\right) \tag{7.42a}$$

To further illustrate the shape of meta-Gaussian copula and its density function, Figure 7.4 graphs the meta-Gaussian copula and its density function with the use of parameters given in this example.

Example 7.9 **(Trivariate meta-Gaussian copula): compute the trivariate meta-Gaussian copula and its density function.**

Compute the copula and its density function with information given as follows:

$$\Sigma = \begin{pmatrix} 1 & 0.2 & 0.6 \\ 0.2 & 1 & 0.4 \\ 0.6 & 0.4 & 1 \end{pmatrix}, \ u_1 = 0.4; u_2 = 0.3; u_3 = 0.8.$$

Also, show the first- and second-order derivatives of the trivariate meta-Gaussian copula.

Applying Equation (7.32) for $d = 3$, we have the following:

$$\begin{aligned}&C(u_1, u_2, u_3; \Sigma)\\ &= \int_{-\infty}^{\Phi^{-1}(u_1)}\int_{-\infty}^{\Phi^{-1}(u_2)}\int_{-\infty}^{\Phi^{-1}(u_d)} \frac{1}{(2\pi)^{\frac{3}{2}}|\Sigma|^{\frac{1}{2}}}\exp\left(-\frac{1}{2}[x_1, x_2, x_3]\Sigma^{-1}\begin{bmatrix} x_1 \\ x_2 \\ x_3 \end{bmatrix}\right)dx_1dx_2dx_3\end{aligned} \tag{7.43}$$

From standard normal distribution, we have the following:

$$\Phi^{-1}(0.4) = -0.2533,\ \Phi^{-1}(0.3) = -0.5244,\ \Phi^{-1}(0.7) = 0.8416$$

$|\Sigma|$ and Σ^{-1} are calculated as follows:

$$|\Sigma| = 0.5360; \Sigma^{-1} = \begin{pmatrix} 1.5672 & 0.0746 & -0.9701 \\ 0.0746 & 1.1940 & -0.5224 \\ -0.9701 & -0.5224 & 1.7910 \end{pmatrix}$$

Integrating Equation (7.43) with the calculated quantity numerically, we have the following:

$$C(0.4, 0.3, 0.8; \Sigma) = 0.1450; c(0.4, 0.3, 0.8; \Sigma) = 0.6309.$$

Applying Equations (7.35) for $d = 3$, we have the first-order derivative of trivariate meta-Gaussian copula as follows:

$$\frac{\partial C(u_1, u_2, u_3)}{\partial u_1} = \frac{1}{\phi(x_1)} \int_{-\infty}^{x_2} \int_{-\infty}^{x_3} g(x_1, w_2, w_3) dw_2 dw_3$$
$$= \frac{1}{\phi(x_1)} \int_{-\infty}^{x_2} \int_{-\infty}^{x_3} \frac{1}{(2\pi)^{\frac{3}{2}} |\Sigma|^{\frac{1}{2}}} \exp\left(-\frac{1}{2} [x_1, w_2, w_3] \Sigma^{-1} \begin{bmatrix} x_1 \\ w_2 \\ w_3 \end{bmatrix} \right) dw_2 dw_3 \tag{7.44}$$

Let $\Sigma = \begin{bmatrix} 1 & \rho_{12} & \rho_{13} \\ \rho_{12} & 1 & \rho_{23} \\ \rho_{13} & \rho_{23} & 1 \end{bmatrix}$, we have the following:

$$\Sigma^{-1} = \frac{1}{|\Sigma|} \begin{bmatrix} 1 - \rho_{23}^2 & \rho_{13}\rho_{23} - \rho_{12} & \rho_{12}\rho_{23} - \rho_{13} \\ \rho_{13}\rho_{23} - \rho_{12} & 1 - \rho_{13}^2 & \rho_{12}\rho_{13} - \rho_{23} \\ \rho_{12}\rho_{23} - \rho_{13} & \rho_{12}\rho_{13} - \rho_{23} & 1 - \rho_{12}^2 \end{bmatrix}$$

where $|\Sigma| = 1 - \rho_{12}^2 - \rho_{13}^2 - \rho_{23}^2 + 2\rho_{12}\rho_{13}\rho_{23}$

The conditional copula defined in Equation (7.44) follows the bivariate normal distribution that is derived in what follows. Under the condition, i.e., $U_1 = u_1$ or equivalently $X_1 = x_1$, we first partition the random variable, Σ, and Σ^{-1} as follows:

$$\begin{bmatrix} x_1 \\ w_2 \\ w_3 \end{bmatrix} = \begin{bmatrix} x_1 \\ \mathbf{w} \end{bmatrix}, \text{ where } \mathbf{w} = \begin{bmatrix} w_2 \\ w_3 \end{bmatrix} \tag{7.44a}$$

$$\Sigma = \begin{bmatrix} 1 & \rho_{12} & \rho_{13} \\ \rho_{12} & 1 & \rho_{23} \\ \rho_{13} & \rho_{23} & 1 \end{bmatrix} = \begin{bmatrix} \Sigma_{11} & \Sigma_{12} \\ \Sigma_{21} & \Sigma_{22} \end{bmatrix} \tag{7.44b}$$

where $\Sigma_{11} = 1$, $\Sigma_{12} = \Sigma_{21}^T = [\rho_{12}, \rho_{13}]$, $\Sigma_{22} = \begin{bmatrix} 1 & \rho_{23} \\ \rho_{23} & 1 \end{bmatrix}$

$$\Sigma^{-1} = \begin{bmatrix} V_{11} & V_{12} \\ V_{21} & V_{22} \end{bmatrix} \tag{7.44c}$$

where $V_{11} = \frac{1}{|\Sigma|}\left(1-\rho_{23}^2\right)$, $V_{12} = V_{21}^T = \frac{1}{|\Sigma|}[\rho_{13}\rho_{23} - \rho_{12} \quad \rho_{12}\rho_{23} - \rho_{13}]$

$$V_{22} = \frac{1}{|\Sigma|}\begin{bmatrix} 1-\rho_{13}^2 & \rho_{12}\rho_{13} - \rho_{23} \\ \rho_{12}\rho_{13} - \rho_{23} & 1-\rho_{12}^2 \end{bmatrix}$$

Substituting Equations (7.44a), (7.44b), and (7.44c) into Equation (7.44), we have the following:

$$[x_1, x_2, x_3]\Sigma^{-1}\begin{bmatrix} x_1 \\ x_2 \\ x_3 \end{bmatrix} = \left[x_1, \mathbf{w}^T\right]\begin{bmatrix} V_{11} & V_{12} \\ V_{21} & V_{22} \end{bmatrix}\begin{bmatrix} x_1 \\ \mathbf{w} \end{bmatrix} = x_1^2 V_{11} + x_1 V_{12}\mathbf{w} + \mathbf{w}^T V_{21} x_1 + \mathbf{w}^T V_{22}\mathbf{w} \tag{7.44d}$$

After some algebra, Equation (7.44d) may be rewritten as follows:

$$[x_1, x_2, x_3]\Sigma^{-1}\begin{bmatrix} x_1 \\ x_2 \\ x_3 \end{bmatrix} = (\mathbf{w}-\mathbf{a})^T V_{22}(\mathbf{w}-\mathbf{a}) + \boldsymbol{b} \tag{7.44e}$$

where $\mathbf{a} = -V_{22}^{-1}V_{21}x_1$, $\boldsymbol{b} = x_1^2\left(V_{11} - V_{21}^T V_{22}^{-1} V_{21}\right)$

Equation (7.44e) can be rewritten as follows:

$$[x_1, x_2, x_3]\Sigma^{-1}\begin{bmatrix} x_1 \\ x_2 \\ x_3 \end{bmatrix} = \left(\mathbf{w} + V_{22}^{-1}V_{21}x_1\right)^T V_{22}\left(\mathbf{w} + V_{22}^{-1}V_{21}x_1\right) + x_1^2\left(V_{11} - V_{21}^T V_{22}^{-1}V_{21}\right) \tag{7.44f}$$

Substituting Equation (7.44f) back into Equation (7.44), we have the following:

$$\frac{\partial C(u_1,u_2,u_3)}{\partial u_1} = \frac{1}{\phi(x_1)}\int_{-\infty}^{x_2}\int_{-\infty}^{x_3}\frac{1}{(2\pi)^{\frac{3}{2}}|\Sigma|^{\frac{1}{2}}}\exp\left(-\frac{1}{2}[x_1, w_2, w_3]\Sigma^{-1}\begin{bmatrix} x_1 \\ w_2 \\ w_3 \end{bmatrix}\right)dw_2 dw_3$$

$$= \frac{1}{\phi(x_1)}\int_{-\infty}^{x_2}\int_{-\infty}^{x_3}\frac{\exp\left(\left(\mathbf{w}+V_{22}^{-1}V_{21}x_1\right)^T V_{22}\left(\mathbf{w}+V_{22}^{-1}V_{21}x_1\right) + x_1^2\left(V_{11} - V_{21}^T V_{22}^{-1}V_{21}\right)\right)}{(2\pi)^{\frac{3}{2}}|\Sigma|^{\frac{1}{2}}}dw_2 dw_3$$

$$\propto \exp\left(-\frac{1}{2}\left(\mathbf{w}+V_{22}^{-1}V_{21}x_1\right)^T V_{22}\left(\mathbf{w}+V_{22}^{-1}V_{21}x_1\right)\right) \sim \mathbf{BVN}\left(-V_{22}^{-1}V_{21}x_1, V_{22}^{-1}\right) \tag{7.45}$$

where $V_{22}^{-1} = \frac{1}{|\Sigma|^2}\begin{bmatrix} 1-\rho_{12}^2 & \rho_{23} - \rho_{12}\rho_{13} \\ \rho_{23} - \rho_{12}\rho_{13} & 1-\rho_{13}^2 \end{bmatrix}$, $-V_{22}^{-1}V_{21}x_1 = \frac{1}{|\Sigma|}\begin{bmatrix} \rho_{12}x_1 \\ \rho_{13}x_1 \end{bmatrix}$.

Similarly, we can derive the second-order derivative of the trivariate meta-Gaussian copula. The second-order derivative of the triavariate meta-Gaussian copula follows the univariate normal distribution, i.e., $\frac{\partial C(u_1, u_2, u_3)}{\partial u_1 \partial u_2} \sim N\left(-\frac{x_1(\rho_{12}\rho_{23} - \rho_{13}) + x_2(\rho_{12}\rho_{13} - \rho_{23})}{1-\rho_{12}^2}, \frac{|\Sigma|}{1-\rho_{12}^2}\right)$.

7.2.2 Meta-Student t Copula

A d-dimensional meta-Student t copula can be expressed as follows:

$$C(u_1,\ldots,u_d;\boldsymbol{\Sigma},v) = T_{\Sigma,v}\left(T_v^{-1}(u_1),\ldots,T_v^{-1}(u_d)\right)$$
$$= \int_{-\infty}^{T_v^{-1}(u_1)}\cdots\int_{-\infty}^{T_v^{-1}(u_d)} \frac{\Gamma\left(\frac{v+d}{2}\right)}{\Gamma\left(\frac{v}{2}\right)} \frac{1}{(\pi v)^{\frac{d}{2}}|\boldsymbol{\Sigma}|^{\frac{1}{2}}}\left(1+\frac{\mathbf{w}^T\boldsymbol{\Sigma}^{-1}\mathbf{w}}{v}\right)^{-\left(\frac{v+d}{2}\right)} d\mathbf{w} \tag{7.46}$$

where

$T_v^{-1}(\cdot)$ represents the inverse of the univariate Student t distribution with v degrees of freedom.
$T_{\Sigma,v}\left(T_v^{-1}(u_1),\ldots,T_v^{-1}(u_d)\right)$ represents the multivariate Student t distribution with correlation matrix $\boldsymbol{\Sigma}$ and v degrees of freedom in which $\boldsymbol{\Sigma} = \begin{bmatrix} 1 & \cdots & \rho_{1d} \\ \vdots & \ddots & \vdots \\ \rho_{d1} & \cdots & 1 \end{bmatrix}$, $\rho_{ij} = \begin{cases} 1, & i=j \\ \rho_{ji}, & i \neq j \end{cases}$.
d represents the dimension of variables; and $\mathbf{w}$ represents the integral matrix: $\mathbf{w} = [w_1,\ldots,w_d]^T$.

Let $g(\mathbf{w}) => \dfrac{\Gamma\left(\frac{v+d}{2}\right)}{\Gamma\left(\frac{v}{2}\right)} \dfrac{1}{(\pi v)^{\frac{d}{2}}|\boldsymbol{\Sigma}|^{\frac{1}{2}}}\left(1+\dfrac{\mathbf{w}^T\boldsymbol{\Sigma}^{-1}\mathbf{w}}{v}\right)^{-\frac{v+d}{2}}$ and $x_1 = T_v^{-1}(u_1), \ldots x_d = T_v^{-1}(u_d)$.

Equation (7.46) can then be rewritten as follows:

$$C(u_1,\ldots,u_d;\boldsymbol{\Sigma},v) = \int_{-\infty}^{x_1}\cdots\int_{-\infty}^{x_d} g(w_1,\ldots,w_d)dw_1\ldots dw_d \tag{7.47}$$

Its copula density function can then be written as follows:

$$c(u_1,\ldots,u_d;\boldsymbol{\Sigma},v) = \frac{\partial^d}{\partial u_1\ldots\partial u_d}C(u_1,\ldots,u_d;\boldsymbol{\Sigma},v)$$
$$= \frac{\partial^d}{\partial u_1\ldots\partial u_d}\int_{-\infty}^{T_v^{-1}(u_1)}\cdots\int_{-\infty}^{T_v^{-1}(u_d)} \frac{\Gamma\left(\frac{v+d}{2}\right)}{\Gamma\left(\frac{v}{2}\right)} \frac{1}{(\pi v)^{\frac{d}{2}}|\boldsymbol{\Sigma}|^{\frac{1}{2}}}\left(1+\frac{\mathbf{w}^T\boldsymbol{\Sigma}^{-1}\mathbf{w}}{v}\right)^{-\frac{v+d}{2}} d\mathbf{w} \tag{7.48}$$

or equivalently

$$c(u_1, \ldots, u_d; \boldsymbol{\Sigma}, v) = \frac{\partial^d}{\partial u_1 \ldots \partial u_d} \int_{-\infty}^{x_1} \cdots \int_{-\infty}^{x_d} g(w_1, \ldots, w_d) dw_1 \ldots dw_d \tag{7.48a}$$

Apply the partial derivative rules of the inverse function:

$$\begin{cases} u_1 = T_v(x_1) \\ \cdots \\ u_d = T_v(x_d) \end{cases} \Rightarrow \begin{cases} \dfrac{du_1}{dx_1} = t_v(x_1) \\ \cdots \\ \dfrac{du_d}{dx_d} = t_v(x_d) \end{cases} \Rightarrow \begin{cases} \dfrac{\partial x_1}{\partial u_1} = \dfrac{dx_1}{du_1} = \dfrac{1}{\dfrac{du_1}{dx_1}} = \dfrac{1}{t_v(x_1)} \\ \cdots \\ \dfrac{\partial x_d}{\partial u_d} = \dfrac{dx_d}{du_d} = \dfrac{1}{\dfrac{du_d}{dx_d}} = \dfrac{1}{t_v(x_d)} \end{cases} \tag{7.49}$$

Now, substituting Equation (7.49) into Equation (7.48) or (7.48a), we can compute the partial derivatives for the d-dimensional meta-Student t copula. Similar to the d-dimensional meta-Gaussian copula, we will calculate the conditional copula by partitioning the random vector $\mathbf{X} = [X_1, \ldots, X_d]^T$, its correlation matrix $\boldsymbol{\Sigma}$, and inverse function $\boldsymbol{\Sigma}^{-1}$ as follows:

- Partitioning $\mathbf{X}$, Σ, $\boldsymbol{\Sigma}^{-1}$ as follows:

$$\mathbf{X} = \begin{bmatrix} \boldsymbol{X}_1 \\ \boldsymbol{X}_2 \end{bmatrix}; \boldsymbol{\Sigma} = \begin{bmatrix} \Sigma_{11} & \Sigma_{12} \\ \Sigma_{21} & \Sigma_{22} \end{bmatrix}; \boldsymbol{\Sigma}^{-1} = \mathbf{V} = \begin{bmatrix} V_{11} & V_{12} \\ V_{21} & V_{22} \end{bmatrix} \tag{7.50}$$

where

$\boldsymbol{X}_1 = [X_1, \ldots, X_{d_1}]^T$ (the conditional m-dimensional vector), $\boldsymbol{X}_2 = [X_{d_1+1}, \ldots, X_d]^T$; $\Sigma_{12} = \Sigma_{21}^T$; $V_{12} = V_{21}^T$;

$$\begin{cases} V_{11} = \left(\Sigma_{11} - \Sigma_{12}\Sigma_{22}^{-1}\Sigma_{21}\right)^{-1}, \quad (d_1 \; by \; d_1 \; matrix) \\ V_{12} = V_{21}^T = -\Sigma_{11}^{-1}\Sigma_{12}\left(\Sigma_{22} - \Sigma_{21}\Sigma_{11}^{-1}\Sigma_{12}\right)^{-1}, \quad (d - d_1) \; by \; d_1 \; matrix) \\ V_{22} = \left(\Sigma_{22} - \Sigma_{21}\Sigma_{11}^{-1}\Sigma_{12}\right)^{-1}, \quad (d - d_1) by (d - d_1) \; matrix) \end{cases} \tag{7.50a}$$

Then, $\mathbf{X}^T\boldsymbol{\Sigma}^{-1}\mathbf{X}$ in Equation (7.48) can be rewritten as follows:

$$\mathbf{X}^T\boldsymbol{\Sigma}^{-1}\mathbf{X} = \boldsymbol{X}_1^T V_{11}\boldsymbol{X}_1 + \boldsymbol{X}_1^T V_{12}\boldsymbol{X}_2 + \boldsymbol{X}_2^T V_{21}\boldsymbol{X}_1 + \boldsymbol{X}_2^T V_{22}\boldsymbol{X}_2$$

$$= \boldsymbol{X}_1^T V_{11}\boldsymbol{X}_1 + 2\boldsymbol{X}_1^T V_{12}\boldsymbol{X}_2 + \boldsymbol{X}_2^T V_{22}\boldsymbol{X}_2 \tag{7.51}$$

Expressing the square in $\boldsymbol{X}_2$, we can compute the conditional distribution as follows:

$$\mathbf{X}^T\boldsymbol{\Sigma}^{-1}\mathbf{X} = (\boldsymbol{X}_2 - \boldsymbol{m})^T\mathbf{M}(\boldsymbol{X}_2 - \boldsymbol{m}) + \mathbf{C} \tag{7.51a}$$

where

$$\mathbf{M} = V_{22}; \mathbf{C} = \boldsymbol{X}_1^T\left(V_{11} - V_{21}^T V_{22}^{-1} V_{21}\right)\boldsymbol{X}_1 = \boldsymbol{X}_1^T\Sigma_{11}^{-1}\boldsymbol{X}_1 \tag{7.51b}$$

$$\boldsymbol{m} = -V_{22}^{-1}V_{21}\boldsymbol{X}_1 = R_{21}R_{11}^{-1}\boldsymbol{X}_1 \tag{7.51c}$$

- Apply the conditional density function $f(\mathbf{X}|X_1) = \dfrac{f(\mathbf{X})}{f(X_1)}$; after some algebra, we have the following:

$$\mathbf{X} \mid X_1 \sim \mathbf{T}\left(X_2; \boldsymbol{\mu}_{2|1}, \boldsymbol{\Sigma}_{2|1}, v_{2|1}\right) \tag{7.52}$$

where

T represents the multivariate (or univariate) Student t distribution;

$$\begin{cases} \boldsymbol{\mu}_{\mathbf{2|1}} = \mathbf{m} = -V_{22}^{-1}V_{21}X_1 = R_{21}R_{11}^{-1}X_1 \\ \boldsymbol{\Sigma}_{\mathbf{2|1}} = \dfrac{v + X_1^T\Sigma_{11}^{-1}X_1}{v + d_1}\left(\Sigma_{22} - \Sigma_{21}\Sigma_{11}^{-1}\Sigma_{12}\right) \\ v_{2|1} = v + d_1 \end{cases} \tag{7.52a}$$

First-Order Partial Derivative

$$\begin{aligned} \frac{\partial C}{\partial u_1} &= \frac{\partial}{\partial u_1}\int_{-\infty}^{x_1}\cdots\int_{-\infty}^{xd} g(w_1,\ldots,w_d)dw_1\ldots dw_d \\ &= \int_{-\infty}^{x_2}\cdots\int_{-\infty}^{x_d}\frac{g(x_1,w_2,\ldots,w_d)}{t_v(x_1)}dw_2\ldots dw_d = \frac{\int_{-\infty}^{x_2}\cdots\int_{-\infty}^{x_d}g(x_1,w_2,\ldots,w_d)dw_2\ldots dw_d}{t_v(x_1)} \end{aligned} \tag{7.53}$$

In Equation (7.53), $\dfrac{g(x_1, w_2, \ldots, w_d)}{f(x_1)}$ is the conditional density function given x_1. Applying Equations (7.50)–(7.52), we have the conditional copula, which follows the $d - 1$ cumulative multivariate (or univariate if d = 2) Student t distribution with the following parameters:

$$\boldsymbol{\Sigma} = \begin{bmatrix} 1 & \cdots & \rho_{1d} \\ \vdots & \ddots & \vdots \\ \rho_{d1} & \cdots & 1 \end{bmatrix}; \Sigma_{11} = 1,\ \Sigma_{12} = \Sigma_{21}^T = [\rho_{12}, \ldots, \rho_{1d}],\ \Sigma_{22} = \begin{bmatrix} 1 & \cdots & \rho_{2d} \\ \vdots & \ddots & \vdots \\ \rho_{d2} & \cdots & 1 \end{bmatrix} \tag{7.54}$$

$$\boldsymbol{\mu}_{\mathbf{2|1}} = (\Sigma_{22} - \Sigma_{21}\Sigma_{12})\left(\Sigma_{12}(\Sigma_{22} - \Sigma_{21}\Sigma_{12})^{-1}\right)^T x_1 \tag{7.54a}$$

$$\boldsymbol{\Sigma}_{\mathbf{2|1}} = \frac{v + x_1^2}{v + 1}(\Sigma_{22} - \Sigma_{21}\Sigma_{12}) \tag{7.54b}$$

$$v_{2|1} = v + 1 \tag{7.54c}$$

Second-Order Partial Derivative

$$\begin{aligned}\frac{\partial^2 C}{\partial u_1 \partial u_2} &= \frac{\partial}{\partial u_2}\left[\frac{1}{t_v(x_1)}\int_{-\infty}^{x_2}\cdots\int_{-\infty}^{x_d} g(x_1, w_2, \ldots, w_d)dw_2 \ldots dw_d\right] \\ &= \frac{1}{t_v(x_1)t_v(x_2)}\int_{-\infty}^{x_3}\cdots\int_{-\infty}^{x_d} g(x_1, x_2, \ldots, w_d)dw_3 \ldots dw_d\end{aligned} \tag{7.55}$$

Similar to the first-order partial derivative for meta-Student t copula, the second-order partial derivative again follows the d-2 cumulative multivariate (or univariate if d = 3) Student t distribution. Based on the derivations given in Equations (7.50)–(7.52), the parameters of the conditional copula are derived in what follows:

Equation (7.50) is rewritten as follows:

$$\mathbf{X} = \begin{bmatrix} \boldsymbol{X}_1 \\ \boldsymbol{X}_2 \end{bmatrix}; \boldsymbol{X}_1 = \begin{bmatrix} x_1 \\ x_2 \end{bmatrix}, \boldsymbol{X}_2 = \begin{bmatrix} x_3 \\ \vdots \\ x_d \end{bmatrix} \tag{7.56}$$

$$\Sigma_{11} = \begin{bmatrix} 1 & \rho_{12} \\ \rho_{12} & 1 \end{bmatrix}, \Sigma_{12} = \Sigma_{21}^T = [\rho_{13}, \ldots, \rho_{1d}], \Sigma_{22} = \begin{bmatrix} 1 & \cdots & \rho_{3d} \\ \vdots & \ddots & \vdots \\ \rho_{d3} & \cdots & 1 \end{bmatrix} \tag{7.56a}$$

Substituting Equation (7.56) back into Equation (7.52), we obtain the parameters for the conditional Student t distribution as follows:

$$\boldsymbol{\mu}_{\mathbf{2|1}} = \left(\Sigma_{22} - \Sigma_{21}\Sigma_{11}^{-1}\Sigma_{12}\right)\left(\Sigma_{12}\left(\Sigma_{22} - \Sigma_{21}\Sigma_{11}^{-1}\Sigma_{12}\right)^{-1}\right)^T \begin{bmatrix} x_1 \\ x_2 \end{bmatrix} \tag{7.56c}$$

$$\boldsymbol{\Sigma}_{\mathbf{2|1}} = \frac{v + [x_1, x_2]\Sigma_{11}^{-1}\begin{bmatrix} x_1 \\ x_2 \end{bmatrix}}{v+2}\left(\Sigma_{22} - \Sigma_{21}\Sigma_{11}^{-1}\Sigma_{12}\right) \tag{7.56d}$$

$$v_{\mathbf{2|1}} = v + 2 \tag{7.56e}$$

d*th-Order Partial Derivative*

Using the same approach, the PDF of d-dimensional meta-Student t copula can be obtained as follows:

$$c(u_1, \ldots, u_d; \boldsymbol{\Sigma}, v) = \frac{\partial^d C(u_1, \ldots, u_d; \boldsymbol{\Sigma}, v)}{\partial u_1 \ldots \partial u_d} = \frac{1}{t_v(x_1) \cdots t_v(x_d)} g(x_1, \ldots, x_d) \tag{7.57}$$

Let $\mathbf{X} = [x_1, \ldots, x_d]^T = \left[T_v^{-1}(u_1), \ldots, T_v^{-1}(u_d)\right]^T$. Then, $g(x_1, \ldots, x_d)$ can be given as follows:

$$g(\mathbf{X}) = g(x_1, \ldots, x_d) = \frac{\Gamma\left(\frac{v+d}{2}\right)}{\Gamma\left(\frac{v}{2}\right)} \frac{1}{(\pi v)^{\frac{d}{2}} |\mathbf{\Sigma}|^{\frac{1}{2}}} \left(1 + \frac{\mathbf{X}^T \mathbf{\Sigma}^{-1} \mathbf{X}}{v}\right)^{-\left(\frac{v+d}{2}\right)} \tag{7.57a}$$

$$c(u_1, \ldots, u_d; \boldsymbol{\Sigma}, v) = \frac{\Gamma\left(\frac{v+d}{2}\right)}{\prod_{i=1}^{d} t_v(x_i) \Gamma\left(\frac{v}{2}\right)} \frac{1}{(\pi v)^{\frac{d}{2}} |\mathbf{\Sigma}|^{\frac{1}{2}}} \left(1 + \frac{\mathbf{X}^T \mathbf{\Sigma}^{-1} \mathbf{X}}{v}\right)^{-\left(\frac{v+d}{2}\right)} \tag{7.57b}$$

Substituting $t_v(x_i) = \dfrac{\Gamma\left(\frac{v+1}{2}\right)}{\Gamma\left(\frac{v}{2}\right)(\pi v)^{\frac{1}{2}}} \left(1 + \dfrac{x_i^2}{v}\right)^{-\frac{v+1}{2}}$ into Equation (7.57b), we have the following:

$$c(u_1, \ldots, u_d) = \frac{\Gamma\left(\frac{v+d}{2}\right) \Gamma^{d-1}\left(\frac{v}{2}\right)}{\Gamma^d\left(\frac{v+1}{2}\right) |\boldsymbol{\Sigma}|^{\frac{1}{2}}} \frac{\left(1 + \frac{\mathbf{X}^T \mathbf{\Sigma}^{-1} \mathbf{X}}{v}\right)^{-\frac{v+d}{2}}}{\prod_{i=1}^{d} \left(1 + \frac{x_i}{v}\right)^{-\frac{v+1}{2}}} \tag{7.57c}$$

Example 7.10 **(Bivariate meta-Student *t* copula): compute the bivariate meta-Student *t* copula and its density function.**

Compute the copula and its density function with the following information:

$$\mathbf{\Sigma} = \begin{bmatrix} 1 & 0.2 \\ 0.2 & 1 \end{bmatrix}, \; v = 2, \; u_1 = 0.4, \; u_2 = 0.3.$$

Also, show the first-order derivative of the bivariate meta-Student *t* copula.

Solution: For the bivariate meta-Student *t* copula, let

$$\boldsymbol{\Sigma} = \begin{bmatrix} 1 & \rho \\ \rho & 1 \end{bmatrix}, \mathbf{X} = [x_1, x_2]^T = \left[T_v^{-1}(u_1), T_v^{-1}(u_2)\right]^T$$

and we have the following:

$$|\boldsymbol{\Sigma}| = 1 - \rho^2; \; |\boldsymbol{\Sigma}|^{-1} = \frac{1}{1-\rho^2} \begin{bmatrix} 1 & -\rho \\ -\rho & 1 \end{bmatrix}$$

$$\begin{aligned} \mathbf{X}^T \boldsymbol{\Sigma}^{-1} \mathbf{X} &= \frac{\left[T_v^{-1}(u_1), T_v^{-1}(u_2)\right] \begin{bmatrix} 1 & -\rho \\ -\rho & 1 \end{bmatrix} \left[T_v^{-1}(u_1), T_v^{-1}(u_2)\right]^T}{1-\rho^2} \\ &= \frac{\left(T_v^{-1}(u_1)\right)^2 - 2\rho T_v^{-1}(u_1) T_v^{-1}(u_2) + \left(T_v^{-1}(u_2)\right)^2}{1-\rho^2} \end{aligned}$$

Then, the bivariate meta-Student t copula and its copula density can be expressed as follows:

$$C(u_1, u_2; \Sigma, v) = T_{\Sigma, v}\left(T_v^{-1}(u_1), T_v^{-1}(u_2)\right)$$
$$= \int_{-\infty}^{T_v^{-1}(u_1)} \int_{-\infty}^{T_v^{-1}(u_2)} \frac{\Gamma\left(\frac{v+2}{2}\right)}{\Gamma\left(\frac{v}{2}\right)} \frac{1}{\pi v |\Sigma|^{\frac{1}{2}}} \left(1 + \frac{\mathbf{w}^T \Sigma^{-1} \mathbf{w}}{v}\right)^{-\frac{v+2}{2}} d\mathbf{w} \tag{7.58}$$

$$c(u_1, u_2; \mathbf{\Sigma}, v) = \frac{\Gamma\left(\frac{v+2}{2}\right)\left(1 + \frac{\mathbf{X}^T \mathbf{\Sigma}^{-1} \mathbf{X}}{v}\right)^{-\frac{v+2}{2}}}{\Gamma\left(\frac{v}{2}\right) \pi v |\mathbf{\Sigma}|^{\frac{1}{2}} t_v(x_1) t_v(x_2)}$$
$$= \frac{\Gamma\left(\frac{v+2}{2}\right)\Gamma\left(\frac{v}{2}\right)}{\Gamma^2\left(\frac{v+1}{2}\right)(1-\rho^2)^{\frac{1}{2}}} \frac{\left(1 + \frac{\left(T_v^{-1}(u_1)\right)^2 - 2\rho T_v^{-1}(u_1) T_v^{-1}(u_2) + \left(T_v^{-1}(u_2)\right)^2}{v(1-\rho^2)}\right)^{-\frac{v+2}{2}}}{\left(1 + \frac{\left(T_v^{-1}(u_1)\right)^2}{v}\right)^{-\frac{v+1}{2}} \left(1 + \frac{\left(T_v^{-1}(u_2)\right)^2}{v}\right)^{-\frac{v+1}{2}}} \tag{7.59}$$

Applying the inverse of univariate Student t distribution with the degrees of freedom (d.f.) = 2, we have the following:

$$x_1 = T_v^{-1}(u_1) = T_2^{-1}(0.4) = -0.2887; x_2 = T_2^{-1}(0.3) = -0.6172;$$

The determinant and the inverse of correlation matrix can be computed as follows:

$$|\mathbf{\Sigma}| = 0.96; \mathbf{\Sigma}^{-1} = \begin{bmatrix} 1.0417 & -0.2083 \\ -0.2083 & 1.0417 \end{bmatrix}.$$

Substituting the computed quantities into Equation (7.58), we have the following:

$$C(u_1, u_2; \mathbf{\Sigma}, v) = T_{\mathbf{\Sigma}, v}(x_1, x_2) = 0.1510.$$

Substituting the computed quantities into Equation (7.59), we can compute the copula density function:

$c(u_1, u_2; \mathbf{\Sigma}, v) = 1.2365.$

Figure 7.5 plots the corresponding copula and its density function.

In what follows, we give the expression for the first-order derivative of the bivariate meta-Student t distribution.

Applying Equation (7.54a), we have the following:

$$\mu_{2|1} = \left(1-\rho^2\right)\left(\rho\left(1-\rho^2\right)^{-1}\right) T_v^{-1}(u_1) = \rho T_v^{-1}(u_1) \tag{7.60}$$

$$\Sigma_{2|1} = \frac{\left(v + \left(T_v^{-1}(u_1)\right)^2\right)}{v+1}\left(1-\rho^2\right) \tag{7.60a}$$

$$v_{2|1} = v + 1 \tag{7.60b}$$

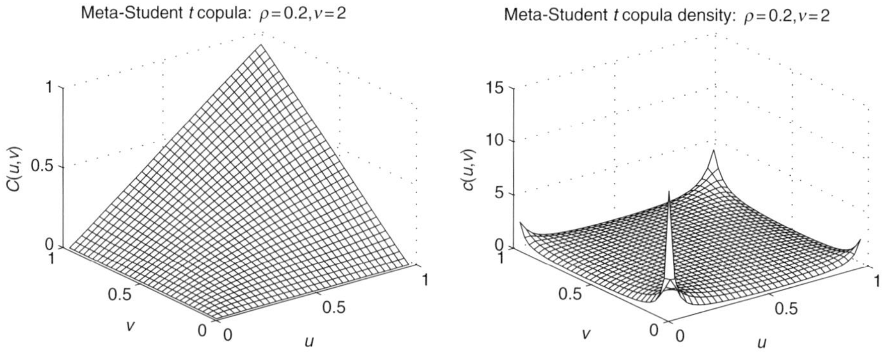

Figure 7.5 Meta-Student *t* copula and its density.

Substituting Equation (7.60) back into Equation (7.52), we have the following:

$$\frac{\partial C(u_1,u_2)}{\partial u_1} = T_{v+1}\left(\frac{\left(T_v^{-1}(u_2) - \rho T_v^{-1}(u_1)\right)}{\sqrt{\frac{\left(v+\left(T_v^{-1}(u_1)\right)^2\right)}{v+1}\left(1-\rho^2\right)}}\right) \tag{7.61}$$

Substituting $v = 2$, $\rho = 0.2$ into Equation (7.61), we have the conditional copula for this example as follows:

$$\frac{\partial C(u_1,u_2)}{\partial u_1} = T_3\left(\frac{\left(T_2^{-1}(u_2) - 0.2T_2^{-1}(u_1)\right)}{\sqrt{\frac{0.96\left(2+\left(T_2^{-1}(u_1)\right)^2\right)}{3}}}\right).$$

Example 7.11 **(Trivariate meta-Student *t* copula): compute the bivariate meta-Student *t* copula and its density function.**

Compute the copula and its density function with the following given information:

$$\boldsymbol{\Sigma} = \begin{bmatrix} 1 & 0.2 & 0.6 \\ 0.2 & 1 & 0.4 \\ 0.6 & 0.4 & 1 \end{bmatrix}, \; v = 2,\; u_1 = 0.4,\; u_2 = 0.3,\; u_3 = 0.8.$$

Also, show the first- and second-order derivative of the trivariate meta-Student *t* copula.

Solution: Applying Equation (7.46) for $d = 3$, we have the following:

$$C(u_1, u_2, u_3) = \int_{-\infty}^{T_v^{-1}(u_1)} \int_{-\infty}^{T_v^{-1}(u_2)} \int_{-\infty}^{T_v^{-1}(u_3)} \frac{\Gamma\left(\frac{v+3}{2}\right)}{\Gamma\left(\frac{v}{2}\right)} \frac{1}{(\pi v)^{\frac{3}{2}} |\mathbf{\Sigma}|^{\frac{1}{2}}} \left(1 + \mathbf{w}^T \mathbf{\Sigma}^{-1} \mathbf{w}\right) d\mathbf{w} \tag{7.62}$$

From the Student t distribution with d.f. = 2, we have the following:

$$T_2^{-1}(0.4) = -0.2887,\ T_2^{-1}(0.3) = -0.6172,\ T_2^{-1}(0.8) = 1.0607.$$

$|\mathbf{\Sigma}|$, Σ^{-1} can be calculated as follows:

$$|\mathbf{\Sigma}| = 0.5360,\ \mathbf{\Sigma}^{-1} = \begin{bmatrix} 1.5672 & 0.0746 & -0.9701 \\ 0.0746 & 1.1940 & -0.5224 \\ -0.9701 & -0.5224 & 1.7910 \end{bmatrix}$$

Integrating Equation (7.62) with the computed quantities, we have the following:

$$C(0.4, 0.3, 0.8; \mathbf{\Sigma}, v) = 0.1445; c(0.4, 0.3, 0.8; \mathbf{\Sigma}, v) = 0.4697.$$

In the following, we will show the first- and second-order derivatives of the trivariate meta-Student t copula.

First-order derivative of the trivariate meta-Student *t* copula:

For the trivariate case, Equation (7.54) can be rewritten for $\mathbf{X} = \begin{bmatrix} \boldsymbol{X}_1 \\ \boldsymbol{X}_2 \end{bmatrix}$, $\boldsymbol{X}_1 = x_1; \boldsymbol{X}_2 = \begin{bmatrix} x_2 \\ x_3 \end{bmatrix}$ as follows:

$$\mathbf{\Sigma} = \begin{bmatrix} 1 & \rho_{12} & \rho_{13} \\ \rho_{12} & 1 & \rho_{23} \\ \rho_{13} & \rho_{23} & 1 \end{bmatrix}; \Sigma_{11} = 1,\ \Sigma_{12} = \Sigma_{21}^T = [\rho_{12}, \rho_{13}],\ \Sigma_{22} = \begin{bmatrix} 1 & \rho_{23} \\ \rho_{23} & 1 \end{bmatrix} \tag{7.63}$$

$$\begin{aligned} \boldsymbol{\mu}_{2|1} &= \begin{bmatrix} 1 - \rho_{12}^2 & \rho_{23} - \rho_{12}\rho_{13} \\ \rho_{23} - \rho_{12}\rho_{13} & 1 - \rho_{13}^2 \end{bmatrix} \left(\frac{[\rho_{12} - \rho_{13}\rho_{23} \quad \rho_{13} - \rho_{12}\rho_{23}]}{|\mathbf{\Sigma}|} \right)^T x_1 \\ &= \begin{bmatrix} (1 - \rho_{12})(\rho_{12} - \rho_{13}\rho_{23}) - (\rho_{23} - \rho_{12}\rho_{13})(\rho_{13} - \rho_{12}\rho_{23}) \\ (\rho_{23} - \rho_{12}\rho_{13})(\rho_{12} - \rho_{13}\rho_{23}) - (1 - \rho_{13}^2)(\rho_{13} - \rho_{12}\rho_{23}) \end{bmatrix} \frac{x_1}{|\mathbf{\Sigma}|} \end{aligned} \tag{7.63a}$$

$$\Sigma_{2|1} = \frac{v + x_1^2}{v + 1} \begin{bmatrix} 1 - \rho_{12}^2 & \rho_{23} - \rho_{12}\rho_{13} \\ \rho_{23} - \rho_{12}\rho_{13} & 1 - \rho_{13}^2 \end{bmatrix} \tag{7.63b}$$

$$v_{2|1} = v + 1 \tag{7.63c}$$

Substituting Equations (7.63a)–(7.63c) into Equation (7.52), we have the first-order derivative for the trivariate meta-Student t copula as follows:

$$C(u_2, u_3 | u_1) = BT\left(\left(\boldsymbol{X}_2 - \boldsymbol{\mu}_{2|1}\right); \Sigma_{2|1}, v_{2|1}\right) \tag{7.63d}$$

where BT represents the bivariate cumulative Student t distribution.

Furthermore, for this example, we have the following:

$$\boldsymbol{\mu}_{2|1} = T_2^{-1}(u_1)\begin{bmatrix}0.2\\0.6\end{bmatrix}, \Sigma_{2|1} = \frac{2+T_2^{-1}(u_1)}{3}\begin{bmatrix}0.96 & 0.28\\0.28 & 0.64\end{bmatrix}, v_{2|1} = 3.$$

$$C(u_2,u_3|u_1) = BT\left(\begin{bmatrix}T_2^{-1}(u_2) - 0.2T_2^{-1}(u_1)\\T_2^{-1}(u_3) - 0.6T_2^{-1}(u_1)\end{bmatrix}; \frac{2+T_2^{-1}(u_1)}{3}\begin{bmatrix}0.96 & 0.28\\0.28 & 0.64\end{bmatrix}, 3\right).$$

Second-order derivative of the trivariate meta-student *t* copula:

In this case, Equation (7.56a) can be rewritten for $\mathbf{X} = \begin{bmatrix}\boldsymbol{X}_1\\\boldsymbol{X}_2\end{bmatrix}, \boldsymbol{X}_1 = \begin{bmatrix}x_1\\x_2\end{bmatrix}; \boldsymbol{X}_2 = x_3$ as follows:

$$\Sigma_{11} = \begin{bmatrix}1 & \rho_{12}\\\rho_{12} & 1\end{bmatrix}, \Sigma_{12} = \Sigma_{21}^T = \begin{bmatrix}\rho_{13}\\\rho_{23}\end{bmatrix}, \Sigma_{22} = 1 \tag{7.64a}$$

$$\mu_{2|1} = \left(\Sigma_{22} - \Sigma_{21}\Sigma_{11}^{-1}\Sigma_{12}\right)\left(\Sigma_{12}\left(\Sigma_{22} - \Sigma_{21}\Sigma_{11}^{-1}\Sigma_{12}\right)^{-1}\right)^T\begin{bmatrix}x_1\\x_2\end{bmatrix}$$

$$= \left(\frac{|\boldsymbol{\Sigma}| + \rho_{12}\rho_{13}\rho_{23}}{1-\rho_{12}^2}\begin{bmatrix}\rho_{13} & \rho_{23}\end{bmatrix}\frac{1-\rho_{12}^2}{|\boldsymbol{\Sigma}| + \rho_{12}\rho_{13}\rho_{23}}\begin{bmatrix}x_1\\x_2\end{bmatrix} = \rho_{13}x_1 + \rho_{23}x_2 \tag{7.64b}$$

$$\Sigma_{2|1} = \frac{v + [x_1, x_2]\Sigma_{11}^{-1}\begin{bmatrix}x_1\\x_2\end{bmatrix}}{v+2}\left(\Sigma_{22} - \Sigma_{21}\Sigma_{11}^{-1}\Sigma_{12}\right)$$

$$= \frac{v + x_1^2 - 2\rho_{12}x_1x_2 + x_2^2}{(v+2)\left(1-\rho_{12}^2\right)^2}\left(|\boldsymbol{\Sigma}| + \rho_{12}\rho_{13}\rho_{23}\right) \tag{7.64c}$$

$$v_{2|1} = v + 2 \tag{7.64d}$$

Substituting Equations (7.64b)–(7.64d) into Equation (7.52), we have the second-order derivative for the trivariate meta-Student *t* copula as follows:

$$C(u_3|u_1,u_2) = T\left(\left(x_3 - \mu_{2|1}\right); \Sigma_{2|1}, v_{2|1}\right) \tag{7.64e}$$

Furthermore, for this example, we have the following:

$$\mu_{2|1} = 0.6T_2^{-1}(u_1) + 0.4T_2^{-1}(u_2)$$

$$\Sigma_{2|1} = 0.584\left(\frac{2 + \left(T_2^{-1}(u_1)\right)^2 - 0.4T_2^{-1}(u_1)T_2^{-1}(u_2) + \left(T_2^{-1}(u_2)\right)^2}{3.6864}\right), v_{2|1} = 4.$$

$$C(u_3|u_1,u_2) = T_4\left(\frac{\left[T_2^{-1}(u_3) - \left(0.6T_2^{-1}(u_1) + 0.4T_2^{-1}(u_2)\right)\right]}{\sqrt{0.584\left(\frac{2 + \left(T_2^{-1}(u_1)\right)^2 - 0.4T_2^{-1}(u_1)T_2^{-1}(u_2) + \left(T_2^{-1}(u_2)\right)^2}{3.6864}\right)}}\right)$$

7.3 Parameter Estimation

7.3.1 Marginal Distributions

Marginal CDF of Symmetric Kotz Type Distribution

From $q_{Kotz} = \frac{2sr^{\frac{N}{s}}}{\pi\Gamma\left(\frac{N}{s}\right)}\int_0^{\infty}\left(t^2+x^2\right)^{N-1}\exp\left(-r\left(t^2+x^2\right)^s\right)dt$, we can use the Gauss–Laguerre numerical integration method to calculate the marginal CDF of the symmetric Kotz type distribution:

$$\int_0^{\infty} f(x)dx = \int_0^{\infty} e^{-x}(e^x f(x))dx \approx \sum_{i=1}^{n} \omega(x_i)e^{x_i}f(x_i) \approx \sum_{i=1}^{n} w(x_i)f(x_i) \tag{7.65}$$

where x_i is the abscissa; $\omega(x_i)$ is the weight of abscissas x_i; $w(x_i)$ is the total weight of abscissa x_i, $w(x_i) = \omega(x_i)e^{x_i}$; and n is the number of integral nodes. For $n = 32$, x_i, $\omega(x_i)$ and $w(x_i)$are given in Table 7.2.

Kotz and Nadarajah (2001) and Nadarajah and Kotz (2005) derived an expression of the hypergeometric function of PDF and CDF of the bivariate symmetric Kotz type distribution and a marginal CDF of the bivariate Pearson type II and VII distributions in the incomplete beta function, respectively.

The PDF and CDF of the bivariate symmetric Kotz type distribution, for $z > 0$, are

$$q_{Kotz}(z) = \frac{r^{\frac{1}{2s}}\exp\left(-rz^{2s}\right)}{\pi\Gamma\left(\frac{N}{s}\right)}\sum_{i=0}^{\infty}(-1)^i r^{\frac{i}{s}}\binom{-\frac{1}{2}}{i} z^{2i}\psi\left(1-\frac{N}{s}+\frac{i}{s}+\frac{1}{2s}, 1-\frac{N}{s}+\frac{i}{s}+\frac{1}{2s}, rz^{2s}\right) \tag{7.66}$$

where ψ is the degenerate hypergeometric function given as follows:

$$\psi(\alpha,\beta;x) = \frac{\Gamma(1-\beta)}{\Gamma(\alpha-\beta+1)}F_1(\alpha,\beta;x) + \frac{\Gamma(\beta-1)}{\Gamma(\alpha)}x^{1-\beta}{}_1F_1(\alpha-\beta+1, 2-\beta;x) \tag{7.66a}$$

$${}_1F_1(\alpha,\beta;x) = 1 + \sum_{i=1}^{\infty}\frac{(a)_i x^i}{(b)_i i!};\ (a)_i = \frac{\Gamma(a+i)}{\Gamma(a)},\ (b)_i = \frac{\Gamma(b+i)}{\Gamma(b)} \tag{7.66b}$$

$$\binom{-\frac{1}{2}}{i} = \frac{-\frac{1}{2}\left(-\frac{1}{2}-1\right)\left(-\frac{1}{2}-2\right)\cdots\left(-\frac{1}{2}-i+1\right)}{i!} = \frac{(-1)^i}{2^{2i}}\binom{2i}{i} \tag{7.66c}$$

The corresponding CDF for $z > 0$

Table 7.2. *Abscissas and weights of Gauss–Laguerre integration.*

No K	Abscissas x_i	Weight $\omega(x_i)$	Total weight $w(x_i)$	No K	Abscissas x_i	Weight $\omega(x_i)$	Total weight $w(x_i)$
1	0.044489	0.109218	0.114187	17	22.63578	4.08E–10	2.764644
2	0.234526	0.210443	0.266065	18	25.62015	2.41E–11	3.228905
3	0.576885	0.235213	0.418793	19	28.87393	8.43E–13	2.920194
4	1.072449	0.195903	0.572533	20	32.33333	3.99E–14	4.392848
5	1.722409	0.129984	0.727649	21	36.1132	8.86E–16	4.279087
6	2.528337	0.070579	0.884537	22	40.13374	1.93E–17	5.204804
7	3.492213	0.031761	1.043619	23	44.52241	2.36E–19	5.114362
8	4.616457	0.011918	1.205349	24	49.20866	1.77E–21	4.155615
9	5.903958	0.003739	1.370222	25	54.35018	1.54E–23	6.198511
10	7.358127	0.000981	1.538776	26	59.87912	5.28E–26	5.347958
11	8.982941	0.000215	1.711646	27	65.98336	1.39E–28	6.283392
12	10.78301	3.92E-05	1.889565	28	72.68427	1.87E–31	6.891983
13	12.76375	5.93E-06	2.073189	29	80.18837	1.18E–34	7.920911
14	14.93091	7.43E-07	2.265901	30	88.73519	2.67E–38	9.204406
15	17.29327	7.63E-08	2.469974	31	98.82955	1.34E–42	11.16374
16	19.85362	6.31E-09	2.642967	32	111.7514	4.51E–48	15.39024

$$Q_{Kotz}(z) = 1 - \frac{1}{\pi\Gamma\left(\frac{N}{s}\right)}\sum\nolimits_{i=0}^{\infty}(-1)^i\binom{-\frac{1}{2}}{i}z^{2i}\left\{\frac{1}{2i+1}\Gamma\left(\frac{N}{s}\right) - r^{\frac{1}{2s}}r^{\frac{i}{s}}z^{2i+1}\Gamma\left(\frac{N}{s}-\frac{i}{s}-\frac{1}{2s}\right)\right.$$
$$\left. + \frac{sr^{\frac{N}{s}}i^{2N}}{N(2N-2i-1)}\,{}_2F_2\left(\frac{N}{s}-\frac{i}{s}-\frac{1}{2s},\frac{N}{s};\frac{N}{s}-\frac{i}{s}-\frac{1}{2s}+1,\frac{N}{s}+1,-rz^{2s}\right)\right\} \tag{7.67}$$

where

$${}_2F_2 = 1 + \sum\nolimits_{i=1}^{\infty}\frac{(a_1)_i(a_2)_i}{(b_1)_i(b_2)_i}\frac{x^i}{i!} \tag{7.67a}$$

Equation (7.67) needs to satisfy $\frac{N}{s}-\frac{i}{s}-\frac{1}{2s}+1 \neq 0$ and $\frac{N}{s}+1 \neq 0$.

Since Equation (7.67) is an expression of hypergeometric function, it needs to satisfy $\frac{N}{s}-\frac{i}{s}-\frac{1}{2s}+1 \neq 0$, and the numerical solution may experience overflow. Therefore, the Gauss–Laguerre integration and multiple complex Gauss–Legendre integral formulae can be used to compute the marginal PDF and CDF of bivariate symmetric Kotz type distribution, respectively.

For the marginal PDF, the Gauss–Laguerre integration can be used as

$$q_{Kotz}(x) \approx \frac{2sr^{\frac{N}{s}}}{\pi\Gamma\left(\frac{N}{s}\right)} \sum_{k=1}^{q} w(t_k)e^{t_k} \sum_{i=1}^{m} \left\{ \left(t_k^2 + x_l^2\right)^{N-1} \exp\left(-r\left(t_k^2 + x_l^2\right)^s\right) \right\} \quad (7.68)$$

where t_k and $w(t_k)$ are the abscissa and the weight of the Gauss–Laguerre integration, respectively; m is the integral node; and q is the node of Gauss–Legendre integration. For CDF, we use multiple complex Gauss–Legendre integral formulae (Zhang, 2000):

$$\int_a^b \left[\int_{\varphi(x)}^{\psi(x)} f(x,y)dy \right] dx \approx \frac{\Delta x}{2} \sum_{i=1}^{q} \alpha_i^{(q)} \sum_{k=1}^{q} \alpha_k^{(q)} \sum_{j=0}^{m} \frac{\Delta y_j}{2} \sum_{l=0}^{nj} f\left(\tilde{x}_{ji}, \tilde{y}_{lk}\right) \quad (7.69)$$

where q is the node of Gauss–Legendre integration; a, b are the upper and lower integral limits of variable x; $\psi(x)$ and $\varphi(x)$ are the upper and lower integral limits of variable y; and m is a positive integer that breaks the interval $[a, b]$ of x into $\boldsymbol{m}$ equal pieces. The width of each piece is $\Delta x = \frac{b-a}{m}$; $x_j = a + j\Delta x$, $j = 0, 1, \ldots, m$; $\tilde{x}_{ji} = x_j + \frac{\Delta x}{2}\left(1 + \tilde{x}_i^{(q)}\right)$, $\tilde{x}_i^{(q)}$ is the abscissa of ith node of the Gauss–Legendre integration; and n_j is a positive integer that breaks the interval $\left[\varphi\left(\tilde{x}_{ji}\right), \psi\left(\tilde{x}_{ji}\right)\right]$ of y into n_j equal pieces, $\Delta y_j = \dfrac{\psi\left(\tilde{x}_{ji}\right) - \varphi\left(\tilde{x}_{ji}\right)}{n_j}$;

$y_l = \varphi\left(\tilde{x}_{ji}\right) + l\Delta y_j$, $l = 0, 1, \ldots, n_j$; $\tilde{y}_{lk} = y_l + \dfrac{1 + \tilde{x}_k^{(q)}\Delta y_j}{2}$; $\alpha_i^{(q)}$ and $\beta^{(q)}$are the abscissas

of the ith and kth nodes of the Gauss–Legendre and the Gauss–Laguerre integration, respectively; $\tilde{x}_k^{(q)}$ is the abscissa of the kth node of the Gauss–Laguerre integration. From Equation (7.69), we know the integral interval is $[0, \infty)$. Using the Gauss–Laguerre integration for y, one can get

$$Q_{Kotz}(x) \approx \frac{1}{2} + \frac{2sr^{\frac{N}{s}}}{\pi\Gamma\left(\frac{N}{s}\right)} \frac{\Delta x \Delta y}{4} \sum_{i=1}^{q} \sum_{k=1}^{q} \alpha_i^{(q)} \beta_k^{(q)} \sum_{j=0}^{m} \sum_{l=0}^{m} \left[\left(\tilde{y}_{lk}^2 + \tilde{x}_{ji}^2\right)^{N-1} \exp\left(-r\left(\tilde{y}_{lk}^2 + \tilde{x}_{ji}^2\right)^s\right) \right] \quad (7.70)$$

Marginal CDF of Symmetric Pearson Type VII Distribution

According to Fang et al. (2002), for $z = [x, y]$ the bivariate symmetric Pearson type VII distribution can be given as follows:

$$q_{PVII}(z) = q_{PVII}(x,y) = \frac{N-1}{\pi m \sqrt{1-\rho^2}} \left[1 + \frac{1}{m(1-\rho^2)} \left(x^2 + y^2 - 2\rho xy\right) \right]^{-N}, \quad N > 1, m > 0 \quad (7.71)$$

Through integration, the marginal CDF of the symmetric Pearson type VII distribution can be written as follows:

$$Q_{p7}(x) = 1 - \frac{\Gamma\left(N - \frac{1}{2}\right)}{\sqrt{\pi m}\Gamma(N-1)} \int_x^{\infty} \left(1 + \frac{y^2}{m}\right)^{-\left(N - \frac{1}{2}\right)} dy$$

$$= \frac{\Gamma\left(N - \frac{1}{2}\right)}{\sqrt{\pi m}\Gamma(N-1)} \int_{-\infty}^{x} \left(1 + \frac{y^2}{m}\right)^{-\left(N - \frac{1}{2}\right)} dy \tag{7.72}$$

On one hand, Equation (7.72) can be solve by applying the Gauss–Laguerre integration to compute the marginal CDF; on the other hand, it can be solved by applying the incomplete beta function (Kotz and Nadarajah, 2001), as follows:

$$Q_{p7}(x) = \begin{cases} \frac{1}{2} I_{\frac{m}{m+x^2}}\left(N - 1, \frac{1}{2}\right), & x \leq 0 \\ 1 - \frac{1}{2} I_{\frac{m}{m+x^2}}\left(N \quad 1, \frac{1}{2}\right), & x > 0 \end{cases} \tag{7.73}$$

where $I_x(a, b)$ is the incomplete beta function, as follows:

$$I_x(a, b) = \frac{1}{B(a, b)} \int_0^x t^{a-1}(1-t)^{b-1} dt \tag{7.73a}$$

$$B(a, b) = \int_0^1 t^{a-1}(1-t)^{b-1} dt \tag{7.73b}$$

Results of the Gauss–Laguerre integration and incomplete beta function results by Kotz and Nadarajah (2001) are very close, as shown in Table 7.3.

Its bivariate copula density can be given as follows:

$$c_{PVII}(u, v) = \frac{q_{PVII}\left(Q_{PVII}^{-1}(u), Q_{PVII}^{-1}(v)\right)}{q_{p7}\left(Q_2^{-1}(u)\right) q_{p7}\left(Q_2^{-1}(v)\right)} = \frac{\Gamma(N-1)\Gamma(N)}{\left(\Gamma\left(N - \frac{1}{2}\right)\right)^2} \frac{\left(1 + \frac{x^2}{m}\right)^{N - \frac{1}{2}} \left(1 + \frac{y^2}{m}\right)^{N - \frac{1}{2}}}{\sqrt{1 - \rho^2}\left(1 + \frac{x^2 + y^2 - 2\rho xy}{m(1 - \rho^2)}\right)^N} \tag{7.74}$$

where $x = Q_{p7}^{-1}(u)$, $y = Q_{p7}^{-1}(v)$.

Marginal CDF of Symmetric Pearson Type II Distribution

Again, based on Fang et al. (2002), the probability density function of symmetric bivariate Pearson II distribution (for $z = [x, y]$) can be given as follows:

$$q_{PII}(z) = q_{PII}(x, y) = \begin{cases} \frac{m+1}{\pi\sqrt{1-\rho^2}} \left[1 - \frac{x^2 + y^2 - 2\rho xy}{1 - \rho^2}\right]^m, & [x, y]\mathbf{R}^{-1}[x, y]^T \leq 1; m > -1 \\ 0, & otherwise \end{cases} \tag{7.75}$$

Table 7.3. *Marginal CDF of the symmetric Pearson type VII distribution (N = 4.0; m = 5.5)*

x	$q_{p7}(x)$	$Q_{p7}(x)^{[1]}$	$Q_{p7}(x)^{[2]}$	x	$q_{p7}(x)$	$Q_{p7}(x)^{[1]}$	$Q_{p7}(x)^{[2]}$
−3.0	0.0134	0.0101	0.0101	0.0	0.3998	0.5000	0.5000
−2.9	0.0155	0.0116	0.0116	0.1	0.3972	0.5399	0.5399
−2.8	0.0180	0.0132	0.0132	0.2	0.3897	0.5793	0.5793
−2.7	0.0208	0.0152	0.0152	0.3	0.3777	0.6177	0.6177
−2.6	0.0242	0.0174	0.0174	0.4	0.3616	0.6547	0.6547
−2.5	0.0280	0.0200	0.0200	0.5	0.3422	0.6899	0.6899
−2.4	0.0326	0.0231	0.0231	0.6	0.3202	0.7230	0.7230
−2.3	0.0378	0.0266	0.0266	0.7	0.2965	0.7539	0.7539
−2.2	0.0439	0.0306	0.0306	0.8	0.2719	0.7823	0.7823
−2.1	0.0509	0.0354	0.0354	0.9	0.2471	0.8083	0.8083
−2.0	0.0590	0.0409	0.0409	1.0	0.2228	0.8317	0.8317
−1.9	0.0684	0.0472	0.0472	1.1	0.1993	0.8528	0.8528
−1.8	0.0790	0.0546	0.0546	1.2	0.1771	0.8716	0.8716
−1.7	0.0912	0.0631	0.0631	1.3	0.1565	0.8883	0.8883
−1.6	0.1049	0.0729	0.0729	1.4	0.1376	0.9030	0.9030
−1.5	0.1204	0.0841	0.0841	1.5	0.1204	0.9159	0.9159
−1.4	0.1376	0.0970	0.0970	1.6	0.1049	0.9271	0.9271
−1.3	0.1565	0.1117	0.1117	1.7	0.0912	0.9369	0.9369
−1.2	0.1771	0.1284	0.1284	1.8	0.0790	0.9454	0.9454
−1.1	0.1993	0.1472	0.1472	1.9	0.0684	0.9528	0.9528
−1.0	0.2228	0.1683	0.1683	2.0	0.0590	0.9591	0.9591
−0.9	0.2471	0.1917	0.1917	2.1	0.0509	0.9646	0.9646
−0.8	0.2719	0.2177	0.2177	2.2	0.0439	0.9694	0.9694
−0.7	0.2965	0.2461	0.2461	2.3	0.0378	0.9734	0.9734
−0.6	0.3202	0.2770	0.2770	2.4	0.0326	0.9769	0.9769
−0.5	0.3422	0.3101	0.3101	2.5	0.0280	0.9800	0.9800
−0.4	0.3616	0.3453	0.3453	2.6	0.0242	0.9826	0.9826
−0.3	0.3777	0.3823	0.3823	2.7	0.0208	0.9848	0.9848
−0.2	0.3897	0.4207	0.4207	2.8	0.0180	0.9868	0.9868
−0.1	0.3972	0.4601	0.4601	2.9	0.0155	0.9884	0.9884

Note: $Q_{PVII}(x)^{[1]}$: Gauss–Laguerre integration; $Q_{PVII}(x)^{[2]}$: Kotz and Nadarajah (2001).

The marginal CDF of symmetric Pearson type II distribution can be expressed as follows:

$$Q_{p2}(x) = \frac{\Gamma(m+2)}{\sqrt{\pi}\Gamma\left(m+\frac{3}{2}\right)} \int_{-1}^{x} \left(1-y^2\right)^{m+\frac{1}{2}} dy; \quad |x| \le 1 \tag{7.76}$$

Table 7.4. *Abscissa and weight of the Gauss–Legendre integration.*

No k	Abscissa x_k	Weight $\omega(x_k)$	No K	Abscissa x_k	Weight $\omega(x_k)$
1	0.99726	0.007018	17	0.048308	0.09654
2	−0.98561	0.016277	18	0.144472	0.095638
3	−0.96476	0.025391	19	0.239287	0.093844
4	−0.93491	0.034275	20	0.331869	0.091174
5	−0.89632	0.042836	21	0.421351	0.087652
6	−0.84937	0.050998	22	0.5069	0.083312
7	−0.79448	0.058684	23	0.587716	0.078194
8	−0.73218	0.065822	24	0.663044	0.072346
9	−0.66304	0.072346	25	0.732182	0.065822
10	−0.58772	0.078194	26	0.794484	0.058684
11	−0.5069	0.083312	27	0.849368	0.050998
12	−0.42135	0.087652	28	0.896321	0.042836
13	−0.33187	0.091174	29	0.934906	0.034275
14	−0.23929	0.093844	30	0.964762	0.025391
15	−0.14447	0.095638	31	0.985612	0.016277
16	−0.04831	0.09654	32	0.997264	0.007018

Applying the Gauss–Legendre integration method, we can compute the marginal CDF of the symmetric Pearson type II distribution using the following:

$$\int_a^b f(x)dx = \frac{b-a}{2}\int_{-1}^{1} f\left(\frac{b-a}{2}\xi + \frac{b+a}{2}\right)d\xi \approx \frac{b-a}{2}\sum\nolimits_{k=1}^{n} w(x_k)f\left(\frac{b-a}{2}x_k + \frac{b+a}{2}\right) \tag{7.77}$$

Table 7.4 lists the abscissa and the weight of the Gauss–Legendre integration.

Similar to the marginal CDF of the symmetric Pearson Type VII distribution, the marginal CDF of symmetric Pearson type II distribution may be solved using the incomplete beta function as follows:

$$Q_{p2}(x) = \begin{cases} \frac{1}{2}I_{1-x^2}\left(m+\frac{3}{2},\frac{1}{2}\right), & -1 \le x \le 0 \\ 1-\frac{1}{2}I_{1-x^2}\left(m+\frac{3}{2},\frac{1}{2}\right), & 0 < x \le 1 \end{cases} \tag{7.78}$$

Comparing the equation of incomplete beta function given by Kotz and Nadarajah (2001), the marginal CDFs computed from the two methods with the given parameter m are very close, as shown in Table 7.5.

Table 7.5. *Marginal CDF of symmetric Pearson type II distribution* (m = *4.5*).

x	$q_{p2}(x)$	$Q_{p2}(x)^{[1]}$	$Q_{p2}(x)^{[2]}$	x	$q_{p2}(x)$	$Q_{p2}(x)^{[1]}$	$Q_{p2}(x)^{[2]}$
–1.0	0.0000	0.0000	0.0000	0.0	1.3535	0.5000	0.5000
–0.9	0.0003	0.0000	0.0000	0.1	1.2872	0.6331	0.6331
–0.8	0.0082	0.0003	0.0003	0.2	1.1036	0.7535	0.7535
–0.7	0.0467	0.0027	0.0027	0.3	0.8446	0.8513	0.8513
–0.6	0.1453	0.0117	0.0117	0.4	0.5661	0.9218	0.9218
–0.5	0.3212	0.0343	0.0343	0.5	0.3212	0.9657	0.9657
–0.4	0.5661	0.0782	0.0782	0.6	0.1453	0.9883	0.9883
–0.3	0.8446	0.1487	0.1487	0.7	0.0467	0.9973	0.9973
–0.2	1.1036	0.2465	0.2465	0.8	0.0082	0.9997	0.9997
–0.1	1.2872	0.3669	0.3669	0.9	0.0003	1.0000	1.0000

Note: $Q_{p2}(x)^{[1]}$: Gauss–Legendre integration; $Q_{p2}(x)^{[2]}$: Kotz and Nadarajah (2001).

7.3.2 Parameter Estimation

Generally speaking, the pseudo-maximum likelihood method may still be used to estimate parameters of meta-elliptical copulas (Nadarajah and Kotz, 2005). Here we will first introduce the pseudo-maximum likelihood function for Kotz and Pearson type meta-elliptical copulas. Then, we again focus on meta-Gaussian and meta-Student *t* copulas with examples.

Bivariate Symmetric Kotz Type Distribution

The joint probability density function of the bivariate symmetric Kotz type distribution can be given as follows:

$$f(x,y)=\frac{sr^{\frac{N}{s}}\left(x^2+y^2-2\rho xy\right)^{N-1}}{\pi\Gamma\left(\dfrac{N}{s}\right)\left(1-\rho^2\right)^{N-\frac{1}{2}}}\exp\left(-r\left(\frac{x^2+y^2-2\rho xy}{1-\rho^2}\right)^{s}\right) \tag{7.79}$$

Then, the log-likelihood function can be given as follows:

$$\begin{aligned} logL(N,r,s,\rho) = {} & \ln s+\frac{N\log r}{s}-\ln\pi-\ln\Gamma\left(\frac{N}{s}\right)+\left(\frac{1}{2}-N\right)\ln\left(1-\rho^2\right) \\ & +(N-1)\ln\left(x^2+y^2-2\rho xy\right)-r\left(\frac{x^2+y^2-2\rho xy}{1-\rho^2}\right)^{s} \end{aligned} \tag{7.79a}$$

Taking the first-order derivative of Equation (7.79a) with respect to parameters N, r, s, ρ, we have the following:

$$\begin{cases} \dfrac{\partial \log L}{\partial N} = \dfrac{\log r}{s} - \dfrac{1}{s}\Psi\left(\dfrac{N}{s}\right) - \ln\left(1-\rho^2\right) + \ln\left(x^2+y^2-2\rho xy\right) \\ \dfrac{\partial \log L}{\partial r} = \dfrac{N}{rs} - \left(\dfrac{x^2+y^2-2\rho xy}{1-\rho^2}\right)^s \\ \dfrac{\partial \log L}{\partial s} = \dfrac{1}{s} - \dfrac{N \ln r}{s^2} + \dfrac{N}{s^2}\Psi\left(\dfrac{N}{s}\right) - r\left(\dfrac{x^2+y^2-2\rho xy}{1-\rho^2}\right)^s \ln\left(\dfrac{\mathrm{x}^2+\mathrm{y}^2-2\rho \mathrm{xy}}{1-\rho^2}\right) \\ \dfrac{\partial \log L}{\partial \rho} = \dfrac{(2N-1)\rho^2}{1-\rho^2} - \dfrac{2(N-1)xy}{x^2+y^2-2\rho xy} - \dfrac{2rs(\rho(x^2+y^2) -\!\!- (1+\rho^2)xy)}{((1-\rho^2))^2}\left(\dfrac{x^2+y^2-2\rho xy}{1-\rho^2}\right)^{s-1} \end{cases} \tag{7.79b}$$

Bivariate Pearson Type VII Distribution

The log-likelihood function of the bivariate Pearson type VII distribution [Equation (7.71)] can be written as:

$$logL(N,m,\rho) = \ln(N-1) - \ln(\pi m) - \frac{1}{2}\ln\left(1-\rho^2\right) - N\ln\left(1+\frac{x^2+y^2-2\rho xy}{m(1-\rho^2)}\right) \tag{7.80}$$

Taking the first-order derivative of Equation (7.80) with respect to parameters N, m, ρ, we have the following:

$$\begin{cases} \dfrac{\partial logL}{\partial N} = \dfrac{1}{(N-1)} - \ln\left(1+\dfrac{x^2+y^2-2\rho xy}{m(1-\rho^2)}\right) \\ \dfrac{\partial logL}{\partial m} = \dfrac{N(x^2+y^2-2\rho xy)}{m^2(1-\rho^2)}\left(1+\dfrac{x^2+y^2-2\rho xy}{m(1-\rho^2)}\right)^{-1} - \dfrac{1}{m} \\ \dfrac{\partial logL}{\partial \rho} = \dfrac{\rho}{1-\rho^2} - \dfrac{2N(\rho(x^2+y^2)-(1+\rho^2)xy)}{m(1-\rho^2)}\left(1+\dfrac{x^2+y^2-2\rho xy}{m(1-\rho^2)}\right)^{-1} \end{cases} \tag{7.80a}$$

Bivariate Pearson Type II Distribution

The log-likelihood function of the bivariate Pearson type II distribution (Equation (7.75)) can be written as follows:

$$\log L(m,\rho) = \ln(m+1) - \ln(\pi) - \frac{1}{2}\ln\left(1-\rho^2\right) + m\ln\left(1-\frac{x^2+y^2-2\rho xy}{1-\rho^2}\right) \tag{7.81}$$

Taking the first-order derivative of Equation (7.81) with respect to parameters m, ρ, we have the following:

$$\begin{cases} \dfrac{\partial logL}{\partial m} = \dfrac{1}{m+1} + \ln\left(1 - \dfrac{x^2 + y^2 - 2\rho xy}{1-\rho^2}\right) \\ \dfrac{\partial logL}{\partial \rho} = \dfrac{\rho}{1-\rho^2} - \dfrac{2m(\rho(x^2+y^2) - (1+\rho^2)xy)}{\left(1 - rho^2\right)^2}\left(1 - \dfrac{x^2 + y^2 - 2\rho xy}{1-\rho^2}\right)^{-1} \end{cases} \tag{7.81a}$$

Setting Equations (7.79b), (7.80a), and (7.81a) to **0**, we can estimate the parameters of the bivariate Kotz, Pearson VII, and Pearson II distributions by solving these equations simultaneously.

Example 7.12 **Estimation of parameters of meta-Gaussian copula with the data given in Table 7.6.**

Table 7.6. *Three-dimensional data sample.*

No.	u_1	u_2	u_3	No.	u_1	u_2	u_3
1	0.8085	0.4026	0.7069	26	0.8044	0.3380	0.9206
2	0.8845	0.9449	0.9775	27	0.8441	0.3217	0.7441
3	0.0483	0.0201	0.0259	28	0.3713	0.5469	0.3967
4	0.5818	0.4478	0.7189	29	0.8165	0.5460	0.6650
5	0.7066	0.6085	0.6556	30	0.0444	0.2351	0.2073
6	0.0543	0.5992	0.0555	31	0.6413	0.8358	0.7090
7	0.4799	0.3308	0.4113	32	0.0675	0.2407	0.1012
8	0.7468	0.6777	0.6236	33	0.0142	0.1737	0.0638
9	0.9989	0.9913	0.9984	34	0.3875	0.7339	0.1912
10	0.9353	0.9649	0.9661	35	0.0237	0.0136	0.0091
11	0.0002	0.0012	0.0033	36	0.7743	0.8217	0.8119
12	0.9388	0.9533	0.9835	37	0.2967	0.8092	0.6397
13	0.8777	0.7798	0.7347	38	0.5267	0.2084	0.1927
14	0.2764	0.7564	0.4758	39	0.8736	0.8376	0.8816
15	0.8212	0.8777	0.7088	40	0.0968	0.0587	0.0861
16	0.4701	0.4711	0.4284	41	0.4120	0.0877	0.4317
17	0.7744	0.1112	0.4433	42	0.3236	0.4496	0.3733
18	0.4937	0.6475	0.8518	43	0.2043	0.7927	0.6416
19	0.7424	0.5635	0.8267	44	0.5628	0.9067	0.5870
20	0.7120	0.9838	0.9100	45	0.1844	0.2117	0.2585
21	0.9757	0.5134	0.7641	46	0.2724	0.5463	0.4876
22	0.3326	0.3769	0.1272	47	0.0737	0.3664	0.3733
23	0.8493	0.6129	0.7113	48	0.5192	0.2766	0.6553
24	0.7328	0.9191	0.9038	49	0.3644	0.6738	0.8504
25	0.5228	0.4322	0.6576	50	0.9005	0.3035	0.8588

Solution: Let $\{x_{1i}, x_{2i}, \ldots, x_{di}\}$ be a d-dimensional sample where $i = 1, \ldots, n$, $u_{1i} = F_1(x_{1i}), \ldots, u_{di} = F_d(x_{di})$. The parameter space is denoted as $\boldsymbol{\theta} = \{\boldsymbol{\Sigma} : \boldsymbol{\Sigma} \in \boldsymbol{\Omega}\}$, where $\boldsymbol{\Sigma}$ is symmetric and a positive definite matrix. Applying Equation (7.39a), the log-likelihood function of the d-dimensional meta-Gaussian copula can be written as follows:

$$\log L(\boldsymbol{\theta}) = logL(\boldsymbol{\Sigma}) = -\frac{n}{2}\ln|\Sigma| - \frac{1}{2}\sum\nolimits_{i=1}^{n} \xi_i^T \left(\boldsymbol{\Sigma}^{-1} - \mathbf{I}\right)\xi_i$$
$$= -\frac{N}{2}\ln|\boldsymbol{\Sigma}| - \frac{1}{2}tr\left(\boldsymbol{\Sigma}^{-1}\sum\nolimits_{i=1}^{n} \xi_i^T \xi_i\right) \tag{7.82}$$

where $\xi_i = [x_{1i}, \ldots, x_{di}]^T = \left[\Phi^{-1}(u_{1i}), \ldots, \Phi^{-1}(u_{di})\right]^T$; $tr(\cdot)$ trace of the matrix.

Assuming Equation (7.82) is differentiable in $\boldsymbol{\theta}$, parameters of the meta-Gaussian copula can be solved for by $\frac{\partial logL}{\partial \boldsymbol{\theta}} = 0$ as follows:

$$\frac{\partial logL}{\partial \boldsymbol{\Sigma}} = -\frac{1}{2}tr\left(-\left(\boldsymbol{\Sigma}^{-1}\right)^2 \sum\nolimits_{i=1}^{n} \xi_i^T \xi_i\right) = -\frac{1}{2}tr\left[\boldsymbol{\Sigma}^{-1}\left(n\mathbf{I}_d - \boldsymbol{\Sigma}^{-1}\sum\nolimits_{i=1}^{n} \xi_i^T \xi_i\right)\right] = 0 \tag{7.82a}$$

From Equation (7.82a), we have the following:

$$n\mathbf{I}_d - \boldsymbol{\Sigma}^{-1}\sum\nolimits_{i=1}^{n} \xi_i^T \xi_i = 0 \Rightarrow \hat{\boldsymbol{\Sigma}} = \frac{1}{n}\sum\nolimits_{i=1}^{n} \xi_i^T \xi_i \tag{7.82b}$$

To estimate the parameters (i.e., covariance matrix) of the meta-Gaussian copula, we first need to compute $\xi_i = \left[\Phi^{-1}(u_{1i}), \Phi^{-1}(u_{2i}), \Phi^{-1}(u_{3i})\right]$; $\Phi(\cdot)$: inverse of $N(0, 1)$, as shown in Table 7.7.

Applying Equation (7.82b), we have $\boldsymbol{\Sigma} = \begin{bmatrix} 1 & 0.6700 & 0.8758 \\ 0.6700 & 1 & 0.7945 \\ 0.8758 & 0.7945 & 1 \end{bmatrix}$.

Table 7.7. *Inverse normal distribution: N(0,1).*

No.	u_1	u_2	u_3	$\Phi^{-1}(u_1)$	$\Phi^{-1}(u_2)$	$\Phi^{-1}(u_3)$
1	0.8085	0.4026	0.7069	0.8723	–0.2466	0.5444
2	0.8845	0.9449	0.9775	1.1977	1.5976	2.0040
3	0.0483	0.0201	0.0259	–1.6612	–2.0523	–1.9443
4	0.5818	0.4478	0.7189	0.2064	–0.1311	0.5797
5	0.7066	0.6085	0.6556	0.5435	0.2754	0.4004
6	0.0543	0.5992	0.0555	–1.6046	0.2512	–1.5941
7	0.4799	0.3308	0.4113	–0.0504	–0.4378	–0.2242
8	0.7468	0.6777	0.6236	0.6643	0.4613	0.3149
9	0.9989	0.9913	0.9984	3.0622	2.3775	2.9566
10	0.9353	0.9649	0.9661	1.5163	1.8109	1.8264
11	0.0002	0.0012	0.0033	–3.4847	–3.0458	–2.7162

Table 7.7. (*cont.*)

No.	u_1	u_2	u_3	$\Phi^{-1}(u_1)$	$\Phi^{-1}(u_2)$	$\Phi^{-1}(u_3)$
12	0.9388	0.9533	0.9835	1.5446	1.6775	2.1332
13	0.8777	0.7798	0.7347	1.1638	0.7714	0.6271
14	0.2764	0.7564	0.4758	–0.5935	0.6947	–0.0606
15	0.8212	0.8777	0.7088	0.9201	1.1633	0.5499
16	0.4701	0.4711	0.4284	–0.0750	–0.0724	–0.1805
17	0.7744	0.1112	0.4433	0.7533	–1.2202	–0.1427
18	0.4937	0.6475	0.8518	-0.0158	0.3787	1.0442
19	0.7424	0.5635	0.8267	0.6508	0.1598	0.9411
20	0.7120	0.9838	0.9100	0.5593	2.1394	1.3409
21	0.9757	0.5134	0.7641	1.9729	0.0335	0.7197
22	0.3326	0.3769	0.1272	–0.4328	–0.3137	–1.1396
23	0.8493	0.6129	0.7113	1.0332	0.2869	0.5573
24	0.7328	0.9191	0.9038	0.6213	1.3990	1.3035
25	0.5228	0.4322	0.6576	0.0573	–0.1707	0.4060
26	0.8044	0.3380	0.9206	0.8574	–0.4179	1.4092
27	0.8441	0.3217	0.7441	1.0114	–0.4630	0.6561
28	0.3713	0.5469	0.3967	–0.3284	0.1179	–0.2619
29	0.8165	0.5460	0.6650	0.9022	0.1157	0.4262
30	0.0444	0.2351	0.2073	–1.7019	–0.7221	–0.8160
31	0.6413	0.8358	0.7090	0.3620	0.9773	0.5504
32	0.0675	0.2407	0.1012	–1.4947	–0.7041	–1.2746
33	0.0142	0.1737	0.0638	–2.1916	–0.9396	–1.5240
34	0.3875	0.7339	0.1912	–0.2859	0.6248	–0.8735
35	0.0237	0.0136	0.0091	–1.9832	–2.2086	–2.3609
36	0.7743	0.8217	0.8119	0.7532	0.9220	0.8849
37	0.2967	0.8092	0.6397	–0.5340	0.8749	0.3576
38	0.5267	0.2084	0.1927	0.0670	–0.8120	–0.8681
39	0.8736	0.8376	0.8816	1.1436	0.9848	1.1832
40	0.0968	0.0587	0.0861	–1.2999	–1.5657	–1.3654
41	0.4120	0.0877	0.4317	–0.2224	–1.3549	–0.1719
42	0.3236	0.4496	0.3733	–0.4576	–0.1266	–0.3232
43	0.2043	0.7927	0.6416	–0.8265	0.8157	0.3627
44	0.5628	0.9067	0.5870	0.1580	1.3205	0.2199
45	0.1844	0.2117	0.2585	–0.8989	–0.8005	–0.6479
46	0.2724	0.5463	0.4876	–0.6056	0.1163	–0.0310
47	0.0737	0.3664	0.3733	–1.4490	–0.3413	–0.3231
48	0.5192	0.2766	0.6553	0.0483	–0.5928	0.3996
49	0.3644	0.6738	0.8504	–0.3468	0.4504	1.0383
50	0.9005	0.3035	0.8588	1.2842	–0.5143	1.0749

Example 7.13 **Show how to estimate parameters of the meta-Student *t* copula.**

Let $\{x_{1i}, x_{2i}, \ldots, x_{di}\}$ be a d-dimensional sample where $i = 1, \ldots, n$, $u_{1i} = F_1(x_{1i}), \ \ldots, \ u_{di} = F_d(x_{di})$. In the case of meta-Student t copula, its parameter space is $\boldsymbol{\theta} = \{(v, \boldsymbol{\Sigma}): v \in (1, \infty), \boldsymbol{\Sigma} \in \boldsymbol{\Omega}\}$. In the same way as in the meta-Gaussian copula, $\boldsymbol{\Sigma}$ is symmetric and positive definite. Applying the meta-Student t copula density function (i.e., Equation (7.57)), the log-likelihood function can be given as follows:

$$logL(v, \boldsymbol{\Sigma}) = n \ln \left(\frac{\Gamma\left(\frac{v+d}{2}\right)}{\Gamma\left(\frac{v+1}{2}\right)} \right) + n(d-1) \ln \left(\frac{\Gamma\left(\frac{v}{2}\right)}{\Gamma\left(\frac{v+1}{2}\right)} \right) - \frac{n}{2} \ln |\boldsymbol{\Sigma}|$$
$$- \frac{v+d}{2} \sum\nolimits_{i=1}^{n} \ln \left(1 + \frac{\xi_i^T \boldsymbol{\Sigma}^{-1} \xi_i}{v}\right) + \frac{v+1}{2} \sum\nolimits_{i=1}^{n} \sum\nolimits_{j=1}^{d} \ln \left(1 + \frac{\xi_{ji}^2}{v}\right) \tag{7.83}$$

where $\xi_i = \left[T_v^{-1}(u_{1i}), \ldots, T_v^{-1}(u_{di})\right]^T$, and v is the degree of freedom.

To estimate the fitted parameters $\hat{\boldsymbol{\theta}} = \left(\hat{v}, \hat{\boldsymbol{\Sigma}}\right)$, we may apply the following two approaches:

1. Optimizing the log-likelihood function (Equation (7.83)) numerically with the constraint of $\boldsymbol{\Sigma}$ being symmetric and with ones on the main diagonal. With this constraint, the MLE estimate of $\hat{\boldsymbol{\Sigma}}$ may not be positive and semidefinite.
2. Estimate $\hat{\boldsymbol{\Sigma}}$ and v separately.
   - $\hat{\boldsymbol{\Sigma}}$ may be estimated from the sample Kendall tau using the following:

$$\hat{\tau}\left(U_i, U_j\right) = \frac{2}{\pi} \arcsin \hat{\rho}_{ij} \Rightarrow \hat{\rho}_{ij} = \sin\left(\frac{\pi}{2}\hat{\tau}_{ij}\right) \tag{7.84}$$

   where $\hat{\tau}_{ij} = \hat{\tau}\left(U_i, U_j\right)$ is the sample Kendall tau between random variable U_i and U_j; and $\hat{\rho}_{ij}$ is the off-diagonal element of correlation matrix $\boldsymbol{\Sigma}$. In the same way as in approach 1, the estimated correlation matrix may not be positive definite.
   - Estimate the single parameter v using MLE (Equation (7.83)) by fixing $\hat{\boldsymbol{\Sigma}}$.

For the estimated $\hat{\boldsymbol{\Sigma}}$ not being positive and semidefinite, we can apply the procedure discussed by McNeil et al. (2005) to convert it into positive definite matrix with the procedure as follows:

i. Compute the eigenvalue decomposition $\boldsymbol{\Sigma} = \boldsymbol{EDE}^T$, where $\boldsymbol{E}$ is an orthogonal matrix that contains eigenvectors, and $\boldsymbol{D}$ is the diagonal matrix that contains all the eigenvalues.
ii. Construct a diagonal matrix $\tilde{\boldsymbol{D}}$ by replacing all negative eigenvalues in $\boldsymbol{D}$ by a small value $\delta > 0$.
iii. Compute $\tilde{\boldsymbol{\Sigma}} = \boldsymbol{E}\tilde{\boldsymbol{D}}\boldsymbol{E}^T$, $\tilde{\boldsymbol{\Sigma}}$ is positive definite but not necessarily a correlation matrix.
iv. Apply the normalizing operator $\mathcal{P}$ to obtain the desired correlation matrix.

Specifically, for the bivariate case, the parameters that need to be estimated are $\boldsymbol{\theta} = (\rho, v)$. Thus, the log-likelihood function (i.e., Equation (7.83)) can be rewritten as follows:

$$\log L(\rho, v) = n \ln \Gamma\left(\frac{v+2}{2}\right) + n \ln \Gamma\left(\frac{v}{2}\right) - 2n \ln \Gamma\left(\frac{v+1}{2}\right) - \frac{n}{2} \ln(1-\rho^2)$$
$$- \frac{v+2}{2} \sum\nolimits_{i=1}^{n} \ln \left(1 + \frac{\left(\xi_{1i}^2 - 2\rho \xi_{1i} \xi_{2i} + \xi_{2i}^2\right)}{(1-\rho^2)v}\right) + \frac{v+1}{2} \sum\nolimits_{i=1}^{n} \sum\nolimits_{j=1}^{2} \ln \left(1 + \frac{\xi_{ji}^2}{v}\right) \tag{7.85}$$

Taking the first-order derivative with respect to ρ, v, we have the following:

$$\frac{\partial logL(\rho, v)}{\partial \rho} = \frac{n\rho}{1-\rho^2} + \frac{v+2}{1-\rho^2}\sum_{i=1}^{n} \frac{(\xi_{1i}-\rho\xi_{2i})(\xi_{2i}-\rho\xi_{1i})}{\left(v(1-\rho^2)+\xi_{1i}^2-2\rho\xi_{1i}\xi_{2i}+\xi_{2i}^2\right)}$$

$$\frac{\partial logL(\rho, v)}{\partial \rho} = -\frac{n}{2}\Psi\left(\frac{v+2}{2}\right)+\frac{n}{2}\Psi\left(\frac{v}{2}\right)-n\Psi\left(\frac{v+1}{2}\right)-\frac{1}{2}\sum_{i=1}^{n}\ln\left(1+\frac{\xi_{1i}^2-2\rho\xi_{1i}\xi_{2i}+\xi_{2i}^2}{v(1-\rho^2)}\right) \tag{7.85a}$$

$$+\frac{v+2}{2v^2}\sum_{i=1}^{n}\frac{v\left(\xi_{1i}^2-2\rho\xi_{1i}\xi_{2i}+\xi_{2i}^2\right)}{v(1-\rho^2)+\xi_{1i}^2-2\rho\xi_{1i}\xi_{2i}+\xi_{2i}^2}$$

$$+\frac{1}{2}\sum_{i=1}^{n}\sum_{j=1}^{2}\ln\left(1+\frac{\xi_{ji}^2}{v}\right)-\frac{v+1}{2v^2}\sum_{i=1}^{n}\sum_{j=1}^{2}\frac{\xi_{ji}^2}{1+\frac{\xi_{ji}^2}{v}}$$

$$-(v+2)\sum_{i=1}^{n}\frac{\xi_{1i}\frac{d\xi_{1i}}{dv}+\xi_{2i}\frac{d\xi_{2i}}{dv}-\rho\left(\xi_{1i}\frac{d\xi_{2i}}{dv}+\xi_{2i}\frac{d\xi_{1i}}{dv}\right)}{v(1-\rho^2)+\xi_{1i}^2-2\rho\xi_{1i}\xi_{2i}+\xi_{2i}^2}+(v+1)\sum_{i=1}^{n}\sum_{j=1}^{2}\frac{\xi_{ji}\frac{d\xi_{ji}}{dv}}{v+\xi_{ji}^2} \tag{7.85b}$$

Example 7.14 Using the data given in Table 7.7, estimate the parameters for the bivariate (using u_1, u_2) and the trivariate meta-Student *t* copula.

Solution:

Bivariate meta-Student *t* copula (using u_1, u_2)

- Approach 1

For the bivariate case, we will apply Equation (7.85), i.e., maximizing the bivariate meta-Student *t* log-likelihood function.

The initial correlation coefficient is set as the sample correlation coefficient computed from the sample Kendall tau ($\hat{\tau}_0 = 0.3812$) using Equation (7.84) as follows:

$$\hat{\rho}_0 = \sin\left(\frac{\pi}{2}\hat{\tau}_0\right) = \sin\left(\frac{0.3812\pi}{2}\right) = 0.5637.$$

The initial degree of freedom (d.f.) is set as the lower limit (i.e., $\hat{v}_0 = 10$).

Then, the final parameter set $\hat{\boldsymbol{\theta}} = \{(\hat{\rho}, \hat{v}) : \rho \in [-1, 1], v > 1\}$ may be estimated using the optimization toolbox (e.g., the fmincon function) by minimizing the negative log-likelihood function (the objective function), which is the dual problem of the MLE estimation. We have the following:

$\hat{\boldsymbol{\theta}} = (\hat{\rho}, \hat{v}) = (0.5591, 6.2531)$. With the estimated correlation coefficient, the correlation matrix is given as follows: $\boldsymbol{\Sigma} = \begin{bmatrix} 1 & 0.5591 \\ 0.5591 & 1 \end{bmatrix}$. The eigenvalue of the correlation matrix is $\boldsymbol{\lambda} = \begin{bmatrix} 0.4409 \\ 1.5591 \end{bmatrix}$, i.e., the correlation matrix is positive definite.

Furthermore, one can use the MATLAB function copulafit to estimate the parameters of the meta-Student t copula using the MLE method. The function is given as follows:

$$\text{MLE}: \left[\hat{\boldsymbol{\Sigma}}, \hat{v}\right] = copulafit\left('t', data\right)$$

Using MLE from MATLAB, we have the following:

$$\hat{\boldsymbol{\Sigma}} = \begin{bmatrix} 1 & 0.5591 \\ 0.5591 & 1 \end{bmatrix}, \hat{v} = 6.2542.$$

- Approach 2
  Fixing $\hat{\rho} = 0.5637$, we have $\hat{v} = 6.4110$.

Trivariate meta-Student *t* copula

- Approach 1
  It is shown for the bivariate case that the parameters estimated using the embedded MATLAB function and those estimated using the fmincon by writing our own objective function are almost the same. So for the trivariate example, we will only show the results obtained from the embedded MATLAB function.
  Applying approach 1 and maximizing the log-likelihood function of the trivariate meta-Student t copula, using the embedded MATLAB function mentioned previously, we have the following ML method:

$$\hat{\boldsymbol{\Sigma}} = \begin{bmatrix} 1 & 0.5831 & 0.8171 \\ 0.5831 & 1 & 0.7518 \\ 0.8171 & 0.7518 & 1 \end{bmatrix}, \hat{v} = 12.8139$$

- Approach 2
  To apply approach 2, we first need to compute the sample correlation matrix from the sample Kendall tau using Equation (7.84) as follows:

$$\boldsymbol{\tau} = \begin{bmatrix} 1 & 0.3812 & 0.6588 \\ 0.3812 & 1 & 0.5102 \\ 0.6588 & 0.5102 & 1 \end{bmatrix}, \boldsymbol{\Sigma} = \begin{bmatrix} 1 & 0.5637 & 0.8598 \\ 0.5637 & 1 & 0.7183 \\ 0.8598 & 0.7183 & 1 \end{bmatrix}.$$

The eigenvalue vector of $\boldsymbol{\Sigma}$ is computed as $\boldsymbol{\lambda} = [0.1116, 0.4537, 2.4347]^T$. Thus, we reach the conclusion that the correlation matrix is positive definite.

Fixing the correlation matrix $\boldsymbol{\Sigma}$, we only have one parameter, i.e., v, that needs to be estimated. Optimizing the log-likelihood equation (i.e., Equation (7.83)), we can estimate v with an initial estimate of $\hat{v}_0 = 2$. Using the fmincon function, we have $\hat{v} = 20.4038$.

It should be noted here for the meta-Student t copula that one can also use the following embedded function: $\left[\hat{\boldsymbol{\Sigma}}, \hat{v}\right] = copulafit\left('t', data,'Method', 'ApproximateML'\right)$. This estimation method is considered as a good estimation only if the sample size is large enough.

7.4 Summary

In this chapter, we have summarized and discussed the properties of meta-elliptical copulas. We have explained the procedures on how to construct and apply the meta-elliptical copulas, especially for the meta-Gaussian and meta-Student *t* copulas. Comparing meta-Gaussian and meta-Student *t* copulas, both copulas may be applied to model the dependence of entire range. The Student *t* copula possesses the symmetric upper (lower) tail dependence, while the meta-Gaussian copula does not possess the tail dependence. The meta-elliptical copula may be applied for the multivariate frequency analysis.

References

Fang, H. B., Fang K. T., and Kotz, S. (2002). The meta-elliptical distributions with given marginals. *Journal of Multivariate Analysis*, 82, 1–16.

Genest, C., Favre, A. C., Be´liveau, J., and Jacques, C. (2007). Meta-elliptical copulas and their use in frequency analysis of multivariate hydrological data. *Water Resources Research*, 43, W09401, doi:10.1029/2006WR005275.

Joe, H. (1997). *Multivariate Models and Dependence Concept*. Chapman & Hall, New York.

Kotz, S. and Nadarajah, S. (2001). Some extreme type elliptical distributions. *Statistics & Probability Letters*, 54, 171–182.

McNeil, A., Frey, R., and Embrechts, P. (2005). *Quantitative Risk Management: Concepts, Techniques, and Tools*. Princeton: Princeton University Press.

Nadarajah, S. (2006). Fisher information for the elliptically symmetric Pearson distributions. *Applied Mathematics and Computation*, 178, 195–206.

Nadarajah, S. (2007). A bivariate gamma model for drought. *Water Resources Research*, 43, W08501, doi:10.1029/2006WR005641.

Nadarajah, S. and Kotz, S. (2005). Information matrices for some elliptically symmetric distribution. *SORT*, 29(1), 43–56.

Zhang, G. (2000). Multiple complex Gauss–Legendre integral formulae and application. *Journal of Lanzhou University (Natural Sciences)*, 36(5), 30–34.

8

Entropic Copulas

ABSTRACT

In previous chapters, we have discussed the Archimedean and non-Archimedean copula families. In this chapter, we will introduce entropic copulas. To be more specific, we will concentrate on the entropic copulas (i.e., most entropic canonical copulas) for the bivariate case. With proper constraints (e.g., the pair rank-based correlation coefficients), the bivariate entropic copula may be easily extended to the higher dimension.

8.1 Entropy Theory and Its Application

Entropy theory has been widely applied to univariate frequency analysis for obtaining the most probable probability distribution of a random variable or the so-called maximum entropy (MaxEnt)–based distribution. The MaxEnt-based distribution is derived with the use of the principle of maximum entropy (Jaynes, 1957a, 1957b), subject to given constraints for the random variable, e.g., first moment, second moment, first moment in logarithm domain, etc. The univariate MaxEnt-based distribution is capable of capturing the shape, mode, as well as the tail of the univariate random variable, since the first four noncentral moments of the random variable almost fully approximate its probability density function. In a similar vein, the entropy theory can also be employed for multivariate hydrological frequency analysis. Conventionally, the MaxEnt-based joint distributions are constructed with the use of covariance (or Pearson's linear correlation coefficient) as constraints (Singh and Krstanovic, 1987; Krstanovic and Singh, 1993a, b; Hao and Singh, 2011; Singh et al., 2012; Singh, 2013, 2015). With the copulas gaining popularity in bivariate/multivariate frequency analysis in hydrology and water resources engineering (Favre, 2004; De Michele, et al., 2005; Kao and Govindaraju, 2007; Vandenberghe et al., 2011; Zhang and Singh, 2012), the entropy theory has been introduced to copula-based bivariate/multivariate frequency analysis. The entropy-based copula modeling may be generalized as follows:

1. The marginal distributions are derived with the use of maximum entropy principle (i.e., MaxEnt-based marginals), and the dependence structure is studied with the use of parametric copulas (e.g., Hao and Singh, 2012; Zhang and Singh, 2012).
2. The dependence function (i.e., copula function) is also derived from the entropy theory (e.g., Chu, 2011).

In the following sections, we will first briefly introduce the Shannon entropy (Shannon, 1948) followed by the derivation of entropic copula.

8.2 Shannon Entropy

In general, entropy is a measure of uncertainty or information of a random variable or its underlying probability distribution, and Shannon entropy (Shannon, 1948) is one measure of uncertainty. The MaxEnt-based distribution may be derived by maximizing the Shannon entropy, subjected to given constraints, which is the least biased and most probable distribution in concert with the principle of maximum entropy. The Shannon entropy for a continuous univariate random variable X can be written as follows:

$$H(X) = -\int f(x) \ln [f(x)]dx \tag{8.1}$$

where H denotes the Shannon entropy, and $f(x)$ denotes the probability density function of random variable X.

The commonly applied constraints to derive the MaxEnt-based distribution from Equation (8.1) may be the following:

$$\int f(x)dx = 1; \int x^i f(x)dx = E(x^i), i = 1, 2, \ldots; \int (\ln x)f(x)dx = E(\ln x) \tag{8.1a}$$

Similarly, the Shannon entropy for the continuous bivariate variables X and Y can be written as follows:

$$H(X, Y) = -\int f(x, y) \ln [f(x, y)]dxdy \tag{8.2}$$

Besides the constraints defined in Equation (8.1a) for a continuous univariate random variable, the other common constraints to derive the MaxEnt-based joint density function $f(x, y)$ are as follows:

$$\iint f(x, y)dxdy = 1; \iint xyf(x, y)dxdy = E(xy) \tag{8.2a}$$

$E(xy)$ in Equation (8.2a) can be written through covariance (i.e., dependence) between random variables X and Y as follows:

$$E(xy) = \mathrm{cov}(x, y) + \mu_X \mu_Y \tag{8.2b}$$

One may refer to Singh (1998, 2013, 2015) in regard to its classical application and parameter estimation. In the section that follows, we will focus on the entropy application to copulas.

8.3 Entropy and Copula

In the previous chapters, we have shown that the joint probability density function may be expressed through the copula density function (i.e., $c(u, v)$) as follows:

$$f(x, y) = f_X(x) f_Y(y) c(u, v), u = F_X(x), v = F_Y(y) \tag{8.3}$$

where f_X, f_Y, F_X, F_y represent, respectively, the probability density function (pdf) and distribution function (cdf) of random variables X and Y; $f(x, y)$ denotes the joint probability density function (jpdf) of random variables X and Y; and $c(u, v)$ denotes the copula density of random variables X and Y.

Equation (8.3) shows that the dependence function and marginal distributions of bivariate random variables can be investigated separately. The Shannon entropy of the copula function may be written as follows:

$$H(u, v) = -\int_0^1 \int_0^1 c(u, v) \ln c(u, v) du dv \tag{8.4}$$

Substituting Equation (8.3) into Equation (8.4), we can show that the Shannon entropy of the copula (i.e., Equation (8.4)) is equivalent to the negative mutual information of random variables X and Y as follows:

$$\begin{aligned} H(u, v) &= -\int_0^1 \int_0^1 c(u, v) \ln c(u, v) du dv \\ &= -\iint \frac{f(x, y)}{f_X(x) f_Y(y)} \ln \left(\frac{f(x, y)}{f_X(x) f_Y(y)} \right) f_X(x) f_Y(y) dx dy \\ &= \iint -f(x, y) \ln \left(\frac{f(x, y)}{f_X(x) f_Y(y)} \right) dx dy = -I(X; Y) \end{aligned} \tag{8.5}$$

Assigning proper constraints, the entropic copula can be derived by maximizing Shannon entropy of the copula (i.e., Equation (8.4)), subject to appropriate constraints. The common constraints for deriving the most entopic copulas are the constraints of total probability of marginals (i.e., for uniform distributed variable on [0, 1]), and measure of dependence (also called association):

$$\int_0^1 \int_0^1 c(u, v) du\, dv = 1 \text{ (total probability)} \tag{8.6a}$$

$$\int_0^1 \int_0^1 u^r c(u, v) du\, dv = E(u^r) = \frac{1}{r+1}, r = 1, 2, \ldots \text{(constraints on } u = F_X(x)) \tag{8.6b}$$

$$\int_0^1 \int_0^1 v^r c(u, v) du\, dv = E(v^r) = \frac{1}{r+1}, r = 1, 2, \ldots \text{(constraints on } v = F_Y(y)) \tag{8.6c}$$

$$\int_0^1 \int_0^1 a_j(u, v) c(u, v) du dv = \Theta_j, j = 1, 2, \ldots \text{(constraints of dependence measure)} \tag{8.6d}$$

In Equation (8.6d), Spearman's rho can be applied as the constraint to measure the dependence if $a_j(u, v) = uv$ with $\Theta_j = \frac{\rho_s + 3}{12}$. From Equation (3.69), it is clear that with $a_j(u, v) = uv$, we have $\int_0^1 \int_0^1 uvc(u, v) du dv = \frac{\rho_s + 3}{12}$. One can also apply other dependence measures, such as Blest's measure and Gini's gamma, discussed in Nelsen (2006) and

Chu (2011). Additionally, Equations (8.6b) and (8.6c) indicate we don't need to know the true underlying marginal distribution to solve for the multipliers of the constraints regarding the marginal variables, since the CDF of any marginal distribution follows the uniform distribution in [0, 1].

Using the constraints (Equations (8.6a)–(8.6d)), the Lagrangian function for the most entropic canonical copula (MECC) can be written as follows:

$$
\begin{aligned}
L = & \int_{[0,1]^2} c(u,v) \ln[c(u,v)] dudv - (\lambda_0 - 1)\left[\int_{[0,1]^2} c(u,v) dudv - 1\right] \\
& - \sum_{i=1}^{m} \lambda_i \left[\int_{[0,1]^2} u^i c(u,v) dudv - \frac{1}{i+1}\right] - \sum_{i=1}^{m} \gamma_i \left[\int_{[0,1]^2} v^i c(u,v) dudv\right] \\
& - \sum_{j=1}^{k} \lambda_{m+j} \left[\int_{[0,1]^2} a_j(u,v) c(u,v) dudv - \hat{\Theta}_j\right]
\end{aligned}
\tag{8.7}
$$

where $\lambda_0, \ldots, \lambda_m, \gamma_1, \ldots, \gamma_m, \lambda_{m+1}, \ldots, \lambda_{m+k}$ are the Lagrange multipliers. $\boldsymbol{\lambda}_U = [\lambda_1, \ldots, \lambda_m], \boldsymbol{\gamma}_V = [\gamma_1, \ldots, \gamma_m]$ are the Lagrange multipliers for the first n noncentral moments of uniformly (0, 1) distributed random variables U and V, respectively. More specifically for MECC, $\boldsymbol{\lambda}_U = \boldsymbol{\gamma}_V$: $\lambda_r = \gamma_r, r = 1, \ldots, m$. $\lambda_{m+1}, \ldots, \lambda_{m+k}$ are the Lagrange multipliers pertaining to the constraints of rank-based dependence measure.

Differentiating Equation (8.7) with respect to $c(u,v)$, we have the following:

$$
c(u,v) = \frac{\exp\left(-\sum_{i=1}^{m} \lambda_i u^i - \sum_{i=1}^{m} \gamma_i v^i - \sum_{j=1}^{k} \lambda_{m+j} a_j(u,v)\right)}{\int_0^1 \int_0^1 \exp\left(-\sum_{i=1}^{m} \lambda_i u^i - \sum_{i=1}^{m} \gamma_i v^i - \sum_{j=1}^{k} \lambda_{m+j} a_j(u,v)\right) dudv}
\tag{8.8}
$$

Similar to the univariate MaxEnt-based distribution, the partition (also called potential) function of the entropic copula can be written as follows:

$$
\begin{aligned}
Z(\Lambda) = & \ln\left[\int_0^1 \int_0^1 \exp\left(-\sum_{i=1}^{m} \lambda_i u^i - \sum_{i=1}^{m} \gamma_i v^i - \sum_{j=1}^{k} \lambda_{m+j} a_j(u,v)\right) dudv\right] \\
& + \sum_{i=1}^{m} \lambda_i \frac{1}{i+1} + \sum_{i=1}^{m} \gamma_i \frac{1}{i+1} + \sum_{j=1}^{k} \lambda_{m+j} \hat{\Theta}_j
\end{aligned}
\tag{8.9a}
$$

or equivalently

$$
Z(\Lambda) = \int_0^1 \int_0^1 \exp\left\{ \begin{array}{c} -\sum_{i=1}^{m} \left[\lambda_i \left(u^i - \frac{1}{i+1}\right)\right] - \sum_{i=1}^{m} \left[\gamma_i \left(v^i - \frac{1}{i+1}\right)\right] \\ -\sum_{j=1}^{k} \left[\lambda_{m+j} \left(a_j(u,v) - \hat{\Theta}_j\right)\right] \end{array} \right\} dudv
\tag{8.9b}
$$

In Equations (8.9a) and (8.9b),

$$
\Lambda = [\lambda_1, \ldots, \lambda_m, \gamma_1, \ldots \gamma_m, \lambda_{m+1}, \ldots, \lambda_{m+k}], [\lambda_1, \ldots, \lambda_m] = [\gamma_1, \ldots, \gamma_m].
$$

To this end, the Lagrange multipliers may be estimated by minimizing the partition function given as Equation (8.9a)–(8.9b).

So far, we have derived the MECC. The MECC may be generalized to most entropic copula (MEC) with respect to a given parametric copula (Chu, 2011). In the case of MEC, Equations (8.8), (8.9a) and (8.9b) can be rewritten as follows:

$$c(u,v) = \frac{\exp\left(-\sum_{i=1}^{m}\lambda_i u^i - \sum_{i=1}^{m}\gamma_i v^i - \sum_{j=1}^{k}\lambda_{m+j}a_j(u,v) - b\tilde{c}(u,v)\right)}{\int_0^1\int_0^1 \exp\left(-\sum_{i=1}^{m}\lambda_i u^i - \sum_{i=1}^{m}\gamma_i v^i - \sum_{j=1}^{k}\lambda_{m+j}a_j(u,v) - b\tilde{c}(u,v)\right)dudv} \tag{8.10}$$

$$Z(\Lambda) = \ln\left[\int_0^1\int_0^1 \exp\left(-\sum_{i=1}^{m}\lambda_i u^i - \sum_{i=1}^{m}\gamma_i v^i - \sum_{j=1}^{k}\lambda_{m+j}a_j(u,v) - b\tilde{c}(u,v)\right) dudv\right] + \sum_{i=1}^{m}\lambda_i\frac{1}{i+1} + \sum_{i=1}^{m}\gamma_i\frac{1}{i+1} + \sum_{j=1}^{k}\lambda_{m+j}\hat{\Theta}_j \tag{8.11a}$$

$$Z(\Lambda) = \int_0^1\int_0^1 \exp\left\{\begin{array}{c} -\sum_{i=1}^{m}\left[\lambda_i\left(u^i - \frac{1}{i+1}\right)\right] - \sum_{i=1}^{m}\left[\gamma_i\left(v^i - \frac{1}{i+1}\right)\right] \\ -b\tilde{c}(u,v) - \sum_{j=1}^{k}\left[\lambda_{m+j}\left(a_j(u,v) - \hat{\Theta}_j\right)\right]\end{array}\right\}dudv \tag{8.11b}$$

In Equations (8.10) and (8.11a), b is a generic constant, and $\tilde{c}(u,v)$ is the given copula. It is seen that the MECC is obtained by setting $b = 0$ (i.e., Equation (8.11b)). In what follows, we will provide examples to illustrate applications of MECC for bivariate cases.

Example 8.1 **Construct the most entropic canonical copula, using the sample dataset listed in Table 8.1 with random variables X and Y sampled from true population X~Gamma (3,4), Y~Gaussian ($5,3^2$).**

The true copula modeling the dependence of random variables X and Y is the Gumbel–Hougaard copula with parameter $\theta = 2.5$:

i. Construct MECC using empirical marginals.
ii. Construct MECC using MaxEnt-based marginals.
iii. Construct MECC using the true underlying population X~Gamma (3,4), Y~Gaussian ($5,3^2$).
iv. Compare the constructed MECC with the underlying copula function as the Gumbel–Hougaard copula with parameter $\theta = 2.5$.

Solution: Before we proceed to build the MECC, we first plot the histograms, the frequency computed from the true population and MaxEnt-based probability distribution in Figure 8.1. The MaxEnt-based univariate distribution (plotted in Figure 8.1) will be further explained in later sections. One purpose of applying empirical, true population and MaxEnt-based univariate distributions is to evaluate the impact of marginals on the derived copula function.

Table 8.1. *Sample dataset for Example 8.1.*

X	Y	X	Y	X	Y	X	Y
22.73	10.53	4.20	4.37	4.42	1.33	16.80	6.38
8.46	1.78	17.27	8.12	26.97	10.26	12.73	6.43
18.68	8.37	17.18	7.41	19.05	5.89	8.77	0.79
11.41	4.85	14.50	7.73	8.80	3.33	5.45	4.26
13.73	5.56	8.11	1.39	11.63	1.24	11.04	4.61
11.74	4.55	26.87	11.63	13.37	5.58	13.68	6.74
3.90	0.15	8.62	1.00	2.46	−0.20	12.40	7.07
14.77	6.12	20.14	4.85	5.73	1.81	19.56	8.62
12.09	5.48	19.97	7.84	4.20	−0.05	9.56	2.44
8.17	3.51	24.13	10.92	26.37	10.13	13.00	6.02
16.60	3.30	11.79	5.13	14.04	6.83	9.92	5.37
16.70	7.21	3.05	0.17	25.73	9.96	9.05	2.16
12.12	6.63	14.30	5.11	15.90	4.24	11.11	2.18
7.73	4.13	12.45	7.83	8.93	2.51	25.19	9.11
13.16	5.71	4.83	0.12	7.34	4.30	6.90	4.31
13.45	2.15	17.13	8.02	11.90	5.78	6.35	4.98
10.96	1.88	22.03	8.55	7.81	5.46	29.83	10.39
6.67	3.24	15.66	6.50	4.39	−2.38	14.50	5.69
19.41	8.85	7.35	5.74	10.07	4.45	11.18	1.80
7.54	2.92	9.00	4.34	9.90	4.13	5.27	0.02
7.54	4.00	3.07	−1.52	10.60	5.43	24.82	10.28
10.79	3.15	7.58	2.90	14.09	3.80	6.67	3.19
14.57	3.08	12.08	5.17	9.59	2.29	19.74	9.12
11.03	5.02	8.57	6.03	4.47	3.02	15.10	4.63
23.81	9.98	7.31	5.73	2.52	7.36	14.24	6.09

Furthermore, throughout the example, the first two noncentral moments of the marginals (Equations (8.12a) and (8.12b)) and $E(UV)$, which is one-to-one related to the rank-based correlation coefficient, Spearman's rho (Equation (8.12c)), will be applied as the constraints for the MECC as follows:

$$\int_0^1\int_0^1 uc(u,v)dudv = E(U) = \int_0^1\int_0^1 vc(u,v)dudv = E(V) = \frac{1}{2} \tag{8.12a}$$

$$\int_0^1\int_0^1 u^2c(u,v)dudv = E\left(U^2\right) = \int_0^1\int_0^1 v^2c(u,v)dudv = E\left(V^2\right) = \frac{1}{3} \tag{8.12b}$$

$$\int_{[0,1]^2} uvc(u,v)dudv = \frac{(\hat{\rho}_s + 3)}{12} = 0.3140; \text{sample } \hat{\rho}_s = 0.7677 \tag{8.12c}$$

In Equation (8.12c), sample $\hat{\rho}_s$ is computed using Equation (3.70).

In what follows, we will proceed with constructing MECC with different marginal distributions.

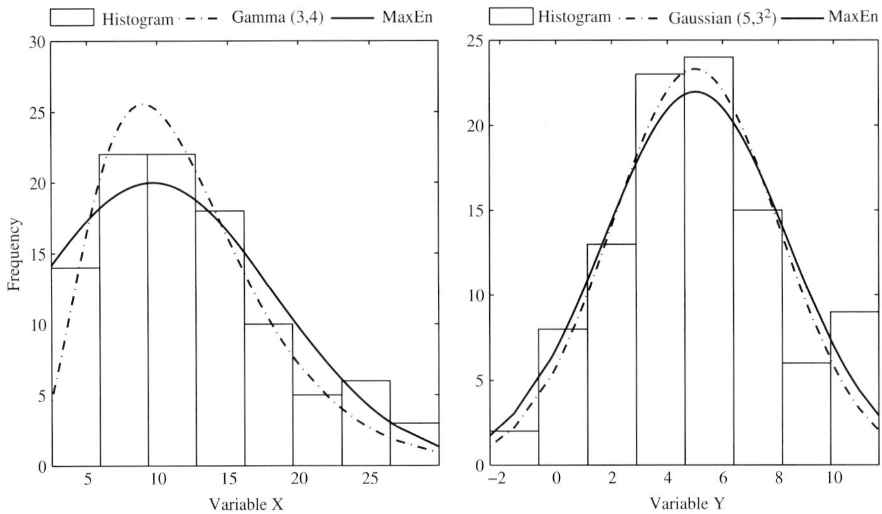

Figure 8.1 Histograms and underlying true probability density functions.

Construct MECC using empirical distribution

The empirical probability is computed with the use of Weibull plotting position formula (Equation (3.103)) that is partially listed in Table 8.2. Minimizing the partition (i.e., objective) function (Equation (8.9a)) using the MATLAB optimization toolbox (e.g., the GA/fminsearch function), Table 8.3 lists the Lagrange multipliers estimated and the relative differences between moment constraints computed from the MECC (Equation (8.12)) and the corresponding sample moments. Figure 8.2 compares the constructed MECC with the empirical copula.

Construct MECC using MaxEnt-based marginal distribution

To apply the MaxEnt-based marginal distribution, we first transform the random variables X and Y into the range (0, 1). To avoid reaching the lower and upper limit, we use the following equation for the monotone transformation:

$$X^t = \frac{x - (1-d) * \min(x)}{(1+d)\max(x) - (1-d)\min(x)}, d = 0.01 \tag{8.13}$$

In Equation (8.13), X denotes the random variable that needs to be transformed, d denotes the threshold ratio to avoid the transformed variable reaching the lower and upper limits, and X^t denotes the variable after transformation.

Strictly from the sample dataset listed in Table 8.1, we evaluate whether the fourth noncentral moment to derive the MaxEnt-based univariate probability distribution by testing whether the sample kurtosis is significantly different from 3 (i.e., kurtosis = 3 for normal distribution) described in Zhang and Singh (2012). The test statistic is computed using the following:

$$T = G_2/SEK \tag{8.14a}$$

$$\gamma_2' = \frac{n\sum_{i=1}^{n}(x_i - \bar{x})^4}{\left[\sum_{i=1}^{n}(x_i - \bar{x})^2\right]^2} - 3 \tag{8.14b}$$

Table 8.2. *Marginal distributions computed by pair.*

No.	Empirical (Weibull)		MaxEnt-based		Underlying population	
	X	Y	X	Y	X~Gamma (3,4)	Y~Gaussian $(5,3^2)$
1	0.901	0.970	0.928	0.972	0.922	0.967
2	0.287	0.139	0.308	0.155	0.354	0.141
3	0.822	0.851	0.823	0.861	0.845	0.869
4	0.485	0.485	0.479	0.483	0.543	0.480
5	0.644	0.584	0.607	0.571	0.667	0.574
6	0.505	0.446	0.498	0.446	0.562	0.441
7	0.050	0.069	0.064	0.059	0.076	0.053
8	0.723	0.693	0.660	0.640	0.713	0.646
9	0.545	0.574	0.518	0.561	0.582	0.563
10	0.277	0.327	0.291	0.321	0.335	0.310
11	0.762	0.307	0.744	0.298	0.783	0.286
12	0.772	0.772	0.748	0.759	0.786	0.769
13	0.554	0.733	0.520	0.698	0.583	0.706
14	0.248	0.366	0.266	0.394	0.305	0.386
15	0.604	0.614	0.577	0.590	0.639	0.594
...	...	...	...	...	...	...
...	...	...	...	...	...	...
...	...	...	...	...	...	...
86	0.396	0.545	0.393	0.547	0.451	0.549
87	0.356	0.188	0.342	0.186	0.394	0.172
88	0.465	0.198	0.462	0.188	0.525	0.174
89	0.941	0.891	0.965	0.911	0.950	0.915
90	0.178	0.406	0.219	0.416	0.250	0.409
91	0.149	0.495	0.189	0.499	0.214	0.498
92	0.990	0.960	0.999	0.968	0.979	0.964
93	0.693	0.604	0.647	0.587	0.702	0.590
94	0.475	0.149	0.466	0.156	0.530	0.143
95	0.119	0.050	0.132	0.054	0.147	0.048
96	0.931	0.950	0.961	0.964	0.947	0.961
97	0.168	0.287	0.206	0.286	0.234	0.273
98	0.861	0.901	0.857	0.911	0.870	0.915
99	0.733	0.465	0.676	0.455	0.727	0.451
100	0.673	0.683	0.633	0.636	0.690	0.642

$$G_2 = \frac{n-1}{(n-2)(n-3}\left[(n+1)\gamma_2' + 6\right] \tag{8.14c}$$

$$SEK = 2\sqrt{\frac{6n(n-1)^2}{(n-2)(n+5)(n^2-9)}} \tag{8.14d}$$

Table 8.3. *Lagrange multipliers estimated for MaxEnt-based univariate distributions as well as the relative difference between computed moment constraints and sample moments.*

Variable		λ_0	λ_1	λ_2	λ_3
X	Parameters	−0.134	−2.666	5.116	0.000
	Relative diff.		3.11E−07	3.73E−07	2.61E−03
Y	Parameters	1.945	−9.615	9.189	0.000
	Relative diff.		2.10E−07	3.68E−07	8.35E−04

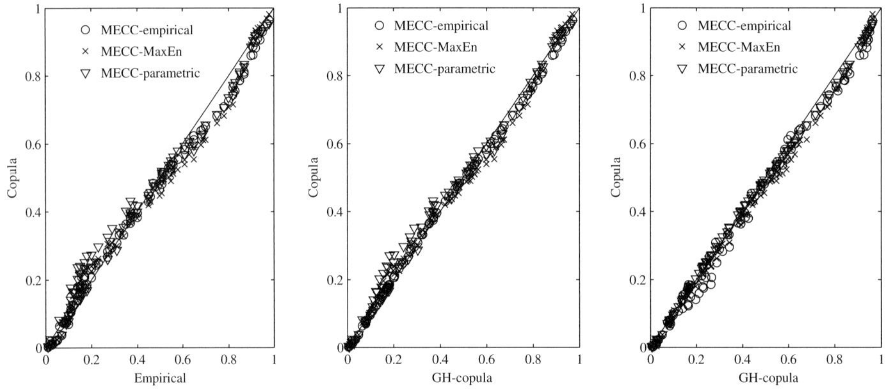

Figure 8.2 Comparison of MaxEnt-based univariate distribution to empirical and underlying distributions.

In Equation (8.14), n is the sample size; γ_2' is the excess kurtosis; G_2 is sample excess kurtosis; and *SEK* is the standard error of kurtosis. The test statistic T follows the standard normal distribution. Applying Equation (8.14), we computed the test statistic (T) for variables X and Y, which were 0.06 and –0.73; and P-values were 0.95 and 0.46, respectively. Thus, the kurtosis was not significantly different from 3 such that we only need to apply the first three noncentral moments to drive the MaxEnt-based distribution with the Lagrange multipliers. The MaxEnt-based univariate distribution for the scaled transformed variable (x^t) is written as follows:

$$f_{X^t}(x^t) = \exp\left(-\lambda_0 - \lambda_1 x^t - \lambda_2 (x^t)^2 - \lambda_3 (x^t)^3\right) \tag{8.14e}$$

$$\text{and } \lambda_0 = \ln\left(\int_0^1 \exp\left(-\sum_{i=1}^{3} \lambda_i (x^t)^i\right) dx\right) \tag{8.14f}$$

The corresponding MaxEnt-based marginal PDF for the observed random variable can be written as follows:

$$f(x) = \frac{1}{A} f(x^t) = \frac{1}{A} \exp\left(-\lambda_0 - \lambda_1 x^t - \lambda_2 (x^t)^2 - \lambda_3 (x^t)^3\right) \tag{8.15}$$

where $A = (1 + d) \max(x) - (1 - d) \min(x)$.

Table 8.4. *Lagrange multipliers estimated for MECC with different consideration of marginals as well as the relative difference between computed moment constraints and sample moments.*

	λ_0	λ_1	λ_2	γ_1	γ_2	λ_3
MECC_emp[a]	−1.450	1.866	8.855	1.866	8.855	−21.441
Relative diff.		1.07E–08	1.45E–08	−4.55E–09	−1.09E–08	1.79E–09
MECC_MaxEn[b]	−1.450	1.866	8.855	1.866	8.855	−21.441
Relative diff.		1.07E–08	1.45E–08	−4.55E–09	−1.09E–08	1.79E–09
MECC_underlying population[c]	−1.450	1.866	8.855	1.866	8.855	−21.441
Relative diff.		1.07E–08	1.45E–08	−4.55E–09	−1.09E–08	1.79E–09

Notes: (a) empirical marginals; (b) MaxEnt-based marginals; (c) true parametric marginals.

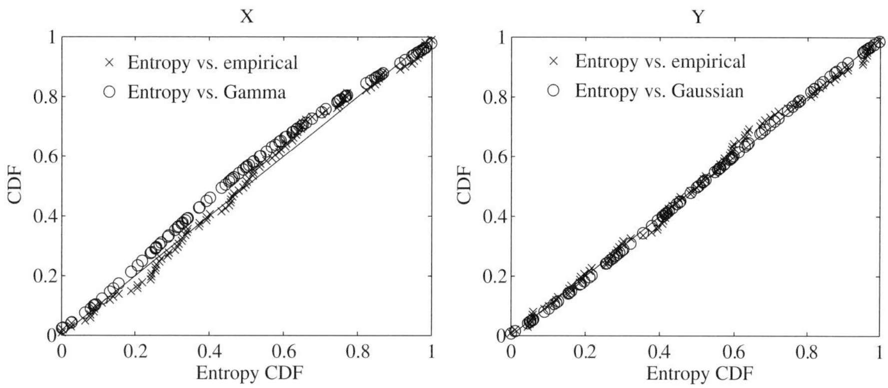

Figure 8.3 Comparison of MECC with the empirical and Gumbel–Hougaard copulas.

The Lagrange multipliers may again be estimated by minimizing the objective function given by Equation (8.14b) using the MATLAB optimization toolbox, as listed in Table 8.4. Results in Table 8.4 indicate that the first three noncentral moments (sample moments) are well preserved. Table 8.2 lists the marginal probabilities computed from the fitted MaxEnt-based distribution. It is worth noting that we may use the transformed variable to compute the marginal probability directly, given the monotone transformation between observed and scale-transformed variables. The MaxEnt-based probability density function is plotted in Figure 8.1, whereas Figure 8.3 compares the MaxEnt-based univariate distribution with the empirical distribution. The comparisons again indicate that the MaxEnt-based distribution matches well the empirical distribution as well as the true population.

Using the CDF computed from the constructed MaxEnt-based univariate distribution, Table 8.4 lists the Lagrange multipliers estimated for the MECC. Figure 8.2 compares the

MECC using the CDF computed from MaxEnt-based univariate distribution with the empirical copulas.

Construct MECC using the underlying population
In this case, gamma (3, 4) and Gaussian (5, 3^2) are applied to random variables X and Y, respectively. The computed CDF is listed in Table 8.2. The MECC is then constructed with the Lagrange multipliers listed in Table 8.4. Figure 8.3 compares the MECC from the underlying population with the empirical copula.

Compare the constructed MECC with the underlying copula function
Applying the Gumbel–Hougaard copula with parameter $\theta = 2.5$ to the marginals computed from the empirical formula, the MaxEnt-based distribution, and underlying populations, Figure 8.3 compares the Gumbel–Hougaard copulas with MECCs. The comparison indicates the following:

a. The Gumbel–Hougaard copula, computed using the empirical distribution, has a better match with the MECC computed from empirical and MaxEnt-based univariate distributions than the MECC computed from underlying univariate populations.
b. The Gumbel-Hougaard copula computed using the underlying univariate populations matches better the MECC computed using the underlying univariate populations than those from empirical and MaxEnt-based univariate distributions.
c. It is understandable that we reach the conclusions in a and b. With the sample data, the MaxEnt-based distribution is derived by equating moment constraints to the sample moments. It is seen from Figure 8.1 that there exists the difference in fitting between MaxEnt-based distributions and true populations. It may be explained with the sample size. It is expected that the MaxEnt-based, true underlying, and empirical distributions should match each other better with the increased sample size.

To summarize this example, we see that using the same moment constraints for the MECC given in Equation (8.12), we obtain exactly the same MECC for the marginals computed from the empirical, MaxEnt-based, and underlying population. It is obvious that with the marginals being uniformly distributed in [0, 1], the moment constraints in Equations (8.12a) and (8.12b) equate the population moments rather than the sample moments and yield $\lambda_1 = \gamma_1; \lambda_2 = \gamma_2$.

In addition, the Lagrange multipliers of the MaxEnt-based univariate distribution and most entropic copula are estimated with the use of MATLAB optimization toolbox in what follows:

MaxEnt-based univariate distribution. According to the principle of maximum entropy for the constraints defined with the noncentral moments, i.e., $E(X^i), i = 1, \ldots, m$; the Lagrange multiplier λ_m for $E(X^m)$ needs to fulfill the condition: $\lambda_m > 0$. To apply the GA function for MaxEnt-based marginal distribution with the first three noncentral moments as constraints and let the objective function (i.e., Equation (8.14b)) be written as a MATLAB function. It is worth noting that the lower and upper bound for the constraints should be set as Lower $= [-\inf, -\inf, 0]$, Upper $= [\inf, \inf, \inf]$.

The Lagrange multipliers can then be estimated using fmincon or GA optimization function. The options may also be stated as optimset for the fmincon function and gaoptimeset for the GA function. It should also be noted that the parameters estimated using GA function may result in different values given the structure of the GA optimization technique; however, parameter values should stay close with each other.

Entropic copula. Similar to the MaxEnt-based univariate distribution, the objective function of entropic copula in MATLAB format can be written using Equation (8.11) and corresponding constraints in Equation (8.12). Again, fmincon, fminsearch, and other optimization MATLAB functions may be applied to estimate the parameters. For example, in this example, we will need to estimate five parameters for entropic copula in theory; however, we will only need to estimate three parameters since $\lambda_1 = \gamma_1,\ \lambda_2 = \gamma_2; i.e., E(U^i) = E(V^i) = 1/(i+1)$.

Example 8.2 **Using the data in Example 8.1, (1) construct MECC by adding Blest (I and II) moment constraints (i.e., Blest's coefficient (Chu, 2011) to MECC, using the empirical marginal distributions; and (2) compare MECC with the additional Blest I and II dependence measure constraints to the Gumbel–Hougaard copula and MECC constructed in Example 8.1 with the empirical marginals.**

Solution:

1. **Construct MECC by adding two Blest moment constraints**

According to Chu (2011), the Blest I and II moment constraints are given as follows:

$$\int_{[0,1]^2} u^2 v c(u,v)dudv = \frac{2\rho_s - \hat{v}_1 + 2}{12} \tag{8.16a}$$

$$\int_{[0,1]^2} v^2 u c(u,v)dudv = \frac{2\rho_s - \hat{v}_2 + 2}{12} \tag{8.16b}$$

In Equations (8.16a) and (8.16b),

$$\text{Sample Blest measure I: } \hat{v}_1 = \frac{2N+1}{N-1} - \frac{12}{N^2-N}\sum_{i=1}^{N}\left(1 - \frac{R_i}{N+1}\right)^2 S_i \tag{8.16c}$$

$$\text{Sample Blest measure II: } \hat{v}_2 = \frac{2N+1}{N-1} - \frac{12}{N^2-N}\sum_{i=1}^{N}\left(1 - \frac{S_i}{N+1}\right)^2 R_i \tag{8.16d}$$

In Equations (8.16c) and (8.16d), $\{R_i, S_i : i = 1 \dots N\}$ is the rank for $\{x_i, y_i : i = 1 \dots N\}$.

By adding the Blest dependence measure, the MECC can be rewritten as folllows:

$$c(u,v) = \exp\left(-\lambda_0 - \lambda_1 u - \lambda_2 u^2 - \gamma_1 v - \gamma_2 v^2 - \lambda_3 uv - \lambda_4 u^2 v - \lambda_5 uv^2\right) \tag{8.17}$$

Table 8.5. *Lagrange multipliers and moment constraints estimated for the MECC with additional Blest I and II dependence measure constraints.*

	λ_0	λ_1	λ_2	γ_1	γ_2	λ_3	λ_4	λ_5
	−0.270	−3.275	11.651	−3.275	11.651	−3.275	−7.745	−7.745
		$E(u)$	$E(u^2)$	$E(v)$	$E(v^2)$	$E(uv)$	$E(u^2v)$	$E(uv^2)$
Moments (M)		0.500	0.333	0.500	0.333	0.314	0.234	0.233
Computed moments (CM)		0.501	0.333	0.501	0.333	0.313	0.233	0.234
Relative diff: (M-CM)/SM		0.001	1.37E−6	0.001	4.94E−7	0.004	0.002	−0.002

The partition (or objective) function (i.e., Equation (8.11a)) can be rewritten as follows:

$$Z(\Lambda) = \ln\left[\int_{[0,1]^2} \exp\left(\sum_{i=1}^{2} \lambda_i u^i - \sum_{i=1}^{2} \gamma_i v^i - \lambda_3 uv - \lambda_4 u^2 v - \lambda_5 v u^2\right) du dv\right]$$
$$+ \sum_{i=1}^{2} \lambda_i \frac{1}{i+1} + \sum_{i=1}^{2} \gamma_i \frac{1}{i+1} + \lambda_3 \hat{E}(uv) + \lambda_4 \hat{E}\left(u^2 v\right) + \lambda_5 \hat{E}\left(uv^2\right) \tag{8.18}$$

In Equation (8.18), $\hat{E}(\cdot)$ denotes the sample moment, and $\lambda_1 = \gamma_1; \lambda_2 = \gamma_2; \lambda_4 = \lambda_5$.

Minimizing Equation (8.18), Table 8.5 lists the Lagrange multipliers estimated, the sample moment constraints, and those computed from the MECC with additional Blest I and II measures. Again, the MATLAB optimization toolbox is applied to minimize the objective function in order to estimate the Lagrange multipliers. Table 8.5 indicates that the moment constraints are preserved reasonably well with the relative error less than 2.5%.

2. **Compare MECC with the additional Blest I and II dependence measure constraints with the Gumbel–Hougaard copula and MECC constructed in Example 8.1**

To compare the MECC with added constraints to the MECC constructed in Example 8.1, Table 8.6 lists the numerical JCDF computed from empirical copula, MECCs in Example 8.1, and the MECC with added Blest I and II constraints of Example 8.2. In this case, we will compare the results from Example 8.2 (i.e., column 7 in Table 8.6) with those from Example 8.1 (i.e., columns 2 and 5 in Table 8.6). Figure 8.4 compares the results graphically for the hypothesized Gumbel–Hougaard copula, MECC (Example 8.1), and MECC (Example 8.2) using empirical marginals. Comparison shows that the MECC constructed in Example 8.1 (i.e., only using Spearman's rho as the constraints for dependence measure) yields better performance than the MECC constructed in Example 8.2 (i.e., with added Blest I and II dependence measure constraints) compared to the hypothesized Gumbel–Hougaard copula. Comparing Table 8.5 to Table 8.3, it is seen that the moment constraints are better preserved by the MECC constructed in Example 8.1 than in Example 8.2. To this end, it is concluded that we need to be cautious when adding more constraints to derive the MECC. In this sample study, the dependence measure through Spearman's rho preserves the dependence structure of datasets well.

Table 8.6. *JCDF computed from Examples 8.1 and 8.2.*

	Example 8.1						Example 8.2
No.	[1]	[2]	[3]	[4]	[5]	[6]	[7]
1	0.91	**0.886**	0.912	0.903	**0.899**	0.919	**0.866**
2	0.11	**0.106**	0.123	0.121	**0.11**	0.122	**0.075**
3	0.83	**0.755**	0.761	0.781	**0.789**	0.814	**0.751**
4	0.4	**0.396**	0.392	0.418	**0.385**	0.41	**0.413**
5	0.52	**0.517**	0.495	0.519	**0.522**	0.525	**0.548**
6	0.4	**0.383**	0.381	0.4	**0.372**	0.39	**0.398**
7	0.03	**0.013**	0.014	0.015	**0.023**	0.026	**0.005**
8	0.65	**0.612**	0.554	0.581	**0.633**	0.596	**0.636**
9	0.5	**0.467**	0.447	0.48	**0.464**	0.479	**0.496**
10	0.17	**0.211**	0.217	0.234	**0.204**	0.224	**0.183**
11	0.31	**0.308**	0.299	0.288	**0.303**	0.283	**0.308**
12	0.75	**0.681**	0.661	0.687	**0.711**	0.717	**0.693**
13	0.55	**0.524**	0.49	0.538	**0.53**	0.546	**0.554**
14	0.17	**0.206**	0.227	0.252	**0.199**	0.241	**0.178**
15	0.51	**0.515**	0.49	0.521	**0.52**	0.526	**0.547**
16	0.18	**0.176**	0.182	0.169	**0.174**	0.167	**0.156**
17	0.15	**0.155**	0.151	0.143	**0.152**	0.141	**0.128**
18	0.12	**0.123**	0.154	0.166	**0.125**	0.164	**0.09**
19	0.85	**0.785**	0.797	0.812	**0.818**	0.843	**0.776**
20	0.11	**0.149**	0.168	0.177	**0.15**	0.174	**0.113**
21	0.14	**0.179**	0.215	0.237	**0.175**	0.227	**0.148**
22	0.22	**0.244**	0.252	0.253	**0.234**	0.244	**0.225**
23	0.27	**0.268**	0.272	0.261	**0.263**	0.256	**0.26**
24	0.38	**0.383**	0.39	0.421	**0.372**	0.413	**0.398**
25	0.91	**0.862**	0.914	0.904	**0.89**	0.925	**0.842**
26	0.06	**0.063**	0.072	0.081	**0.065**	0.083	**0.045**
27	0.82	**0.743**	0.715	0.743	**0.777**	0.776	**0.742**
...	...	...	...	...	...	...	...
...	...	...	...	...	...	...	...
93	0.56	**0.547**	0.52	0.541	**0.557**	0.55	**0.578**
94	0.14	**0.14**	0.147	0.138	**0.138**	0.136	**0.115**
95	0.05	**0.021**	0.024	0.024	**0.031**	0.034	**0.01**
96	0.92	**0.899**	0.934	0.92	**0.921**	0.938	**0.878**
97	0.12	**0.127**	0.152	0.164	**0.129**	0.162	**0.094**
98	0.87	**0.812**	0.814	0.826	**0.843**	0.856	**0.797**
99	0.45	**0.452**	0.436	0.44	**0.451**	0.437	**0.478**
100	0.61	**0.583**	0.54	0.569	**0.599**	0.583	**0.611**

Note: [1] Empirical copula; [2] MECC with empirical marginals; [3] MECC with MaxEnt marginal distributions; [4] MECC with true underlying marginal distributions; [5] True Gumbel–Hougaard copula with empirical marginals; [6] True Gumbel–Hougaard copula with true underlying marginal distributions; [7] MECC with empirical marginals.

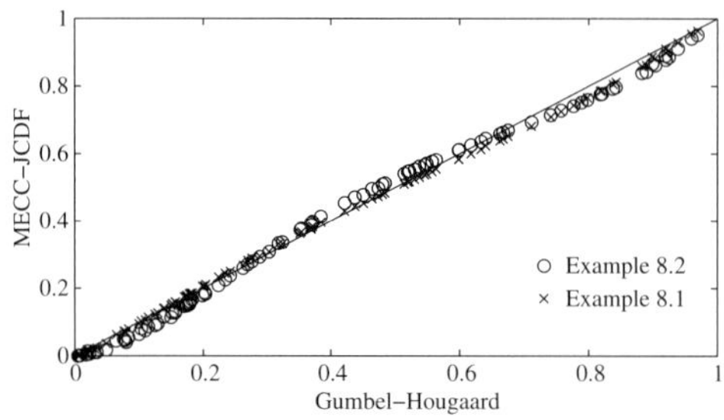

Figure 8.4 Comparison of the MECC constructed in Examples 8.1 and 8.2 with the hypothesized Gumbel–Hougaard copula in Example 8.1.

Until now, we have concentrated on the MECC construction. In what follows, we will show its real-world application using flood data from the Walnut Gulch Experimental Watershed (Flume 1).

Example 8.3 **Use actual flood data from the Walnut Gulch Experimental Watershed (Flume 1) to construct and compare MECC and Gumbel–Hougaard copulas.**

For a real-world example using flood data from the Walnut Gulch Experimental Watershed (Flume 1) given in Table 8.7, do the following:

1. Construct the MaxEnt-based marginal distributions using the first three noncentral moments as constraints.
2. Construct the MECC using Equation (8.12) as the constraint with the MaxEnt-based marginals from step 1. Then compare the MECC constructed with the Gumbel–Hougaard copula with the same marginals.
3. Construct the MECC and fit the Gumbel–Hougaard copula to the flood data with empirical marginal distributions.
4. Compare the MECC and Gumbel–Hougaard copulas fitted in steps 2 and 3 with empirical copulas.

Solution: Flume 1 is located at the most downstream point of the Walnut Gulch Experimental Watershed (i.e., 31°43'45.32" N and 110°9'12.06" W). It covers an area of about 150 km^2. The annual maximum series (AMS) are extracted from the event-based dataset (1957–2012). In this example, flood data of the year 1979 were not used in analysis to avoid uncertainty (i.e., from the dataset, there was no obvious runoff for the entire year).

1. **Construct the MaxEnt-based univariate distributions.**

Using the first three noncentral moments as the constraints, we have the constraint equation for the univariate density function $f(x)$:

$$\int f(x)dx = 1; \int x^i f(x)dx = E\left(x^i\right) \approx \overline{x^i}, i = 1, \ldots, 3.$$

and the MaxEnt-based univariate density function is given as follows:

Table 8.7. *Annual maximum flood data (Flume 1 at the Walnut Gulch Watershed).*

Year	Volume (ft^3)	Discharge (cfs)	Year	Volume (ft^3)	Discharge (cfs)
1957	34,530,000	11,250	1985	1,427,000	233.9
1958	13,960,000	3,388	1986	2,800,000	751.2
1959	14,450,000	2,767	1987	798,600	387.6
1960	4,0180	53.95	1988	475,500	184
1961	12,200,000	3929	1989	409,900	128
1962	3,094,000	850.6	1990	12,170,000	1841
1963	15,950,000	2,709	1991	1,829,000	710.9
1964	15,520,000	4,290	1992	727,000	196.8
1965	4,326,000	841.3	1993	1,607,000	397.8
1966	5,920,000	1,574	1994	2,780,000	477
1967	15,930,000	4,681	1995	631,700	146.6
1968	4,447,000	807.4	1996	4,397,000	1011
1969	3,429,000	1679	1997	5,495,000	800.8
1970	2,076,000	710.3	1998	8,037,000	1116
1971	14,890,000	3,615	1999	12,890,000	2566
1972	16,010,000	6,057	2000	21,740,000	5456
1973	7,211,000	2,978	2001	1,183,000	503.5
1974	3,262,000	6,38.9	2002	3,404,000	1175
1975	10,300,000	2,071	2003	5,210,000	1184
1976	6,398,000	883	2004	42,260	25.29
1977	11,660,000	2,852	2005	762,100	368.2
1978	3,228,000	1,205	2006	8,102,000	1570
1979	776.9	0.6325	2007	1,801,000	901.2
1980	415,200	371.8	2008	13,570,000	2700
1981	10,830,000	1,036	2009	519,700	211.2
1982	5,491,000	1,939	2010	4,124,000	1203
1983	9,543,000	1,068	2011	1,998,000	737.4
1984	2,342,000	437.6	2012	19,700,000	3190

$$f(x) = \exp\left(-\lambda_0 - \lambda_1 x - \lambda_2 x^2 - \lambda_3 x^3\right)$$

where $\lambda_0 = g(\lambda_1, \lambda_2, \lambda_3) = \ln \int \exp\left(-\sum_{i=1}^{3} \lambda_i x^i\right) dx; \lambda_3 > 0.$

As discussed in the previous examples, parameters $\Lambda = (\lambda_1, \lambda_2, \lambda_3)$ can be estimated by minimizing the partition function: $Z(\Lambda) = \lambda_0 + \sum_{i=1}^{3} \lambda_i \overline{x^i}$. Similar to the previous example, the flood variables are transformed from $(0, +\infty)$ to (0, 1) using Equation (8.13) with $d = 0.1$.

Applying the Shannon entropy, the Lagrange multipliers estimated for the peak discharge and flood volume are listed in Table 8.8, and comparison of MaxEnt-based PDF and CDF with their empirical form is plotted in Figure 8.5. Results from Table 8.8 indicate that the first three noncentral moments for the transformed flood variables are well preserved. Comparison in Figure 8.5 graphically confirms the good fit between empirical and MaxEnt-based distributions. The MaxEnt-based univariate distributions may be written for discharge and flood volume without transformation as follows:

Table 8.8. *Results of the MaxEnt-based univariate distributions for the transformed discharge and flood volume variables.*

	λ_0	$\lambda_1 : [E(x)]$	$\lambda_2 : [E(x^2)]$	$\lambda_3 : [E(x^3)]$
		Transformed discharge variables		
Multipliers estimated	−2.137	9.584	−4.146	4.499E−04
$\overline{x^i} : i = 1, 2, 3$		0.138	0.043	0.022
$\frac{E(x^i) - \overline{x^i}}{\overline{x^i}}$		−2.50E−04	−1.58E−04	1.12E−03
		Transformed volume variables		
Multipliers estimated	−1.695	5.456	−0.170	0.139
$\overline{x^i} : i = 1, 2, 3$		0.184	0.067	0.033
$\frac{E(x^i) - \overline{x^i}}{\overline{x^i}}$		3.58E−03	4.31E−03	3.08E−03

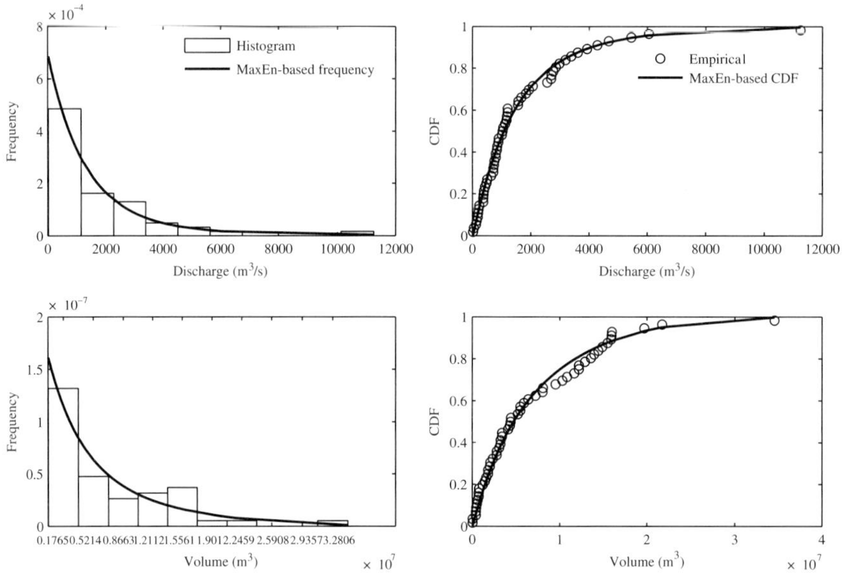

Figure 8.5 Comparison of MaxEnt-based univariate distribution with empirical frequency and CDFs.

$$f(dis) = \frac{1}{1.235 \cdot 10^4} \exp\left(2.137 - 9.584 dis^T + 4.146\left(dis^T\right)^2 - 4.499 \cdot 10^{-4}\left(dis^T\right)^3\right)$$

$$f(vol) = \frac{1}{3.975 \cdot 10^7} \exp\left(1.695 - 5.456 vol^T + 0.17\left(vol^T\right)^2 - 0.139\left(vol^T\right)^3\right)$$

$$dis^T = \frac{dis - 22.76}{1.235 \cdot 10^4}, \quad vol^T = \frac{vol - 36162}{3.975 \cdot 10^7}; \quad dis^T, vol^T \in (0, 1).$$

2. Construct MECC using Equation (8.12) with given constraints and marginal distributions estimated in step 1.

Using the Lagrange multipliers estimated in step 1, the cumulative probability distributions of discharge and flood volume are computed, as listed in Table 8.9.
To apply Equation (8.12) as the constraints to construct MECC, the Spearman's rho rank-based correlation coefficient is computed as $\hat{\rho}_s = 0.9387$. Equating the sample moment to the dependence measure constraint in Equation (8.12), we have

$$E(uv) \approx \overline{uv} = \frac{\hat{\rho}_s + 3}{12} = 0.3282.$$

Table 8.9. *MaxEnt-based and empirical CDF.*

	MaxEnt-based		Empirical (Weibull formula)	
Year	Disch.	Vol.	Disch.	Vol.
1957	0.996	0.997	0.982	0.982
1958	0.861	0.866	0.857	0.821
1959	0.809	0.875	0.786	0.839
1960	0.021	0.001	0.036	0.018
1961	0.893	0.827	0.893	0.768
1962	0.421	0.355	0.429	0.375
1963	0.803	0.900	0.768	0.911
1964	0.909	0.894	0.911	0.875
1965	0.418	0.460	0.411	0.482
1966	0.629	0.570	0.643	0.589
1967	0.924	0.900	0.929	0.893
1968	0.405	0.469	0.393	0.518
1969	0.651	0.385	0.661	0.446
1970	0.367	0.254	0.304	0.304
1971	0.876	0.883	0.875	0.857
1972	0.957	0.901	0.964	0.929
1973	0.829	0.643	0.821	0.625
1974	0.337	0.370	0.286	0.411
1975	0.722	0.772	0.714	0.696
1976	0.433	0.599	0.446	0.607
1977	0.818	0.813	0.804	0.732
1978	0.536	0.367	0.607	0.393
1980	0.210	0.053	0.179	0.071
1981	0.485	0.789	0.500	0.714
1982	0.700	0.543	0.696	0.554
1983	0.495	0.745	0.518	0.679
1984	0.244	0.282	0.232	0.321
1985	0.134	0.181	0.143	0.214
1986	0.383	0.327	0.357	0.357

Table 8.9. (*cont.*)

Year	MaxEnt-based Disch.	MaxEnt-based Vol.	Empirical (Weibull formula) Disch.	Empirical (Weibull formula) Vol.
1987	0.218	0.104	0.196	0.179
1988	0.104	0.061	0.089	0.089
1989	0.069	0.052	0.054	0.054
1990	0.682	0.826	0.679	0.750
1991	0.367	0.227	0.321	0.268
1992	0.112	0.094	0.107	0.143
1993	0.224	0.202	0.214	0.232
1994	0.263	0.325	0.250	0.339
1995	0.081	0.082	0.071	0.125
1996	0.477	0.465	0.482	0.500
1997	0.403	0.543	0.375	0.571
1998	0.510	0.683	0.536	0.643
1999	0.788	0.843	0.732	0.786
2000	0.945	0.959	0.946	0.964
2001	0.276	0.152	0.268	0.196
2002	0.528	0.383	0.554	0.429
2003	0.530	0.524	0.571	0.536
2004	0.002	0.001	0.018	0.036
2005	0.208	0.099	0.161	0.161
2006	0.628	0.686	0.625	0.661
2007	0.440	0.224	0.464	0.250
2008	0.802	0.858	0.750	0.804
2009	0.120	0.067	0.125	0.107
2010	0.536	0.444	0.589	0.464
2011	0.378	0.245	0.339	0.286
2012	0.847	0.943	0.839	0.946

Now the objective function (i.e., Equation (8.9)) can be rewritten as follows:

$$Z(\Lambda) = \ln\left[\int_{[0,1]^2} \exp\left(-\sum_{i=1}^{2} \lambda_i u^i - \sum_{i=1}^{2} \gamma_i v^i - \lambda_3 uv\right) dudv\right] + \sum_{i=1}^{2} \lambda_i \frac{1}{i+1} + \sum_{i=1}^{2} \gamma_i \frac{1}{i+1} + \lambda_3 E(uv)$$

or equivalently

$$Z(\Lambda) = \int_{[0,1]^2} \exp\left\{-\sum_{i=1}^{2}\left[\lambda_i\left(u^i - \frac{1}{i+1}\right)\right] - \sum_{i=1}^{2}\left[\gamma_i\left(v^i - \frac{1}{i+1}\right)\right] - \lambda_3[uv - E(uv)]\right\} dudv$$

As discussed earlier, we have $\lambda_1 = \gamma_1; \lambda_2 = \gamma_2$. Minimizing the objective function, we list the estimated Lagrange multipliers in Table 8.10.

Table 8.10. *Lagrange multipliers estimated for MECC and comparison with Gumble–Hougaard copula.*

Marginals		λ_0	λ_1 $E(U)$	λ_2 $E(U^2)$	γ_1 $E(V)$	γ_2 $E(V^2)$	λ_3 $E(UV)$
	MECC	–1.803	1.163	42.331	1.163	42.331	–86.990
	Moment constraint (comp.)		0.50	0.333	0.50	0.333	0.328
	Relative differences		–3.21E–09	–3.44E–09	4.59E–09	9.14E–09	2.70E–09
	Gumbel	$\hat{\theta} = 3.926; \rho_s^{comp} = 0.909$					
	MECC vs. Gumbel[a]	$k = 0.9978; R_{square} = 0.9969$					
	MECC	–1.803	1.163	42.331	1.163	42.331	–86.990
Empirical	Moment constraint (comp.)		0.50	0.333	0.50	0.333	0.328
	Relative differences		–3.21E–09	–3.44E–09	4.59E–09	9.14E–09	2.70E–09
	Gumbel	$\hat{\theta} = 4.335; \rho_s^{comp} = 0.925$					
	MECC vs. Gumbel[a]	$k = 0.9954; R_{square} = 0.9975$					

Notes: [a]Regression MECC on Gumbel using $y = kx$.

As shown in Table 8.10, the moment constraints are well preserved with the relative difference of less than 10^{-8}. Figure 8.6 compares the MECC and Gumbel–Hougaard copula with the use of MaxEnt-based marginals. Comparison shows that the Gumbel–Hougaard copula and MECC yield very similar results. As shown in the scatter plot, the joint CDF computed from two copulas closely follows a 45° line. Numerical regression ensures a very similar performance of MECC and Gumbel–Hougaard copulas.

3. Construct the MECC and fit the Gumbel–Hougaard copula to the flood data with empirical marginal distributions.

Using the same moment constraints as in step 2, we will obtain exactly the same MECC copula for the MaxEnt-based and empirical CDFs (i.e., Table 8.10). Fitting the Gumbel–Hougaard copula with the use of empirical marginals listed in Table 8.9, the parameters estimated are higher than those from MaxEnt-based marginals. The values of Spearman's rho computed using the estimated parameters (the Gumbel–Hougaard copulas) are close to each other. The relative differences to the sample Spearman's rho are about 0.032 and 0.015 for the Gumbel–Hougaard copula with MaxEnt-based and empirical marginals, respectively. This information indicates the advantage of applying empirical marginals to construct copulas, i.e., better avoiding the misidentification of the univariate distributions.

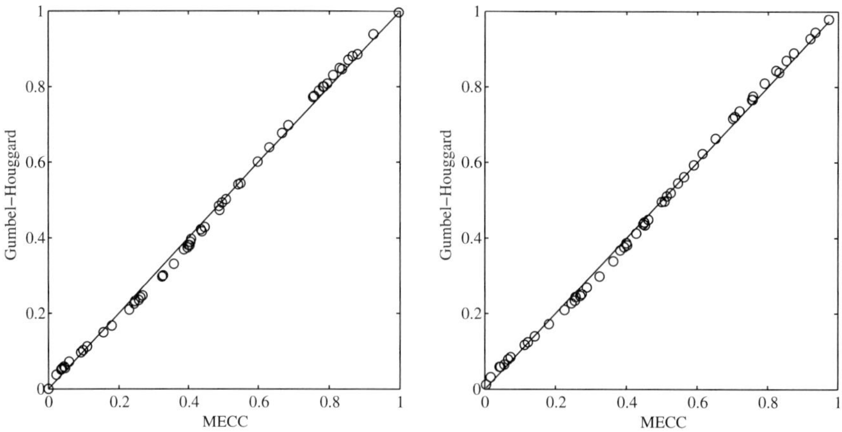

Figure 8.6 Comparison of MECC with Gumbel–Hougaard copulas.

Again, comparing the MECC with the Gumbel–Hougaard copulas fitted to empirical marginals, two copulas yield very similar results, as shown in Table 8.10 and Figure 8.6.

4. Compare the MECC and Gumbel–Houggard copulas fitted in steps 2 and 3 with empirical copulas.

The empirical copula for the bivariate flood variable is computed using Equation (3.65). Table 8.11 lists the JCDF computed from steps 2 and 3, as well as the empirical copulas. Table 8.12 lists the regression results with the use of simple linear regression $y = kx$, in which the empirical copula is considered as the independent variable x, with a visual comparison plotted in Figure 8.7. As shown in Table 8.12 and Figure 8.7, the MECC and Gumbel–Hougaard copulas (fitted to the MaxEnt-based and empirical marginals) indicate good fit to the empirical copulas.

Table 8.11. *JCDF computed from MECC, Gumbel–Hougaard, and empirical copulas.*

MECC		Gumbel–Hougaard		Empirical copula
MaxEnt marginals	Empirical marginals	MaxEnt Marginals	Empirical marginals	
0.995	0.974	0.996	0.979	1.000
0.817	0.790	0.837	0.810	0.818
0.785	0.757	0.802	0.776	0.782
0.000	0.004	0.001	0.013	0.018
0.802	0.754	0.817	0.767	0.782
0.345	0.363	0.317	0.339	0.364
0.787	0.757	0.800	0.767	0.782

Table 8.11. (*cont.*)

MECC		Gumbel–Hougaard		Empirical copula
MaxEnt marginals	Empirical marginals	MaxEnt Marginals	Empirical marginals	
0.862	0.853	0.879	0.870	0.891
0.399	0.402	0.371	0.381	0.382
0.544	0.563	0.536	0.562	0.564
0.874	0.875	0.889	0.889	0.909
0.393	0.394	0.367	0.376	0.382
0.391	0.450	0.379	0.442	0.455
0.252	0.270	0.231	0.247	0.255
0.836	0.822	0.855	0.843	0.855
0.890	0.921	0.896	0.928	0.945
0.628	0.615	0.636	0.623	0.636
0.318	0.288	0.288	0.269	0.291
0.687	0.652	0.699	0.663	0.691
0.434	0.447	0.420	0.437	0.455
0.762	0.706	0.781	0.722	0.727
0.371	0.399	0.353	0.387	0.400
0.042	0.055	0.050	0.065	0.073
0.485	0.499	0.483	0.497	0.509
0.533	0.545	0.530	0.545	0.564
0.495	0.513	0.492	0.510	0.509
0.222	0.226	0.201	0.210	0.236
0.104	0.122	0.106	0.124	0.145
0.315	0.324	0.286	0.299	0.345
0.088	0.142	0.093	0.140	0.182
0.033	0.042	0.047	0.059	0.073
0.020	0.016	0.035	0.032	0.055
0.667	0.652	0.677	0.664	0.691
0.227	0.255	0.210	0.234	0.255
0.053	0.072	0.066	0.084	0.109
0.169	0.182	0.157	0.172	0.218
0.251	0.245	0.227	0.227	0.255
0.035	0.044	0.050	0.059	0.073
0.431	0.451	0.406	0.434	0.436
0.404	0.382	0.384	0.368	0.382
0.506	0.524	0.500	0.520	0.509
0.758	0.701	0.775	0.717	0.745
0.929	0.935	0.940	0.945	0.964
0.142	0.182	0.137	0.172	0.200
0.385	0.428	0.365	0.413	0.418
0.483	0.507	0.465	0.498	0.491

Table 8.11. *(cont.)*

MECC		Gumbel–Hougaard		Empirical copula
MaxEnt marginals	Empirical marginals	MaxEnt Marginals	Empirical marginals	
0.000	0.004	0.000	0.013	0.018
0.083	0.113	0.088	0.117	0.145
0.598	0.590	0.599	0.593	0.600
0.227	0.257	0.215	0.244	0.255
0.774	0.719	0.791	0.736	0.764
0.040	0.065	0.053	0.079	0.091
0.435	0.461	0.414	0.449	0.436
0.246	0.273	0.226	0.251	0.273
0.838	0.832	0.846	0.839	0.855

Table 8.12. *Regression comparison results using empirical copula as independent variables.*

	MaxEnt-based marginals		Empirical marginals	
	MECC	GH[a]	MECC	GH
K	0.978	0.98	0.972	0.976
R-square	0.989	0.987	0.996	0.997

Note: [a] GH denotes Gumbel–Hougaard copula.

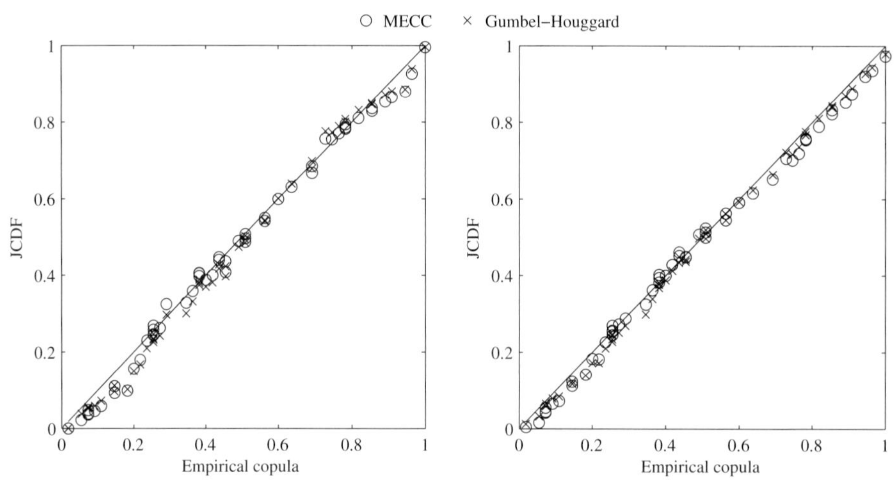

Figure 8.7 Comparison of MECC and Gumbel–Hougaard copulas to empirical copulas.

8.4 Summary

In this chapter, we introduce the entropy theory to study bivariate frequency analysis. The entropy-copula modeling discussed here may also be called most entropic canonical copula (MECC). Through examples, we have shown the following:

1. MECC construction only depends on the assigned constraints, that is, the Lagrange multipliers will not change in regard to different marginal distributions to be imposed. This is because (i) the marginals (i.e., CDFs) are uniformly distributed and $E(U^i) = \frac{1}{i+1}, U \sim uniform\ (0, 1)$; and (ii) the rank-based dependence measure does not depend on the marginal distributions.
2. $E(U^i), E(V^i), i = 1, 2$ may be enough in regard to the marginal constraints to construct MECC (i.e., Equations (8.12a) and (8.12b)).
3. As shown in Example 8.2, the performance is not significantly improved by adding more constraints in dependence measure besides $E(uv)$ rather than making the optimization more complex.
4. In general, it is good enough to preserve the dependence measure through $E(uv)$. $E(uv)$ directly corresponds to the rank-based Spearman correlation coefficient (ρ_s) (i.e., Equation (8.12c)). This is not unusual, since ρ_s is a popular nonparametric dependence measure used for parameter estimation besides Kendall's tau (τ).
5. The MECC constructed yields very similar performance, compared with the parametric copula with the same marginal distributions (e.g., the Gumbel–Hougaard copula applied in this chapter).
6. As with other parametric or nonparametric copulas, the marginal distributions and MECC can be investigated separately.
7. The overall advantage of MECC is that we obtain a unique Shannon entropy–based copula function with the given constraints. The parameters will not change with different marginal distribution candidates; however, parameters of parametric copulas do change if different marginal distribution candidates are used for parameter estimation. To some degree, the MECC minimizes the risk of improper choice of parametric copulas.
8. The MECC may be easily extended to a higher dimension with the use of a pairwise rank-based dependence structure.

References

Chu, B. (2011). Recovering copulas from limited information and an application to asset allocation. *Journal of Banking and Finance*, 35, 1824–1842. doi:10.1016/j.jbankfin.2010.12.011.

De Michele, C., Saladori, G., Canossi, M., Petaccia, A., and Rosso R. (2005). Bivariate statistical approach to check adequacy of dam spillway. *Journal of Hydrological Engineering ASCE*, 10(1), 50–57.

Favre, A.-C., El Adlouni, S., Perreault, L., Thiemonge, N., and Bebee, B. (2004). Multivariate hydrological frequency anlaysis using copulas. *Water Resources Research*, 40, W01101.

Hao, Z. and Singh, V. P. (2011). Single-site monthly streamflow simulation using entropy theory. *Water Resource Research*, 47, W09528, doi:10.1029/2100WR011419.
Hao, Z. and Singh, V. P. (2012). Entropy-copula method for single-site monthly streamflow simulation. *Water Resources Research,* 48, W06604, doi:10.1029/WR011419.
Jaynes, E. T. (1957). Information theory and statistical mechanics. *Physical Review. Series II,* 106(4), 620–630.
Jaynes, E. T. (1957). Information theory and statistical mechanics II. *Physical Review. Series II,* 108(2), 171–190.
Kao, S.-H. and Govindaraju, R. S. (2007). A bivariate frequency analysis of extreme rainfall with implications for design. *Journal of Geophysical Research,* 112, D13119, doi:10.1029/2007JD008522.
Krstanovic, P. F. and Singh, V. P. (1993a). A real-time flood forecasting model based on maximum-entropy spectral analysis: I. Development. *Water Resources Management*, 7(2), 109–129.
Krstanovic, P. F. and Singh, V. P. (1993b). A real-time flood forecasting model based on maximum-entropy spectral analysis: II. Application. *Water Resources Management*, 7(2), 131–151.
Nelsen, R. B. (2006). *An Introduction to Copulas*. 2nd edition. Springer, New York.
Shannon, C. E. (1948). A mathematical theory of communication. *Bell System Technical Journal,* 27, 379–423.
Singh, V.P. (1998). *Entropy-Based Parameter Estimation in Hydrology*. Kluwer Academic Publishers, Boston.
Singh, V. P. (2013). *Entropy Theory in Environmental and Water Engineering*. John Wiley, Sussex.
Singh, V. P. (2015). *Entropy Theory in Hydrologic Science and Engineering*. McGraw-Hill Education, New York.
Singh, V. P. and Krstanovic, P. F. (1987). A stochastic model for sediment yield using the principle of maximum entropy. *Water Resources Research*, 23(5), 781–793.
Singh, V. P., Zhang, L., and Rahimi, A. (2012). Probability distribution of rainfall-runoff using entropy theory. *Transactions of the ASABE*, 55(5), 1733–1744.
Vandenberghe, S., Verhoest, N. E. C., Onof, C., and De Baets, B. (2011) A comparative copula-based bivariate frequency analysis of observed and simulated storm events: a case study on Bartlett-Lewis modeled rainfall. *Water Resources Research*, 47, W07529, doi:10.1029/2009WR008388.
Zhang, L. and Singh, V. P. (2012). Bivariate rainfall and runoff analysis using entropy and copula theories. *Entropy*, 14, 1784–1812. doi:10.3390/e14091784.

9

Copulas in Time Series Analysis

ABSTRACT

In previous chapters, we have mainly discussed copula models for bivariate/multivariate random variables. Now we ask two other questions that usually arise in hydrology and water resources engineering. Can we use the stochastic approach to predict streamflow at a downstream location using streamflow at the upstream location? If streamflow is time dependent, then it cannot be considered as a random variable as is done in frequency analysis. Can we model the temporal dependence of an at-site streamflow sequence (e.g., monthly streamflow) more robustly than with the classical time series and Markov modeling approach (e.g., modeling the nonlinearity of time series freely)? This chapter attempts to address these questions and introduces how to model a time series with the use of copula approach.

9.1 General Concept of Time Series Modeling

In this section, we briefly introduce time series modeling. The reader may refer to Box et al. (2008) for a complete discussion. A time series (more specifically with even time intervals) may be stationary, nonstationary, or long-memory; linear or nonlinear. Following Box et al. (2008), a general form of a linear time series Y_t may be written as follows:

$$\phi(B)(1-B)^d Y_t = c + \theta(B)a_t \tag{9.1}$$

where Y_t is the time series; B is the backward operator; d is the differencing operator; $d=0$ for stationary; d is a positive integer (usually $d=1$ or 2) for nonstationary; $d \in (0,1)$ for long memory time series; $\phi(B) = 1 - \phi_1 B - \phi_2 B^2 - \cdots - \phi_p B^p$ is the autoregressive term; $\theta(B) = 1 + \theta_1 B + \theta_2 B^2 + \cdots + \theta_q B^q$ is the moving average term; and $a_t$is the innovation (i.e., white noise and more specifically white Gaussian noise).

The classic time series model given in Equation (9.1) may be identified with the following procedures:

1. Graph the sample autocorrelation (ACF) and partial autocorrelation (PACF) function for time series $\{X_t\colon t = 1, \ldots, n\}$.

2. Identify the possible model order from sample ACF and PACF, if the visual evidences are observed:
   i. If sample ACF falls into the 95% confidence bound quickly, then the time series X_t may be considered stationary (shown in Figure 9.1(a)); otherwise, the time series is nonstationary or long memory (Figure 9.1(b)), and differencing is needed to convert a nonstationary time series into the stationary time series (Figure 9.1(c)).
   ii. With the stationary time series, the model order may then be estimated from the sample ACF and PACF as follows: (a) if the cutoff point in ACF with the PACF falls into the 95% confidence bound, we will have moving average (MA) time series model (Figure 9.2(a)); (b) if the cutoff point in PACF with the ACF falls into the 95% confidence bound, we will have autoregressive (AR) time series model (Figure 9.2(b)); and (c) if both ACF and PACF fall into the 95% confidence bound, we will have an autoregressive and moving average (ARMA) time series model (Figure 9.2(c)).
3. Estimate the model parameters for the stationary time series with the assumption of model residual: $a_t \sim N(0, \sigma_a^2)$.

With the preceding initial introduction, we will now further illustrate Equation (9.1) using streamflow as an example. It is supposed that the differencing order, $d = 0$, occurs most likely for a watershed before experiencing climate change and/or alteration by human activities; $d = 1$ occurs most likely for the watershed with these impacts; and $d \in (0, 0.5)$ occurs usually for reservoir operations. In other words, the original stationary streamflow series (or the stationary streamflow series after necessary differencing) at time t is dependent on the value at previous p times (i.e., it depends on the streamflow at $t-1, t-2, \ldots, t-p$). Constant c relates to the long-term average of the stationary series given in Equation (9.2b). $\theta(B) = 1 + \theta_1 B + \theta_2 B^2 + \cdots + \theta_q B^q$ represents the moving

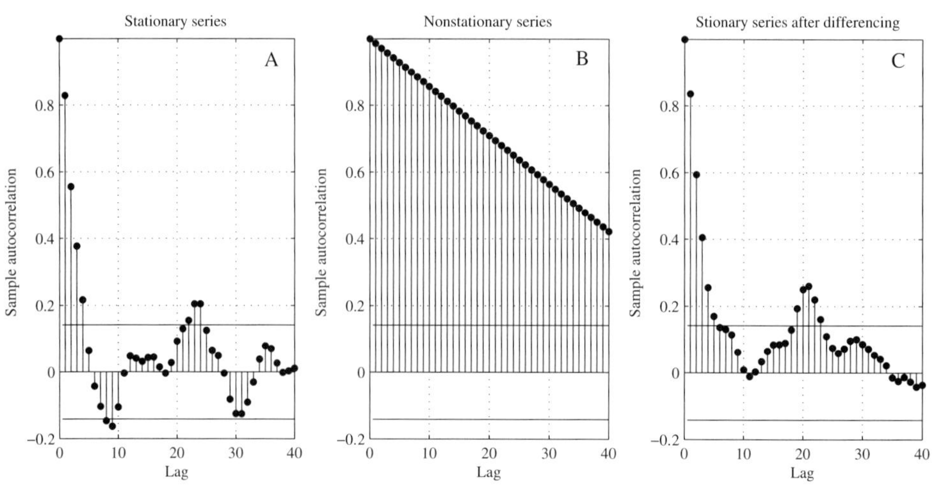

Figure 9.1 Sample autocorrelation function illustration plots.

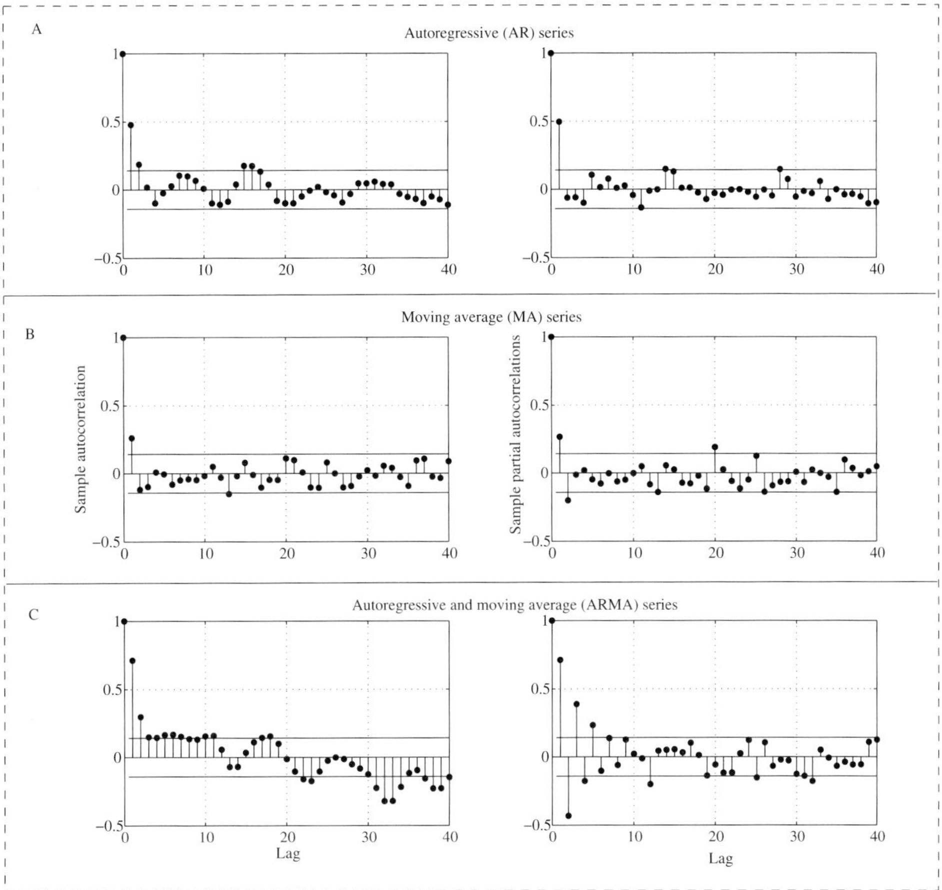

Figure 9.2 Sample ACF and PACF for the simulated stationary time series.

average term. Replacing $w_t = (1-B)^d Y_t$ such that Y_t is now written as stationary time series after necessary differencing, Equation (9.1) may be rewritten as follows:

$$\begin{aligned} w_t &= c + \varphi_1 w_{t-1} + \varphi_2 w_{t-2} + \cdots + \varphi_p w_{t-p} + a_t + \theta_1 a_{t-1} + \theta_2 a_{t-2} + \cdots + \theta_q a_{t-q} \\ &= c + \sum_{i=1}^{p} \varphi_i w_{t-i} + a_t + \sum_{j=1}^{q} \theta_j a_{t-j};\, a_t \sim i.i.d.\left(0, \sigma_a^2\right) \end{aligned} \tag{9.2}$$

Taking the expectation of Equation (9.2), we have the following:

$$E(w_t) = c + \sum_{i=1}^{p} \phi_i E(w_{t-i}) + E(a_t) + \sum_{j=1}^{q} \theta_j E\left(a_{t-j}\right) \tag{9.2a}$$

Substituting $E(w_t) = E(w_{t-i}), i = 1, \ldots, p,$ and $E(a_t) = E\left(a_{t-j}\right) = 0$ for the stationary time series into Equation (9.2a), we have the following:

$$E(w_t) = c + \sum_{i=1}^{p} \phi_i E(w_t) \Rightarrow E(w_t) = \frac{c}{1 - \sum_{i=1}^{p} \phi_i} \tag{9.2b}$$

To further evaluate if differencing is necessary, two statistical tests can help make a reasonable and formal decision. The Kwiatkowski–Phillips–Schmidt–Shin (KPSS) test (1992) has the null hypothesis of time series being stationary, while the augmented Dickey–Fuller (ADF) test (Dickey and Fuller, 1979) has the null hypothesis of time series as a unit root process (or is simply called nonstationary). The KPSS and ADF tests are complementary to each other (Arya and Zhang, 2015) as follows:

i. Time series being stationary: *acceptance* by KPSS test while *rejection* by ADF test.
ii. Time series being nonstationary: *rejection* by KPSS test while *acceptance* by ADF test.
iii. Time series belonging to a long memory process: *rejection* by both KPSS and ADF tests. In this case, the Hurst coefficient (Hurst, 1951) is applied to evaluate the necessary fractional differencing order.
iv. Not enough evidence to decide whether the time series is stationary or nonstationary: *acceptance* by both KPSS and ADF tests.

Furthermore, if there exists heteroscedasticity (i.e., changing variance) in the time series such that the time series tends to have a large value following a large value and a small value following a small value as a simple illustration. Then, for the time series with heteroscedasticity, the model error of Equation (9.1) needs to be further revised using (Generalized) Autoregressive Conditional Heteroscedastic (G) (ARCH) models. A (G) ARCH model indicates a second-order dependent time series. In other words, the conditional variability depends on the past history of the time series. An ARCH model can be written as follows:

$$h_t \equiv \text{var}(a_t | a_{t-1}, a_{t-2} \ldots) = E\left(a_t^2 | a_{t-1}, a_{t-2} \ldots\right) = w_0 + \sum_{i=1}^{s} w_i a_{t-i}^2 \tag{9.3}$$

and a Generalized ARCH (Bollerslev, 1986) model can be written as follows:

$$\begin{aligned} h_t &\equiv \text{var}(a_t | a_{t-1}, a_{t-2} \ldots) = E\left(a_t^2 | a_{t-1}, a_{t-2} \ldots\right) \\ &= w_0 + \sum_{i=1}^{s} w_i a_{t-i}^2 + \sum_{j=1}^{r} q_j h_{t-j} \end{aligned} \tag{9.4}$$

In Equations (9.3) and (9.4), h_t denotes the conditional variance (variability) of a_t given $a_{t-1}, a_{t-2}, \ldots$; $w_0 > 0, w_i \geq 0, i = 1, \ldots, s, q_j \geq 0, j = 1, \ldots, r$; w_i denotes the coefficients of the ARCH effects (i.e., for the correlated squared model errors a_t); and q_i denotes the coefficients of the correlated conditional variance h_t.

In addition, there exists a relation among conditional variance (h_t), innovation (i.e., model residual a_t) and standard white Guassian noise $(e_t, e_t \sim N(0, 1))$ as follows:

$$a_t = \sqrt{h_t} e_t \tag{9.5}$$

The parameters of the time series model may be estimated with the use of the maximum likelihood method.

Example 9.1 **Fit an autoregressive time series model with order 1 (i.e., AR(1))$Y_t = c + \phi_1 Y_{t-1} + e_t, e_t \sim N(0, \sigma_e^2)$ to annual streamflow data given in Table 9.1.**

Plot the original time series and residual sequence. Is the residual sequence a white Gaussian noise?

Table 9.1. *Annual streamflow data.*

Year	Flow (cfs)	Year	Flow (cfs)
1960	517.9	1988	280.8
1961	367.8	1989	500.9
1962	252.2	1990	528.9
1963	258.6	1991	602.8
1964	281.3	1992	386
1965	308.4	1993	627.2
1966	317.3	1994	520
1967	372.4	1995	345.3
1968	349.1	1996	575
1969	504.3	1997	663.2
1970	330.9	1998	412.4
1971	413.3	1999	311
1972	461.7	2000	385.3
1973	567	2001	299.3
1974	500.6	2002	417.5
1975	654.9	2003	578.5
1976	550.3	2004	715.8
1977	401	2005	676.3
1978	593.8	2006	507
1979	508	2007	736
1980	543.5	2008	677.4
1981	442.3	2009	508.9
1982	477.3	2010	418.3
1983	473	2011	721.2
1984	548.3	2012	609
1985	467.7	2013	536.3
1986	539.3	2014	608.1
1987	431.7	2015	552.4

Solution: To fit AR(1) to the time series listed in Table 9.1, we can simply use the MATLAB function as follows:

1. Assign the TS as the time series listed in Table 9.1.
2. Set up the AR(1) model, where we need to fit using arima:

```
model=arima(1,0,0); % ARIMA (P, D, Q): P=1, AR term;
D=0, stationary series; Q=0, MA term.
```

3. Apply the estimate function (through MLE):

```
[param,Var,LogL]=estimate(model,TS);
% param: estimated parameter for the model defined above.
% Var: variance-covariance matrix for the parameter
estimated. Here, we have 3 parameters: constant-C,
autoregressive parameter, and variance of model residual.
% LogL: the loglikelihood of the objective function after
optimization.
```

Using the preceding functions, we get the results listed in Table 9.2.
The fitted AR(1) time series model is now written as follows:

$$Z_t = 255.13 + 0.475Z_{t-1} + e_t \tag{9.6}$$

4. Apply the infer function to compute the model residual sequence listed in Table 9.3:

```
res=infer(param,TS);
```

Table 9.2. *Parameter estimated for the AR(1) model.*

ARIMA (1,0,0) model (AR(1) model)			
Conditional probability distribution: Gaussian			
Parameter	Value	Standard error	*T*-statistics
Constant (cfs)	255.13	71.80	3.55
AR{1}	0.475	0.15	3.18
Variance (cfs^2)	12751.6	3141.41	4.06
LogL = –434.18			

Table 9.3. *Fitted model residual.*

Year	Residual (cfs)	Year	Residual (cfs)
1960	24.90	1988	–179.29
1961	–133.22	1989	112.45
1962	–177.55	1990	35.95
1963	–116.27	1991	96.56
1964	–96.61	1992	–155.32
1965	–80.29	1993	188.81
1966	–84.25	1994	–32.91

Table 9.3. (*cont.*)

Year	Residual (cfs)	Year	Residual (cfs)
1967	–33.38	1995	–156.71
1968	–82.84	1996	155.93
1969	83.43	1997	135.07
1970	–163.66	1998	–157.60
1971	1.07	1999	–139.93
1972	10.34	2000	–17.49
1973	92.67	2001	–138.76
1974	–23.73	2002	20.27
1975	162.10	2003	125.15
1976	–15.76	2004	186.01
1977	–115.40	2005	81.33
1978	148.28	2006	–69.22
1979	–29.05	2007	240.16
1980	47.18	2008	72.84
1981	–70.87	2009	–67.84
1982	12.18	2010	–78.44
1983	–8.74	2011	267.47
1984	68.60	2012	11.46
1985	–47.75	2013	–7.97
1986	62.12	2014	98.35
1987	–79.48	2015	8.56

Figure 9.3 plots the original time series, fitted model residuals, and the histogram compared to the hypothesized white Gaussian noise. From the histogram plot, it seems that the hypothesized distribution may properly represent the distribution of the fitted model residuals. To formally assess whether the fitted model residuals are a white Gaussian noise, we apply the Kolmogorov–Smirnov (KS) test. The KS test evaluates the maximum distance of empirical and parametric CDFs. Its test statistic D_n can be written as follows:

$$D_n = \sup_x \mid F_n(x) - F(x) \mid \tag{9.7}$$

The null hypothesis (H_0) is that the fitted model residuals follow $N(0, \sigma_e^2)$, *i.e.*, $N(0, 12751.6)$. With this null hypothesis, we can either use the parametric bootstrap method or MATLAB function kstest directly. Here we will simply use the MATLAB function kstest. Table 9.4 lists the empirical and parametric CDFs for the fitted model residuals.

Applying kstest as follows

```
[H,Pvalue,stat]=kstest(res,[res,normcdf(res,0,param.Vari-
ance^0.5)],0.05)
```

Table 9.4. *Empirical and parametric CDFs of the fitted model residuals.*

Residual	Parametric CDF	Empirical CDF	Residual	Parametric CDF	Empirical CDF
24.90	0.59	0.63	−179.29	0.06	0.02
−133.22	0.12	0.16	112.45	0.84	0.82
−177.55	0.06	0.04	35.95	0.62	0.65
−116.27	0.15	0.18	96.56	0.80	0.79
−96.61	0.20	0.21	−155.32	0.08	0.11
−80.29	0.24	0.26	188.81	0.95	0.95
−84.25	0.23	0.23	−32.91	0.39	0.40
−33.38	0.38	0.39	−156.71	0.08	0.09
−82.84	0.23	0.25	155.93	0.92	0.89
83.43	0.77	0.75	135.07	0.88	0.86
−163.66	0.07	0.05	−157.60	0.08	0.07
1.07	0.50	0.53	−139.93	0.11	0.12
10.34	0.54	0.56	−17.49	0.44	0.46
92.67	0.79	0.77	−138.76	0.11	0.14
−23.73	0.42	0.44	20.27	0.57	0.61
162.10	0.92	0.91	125.15	0.87	0.84
−15.76	0.44	0.47	186.01	0.95	0.93
−115.40	0.15	0.19	81.33	0.76	0.74
148.28	0.91	0.88	−69.22	0.27	0.33
−29.05	0.40	0.42	240.16	0.98	0.96
47.18	0.66	0.67	72.84	0.74	0.72
−70.87	0.27	0.32	−67.84	0.27	0.35
12.18	0.54	0.60	−78.44	0.24	0.30
−8.74	0.47	0.49	267.47	0.99	0.98
68.60	0.73	0.70	11.46	0.54	0.58
−47.75	0.34	0.37	−7.97	0.47	0.51
62.12	0.71	0.68	98.35	0.81	0.81
−79.48	0.24	0.28	8.56	0.53	0.54

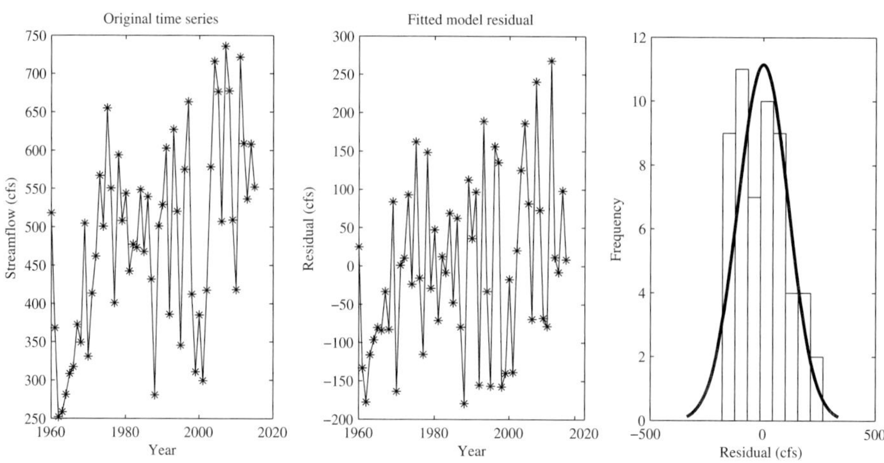

Figure 9.3 Original time series, fitted model residual plots, and histogram.

we have $H = 0$, Pvalue = 0.803, and test statistic = 0.083. With null hypothesis being accepted and Pvalue > 0.05, we show that the fitted model residual is a white Gaussian noise.

9.2 Spatially Dependent Bivariate or Multivariate Time Series

In stock exchanges, the stock values among major exchanges (e.g., London, Hong Kong, New York, Tokyo) always impact one another. In other words, these major exchanges have a tendency to follow each other. In the field of hydrology and water resources engineering, there also exists a similar tendency (or spatial dependence). For example, streamflow (or flood) at a downstream location is generally positively dependent on that at the upstream location. In this section, we will show how to evaluate spatially dependent time series.

As discussed in the previous chapters, copulas are applied to bivariate/multivariate independent random variables. Thus, to employ the copula theory for a bivariate (multivariate) time-dependent sequence (e.g., spatial dependence of bivariate/multivariate time series), we need to investigate each individual time series first and fit the time series with proper models (e.g., the equations in Section 9.1). Three steps are needed for bivariate/multivariate time series analysis using copulas:

1. Investigate each univariate time series separately, including the assessment of stationarity, time series model identification, and estimation of model parameters;
2. Compute model residuals from the fitted univariate time series model.
3. Apply copulas to the model residuals obtained from step 2.

In what follows, we will use a simple example to illustrate how to model a bivariate time series.

Example 9.2 **Perform a dependence study using the daily time series data given in Table 9.5.**

Table 9.5. *Bivariate time series.*

	TS1	TS2		TS1	TS2
1	60.25	476.35	26	45.24	475.13
2	55.87	475.48	27	11.28	475.01
3	76.18	476.02	28	30.96	475.44
4	74.84	476.75	29	78.03	475.49
5	84.79	476.89	30	97.69	475.91
6	68.39	477.27	31	90.31	476.52
7	73.14	477.51	32	60.44	475.51
8	63.01	476.50	33	52.28	474.24

Table 9.5. (*cont.*)

	TS1	TS2		TS1	TS2
9	63.28	475.87	34	35.00	473.57
10	80.17	475.74	35	38.98	473.33
11	72.42	475.29	36	49.74	473.42
12	57.82	476.75	37	75.43	473.87
13	86.49	476.60	38	81.49	471.65
14	99.88	475.69	39	56.48	470.84
15	75.34	474.94	40	51.48	472.87
16	66.07	473.83	41	50.26	473.17
17	56.61	473.63	42	52.20	474.16
18	69.76	473.84	43	76.67	475.32
19	94.75	475.58	44	83.07	475.63
20	62.69	476.17	45	79.80	475.90
21	64.71	475.11	46	81.26	474.64
22	88.54	476.29	47	73.69	474.19
23	84.75	478.09	48	72.70	476.01
24	94.86	478.04	49	77.68	476.85
25	81.05	476.20	50	89.06	476.79

Solution: Applying the procedure similar to Example 9.1, the time series TS1 and TS2 are fitted with ARIMA (1,0,1) (i.e., ARMA(1,1)) and ARIMA(2,0,0) (i.e., AR(2)), respectively. The fitted parameters for the time series are given in Table 9.6. Figure 9.4 plots the original time series and empirical frequencies (histogram) for the model residuals.

The acceptance by the KS test for the fitted model residuals indicates that the model residuals belong to the white noise.

Table 9.6. *Fitted time series model and parameter estimated.*

	TS1-ARIMA(1,0,1) model: conditional probability distribution: Gaussian		
Parameter	Value	Standard error	T statistics
Constant	55.24	13.18	4.19
AR{1}	0.21	0.19	1.09
MA{1}	0.66	0.15	4.34
Variance	185.21	42.53	4.36
	TS2-ARIMA(2,0,0) model: conditional probability distribution: Gaussian		
Parameter	Value	Standard error	T statistics
Constant	125.18	38.95	3.16
AR{1}	1.1	0.15	7.15
AR{2}	–0.36	0.16	–2.3
Variance	0.7	0.12	5.63

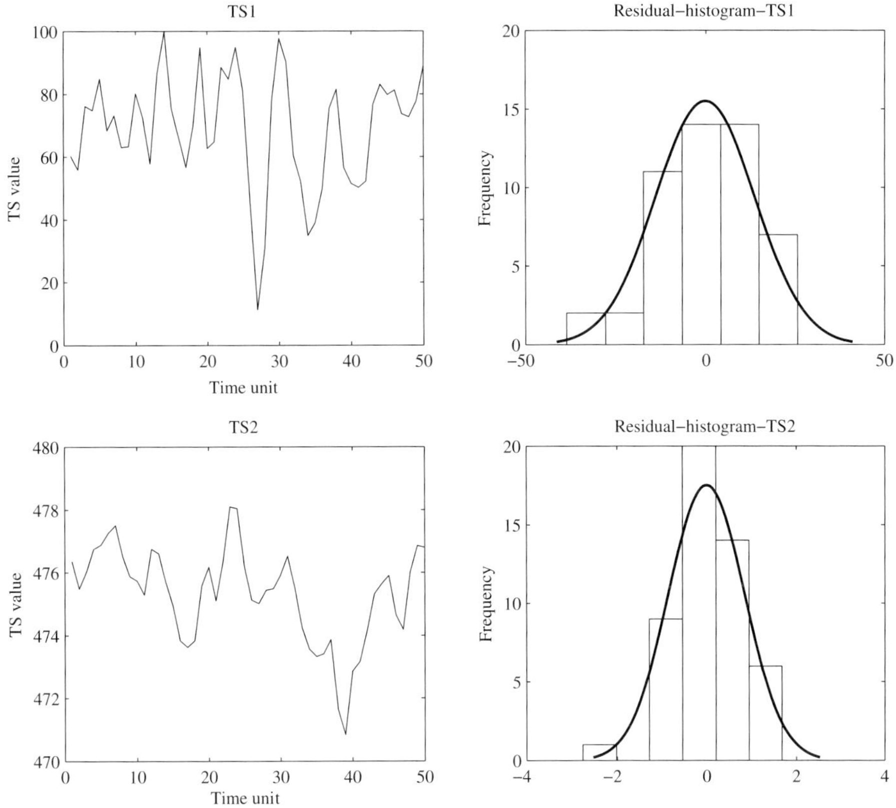

Figure 9.4 Plots of original time series and histograms of the fitted model residuals.

We can write the residuals as a function of time series:

$$TS1:\ res_t^{TS1} = -55.24 + TS1_t - 0.21TS1_{t-1} - 0.65res_{t-1}^{TS1} \quad (9.8a)$$

$$TS2:\ res_t^{TS2} = -123.18 + TS2_t - 1.10TS2_{t-1} + 0.36TS2_{t-2} \quad (9.8b)$$

Now, we may apply copula to res_t^{TS1}, res_t^{TS2} that are considered as independent random variables. Figure 9.5 shows the scatter plot of the random variables.

From Figure 9.5, it is seen the fitted model residuals are positively correlated. Using Equations (3.70) and (3.73), the empirical rank-based correlation coefficients, Spearman's ρ, and Kendall's τ are computed as $\rho_n \approx 0.28,\ \tau_n \approx 0.16$. To this end, we apply the Archimedean copulas (i.e., Gumbel–Hougaard, Clayton, and Frank copulas presented in Chapter 4) and meta-elliptical copulas (i.e., the Gaussian and Student t copulas presented in Chapter 7). Similar to the discussion in the previous chapters, we apply the pseudo (i.e., semiparametric) and two-stage maximum likelihood methods for parameter estimation. Table 9.7 lists the empirical and parametric CDFs of the fitted model residuals. Table 9.8 lists the parameters and corresponding estimated likelihood values. The likelihood values in Table 9.8 suggest that the

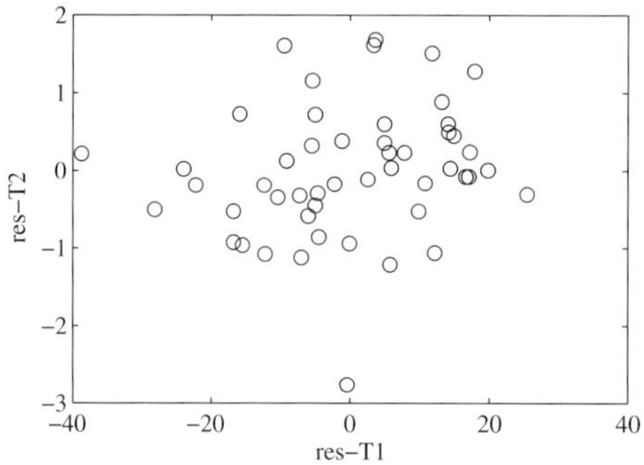

Figure 9.5 Scatter plot of the fitted model residuals: $\boldsymbol{res}_t^{T1}$ and $\boldsymbol{res}_t^{T2}$.

Table 9.7. *Fitted model residuals, empirical and parametric CDFs computed.*

		Empirical CDF		Parametric CDF	
res_t^{T1}	res_t^{T2}	res_t^{T1}	res_t^{T2}	res_t^{T1}	res_t^{T2}
–9.03	0.13	0.25	0.59	0.25	0.56
–5.91	–0.59	0.31	0.20	0.33	0.24
13.25	0.89	0.78	0.86	0.83	0.86
–4.93	0.72	0.39	0.82	0.36	0.80
17.28	0.24	0.92	0.67	0.90	0.61
–15.82	0.73	0.14	0.84	0.12	0.81
14.15	0.60	0.80	0.80	0.85	0.76
–16.72	–0.53	0.12	0.24	0.11	0.26
5.99	0.03	0.67	0.57	0.67	0.51
7.86	0.23	0.69	0.65	0.72	0.61
–4.62	–0.29	0.41	0.35	0.37	0.36
–9.39	1.60	0.24	0.94	0.24	0.97
25.46	–0.31	0.98	0.33	0.97	0.36
9.94	–0.53	0.71	0.22	0.77	0.26
–7.15	–0.33	0.27	0.31	0.30	0.35
–0.08	–0.94	0.51	0.14	0.50	0.13
–12.28	–0.19	0.18	0.37	0.18	0.41
10.88	–0.16	0.73	0.43	0.79	0.42
17.88	1.28	0.94	0.90	0.91	0.94
–23.98	0.02	0.06	0.53	0.04	0.51
12.28	–1.06	0.76	0.10	0.82	0.10
11.79	1.51	0.75	0.92	0.81	0.96
3.38	1.62	0.55	0.96	0.60	0.97
19.83	0.01	0.96	0.51	0.93	0.50

Table 9.7. (*cont.*)

		Empirical CDF		Parametric CDF	
res_t^{T1}	res_t^{T2}	res_t^{T1}	res_t^{T2}	res_t^{T1}	res_t^{T2}
–6.93	–1.12	0.29	0.06	0.31	0.09
–22.24	–0.19	0.08	0.39	0.05	0.41
–38.68	0.22	0.02	0.61	0.00	0.60
–1.12	0.38	0.47	0.73	0.47	0.68
17.11	–0.08	0.90	0.47	0.90	0.46
14.99	0.45	0.86	0.75	0.86	0.70
4.93	0.60	0.59	0.78	0.64	0.76
–16.78	–0.93	0.10	0.16	0.11	0.13
–4.43	–0.85	0.43	0.18	0.37	0.15
–28.16	–0.51	0.04	0.25	0.02	0.27
–4.96	–0.45	0.37	0.27	0.36	0.30
–10.31	–0.35	0.22	0.29	0.22	0.34
16.67	–0.08	0.88	0.49	0.89	0.46
–0.37	–2.76	0.49	0.02	0.49	0.00
–15.41	–0.97	0.16	0.12	0.13	0.12
–5.32	1.16	0.35	0.88	0.35	0.92
–12.14	–1.08	0.20	0.08	0.19	0.10
–5.46	0.32	0.33	0.69	0.34	0.65
14.21	0.49	0.82	0.76	0.85	0.72
2.57	–0.11	0.53	0.45	0.57	0.45
5.65	0.23	0.63	0.63	0.66	0.61
5.76	–1.21	0.65	0.04	0.66	0.07
–2.19	-0.18	0.45	0.41	0.44	0.42
3.63	1.68	0.57	0.98	0.61	0.98
4.98	0.36	0.61	0.71	0.64	0.67
14.43	0.03	0.84	0.55	0.86	0.51

Frank copula attains the best overall performance, followed by Gaussian copula. Figure 9.6 compares the CDF of the model residuals to the simulated random variates from the fitted parametric copulas. Figure 9.7 compares the model residuals to that computed from the two-stage estimation. Comparison indicates a similar performance between the Frank and Gaussian copulas.

From this example, we also note that the rank-based correlation coefficient of the model residuals may be different from that of the original time series. In this example, we have $\tau_n \approx 0.16$ for the model residuals, while $\tau_n \approx 0.35$ for the original time series. The reduction of the degree of association for the model residuals may be due to the autoregressive component of time series modeling.

Table 9.8. *Estimated copula parameters and estimated log-likelihood values.*

Copula	Pseudo	MLE	Semiparametric	MLE
Gumbel–Hougaard	1.15	0.75	1.13	0.61
Clayton	0.31	0.88	0.08	0.15
Frank	1.63	1.79	1.56	1.69
Gaussian	0.26	1.30	0.20	0.99
Student *t*	$[0.26, 1.38 \times 10^7]$	1.30	$[0.20, 1.38 \times 10^7]$	0.99

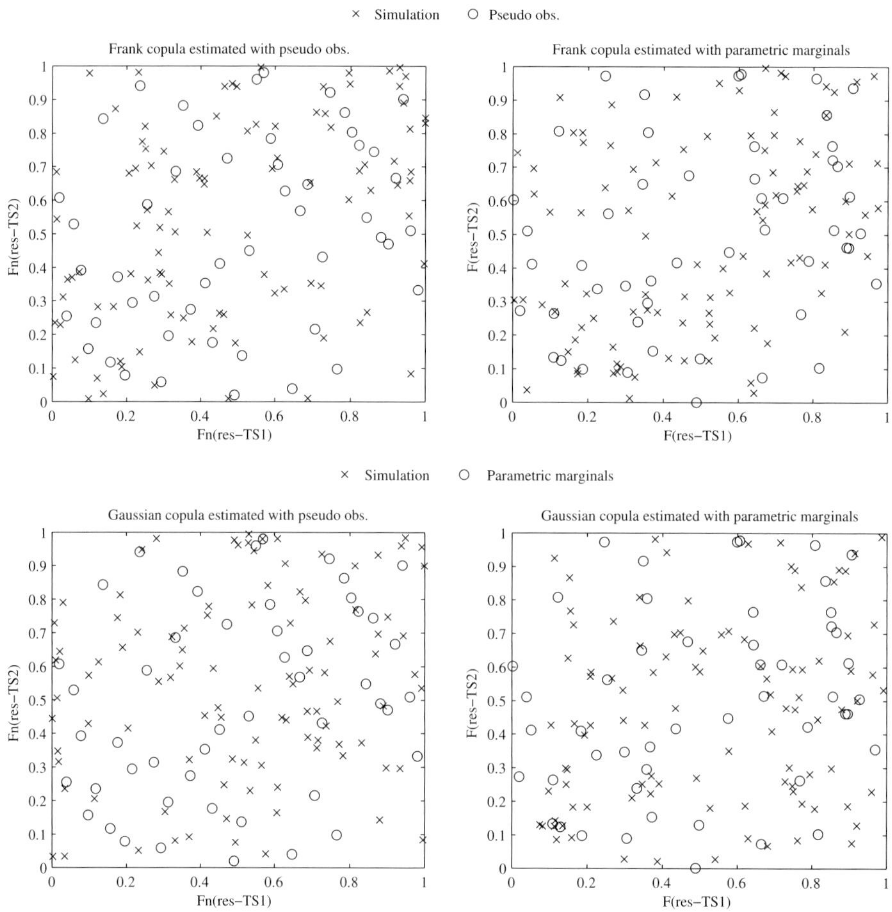

Figure 9.6 Comparison of simulated random variates to pseudo-observations and fitted parametric marginals.

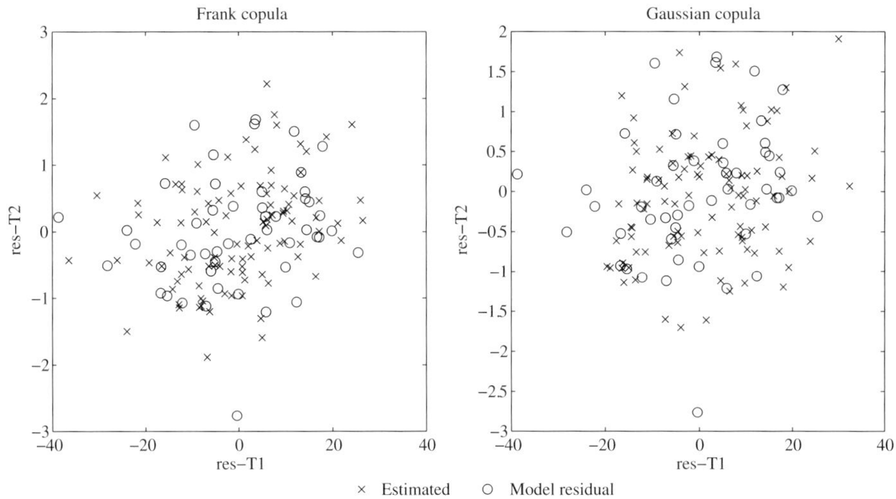

Figure 9.7 Comparison of fitted model residuals to those computed from copula with parameter estimated using two-stage MLE.

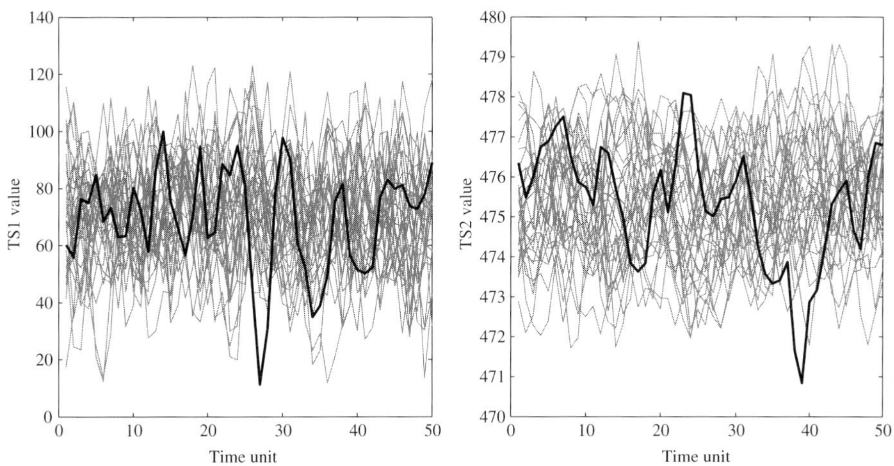

Figure 9.8 Reconstructed time series using the copula estimated with two-stage MLE.

From Equation (9.8), we can also reconstruct the time series with the use of random variates from the simulated copula. Here, we will again use copulas with two-stage MLE as an example. Additionally, we will use the last two values of the time series (i.e., Table 9.5) as initial estimates. Figure 9.8 plots the reconstructed time series, which shows that the reconstructed time series reasonably follows the same pattern as does the original time series.

Now, we have explained how to study the spatial dependence for the sequence with time dependence. In the previous example, we studied time-dependent sequences and spatial dependence of the sequences separately. We first built the time series (i.e., the autoregressive and moving average) model for each univariate time-dependent sequence. Then, we built the copula model on the residual (also called innovation) of the time series model, since the residuals are now random variables.

Copula modeling can also be applied to study the serial dependence of univariate time series, in addition to studying the previously discussed bivariate/multivariate spatial-temporal dependent time series (i.e., spatial dependence for the time-dependent sequence). In the following section, we will introduce how to model the serial dependence of univariate time series.

9.3 Copula Modeling for Univariate Time Series with Serial Dependence: General Discussion

Darsow, et al. (1992) introduced the condition (i.e., equivalent to Chapman–Kolmogorov equations) for a copula-based time series to be a Markov process. Joe (1997) introduced a class of parametric stationary Markov models based on parametric copulas and parametric marginal distributions. Similar to the copula application in bivariate or multivariate frequency analysis discussed previously, a copula-based time series model also allows one to consider serial dependence and marginal behavior of the time series investigated separately.

Following Joe (1997) and Chen and Fan (2006), copulas can be applied to the stationary time series of (i) Markov chain models (both discrete and continuous, including autoregressive models); (ii) K-dependent time series models (i.e., moving average models with order k); (iii) convolution-closed infinitely divisible univariate marginal models. For stationary time series $\{Z_t\text{: } t = 1, 2, \ldots\}$.

Let $\{e_t\}$ be i.i.d. random variables that are independent of $\{Z_{t-1}, Z_{t-2} \ldots\}$ (i.e., the innovation of the time series $\{Z_t : t = 1, 2, \ldots\}$). We may express the preceding three cases by using one of the following models (Joe, 1997).

Markov Chain models:

➢ Kth-order autoregressive model:

$$Z_t = \alpha_1 Z_{t-1} + \alpha_t Z_{t-2} + \cdots + \alpha_K Z_{t-K} + e_t \tag{9.9}$$

where $\alpha_1, \alpha_2, \ldots, \alpha_K$ are the scalars.

➢ Kth-order Markov chain:

$$Z_t = g(Z_{t-1}, Z_{t-2}, \ldots, Z_{t-k}, e_t) \tag{9.10}$$

where g is a real-valued function.

➢ First-order convolution-closed infinitely divisional univariate margin model:

$$Z_t = S_t(Z_{t-1}) + e_t \tag{9.11}$$

where S_t is an independent realization of the stochastic operator.

***K*-dependent time series models:**

➢ *K*th-order Moving Average model:

$$Z_t = e_t + \beta_1 e_{t-1} + \beta_2 e_{t-2} + \cdots + \beta_K e_{t-K} \quad (9.12)$$

where $\beta_1, \beta_2, \ldots, \beta_K$ are the scalars.

➢ *K*-dependent model:

$$Z_t = h(e_t, e_{t-1}, \ldots, e_{t-K}) \quad (9.13)$$

where h is a real-valued function.

➢ One-dependent convolution-closed infinitely divisible univariate marginal model:

$$Z_t = e_t + S_t(e_{t-1}) \quad (9.14)$$

where S_t is the independent realization of the stochastic operator.

Now, with the classified models given in Equations (9.10)–(9.14), we will focus on the continuous Markov chain (also called Markov process) models for the rest of the chapter. We will introduce the simple first-order Markov models first, followed by the Kth-order Markov models.

9.4 First-Order Copula-Based Markov Model

9.4.1 General Concept of the First-Order Copula-Based Continuous Markov Model

For continuous time series $\{Z_t: t = 1, 2, \ldots\}$ modeled with the first-order Markov process, its transition probability can be expressed as follows:

$$P(Z_t \le z_t | Z_{t-1} = z_{t-1}, Z_{t-2} = z_{t-2}, \ldots, Z_1 = z_1) = P(Z_t \le z_t | Z_{t-1} = z_{t-1}) \quad (9.15)$$

Equation (9.15) means that the probabilistic behavior of the time series $\{Z_t\}$ is fully governed by the joint distribution of $\{Z_t, Z_{t-1}\}$. We can apply copula modeling as a robust and powerful representation of Equation (9.15) as follows:

$$P(Z_t \le z_t | Z_{t-1} = z_{t-1}, Z_{t-2} = z_{t-2}, \ldots, Z_1 = z_1)$$

$$= P(Z_t \le z_t | Z_{t-1} = z_{t-1}) = \frac{\partial C(F(z_t), F(z_{t-1}))}{\partial (F(z_{t-1}))} \quad (9.16a)$$

and the conditional density of Z_t given Z_{t-1} can be expressed using copula as follows:

$$h(z_t | z_{t-1}) = f(z_t) c(F(z_t), F(z_{t-1})) \quad (9.16b)$$

where C and c represent the copula and its density function of (z_t, z_{t-1}), and F and f represent the marginal distribution and the density function of z_t, respectively

9.4.2 Parameter Estimation of the First-Order Copula-Based Continuous Markov Model

Chen and Fan (2006) proposed an estimation method similar to the semiparametric maximum likelihood estimation method discussed in the previous chapters. Following Chen and Fan (2006) and Equations (9.16a) and (9.16b), we see the time series is fully determined by the true unknown marginal distribution F^* and a copula function with parameter $\boldsymbol{\alpha}^*$ (or simply written as $(F^*, \boldsymbol{\alpha}^*)$). Here we again note the advantage of investigating the marginal distribution and copula separately.

To evaluate the copula parameter, we may first apply the empirical marginal to time series $\{Z_t\}$ with the Weibull plotting-position formula in the same fashion as Equation (3.103):

$$F_n(z) = \frac{1}{m+1}\sum\nolimits_{t=1}^{m} 1(Z_t \le z) \tag{9.17}$$

where: m is the length of time series (or simply called the sample size).

Replacing Equation (9.17) with the true unknown marginal and its density function, and true copula density function, the log-likelihood function for the first-order Markov model can be expressed as follows:

$$L(\boldsymbol{\alpha}) = \frac{1}{m}\sum\nolimits_{t=1}^{m} \log f^*(z_t) + \frac{1}{n}\sum\nolimits_{t=2}^{m} \log c(F^*(z_t), F^*(z_{t-1}); \boldsymbol{\alpha}) \tag{9.18a}$$

Equation (9.18a) can be simplified using the empirical distribution F_n as follows:

$$\hat{L}(\boldsymbol{\alpha}) = \frac{1}{n}\sum\nolimits_{t=2}^{m} \log c(F_n(z_t), F_n(z_{t-1}); \boldsymbol{\alpha}) \tag{9.18b}$$

Equation (9.18b) is in the same form as Equation (3.104).

9.4.3 Simulation (Realizations) of the Time Series from the First-Order Copula-Based Markov Process

The univariate time series from the first-order copula-based Markov process can be simulated with a similar approach discussed in Section 3.7. With little modification, the simulation procedure is presented as follows:

i. Generate i.i.d. uniformly distributed random variables $U = \{u_i: i = 1, 2, 3, \ldots, N\}$.
ii. Set $y_1 = u_1$.
iii. Set $u_2 = C(y_2|y_1) = C(y_2|u_1) = \frac{\partial C(y_2, u_1)}{\partial u_1} \Rightarrow y_2 = h^{-1}(u_2, y_1; \boldsymbol{\alpha})$, in which the h function is defined as the conditional copula.
iv. Continue until we obtain $y_N = h^{-1}(u_n, y_{n-1}; \boldsymbol{\alpha})$.

It should be noted that $\{y_1, y_2, \ldots, y_n\}$ simulated from steps i to iv are the time series in the frequency domain (i.e., marginals), and we will need to perform the one-to-one transformation to obtain the corresponding time series simulated in the real domain (e.g., through parametric distribution, empirical distribution, or kernel density based on the observed time series).

9.4.4 Forecast and Quantile Estimation of the First-Order Markov Process

As in economics and finance, it is our interest to forecast the future behavior of the time series, or in other words median forecast and conditional quantile estimation. For given quantile, the conditional quantile estimation may also be called value-at-risk (VaR). From the transitional probability of the first-order Markov process (i.e., Equation (9.16a)), we have the median forecast expressed as follows:

$$\begin{aligned} E[Z_t|Z_{t-1} = z_{t-1}] &= \int z_t h^*(z_t|z_{t-1})dz_t \\ &= \int z_t f^*(z_t)c(F^*(z_t), F^*(z_{t-1}); \boldsymbol{\alpha})dz_t \\ &= \int z_t c(F^*(z_t), F^*(z_{t-1}); \boldsymbol{\alpha})dF^*(z_t) \end{aligned} \tag{9.19a}$$

Again, replacing the unknown true marginal distribution by its empirical distribution, and the true copula parameter by its estimated parameter ($\hat{\boldsymbol{\alpha}}$) from Equation (9.18b), Equation (9.19a) can be rewritten as follows:

$$E[Z_t|Z_{t-1} = z_{t-1}] = \int z_t c(F_n(z_t), F_n(z_{t-1}); \hat{\boldsymbol{\alpha}})dF_n(z_t) \tag{9.19b}$$

Equation (9.19b) implies that the conditional probability (i.e., conditional copula) of $Z_t \mid Z_{t-1}$ equals 0.5 (also called 50% conditional quantile) as follows:

$$\begin{aligned} &C(F_n(z_t)|F_n(z_{t-1}); \hat{\boldsymbol{\alpha}}) = 0.5 \\ &\Rightarrow F_n(z_t) = C^{-1}_{F_n(z_t)|F_n(z_{t-1})}(0.5|F_n(z_{t-1}); \hat{\boldsymbol{\alpha}}) \end{aligned} \tag{9.20a}$$

From Equation (9.20a), we can further forecast the behavior of z_t as follows:

$$\hat{z}_t = F_n^{-1}\left(C^{-1}_{F_n(z_t)|F_n(z_{t-1})}(0.5|F_n(z_{t-1}); \hat{\boldsymbol{\alpha}})\right) \tag{9.20b}$$

Similarly, Equations (9.20a) and (9.20b) can be easily reformulated for the estimation of any given conditional quantile q as follows:

$$F_n(z_t) = C^{-1}_{F_n(z_t)|F_n(z_{t-1})}(q|F_n(z_{t-1}); \hat{\boldsymbol{\alpha}}) \Rightarrow \hat{z}_t^q = F_n^{-1}\left(C^{-1}_{F_n(z_t)|F_n(z_{t-1})}(q|F_n(z_{t-1}); \hat{\boldsymbol{\alpha}})\right) \tag{9.21}$$

Example 9.3 **Rework Example 9.1 using the Gumbel–Hougaard and Gaussian copula-based first-order Markov model. Also, compare the one-step ahead forecast (i.e., forecasting the annual flow for water year 2016) with both the classic AR(1) model and copula-based first-order Markov model.**

Solution: In Example 9.1, we applied the AR(1) model to investigate the behavior and annual flow listed in Table 9.1. From Example 9.1 we conclude that statistically, we can apply AR(1) model to the annual flow under the assumption that annual flow shows a linear temporal dependence; however, in reality the dependence is usually nonlinear. Without imposing more complex (G)ARCH model, the copula-based Markov model is an excellent alternative approach to solve this issue. In addition, the Gaussian process assumption is also relaxed when applying the copula-based Markov model.

In this example, we will use semiparametric estimation such that the empirical distribution is applied for the marginals. The following steps are needed to model the temporal dependence with copulas:

1. With Equation (9.17), the Weibull plotting position formula and kernel density function are employed to compute the empirical marginal distribution (Table 9.9). It is worth noting that

Table 9.9. *Empirical marginals using the Weibull plotting position formula and kernel density function.*

(1)	(2)	(3)	(4)	(1)	(2)	(3)	(4)
1960	517.9	0.58	0.58	1988	280.8	0.05	0.08
1961	367.8	0.21	0.23	1989	500.9	0.49	0.53
1962	252.2	0.02	0.05	1990	528.9	0.61	0.61
1963	258.6	0.04	0.06	1991	602.8	0.81	0.79
1964	281.3	0.07	0.08	1992	386	0.26	0.26
1965	308.4	0.11	0.12	1993	627.2	0.86	0.84
1966	317.3	0.14	0.13	1994	520	0.60	0.59
1967	372.4	0.23	0.23	1995	345.3	0.18	0.18
1968	349.1	0.19	0.19	1996	575	0.75	0.73
1969	504.3	0.51	0.54	1997	663.2	0.89	0.89
1970	330.9	0.16	0.16	1998	412.4	0.30	0.32
1971	413.3	0.32	0.32	1999	311	0.12	0.12
1972	461.7	0.40	0.43	2000	385.3	0.25	0.26
1973	567	0.74	0.71	2001	299.3	0.09	0.11
1974	500.6	0.47	0.53	2002	417.5	0.33	0.33
1975	654.9	0.88	0.88	2003	578.5	0.77	0.74
1976	550.3	0.70	0.67	2004	715.8	0.95	0.95
1977	401	0.28	0.29	2005	676.3	0.91	0.91
1978	593.8	0.79	0.77	2006	507	0.53	0.55
1979	508	0.54	0.55	2007	736	0.98	0.96
1980	543.5	0.67	0.65	2008	677.4	0.93	0.91
1981	442.3	0.39	0.39	2009	508.9	0.56	0.56
1982	477.3	0.46	0.47	2010	418.3	0.35	0.33
1983	473	0.44	0.46	2011	721.2	0.96	0.95
1984	548.3	0.68	0.66	2012	609	0.84	0.80
1985	467.7	0.42	0.45	2013	536.3	0.63	0.63
1986	539.3	0.65	0.64	2014	608.1	0.82	0.80
1987	431.7	0.37	0.36	2015	552.4	0.72	0.67

Note: (1): year; (2): annual flow (cfs); (3): empirical CDF; (4): CDF computed through kernel density.

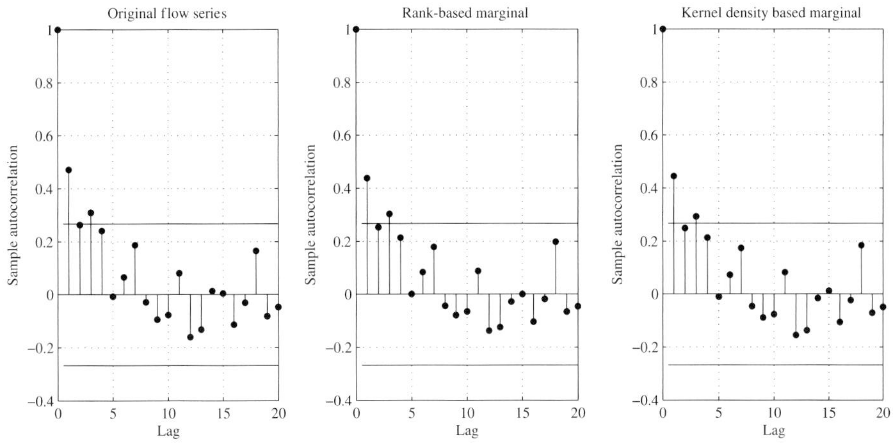

Figure 9.9 Sample autocorrelation function plots for original Flow series, rank-based and kernel density based marginals.

the marginal estimated nonparametrically will not change the structure of the time series dataset, as shown in Figure 9.9 (through the sample autocorrelation function plot).

2. Estimate the copula parameter for the first-order Markov model using Equation (9.17b).
3. Estimate the copula parameter for the first-order Markov model using Equation (9.17b).

As discussed in the previous chapters, we can first estimate the rank-based Kendall's tau for lag-1 temporal dependence. We computed $\tau \approx 0.31$. As the autoregressive coefficient estimated in Example 9.1 ($\varphi \approx 0.47$), the annual flow at the current time t is positively dependent on that at the previous time of $t-1$. Using computed sample $\tau \approx 0.31$, we obtain the initial parameter estimate:

Gumbel–Hougaard copula: $\theta_{GH}^{ini} = \frac{1}{1-0.31} \approx 1.44$

Meta-Gaussian copula: $\theta_{GAU}^{ini} = \rho^{ini} = \sin\left(\frac{\pi}{2}(0.31)\right) \approx 0.463.$

Now maximizing Equation (9.18b) for the log-likelihood functions of Gumbel–Hougaard and meta-Gaussian copulas, we have $\theta_{GH} = 1.38; \theta_{GAU} = 0.51$.

Comparing to the parameter estimated from the AR(1) model and that estimated from meta-Gaussian copula-based first-order Markov model, there is minimal difference in regard to the parameter estimated. To further compare AR(1) model and copula-based first-order Markov model, we simulate the time series with 100 realizations. We directly simulate the realizations in the real domain for the AR(1) model (with the parameters estimated in Example 9.1). For the copula-based first-order Markov process, we first simulate the marginals, and then we perform the inverse transformation with the kernel density function approach. Figure 9.10 plots the realizations from all three approaches.

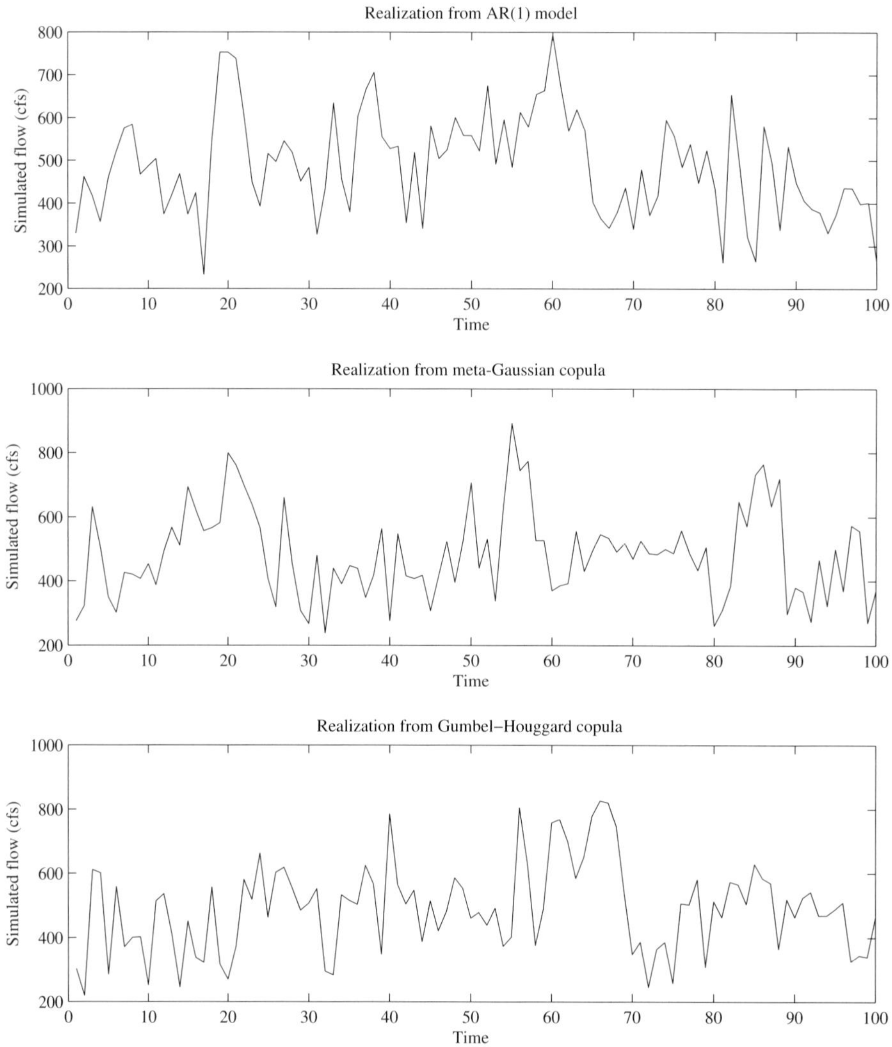

Figure 9.10 Realization from the classic AR(1) model and copula-based first-order Markov model.

One-step ahead forecast with the AR(1) model: From Equation (9.6), the corresponding one-step ahead forecast is written using the difference equation as $[Z_{t+1}] = 255.132 + 0.475[Z_t] + [\epsilon_{t+1}]$.

Substituting $[\epsilon_{t+1}] = 0$ and $z_{2015} = 552.4$ *cfs* into the forecast equation, we have the following:

$$[z_{2016}] = 255.132 + 0.475(552.4) = 517.52 \ cfs.$$

One-step ahead forecast from copula-based first-order Markov models: First, we can rewrite Equation (9.20b) in a similar fashion as the preceding AR(1) forecast equation:

$$[z_{t+1}] = F_n^{-1}\left(C^{-1}_{F_n(z_{t+1})|F_n(z_t)}(0.5|F_n(z_t);\hat{\boldsymbol{\alpha}}\right) \tag{9.22}$$

In Equation (9.22) F_n represents the marginal estimated nonparametrically through the kernel density function.

i. Gumbel–Hougaard copula: Substituting $\hat{\alpha} = 1.38$ and $F_n(z_{2015}) = 0.67$ into Equation (9.22) for Gumbel–Hougaard copula, we obtain the estimated marginal for 2016 and the corresponding forecasted annual flow as follows:

$$[F_n(z_{2016})]^{GH} = 0.56, \textit{and } [z_{2016}]^{GH} = 509.36 \textit{ cfs}$$

ii. Meta-Gaussian copula: Substituting $\hat{\alpha} = 0.514$ and $F_n(z_{2015}) = 0.67$ into Equation (8.22) for the meta-Gaussian copula, we obtain the estimated marginal for 2016 and the corresponding forecasted annual flow as follows:

$$[F_n(z_{2016})]^{GAU} = 0.591, \textit{and } [z_{2016}]^{GAU} = 521.62 \textit{ cfs}$$

Comparing the one-step ahead forecast with the AR(1) and meta-Gaussian copula-based first-order Markov model, it is seen that the relative difference of forecast results is less than 1% with the meta-Gaussian copula reaching a slightly better forecasting result.

9.5 Kth-Order Copula-Based Markov Models ($K \geq 2$)

Similar to the discussion in Section 9.4; the continuous time series $\{Z_t : t = 1, 2, \ldots\}$, modeled by the Kth-order Markov process, is fully governed by the joint distribution of $\{Z_t, Z_{t-1}, \ldots, Z_{t-k}\}$. The transition probability for the first-order Markov process (i.e., Equation (9.15)) can be rewritten as follows:

$$P(Z_t \leq z_t | Z_{t-1} = z_{t-1}, Z_{t-2} = z_{t-2}, \ldots Z_{t-K} = z_{t-K}) \tag{9.23}$$

Similar to the application of bivariate copula to study the first-order Markov models, the serial dependence of *K*th order copula-based Markov process may be fully assessed using (*K*+1)-dimensional copulas.

9.5.1 Building Copula Structure for Kth-Order Markov Models

Given the property of serial dependence for a higher-order Markov process, we need to construct a D-vine copula for the modeling purpose.

We will use a third-order Markov process as an example (Figure 9.11 similar to Figure 5.11).

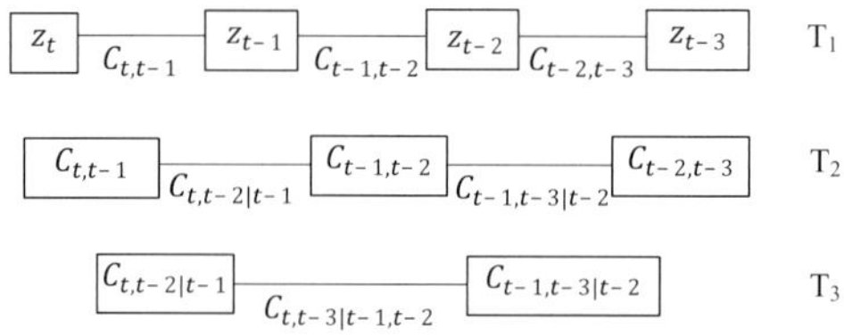

Figure 9.11 D-vine copula for a third-order Markov process.

To comply with the properties of the Markov process, Figure 9.11 shows the following:

a. The same as the first-order Markov model, z_t, z_{t-1}, z_{t-2}, and z_{t-3} have the same marginal distribution.
b. T_1 directly represents the lag-1 serial dependence, i.e., the same copula applies to $\{z_t, z_{t-1}\}, \{z_{t-1}, z_{t-2}\}, \{z_{t-2}, z_{t-3}\}$, i.e., $C_{t,t-1} \equiv C_{t-1,t-2} \equiv C_{t-2,t-3}$.
c. For the lag-2 serial dependence, we also have the same copula applying to $\{z_t, z_{t-1}, z_{t-2}\}, \{z_{t-1}, z_{t-2}, z_{t-3}\}$, i.e., $C_{t,t-1,t-2} = C_{t-1,t-2,t-3}$.
d. The copulas in b and c are differentiable.

With the same philosophy, we can model the given Kth-order Markov process using copulas.

9.5.2 Order Identification for the Markov Process

We have explained how to build a copula structure for higher-order Markov process. Here we further explain how to identify the order of Markov process for the time series with the following procedure:

i. Study the lag-1 dependence by evaluating the dependence of $\{F_{z_t}, F_{z_{t-1}}\}$. If F_{z_t} and $F_{z_{t-1}}$ are stochastically independent, we say there is no lag-1 dependence (or say z_t and z_{t-1} are independent).
ii. If step i indicates the lag-1 dependence being statistically significant, we evaluate the dependence of $\{F_{z_t}, F_{zt-2}\}$ for lag-2 dependence of z_t and z_{t-2} through the evaluation of $F_{t|t-1}(z_t|z_{t-1})$ and $F_{t-2|t-1}(z_{t-2}|z_{t-1})$ (i.e., the copula form of $C_{t|t-1}(F(z_t)|F(z_{t-1}))$ and $C_{t-2|t-1}(F(z_{t-2})|F(z_{t-1}))$.
iii. We move sequentially to higher orders until we identify that $F_{t|t-1,\ldots,t-k}$ and $F_{t-(k+1)|t-1,\ldots,t-k}$ (i.e., $C_{t|t-1,\ldots,t-k}$ and $C_{t-k-1|t-1,\ldots,t-k}$) are stochastically independent.
iv. Until now, we have successfully identified the order of Markov process, i.e., order k.

It is worth noting that Equation (9.24) will be applied to compute the conditional CDF needed. Also, similar to the first-order Markov process, we may apply the empirical marginal to the univariate time series.

9.5.3 Parameter Estimation for Kth-Order Copula-Based Markov Models

The parameters of the D-vine copula may be again estimated semiparametrically. Similar to Equation (9.16a), the transitional probability and its density function can be written as follows:

$$P(Z_t \leq z_t | Z_{t-1} = z_{t-1}, \ldots, Z_{t-k} = z_{t-k}) = C(F(z_t)|F(z_{t-1}), \ldots F(z_{t-k})) = \frac{A_k}{B_k} \quad (9.24)$$

where

$$A_k = \frac{\partial C^k(F(z_t), F(z_{t-1}), \ldots, F(z_{t-k}))}{\partial F(z_{t-1}) \ldots \partial F(z_{t-k})} \quad (9.24a)$$

$$B_k = \frac{\partial C^k(F(z_{t-1}), \ldots, F(z_{t-k}))}{\partial F(z_{t-1}) \ldots \partial F(z_{t-k})} = c(F(z_{t-1}), \ldots, F(z_{t-k})) \quad (9.24b)$$

$$h(z_t | Z_{t-1} = z_{t-1}, \ldots, Z_{t-k} = z_{t-k}) = f(z_t)c(F(z_t), \ldots, F(z_{t-k})) \quad (9.25)$$

In Equations (9.24) and (9.25), $C(.|.)$ represents the conditional copula; c represents the copula density function; and F and f represent the marginal distribution and marginal density function, respectively.

Similar to the first-order Markov process (i.e., Equation (9.18b)), the semiparametric log-likelihood function for the (k+1)-dimensional D-vine copula can be written as follows:

$$L(\boldsymbol{\alpha}) = \frac{1}{n} \sum_{t=k+1}^{m} \ln c(F_n(z_t), F_n(z_{t-1}), \ldots, F_n(z_{t-k}); \boldsymbol{\alpha}) \quad (9.26)$$

Looking at Equation (9.26), it is shown that the algebra of the copula density function may be getting complicated when the order of the Markov model needed is high. Thus, to estimate the parameters, we can proceed with two approaches: (i) sequential estimation or (ii) simultaneous estimation.

i. Sequential Estimation Approach

1. Choose and estimate the parameters of the copula candidate for the first level.
2. Compute the conditional copulas using the fitted parametric copula for the first level.
3. Choose and estimate the parameters of the copula candidates for the second level with the use of conditional copulas computed in step 2.
4. Continue these steps sequentially, until we reach the top level of the copula structure.

ii. Simultaneous Estimation Approach

Unlike the sequential approach, where we estimate the copula parameters for each level separately using the fitted copulas from the previous level, we may estimate the copula parameters of all levels simultaneously using the full semiparametric log-likelihood function of the entire vine structure as the objective function.

9.5.4 Simulation (Realizations) of the Time Series from Kth-Order Copula-Based Markov Models

Similar to the simulation for first-order copula-based Markov models discussed in Section 9.4.3, the D-vine copula simulation algorithm (i.e., algorithm 4 in Aas et al., 2009) may be modified and applied as follows:

i. Generate i.i.d. uniformly distributed random variables $U = \{u_i : i = 1, 2, \ldots N\}$, where N is the length of time series that needs to be simulated.
ii. Set $y_1 = u_1$.
iii. Set $u_2 = C(y_2|y_1) = C(y_2|u_1) = \frac{\partial C(y_2, u_1)}{\partial u_1} \Rightarrow y_2 = h^{-1}(u_2, u_1; \boldsymbol{\alpha}_{12})$.
iv. Based on the general formula (i.e., Equation 5.24), set

$$u_3 = C(y_3|y_1, y_2) = \frac{\partial C_{13|2}(C(y_3|y_2; \boldsymbol{\alpha}_{12}), C(y_1|y_2; \boldsymbol{\alpha}_{12}))}{\partial C(y_1|y_2; \boldsymbol{\alpha}_{12})}$$

v. Continue until we obtain y_N by setting $u_N = C(y_N|y_{N-1}, y_{N-2}, \ldots y_{N-K}; \boldsymbol{\alpha})$.

Now, we have simulated the desired Kth-order Markov process in the frequency domain. Again, similar to the simulation of the first-order Markov model, the sequence simulated in the frequency domain needs to be transformed into the real-domain through either parametric marginal distribution or nonparametric marginal distribution (e.g., empirical distribution using plotting-position formulas and kernel density) with proper interpolation.

9.5.5 Forecast and Quantile Estimation of Kth-order Copula-Based Markov Models

For the Kth-order copula-based Markov process, the first-order median forecast formula (i.e., Equation (9.19)) can be rewritten as follows:

$$\begin{aligned} E[Z_t|Z_{t-1} = z_{t-1}, \ldots, Z_{t-K} = z_{t-K}] &= \int z_t f^*(z_t) c(F^*(z_t), \ldots, F^*(z_{t-K}), \boldsymbol{\alpha}) dz_t \\ &= \int z_t c(F^*(z_t), \ldots, F^*(z_{t-K})) dF^*(z_t) \end{aligned} \quad (9.27)$$

Replacing the unknown true marginal distribution (F^*) by its nonparametric marginal distribution (F_n), and true copula parameter vector $\boldsymbol{\alpha}$ **by** the estimated parameter $\hat{\boldsymbol{\alpha}}$, Equation (9.27) can be rewritten as follows:

$$E[Z_t|Z_{t-1} = z_{t-1}, \ldots, Z_{t-K} = z_{t-K}] = \int z_t c(F_n(z_t), \ldots, F_n(z_{t-K}); \hat{\boldsymbol{\alpha}}) dF_n(z_t) \quad (9.27a)$$

Similar to Equation (9.19a), Equation (9.27a) of the conditional probability (also called the conditional copula) of $Z_t| Z_{t-1} = z_{t-1}, \ldots, Z_{t-K} = z_{t-K}$ is equal to 0.5 (also called the 50% conditional quantile). The median forecast can be computed using the following:

$$\hat{z}_t = F_n^{-1}(C_{F_n(z_t)|F_n(z_{t-1}, \ldots, z_{t-K})}(0.5|F_n(z_{t-1}), \ldots, F_n(z_{t-K}); \hat{\boldsymbol{\alpha}}) \quad (9.28)$$

Furthermore, for any given conditional quantile q, its associated time series value may be computed using the following:

$$\hat{z}_t^q = F_n^{-1}\left(C^{-1}_{F_n(z_t)|F_n(z_{t-1}),\ldots,F_n(z_{t-K})}(q|F_n(z_{t-1}),\ldots,F_n(z_{t-K});\hat{\boldsymbol{\alpha}})\right) \tag{9.29}$$

Example 9.4 **Rework TS2 series in Example 9.2 using (i) meta-Gaussian and (ii) Frank copulas. Also, compare the results with those from AR(2) model in Example 9.2.**

Solution: According to Example 9.2, the time series TS2 is fitted with the classic AR(2) model. We will proceed with the following procedure:

i. Identify the Markov order for the time series:
- Lag-1 dependence: To assess the lag-1 dependence (i.e., $TS2_t$ and $TS2_{t-1}$), we will simply compute the rank-based Kendall correlation coefficient and assess its significance (using the critical value of $\alpha = 0.05$).

  Using Equation (3.68) (or simply using the MATLAB function corr), we compute the following:

$$\tau_n^{Lag1} = 0.5765, Pvalue \leq 10^{-4}$$

  Results show that the lag-1 dependence is significant.
- Lag-2 dependence: To assess the lag-2 dependence (i.e., $TS2_t$ *and* $TS2_{t-2}$), we need to evaluate the dependence of $F_{t|t-1}(z_t|z_{t-1})$ and $F_{t-2|t-1}(z_{t-2}|z_{t-1})$. Thus, we first need to estimate the conditional distribution (or simply conditional copula) with the understanding that $\{z_t, z_{t-1}\}$ and $\{z_{t-1}, z_{t-2}\}$ have the same copula. Using meta-Gaussian copula, we can simply estimate the copula parameter through Kendall's tau, previously computed as follows:

$$\rho = \sin\left(\frac{\pi}{2}\tau_n^{lag_1}\right) = 0.7868$$

  We now build the bivariate Gaussian copula for the lag-1 sequence and compute the corresponding conditional copulas. The conditional copula results are listed in Table 9.9. The Kendall correlation coefficient is computed as follows:

$$\tau_n[F(z_t|z_{t-1}), F(z_{t-2}|z_{t-1})] = -0.2748, Pvalue \approx 0.006$$

  Now, we conclude that lag-2 dependence is still significant. We will need to move on to the evaluation of lag-3 dependence.
- Lag-3 dependence: Similar to the lag-2 dependence assessment, we will need to evaluate the rank-based conditional dependence of the following:

$$\{F(z_t|z_{t-1}, z_{t-2}),\ F(z_{t-3}|z_{t-1}, z_{t-2})\}$$

  In the preceding formulation, we can further write the two components using Equation (5.24) as follows:

Table 9.10. *Markov order identification results table.*

Time series		Lag-2		Lag-3					
TS2	CDF	(1)	(2)	(3)	(4)	(5)	(6)	(7)	(8)
		$\tau \approx -0.27$						$\tau \approx 0.1674$	
		$P \approx 0.0056$						$P \approx 0.099$	
476.35	0.72								
475.48	0.49								
476.02	0.64	0.72	0.84						
476.75	0.81	0.84	0.32	0.84	0.32	0.72	0.84	0.81	0.91
476.89	0.84	0.68	0.29	0.68	0.29	0.84	0.32	0.60	0.47
477.27	0.90	0.79	0.57	0.79	0.57	0.68	0.29	0.83	0.34
477.51	0.93	0.76	0.50	0.76	0.50	0.79	0.57	0.79	0.71
476.50	0.76	0.24	0.59	0.24	0.59	0.76	0.50	0.25	0.63
475.87	0.60	0.31	0.93	0.31	0.93	0.24	0.59	0.57	0.47
475.74	0.56	0.47	0.80	0.47	0.80	0.31	0.93	0.63	0.91
475.29	0.45	0.34	0.58	0.34	0.58	0.47	0.80	0.36	0.81
476.75	0.81	0.95	0.66	0.95	0.66	0.34	0.58	0.98	0.51
476.60	0.78	0.55	0.09	0.55	0.09	0.95	0.66	0.30	0.89
475.69	0.55	0.21	0.67	0.21	0.67	0.55	0.09	0.25	0.08
474.94	0.37	0.24	0.86	0.24	0.86	0.21	0.67	0.41	0.55
473.83	0.19	0.15	0.73	0.15	0.73	0.24	0.86	0.20	0.81
473.63	0.16	0.31	0.72	0.31	0.72	0.15	0.73	0.40	0.59
473.84	0.19	0.43	0.43	0.43	0.43	0.31	0.72	0.39	0.67
475.58	0.52	0.89	0.31	0.89	0.31	0.43	0.43	0.87	0.39
476.17	0.68	0.75	0.07	0.75	0.07	0.89	0.31	0.50	0.51
475.11	0.40	0.16	0.31	0.16	0.31	0.75	0.07	0.09	0.09
476.29	0.71	0.88	0.85	0.88	0.85	0.16	0.31	0.97	0.16
478.09	0.97	0.99	0.14	0.99	0.14	0.88	0.85	0.98	0.96
478.04	0.97	0.72	0.07	0.72	0.07	0.99	0.14	0.46	0.45
476.20	0.68	0.06	0.76	0.06	0.76	0.72	0.07	0.08	0.08
475.13	0.41	0.16	0.99	0.16	0.99	0.06	0.76	0.53	0.52
475.01	0.38	0.43	0.86	0.43	0.86	0.16	0.99	0.63	0.98
475.44	0.48	0.62	0.50	0.62	0.50	0.43	0.86	0.63	0.86
475.49	0.49	0.51	0.34	0.51	0.34	0.62	0.50	0.43	0.56
475.91	0.61	0.68	0.48	0.68	0.48	0.51	0.34	0.69	0.33
476.52	0.76	0.79	0.35	0.79	0.35	0.68	0.48	0.76	0.56
475.51	0.50	0.18	0.32	0.18	0.32	0.79	0.35	0.10	0.48
474.24	0.25	0.13	0.88	0.13	0.88	0.18	0.32	0.25	0.18
473.57	0.15	0.21	0.81	0.21	0.81	0.13	0.88	0.32	0.78
473.33	0.12	0.28	0.58	0.28	0.58	0.21	0.81	0.30	0.72
473.42	0.13	0.37	0.42	0.37	0.42	0.28	0.58	0.32	0.49
473.87	0.19	0.50	0.32	0.50	0.32	0.37	0.42	0.41	0.36

Table 9.10. (*cont.*)

Time series		Lag-2		Lag-3					
TS2	CDF	(1) $\tau \approx -0.27$ $P \approx 0.0056$	(2)	(3)	(4)	(5)	(6)	(7) $\tau \approx 0.1674$ $P \approx 0.099$	(8)
471.65	0.03	0.03	0.24	0.03	0.24	0.50	0.32	0.01	0.31
470.84	0.01	0.11	0.84	0.11	0.84	0.03	0.24	0.19	0.05
472.87	0.08	0.71	0.42	0.71	0.42	0.11	0.84	0.70	0.70
473.17	0.10	0.41	0.03	0.41	0.03	0.71	0.42	0.12	0.51
474.16	0.23	0.66	0.24	0.66	0.24	0.41	0.03	0.55	0.02
475.32	0.45	0.77	0.13	0.77	0.13	0.66	0.24	0.60	0.28
475.63	0.53	0.61	0.15	0.61	0.15	0.77	0.13	0.42	0.19
475.90	0.60	0.63	0.38	0.63	0.38	0.61	0.15	0.58	0.16
474.64	0.31	0.13	0.42	0.13	0.42	0.63	0.38	0.08	0.43
474.19	0.24	0.30	0.85	0.30	0.85	0.13	0.42	0.48	0.23
476.01	0.63	0.93	0.54	0.93	0.54	0.30	0.85	0.95	0.82
476.85	0.83	0.87	0.06	0.87	0.06	0.93	0.54	0.68	0.79
476.79	0.82	0.60	0.25	0.60	0.25	0.87	0.06	0.48	0.11

Note: Lag-2: (1) $F(z_t|z_{t-1})$; (2) $F(z_{t-1}|z_{t-2})$
Lag-3: (3) $F(z_t|z_{t-1})$; (4) $F(z_{t-2}|z_{t-1})$; (5) $F(z_{t-1}|z_{t-2})$; (6) $F(z_{t-3}|z_{t-2})$
(7) $F(z_t|z_{t-1}, z_{t-2})$; (8) $F(z_{t-3}|z_{t-1}, z_{t-2})$

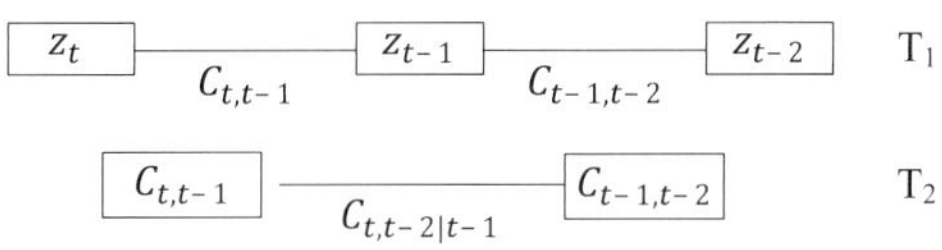

Figure 9.12 Vine-copula structure for the second-order copula-based Markov process.

$$F(z_t|z_{t-1}, z_{t-2}) = \frac{\partial C_{z_t, z_{t-2}|z_{t-1}}\left(C_{z_t|z_{t-1}}, C_{z_{t-2}|z_{t-1}}\right)}{\partial C_{z_{t-2}|z_{t-1}}} \tag{9.30a}$$

$$F(z_{t-3}|z_{t-1}, z_{t-2}) = \frac{\partial C_{z_{t-3}, z_{t-1}|z_{t-2}}\left(C_{z_{t-3}|z_{t-2}}, C_{z_{t-1}|z_{t-2}}\right)}{\partial C_{z_{t-1}|z_{t-2}}} \tag{9.30b}$$

To compute the conditional probability for Equations (9.30a) and (9.30b), we apply the meta-Gaussian copula first using the Gaussian copula fitted to the lag-2 dependence assessment. The conditional distribution for $C_{z_t|z_{t-1}}, C_{z_{t-2}|z_{t-1}}, C_{z_{t-1}|z_{t-2}}, C_{z_{t-3}|z_{t-2}}$ is also listed in Table 9.10.

Table 9.11. *Parameter estimation results.*

Time series			Meta-Gaussian		Frank	
Lag-2	Lag-1	TS2	$C_{t\|t-1}$	$C_{t-2\|t-1}$	$C_{t\|t-1}$	$C_{t-2\|t-1}$
$(t-2)$	$(t-1)$	t	$\rho^{lag1} = 0.8265$		$\theta^{lag1} = 7.9422$	
NaN	NaN	0.72				
NaN	0.72	0.49				
0.72	0.49	0.64	0.74	0.86	0.77	0.87
0.49	0.64	0.81	0.86	0.29	0.84	0.24
0.64	0.81	0.84	0.68	0.25	0.63	0.21
0.81	0.84	0.90	0.79	0.55	0.74	0.51
0.84	0.90	0.93	0.76	0.46	0.74	0.47
0.90	0.93	0.76	0.19	0.55	0.24	0.59
0.93	0.76	0.60	0.28	0.94	0.22	0.89
0.76	0.60	0.56	0.46	0.81	0.43	0.81
0.60	0.56	0.45	0.32	0.58	0.28	0.58
0.56	0.45	0.81	0.96	0.68	0.96	0.72
0.45	0.81	0.78	0.53	0.06	0.49	0.05
0.81	0.78	0.55	0.18	0.67	0.14	0.62
0.78	0.55	0.37	0.22	0.88	0.19	0.88
0.55	0.37	0.19	0.14	0.76	0.15	0.81
0.37	0.19	0.16	0.32	0.76	0.37	0.80
0.19	0.16	0.19	0.45	0.45	0.49	0.49
0.16	0.19	0.52	0.92	0.32	0.93	0.36
0.19	0.52	0.68	0.77	0.05	0.79	0.05
0.52	0.68	0.40	0.14	0.28	0.10	0.22
0.68	0.40	0.71	0.91	0.88	0.93	0.90
0.40	0.71	0.97	0.99	0.11	0.97	0.08
0.71	0.97	0.97	0.69	0.04	0.81	0.12
0.97	0.97	0.68	0.03	0.74	0.10	0.83
0.97	0.68	0.41	0.13	0.99	0.10	0.98
0.68	0.41	0.38	0.43	0.88	0.44	0.91
0.41	0.38	0.48	0.64	0.51	0.69	0.54
0.38	0.48	0.49	0.52	0.32	0.52	0.30
0.48	0.49	0.61	0.69	0.48	0.72	0.47
0.49	0.61	0.76	0.81	0.33	0.80	0.29
0.61	0.76	0.50	0.15	0.29	0.11	0.23
0.76	0.50	0.25	0.11	0.90	0.10	0.90
0.50	0.25	0.15	0.20	0.84	0.24	0.88
0.25	0.15	0.12	0.29	0.62	0.33	0.65
0.15	0.12	0.13	0.39	0.45	0.41	0.47
0.12	0.13	0.19	0.53	0.33	0.55	0.36

Table 9.11. (*cont.*)

Time series			Meta-Gaussian		Frank	
Lag-2	Lag-1	TS2	$C_{t\|t-1}$	$C_{t-2\|t-1}$	$C_{t\|t-1}$	$C_{t-2\|t-1}$
0.13	0.19	0.03	0.02	0.24	0.05	0.29
0.19	0.03	0.01	0.11	0.89	0.08	0.74
0.03	0.01	0.08	0.78	0.47	0.44	0.19
0.01	0.08	0.10	0.44	0.03	0.41	0.05
0.08	0.10	0.23	0.71	0.25	0.70	0.27
0.10	0.23	0.45	0.80	0.12	0.85	0.17
0.23	0.45	0.53	0.63	0.13	0.66	0.13
0.45	0.53	0.60	0.64	0.37	0.65	0.34
0.53	0.60	0.31	0.10	0.40	0.08	0.36
0.60	0.31	0.24	0.29	0.88	0.32	0.91
0.31	0.24	0.63	0.95	0.57	0.96	0.62
0.24	0.63	0.83	0.89	0.04	0.87	0.04
0.63	0.83	0.82	0.59	0.21	0.55	0.18

With these conditional probabilities computed, we estimate the copula parameters for $C_{z_t, z_{t-2}|z_{t-1}}, C_{z_{t3}, _{t1|t2}}$ that are –0.45 and –0.42 respectively. Now, we can compute $F(z_t|z_{t-1}, z_{t-2}), F(z_{t-3}|z_{t-1}, z_{t-2})$, which are again listed in Table 9.10. Now Kendall's correlation coefficient is computed as follows:

$$\tau_n[F(z_t|z_{t-1}, z_{t-2}), F(z_{t-3}|z_{t-1}, z_{t-2})] = 0.1674, Pvalue \approx 0.099 > 0.05$$

It is then concluded that it is reasonable to apply the second-order copula-based Markov process for the dataset.

ii. Estimate the parameters for the second-order copula-based Markov model.

From the results obtained in step i, we know that a trivariate copula will be needed to model the second-order Markov process for time series TS2, for which the schematic for the trivariate D-vine copula is shown in Figure 9.12. As given in the problem statement, the parameters of the meta-Gaussian, the Student t, and the Frank-vine copula structures will be estimated with the use of sequential parameter estimation.

Meta Gaussian vine copula structure:
In this case, we apply the meta-Gaussian copulas (Equation (7.40)) for T_1 and T_2. We will use the parameters estimated nonparametrically for lag-dependence evaluation as initial parameters.

T_1: using $\rho^{in} = 0.7868$ for $\{z_t, z_{t-1}\}$ & $\{z_{t-1}, z_{t-2}\}$ and applying the semiparametric MLE, we have the parameter estimated for the lag-1 serial dependence of T_1 as $\rho_{T_1} = 0.8265$.

T_2. Fix the parameter estimated for T_1 to compute the conditional copula of $C_{t|t-1}, C_{t-1|t-2}$ (listed in Table 9.11). Finally, using the computed conditional copula of bivariate variables, we estimate the parameter for T_2 as $\rho_{T_2} = -0.3961$.

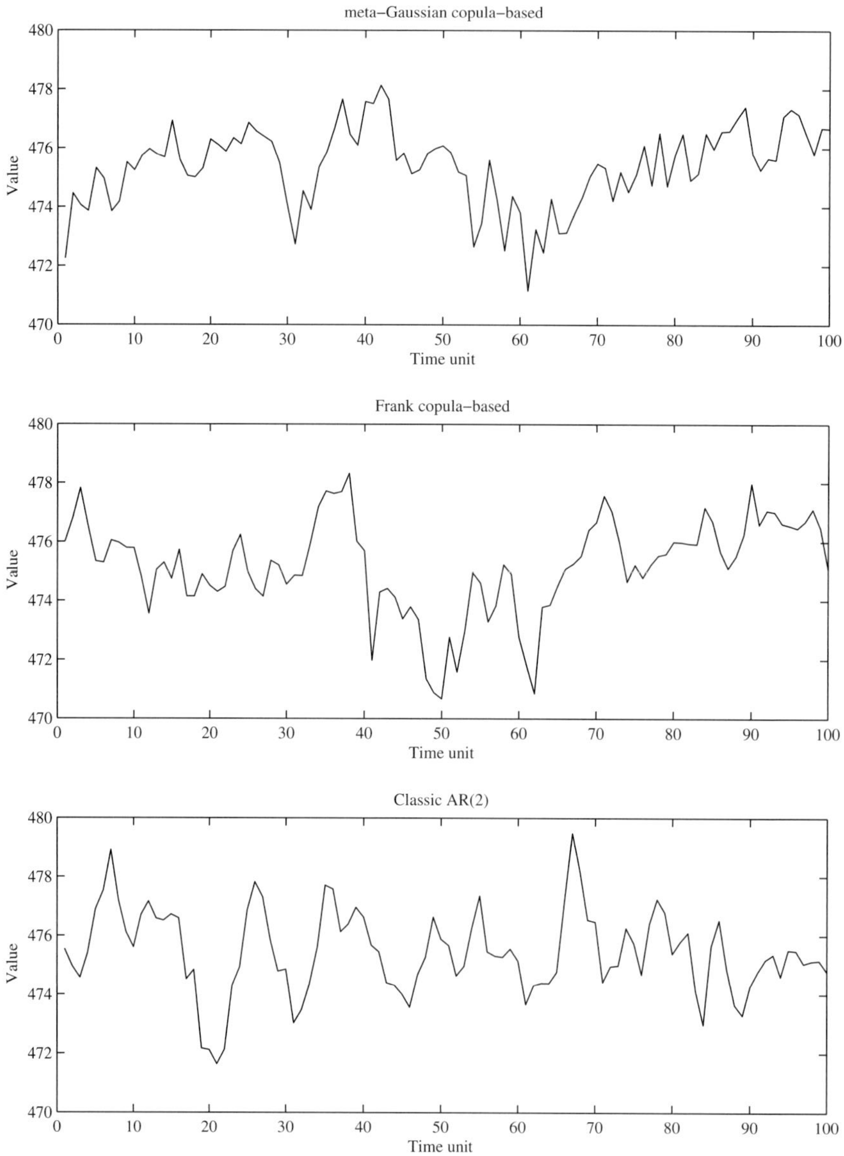

Figure 9.13 Simulations from AR(2) and copula-based second-order Markov models.

Frank vine copula structure:
Using the same procedure as that for the meta-Gaussian copula, we again estimate the parameters for the Frank copula (Copula No. 5, Table 3.1) using the semiparametric MLE as follows:

$$\alpha_{T_1} = 7.9422; \quad \alpha_{T_2} = -2.4461.$$

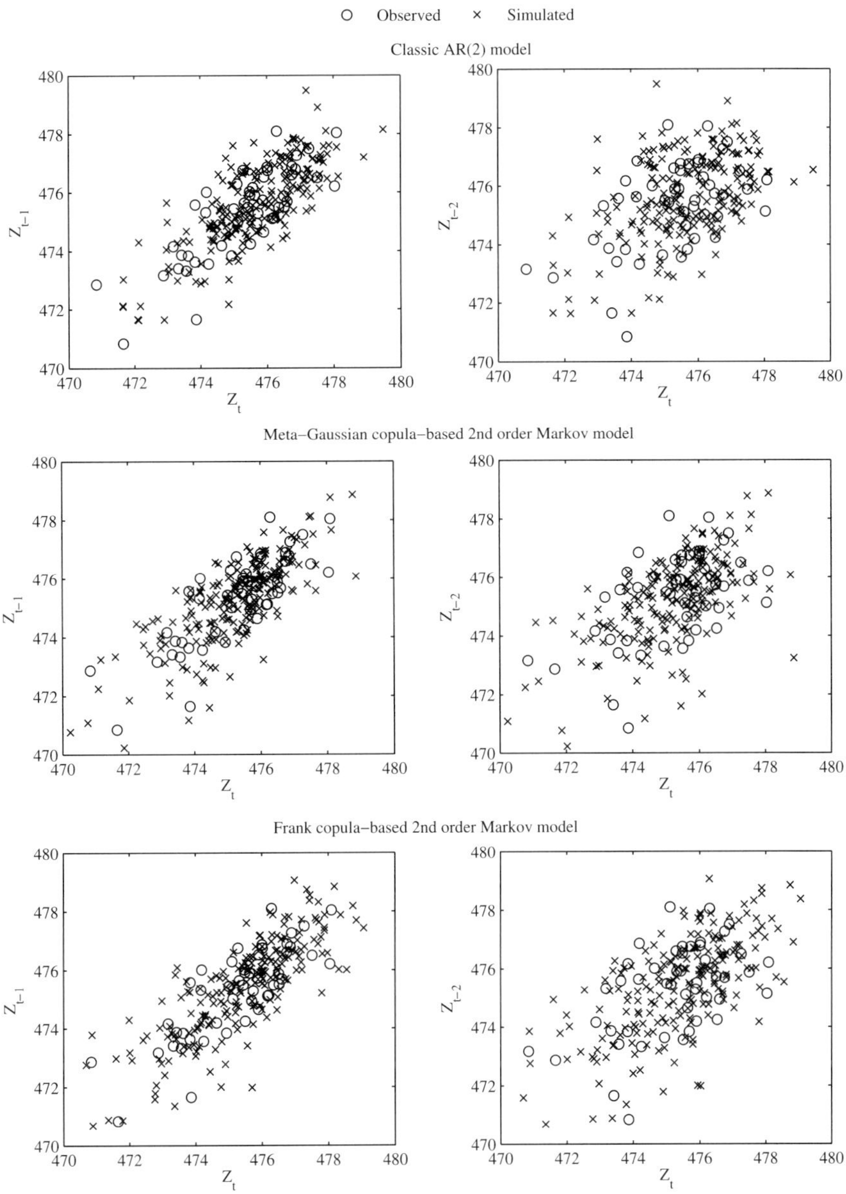

Figure 9.14 Lag-1 and lag-2 dependence comparison of simulated time series to the orginal time series TS2.

Again, the conditional copula needed for the parameter estimation for T_2 is listed in Table 9.11.

iii. Simulate the univariate time series.

Compared to the simulation for the first-order copula-based Markov process, we will need to simulate variate y_3 through Equation (9.30b). Following the simulation discussed in Section 9.5.4, Figure 9.13 shows simulations from the copula-based model as well as the

simple AR(2) model. To further compare the classic AR(2) model to the second-order copula-based Markov model, we perform the one-step ahead forecast using exactly the same rationale as in Example 9.3.
One-step ahead forecast from AR(2): $\widehat{TS2}_{51} \approx 476.40$.
One-step ahead forecast from the Gaussian copula-based second-order Markov model: $\hat{F}(TS2_{51}) = 0.7512; \widehat{TS2}_{51}^{Gaussian} = 476.472$
One-step ahead forecast from the Frank copula-based second-order Markov model: $\hat{F}(TS2_{51}) = 0.7719; \widehat{TS2}_{51}^{Gaussian} = 476.562$

From the forecast results, it is seen that there is minimal difference between the classic AR(2) model and copula-based models. Given the time series data applied in Examples 9.2 and 9.4 as the synthetic time series generated from the AR(2) model, it is no surprise that overall the Gaussian copula-based model performs more similarly to the AR(2) model. Figure 9.14 plots the scatter plots for the lag-1 and lag-2 dependences. Figure 9.14 shows that the copula-based Markov model captures the serial dependence well.

9.6 Summary

This chapter further reveals the advantages of the copula theory not only in traditional frequency analysis but also in time series analysis:

i. It allows the investigation of spatial and temporal dependences separately from their marginals and their effect.
ii. It is more robust for to modeling any type of temporal (serial dependence) and avoids the Gaussian process assumption of the time series modeling approach.
iii. It provides a better approach to identify the necessary order for the Markov process.
iv. Vine copula may be easily applied to model a higher-order Markov process.
v. These advantages are very important for the hydrological analysis under the impact of climate change and land use/land cover (LULC) when the univariate hydrological variables may no longer be considered as independent random variables.

References

Aas, K., Czado, C., Frigessi, A., and Bakken, H. (2009). Pair-copula constructions of multiple dependence. *Insurance: Mathematics and Economics*, 44, 182–198. doi:10.1016/j.insmatheco.2007.02.001.

Arya, F. K. and Zhang, L. (2015). Time series analysis of water quality parameters at Stillaguamish River using order series method. *Stochastic Environmental Research and Risk Assessment*, 29, 227. doi:10.1007/s00477–014–0907–2.

Bollerslev, T. (1986). Generalized autoregressive conditional heteroskedasticity. *Journal of Econometrics*, 31, 307–327.

Box, G. E. P., Jenkins, G. M., and Reinsel, G. C. (2008). *Time Series Analysis: Forecasting and Control*. John Wiley & Sons, Inc., Hoboken.

Chen, X. and Fan, Y. (2006). Estimation and model selection of semiparametric copula-based multivariate dynamic models under copula misspecification. *Journal of Econometrics*, 135, 125–154.

Darsow, W., Nguyen, B., and Olsen, E. (1992). Copulas and Markov processes. *Illinois Journal of Mathematics,* 36, 600–642.

Dickey, D. A. and Fuller, W. A. (1979). Distribution of the estimates for autoregressive time series with a unit root. *Journal of American Statistical Association*, 74, 427–431.

Hurst, H. E. (1951). Long-term storage capacity of reservoirs. *Transactions of American Society of Civil Engineers*, 116, 770.

Kwiatkowski, D., Phillips, P. C. B., Schmidt, P., and Shin, Y. (1992). Testing the null hypothesis of stationarity against the alternative of a unit root. *Journal of Econometrics*, 54, 159–178.

Joe, H. (1997). *Multivariate Models and Multivariate Dependence Concepts*. Chapman & Hall/CRC, New York.

Part Two

Applications

10

Rainfall Frequency Analysis

ABSTRACT

In this chapter, we will illustrate the application of copulas in rainfall frequency analysis. This chapter is divided into two parts: (1) rainfall depth-duration frequency (DDF) analysis; and (2) multivariate rainfall frequency (i.e., four-dimensional) analysis. The rainfall data from the watersheds in the United States are collected and applied for analyses. The Archimedean, meta-elliptical, and vine copulas are applied to model the dependence among rainfall variables. Application shows that the DDF may be modeled by the Gumbel–Hougaard copula. Both vine and meta-elliptical copulas may be applied to model the spatial dependence of rainfall variables. Compared to the vine copula, modeling is easier to do when applying the meta-elliptical copula.

10.1 Introduction

Rainfall frequency analysis is of fundamental importance for hydrologic and hydraulic engineering design. In what follows, we will first introduce some examples with regard to rainfall analysis. Rainfall intensity-duration-frequency (IDF) or rainfall depth-duration frequency (DDF) curves published by National Oceanographic Atmospheric Administration (NOAA) are classic examples of rainfall frequency analysis. The IDF (or DDF) curves are derived first by separating rainfall events based on their durations (e.g., 15 minutes, 30 minutes, one hour, etc.) and then by fitting a univariate probability distribution to the rainfall depth or intensity data of a certain duration. The fitted univariate distribution is applied to produce a family of rainfall depth-frequency curves. In this manner, the two-dimensional depth-duration analysis is reduced to a one-dimensional analysis, involving only intensity (or depth) corresponding to a fixed duration. As described by the NOAA documents (e.g., TP-40), the IDF (or DDF) curves may be estimated from either annual maximum series or partial duration series. The IDF (or DDF) curves are widely applied in hydrological and hydraulic engineering design.

The rational method relates rainfall intensity (I) of a given duration (normally equal to the time of concentration) of a certain return period to peak runoff (discharge) (Q), where the peak runoff is assumed as a linear function of rainfall ($Q = CIA$), where A is the area of the drainage basin. In this method, rainfall of a certain return period results in the runoff peak of exactly the same return period. To date, the rational method is commonly applied in urban hydrology (e.g., urban rainfall and runoff analysis) and urban hydraulic engineering design (e.g., detention/retention basin design, storm sewer design, and highway drainage design).

The SCS method, developed by Soil Conservation Service (now, the Natural Resources Conservation Service), may be applied to larger areas compared to the rational method (usually less than 60 acres [about 25 hectares]) for estimating runoff of a given rainfall amount. This method estimates the amount of surface runoff (or excess rainfall) through what is called the Curve Number (CN), which is related to land use and land cover, antecedent soil moisture, hydrologic condition, and soil moisture retention capacity.

The probable maximum precipitation (PMP) method, which does not rely on the IDF (DDF) curve, estimates the maximum amount of precipitation that may probably occur. The PMP analysis is required for the design of dams, dam breach analysis, spillway analysis, design of nuclear power plants, etc. These examples may be considered to illustrate applications of univariate rainfall analysis in hydrologic and hydraulic engineering design.

In the past three decades, bivariate (and multivariate) rainfall frequency analysis has attracted significant attention, because rainfall variables may be correlated and may significantly affect surface runoff (Cordova and Rodriguez-Iturbe, 1985). In the early days, the bivariate exponential distribution was applied to model the correlation structure of extreme rainfall variables (e.g., Hashino, 1985; Singh and Singh, 1991; Bacchi et al., 1994). Later, other bivariate rainfall models were investigated to model the relation between rainfall intensity and rainfall duration, for example, improved derived flood frequency distribution (DFFD) model by Kurothe et al. (1997) and Goel et al. (2000); Yue (2000a, 2000b, 2000c) investigated the applicability of bivariate normal, Gumbel logistic, and Gumbel mixed distributions. Besides the application to river discharge (Favre et al., 2004), the copula theory has been applied to bivariate and multivariate rainfall analysis (Grimaldi et al., 2005; Zhang and Singh, 2007a, 2007b, 2007c; Kao and Govindaraju, 2007, 2008; Cong and Brady, 2012; Zhang et al., 2012; Hao and Singh, 2013; Zhang et al., 2013; Abdul Rauf and Zeephongsekul, 2014; Cantet and Arnaud, 2014; Khedun et al., 2014; Moazami et al., 2014; Vernieuwe et al., 2015; among others).

With the advantages of the copula theory discussed in the preceding chapters, we will illustrate the application of copula theory to bivariate (or multivariate) rainfall frequency analysis. It is assumed that rainfall variables are continuous variates. However, rainfall variables may actually be discrete in nature.

10.2 Rainfall Depth-Duration Frequency (DDF) Analysis

Many studies have employed copulas for bivariate (multivariate) rainfall analysis based on annual maximum series (AMS). In this section, we will use Partial Duration Series (PDS) to illustrate the copula application to derive the DDF curves. The rainfall data with a 15-minute interval were collected for the rain gauge station: coop-166394 near Morgan City, Louisiana. The recorded data cover a period from May 8, 1971, to January 1, 2014. The rainfall data are available upon request from National Climate Data Center (NCDC).

The general procedure for DDF analysis includes the following steps:

1. Separate the rainfall records collected into independent rainfall events. Extract the rainfall depth and rainfall duration from these independent rainfall events obtained.
2. Evaluate the marginal rainfall depth and rainfall duration variables and corresponding marginal distributions.
3. Evaluate the rank-based correlation of rainfall depth and rainfall duration. Choose the possible copula candidates.
4. Perform the rainfall depth and rainfall duration analysis with the use of the possible copula candidates. Select the best-fitted copula functions.
5. Estimate the rainfall depth of given rainfall duration for a given return period.

In what follows, we will discuss how to perform the DDF analysis in detail.

10.2.1 Rainfall Data Processing

Before analyzing bivariate rainfall variables (i.e., rainfall depth and duration), we need to separate the rainfall data into individual rainfall events first. As commonly done, a six-hour duration of no rain is considered as the criterion to separate any two events. From a total of 12,089 available rainfall records for rain gage coop-166394, a total of 2,816 events were identified for the 43-year duration. Table 10.1 illustrates the rainfall event separation year 1971 as an example. From Table 10.1, it can be seen that there are nine independent rainfall events identified from May 22, 1971, to June 29, 1971. Table 10.2 lists the nine rainfall events separated. As an example, we consider the No. 5 event, which started on June 20, 1971, at 14:15 and ended on 15:15 on the same day. Summing up the incremental rainfall depths within this time window, we have the following:

$$\begin{aligned}&\text{depth} = 2.54 + 2.54 + 10.16 + 7.62 = 22.86 \text{ mm}.\\&\text{maximum rainfall intensity} = 10.16/0.25 = 40.64 \text{ mm/h}.\\&\text{duration} = 1 \text{ h}.\end{aligned}$$

Similarly, all rainfall events may be separated based on the six-hour duration of no rain criterion. Setting the threshold for the identified rainfall events as follows:

$$Threshold = median\ (rainfall\ depth) + std\ (rainfall\ depth) \tag{10.1}$$

From the record, we have median = 7.62 mm and standard deviation = 23.87 mm, which yield the threshold = 31.49 mm. Applying this threshold, we reduced the number of rainfall events to 378 that is roughly about nine events per year. With the partial duration rainfall series thus identified, we can then start to investigate bivariate rainfall characteristics through (i) the investigation of the marginal distribution and (ii) the investigation of dependence.

10.2.2 Investigation of Marginal Distributions: Depth and Duration

Before we investigate marginal distributions, we will first look at a scatter plot of rainfall variables (Figure 10.1(a)). Zooming in on the lower-right corner (Figure 10.1(b)), we see there are ties in both rainfall depth and rainfall duration variables. The kernel density

Table 10.1. *Illustration of rainfall event separation.*

Event no.	Date	Rainfall amount (mm)[a]	Interarrival time (h)[b]
	19710522 15:30	7.62	—
	19710601 00:15	0	—
1	19710605 16:15	2.54	336.75
2	19710616 13:45	2.54	261.50
3	19710618 16:00	5.08	50.25
	19710618 17:30	2.54	1.50
4	19710619 13:30	2.54	20.00
	19710619 13:45	2.54	0.25
	19710619 14:30	2.54	0.75
5	19710620 14:30	2.54	24.00
	19710620 14:45	2.54	0.25
	19710620 15:00	10.16	0.25
	19710620 15:15	7.62	0.25
6	19710621 10:00	2.54	18.75
	19710621 10:15	7.62	0.25
7	19710622 20:00	12.7	33.75
	19710622 20:15	5.08	0.25
	19710622 20:30	5.08	0.25
	19710622 20:45	5.08	0.25
	19710622 21:00	2.54	0.25
	19710622 21:15	2.54	0.25
	19710622 22:00	2.54	0.75
	19710622 22:30	2.54	0.50
8	19710624 15:30	2.54	41.00
9	19710629 12:00	5.08	116.50

Notes: [a] Incremental rainfall depth with 15-minute interval until the time stated.
[b] Difference between current day and time with the previous day and time.

(nonparametric probability density estimation (Wand and Jones, 1995) is applied to approximate the nonparametric probability density and distribution function for univariate rainfall variables. The kernel density function is given as follows:

$$\hat{f}(x) = \frac{1}{nh}\sum_{i=1}^{n} K\left(\frac{x - x_i}{h}\right) \tag{10.2}$$

In Equation (10.2), $K(.)$ is the kernel function. Here we use the commonly applied $K(x) = \phi(x)$, i.e., the normal kernel (the normal density function); h is the smoothing

Table 10.2. *Nine rainfall events separated based on Table 10.1.*

No	Depth (mm)	Duration (h)	Max. Intensity (mm/h)[a]	Start	End
1	2.54	0.25	10.16	6/5/71 16:00	6/5/71 16:15
2	2.54	0.25	10.16	6/16/71 13:30	6/16/71 13:45
3	7.62	1.75	20.32	6/18/71 15:45	6/18/71 17:30
4	7.62	1.25	10.16	6/19/71 13:15	6/19/71 14:30
5	22.86	1.00	40.64	6/20/71 14:15	6/20/71 15:15
6	10.16	0.50	30.48	6/21/71 9:45	6/21/71 10:15
7	38.1	2.75	50.8	6/22/71 19:45	6/22/71 22:30
8	2.54	0.25	10.16	6/24/71 15:14	6/24/71 15:30
9	5.08	0.25	20.32	6/29/71 11:45	6/29/71 12:00

Note: [a] Maximum average intensity of 15-minute interval.

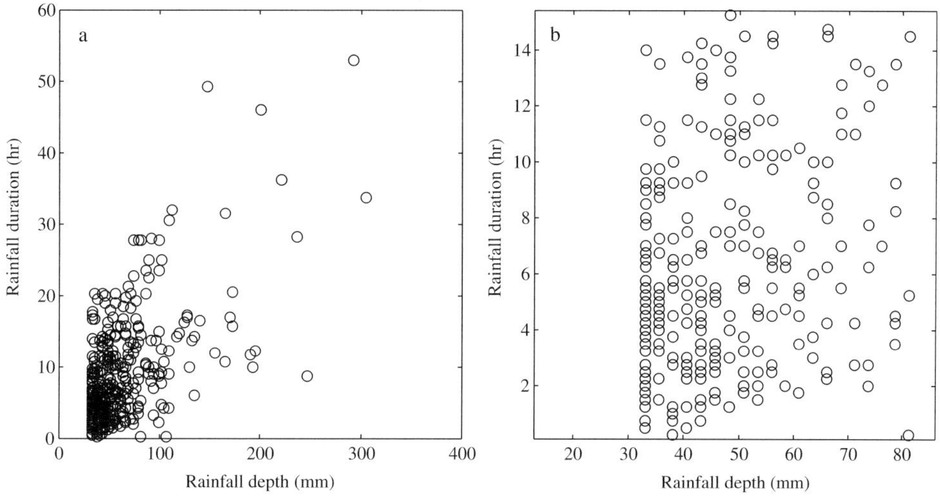

Figure 10.1 Scatter plot for rainfall depth and rainfall duration: (a) original; (b) zoomed in at lower-right corner.

parameter, which is also called bandwidth ($h = 6.086$ mm,1.797 hr for rainfall depth and rainfall duration respectively); and n is the sample size.

To compute the probability density and marginal probability using the kernel density, the MATLAB function is applied as follows:

$$pdf = ksdensity(x, x_1, 'support', 'positive') \tag{10.2a}$$

$$cdf = ksdensity(x, x_1, 'function', 'cdf', 'support', 'positive') \tag{10.2b}$$

In Equations (10.2a) and (10.2b), x and x_1 represent the random variable and the data points where the nonparametric *pdf* and *cdf* need to be evaluated. Figure 10.2 plots the

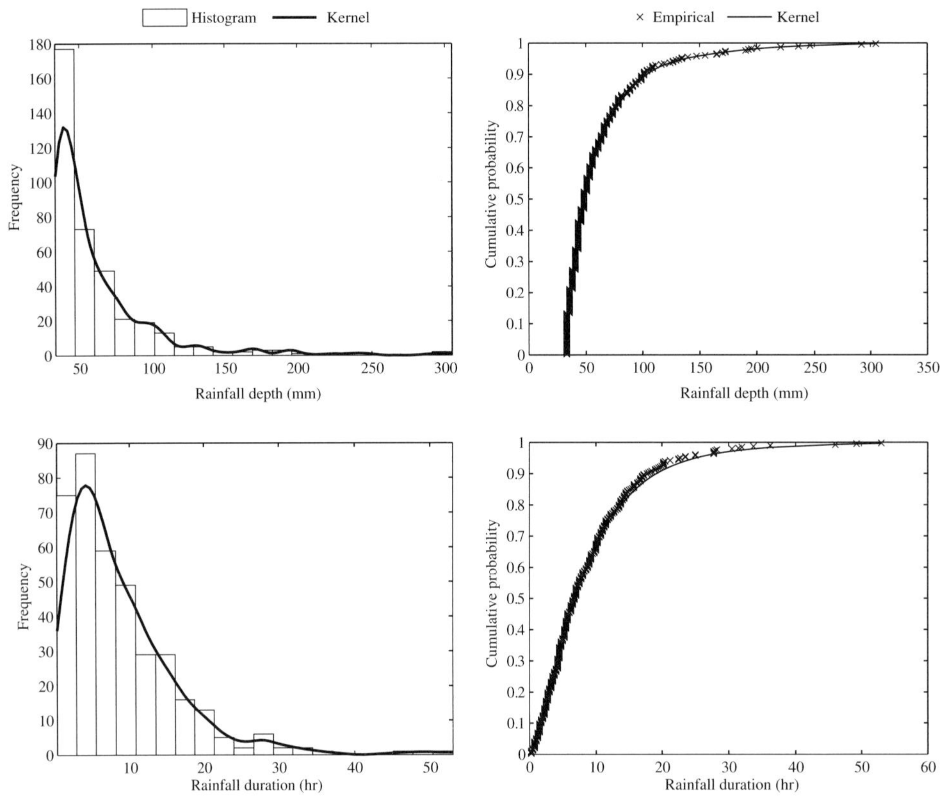

Figure 10.2 Frequency and cumulative probability plots with kernel density function for rainfall depth and rainfall duration series.

density function as well as the cumulative probabilities for both rainfall variables. The CDF estimated from the kernel density is applied for bivariate analysis using copulas.

10.2.3 Bivariate Rainfall Frequency Analysis

The scatter plot in Figure 10.1 indicates positive dependence between rainfall depth and rainfall duration. The rank-based Kendall correlation coefficient is computed as $\tau_n \approx 0.32$. Among the copula candidates (i.e., Gumbel–Hougaard, Clayton, Frank, Gaussian, and Student's *t* copulas), the Frank copula is found to better model the bivariate rainfall characteristics. Figure 10.3 compares the empirical CDF estimated from the kernel density with the bivariate random variables simulated from the fitted Frank copula with its parameter value of 3.529. Comparison shows that (i) the simulated random variates cover the overall dependence fairly well; and (ii) the tie existing in both rainfall depth and rainfall duration variables may impact the concordance of the bivariate rainfall variables.

However, with the continuous assumption, we will proceed to estimate the rainfall depth for a given duration of a given return period. The exceedance probability (P_{ex}) corresponding to a given return period (T) for the partial duration series may be written as follows:

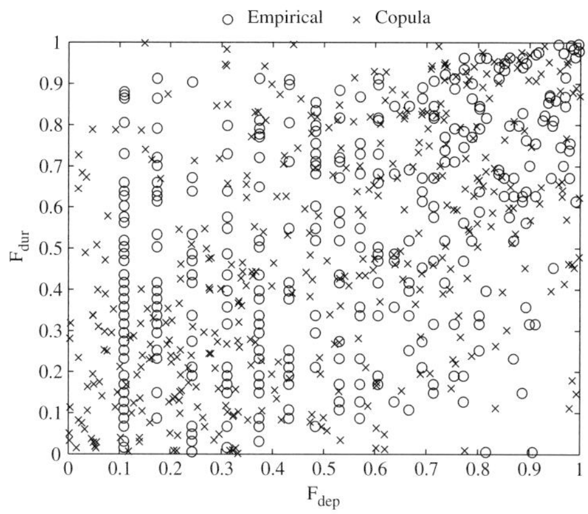

Figure 10.3 Comparison of bivariate empirical distribution using kernel density with the random variables simulated from the fitted Frank copula.

$$P_{ex} = \frac{1}{\mu T} \tag{10.3}$$

In Equation (10.3), $\mu \approx 9$, the average number of events per year.

Equating Equation (10.3) to the exceedance probability of rainfall depth of a given rainfall duration, we have the following:

$$P\left(rain_{dep} > x | Rain_{dur} = d\right) = \frac{1}{\mu T} \tag{10.4}$$

Equation (10.4) is equivalent to the following:

$$C^{Frank}\left(F_{dep} \le F_{dep}(x) | F_{dur} = F_{dur}(d)\right) = 1 - \frac{1}{\mu T} \tag{10.5}$$

In Equation (10.5), $C^{Frank}\left(F_{dep} \le F_{dep}(x) | F_{dur} = F_{dur}(d)\right) = P(dep \le x | dur = d)$. The conditional copula in Equation (10.5) is listed as #5 in Table 4.2. Applying the kernel density to the given durations of 1, 2, 3, 6, 12, and 24 fours, we have $F_{dur}(d)$ computed as follows:

$$F_{dur}(d) = [0.0818, 0.1385, 0.2079, 0.4362, 0.7480, 0.9551].$$

For the return period of 1, 2, 5, 10, 25, 50, and 100 years, we have the exceedance probability computed using Equation (10.3) directly as follows:

$$P_{ex} = [0.8862, 0.9431, 0.9772, 0.9886, 0.9954, 0.9997, 0.9989].$$

Substituting $F_{dur}(d), P_{ex}$ into Equation (10.5), we can compute $F_{dep}(x)$ numerically using the bisection method. Finally, we can estimate the corresponding rainfall depth using the inverse of the kernel density (fitted to the observed rainfall depth) with the computed $F_{dep}(x)$. Table 10.3 lists the estimated $F_{dep}(x)$ and the corresponding estimated rainfall depth. Figure 10.4 compares the rainfall depth estimated from copula-based analysis with the published DDF of partial duration for Morgan City, Louisiana (http://hdsc.nws.noaa.gov/hdsc/pfds/pfds_map_cont.html?bkmrk=la). Comparison shows that (i) for the storms

Table 10.3. *Estimated probability distribution of rainfall depth and estimated rainfall depth of given duration with given return period.*

	1-yr	2-yr	5-yr	10-yr	25-yr	50-yr	100-yr
				$F_{dep}(x)$			
1-hr	0.5988	0.7364	0.8656	0.9252	0.9677	0.9834	0.9916
2-hr	0.6512	0.7793	0.8921	0.9413	0.9751	0.9873	0.9936
3-hr	0.7034	0.8195	0.9153	0.9548	0.9811	0.9904	0.9952
6-hr	0.8296	0.9064	0.9599	0.9794	0.9916	0.9958	0.9979
12-hr	0.9250	0.9622	0.9848	0.9924	0.9969	0.9985	0.9992
24-hr	0.9594	0.9802	0.9922	0.9961	0.9984	0.9992	0.9996
				Rainfall depth (mm)			
1-hr	55.30	68.64	92.78	116.32	166.01	206.38	246.33
2-hr	59.54	74.73	101.12	128.39	182.26	221.99	261.82
3-hr	64.72	81.89	110.86	143.97	198.76	238.62	277.24
6-hr	84.01	106.74	151.78	193.73	246.69	284.32	314.98
12-hr	116.25	155.60	211.31	251.90	299.25	327.08	350.48
24-hr	150.98	195.86	250.48	287.99	326.54	350.01	370.53

with shorter duration and return periods less than 10 years, the copula estimates are either closely following the NOAA estimates or well within NOAA 90% bounds; (ii) for short durations (i.e., $D = 1$ and 2 hours) and higher return periods ($T \geq 25\ yr$), the copula estimates are higher than the NOAA upper 90% bounds; and (iii) as the storm duration increases, the copula estimates for higher return periods get closer to either NOAA upper 90% bounds or actually closely follow the NOAA estimates.

The differences between the NOAA-DDF and the copula-based DDF curves may be due to the following:

i. The NOAA-DDF analysis only extracts rainfall events for certain durations. These extracted events are then treated as univariate random variables and are fitted by univariate probability distributions.
ii. In the copula-based DDF analysis, on the other hand, rainfall events extracted may yield different rainfall durations. The bivariate rainfall depth-duration model is then constructed, and the rainfall depth of a given duration is estimated from the conditional probability function of $f(depth < depth^* | duration = duration^*)$. In this analysis, the duration can take on any value.
iii. The ties that may exist in the NOAA-DDF extracted events may not have the same degree of impact as that of copula-based DDF events. As discussed earlier, there may be many ties in the rainfall depth and duration of the extracted rainfall events (partial duration or annual maximum series), and these tied values may distort the concordance of the bivariate rainfall analysis. Additionally, the rainfall variables (especially rainfall duration) may be discrete in nature.

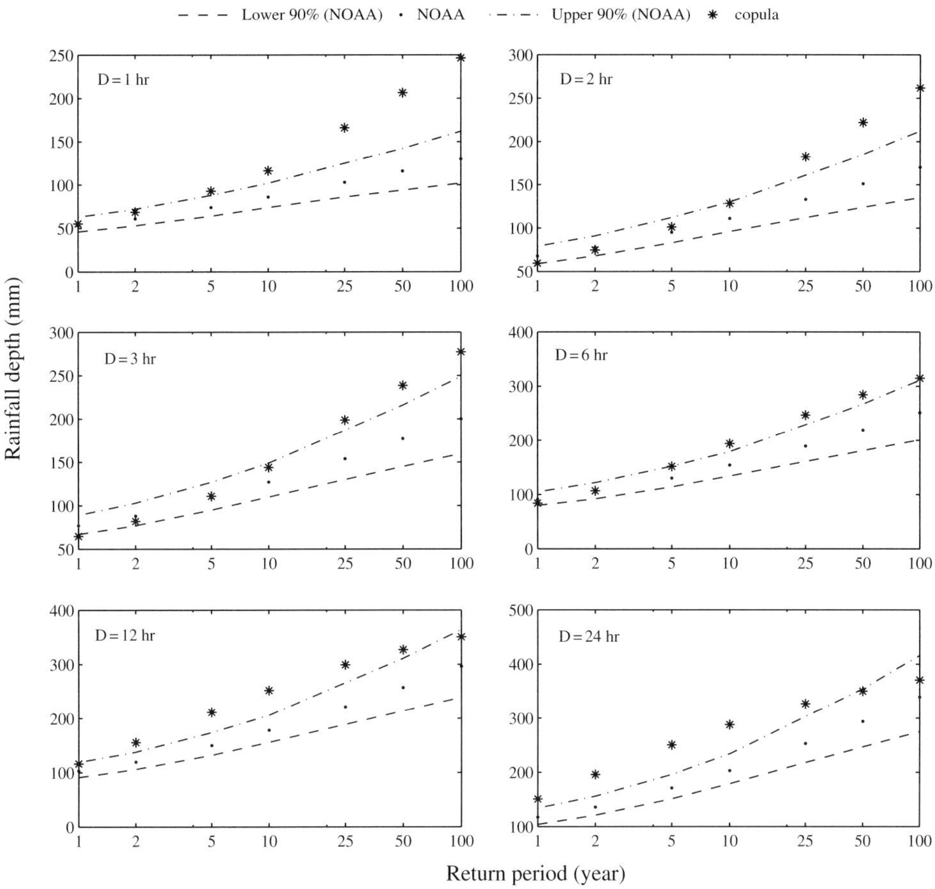

Figure 10.4 Comparison of copula estimates with the NOAA estimations with a 90% confidence bound.

Even with the differences between the NOAA and copula-based DDF curves constructed for the partial duration time series, the copula-based method may be considered as a rational alternative for rainfall DDF (or IDF) construction with simpler and faster rainfall separation (events regardless of the length of rainfall duration) compared to that of NOAA analysis (rainfall duration–based directly).

10.3 Spatial Analysis of Annual Precipitation

With the assumption of annual precipitation amount as a random variable, the general procedure for spatial analysis of annual precipitation includes the following steps:

1. Select the region of interest, identify the rain gauges, and collect the annual precipitation records.
2. Evaluate the pairwise rank-based correlation coefficient of annual precipitation.

3. Identify the possible vine structure based on the rank-based correlation coefficients computed, and select possible copula candidates for T1 first, and then proceed with the analysis for the rest of the tree structure as discussed in Chapter 5.
4. Identify the proper tree structure for the asymmetric Archimedean copula and then proceed with the analysis as discussed in Chapter 5 for the asymmetric Archimedean copula.
5. Construct the meta-elliptical copula for the multivariate precipitation variables.
6. Compare the performance of different copula construction approaches.

To illustrate the spatial analysis of annual precipitation (rainfall), we will use four NOAA rainfall stations located in the Cuyahoga River Watershed, Ohio (see Table 10.4). The copula model is constructed from the annual rainfall data collected from 1953 to 2012 from NCDC. In this case study, we will apply D-vine, meta-elliptical copulas (i.e., meta-Gaussian and meta-Student T) and asymmetric Archimedean copulas. The reason that a D-vine copula is chosen from the pair copula construction is that there is no obvious center variable governing the dependence structure among all four rainfall stations (see the rank-based Kendall correlation coefficient listed in Table 10.5).

10.3.1 Application of D-Vine Copula to Four-Dimensional Rainfall Variables

Copula Identification for T_1

According to Kendall's tau correlation coefficient matrix, the proper structure for T_1 is as follows: $R330058 - R333780 - R336949 - R331458$ (i.e., the bivariate pairs for T_1 are [R330058, R333780]; [R333780, R336949]; [R336949, R331458]). Using the empirical marginals (Weibull plotting position formula), let U_1, U_2, U_3, U_4 represent the empirical marginals as follows:

$$U_1 = \widehat{F}_n(R330058); U_2 = \widehat{F}_n(R333780); U_3 = \widehat{F}_n(R336949); and\ U_4 = \widehat{F}_n(R331458).$$

The D-vine structure for this example is the same as in Figure 10.5. In this case study, we choose Archimedean copulas for dealing with the positive dependence (Gumbel–Hougaard, Clayton, Frank, Joe, and BB1 copulas) as the candidates. Chapter 4 listed the one-parameter Archimedean copulas candidates. Hence we only give the formula for BB1 copula, which is a two-parameter Archimedean copula with the limiting conditions of either the Clayton or Gumbel–Hougaard copula. The BB1 copula (Joe, 1997) can be formulated as follows:

$$C(u, v; \theta_1, \theta_2) = \left\{1 + \left[\left(u^{-\theta_1} - 1\right)^{\theta_2} + \left(v^{-\theta_1} - 1\right)^{\theta_2}\right]^{\frac{1}{\theta_2}}\right\}^{-\frac{1}{\theta_1}}; \theta_1 > 0; \theta_2 \geq 1 \quad (10.6)$$

The BB1 copula converges to (i) the Gumbel–Hougaard copula if $\theta_1 \to 0$; and (ii) the Clayton copula if $\theta_2 = 1$.

Table 10.4. *Annual rainfall amount (mm) at four rain gauges.*

	Rain gauges			
Year	R330058	R336949	R333780	R331458
1953	668.528	634.746	536.702	744.728
1954	855.726	970.788	915.416	970.28
1955	705.866	943.61	943.864	872.998
1956	1,071.88	1,110.996	1,197.61	950.468
1957	735.584	954.532	859.79	839.47
1958	838.962	996.95	1,021.588	923.798
1959	1,025.906	1,334.008	1,240.282	1,075.182
1960	504.952	524.51	791.464	603.504
1961	716.026	773.43	909.828	930.656
1962	567.944	655.828	641.096	678.18
1963	437.134	535.94	499.11	503.428
1964	895.096	817.88	784.352	721.614
1965	754.38	757.682	912.876	757.682
1966	621.792	657.352	787.654	745.236
1967	676.656	684.784	822.452	917.194
1968	796.544	852.932	1,136.142	817.372
1969	738.632	786.384	1,152.652	768.604
1970	879.348	929.132	926.338	754.634
1971	684.022	826.77	786.638	628.142
1972	1,003.554	937.514	1,070.864	934.72
1973	841.248	940.054	1,013.714	1041.146
1974	899.922	1,041.146	1,021.588	1048.004
1975	933.196	1,049.782	998.474	1032.764
1976	759.714	892.556	767.588	1045.21
1977	54.864	957.072	1,027.684	1169.924
1978	699.262	829.31	910.336	805.434
1979	876.046	1,065.53	1,082.294	1132.586
1980	854.71	875.538	863.092	848.106
1981	881.38	914.4	970.534	897.89
1982	733.806	881.38	833.12	845.058
1983	885.19	984.504	1,007.11	851.916
1984	753.364	833.12	827.532	1026.922
1985	852.424	939.8	923.29	1,005.586
1986	721.106	850.646	934.466	1,125.474
1987	618.744	670.56	807.212	897.382
1988	735.33	721.614	744.728	773.938
1989	884.428	897.128	811.022	993.14
1990	1,592.834	1,193.546	1,251.712	1,347.216
1991	530.86	628.142	680.212	716.28
1992	1,019.048	1,010.412	1,069.34	1,020.064

Table 10.4. (*cont.*)

	Rain gauges			
Year	R330058	R336949	R333780	R331458
1993	898.652	915.416	811.276	805.434
1994	909.066	867.41	852.17	788.924
1995	813.308	918.718	799.592	830.072
1996	1,128.014	1,097.28	979.424	1,168.146
1997	797.56	861.822	875.284	968.248
1998	978.662	970.788	869.442	963.168
1999	751.586	806.45	791.972	987.044
2000	1,013.968	932.434	969.518	887.222
2001	764.54	771.906	778.256	774.192
2002	931.926	744.982	891.54	837.946
2003	1,149.604	1,142.492	1,205.484	1,036.828
2004	1,049.274	1,088.136	1,056.386	956.818
2005	897.382	982.726	908.812	910.59
2006	1,053.084	1,087.882	1,160.78	1,362.964
2007	918.21	970.534	985.266	1,145.794
2008	887.984	1,012.19	963.422	1,165.606
2009	781.558	899.668	941.832	965.708
2010	774.446	846.328	747.776	927.862
2011	1,360.678	406.908	1,551.432	315.722
2012	782.574	664.718	849.122	782.828

Table 10.5. *Kendall's tau correlation coefficient matrix.*

	R330058	R336949	R333780	R331458
R330058	1	0.6418	0.5064	0.4151
R336949	0.6418	1	0.5631	0.5300
R333780	0.5064	0.5631	1	0.4490
R331458	0.4151	0.5300	0.4490	1

In addition, the BB1 copula has both upper- and lower-tail dependence coefficients, as follows:

$$\lambda_L = 2^{-\frac{1}{\theta_1 \theta_2}}, \lambda_U = 2 - 2^{\frac{1}{\theta_2}}$$

The parameters of T_1 are estimated with the pseudo-MLE through the empirical marginals for all the copula candidates (Table 10.5). Table 10.6 also lists the log-likelihood, AIC, and BIC values with the best-fitted copula highlighted. From Table 10.6, we see that the two-parameter BB1 copula is the best-fitted copula for stations (R330058, R333780, R333780, and R336949), and the Gumbel–Hougaard copula is the best-fitted copula for stations R336949 and R331458.

Table 10.6 *Estimation results for copula candidates.*

Variables	Copulas	θ	L	AIC	BIC
	Gumbel-Hougaard (*GH*)	2.7782	35.0603	−68.1206	−66.0601
	Clayton (*C*)	2.9226	33.9990	−65.9979	−63.9375
U_1 *v.s.* U_2	Frank (*F*)	8.8627	31.7661	−61.5322	−59.4718
	Joe (*J*)	3.3021	28.8828	−55.7655	−53.7051
	BB1	**[1.0203, 1.9788]**	**39.0411**	**−74.0821**	**−69.9612**
	Gumbel-Hougaard (*GH*)	2.3336	26.0645	−50.1290	−48.0685
	Clayton (*C*)	1.8549	20.2556	−38.5113	−36.4508
U_2 *v.s.* U_3.	Frank (*F*)	6.8861	22.8811	−43.7622	−41.7017
	Joe (*J*)	2.8627	23.0147	−44.0294	−41.9689
	BB1	**[0.3841,2.0196]**	**26.7981**	**−49.5963**	**−45.4754**
	Gumbel-Hougaard (*GH*)	**1.7527**	**22.9526**	**−43.9052**	**−41.8447**
	Clayton (*C*)	1.2274	12.7401	−23.4801	−21.4197
U_3 *v.s.* U_4	Frank (*F*)	4.9432	14.3614	−26.7228	−24.6623
	Joe (*J*)	1.9670	10.9197	−19.8394	−17.7790
	BB1	[0.4963,1.4682]	15.1711	−26.3423	−22.2214

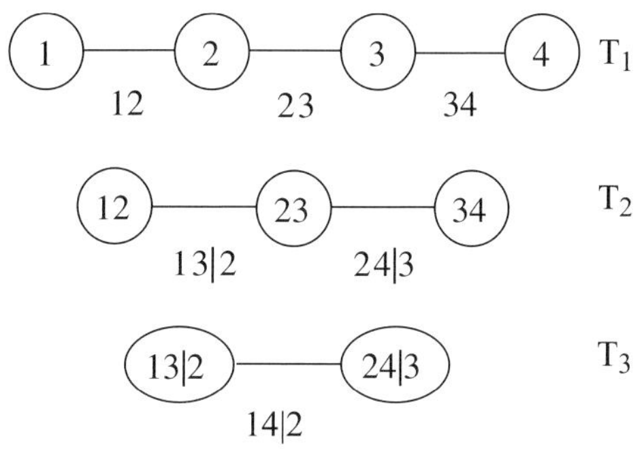

Figure 10.5 D-vine structure for four-dimensional rainfall variables: (1) R330058, (2) R336949, (3) R333780, and (4) R331458.

Based on the AIC/BIC model selection criteria, we again find that (1) the BB1 copula reaches the lowest AIC/BIC values for pairs (U_1,U_2); (2) the BB1 copula is also selected to model the pairs (U2 and U3) since it yields the compariable AIC/BIC and may capture the lower tail dependence, compared with Gumbel–Houggard copula and (3) the Gumbel–Hougaard copula reaches the lowest AIC/BIC for pair (U_3,U_4).

Copula Identification for T_2

Using the best-fitted copulas for T_1, Table 10.7 lists the conditional probability computed for T_2.

Table 10.7. *Conditional probability needed for T_2.*

No.	(1)	(2)	(3)	(4)	No.	(1)	(2)	(3)	(4)
1	0.748	0.290	0.588	0.302	30	0.100	0.544	0.877	0.445
2	0.199	0.578	0.803	0.820	31	0.187	0.611	0.611	0.066
3	0.005	0.551	0.664	0.384	32	0.287	0.483	0.506	0.791
4	0.884	0.664	0.962	0.550	33	0.282	0.591	0.716	0.730
5	0.012	0.558	0.872	0.592	34	0.115	0.411	0.189	0.878
6	0.719	0.441	0.149	0.609	35	0.195	0.363	0.057	0.354
7	0.106	0.698	0.997	0.770	36	0.800	0.345	0.273	0.647
8	0.594	0.275	0.003	0.252	37	0.803	0.476	0.730	0.885
9	0.717	0.380	0.070	0.268	38	0.946	0.684	0.525	0.428
10	0.173	0.325	0.325	0.316	39	0.049	0.315	0.315	0.301
11	0.358	0.256	0.123	0.078	40	0.727	0.638	0.356	0.249
12	0.957	0.433	0.714	0.202	41	0.878	0.537	0.778	0.437
13	0.638	0.388	0.076	0.185	42	0.925	0.448	0.522	0.100
14	0.502	0.303	0.164	0.559	43	0.242	0.531	0.789	0.602
15	0.606	0.335	0.511	0.818	44	0.752	0.671	0.830	0.924
16	0.445	0.455	0.200	0.252	45	0.243	0.419	0.419	0.757
17	0.412	0.403	0.006	0.091	46	0.816	0.497	0.828	0.398
18	0.587	0.504	0.281	0.087	47	0.831	0.396	0.490	0.674
19	0.089	0.462	0.439	0.039	48	0.960	0.571	0.458	0.406
20	0.894	0.564	0.247	0.228	49	0.676	0.354	0.394	0.072
21	0.354	0.517	0.972	0.878	50	0.987	0.371	0.099	0.133
22	0.410	0.658	0.864	0.595	51	0.690	0.691	0.424	0.196
23	0.425	0.631	0.803	0.523	52	0.572	0.678	0.563	0.266
24	0.259	0.517	0.858	0.979	53	0.549	0.618	0.778	0.851
25	0.316	0.604	0.689	0.966	54	0.747	0.624	0.395	0.985
26	0.382	0.426	0.506	0.330	55	0.855	0.517	0.222	0.909
27	0.105	0.651	0.698	0.535	56	0.468	0.644	0.474	0.939
28	0.695	0.469	0.492	0.175	57	0.328	0.490	0.335	0.592
29	0.443	0.584	0.359	0.232	58	0.164	0.598	0.930	0.942

$\tau((1),(2)) = 0.1446; \tau((3),(4)) = 0.2922.$

Note: (1) $C^{BB1}(U_1|U_2 = u_2; \boldsymbol{\theta}_{12})$; (2) $C^{BB1}(U_3|U_2 = u_2; \boldsymbol{\theta}_{23})$; (3) $C^{BB1}(U_2|U_3 = u_3; \boldsymbol{\theta}_{23})$; (4) $C^{GH}(U_4|U_3 = u_3; \theta_{34})$.

From Kendall's correlation coefficient estimated in Table 10.7, we again have the positive dependence for $[U_1|U_2, U_3|U_2]$ and $[U_2|U_3, U_4|U_3]$. Using all copula candidates for T_1, Table 10.8 lists the results from pseudo-MLE for T_2. Based on AIC and BIC, Frank copula is found as the best fitted copula for T_2 variables as shown in Table 10.8. However the goodness-of-fit study shows that BB1 copula should be applied to model the dependence at T_2 (Table 10.8).

Table 10.8 *Results of pseudo-maximum likelihood estimation for T2.*

Variables	Copulas	θ	L	AIC	BIC
	Gumbel-Hougaard (*GH*)	1.1059	0.5369	0.9262	2.9867
	Clayton (*C*)	0.2371	1.0970	−0.1937	1.8667
	Frank (*F*)	1.3126	1.4286	−0.8571	1.2033
	Joe (*J*)	1.0796	0.1082	1.7835	3.8439
$U_1\|U_2$ *v.s.* $U_3\|U_2$	BB1	[0.2336, 1.0001]	1.0967	1.8069	5.9278
	Gumbel-Hougaard (*GH*)	1.2934	3.6637	−5.3274	−3.2670
	Clayton (*C*)	0.5281	2.9222	−5.9477	−3.8873
$U_2\|U_3$ *v.s.* $U_4\|U_3$	Frank (*F*)	2.7826	5.8333	−9.6668	−7.6064
	Joe (*J*)	1.3356	2.1710	−2.3419	−0.2815
	BB1	[0.3112, 1.1461]	4.3801	−4.7604	−0.6396

Copula Identification for T_3

Now we can move on to T_3. Similar to T_2, we first need to compute the conditional copula through the BB1 copula as follows:

$$F(U_1 \le u_1|U_2 = u_2, U_3 = u_3) = \frac{\partial C_{13|2}(F(u_1|u_2)|F(u_3|u_2))}{\partial F(u_3|u_2)}$$

$$F(U_4 \le u_4|U_2 = u_2, U_3 = u_3) = \frac{\partial C_{24|3}(F(u_4|u_3)|F(u_2|u_3))}{\partial F(u_2|u_3)}$$

The computed conditional probability distribution using the selected fitted copulas in T_1 and T_2 is listed in Table 10.9.

Using the conditional probability listed in Table 10.9, Kendall's correlation coefficient is computed as $\tau = -0.06$, and we have negative dependent variables for T_3. Applying the Frank copula, we have $\theta = -0.5062$.

Applying the goodness-of-fit study for the fitted vine copula, the Rosenblatt transform approach is applied. The test results (Table 10.10) further confirmed the selected vine copula may properly study the dependence structure of the four-dimensional rainfall dataset.

Applying the goodness-of-fit study, we have $S_n^B = 0.046, P = 0.95$. Thus, we have the four-dimensional fitted D-vine copula as BB1-BB1-GH (T_1), BB1-BB1 (T_2) and Frank (T_3) as shown in Figure 10.6.

With the fitted D-vine copula in Figure 10.6, we can simulate the four-dimensional pseudo-rainfall variables (i.e., the marginal CDF of rainfall variables) as shown in Figure 10.7. Here, we will show how to simulate the random variates from the fitted D-vine copula given in Figure 10.6 with a simple example:

Table 10.9. *Conditional probability distribution computed for parameter estimation of T_3.*

No.	$C(U_1\|U_2, U_3)$	$C(U_4\|U_2, U_3)$	No.	$C(U_1\|U_2, U_3)$	$C(U_4\|U_2, U_3)$
1	0.8222	0.2421	30	0.1612	0.2798
2	0.2184	0.7546	31	0.1519	0.0328
3	0.0030	0.3020	32	0.2693	0.8071
4	0.9089	0.3088	33	0.2700	0.6734
5	0.0191	0.4304	34	0.0798	0.9377
6	0.6749	0.7570	35	0.1388	0.6254
7	0.2114	0.3717	36	0.7956	0.7325
8	0.5270	0.7909	37	0.8355	0.8653
9	0.6707	0.5060	38	0.9368	0.3979
10	0.1761	0.3532	39	0.0476	0.3403
11	0.3406	0.1462	40	0.6812	0.2610
12	0.9673	0.1203	41	0.8936	0.3145
13	0.5824	0.3821	42	0.9276	0.0656
14	0.4791	0.7055	43	0.2885	0.4868
15	0.6624	0.8341	44	0.7448	0.8926
16	0.3849	0.3462	45	0.2316	0.7938
17	0.3342	0.5159	46	0.8633	0.2583
18	0.5362	0.0943	47	0.8433	0.6859
19	0.0726	0.0241	48	0.9543	0.3989
20	0.8738	0.2856	49	0.6909	0.0563
21	0.6729	0.6668	50	0.9843	0.2652
22	0.4138	0.4388	51	0.6391	0.1787
23	0.4219	0.3929	52	0.5194	0.2126
24	0.3643	0.9708	53	0.5467	0.8060
25	0.2926	0.9680	54	0.7068	0.9923
26	0.3924	0.2969	55	0.8297	0.9524
27	0.0793	0.4514	56	0.4078	0.9552
28	0.6891	0.1395	57	0.2790	0.6537
29	0.3809	0.2404	58	0.2531	0.8725

Table 10.10. *Goodness-of-fitted test for T_1, T_2, and T_3.*

	Copula	P	S_{nb}		Copula	P	S_{nb}
	BB1 (U_1,U_2)	0.59	0.23	T_2	BB1(U1|U2, U3|U2)	0.31	0.35
T_1	BB1(U_2,U_3)	0.62	0.21		BB1(U2|U3, U4|U3)	0.58	0.23
	GH (U_3,U_4)	0.41	0.12	T_3	Frank $(U_1\|U_2,U_3, U_4\|U_2,U_3)$	0.046	0.95

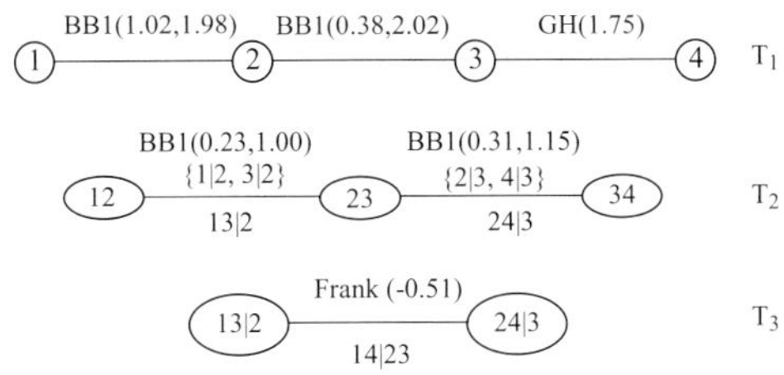

Figure 10.6 Fitted D-vine copula for four-dimensional rainfall variables.

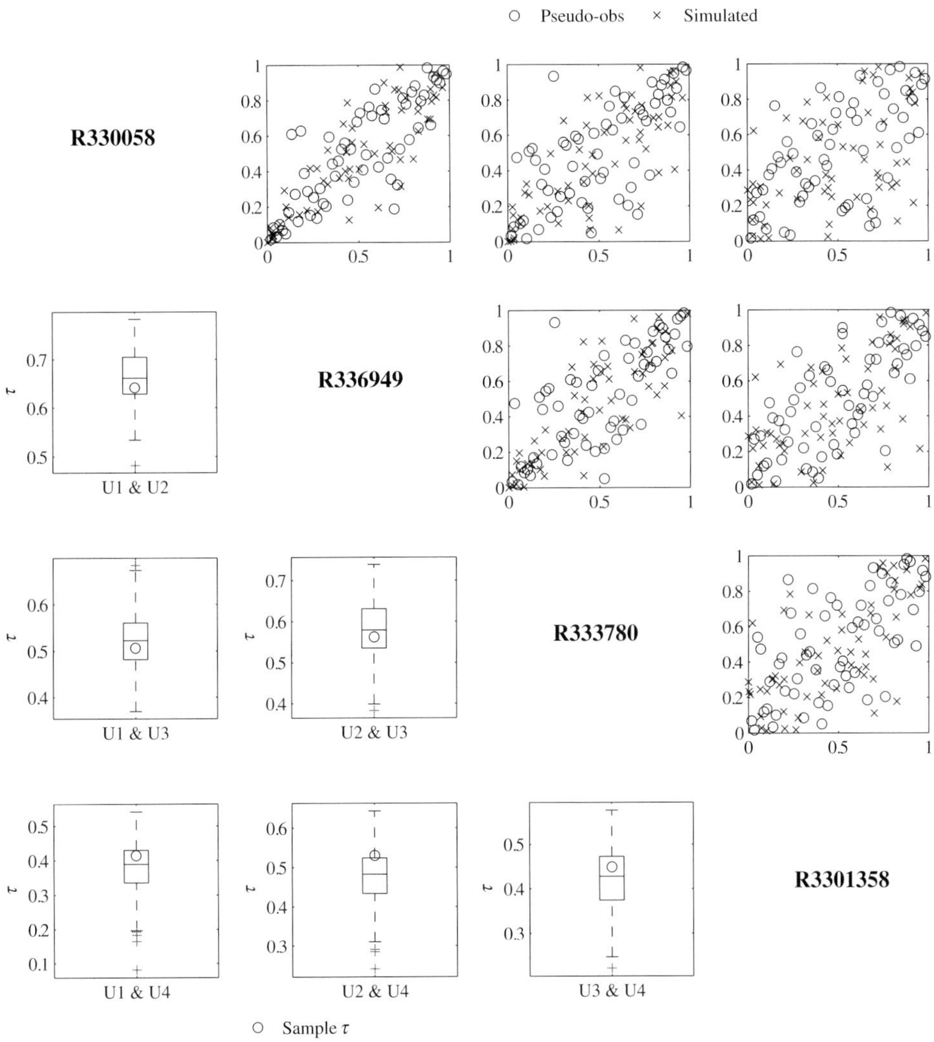

Figure 10.7 Comparison of simulated random variables with the pseudo-rainfall observations and simulated rank-based Kendall correlation coefficient with sample Kendall correlation coefficient.

1. Generate four independent uniform random variables in [0,1]

$$W = [0.7582, 0.6289, 0.9611, 0.2743].$$

For the independent random variables generated, we set the following:

$$\begin{aligned}
\boldsymbol{u}_1 &= \boldsymbol{W}(\mathbf{1}) = \mathbf{0.7582} \\
C(u_2|U_1 = u_1) &= 0.6289 \\
C(u_3|U_1 = u_1, U_2 = u_2) &= 0.9611; \\
C(U_4|U_1 = u_1, U_2 = u_2, U_3 = u_3) &= 0.2743.
\end{aligned}$$

2. Simulate $\boldsymbol{u}_2$ from $\boldsymbol{C}(\boldsymbol{U}_2|\boldsymbol{U}_1 = \boldsymbol{u}_1) = 0.6289$.

According to the fitted D-vine copula, we know $[U_1, U_2]$ is fitted with the BB1 copula (i.e., Equation (9.1)) with parameters [1.0203,1.9788]. Its conditional copula is then written as follows:

$$C(u_2|U_1 = u_1) = \left.\frac{\partial C(u_1, u_2)}{\partial u_1}\right|_{u_1=0.6478} = \frac{S^{\frac{1}{\theta_2}-1}\left(u_1^{-\theta_1} - 1\right)^{\theta_2 - 1}}{u_1^{\theta_1+1}\left(S^{\frac{1}{\theta_2}} + 1\right)^{\frac{1}{\theta_1}+1}} \tag{10.7}$$

where: $S = \left(\frac{1}{u_1^{\theta_1}} - 1\right)^{\theta_2} + \left(\frac{1}{u_2^{\theta_1}} - 1\right)^{\theta_2}$.

Substituting $u_1 = 0.7582$ and $C(u_2|U_1 = u_1) = 0.6289$ into Equation (10.7), we can solve for u_2 numerically and obtain the following:

$$\boldsymbol{u}_2 = \mathbf{0.7755}.$$

3. Simulate $\boldsymbol{u}_3$ from $\boldsymbol{C}(\boldsymbol{u}_3|\boldsymbol{U}_1 = 0.7582, \boldsymbol{U}_2 = 0.7755)$.

According to the fitted D-vine copula, we know U_2 is the one of the center variables, and from the probability density composition discussed in Chapter 5, we have the following:

$$C(u_3|U_1 = u_1, U_2 = u_2) = \frac{\partial C_{13|2}\left(C_{3|2}, C_{1|2}\right)}{\partial C_{1|2}} \tag{10.8}$$

As seen in Equation (10.8), we may simulate u_3 with the following two steps:

i. Compute $C_{3|2}$ from $C_{13|2}$. According to Figure 10.6, we know that the BB1 copula with parameter [0.2336, 1.0001] properly models $C_{13|2}\left(C_{3|2}, C_{1|2}\right)$. With this in mind and after computing $C_{1|2}$ using the BB1 conditional copula (i.e., Equation (10.7)), we immediately have the following: $C(U_1 \le 0.7582|U_2 = 0.7755) = 0.5465$. Given $\left\{C_{3|2}, C_{1|2}\right\}$ (i.e., one of the bivariate copulas at T_2) again modeled by the BB1 copula, $C_{3|2}$ can then be computed by substituting $C_{1|2} = 0.5465$ as u_1 and $C_{3|2}$ as u_2, and by equating Equation (10.8) to 0.9383. We can solve for $C_{3|2}$ numerically as $C_{3|2} = 0.9636$.

ii. Compute u_3 from $C_{3|2}$. From Figure 10.6, $\{U_2, U_3\}$ is also modeled with the BB1 copula; u_3 can then be solved for numerically by substituting $u_2 = 0.7755$ as u_1, and u_3 as u_2 into Equation (10.7), and by setting the equation equal to $C_{3|2} = 0.9636$. We then have the following:

$$u_3 = \mathbf{0.9383}.$$

4. Simulate $\boldsymbol{u}_4$ from $\boldsymbol{C}(\boldsymbol{u}_4|\boldsymbol{U}_1 = 0.7582, \boldsymbol{U}_2 = 0.7755, \boldsymbol{U}_3 = 0.6865)$.

From the fitted D-vine copula, we know the conditional copula $C(u_4|U_1 = u_1, U_2 = u_2, U_3 = u_3)$ may be written using the probability function decomposition discussed in Chapter 5 as follows:

$$C(u_4|U_1 = u_1, U_2 = u_2, U_3 = u_3) = \frac{\partial C_{14|23}(C_{1|23}, C_{4|23})}{\partial C_{1|23}} \tag{10.9}$$

where:

$$C_{1|23} = \frac{\partial C_{13|2}(C_{1|2}, C_{3|2})}{\partial C_{3|2}}; C_{4|23} = \frac{\partial C_{24|3}(C_{4|3}, C_{2|3})}{\partial C_{2|3}} \tag{10.9a}$$

We know from Equations (10.9) and (10.9a) that $C_{14|23}$, $C_{1|23}$, and $C_{4|23}$ are modeled by bivariate Frank, BB1, and BB1 copulas, respectively (Figure 10.6). To this end, we can simulate u_4 with the steps given in what follows:

i. Simulate $C_{4|23}$ using $C(u_4|U_1 = 0.7582_1, U_2 = 0.7755, U_3 = 0.9383) = 0.2743$.

With the previously simulated u_1, u_2, u_3, we first compute the conditional copula $C_{1|23}$ in Equation (10.9a). Applying the corresponding fitted BB1 copulas, we compute the conditional copula as follows: $C_{1|2} = 0.5465, C_{3|2} = 0.9636, C_{1|23} = 0.4764$. The given Frank copula may be applied to model $C_{14|23}$ of T_3, and Equation (10.9) may be rewritten using the conditional Frank copula as follows:

$$C(u_4|U_1 = u_1, U_2 = u_2, U_3 = u_3) = \frac{e^{-\theta C_{1|23}}\left(e^{-\theta C_{4|23}} - 1\right)}{e^{-\theta\left(C_{1|23}+C_{4|23}\right)} - e^{-\theta C_{1|23}} - e^{-\theta C_{4|23}} + e^{-\theta}} \tag{10.10}$$

Substituting

$$\theta = -0.5062, C_{1|23} = 0.4764, C(u_4|U_1 = u_1, U_2 = u_2, U_3 = u_3) = 0.2743$$

into Equation (10.10), $C_{4|23}$ is solved for numerically as follows: $C_{4|23} = 0.2777$.

ii. Simulate $C_{4|3}$ using $C(u_4|U_2 = 0.7755, U_3 = 0.9383) = 0.2777$.

$C_{4|3}$ can be simulated from the conditional copula $C_{4|23}$ through $C_{24|3}$, given as Equation (10.9a). $C_{24|3}$ may be modeled with the bivariate BB1 copula through $C_{4|3}, C_{2|3}$. Applying the BB1 copula to $\{U_2, U_3\}$, we can easily compute $C_{2|3} = C(U_2 \leq 0.7755|U_3 = 0.9383) = 0.1736$. In model construction (e.g., Figure 10.6), $C_{24|3}$ is also modeled by the BB1 copula. Thus, we can solve for $C_{4|3}$ numerically as $C_{4|3} = 0.1867$.

iii. Simulate u_4 using $C_{4|3} = 0.1867$.

We know that $\{U_3, U_4\}$ is modeled with the Gumbel–Hougaard copula (Figure 10.6), as shown in Chapter 4; the conditional Gumbel–Hougaard copula can be written as follows:

$$C(u_4|U_3 = u_3) = \frac{(-\ln u_3)^{-1+\theta}\left[(-\ln u_3)^{\theta} + (-\ln u_4)^{\theta}\right]^{-1+\frac{1}{\theta}}}{u_3 \exp\left[(-\ln u_3)^{\theta} + (-\ln u_4)^{\theta}\right]} \tag{10.11}$$

Substituting $u_3 = 0.9383, C_{4|3} = 0.1867$ into Equation (10.11), we have the following:

$$u_4 = \mathbf{0.6865}$$

Finally, with four independent uniform random variables $W = [0.7582, 0.6289, 0.9611, 0.2743]$; we successfully simulate the pseudo-rainfall variables from the fitted D-vine copula as follows:

$$\boldsymbol{U} = [0.7582, 0.7755, 0.9383, 0.6865].$$

Comparison of the simulated copula random variables with the pseudo-rainfall variables (the upper triangle of Figure 10.7) shows that the fitted D-vine copula reasonably preserves the overall dependence. With the use of 200 simulations, the lower triangle of Figure 10.7 compares the Kendall correlation coefficient computed from the simulations with the sample Kendall correlation coefficient computed from the observed four-dimensional rainfall variables. Comparison through the Kendall's correlation coefficient indicates the following:

1. The sample correlation coefficient is within 50% bound for all free bivariate variates in T_1, i.e.,

$$(U_1, U_2) : (R330058, R336949), (U_2, U_3) : (R336949, R333780), (U_3, U_4) : (R333780, R331358);$$

2. The sample correlation coefficient is also within 50% bound for the bivariate variates through conditioning, i.e., $(U_1, U_3) : (R330058, R333780), (U_1, U_4) : (R330058, R331358)$.
3. The sample correlation coefficient is very close to the 50% bound for the last pair of the bivariate variate through conditioning: $(U_2, U_4) : (R336949, R331358)$.

The preceding comparison ensures the appropriateness of applying the fitted D-vine copula model to investigate the four-dimensional rainfall variables. In addition, with the closeness of rain gauges, it is reasonable to assume that there may exist the tail dependence among the rainfall variables (i.e., there is the concurrent tendency of extreme weather events, e.g., storm events). The possible tail dependence makes the BB1 copula the best choice for a majority of cases. We will provide a detailed discussion in this regard when we compare the fitted vine copula to meta-elliptical and asymmetric Archimedean copulas later in the chapter.

10.3.2 Application of Meta-Elliptical Copula to Four-Dimensional Rainfall Variables

In this section, we will apply the meta-Gaussian and meta-Student t copula to model the four-dimensional rainfall variables. Using the same empirical marginals as those in Section 10.3.1, Table 10.11 lists the parameters (i.e., the correlation matrix for meta-Gaussian

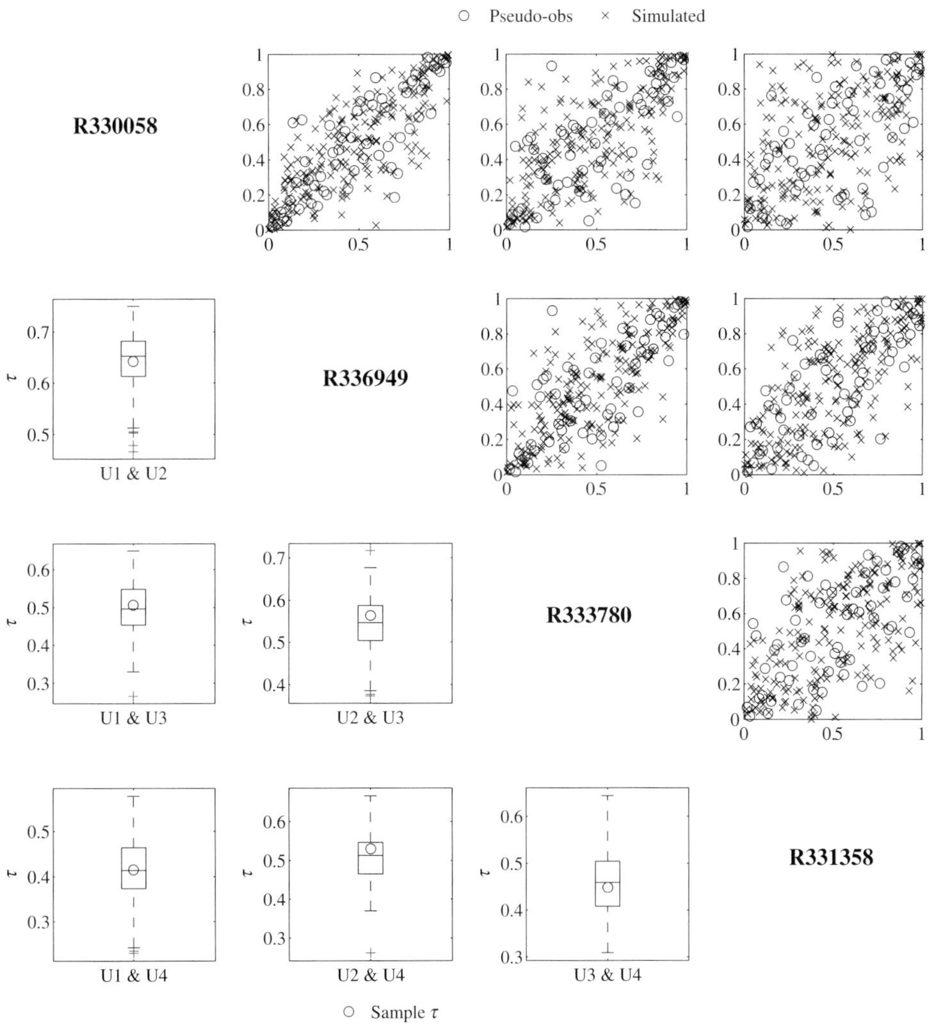

Figure 10.8 Comparison with the fitted meta-Gaussian copula.

Table 10.11. *Parameters estimated for meta-Gaussian and meta-Student t copulas.*

	R330058	R336949	R333780	R331358	R330058	R336949	R333780	R331358
Stations	Meta-Gaussian copula				Meta-Student t copula			
R330058	1	0.85	0.71	0.61	1	0.87	0.74	0.64
R336949	0.85	1	0.75	0.72	0.87	1	0.78	0.74
R333780	0.71	0.75	1	0.65	0.74	0.78	1	0.68
R331358	0.61	0.72	0.65	1	0.64	0.74	0.68	1
					d.f. $v = 62.16$			

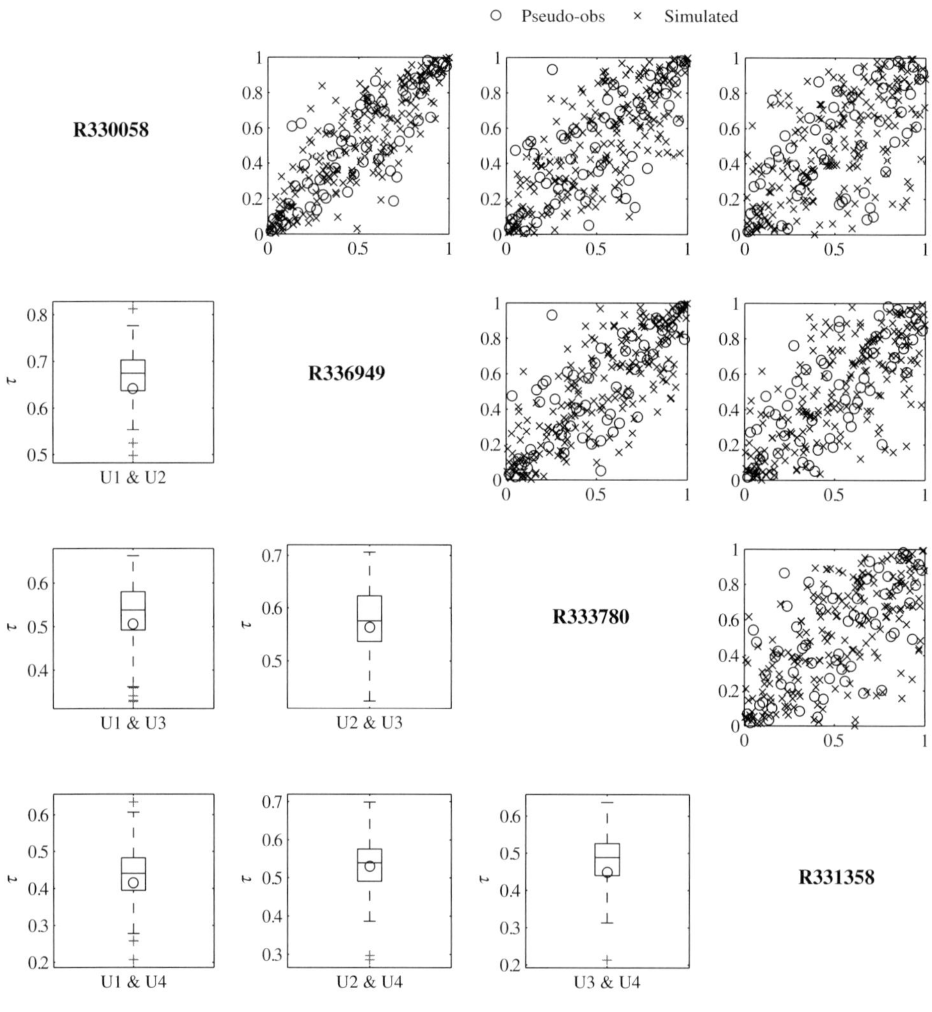

Figure 10.9 Comparison with the fitted meta-Student t copula.

copula, and the correlation matrix and degree of freedom for meta-Student t copulas). With the estimated parameters, Figures 10.8 and 10.9 compare the simulated copula random variables with the pseudo-rainfall random variables as well as the simulated Kendall correlation coefficient with the sample Kendall correlation coefficient.

Simulations shown in Figures 10.8 and 10.9 indicate that the overall dependence structure of rainfall variables is very well preserved. In the case of overall dependence structure, the meta-Gaussian and meta-Student *t* copula visually perform better than the previously fitted D-vine copula, e.g., all sample Kendall correlation coefficients are within 50% bounds of the simulated Kendall correlation coefficients (200 simulations). Furthermore, the goodness-of-fit studies using the Rosenblatt transform yield the following:

Meta-Gaussian copula: $SnB = 0.0245, P = 0.964$.

Meta-Student t copula: $SnB = 0.094, P = 0.785$.

10.3.3 Application of the Asymmetric Archimedean Copula to Four-Dimensional Rainfall Variables

In this section, we will evaluate the performance of asymmetric Archimedean copulas. Here we will choose the following two types of asymmetric structures (Figure 10.10). In Figure 10.10, U_1, U_2, U_3, and U_4 represent R330058, R336949, R333780, and R331358, respectively, as that applied for the D-vine and meta-elliptical copulas.

As seen in Section 10.3.1, the BB1 and Gumbel–Hougaard copulas are found to properly model the bivariate random variables in T_1. From Table 10.6, we see that the Gumbel–Hougaard copula comes to the second place to model $(R330058, R336949)$, $(R336949, R333780)$, and the BB1 copula comes as the second place to model $(R333780, R331358)$. Given the possible difficulties to assess that the parameter of the higher level are lower than the parameter in the lower level (i.e., the parameter of C2 should be less than that of C1) for the two-parameter copulas, we will apply the second-best Gumbel–Hougaard copula for analysis. Applying the Gumbel–Hougaard copula and letting θ_1,θ_2,θ_3 represent the parameter for C_1,C_2,C_3 respectively, the nested asymmetric Gumbel–Hougaard copula for the four-dimensional case can be written as follows:

$$C(u_1, u_2, u_3, u_4; \theta_1, \theta_2, \theta_3) = C_3^{GH}\left(C_2^{GH}\left(C_1^{GH}(u_1, u_2; \theta_1), u_3; \theta_2\right), u_4; \theta_3\right) \tag{10.12}$$

where $\theta_1 \geq \theta_2 \geq \theta_3$

Table 10.12 lists the parameters as well as the Kendall correlation coefficient estimated for each level. The parameters listed in Table 10.12 fulfills the conditions for the nested asymmetric Archimedean copula (i.e., given as part of Equation (10.12)). Applying the

Table 10.12. *Results from the nested asymmetric Gumbel–Hougaard copula.*

Parameters	C1	C2	C3
θ	2.78	2.22	1.65
τ	0.64	0.56	0.51

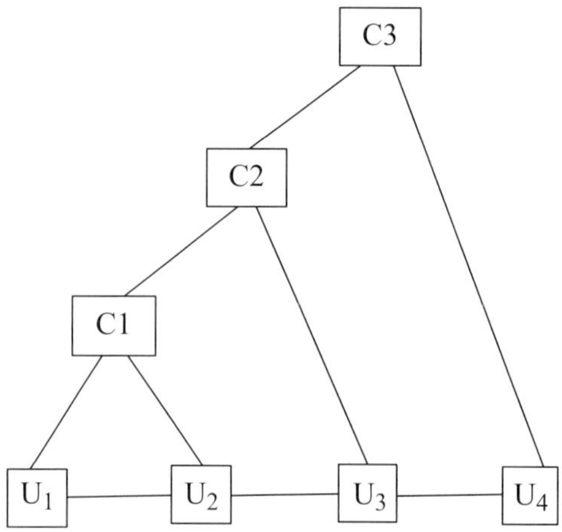

Figure 10.10 Asymmetric Archimedean copula structure.

goodness-of-fit study through the Rosenblatt transform, we obtain $SnB = 0.048$, $P = 0.323$. The goodness of fit results indicate the appropriateness of the fitted four-dimensional asymmetric Gumbel–Hougaard copula.

Furthermore, according to the discussion in Chapter 5, we may conclude that (i) pairs (R330058, R333780) and (R336949, R333780) should follow the Gumbel–Houggard copula with parameter θ_2; (ii) and pairs (R330058, R331358), (R336949, R331358), and (R333780, R331358) should all follow the Gumbel–Hougaard copula with parameter θ_3. Figure 10.11 compares the asymmetric Gumbel–Hougaard copula and bivariate Gumbel–Hougaard copula with the pseudo-observations. The scatter plots in Figure 10.11 show that the overall positive dependence may be captured; however, the box plots for the Kendall correlation coefficient indicate that (i) the sample Kendall correlation coefficient obviously falls out of the upper 50% bound for (R336949, R331358), and (R333780, R331358); (ii) the sample Kendall correlation coefficient is slightly higher than the upper 50% bound for (R336949, R333780); and (iii) the sample Kendall correlation coefficient is slightly higher than the 75% bound for (R336949, R331358).

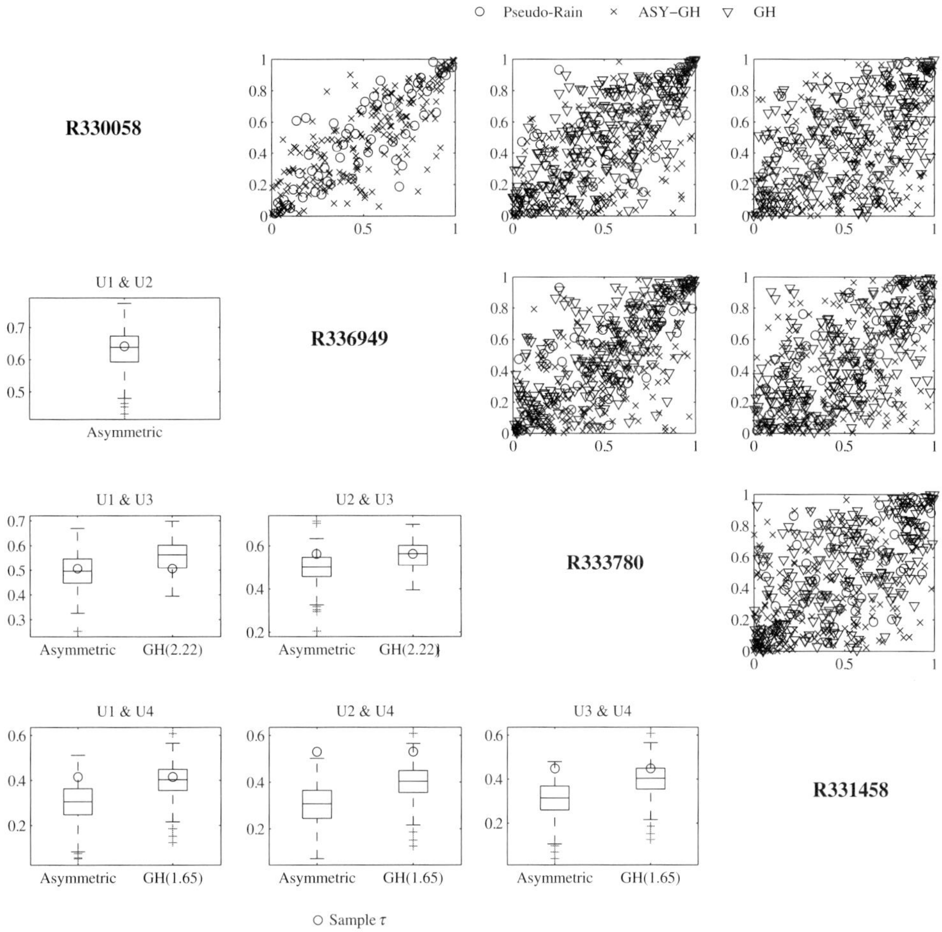

Figure 10.11 Comparison of the asymmetric Gumbel–Hougaard copula with the pseudo-rainfall variables.

10.3.4 Comparison of D-vine, Meta-Elliptical, and Asymmetric Archimedean Copulas

In this section, we will compare the performances of fitted D-vine, meta-elliptical, and asymmetric Archimedean copulas for modeling the dependence in higher dimensions (i.e., four dimensions in this case study). Given all three types of fitted copulas passing the goodness-of-fit study, we will focus on the performance, freedom, and complexity of copula functions.

Flexibility and Complexity of Copula Functions

First, as discussed in Chapter 5, the vine copula is constructed, based on the probability density function decomposition, such that only bivariate copulas are considered as the building blocks either unconditionally (base level T_1) or conditionally (upper levels). The bivariate copulas (i.e., the building blocks) are allowed for free specification (i.e., the copulas do not need to belong to the same family at all). Additionally, there are many choices for model construction. For example, in our four-dimension example illustrated here, we may be able to build 24 different D-vine copula structures through different pairing schemes.

Second, the meta-elliptical copula is only dependent on the correlation matrix for the meta-Gaussian copula, and the correlation matrix and degree of freedom for meta-Student t copula. In addition, its parameter estimation is easier than that of a vine copula.

Third, there are constraints on the asymmetric Archimedean copula. In addition, there are implications for the dependence for indirectly connected bivariate random variates (as discussed in the previous section).

Overall, the vine copula is most complex with the most flexibility of model construction. The meta-elliptical copula may always be able to capture the overall dependence. The asymmetric Archimedean copula has the least flexibility for model construction, and the dependence structure may not be properly captured due the theoretical constraints of the asymmetric copula function.

Comparison of Copula Performances

Applying Equation (5.61) in Chapter 5 to the fitted D-vine copula in Section 10.3.1, we will be able to compute the joint CDF for the four-dimensional rainfall variables. Similar to application of Equation (7.32) and Equation (7.46) in Chapter 7 and Equation (10.12), we can compute the joint CDF fitted by the meta-Gaussian, meta-Student t and asymmetric Gumbel–Hougaard copulas for the four-dimensional rainfall variables. Figure 10.12 compares the fitted parametric four-dimensional copula function with the nonparametric empirical copulas. Table 10.13 lists the RMSE computed between the parametric and empirical copulas. Figure 10.12 shows that (1) there is minimal visual difference between the performance of meta-Gaussian and that of meta-Student T copulas; (2) there is visual difference between the performance of fitted D-vine copula and asymmetric GH copula; and (3) the fitted D-vine copula may underestimate the JCDF for higher orders (>35) more than the asymmetric GH copula. The RMSE results listed in Table 10.13 further confirm the findings visually seen in Figure 10.12.

Table 10.13. *RMSE computed between parametric and empirical copulas.*

Copula	D-vine	Asymmetric GH	Meta-Gaussian	Meta-Student t
RSME	0.040	0.032	0.029	0.027

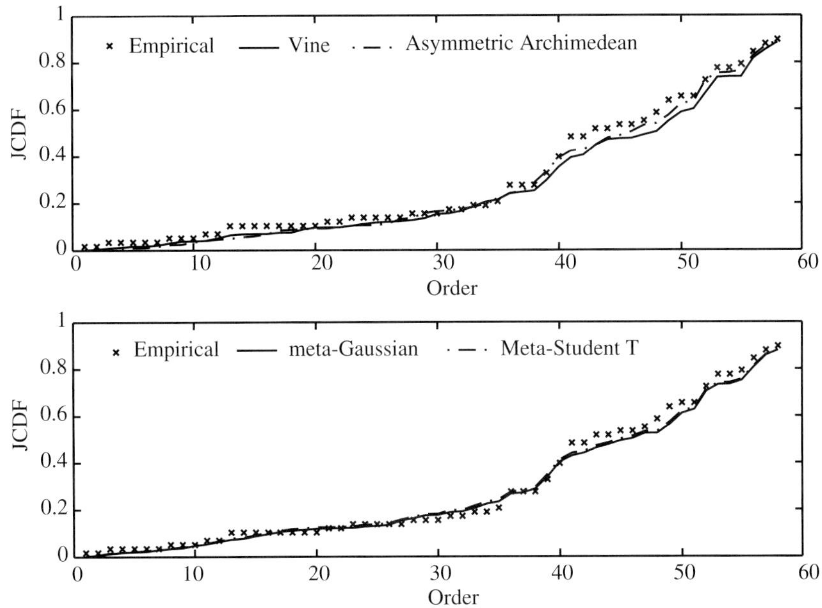

Figure 10.12 Comparison of vine, meta-elliptical, and asymmetric Archimedean copulas with empirical copula.

Comparing all three types of the copulas, one may directly apply the meta-elliptical copula for higher dimensions as the following:

1. The variance–covariance structure may be very well preserved (Figures 10.8 and 10.9).
2. A meta-elliptical copula is easy to construct, compared to both vine and asymmetric Archimedean copulas.
3. A meta-elliptical copula yields the overall best performance.

10.4 Summary

In this chapter, we discussed the application of copula to (1) the partial duration rainfall sequences to construct the DDF curve, and (2) the spatial dependence of precipitation measured from multiple rain gauge stations (i.e., four stations are selected in the case study).

The study shows the following:

i. Even with the differences between the NOAA and copula-based DDF curves constructed for the partial duration time series, the copula-based method may be considered as a rational alternative for rainfall DDF (or IDF) construction with simpler and faster rainfall separation (events regardless of the length of rainfall duration) compared to that of NOAA analysis (rainfall duration based directly).

ii. Applying vine, meta-elliptical, and asymmetric copulas to model the spatial dependence, we have found that the vine copula is most complex and most flexible at the same time. In regard to the copula performance, one may directly apply the meta-elliptical copula, given the simplicity of the parameter estimation and the capture of pairwise dependence structure for all correlated random variables.

References

Abdul Rauf, U. F. A. and Zeephongsekul, P. (2014). Copula based analysis of rainfall severity and duration: a case study. *Theoretical and Applied Climatology,* 115(1–2), 153–166.

Bacchi, B., Becciu, G., and Kottegoda, N. T. (1994). Bivariate exponential model applied to intensities and durations of extreme rainfall. *Journal of Hydrology,* 155, 225–236.

Cantet, P. and Arnaud, P. (2014). Extreme rainfall analysis by a stochastic model: impact of the copula choice on the sub-daily rainfall generation. *Stochastic Environmental Research and Risk Assessment,* 28, 1479–1492.

Cong, R.-G. and Brady, M. (2012). The interdependence between rainfall and temperature: copula analysis. *Scientific World Journal,* 405675, doi:10.1100/2012/405675.

Cordova, J. R. and Rodriguez-Iturbe, I. (1985). On probabilistic structure of storm surface runoff. *Water Resources Research,* 21(5), 755–763.

Favre, A.-E., El Adlouni, S., Perreault, L, Thiemonge, N., and Bobee, B. (2004). Multivariate hydrological frequency analysis using copulas. *Water Resources Research,* 40, W01101.

Goel, N. K., Kurothe, R. S., Mathur, B. S., and Vogel, R.M. (2000). A derived flood frequency distribution for correlated rainfall intensity and duration. *Journal of Hydrology,* 228, 56–67.

Grimaldi, S., Serinaldi, F., Napolitano, F., and Ubertine, L. (2005). A 3-copula function application for design hyetograph analysis. *IAHS Publication,* 293, 1–9.

Hao, Z. and Singh, V. P. (2013). Entropy-based method for extreme rainfall analysis in Texas. *Journal of Geophysical Research*, 118, 263–273, doi:10.1029/2011JD017394.

Hashino, M. (1985). Formulation of the joint return period of two hydrologic variates associated with a poisson process. *Journal of Hydroscience and Hydraulic Engineering,* 3(2), 73–84.

Joe, H. (1997). *Multivariate Models and Multivariate Dependence Concepts*. Chapman & Hall/CRC, New York.

Kao, S.-C. and Govindaraju, R. S. (2007). A bivariate frequency analysis of extreme rainfall with implications for design. *Water Resources Research,* 112, D13119.

Kao, S.-C. and Govindaraju, R. S. (2008). Tivariate statistical analysis of extreme rainfall events via Plackett family of copulas. *Water Resources Research,* 44, W02415.

Khedun, C. P., Mishra, A. K., Singh, V. P., and Giardino, J. R. (2014). A copula-based precipitation model: investigating the interdecadal modulation of ENSO's impacts on monthly precipitation. *Water Resources Research*, 50, 1–20, doi:10.1002/2013WR013763.

Kurothe, R. S., Goel, N. K., and Mathur, B. S. (1997). Derived flood frequency distribution of negatively correlated rainfall intensity and duration. *Water Resources Research,* 33(9), 2103–2107.

Moazami, S., Golian, S., Kavianpour, M. R., and Hong, Y. (2014). Uncertainty analysis of bias from satellite rainfall estimates using copula method. *Atmospheric Research*, 137, 145–166

Singh, K. and Singh, V. P. (1991). Derivation of bivariate probability density functions with exponential marginals. *Stochastic Hydrology and Hydraulics,* 6(1), 47–54.

Vernieuwe, H., Vandenberghe, S., De Bates, B., and Verhoest, N. E. C. (2015). A continuous rainfall model based on vine copulas. *Hydrology and Earth System Sciences,* 19, 2685–2699. doi:10.5194/hess-19-2685-2015.

Yue, S. (2000a). Joint probability distribution of annual maximum storm peaks and amounts as represented by daily rainfalls. *Hydroscience Journal,* 45(2), 315–326.

Yue, S. (2000b). The Gumbel logistic model for representing a multivariate storm event. *Advances in Water Resources,* 24(2), 179–185.

Yue, S. (2000c). The Gumbel mixed model applied to storm frequency analysis. *Water Resources Management.* 14(5), 377–389.

Zhang, L. and Singh, V. P. (2007a). Bivariate rainfall frequency analysis using Archimedean copulas. *Journal of Hydrology,* 332, 93–109.

Zhang, L. and Singh, V. P. (2007b). IDF curves using Frank Archimedean copula. *Journal of Hydrologic Engineering,* 12(6), 651–662.

Zhang, L. and Singh, V. P. (2007c). Gumbel–Houggard copula for trivariate rainfall frequency analysis. *Journal of Hydrologic Engineering* 12(4), 409–419.

Zhang, Q., Li, J., and Singh, V. P. (2012). Application of Archimedean copulas in the analysis of the precipitation extremes: effects of precipitation change. *Theoretical and Applied Climatology*, 107(1–2), 255–264.

Zhang, Q., Li, J., Singh, V. P., and Xu, C.-Y. (2013). Copula-based spatio-temporal patterns of precipitation extremes in China. *International Journal of Climatology*, 33(5), 1140–1152.

11

Flood Frequency Analysis

ABSTRACT

In this chapter, copula modeling is applied to flood analysis with the use of real-world flood data. The chapter is structured in the following sections: (i) an introduction; (ii) at-site flood frequency analysis; (iii) spatial dependence for flood variables; and (iv) concluding remarks.

11.1 Introduction

Univariate flood frequency analysis has long been done for design of hydraulic structures, such as levees, flood walls, spillways, dams, culverts, drainage structures, and reservoirs, as well as for risk and uncertainty analysis. In the past decade, hydrologists have employed the copula theory for bivariate/multivariate flood frequency analyses. The advantages of applying the copula theory are that (i) it allows for separate consideration of marginal distributions and the joint distribution (i.e., copulas); (ii) it allows one to investigate both linear and nonlinear dependence structures; (iii) the tail dependence may be better captured; and (iv) it is easier to extend to higher dimensions through the vine copula or meta-elliptical copulas. The copula methodology has been applied to model the bivariate and multivariate flood frequency analysis (Chowdhary et al., 2011; Chen et al., 2012, 2013; Bezak et al., 2014; Sraj et al., 2015; Durocher et al., 2016; Requena et al., 2016; among others).

11.2 At-Site Flood Frequency Analysis

Univariate flood frequency analysis (e.g., using annual peak discharge) has long been a standard hydrological design method. In the United States, the log-Pearson type III distribution is still the standard distribution for flood frequency analysis, even though it is known that annual peak discharge by itself is not sufficient to account for flood risk.

A given flood event may be characterized by three important characteristics, i.e., peak discharge, volume, and duration. These three characteristics interact with one another when assessing flood risk or flood damage. As an example, a flood event with a longer duration may breach a levee due to long inundation time and possibly a large flood volume, while the peak discharge in this case may not be high. Another example is when a flood event

Table 11.1. *Flood data (Yue 1999).*[a]

Year	Q (cms)	V (day.cms)	D (days)	Year	Q (cms)	V (day.cms)	D (days)
1963	968	58,538	111	1980	949	33,010	69
1964	1780	68,828	98	1981	1,500	64,631	114
1965	1330	38,682	73	1982	1,920	50,525	77
1966	1650	54,139	78	1983	1,590	67,223	80
1967	934	39,744	75	1984	1,460	57,769	96
1968	1100	37,213	84	1985	1,210	47,627	80
1969	1380	50,895	80	1986	1,690	46,735	74
1970	1780	66,879	96	1987	610	35,600	96
1971	1420	38,634	66	1988	993	36,882	80
1972	1160	42,497	79	1989	1,490	41,943	63
1973	1470	55,766	78	1990	1,570	38,568	59
1974	2400	84,198	80	1991	1,130	49,226	93
1975	1260	48,790	83	1992	1,820	51,752	77
1976	1490	60,767	84	1993	1,360	45,263	83
1977	1370	60,824	92	1994	1,170	74,840	126
1978	1530	63,663	102	1995	1,550	51,853	80
1979	2040	59,254	76				

Note: [a] In this dataset, discharge (Q), flood volume (V), and flood duration (D) are considered independent identically distributed (i.i.d.) random variables.

with a higher peak discharge and a shorter duration may overtop a flood wall, causing flood damage. To further explain how to do flood frequency analysis considering all three characteristics, we will use the flood data listed in Table 11.1 (Yue, 1999) as an illustrative example.

The at-site trivariate flood frequency analysis in this chapter follows this procedure:

1. Collect the streamflow sequence and separate the streamflow sequence into peak discharge, flood duration, and flood volume variable.
2. Assess the pairwise overall dependence nonparametrically with the use of the Kendall rank-based correlation coefficient.
3. Apply the vine copula approach to study the dependence structure. The bivariate copula (building block) candidates are selected based on the nonparametric tail dependence coefficient and Kendall correlation coefficient.
4. Perform the risk analysis through the joint and conditional return period.

11.2.1 Brief Discussion of Dataset

As stated in Yue (1999), the flood dataset was collected for Asuapmushuan River basin in Quebec, Canada. Due to the constraints of the dataset, Yue (1999) applied the maximum annual daily discharge (Q [m^3/s]), the corresponding flood volume (V [day m^3/s]), and

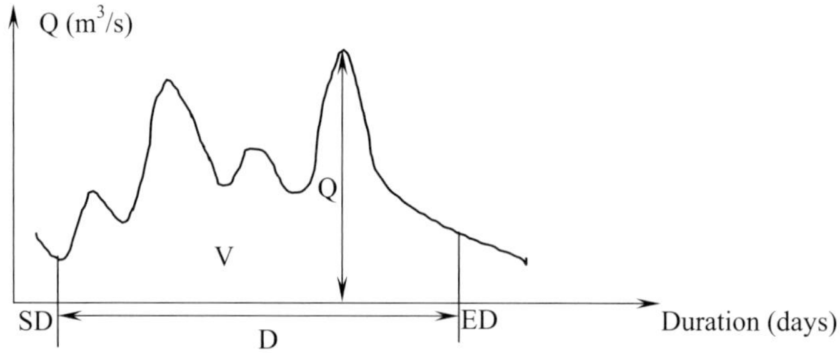

Figure 11.1 Schematic for a given flood event.

duration (D [day]) for frequency analysis. According to Yue (1999), the values of flood volume and duration were determined from the schematic (Figure 11.1) and Equation (11.1) as follows:

$$D = ED - SD; V = \sum_{i=SD}^{ED} q_i - \frac{1}{2}(q_{SD} + q_{ED}) \tag{11.1}$$

In Equation (11.1), SD and ED represent the starting time and the ending time of the flood event, respectively; D represents the duration of the flood event; q_i represents the discharge of day-i during the flood event; and V represents the flood volume.

11.2.2 Dependence Measure of Flood Variables: Nonparametric Assessment

Before we apply copulas to at-site flood frequency analysis, we compute the sample Kendall's correlation coefficient using Equation (3.73), as listed in Table 11.2. Using Equations (3.76)–(3.79) and Equation (3.80) presented in Sections 3.4.3 and 3.4.4, Figure 11.2 graphs the chi- and K-plots to assess the dependence among the flood random variables. Table 11.2 and Figure 11.2 clearly indicate the positive dependence between $\boldsymbol{Q}$ and $\boldsymbol{V}$ as well as between $\boldsymbol{V}$ and $\boldsymbol{D}$, while the negative dependence is detected between $\boldsymbol{Q}$ and $\boldsymbol{D}$. Physically, the dependence structure implies that (i) high flow tends to result in high flood volume; (ii) long flood duration (or long inundation time) also tends to lead to high flood volume (e.g., flood events due to slow moving storms); (iii) a high flow event may lead to a short-duration flood event (e.g., flash flooding caused by short-duration, high-intensity storms). Thus, it is more advantageous to take all three flood characteristics into consideration than assuming peak flow and flood duration being independent, as is usually done in conventional at-site bivariate/multivariate flood frequency analysis (e.g., Yue (1999)). In addition, the K-plots in the upper triangle of Figure 11.2 confirm the positive dependence of Q and V and of V and D and close to the independence of Q and D, the same as the chi-plots in the lower triangle and scatter plots of (Q, V), (V, D), and (Q, D) placed diagonally.

With the initial dependence assessment, we will apply the vine copula, meta-Gaussian, and meta-Student t copulas to model the dependence structure for trivariate flood variables.

Table 11.2. *Sample Kendall's tau correlation coefficient.*

	Q	V	D
Q	1	0.41	–0.13
V	0.41	1	0.42
D	–0.13	0.42	1

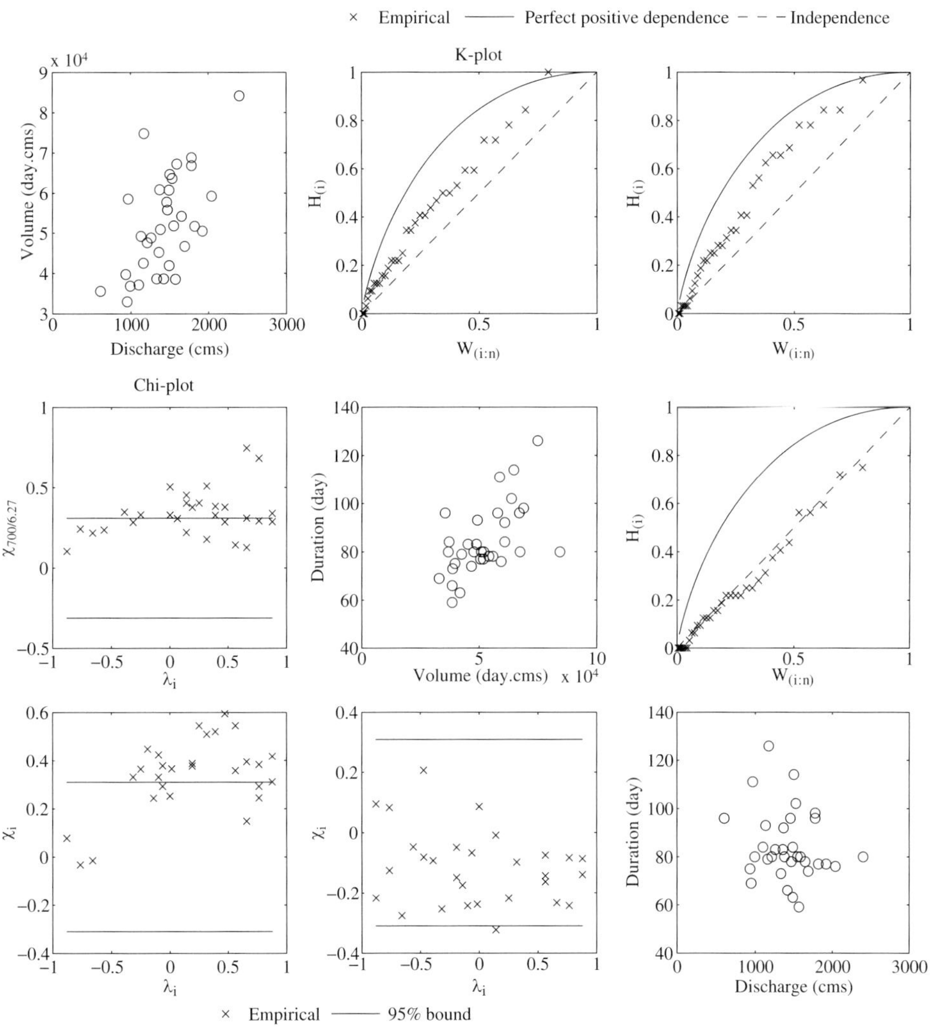

Figure 11.2 K-plots, chi-plots, and scatter plots for flood variables.

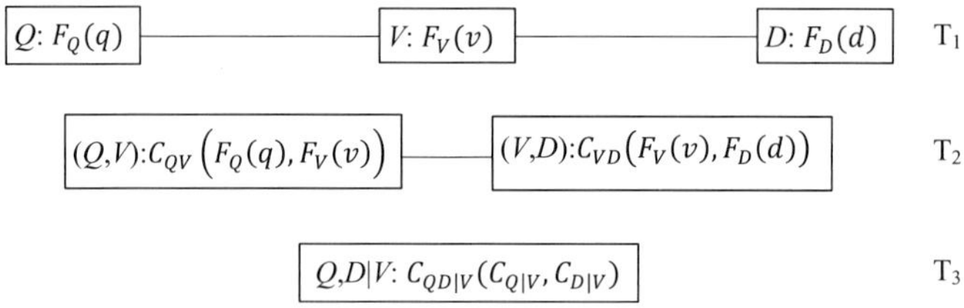

Figure 11.3 Vine-copula schematic for at-site trivariate flood variables.

11.2.3 Vine Copula–Based at-Site Flood Frequency Analysis

As discussed in Chapter 5, vine copulas belong to the asymmetric copula family. Given that flood volume has a higher degree of association with both flood peak flow and flood duration, we choose volume as the center variable to build the vine structure, as shown in Figure 11.3. As shown in Figure 11.3 and the discussions in Chapter 5, the bivariate copula is the building block for the entire structure. More specifically, we have full freedom to choose the best-fitted copula for (*Q*, *V*) and (*V*, *D*) separately in T_1. Then, based on the best-fitted copula in T_1 we will be able to choose the best-fitted copula in T_2.

Copula Candidates for T_1

As shown in Figure 11.3 and the discussions in the previous chapters, we need to first compute the marginal distributions nonparametrically (e.g., Weibull plotting position formula Equation (3.103), or kernel density) or parametrically with fitted marginal distributions. Here we will use the Weibull plotting position formula to compute the marginals, as shown in Table 11.3. Before we choose the copula candidate, we assess the tail dependence of (*Q*, *V*) and (*V*, *D*) such that we can make a better judgment to choose the candidate. Figure 11.4 shows the scatter plot using the empirical marginals of each of (*Q*, *V*) and of each of (*V*, *D*). Compared to the left tail (i.e., the lower tail) dependence, we are usually more interested in the right tail (i.e., the upper tail) dependence for these extreme events. Based on the tail dependence concept discussed in Chapter 3, we will first introduce how to evaluate the empirical tail dependence coefficient in what follows.

The tail dependence may be evaluated either graphically (Abberger, 2005) or numerically (Frahm et al., 2005; Schmidt and Stradtmuller, 2006). Here the nonparametric estimation is discussed in detail. Following Frahm et al. (2005), the nonparametric estimation is based on the empirical copula (i.e., Equation (3.64)) without any assumption on either parametric copula or marginals. In general, there are three types of nonparametric estimation (i.e., log-estimator [LOG], secant of the copula's diagonal [SEC], and CFG; Poulin et al., 2007) for the upper-tail dependence ($\hat{\lambda}_U$) that can be expressed as follows:

$$\hat{\lambda}_U^{LOG} = 2 - \frac{\log C_m\left(\frac{n-k}{n},\frac{n-k}{n}\right)}{\log\left(\frac{n-k}{n}\right)}, 0 < k < n \tag{11.2}$$

Table 11.3. *Marginal distributions computed using the Weibull plotting position formula.*

Year	Q (cms)	V (day cms)	D (days)	$F(Q)$	$F(V)$	$F(D)$
1963	968	58,538	111	0.12	0.68	0.91
1964	1,780	68,828	98	0.84	0.91	0.85
1965	1,330	38,682	73	0.35	0.21	0.15
1966	1,650	54,139	78	0.76	0.59	0.34
1967	934	39,744	75	0.06	0.24	0.21
1968	1,100	37,213	84	0.18	0.12	0.66
1969	1,380	50,895	80	0.44	0.50	0.49
1970	1,780	66,879	96	0.84	0.85	0.79
1971	1,420	38,634	66	0.47	0.18	0.09
1972	1,160	42,497	79	0.24	0.29	0.38
1973	1,470	55,766	78	0.53	0.62	0.34
1974	2,400	84,198	80	0.97	0.97	0.49
1975	1,260	48,790	83	0.32	0.41	0.60
1976	1,490	60,767	84	0.57	0.74	0.66
1977	1,370	60,824	92	0.41	0.76	0.71
1978	1,530	63,663	102	0.65	0.79	0.88
1979	2,040	59,254	76	0.94	0.71	0.24
1980	949	33,010	69	0.09	0.03	0.12
1981	1,500	64,631	114	0.62	0.82	0.94
1982	1,920	50,525	77	0.91	0.47	0.28
1983	1,590	67,223	80	0.74	0.88	0.49
1984	1,460	57,769	96	0.50	0.65	0.79
1985	1,210	47,627	80	0.29	0.38	0.49
1986	1,690	46,735	74	0.79	0.35	0.18
1987	610	35,600	96	0.03	0.06	0.79
1988	993	36,882	80	0.15	0.09	0.49
1989	1,490	41,943	63	0.57	0.26	0.06
1990	1,570	38,568	59	0.71	0.15	0.03
1991	1,130	49,226	93	0.21	0.44	0.74
1992	1,820	51,752	77	0.88	0.53	0.28
1993	1,360	45,263	83	0.38	0.32	0.60
1994	1,170	74,840	126	0.26	0.94	0.97
1995	1,550	51,853	80	0.68	0.56	0.49

$$\widehat{\lambda}_U^{SEC} = 2 - \frac{1 - C_m\left(\frac{n-k}{n}, \frac{n-k}{n}\right)}{1 - \frac{n-k}{n}}, 0 < k < n \tag{11.3}$$

$$\lambda_U^{CFG} = 2 - 2\exp\left(\frac{1}{n}\sum_{i=1}^{n} \log\left(\sqrt{\log\left(\frac{1}{U_i}\right)\log\left(\frac{1}{V_i}\right)} \Big/ \log\left(\frac{1}{\max\left(U_i, V_i\right)}\right)\right)\right) \tag{11.4}$$

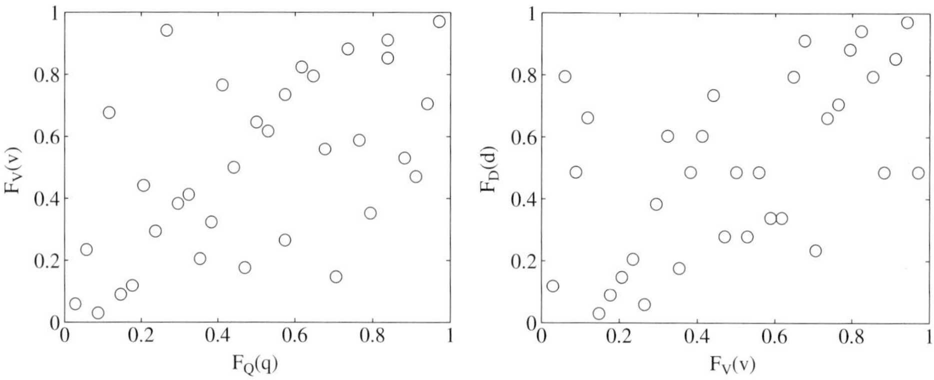

Figure 11.4 Scatter plots for the marginal of (Q, V) and of (V, D).

In Equations (11.2)–(11.4), n is the sample size; U_i, V_i are the marginal variables; and k is the chosen threshold of the LOG and SEC methods.

The LOG method was proposed by Coles et al. (1999). The SEC method first appeared in Joe (1997). The threshold k can be estimated using the heuristic plateau-finding algorithm proposed by Frahm et al. (2005), which can be formulated as follows:

1. Smooth using the box kernel with bandwidth $b \in \mathbb{N}$ (usually each moving average window should maintain 1% data) to compute the average of $(2b+1)$ successive points from $\hat{\lambda}_1, \ldots, \hat{\lambda}_n$ (i.e., mapping $k \mapsto \hat{\lambda}_k, k = 1,2,\ldots,n$) to obtain $\bar{\lambda}_1, \ldots, \bar{\lambda}_{n-2b}$.
2. Set plateau length $m = \lfloor \sqrt{n-2b} \rfloor$ and define a vector: $p_k = \left(\bar{\lambda}_k, \ldots, \bar{\lambda}_{k+m-1}\right), k = 1, \ldots, n-2b-m+1$.
3. Set the stopping criteria using the standard deviation of $\bar{\lambda}_1, \ldots, \bar{\lambda}_{n-2b}$. The threshold k can then be estimated from the first plateau p_k that satisfies the condition:

$$\sum_{i=k+1}^{k+m-1} \left|\bar{\lambda}_i - \bar{\lambda}_k\right| \leq 2\sigma \tag{11.5}$$

   If k is un-identified, $\hat{\lambda}_U$ is set as 0; otherwise, move on to step 4.
4. Estimate the upper-tail dependence coefficient for threshold k as follows:

$$\hat{\lambda}_U(k) = \frac{1}{m}\sum_{i=1}^{m} \bar{\lambda}_{k+i-1} \tag{11.6}$$

The CFG method (i.e., Equation (11.4)) first appeared in Capéraà et al. (2007) that does not require the estimation of a threshold. However, there exists a strong underlying assumption: the empirical copula may be approximated by the extreme value (EV) copula (e.g., the Gumbel–Hougaard copula as an example). It is worth noting that the lower-tail dependence is the same as the upper-tail dependence of the survival copula.

The empirical upper-/lower-tail dependence coefficient is computed, as listed in Table 11.4. To illustrate the procedure, the empirical upper-tail dependence coefficient is further explained using Q and V with the LOG method. From the sample data listed in

Table 11.4. *Upper- and lower-tail dependence coefficients for* (Q, V) *and* (V, D).

	Upper			Lower	
	LOG	SEC	CFG	LOG	SEC
Q & *V*	0.29	0.38	0.43	0.74	0.95
V & *D*	0.49	0.60	0.51	0.60	0.92

Table 11.5. $\hat{\lambda}_k$ *computed using Equation (11.2).*

k	C_m	$\hat{\lambda}_k$	k	C_m	$\hat{\lambda}_k$
1	0.9697	1.0000	17	0.3636	0.6026
2	0.9091	0.4755	18	0.3333	0.6066
3	0.8485	0.2761	19	0.3030	0.6076
4	0.7879	0.1549	20	0.2727	0.6053
5	0.7576	0.3102	21	0.2121	0.4672
6	0.7273	0.4131	22	0.1818	0.4483
7	0.6667	0.2993	23	0.1818	0.5721
8	0.6061	0.1963	24	0.1515	0.5476
9	0.5758	0.2664	25	0.1515	0.6683
10	0.5455	0.3210	26	0.1212	0.6391
11	0.4848	0.2146	27	0.1212	0.7622
12	0.4545	0.2556	28	0.0909	0.7293
13	0.4242	0.2878	29	0.0606	0.6715
14	0.3939	0.3126	30	0.0606	0.8309
15	0.3939	0.4631	31	0.0303	0.7527
16	0.3939	0.5956	32	0	

Table 11.1, the sample size is $n = 33$. Applying Equation (11.2), we compute $\hat{\lambda}_k$ for $k = 1, 2, \ldots, 32$, as listed in Table 11.5. With the initial $\hat{\lambda}_k$s estimated for the LOG method, we can now move on to evaluate the tail dependence. With the sample size of 33, we set the bandwidth $b = 0$. With $b = 0$, we have $\hat{\lambda}_k = \bar{\lambda}_k$, and the standard deviation of vector $\bar{\lambda}$s is 0.2114. The plateau length $m = 5$ yields the vector with size of 27 by 5 for the non-NaN values that are listed in Table 11.6. Finally, applying Equation (11.5), we obtain the first p vector that satisfies the condition that index $k = 3$ that results in the following:

$$\sum\nolimits_{i=4:7} \left|\bar{\lambda}_i - \bar{\lambda}_3\right| = 0.3155 < 2(0.2114) = 0.4229.$$

We obtain the upper tail dependence as $\lambda_U^{LOG} \approx 0.29$.

From the tail dependence coefficients evaluated and listed in Table 11.4, it is seen that there exist both upper and lower tail dependences for the bivariate (Q, V) and

Table 11.6. *Vector p with the plateau length* m = *5.*

k	$\bar{\lambda}_k$	$\bar{\lambda}_{k+1}$	$\bar{\lambda}_{k+2}$	$\bar{\lambda}_{k+3}$	$\bar{\lambda}_{k+4}$
1	1.0000	0.4755	0.2761	0.1549	0.3102
2	0.4755	0.2761	0.1549	0.3102	0.4131
3	0.2761	0.1549	0.3102	0.4131	0.2993
4	0.1549	0.3102	0.4131	0.2993	0.1963
5	0.3102	0.4131	0.2993	0.1963	0.2664
6	0.4131	0.2993	0.1963	0.2664	0.3210
7	0.2993	0.1963	0.2664	0.3210	0.2146
8	0.1963	0.2664	0.3210	0.2146	0.2556
9	0.2664	0.3210	0.2146	0.2556	0.2878
10	0.3210	0.2146	0.2556	0.2878	0.3126
11	0.2146	0.2556	0.2878	0.3126	0.4631
12	0.2556	0.2878	0.3126	0.4631	0.5956
13	0.2878	0.3126	0.4631	0.5956	0.6026
14	0.3126	0.4631	0.5956	0.6026	0.6066
15	0.4631	0.5956	0.6026	0.6066	0.6076
16	0.5956	0.6026	0.6066	0.6076	0.6053
17	0.6026	0.6066	0.6076	0.6053	0.4672
18	0.6066	0.6076	0.6053	0.4672	0.4483
19	0.6076	0.6053	0.4672	0.4483	0.5721
20	0.6053	0.4672	0.4483	0.5721	0.5476
21	0.4672	0.4483	0.5721	0.5476	0.6683
22	0.4483	0.5721	0.5476	0.6683	0.6391
23	0.5721	0.5476	0.6683	0.6391	0.7622
24	0.5476	0.6683	0.6391	0.7622	0.7293
25	0.6683	0.6391	0.7622	0.7293	0.6715
26	0.6391	0.7622	0.7293	0.6715	0.8309
27	0.7622	0.7293	0.6715	0.8309	0.7527

(V, D) flood variables. To this end, we will have the following choices to investigate the dependence:

i. Use a mixed copula to model the bivariate flood variables.
ii. Use two-parameter copulas (Joe, 1997) to model the bivariate flood variables.
iii. Use copulas with upper-tail dependence to model the bivariate flood variables.

In theory, (a) all three approaches should be able to capture the overall dependence structure; (b) compared with approaches ii and iii, approach i may better capture both upper and tail dependences; (c) among the three approaches, parameter estimation for approach i is most complex; and (d) if we are only concerned with the upper-tail dependence, we may prefer approach iii. In what follows, we will discuss the copula candidates for all three approaches.

Approach i: Mixture Copula for Bivariate Variables Following the discussion in Chapter 4, we introduce the Archimedean copula class. In this class, the Gumbel–Hougaard copula possesses the upper-tail dependence only $(\lambda_U = 2 - 2^{\frac{1}{\theta_{GH}}})$, while its survival copula possesses lower-tail dependence $(\lambda_L = 2 - 2^{\frac{1}{\theta_{SGH}}})$; and the Clayton copula only possesses the lower-tail dependence $(\lambda_L = 2^{-\frac{1}{\theta_C}})$. In addition, the Gumbel–Hougaard copula may only model the positive dependence, while the Clayton copula may model both positive and negative dependences. Following the discussion in Chapter 7, the meta-Gaussian copula, which is elliptical, has no tail dependence. Now through this approach, we will choose two candidates:

✓ Gumbel–Hougaard + meta-Gaussian + survivial Gumbel–Hougaard copulas
✓ Gumbel–Hougaard + meta-Gaussian + Clayton copulas

The Gumbel–Hougaard and Clayton copulas are listed in Chapter 4. The bivariate meta-Gaussian copula is expressed in Chapter 7. The survival Gumbel–Hougaard copula (C^{SGH}) and its density function (c^{SGH}) can be written as follows:

$$C^{SGH}(u_1, u_2; \theta) = u_1 + u_2 - 1 + C^{GH}(1 - u_1, 1 - u_2; \theta) \tag{11.7a}$$

$$c^{SGH}(u_1, u_2; \theta) = c^{GH}(1 - u_1, 1 - u_2; \theta) \tag{11.7b}$$

In Equations (11.7a) and (11.7b), $\theta\colon \theta \geq 1$ represents the copula parameter to be estimated.

The corresponding mixture copula model may then be written as follows:

$$(\text{A})\colon C(u_1, u_2; \boldsymbol{\theta}) = a_1 C^{GH}(u_1, u_2; \theta_1) + a_2 C^{SGH}(u_1, u_2; \theta_2) + a_3 C^{Normal}(u_1, u_2; \theta_3) \tag{11.8a}$$

$$(\text{B})\colon C(u_1, u_2; \boldsymbol{\theta}) = a_1 C^{GH}(u_1, u_2; \theta_1) + a_2 C^{clayton}(u_1, u_2; \theta_2) + a_3 C^{Normal}(u_1, u_2; \theta_3) \tag{11.8b}$$

where $\boldsymbol{\theta} = [\theta_1,\ \theta_2,\ \theta_3]$. $a_1, a_2, a_3 \in [0, 1]\colon a_1 + a_2 + a_3 = 1;$ are the weight factors.

Approach ii: Two-Parameter Copulas for Bivariate Variables As discussed in Joe (1997), the two-parameter Archimedean copulas may be capable of capturing both the overall dependence and the tail dependence. Following Joe (1997), we will briefly introduce BB1 BB4, and BB7 copulas.

BB1 Copula

$$\begin{aligned} C(u, v; \theta_1, \theta_2) &= \left\{1 + \left[\left(u^{-\theta_1} - 1\right)^{\theta_2} + \left(v^{-\theta_1} - 1\right)^{\theta_2}\right]^{\frac{1}{\theta_2}}\right\}^{-\frac{1}{\theta_1}} \\ &= \phi^{-1}\left(\phi_{\theta_1,\theta_2}(u) + \phi_{\theta_1,\theta_2}(v)\right); \theta_1 > 0, \theta_2 \geq 1 \end{aligned} \tag{11.9}$$

Its generating function and tail dependence function can be written as follows:

$$\phi_{\theta_1,\theta_2}(t) = \left(t^{-\theta_1} - 1\right)^{\theta_2},\ \lambda^U = 2 - 2^{\frac{1}{\theta_2}},\ \lambda^L = 2^{-\left(\frac{1}{(\theta_1\theta_2)}\right)} \tag{11.9a}$$

The BB1 copula can only be applied to model the positive dependence and may be considered as a two-parameter Archimedean copula. It possesses both upper- and lower-tail dependences. The limiting copulas are Gumbel–Hougaard copula ($\theta_1 \to 0$) and Clayton copula ($\theta_2 = 1$). With the combination of the Gumbel–Hougaard and Clayton copulas, the BB1 copula is able to capture both upper- and lower-tail dependences in which the upper-tail dependence is independent of parameter θ_1.

BB4 Copula

$$C(u, v; \theta_1, \theta_2) = \left(u^{-\theta_1} + v^{-\theta_1} - 1 - \left[\left(u^{-\theta_1} - 1\right)^{-\theta_2} + \left(v^{-\theta_1} - 1\right)^{-\theta_2}\right]^{-\frac{1}{\theta_2}}\right)^{-\frac{1}{\theta_1}} \tag{11.10}$$

where $\theta_1 \geq 0, \theta_2 > 0$. Its tail dependence functions can be written as follows:

$$\lambda^U = 2^{-\frac{1}{\theta_2}},\ \ \lambda^L = \left(2 - 2^{-\frac{1}{\theta_2}}\right)^{-\frac{1}{\theta_1}} \tag{11.10a}$$

Unlike the BB1 copula, the BB4 copula is not a two-parameter Archimedean copula. Its limiting copulas are the Clayton copula when $\theta_2 \to 0$ and the Galambos copula when $\theta_1 \to 0$. The Glambos copula belongs to an extreme value copula given as follows:

$$C(u, v, \delta) = uv \exp\left(\left((-\log u)^{-\delta} + (-\log v)^{-\delta}\right)^{-\frac{1}{\delta}}\right),\ \delta > 0 \tag{11.10b}$$

As seen from Equation (11.10a), the upper-tail dependence of the BB4 copula is independent of parameter θ_1.

BB7 Copula

$$C(u, v; \theta_1, \theta_2) = 1 - \left(1 - \left(\left(1 - (1-u)^{\theta_1}\right)^{-\theta_2} + \left(1 - (1-v)^{\theta_1}\right)^{-\theta_2} - 1\right)^{-\frac{1}{\theta_2}}\right)^{\frac{1}{\theta_1}} \tag{11.11}$$

where $\theta_1 \geq 1, \theta_2 > 0$.

BB7 is the same as the BB1 copula and is also a two-parameter Archimedean copula. Its generating function and tail dependence functions can be expressed as follows:

$$\phi_{\theta_1,\theta_2}(t) = \left(1 - (1-t)^{\theta_1}\right)^{-\theta_2} - 1;\ \lambda^U = 2 - 2^{\frac{1}{\theta_1}};\ \lambda^L = 2^{-\frac{1}{\theta_2}} \tag{11.11a}$$

The limiting copulas for the BB7 copula are the Clayton copula when $\theta_1 = 1$ and the Joe copula when $\theta_2 \to 0$.

Approach iii: Choosing Copulas with Upper-Tail Dependence The copulas are chosen from the Archimedean, extreme, and elliptical copula families as follows:

Archimedean family: Gumbel–Hougaard and Joe copulas
Extreme copula family: Galambos copula
Elliptical copula family: meta-Student t copula, $\lambda^U = \lambda^L = 2t_{v+1}\left(\frac{-\sqrt{v+1}\sqrt{1-\rho}}{\sqrt{1+\rho}}\right)$.

Among the copulas listed in approach iii, all four copulas possess the upper-tail dependence. In addition, only the meta-Student t copula also possesses the symmetric lower-tail dependence.

Parameter Estimation and the Best-Fitted Copula for T_l

Parameter Estimation for Approach i: Mixture Copula The pseudo-MLE discussed in Chapter 4 is applied to estimate the parameters of the mixture copula. The initial parameters are set as follows:

✓ Each copula is of equal weight.
✓ The initial copula parameters are represented as the random variables, which may be modeled by one copula.

For each case, we have the following:

Q & V

Case (A):

$a_1 = 0.1652, \theta_{GH} = 4.0597;\ a_2 = 0.8348, \theta_{SGH} = 1.6549, a_3 = 0, \theta_{normal} = 0.5955.$
$LL = 8.5657,\ \boldsymbol{AIC} = -\mathbf{7.131};\ \lambda^U = 0.13;\ \lambda^L = 0.40.$

Case (B):

$a_1 = 0.2295, \theta_{GH} = 3.7289;\ a_2 = 0.7705, \theta_{clayton} = 1.1227;\ a_3 = 0;\ \theta_{normal} = 0.5955$
$LL = 8.8470,\ \boldsymbol{AIC} = -\mathbf{7.694};\ \lambda^U = 0.1826;\ \lambda^L = 0.4156$

V & D

Case (A):

$a_1 = 0.7482, \theta_{GH} = 2.0434;\ a_2 = 0.2518, \theta_{SGH} = 1.1900, a_3 = 0, \theta_{normal} = 0.5845.$
$LL = 6.7587,\ \boldsymbol{AIC} = -\mathbf{3.5174};\ \lambda^U = 0.446;\ \lambda^L = 0.0528.$

Case (B):

$a_1 = 0.7628, \theta_{GH} = 2.0164;\ a_2 = 0.2372, \theta_{clayton} = 0.3963, a_3 = 0, \theta_{normal} = 0.5845.$
$LL = 6.7627,\ \boldsymbol{AIC} = -\mathbf{3.525};\ \lambda^U = 0.4499;\ \lambda^L = 0.0413.$

Parameter Estimation for Copula Candidates in Approaches ii and iii Similar to approach i, the pseudo-MLE is applied to estimate the parameters for the copula candidates presented in approaches ii and iii, which are listed in Table 11.7.

Considering the number of parameters, the overall dependence, and tail dependence, the BB7 copula is selected to model the dependence of Q & V as well as V & D for T_1. To further ensure the appropriateness of the selected copula, the formal S_n^B test is performed. Based on Rosenblatt's transform, the S_n^B test was introduced in Section 3.8.3. Hence, we have the following:

$$\text{Q \& V: } S_n^B = 0.0215, P = 0.897; \quad \text{V \& D: } S_n^B = 0.0351, P = 0.52.$$

With the confirmation from the formal goodness-of-fit statistical test, we fix the copula in T_1 and move on to T_2.

Copula Selection for T_2

To select the copula and estimate its parameters for T_2, we first compute the conditional copula of $Q|V$ and $D|V$ using the fitted BB7 copula for T_1. The conditional copulas of Q|V and D|V are obtained by taking the partial derivative with respect to $F(v)$ that are listed in Table 11.8.

Computing Kendall's correlation coefficient, we have $\tau = -0.5265$. With the negative correlation, we will choose the Frank, meta-Gaussian, and meta-Student t copulas as the candidates for modeling. Applying pseudo-MLE to the marginals estimated from the conditional copulas of $Q|V$ and $D|V$ (i.e., columns 4 and 5) with the initial parameter estimated from the estimated Kendall's correlation coefficient, we obtain

Frank: $\theta = -5.655, LL = 10.705, AIC = -19.411$
Meta-Gaussian: $\rho = -0.6898, LL = 10.656, AIC = -19.3127$
Meta-Student t: $\rho = -0.7034, v = 17.4524; LL = 10.776, AIC = -17.553$

Based on the log-likelihood and AIC values, we choose Frank copula to model T_2. Again applying the formal S_n^B goodness-of-fit test, we have the following: $S_n^B = 0.0456, P = 0.207$.

Now, we have finished building the vine-copula structure for the flood peak (Q), flood duration (D), and flood volume (V) variables in which the BB7 copula is applied to model the dependence of Q&V as well as of D&V in T_1. The reasons we choose the BB7 copula are that (a) compared to the five-parameter mixture copula, the two-parameter BB7 copula reaches the smallest AIC value; and (b) the BB7 copula reasonably captures the tail dependence of Q&V as well as D&V. Figure 11.5 compares simulated random variates with pseudo-observations. In Figure 11.5 Kendall's tau computed from simulation is also compared with sample Kendall's tau: $\tau_{Q,V} = 0.41, \tau_{V,D} = 0.42; \tau_{Q,D} = -0.13$. As seen in the box plots for (Q, V) and (V, D), the random variables simulated from the fitted BB7 copula well represent their dependence structure. Even though we did not directly investigate the copula function of (Q, D), the random variables simulated from the fitted BB7–BB7–Frank vine copula again reasonably represent the dependence structure of (Q, D).

Table 11.7. *Results of two-parameter and one-parameter copula candidates for T_1.*

Approach	Copula Family	Copulas	*Q&V*			*D&V*		
			θ, *LL*	λ^U	λ^L	θ,*LL*	λ^U	λ^L
ii	Two-parameter	BB1	[0.829, 1.235]	0.303	0.529	[0.1298, 1.6336]	0.472	0.038
			LL = 8.696, **AIC = –13.39**			LL = 6.654, **AIC = –11.308**		
		BB4	N/A	N/A	N/A	N/A	N/A	N/A
		BB7	[1.528,1.235]	0.426	0.57	[1.828, 0.535]	0.539	0.273
			LL = 9.024, **AIC = –14.048**			LL = 6.571, **AIC = –11.142**		
iii	One-parameter Archimedean	Gumbel–Hougaard	1.7508	0.514		1.717	0.503	
			LL = 7.047, **AIC = –12.094**			LL = 6.60, **AIC = –11.2**		
		Joe	1.957	0.575		2.021	0.5909	
			LL = 5.376, **AIC = –8.75**			LL = 5.935, **AIC = –9.87**		
	Extreme copula	Galambos	1.027	0.509		0.9976	0.499	
			LL = 6.947, **AIC = –11.894**			LL = 6.477, **AIC = –10.954**		
	Elliptical	Meta-Student *t*	[0.594, 2.438]	0.205	0.205	[0.574, 4.989]	0.125	0.125
			LL = 8.223, **AIC = –12.446**			LL = 6.3999, **AIC = –8.799**		

Table 11.8. *Conditional copula computed using the fixed BB7 copula in T_1.*

$F_n(Q)$	$F_n(V)$	$F_n(D)$	$F(Q\|V) = \frac{\partial C(F_n(q),F_n(v))}{\partial F_n(v)}$	$F(D\|V) = \frac{\partial C(F_n(d),F_n(v))}{\partial F_n(v)}$
0.12	0.68	0.91	0.02	0.95
0.84	0.91	0.85	0.57	0.54
0.35	0.21	0.15	0.59	0.22
0.76	0.59	0.34	0.78	0.26
0.06	0.24	0.21	0.04	0.30
0.18	0.12	0.66	0.47	0.90
0.44	0.50	0.49	0.39	0.50
0.84	0.85	0.79	0.69	0.58
0.47	0.18	0.09	0.77	0.14
0.24	0.29	0.38	0.27	0.50
0.53	0.62	0.34	0.42	0.24
0.97	0.97	0.49	0.78	0.05
0.32	0.41	0.60	0.30	0.71
0.57	0.74	0.66	0.39	0.54
0.41	0.76	0.71	0.18	0.58
0.65	0.79	0.88	0.44	0.85
0.94	0.71	0.24	0.96	0.11
0.09	0.03	0.12	0.67	0.49
0.62	0.82	0.94	0.37	0.94
0.91	0.47	0.28	0.96	0.25
0.74	0.88	0.49	0.45	0.17
0.50	0.65	0.79	0.36	0.82
0.29	0.38	0.49	0.28	0.58
0.79	0.35	0.18	0.90	0.18
0.03	0.06	0.79	0.11	0.97
0.15	0.09	0.49	0.50	0.80
0.57	0.26	0.06	0.77	0.06
0.71	0.15	0.03	0.94	0.05
0.21	0.44	0.74	0.13	0.84
0.88	0.53	0.28	0.93	0.23
0.38	0.32	0.60	0.47	0.75
0.26	0.94	0.97	0.03	0.89
0.68	0.56	0.49	0.67	0.46

11.2.4 At-Site Flood Risk Analysis

Similar to Yue et al. (1999), the Gumbel distribution is applied to model the marginal distributions for flood peak, volume, and duration to assess and compare the risk measure. The Gumbel distribution (also called the EV1 distribution) can be given as follows:

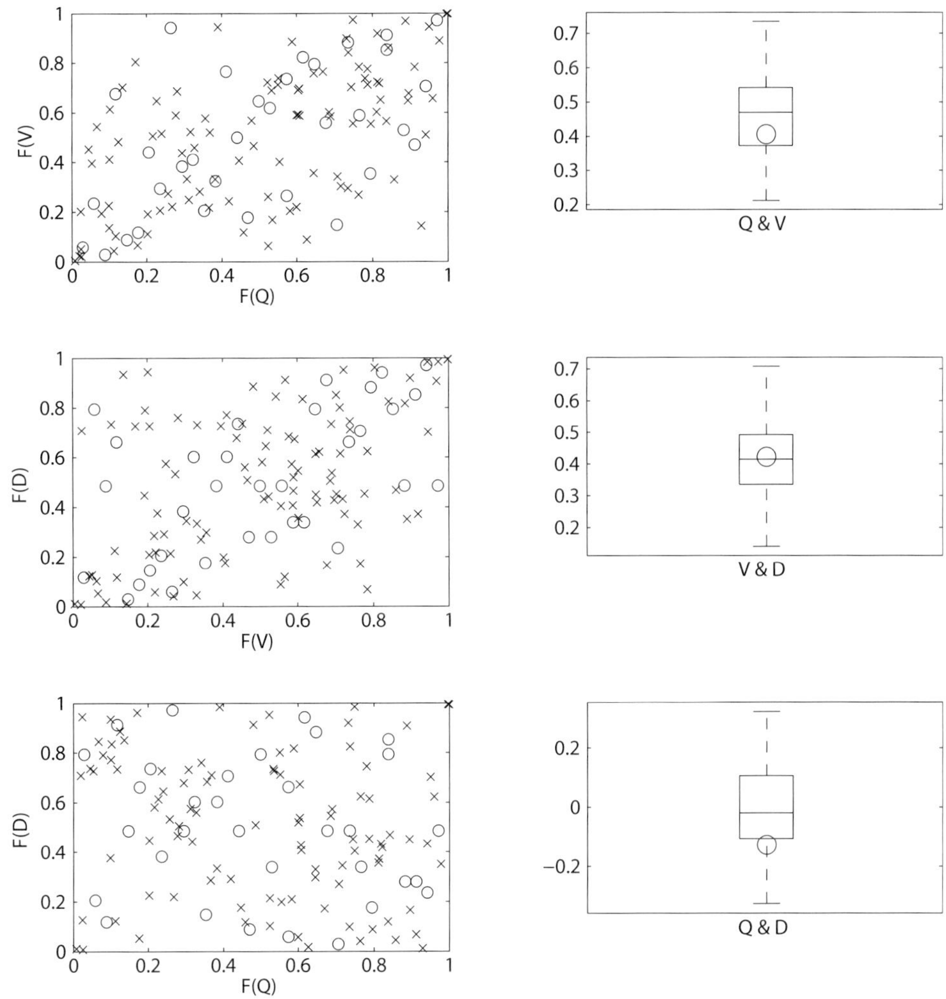

Figure 11.5 Comparison of simulated variables with pseudo-observations.

$$F(x) = \exp\left[-exp\left(\frac{x-\mu_x}{\alpha_x}\right)\right] \tag{11.12}$$

where μ_x, α_x are, respectively, the location and the scale parameters for random variable X.

Using MLE, parameters of the marginal distributions are listed in Table 11.9. Figure 11.6 plots the frequency histograms for flood variables. Figure 11.6 shows that Gumbel distribution may not be the proper choice for flood duration. We further choose the log-normal distribution for flood duration. The parameters are also listed in Table 11.9.

Table 11.9. *Fitted parametric marginal distributions and KS test results.*

	Discharge		Volume		Duration	
Distribution	Parameters	KS test[a]	Parameters	KS test	Parameters	KS test
Gumbel	**[1608.5, 383.9]**	**[0.14, 0.47]**	**[58591, 13148]**	**[0.13, 0.56]**	[92.04, 16.89]	[0.23, 0.04]
Log-normal					**[4.42, 0.17]**	**[0.17, 0.24]**

Note: [a] In the KS test, the first column is the test statistics, the second column is the *P*-value.

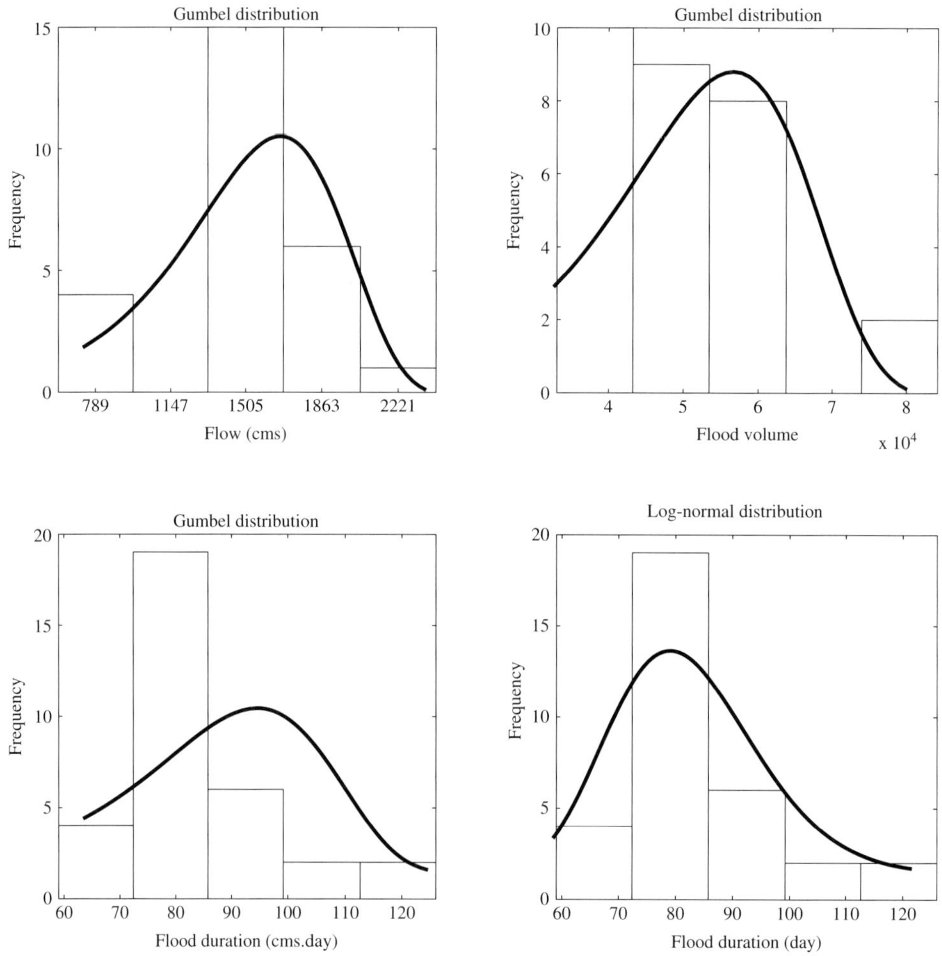

Figure 11.6 Frequency histograms for the fitted Gumbel and log-normal distributions.

Joint and Conditional Return Periods for Bivariate Cases of Discharge and Flood Volume, and Flood Volume and Duration

In this section, we will discuss the important risk measure by using joint and conditional return periods for the bivariate case. For the joint return period, we will consider the "AND" and "OR" cases. For the conditional return period, we will consider the $X > x \mid Y > y$ (e.g., $Q > q \mid V > v$) and $X > x \mid Y = y$ (e.g., the $Q > q \mid V = v$) cases.

Joint Return Period of Discharge and Flood Volume, and Flood Volume and Duration

As discussed in Section 3.10.2, the joint return periods are computed for two cases: the "AND" case and "OR" case. In the "AND" case, the critical values set for both variables are exceeded. In the "OR" case, the critical value for at least one variable is exceeded. Using the 5-, 10-, 25-, 50-, 100-year discharge and flood volume events as criteria, we can easily compute the joint return period with the use of the BB7 copula fitted to the flood discharge and flood volume. The design events are listed in Table 11.10 for flood discharge (*Q*) and flood volume (*V*) with given return periods using the fitted parametric Gumbel distribution.

"AND" Case: $T(V > v \cap D > d)$ From Equation (3.136), the "AND" case implies to compute the survival copula of the bivariate random variable. using the five-year design discharge and the five-year design flood volume as an example, we can write the following:

$$\begin{aligned} F(Q > Q_5 \ \& \ V > V_5) &= C(F_Q > 0.8 \ \& \ F_V > 0.8) \\ &= 1 - F_Q - F_v + C(F_Q, F_V) \\ &= 1 - 0.8 - 0.8 + C_{BB7}(0.8, 0.8) \\ &= 1 - 0.8 - 0.8 + 0.7033 = 0.1033 \end{aligned}$$

$$T(Q > Q_5 \ \& \ V > V_5) = \frac{1}{F(Q > Q_5 \ \& \ V > V_5)} = \frac{1}{0.1033} = 9.68 \ yr.$$

With the same logic, the joint return periods of the "AND" case for *Q*&*V* and *V*&*D* are computed, as listed in Table 11.11.

Table 11.10. *Marginal design events with given return periods.*

Variables	Marginal	Return period (years) 5	10	25	50	100
Discharge (cms)	Gumbel	1,791.14	1,928.62	2,057.21	2,132.07	2,194.69
Volume (cms·day)	Gumbel	64,848.11	69,557.12	73,961.78	76,525.99	78,670.80
Duration (day)	Log-normal	95.57	102.78	111.06	116.76	122.14
	Gumbel	100.08	106.13	111.79	115.08	117.84

Table 11.11. *Joint return period of the "AND" case.*

Return period (years)		V (cms·day) 64848.11	69557.12	73961.78	76525.99	78670.80
	1,791.14	9.68	15.54	32.33	59.40	112.37
Q (cms)	1,928.62	15.54	21.78	39.29	67.40	122.04
	2,057.21	32.33	39.29	57.56	86.74	143.61
	2,132.07	59.40	67.40	86.74	116.59	174.80
	2,194.69	112.37	122.04	143.61	174.80	234.21
		D (day) 95.57	102.78	111.06	116.76	122.14
	64,848.11	8.76	13.68	28.78	54.03	104.39
	69,557.12	13.68	18.25	32.93	58.10	108.59
V (cms·day)	73,961.78	28.78	32.93	46.24	70.38	120.45
	76,525.99	54.03	58.10	70.38	92.69	140.92
	78,670.80	104.39	108.59	120.45	140.92	185.50

"OR" Case: $T(Q > q \cup V > v)$ The "OR" case implies that at least one variable exceeds the critical design value. The return period of the "OR" case is given in Equation (3.137). Using the five-year design discharge and the five-year design flood volume as an example, the exceedance probability of the "OR" case can be written as follows:

$$F(Q > Q_5 \text{ or } V > V_5) = 1 - C(F_Q \leq 0.8, F_V \leq 0.8) = 1 - 0.7033 = 0.2967.$$

$$T(Q > Q_5 \text{ or } V > V_5) = \frac{1}{F(Q > Q_5 \text{ or } V > V_5)} \approx 3.37 \text{ yr}.$$

The rest of the "OR' case computations are listed in Table 11.12.

Compared with the "AND" case, the return period of the "OR" case is less than that of the "AND" case. It is obviously in agreement with reality. As an example, the discharge may be exceeded, while the volume does not exceed the design volume and vice versa.

Conditional Return Period for Flood Discharge and Flood Volume, and Flood Volume and Flood Duration

In this section, we will discuss two cases of conditional return periods: (i) $X > x|Y > y$ and (ii) $X > x|Y = y$.

Case i: $T(X > x|Y > y)$ Following Nelsen (2006) as well as the discussion in Chapter 3, the conditional probability of $P(X > x|Y > y)$ *or* $C(F_X > u|F_Y > v)$ may lead to the right tail increasing (RTI) property, if $\frac{1-u-v+C(u,v)}{1-v}$ is nondecreasing in u.

Table 11.12. *Joint return period of "OR" case.*

Return period (years)		V (cms·day) 64848.11	69557.12	73961.78	76525.99	78670.80
	1,791.14	3.37	4.24	4.78	4.92	4.97
Q (cms)	1,928.62	4.24	6.49	8.73	9.51	9.82
	2,057.21	4.78	8.73	15.97	20.63	23.24
	2,132.07	4.92	9.51	20.63	31.82	41.19
	2,194.69	4.97	9.82	23.24	41.19	63.57
		D (day) 95.57	102.78	111.06	116.76	122.14
	6,4848.11	3.50	4.41	4.87	4.96	4.99
	69,557.12	4.41	6.89	9.12	9.73	9.92
V (cms.day)	73,961.78	4.87	9.12	17.13	21.84	23.98
	76,525.99	4.96	9.73	21.84	34.23	43.66
	78,670.80	4.99	9.92	23.98	43.66	68.45

The return period is written with $\mu = 1$ as follows:

$$T(X > x|Y > y) = \frac{1}{(1-v)(1-u-v+c(u,v))} \tag{11.13}$$

Using the flood volume as a conditioning variable, the conditional distribution and conditional return period are computed, as listed in Table 11.13. Figure 11.7 plots the conditional probability given $V > v$ for flood discharge and flood duration computed using the copula. Table 11.13 and Figure 11.7 show that the RTI property does exist for Q & V as well as V & D. The existence of RTI also implies the right tail dependence.

Case (ii): $\boldsymbol{T(X > x|Y = y)}$. Following Nelsen (2006) and the discussion in Chapter 3, the conditional probability of $P(X > x|Y = y)$ or equivalently $C(U > u|V = v)$ may lead to stochastic monotonicity (or stochastic increasing of X in Y), i.e., $\partial C(u,v)/\partial v$ is a non-increasing function in v. Or in other words, $1 - \partial C(u,v)/\partial v$ is a nondecreasing function in v. For the chosen BB7 copula (i.e., Equation (11.11)), its partial derivative can be written as follows:

$$\frac{\partial C(u,v)}{\partial v} = \frac{(1-v)^{\theta_1-1}\left(1-\frac{1}{S^{\frac{1}{\theta_2}}}\right)^{\frac{1}{\theta_1}-1}}{\left(1-(1-v)^{\theta_1}\right)^{\theta_2+1} S^{\frac{1}{\theta_2}+1}}; S = \left(1-(1-u)^{\theta_1}\right)^{-\theta_2} + \left(1-(1-v)^{\theta_1}\right)^{-\theta_2} - 1 \tag{11.14}$$

Figure 11.8 plots the conditional probability of discharge given flood volume as well as flood duration given flood volume. Figure 11.8 clearly shows that discharge and duration

Table 11.13. *Conditional return periods of* $Q \mid V > v$ and V| $D > d$.

Return period (years)		Given $V > v$ (cms·day)				
		64848.11	69557.12	73961.78	76525.99	78670.80
	1,791.14	48.41	77.72	161.63	297.02	561.84
Q (cms)	1,928.62	155.44	217.83	392.86	673.97	1,220.36
	2,057.21	808.16	982.16	1,439.00	2,168.49	3,590.12
	2,132.07	2,970.17	3,369.86	4,336.98	5,829.03	8,739.74
	2,194.69	11,236.71	12,203.58	14,360.48	17,479.48	23,419.82
		Given $D > d$ (day)				
		95.57	102.78	111.06	116.76	122.14
	64,848.11	95.57	102.78	111.06	116.76	122.14
	69,557.12	43.81	68.38	143.90	270.13	521.96
V (cms·day)	73,961.78	136.75	182.47	329.27	580.97	1,085.94
	76,525.99	719.48	823.17	1,155.97	1,759.44	3,011.34
	78,670.80	2,701.34	2,904.83	3,518.89	4,634.37	7,046.14

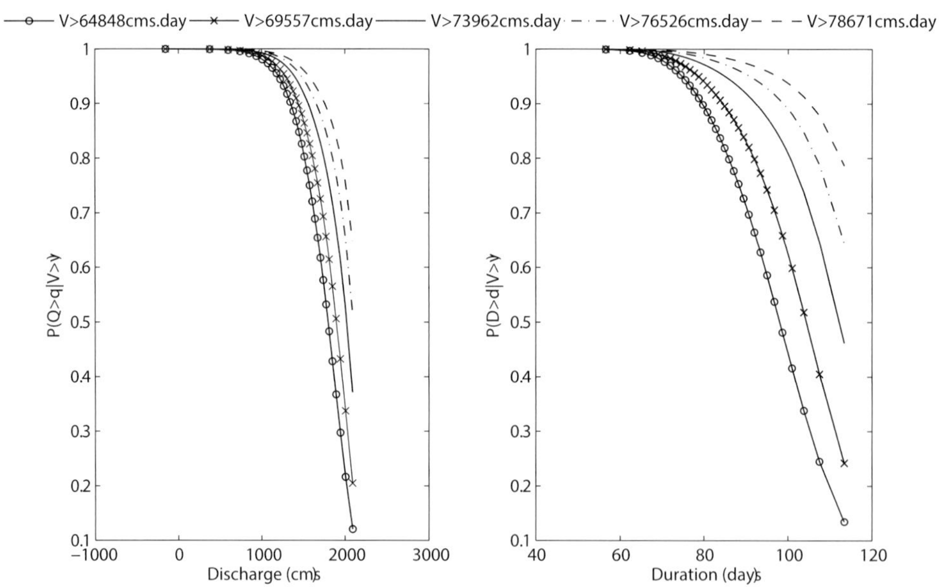

Figure 11.7 Conditional probability plot for discharge and duration given that the flood volume is greater than the given threshold.

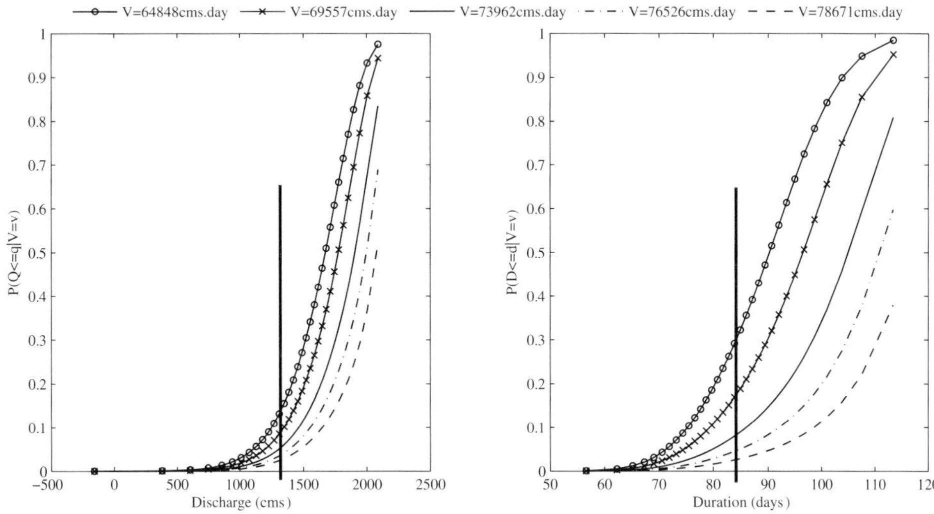

Figure 11.8 Conditional probability of flood discharge and duration for the given flood volume.

are nonincreasing in flood volume. In other words, discharge and duration are stochastically increasing on flood volume and vice versa. As an example, under the conditions of $V = \{64848, 69557, 73926, 76526, 78670\}$ *cmsday*, the conditional probability of $P(Q > 1500\ cms|V = v_i)$ and $P(D > 90\ day|V = v_i)$ decreases as V increases. Figure 11.9 plots the conditional return period for given flood volume of Case ii using the following:

$$T(X \geq x|Y = y) = \frac{1}{P(X > x|Y = y)} = \frac{1}{1 - C(U \leq u|V = v)} = \frac{1}{1 - \dfrac{\partial C(u, v)}{\partial v}} \quad (11.15)$$

Similar to Figure 11.8, Figure 11.9 also shows that under given flood volume (i.e., V = v), the higher discharge and longer duration result in a shorter return period and vice versa.

Comparing the results of the univariate return period, the joint return period ("OR" and "AND" cases), and the conditional return periods ($Q > q|V > v, V > v|D > d$), the same conclusion (Serinaldi, 2015) is obtained, as follows:

$$T_{OR}(Q, V) \leq \min(T_Q, T_V) \leq \max(T_Q, T_V) \leq T_{AND}(Q, V) \leq T_{COND}(Q|V > v)$$

$$T_{OR}(V, D) \leq \min(T_V, T_D) \leq \max(T_V, T_D) \leq T_{AND}(V, D) \leq T_{COND}(V|D > d)$$

11.2.5 Joint and Conditional Return Periods of Flood Discharge, Flood Volume and Flood Duration (Trivariate Case)

Similar to the bivariate case discussed in the previous sections, we will again consider the "AND" and "OR" cases for the joint return period. We will consider the following cases for the conditional return periods: (i) $X > x \cup Y > y|Z > z$ (ii) $X > x \cup Y > y|Z = z$,

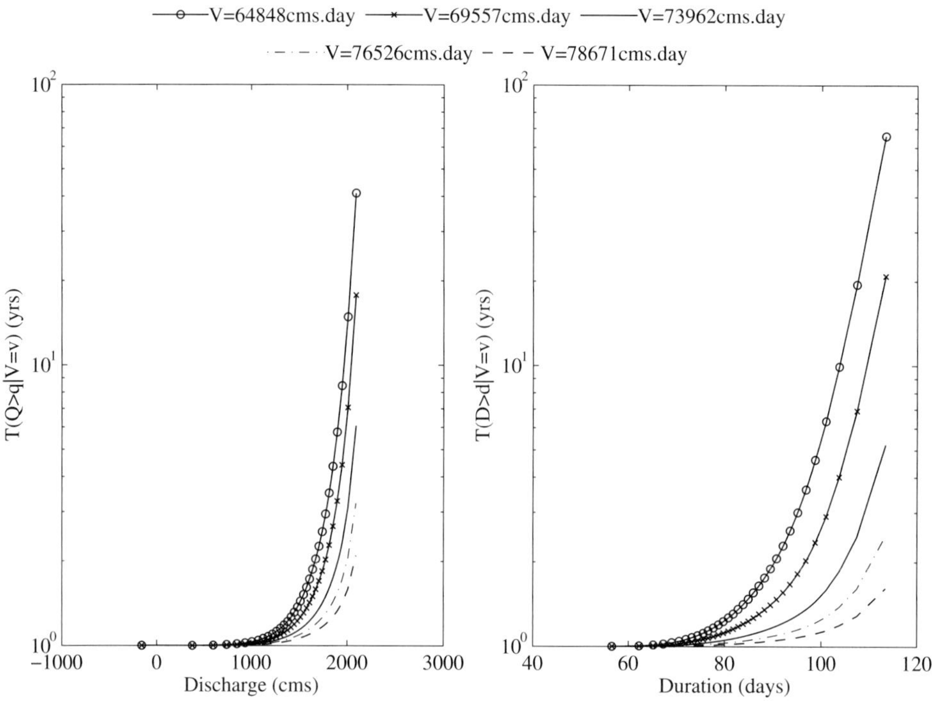

Figure 11.9 Conditional return period of flood discharge and duration for the given flood volume.

(iii) $X > x \cap Y > y|Z > z$, (iv) $X > x \cap Y > y|Z = z$, (v) $X > x|Y > y, Z > z$, (vi) $X > x|Y = y, Z = z$ As shown in Equation (5.60), the joint probability distribution of flood discharge (Q), flood volume (V), and flood duration (D) may be expressed through the conditional probability distribution as follows:

$$\begin{aligned} F(Q \leq q, V \leq v, D \leq d) &= C(F_Q, F_V, F_D) \\ &= F_V \cdot C_{QD|V}\Big(C_{Q|V}\Big(F_Q \leq F_q^* | F_V \leq F_v^*\Big), C_{D|V}\big(F_D \leq F_d^* | F_V \leq F_v^*\big)\end{aligned} \tag{11.16}$$

In Equation (11.16), $C_{QD|V}, C_{Q|V}, C_{D|V}$ are fitted using Frank, BB7, and BB7 copulas, respectively. In Section 11.2.3, we have shown that such a fitted vine copula may properly represent the trivariate dependence structure for the trivariate flood variables using a formal goodness-of-fit test. Figure 11.10 graphically illustrates the appropriateness through the joint probability plot by ordered pair. In what follows, we will discuss the joint return periods first, followed by the conditional return periods.

Joint Return Period of Flood Discharge, Flood Volume, and Flood Duration

"AND" Case: $T(Q > q \cap V > v \cap D > d)$ As introduced in Chapter 3, the joint return period of the "AND" case may be expressed using Equation (3.149), which implies that

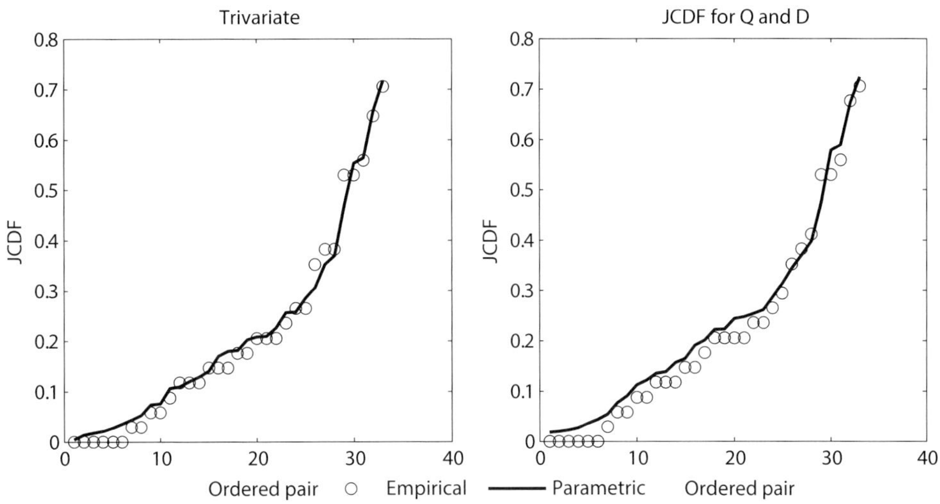

Figure 11.10 Joint CDF plot for flood variables.

flood discharge, flood volume, and flood duration all exceed their threshold values. To estimate the joint return period for the "AND" case, we need to know the bivariate joint distribution of flood discharge and flood duration. From the fitted vine copula structure, there does not exist a direct connection between flood discharge and flood duration; however, they are indirectly connected through flood volume. From Nelsen (2006) and the copula properties discussed in Chapter 3, we evaluate the joint distribution of flood discharge and duration by setting the marginal CDF for flood volume as 1, i.e.,

$$C(F_Q, F_D) = C(F_Q, 1, F_D) = \int_0^1 C(F_Q, F_D|t)dt \tag{11.17}$$

Using the fitted BB7–BB7–Frank vine copula, Equation (11.17) is further reduced to integrating the conditional frank copula. Figure 11.10 also compares the empirical distribution with the parametric distribution derived from the fitted vine copula.

Table 11.14 shows the joint return period for the "AND" case using $D = 90$ days as the threshold of flood duration for 5-, 10-, 25-, 50-, and 100-year design flood discharges and flood volumes.

"OR" Case: $\boldsymbol{T(Q > q \cup V > v \cup D > d)}$ As discussed in Chapter 3, at least one variable exceeds the threshold value. The joint return period is computed using Equation (3.150) for the "OR" case, that is, $Q > q \cup V > v \cup D > d$. As in the "AND" case, $D = 90$ days is applied as the fixed threshold for flood duration. Table 11.14 also lists the computed "OR" case joint return period using the 5-, 10-, 25-, 50-, and 100-year design flood discharge and flood volume values as threshold values.

Figure 11.11 plots the joint return periods for the "AND" and "OR" cases. Figure 11.11 and Table 11.14 indicate that the risk of all three flood variables exceeding the threshold

Table 11.14. *Joint return period for trivariate flood variables* (D = *90 days*).

Return period (years)		V (cms·day) *"AND" case*				
		64,848.11	69,557.12	73,961.78	76,525.99	78,670.80
	1,791.14	15.96	16.30	16.77	17.03	17.20
Q (cms)	1,928.62	27.02	27.32	27.79	28.09	28.29
	2,057.21	52.95	53.23	53.74	54.14	54.48
	2,132.07	89.73	90.01	90.55	91.05	91.55
	2,194.69	156.13	156.42	157.01	157.61	158.31
		V (cms·day) ***"OR" CASE***				
		64,848.11	69,557.12	73,961.78	76,525.99	78,670.80
	1,791.14	2.13	2.14	2.06	2.02	1.99
	1,928.62	2.45	2.59	2.55	2.50	2.46
Q (cms)	2,057.21	2.61	2.88	2.93	2.90	2.87
	2,132.07	2.65	2.96	3.05	3.04	3.02
	2,194.69	2.67	2.98	3.10	3.11	3.10

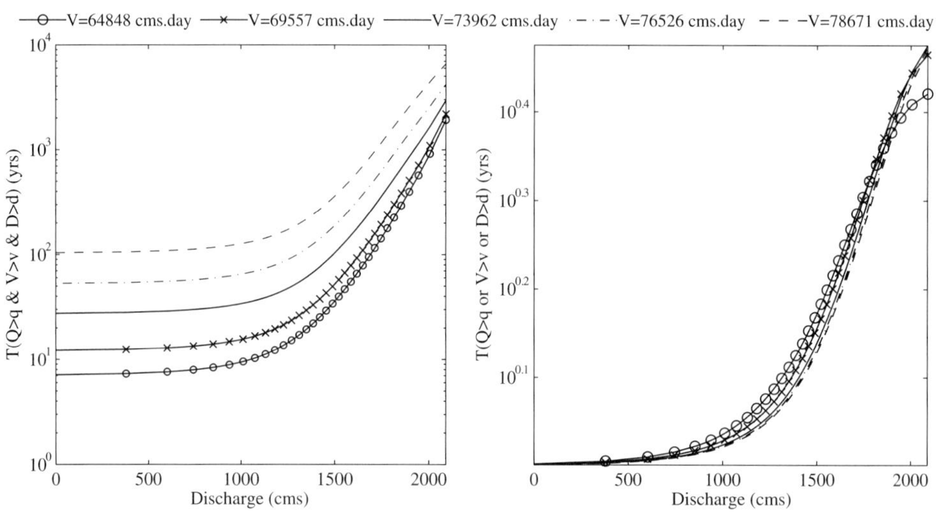

Figure 11.11 Joint return periods for trivariate flood variables: "AND" and "OR" cases.

values is significantly smaller than at least one of the variables exceeding its threshold value.

Conditional Return Periods of Flood Discharge, Volume, and Duration

As stated earlier, we are going to evaluate six different types of conditional return periods for flood discharge, volume, and duration. In traditional flood frequency analysis, the standard approach is to investigate the discharge variable only. Thus, in all six cases, we will consider discharge as one conditional variable.

Cases I and II: $\boldsymbol{T(Q > q \cup V > v|D > d)}$; $\boldsymbol{T(Q > q \cup V > v|D = d)}$ For case I, its conditional probability $P(Q > q \cup V > v|D > d)$ can be derived as follows:

$$P(Q > q \cup V > v|D > d) = 1 - P(Q \leq q \cap V \leq v|D > d) \tag{11.18a}$$

$$P(Q \leq q \cup V \leq v|D > d) = \frac{P(Q \leq q, V \leq v, D > d)}{1 - P_d(D \leq d)} \tag{11.18b}$$

$$P(Q \leq q \cup V \leq v|D > d) = \frac{C_{QV}(F_Q(q), F_V(v)) - C(F_Q(q), F_V(v), F_D(d))}{1 - F_D(D \leq d)} \tag{11.18c}$$

Following the same logic as that discussed for the bivariate case in Serinaldi (2015), the conditional return period of $T(Q > q \cup V > v|D > d)$ can be written as follows:

$$T(Q > q \cup V > v|D > d) = \frac{1}{(1 - F_D(d))(1 - C_{QV}(F_Q(q), F_V(v)) + C(F_Q(q), F_V(v), F_D(d)))} \tag{11.18d}$$

For case II, i.e., $T(Q > q \cup V > v|D = d)$, its conditional probability of $Q > q \cup V > v|D = d$ can be written as follows:

$$P(Q > q \cup V > v|D = d) = 1 - P(Q \leq q, V \leq v|D = d) = 1 - \left.\frac{\partial C(F_Q(q), F_V(v), F_D(d))}{\partial d}\right|_{D=d} \tag{11.19a}$$

and

$$T(Q > q \cup V > v|D = d) = \frac{1}{P(Q > q \cup V > v|D = d)} \tag{11.19b}$$

Applying the BB7–BB7–Frank copula to Equations (11.18) and (11.19), the conditional return periods are computed, as listed in Table 11.15, using five design flood discharge values and flood volume values as threshold values with the flood duration threshold value set as 90 days for exceedance (case I) and conditioning (case II). As shown in the preceding equations, in both of the cases at least one of the flood discharge or flood volume values exceeds its threshold value. Table 11.15 shows that higher conditional periods are obtained for case I than those for case II. Using the fitted log-normal distribution, the marginal probability $F_D(D \leq 90) = 0.68$. In general, the flood event with this duration occurs once in about three years. It is more likely for the large discharge or flood volume to occur for case I compared to case II. Figure 11.12 shows the conditional return periods for cases I and II of trivariate flood variables.

Table 11.15. *Conditional return period for cases I and II.*

Return period (years)		V (cms·day) $Q > q \cup V > v \mid D > d$ (90 days) 64,848.11	69,557.12	73,961.78	76,525.99	78,670.80
	1,791.14	3.82	4.11	4.36	4.47	4.52
Q (cms)	1,928.62	3.82	4.12	4.38	4.49	4.55
	2,057.21	3.82	4.12	4.39	4.49	4.55
	2,132.07	3.82	4.12	4.39	4.50	4.56
	2,194.69	3.82	4.12	4.39	4.50	4.56
		V (cms·day) $Q > q \cup V > v \mid D = d$ (90 days) 64,848.11	69,557.12	73,961.78	76,525.99	78,670.80
	1,791.14	4.92	12.29	23.65	25.67	25.18
	1,928.62	5.09	14.65	45.16	63.63	67.72
Q (cms)	2,057.21	5.15	15.66	67.30	143.75	193.39
	2,132.07	5.16	15.88	75.66	207.61	370.12
	2,194.69	5.16	15.96	79.23	251.33	603.00

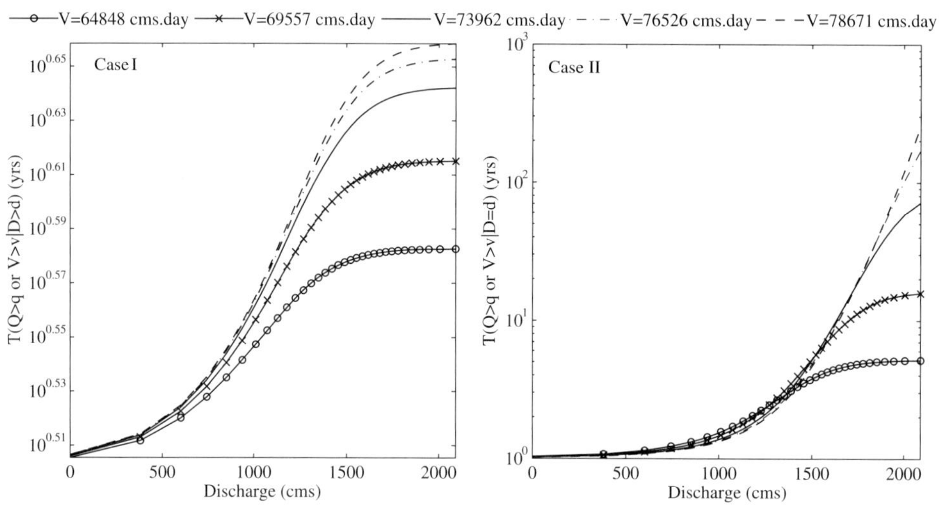

Figure 11.12 Conditional return periods for cases I and II of trivariate flood variables.

Cases III and IV: $T(Q > q \cap V > v|D > d)$; $T(Q > q \cap V > v|D = d)$ For case III, i.e., $T(Q > q \cap V > v|D > d)$; its corresponding exceedance conditional probability can be written as follows:

$$P(Q > q \cap V > v|D > d) = \frac{P(Q > q \cap V > v \cap d > d)}{P(D > d)} \tag{11.20a}$$

Substituting Equation (3.136) with the copula from Chapter 3 into Equation (11.20a), we can rewrite Equation (11.20a) as follows:

$$P(Q > q \cap V > v|D > d) = \frac{1 - F_Q(q) - F_V(v) - F_D(d) + C_{QV} + C_{VD} + C_{QD} - C_{QVD}}{1 - F_D(d)} \tag{11.20b}$$

Again, following the logic in Serinaldi (2015), the conditional return period can then be given as follows:

$$T(Q>q\cap V>v|D>d) = \frac{1}{(1-F_d)(1-F_Q(q)-F_V(v)-F_D(d)+C_{QV}+C_{VD}+C_{QD}-C_{QVD})} \tag{11.20c}$$

For case IV, i.e. $T(Q > q\cap V > v|D=d)$; its corresponding exceedance conditional probability can be written as follows:

$$\begin{aligned} &P(Q > q \cap V > v|D = d) \\ &= 1 - P(Q \leq q|D = d) - P(V \leq v|D = d) + P(Q \leq q, V \leq v|D = d) \end{aligned} \tag{11.21a}$$

The conditional return period can then be given as follows:

$$\begin{aligned} &T(Q > q \cap V > v|D = d) \\ &= \frac{1}{1 - P(Q \leq q|D = d) - P(V \leq v|D = d) + P(Q \leq q, V \leq v|D = d)} \end{aligned} \tag{11.21b}$$

In Equation (11.21), $P(Q \leq q|D = d) = \frac{\partial C_{QD}}{\partial(F_D(d))}$ with the joint distribution of flood discharge and duration derived in Equation (11.17).

Applying the fitted BB7–BB7–Frank vine copula, we compute the conditional return periods for the design events of discharge and flood volume using D = 90 days as the threshold value for flood duration. Table 11.16 lists the conditional return period computed for cases III and IV, and Figure 11.13 plots the conditional return periods. Compared to cases I and II, it is seen that the conditional return period computed for cases III and IV is much higher. The results confirm the real-world situation, that is, it is much harder for both flood discharge and flood volume to exceed the threshold values concurrently.

Cases V and VI: $T(Q > q|V > v, D > d); T(Q > q|V = v, D = d)$ For case V, the conditional probability may be written as follows:

$$P(Q > q|V > v, D > d) = \frac{P(Q > q, V > v, D > d)}{P(V > v, D > d)} \tag{11.22a}$$

Using the same approach as described in Serinaldi (2015), its conditional return period can be given as follows:

Table 11.16. *Conditional return period for cases III and IV.*

Return period (years)		V (cms·day) $Q > q \cap V > v \mid D > d$ (90 days) 64,848.11	69,557.12	73,961.78	76525.99	78670.80
	1791.14	606.85	811.67	1,459.08	2,535.62	4,663.29
Q (cms)	1928.62	1,584.71	1,993.04	3,273.31	5,396.37	9,570.15
	2057.21	4,486.15	5,217.81	7,385.68	10,931.92	17,886.38
	2132.07	9,212.59	10,234.36	13,039.06	17,435.98	26,007.61
	2194.69	18,526.45	19,913.44	23,434.22	28,573.04	38,283.46
		V (cms·day) $Q > q \cap V > v \mid D = d$ (90 days) 64,848.11	69557.12	73961.78	76525.99	78,670.80
	1791.14	30.17	42.27	79.65	141.22	262.49
	1928.62	76.89	99.76	168.91	282.02	503.34
Q (cms)	2057.21	214.06	253.33	363.83	541.51	888.36
	2132.07	436.19	489.90	628.74	842.38	1,257.44
	2194.69	872.64	944.54	1,115.45	1,360.10	1,821.49

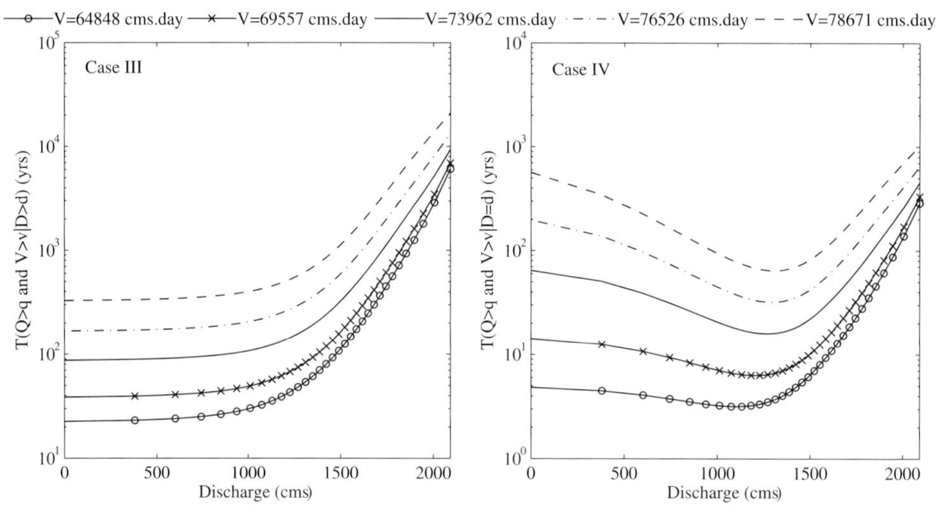

Figure 11.13 Conditional return period plots for cases III and IV.

$$T(Q > q|V > v, D > d) = \frac{1}{P(V > v, D > d) * P(Q > q, V > v, D > d)} \tag{11.22b}$$

For case VI, its conditional probability may be written as follows:

$$P(Q > q|V = v, D = d) = 1 - P(Q \leq q|V = v, D = d) \tag{11.23a}$$

$$P(Q \le q|V = v, D = d) = \frac{\partial C_{QD|V}\left(C_{Q|V}, C_{D|V}\right)}{\partial C_{D|V}} \tag{11.23b}$$

The conditional return periods computed for cases V and VI are tabulated and plotted in Table 11.17 and Figure 11.14, respectively. Table 11.17 indicates that higher conditional return periods are obtained for case V under the condition that $V > v_i \cap D > 90$ days than those for case VI under the condition that $V = v_i \cap D = 90$ days. It is also seen that the

Table 11.17. *Conditional return period for cases V and VI.*

		V (cms·day) $Q > q \mid V > v, D > d$ (90 days)				
Return period (years)		64,848.11	69,557.12	73,961.78	76,525.99	78,670.80
	1,791.14	1,344.57	3,090.00	1.25E+04	4.20E+04	1.51E+05
Q (cms)	1,928.62	3,511.13	7,587.46	2.82E+04	8.94E+04	3.11E+05
	2,057.21	9,939.68	1.99E+04	6.35E+04	1.81E+05	5.81E+05
	2,132.07	2.04E+04	3.90E+04	1.12E+05	2.89E+05	8.44E+05
	2,194.69	4.10E+04	7.58E+04	2.02E+05	4.73E+05	1.24E+06
		V (cms·day) $Q > q \mid V = vD = d$ (90 days)				
		64,848.11	69,557.12	73,961.78	76,525.99	78,670.80
	1,791.14	4.04	1.38	1.05	1.02	1.01
	1,928.62	14.66	2.70	1.17	1.05	1.02
Q (cms)	2,057.21	68.92	10.09	1.89	1.20	1.06
	2,132.07	206.03	29.26	3.94	1.66	1.17
	2,194.69	601.85	84.80	10.04	3.12	1.55

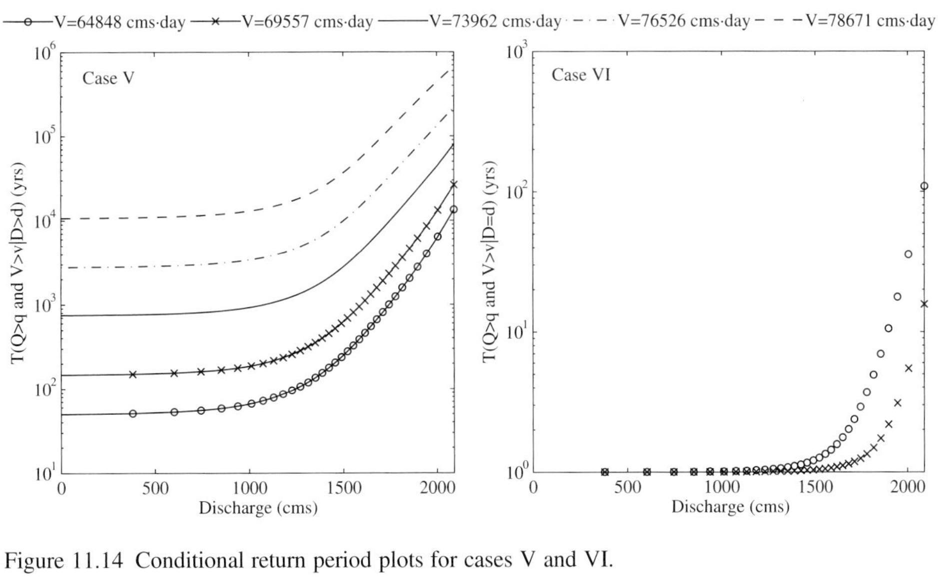

Figure 11.14 Conditional return period plots for cases V and VI.

conditional return period decreases for $Q > q_i$ with the increase of flood volume for case VI. This result again agrees with the right tail dependence between flood discharge and flood volume. Compared to low discharge, high discharge is more likely to occur under the condition of high flood volume.

11.2.6 Comparison with the Yue et al. (1999) Results

Compared with the results in Yue et al. (1999), there are some major differences for the case study presented using the same data listed in Yue et al. (1999). First, Yue, et al. (1999) applied the Gumbel (EV1) distribution as the marginal distribution for flood discharge, volume, and duration. According to the univariate goodness-of-fit test (i.e., the KS test), Table 11.9 shows the Gumbel distribution is proper for flood discharge and flood volume; however, it may not be a proper model flood duration (KS statistics = 0.23 with P-value = $0.04 < 0.05$). Rather than the Gumbel distribution, the log-normal distribution is shown to be a proper marginal distribution for flood duration. Hence, both Gumbel and log-normal distributions are applied to flood discharge, volume, and duration, respectively, rather than applying the Gumbel distribution to all three flood variables.

Second, the bivariate Gumbel mixed distribution is applied to model flood discharge and flood volume, and flood volume and flood duration, in Yue et al. (1999). Given the limitations of conventional bivariate flood frequency analysis, the Gumbel distribution is applied as the marginal distribution for all three flood variables. The Gumbel distribution is proper for flood discharge and flood volume; however, it is not proper to model flood duration based on the KS goodness-of-fit test. Instead, the log-normal distribution may be properly applied to model the flood duration (Table 11.9). As seen in this section, the proper marginal distribution is applied to each flood variable for the case study presented here using the same flood data as recorded in Yue et al. (1999).

Third, only bivariate flood frequency analysis was performed in Yue et al. (1999), but the trivariate flood frequency analysis is presented here using the vine copula. In the vine copula structure, flood discharge and flood volume, and flood volume and flood duration, are modeled with the unconditional BB7 copula. Even though Yue et al. (1999) did not specifically state the tail dependence, right tail increasing, and stochastic monotonic properties for the bivariate conditional return period, both the results in Yue et al. (1999) and the results obtained here clearly indicate these interesting properties, which are in line with the physical world. In addition to the bivariate analysis, the case study in this section also computes the joint and conditional return periods, based on trivariate frequency analysis. As shown in Section 11.2.5, the trivariate joint and conditional return periods also reveal the interactions among the three flood variables.

11.3 Spatially Dependent Discharge Analysis

Similar to the spatial rainfall frequency analysis, the spatial discharge (streamflow) frequency analysis involves the following procedure:

Table 11.18. *Monthly (May) discharge at the Yampa and Colorado rivers (cfs).*

Year	USGS9239500	USGS9251000	USGS9070500	USGS9095500	USGS9163500	USGS9180500
1951	1,746	5,356	5,618	8,725	12,340	12,330
1952	2,139	8,394	8,693	15,910	30,500	35,000
1953	1,064	3,602	3,430	5,634	8,905	9,857
1954	1,082	3,398	2,248	4,807	6,256	7,089
1955	1,579	4,881	3,119	6,248	10,130	12,230
1956	2,058	6,518	6,606	11,140	15,640	16,350
1957	1,786	7,156	5,458	9,616	18710	22,360
1958	2,681	8,931	7,917	13,780	28820	33,050
1959	1,372	4,306	3,979	6,375	8,337	8,710
1960	1,543	4,675	4,325	7,028	11,170	12,330
1961	1,261	3,790	3,158	5,767	9,300	11,010
1962	2,207	7,145	8,600	14,520	23,650	26,070
1963	1,261	4,081	2,460	5,245	7,579	8,402
1964	1,376	5,428	3,180	6,560	12,520	14,000
1965	1,494	6,280	4,123	7,763	16,890	20,680
1966	1,256	3,858	2,760	6,068	8,995	11,330
1967	1,122	4,063	2,641	5,342	6,899	7,506
1968	1,402	5,584	2,660	5,302	8,895	10,850
1969	1,752	6,510	4,062	9,121	13,490	16,060
1970	2,378	8,302	8,513	13,600	19,720	22,520
1971	1,707	6,401	5,533	8,473	11,570	12,490
1972	1,496	4,248	3,721	6,409	7,386	7,366

Table 11.18. (*cont.*)

Year	USGS9239500	USGS9251000	USGS9070500	USGS9095500	USGS9163500	USGS9180500
1973	2,216	7,689	5,163	9,630	17,710	25,320
1974	2,862	9,695	7,890	11,540	15,230	16,530
1975	1,276	5,439	3,528	6,331	13,150	16,380
1976	1,498	5,011	3,547	6,520	8,843	10,400
1977	702.4	1,850	1,436	2,536	2,283	2,322
1978	1576	6,470	4,177	7,018	11,540	15,560
1979	1825	7,784	5,413	9,865	18,650	24,610
1980	1909	8,321	5,682	10,420	20,300	26,920
1981	896.2	3,031	1,735	3,259	4,600	4,821
1982	1,702	6,866	3,411	6,857	12,340	14,530
1983	1,405	6,068	4,279	8,783	17,540	25,420
1984	3,350	14,000	10,770	20,290	37,960	42,090
1985	2,203	9,518	7,635	16,440	28,570	31,970
1986	1,867	7,456	7,024	12,700	22,370	24,360
1987	1,356	4,409	3,931	8,229	15,520	20,830
1988	1,486	5,430	3,812	6,337	8,551	8,788
1989	1,135	3,310	2,974	5,287	6,651	7,011
1990	905.7	2,642	1,823	3,085	4,078	4,070
1991	1,422	5,170	3,650	6,449	10,610	10,860
1992	1,437	3,985	2,995	5,874	10,170	11,330
1993	1,772	7,964	6,371	13,680	27,350	32,030
1994	1,368	4,205	3,103	6,203	9,912	11,200

1995	869.3	5,965	2,657	5,611	15,040	18,450
1996	2,458	9,091	8,061	12,570	18,460	18,840
1997	2,420	9,921	7,875	13,830	22,500	26,960
1998	1,893	8,196	4,576	10,540	18,470	22,280
1999	1,341	5,568	3,093	5,665	9,775	11,600
2000	2,079	6,285	4,785	7,986	10,940	12,360
2001	1,775	5,250	3,112	6,301	9,017	9,780
2002	742.8	2,007	1,254	2,683	2,640	2,696
2003	1,730	6,358	3,538	6,855	9,043	9,027
2004	1,211	4,031	2,011	4,571	6,615	7,255
2005	1,502	6,596	3,276	8,059	16,110	20,690
2006	2,236	7,115	5,008	9,854	13,140	12,840
2007	1,540	4,545	3,918	7,200	10,200	10,500
2008	1,796	9,000	6,600	10,950	22,020	23,380
2009	2,105	8,248	6,937	12,960	20,390	22,010
2010	1,137	5,225	3,375	6,072	9,452	10,710
2011	1,818	8,905	7,568	11,480	18,210	18,220
2012	929.1	2,377	1,566	3,446	3,836	4,112
2013	1,622	4,925	3,222	5,558	6,959	7,197
2014	2,117	7,092	8,014	11,230	14,850	13,900
2015	2,051	5,186	4,586	7,030	10,660	10,370

Table 11.19. *Kendall correlation coefficient.*

	USGS9239500	USGS9251000	usgs9070500	USGS9095500	USGS9163500	USGS9180500
USGS9239500	1	0.70	0.72	0.73	0.60	0.54
USGS9251000	0.70	1	0.67	0.74	0.73	0.69
USGS9070500	0.72	0.67	1	0.84	0.69	0.60
USGS9095500	0.73	0.74	0.84	1	0.79	0.71
USGS9163500	0.60	0.73	0.69	0.79	1	0.89
USGS9180500	0.54	0.69	0.60	0.71	0.89	1

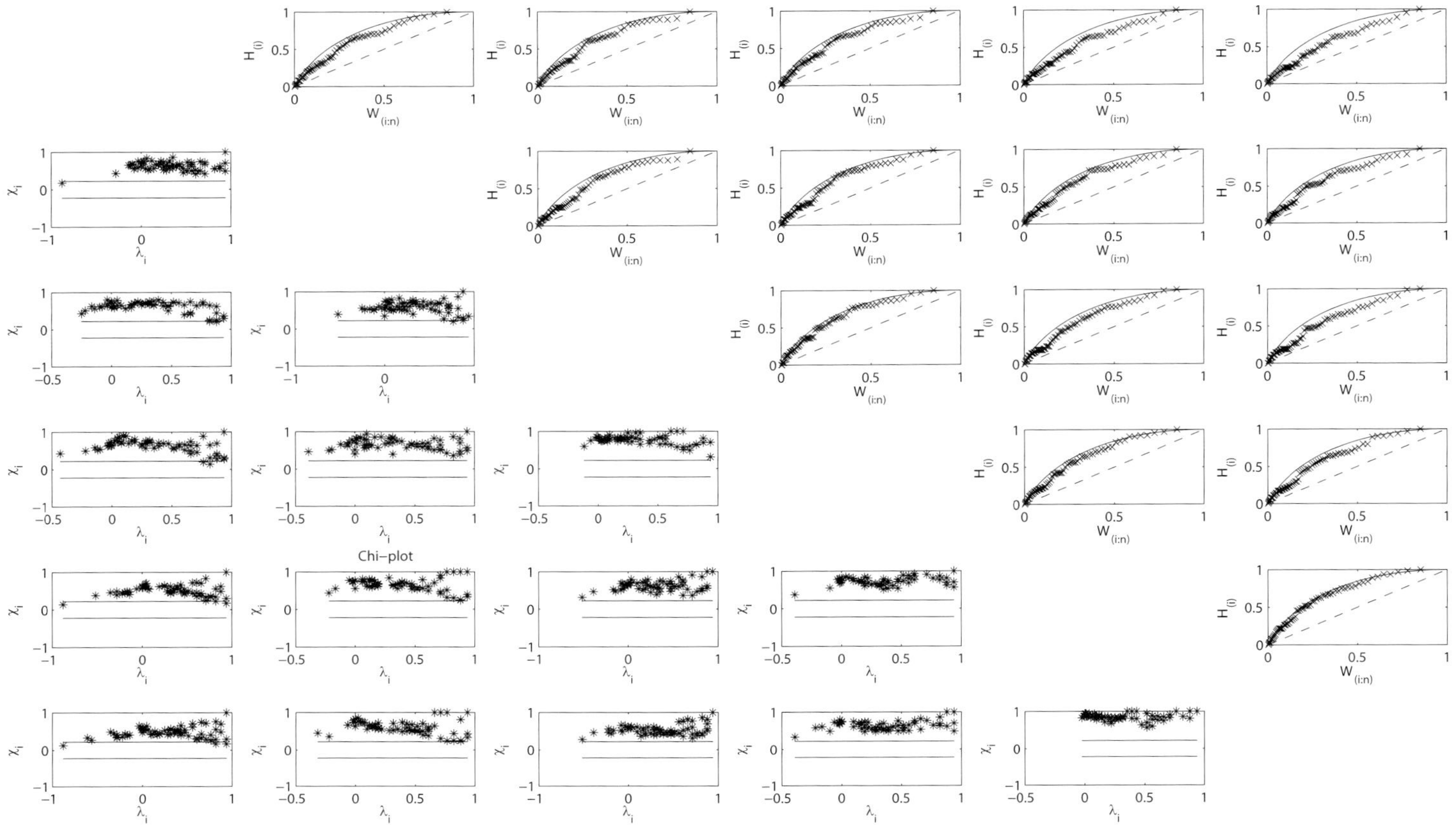

Figure 11.15 K-plots and chi-plots for monthly discharge (May).

Table 11.20. *Parameters estimated for the meta-Gaussian copula.*

	USGS9239500	USGS9251000	USGS9070500	USGS9095500	USGS9163500	USGS9180500
USGS9239500	1	0.88	0.90	0.90	0.79	0.74
USGS9251000	0.88	1	0.87	0.91	0.90	0.87
USGS9070500	0.90	0.87	1	0.96	0.88	0.82
USGS9095500	0.90	0.91	0.96	1	0.94	0.90
USGS9163500	0.79	0.90	0.88	0.94	1	0.98
USGS9180500	0.74	0.87	0.82	0.90	0.98	1

Notes: SnB goodness-of-fit test: test statistics = 0.011; *P*-value = 0.67.

Table 11.21. *Parameters estimated for the meta-Student t copula.*

	USGS9239500	USGS9251000	USGS9070500	USGS9095500	USGS9163500	USGS9180500
USGS9239500	1	0.89	0.90	0.90	0.81	0.77
USGS9251000	0.89	1	0.87	0.91	0.90	0.88
USGS9070500	0.90	0.87	1	0.96	0.88	0.83
USGS9095500	0.90	0.91	0.96	1	0.95	0.91
USGS9163500	0.81	0.90	0.88	0.95	1	0.98
USGS9180500	0.77	0.88	0.83	0.91	0.98	1
	$v = 17.04$					

Notes: SnB goodness-of-fit test: test statistics = 0.016; *P*-value = 0.42.

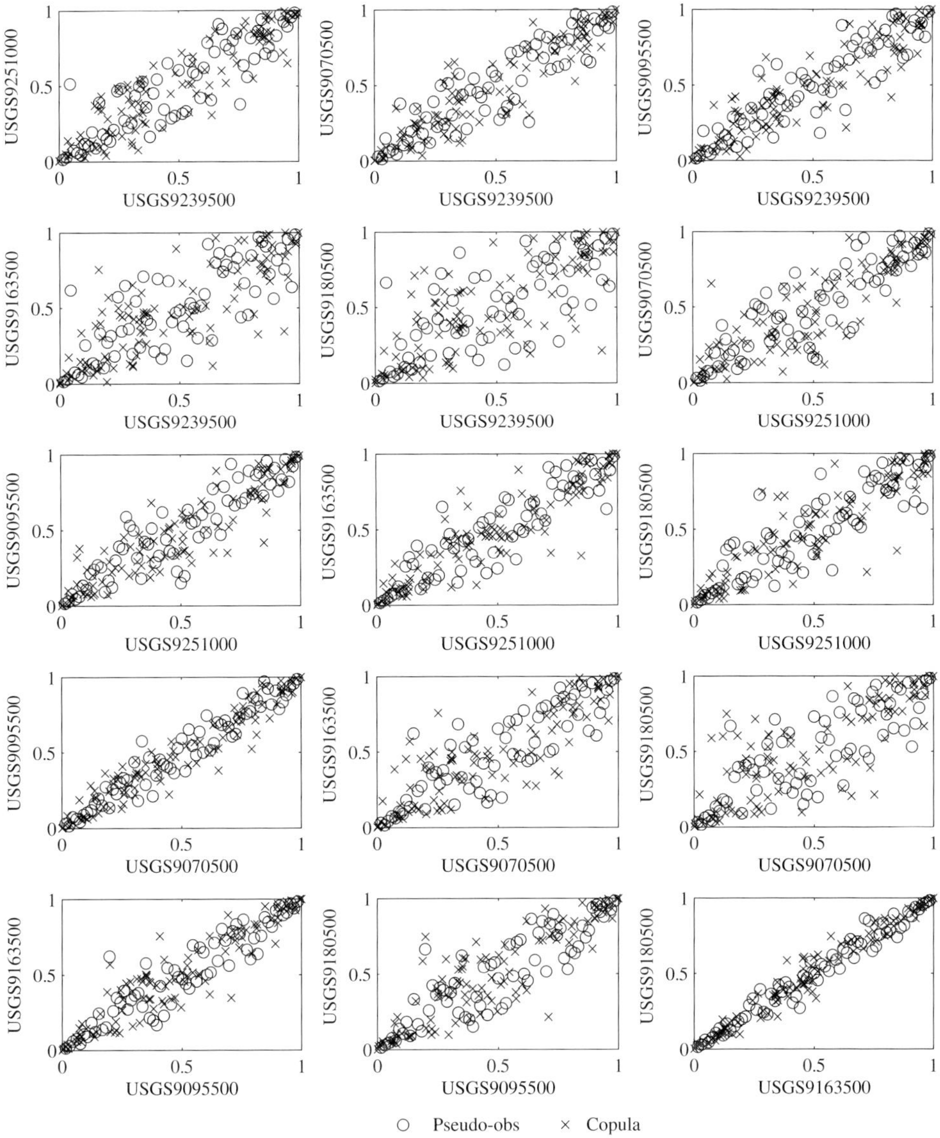

Figure 11.16 Comparison of variates simulated from the meta-Gaussian copula with pseudo-observations.

1. Select the gauging stations and collect the streamflow time series.
2. Compute the Kendall correlation coefficient matrix.
3. Apply the meta-elliptical copula to study the spatial dependence.

To illustrate the spatial dependence of discharge, we will use monthly streamflow of May along the Yampa River and the upper stream of the Colorado River. Six gauging stations are selected for analysis, as listed in Table 11.18.

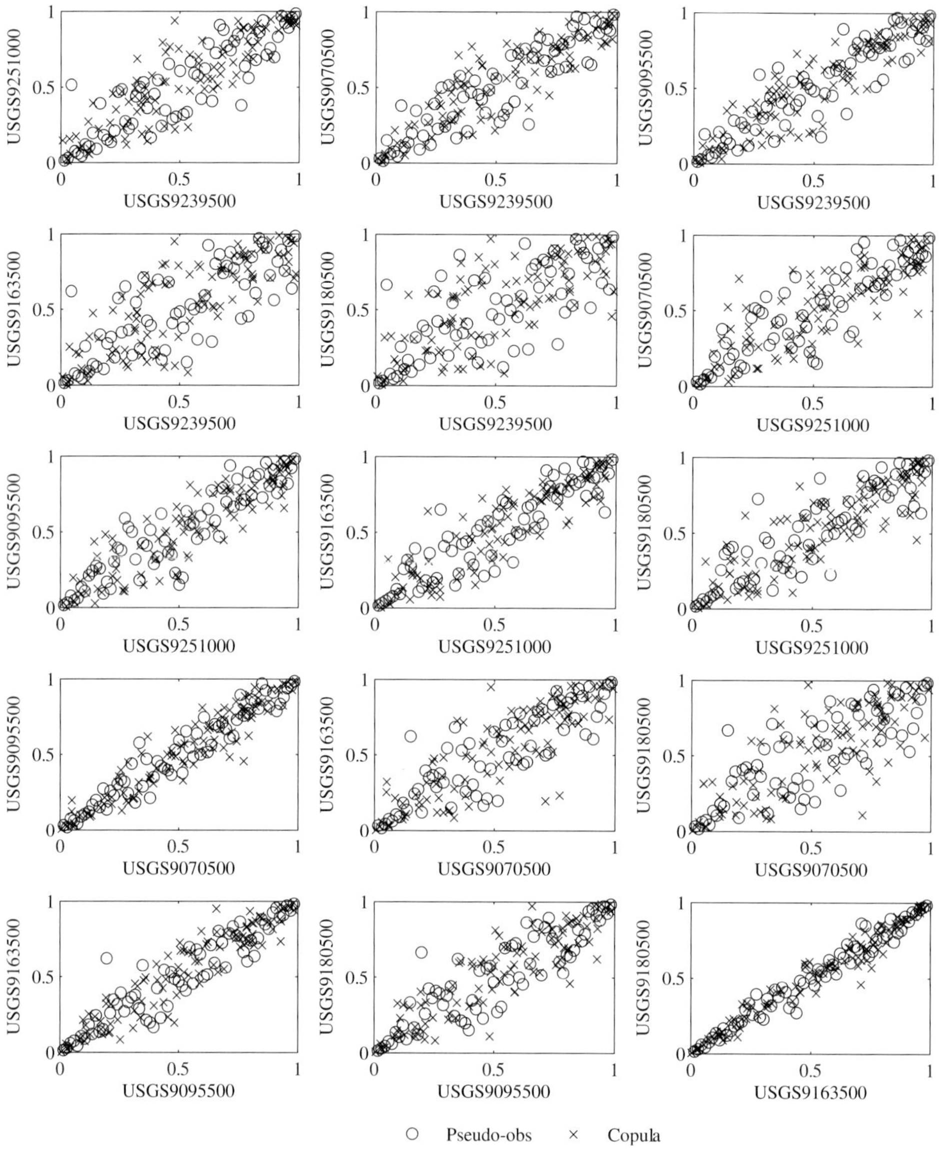

Figure 11.17 Comparison of variates simulated from the meta-Student t copula with pseudo-observations.

In this case study, we assume discharges (the month of May) at all six sites as random variables. In addition, the most commonly applied meta-elliptical copulas discussed in Chapter 7 (i.e., meta-Gaussian and meta-Student *t*) are applied to model the spatial dependence. Table 11.19 lists the Kendall correlation coefficient. It is seen that monthly discharge is positively correlated. Figure 11.15 graphs the K-plots and chi-plots. The

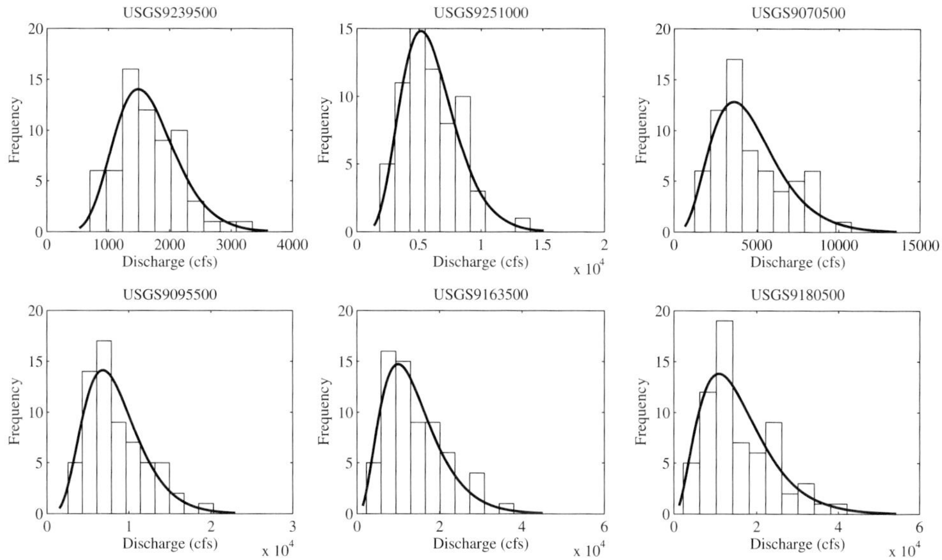

Figure 11.18 Histogram and fitted gamma distribution for all six locations.

K-plots of each pair are shown in the upper triangle, and the chi-plots of each pair are shown in the lower triangle. The K-plots and chi-plots again show that monthly discharge variables are highly positively dependent.

With the use of the Weibull plotting-position formula to compute the empirical distribution (i.e., pseudo-observations) and applying pseudo-MLE for the meta-elliptical Gaussian copula, Table 11.20 lists the estimated parameters, i.e., the correlation coefficient matrix. Similarly, applying pseudo-MLE parameters of the meta-Student *t* copula (i.e., the correlation matrix and degree of freedom) are estimated, as listed in Table 11.21. To assess the fitness of the meta-Gaussian and meta-Student t copulas, the SnB goodness-of-fit test is applied and the test results are listed in Tables 11.20 and 11.21 for the fitted meta-Gaussian and meta-Student *t* copulas, respectively. The test results indicate that both copulas may properly model the monthly discharge. In addition, the test statistic of the meta-Gaussian copula is less than that of the meta-Student *t* copula.

Using the parameters listed in Tables 11.20 and 11.21, we then simulate the pseudo-observations from meta-Gaussian and meta-Student *t* copulas; comparison with the meta-Gaussian copula is shown in Figure 11.16, and comparison with the meta-Student *t* copula is shown in Figure 11.17. From Figures 11.16 and 11.17, we notice that the two gauging stations on the Colorado River (i.e., USGS 9163500 and USGS 9180500) are almost perfectly correlated, with a correlation coefficient very close to 1.

Until now, we have successfully fitted meta-Gaussian and meta-Student *t* copulas to monthly discharge in the frequency domain. Next we will assess the fit in the real domain. Figure 11.18 plots the histogram as well as the fitted gamma distribution. As shown in

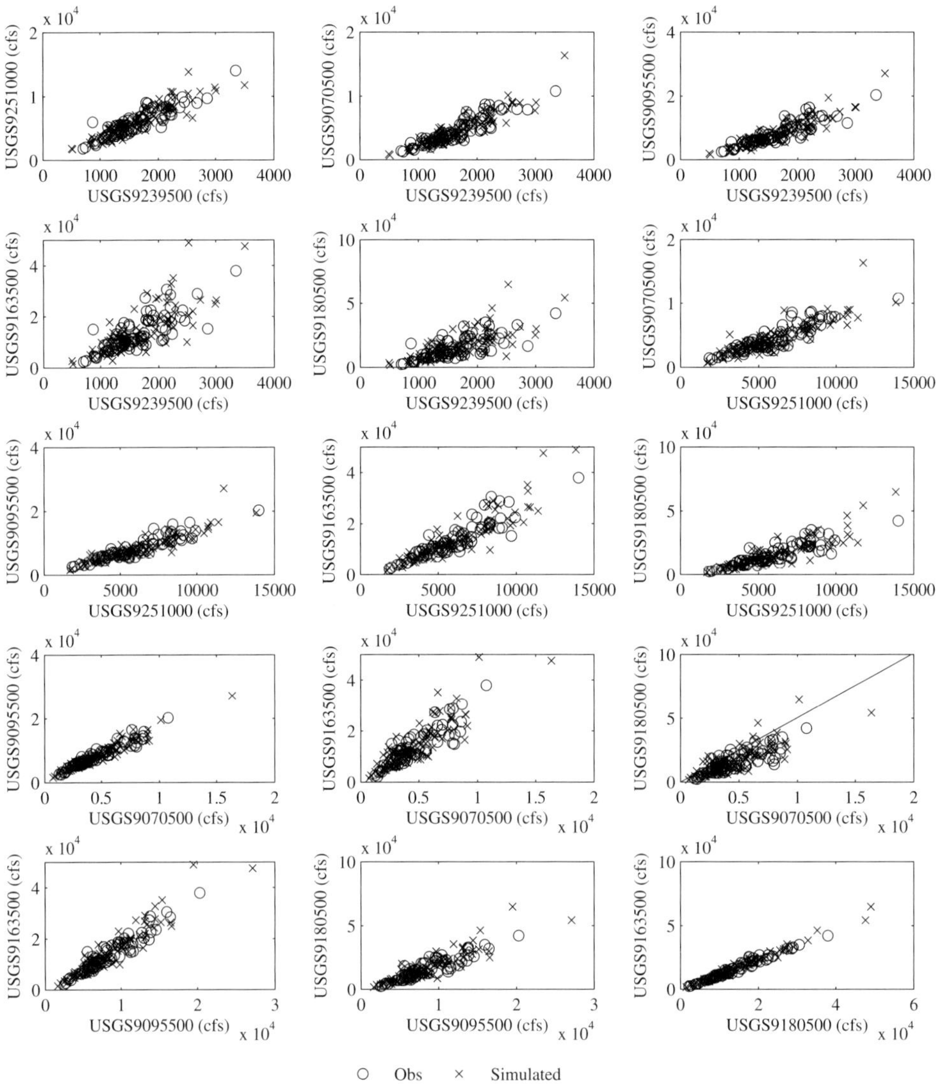

Figure 11.19 Comparison of observed monthly discharge with simulated monthly discharge from the meta-Gaussian copula.

Figure 11.18, the gamma distribution may be applied to model the univariate monthly discharge with the KS goodness-of-fit test results listed in Table 11.22. Table 11.22 shows that the gamma distribution can be applied to model univariate monthly discharge. With the fitted gamma distribution, Figures 11.19 and 11.20 present the comparison in the real domain. These comparisons again confirm the appropriateness of meta-Gaussian and meta-Student *t* copulas, as well as the fitted univariate gamma distribution.

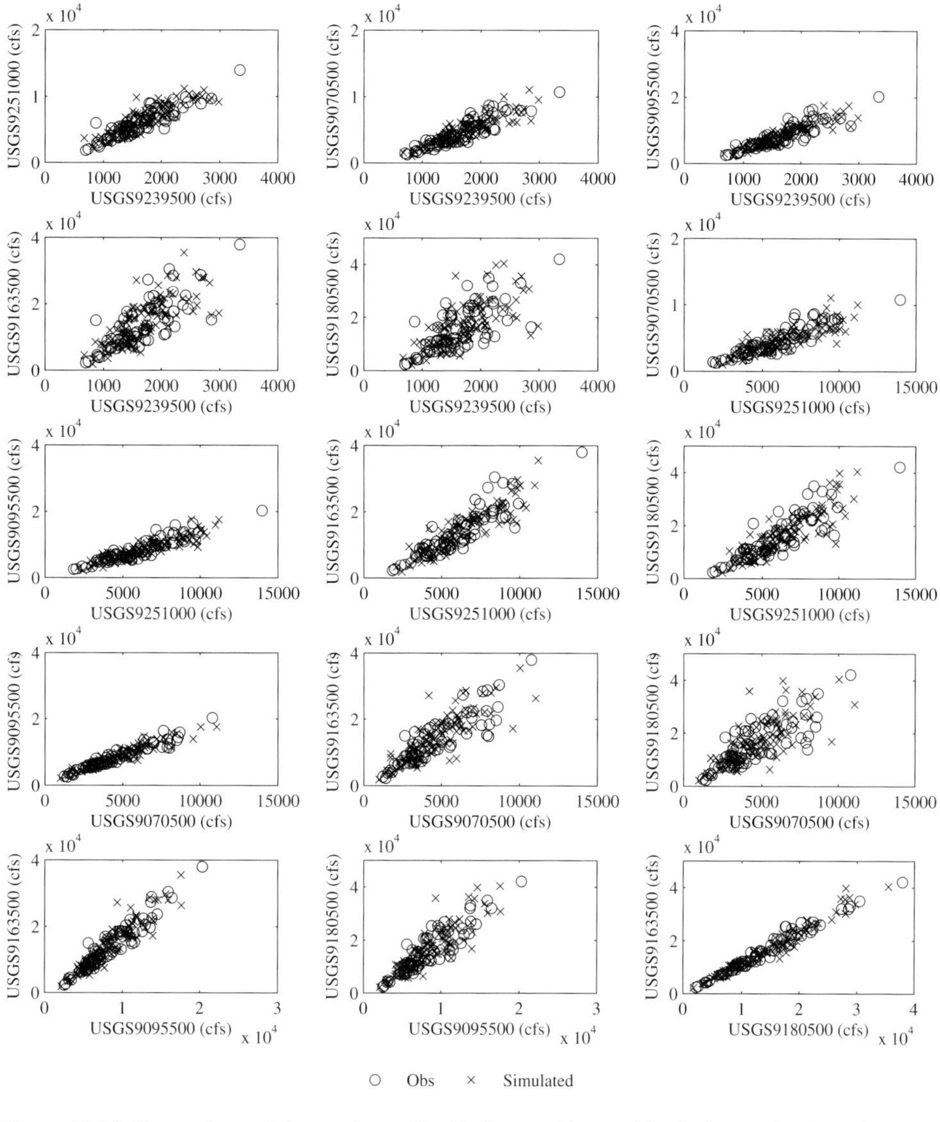

Figure 11.20 Comparison of observed monthly discharge with monthly discharge simulated from the meta-Student t copula.

In this case study, we show how to model the spatial dependence when the variables may be considered as random variables. With the highly positively correlated discharge variables, we may expect high/low flow across the region at the same time. Additionally, the spatial dependence will allow us to investigate the flow pattern and aid us with hydrological design.

Table 11.22. *Estimated parameters for univariate discharge (gamma) and KS goodness-of-fit test results.*

	USGS9239500	USGS9251000	USGS9070500	USGS9095500	USGS9163500	USGS9180500
Parameter	[19.48, 157.46]	[7.06,851.76]	[4.67,982.43]	[5.55,1501.2]	[3.68, 3720.5]	[3.29,4759.9]
KS statistics	0.047	0.056	0.088	0.119	0.062	0.090
P-value	0.998	0.979	0.660	0.289	0.949	0.636

11.4 Summary

In this chapter, we introduce case studies of copula application for both at-site and spatial flood frequency analyses. The case studies indicate the following:

I. Compared with conventional approaches, the copula approach indeed offers the advantage to better capture the dependence structure among flood variables as well as to minimize the impact of marginal distribution misidentification with the use of the empirical marginals for copula construction and parameter estimation.
II. For at-side flood frequency analysis, the overall dependence structure may be well captured by different copulas that may or may not capture the tail dependence. Given the characteristics of flood variables (e.g., flood peak vs. flood volume; flood volume vs. flood duration), it is recommended to choose the copulas at least handling the upper-tail dependence (e.g., the Gumbel–Hougaard copula) or mixed copulas to capture the important upper-tail dependence. Better capturing the upper-tail dependence may directly yield better engineering design by minimizing flood risk.
III. Spatial flood frequency analysis, in general, provides a pattern of spatial distribution. The complexity of constructing the proper vine copula will increase significantly with the increase of dimension (i.e., the number of gauging stations considered within the watershed or region). Thus, it is recommended to apply the meta-elliptical copulas to spatial frequency analysis. Similar to other copula families, the meta-elliptical copula is capable of capturing the overall dependence well, in addition to its relatively simple and easy parameter estimation. This simple construction may allow the water resources engineer to better implement the methodology and make viable watershed management decisions.

References

Abberger, K. (2005). A simple graphical method to explore tail-dependence in stock return pairs. *Applied Financial Economics*, 15(1), 43–51.

Bezak, N., Mikos, M., and Sraj, M. (2014). Trivariate frequency analysis of peak discharge, hydrograph volume and suspended sediment concentration data using copulas. *Water Resources Management*, 28, 2195–2212. doi:10.1007/s11269-014-0606-2.

Capéraà, P., Fougeres, A.-L., and Genest, C. (1997). A nonparametric estimation procedure for bivariate exteme value copulas. *Biometrika*, 84, 567–577.

Chen, L., Singh, V. P., and Guo, S. (2013). Measure of correlation between river flows using the copula-entropy theory. *Journal of Hydrologic Engineering*, 18(12), 1591–1608.

Chen, L., Singh, V. P., Guo, S., Hao, Z., and Li, T. (2012). Flood coincidence risk analysis using multivariate copula functions. *Journal of Hydrologic Engineering*, 17(6), 742–755.

Chowdhary, H., Escobar, L. A., and Singh, V. P. (2011). Identification of suitable copulas for bivariate frequency analysis of flood peak and flood volume data. *Hydrology Research*, 42(2–3), 193–216.

Coles, S. G., Heffernan, J. E., and Tawn, J. A. (1999). Dependence measures for extreme value analyses. *Extremes*, 2, 339–365.

Durocher, M., Chebana, F., and Ouarda, T. B. M. J. (2016). On the prediction of extreme flood quantiles at ungauged locations with spatial copula. *Journal of Hydrology*, 533, 523–532. doi:10.1016/j.jhydrol.2015.12.029.

Frahm, G., Junker, M., and Schmidt, R. (2005). Estimating the tail dependence coefficient: properties and pitfalls, *Insurance Mathematics & Economics*, 37, 80–100.

Joe, H. (1997). *Multivariate Models and Dependence Concepts*. Chapman & Hall, New York.

Nelsen, R. B. (2006). *An Introduction to Copulas*, 2nd edition. Springer, U.S.A.

Poulin, A., Huard, D., Favre, A.-C., and Pugin, S. (2007). Importance of tail dependence in bivariate frequency analysis. *Journal of Hydrologic Engineering*, 12(4), 394–403, doi:10.1061/(ASCE)1084–0699(2007).

Requena, A. I., Chebana, F., and Mediero, L. (2016). A complete procedure for multivariate index-flood model application. *Journal of Hydrology*, 535, 559–580. doi:10.1016/j.jhydrol.2016.02.004.

Schmidt, R. and Stadtmuller, U. (2006). Non-parametric estimation of tail dependence. Scandinavian Journal of Statistics: Theory and Applications, 33(2), 307–335.

Serinaldi, F. (2015). Dismissing return periods. *Stochastic Environmental Research and Risk Assessment*, 29, 1179–1189, doi:10.1007/s00477–014–0916–1.

Sraj, M., Bezak, N., and Brilly, M. (2015). Bivariate flood frequency analysis using the copula function: a case study of the Litija station on the Sava River. *Hydrological Processes*, 29, 225–238. doi:10.1002/hyp.10145.

Yue, S., Ouarda, T. B. M. J., Bobee, B., Legendre, P., and Bruneau, P. (1999). The Gumbel mixed model for flood frequency analysis. *Journal of Hydrology*, 226, 88–100.

12

Water Quality Analysis

ABSTRACT

This chapter discusses how to apply copulas in water quality analysis. For monthly water quality observations, applications will include (i) a copula-based Markov process to study the water quality sequence with temporal dependence; and (ii) a copula-based multivariate water quality time series analysis. This chapter is in line with Chapter 9.

12.1 Case-Study Sites

According to the availability of water quality data, two watersheds are used as a case study. One watershed is a natural watershed, while the other is an urban watershed. The watershed boundaries and streams data were retrieved from the NHDPlus High Resolution National Hydrography Dataset and Watershed Boundary Dataset (https://nhd.usgs.gov/). The land use and land cover (LULC) were retrieved from the National Land Cover Database (www.nrlc.gov).

12.1.1 Snohomish River Watershed

According to the Department of Ecology of the State of Washington (ecy.wa.gov), the Snohomish River is formed near the city of Monroe, where the Skykomish and Snohomish rivers meet. The Snohomish River continues its way through the estuary of the city of Snohomish before entering into Puget Sound. The Snohomish watershed covers an area of 1,978 square miles (about 5,123 square kilometers) and provides important water recreation activities. In the past, agriculture and forest were two main LULC within the watershed; however, throughout the last century, more human activity has been introduced into the watershed. The Department of Ecology clearly stated (ecy.wa.gov): "over the last century, diking and other engineering activities in the lower part of the basin greatly changed how water is stored and managed in floodplain areas. More recently, cities and suburban areas have grown rapidly, creating more change to the natural water cycle."

Besides the change in the natural water cycle induced by human activities, water quality issues (including but not limited to bacteria, dissolved oxygen (DO), temperature, and pH) have also been identified for some areas. Four stations located in the Snohomish watershed

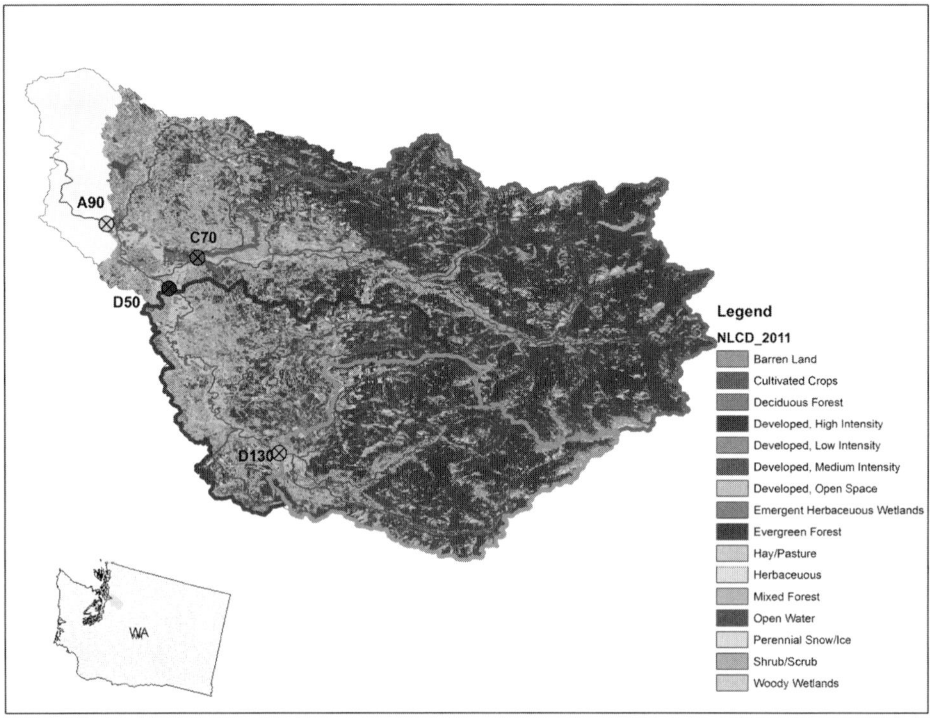

Figure 12.1 Snohomish watershed map and its LULC in 2011(retrieved from USGS and NLCD). A black and white version of this figure will appear in some formats. For the color version, please refer to the plate section.

are selected for the case study: A90, C70, D50, and D130 (shown in Figure 12.1). The total persulfate nitrogen (TPN) and DO at C70 are chosen as the targeting water quality parameters for the temporal dependence case study. DO at all four stations is chosen for the spatial dependence study.

12.1.2 Chattahoochee River Watershed

As a tributary of Apalachicola River, the Chattahoochee River originates south of the Alabama and Georgia border and joins the Apalachicola River at the Georgia and Florida border. The Chattahoochee River watershed is the largest subwatershed of Apalachicola–Chattahoochee–Flint river basin.

The city of Atlanta is located within the watershed. There are gauging stations upstream and downstream of metropolis, i.e., the Belton Bridge station (USGS02332017, upstream) and Whitesburg station (USGS02338000, downstream). The subwatershed upstream of the Bridge station may be classified as the forest watershed. With the major metropolitan area – the city of Atlanta – the subwatershed upstream of Whitesburg is more developed (by 2011, the developed land alone accounts for about 34%) and may be considered the

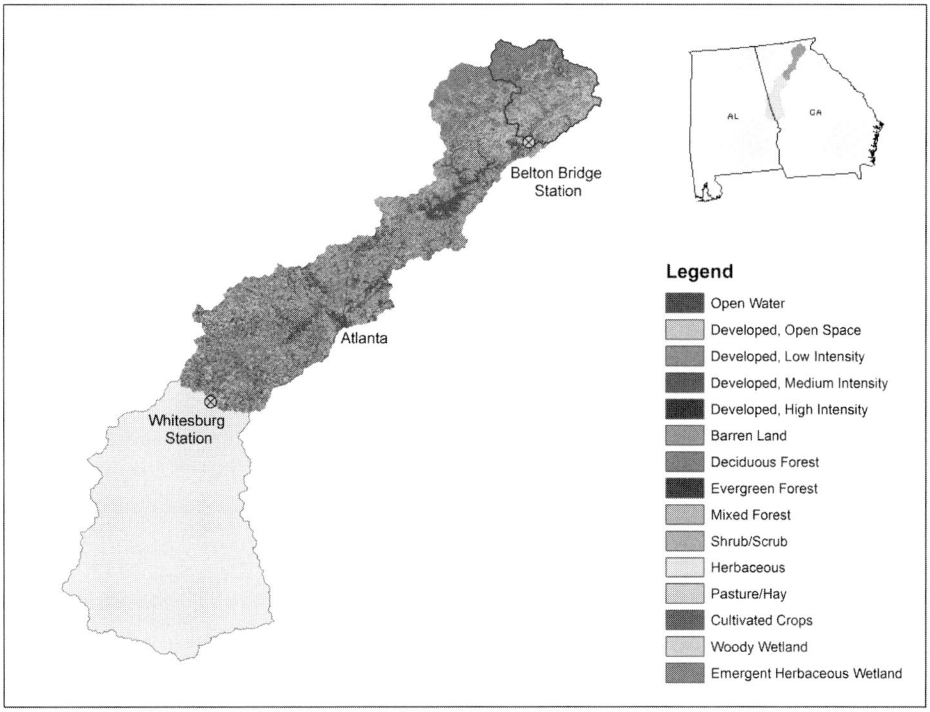

Figure 12.2 Chattahoochee River watershed upstream of the Whitesburg station and its LULC in 2011 (retrieved from USGS and NLCD). A black and white version of this figure will appear in some formats. For the color version, please refer to the plate section.

urban watershed (shown in Figure 12.2). The targeting water quality parameters are total nitrogen (TKN, mg/L), DO, and phosphorus (mg/L).

12.2 Dependence Study at the Snohomish River Watershed

In this section, we will investigate the temporal and spatial–temporal dependence for the water quality parameters at the case-study site of the Snohomish River watershed. For the case study of the Snohomish River watershed, monthly TPN (C70 only) and DO (C70, D50, D130, and A90) are applied for the study. The monthly TPN and DO at station C70 are used for the study of temporal dependence. The monthly DO at all four of these stations is applied to study the spatial–temporal dependence.

12.2.1 Study of Temporal Dependence Using Copulas

Temporal Dependence of Monthly TPN and DO at Station C70

The TPN and DO at station C70 are chosen for the study of temporal dependence. Table 12.1 lists the dataset for both temporal and spatial dependence study. We will use

Table 12.1. *TPN and DO monthly dataset from the Snohomish River watershed.*

Dates	TPN (C70)	DO (C70)	DO (D50)	DO (D130)	DO (A90)
Oct-94	0.116	11.3	10.8	10.5	10.9
Nov-94	0.312	12	11.7	11.9	11.9
Dec-94	0.287	12.7	12.6	12.6	12.6
Jan-95	0.285	12.8	11.9	12.3	12.15
Feb-95	0.21	12.4	12.4	12.8	12.8
Mar-95	0.212	12.1	11.6	11.8	11.7
Apr-95	0.141	12.2	11.4	11.9	11.6
May-95	0.081	11.6	10.4	11.1	10.8
Jun-95	0.086	11.2	10.2	10.9	10.6
Jul-95	0.066	10.2	9.2	9.4	9.7
Aug-95	0.117	10.5	9.6	9.8	9.8
Sep-95	0.125	10.3	9.6	9.3	9
Oct-95	0.253	11	10.4	10.8	10.6
Nov-95	0.178	12.2	10.9	12	11.5
Dec-95	0.203	12.5	11.4	12.1	11.8
Jan-96	0.223	12.8	11.9	12.4	12.3
Feb-96	0.155	12.4	12.3	12.2	12.2
Mar-96	0.119	12.6	11.7	12.2	12
Apr-96	0.114	11.7	10.8	11.3	11.1
May-96	0.09	12	11.3	11.7	11.6
Jun-96	0.043	11.3	10.1	10.8	10.7
Jul-96	0.045	10.5	9.7	9.7	10
Aug-96	0.107	10.1	9.4	9.6	9.8
Sep-96	0.194	10.7	9.9	10.5	10.2
Oct-96	0.331	11.6	11.4	11.7	11.3
Nov-96	0.237	12.3	11.8	12.2	12
Dec-96	0.286	12.7	12	12.5	12.2
Jan-97	0.201	13	12.3	12.6	12.4
Feb-97	0.232	12.5	12	12.3	12
Mar-97	0.299	12.5	11.9	12.3	12.2
Apr-97	0.218	12.7	12.7	12.5	12.6
May-97	0.077	11.9	10.9	11.8	11.2
Jun-97	0.076	11.6	10.8	11.2	11.1
Jul-97	0.058	10.2	9	9.3	9.3
Aug-97	0.055	10.2	9.3	9.4	9.2
Sep-97	0.204	10.4	9.9	10.3	10
Oct-97	0.163	11.1	10.7	11	11
Nov-97	0.187	11.8	11.3	11.3	11.4
Dec-97	0.282	12.6	11.7	11.6	11.8
Jan-98	0.387	12.2	11.6	11	11.7
Feb-98	0.258	12.4	11.5	11.7	11.8
Mar-98	0.217	12.1	11.5	12.2	11.8

Table 12.1. (*cont.*)

Dates	TPN (C70)	DO (C70)	DO (D50)	DO (D130)	DO (A90)
Apr-98	0.165	12	10.9	11.5	11.5
May-98	0.14	12	11.1	11.9	11.4
Jun-98	0.065	10.8	9.7	10.3	10.1
Jul-98	0.077	10.4	9	9.9	9.3
Aug-98	0.075	10.1	8.3	9.4	9.1
Sep-98	0.097	10.3	9.6	9.4	9.4
Oct-98	0.23	12.1	11.7	11.5	11.2
Nov-98	0.378	11.7	11.2	11.5	11.4
Dec-98	0.392	12.6	12.4	12.6	12.2
Jan-99	0.333	12.2	11.9	12.1	12
Feb-99	0.275	12.8	11.6	12.4	12.4
Mar-99	0.19	12.3	12	12.5	12
Apr-99	0.162	12.4	12.1	12.7	11.9
May-99	0.146	12.2	10.9	12.2	11.5
Jun-99	0.064	11.5	11.1	11.5	11.5
Jul-99	0.057	11.3	10.3	11.1	10.6
Aug-99	0.084	11.3	9.8	10.5	10.5
Sep-99	0.124	9.8	9.8	9.3	9
Oct-99	0.189	11.3	11	11.2	10.9
Nov-99	0.29	12	11.7	11.9	11.7
Dec-99	0.242	12	11.3	11.7	11.6
Jan-00	0.271	13.1	12	12.4	12.3
Feb-00	0.481	12.8	12.1	12.3	12.2
Mar-00	0.195	13	12.2	12.7	12.5
Apr-00	0.141	12.1	11.7	12.1	11.8
May-00	0.146	11.9	10.7	11.7	11.1
Jun-00	0.061	12.2	11.1	11.6	11.4
Jul-00	0.049	10.9	9.8	10.5	10.5
Aug-00	0.082	11	9.5	9.69	9.69
Sep-00	0.082	10.4	9.79	9.89	9.59
Oct-00	0.148	11.1	11	11.2	10.7
Nov-00	0.149	13.36	12.57	12.57	12.67
Dec-00	0.229	12.86	12.46	12.48	12.46
Jan-01	0.229	13.06	12.44	13.36	12.85
Feb-01	0.224	13.57	12.95	13.06	13.06
Mar-01	0.257	12.62	13.03	13.73	12.42
Apr-01	0.187	12.32	11.81	11.81	11.71
May-01	0.11	12.18	11.97	12.08	11.47
Jun-01	0.074	11.4	11.6	11.7	11
Jul-01	0.089	10.71	9.89	10.51	11.22
Aug-01	0.135	9.89	10	10	9.28
Sep-01	0.181	9.9	10.1	10	9.2

Table 12.1. (*cont.*)

Dates	TPN (C70)	DO (C70)	DO (D50)	DO (D130)	DO (A90)
Oct-01	0.309	12.22	11.81	12.42	11.81
Nov-01	0.193	12.09	11.29	11.49	11.49
Dec-01	0.305	12.82	12.32	11.91	13.33
Jan-02	0.22	13.26	12.07	12.57	12.57
Feb-02	0.207	14.03	12.74	13.83	13.33
Mar-02	0.245	13.83	12.53	13.23	12.83
Apr-02	0.149	12.9	12.31	12.51	12.21
May-02	0.099	11.96	11.47	11.76	11.47
Jun-02	0.077	11.73	11.25	11.35	11.35
Jul-02	0.068	11	9.69	10.8	10.1
Aug-02	0.046	9.8	9.19	9.8	9.69
Sep-02	0.096	11.34	9.75	10.34	9.95
Oct-02	0.08	10.65	10.85	10.25	10.65
Nov-02	0.349	12.32	11.51	11.71	11.81
Dec-02	0.205	12.69	12.18	11.97	12.48
Jan-03	0.21	13.16	13.06	12.65	12.55
Feb-03	0.219	13.6	12.89	13.6	12.99
Mar-03	0.18	12.5	11.4	12.4	11.8
Apr-03	0.153	12.4	11	11.2	10.4
May-03	0.105	12.08	11.67	12.18	11.26
Jun-03	0.054	11.06	9.64	10.86	10.15
Jul-03	0.11	10.3	8.97	10.2	8.87
Aug-03	0.094	9.2	9.6	9	8.8
Sep-03	0.19	10.6	10	10.1	9.19
Oct-03	0.25	10.9	10.7	11.11	10.8
Nov-03	0.2745	12.46	11.71	11.91	11.91
Dec-03	0.23	12.56	12.16	12.56	12.16
Jan-04	0.273	13.06	12.36	12.56	12.46
Feb-04	0.193	12.7	11.7	12.5	12.1
Mar-04	0.15	12.3	11.7	11.7	11.8
Apr-04	0.12	11.9	11.2	11.8	11.1
May-04	0.08	11.7	10.8	12	11.2
Jun-04	0.073	10.6	9.69	10.5	9.8
Jul-04	0.076	11.1	8.8	9.4	8.8
Aug-04	0.12	9.4	8.69	8.69	8.6
Sep-04	0.15	11.11	10.5	11.11	10.7
Oct-04	0.18	11.2	11.2	11.3	11.2
Nov-04	0.2	11.7	11	11.3	11.3
Dec-04	0.24	12.2	12.53	12.1	11.8
Jan-05	0.15	12.5	11	12.3	11.6
Feb-05	0.19	13.2	12.1	12.3	12.5
Mar-05	0.251	12.2	11.8	12.1	11.3

Table 12.1. (*cont.*)

Dates	TPN (C70)	DO (C70)	DO (D50)	DO (D130)	DO (A90)
Apr-05	0.2	12	11.7	12	11.5
May-05	0.094	11.5	11.4	12	10.7
Jun-05	0.087	11.4	10.5	11.1	10.5
Jul-05	0.13	10.19	9.3	9.69	8.9
Aug-05	0.13	9.31	8.81	9.21	8.51
Sep-05	0.13	10.19	9.69	9.6	9.5
Oct-05	0.19	10.8	10.4	10.7	10.19
Nov-05	0.329	12.6	12	12.3	12.4
Dec-05	0.22	13.3	12.9	13	13.1
Jan-06	0.263	12.8	11.7	12.4	12.3
Feb-06	0.24	13.3	11.9	12.7	12.4
Mar-06	0.253	13.1	11.6	12.4	12.3
Apr-06	0.2	12.5	11.9	12.5	12.2
May-06	0.1	12.2	10.8	11.8	11.2
Jun-06	0.052	11.9	10.8	11.6	11.2
Jul-06	0.067	11.1	9.5	10.4	9.9
Aug-06	0.094	9.5	9.19	8.9	9.9
Sep-06	0.11	10.4	10	9.8	9.5
Oct-06	0.308	10.7	10.6	10.6	10.3
Nov-06	0.2645	12.61	11.2	11.8	11.4
Dec-06	0.23	12.19	13	12.9	12.8
Jan-07	0.24	13	12.3	12.7	12.8
Feb-07	0.14	12.9	11.8	12.7	12.2
Mar-07	0.12	12.8	12.4	12.8	12.3
Apr-07	0.099	12.6	11.6	12.4	11.9
May-07	0.061	12.26	11.25	11.85	11.95
Jun-07	0.073	11.7	11.2	11.2	11.5
Jul-07	0.1	10.5	9.4	9.8	9.9
Aug-07	0.098	10	8.9	9.4	9.19
Sep-07	0.12	11.18	10.29	9.9	9.5
Oct-07	0.25	11.8	11.9	11.8	11.5
Nov-07	0.2	12.63	12.13	12.33	12.33
Dec-07	0.24	12.5	11.6	11.9	11.9
Jan-08	0.28	13.23	12.33	12.33	12.63
Feb-08	0.2	12.95	12.04	12.44	12.34
Mar-08	0.2	13.1	12	12.4	12.3
Apr-08	0.215	12.63	11.94	12.33	12.03
May-08	0.13	12.7	12.1	12.5	12.3
Jun-08	0.098	12.1	11	12	11.4
Jul-08	0.06	11.1	10.1	10.4	10.7
Aug-08	0.055	10	8.69	9.1	9.19
Sep-08	0.12	11.1	10	10.19	9.8

Table 12.1. (*cont.*)

Dates	TPN (C70)	DO (C70)	DO (D50)	DO (D130)	DO (A90)
Oct-08	0.18	12.3	11.8	11.8	11.7
Nov-08	0.21	9.6	11.7	11.4	10.5
Dec-08	0.216	13.6	13.4	13.6	13.6
Jan-09	0.212	13.5	12.4	12.9	12.9
Feb-09	0.19	12.3	11.5	11.4	12.4
Mar-09	0.214	13.4	12.8	13.5	12.8
Apr-09	0.227	12.2	11.6	12.6	11.6
May-09	0.091	12.5	12	12	12
Jun-09	0.04	11.3	10.4	10.8	10.6
Jul-09	0.083	10.1	9	9.1	9.19
Aug-09	0.067	10	9.5	9.3	9.5
Sep-09	0.159	9.4	9.3	9.19	8.19
Oct-09	0.327	10.6	10.19	10.6	10
Nov-09	0.163	12.4	12	12.3	12
Dec-09	0.279	13	12.6	12.6	12.7
Jan-10	0.276	12.5	11.8	12.1	12.1
Feb-10	0.182	12.8	12.6	12.1	12.6
Mar-10	0.161	12.4	11.6	12.2	12
Apr-10	0.14	12.3	11.9	12.4	11.6
May-10	0.076	12.1	11.3	11.9	10.8
Jun-10	0.049	11.3	10.6	11.1	11.1
Jul-10	0.078	10.3	9	9.4	8.8
Aug-10	0.083	10.4	9.59	9.89	9.49
Sep-10	0.109	11.1	10	10.4	10.5
Oct-10	0.17	10.94	10.74	10.94	11.34
Nov-10	0.162	11.85	11.75	11.75	11.85
Dec-10	0.267	13.13	11.6	12.2	11.9
Jan-11	0.217	13.1	12.6	12.8	13.2
Feb-11	0.161	12.6	12	12.2	12.6
Mar-11	0.224	12.3	11.9	11.7	11.8
Apr-11	0.154	12	11.8	11.7	11.9
May-11	0.095	12.6	12.3	12.2	12.2
Jun-11	0.042	11.87	11.16	11.47	11.47
Jul-11	0.04	11.4	10.8	10.6	10.8
Aug-11	0.05	10.65	9.44	9.64	10.05
Sep-11	0.149	10.02	9.12	9.62	9.82
Oct-11	0.188	11.12	10.92	11.22	10.5
Nov-11	0.233	13.7	12.8	13.2	12.7
Dec-11	0.255	13.6	13.1	13	12.8
Jan-12	0.356	13.2	12.6	12.8	12.8
Feb-12	0.191	12.8	12	12.3	12.8
Mar-12	0.233	12.8	12.2	12.6	13.2

Table 12.1. (*cont.*)

Dates	TPN (C70)	DO (C70)	DO (D50)	DO (D130)	DO (A90)
Apr-12	0.123	12.83	12.42	13.03	12.73
May-12	0.115	12.73	11.52	12.73	12.02
Jun-12	0.094	12.2	11.9	11.9	11.5
Jul-12	0.041	11.2	10.1	10.5	10
Aug-12	0.061	10.8	9	9.5	9.6
Sep-12	0.081	10.8	10.1	9.5	10.16
Oct-12	0.21	11.6	11.3	11.3	11.2
Nov-12	0.227	12.6	11.5	12.2	11.7
Dec-12	0.42	12	11.8	11.9	12.1
Jan-13	0.337	13.3	12.9	12.7	12.7
Feb-13	0.244	12.5	11.7	12.1	11.3
Mar-13	0.175	12.8	12.2	12.6	11.7
Apr-13	0.171	12.2	11.6	12	11.8
May-13	0.067	11.8	11.2	11.6	11.3
Jun-13	0.058	11.4	10.2	11	10.3
Jul-13	0.084	9.8	8.9	9.2	9
Aug-13	0.226	9.6	8.7	8.7	8.9
Sep-13	0.177	10	9.2	10.3	9.3

the water quality data before 2012 to build a copula-based Markov process model, and the water quality data of 2012 and 2013 will be used for model validation purpose.

In general, before we proceed to investigate the temporal dependence using copulas, we first evaluate whether there exists periodicity (or seasonality) in the sequence. For monthly TPN and DO, we suspect that there should exist seasonality. We can use the sample autocorrelation function plot or cumulative periodogram through spectral analysis (Box et al., 2007) to assess the seasonality.

The sample autocorrelation coefficient $[\gamma_k]$ for time series x_t at lag k can be written as follows:

$$c_k = \frac{1}{N}\sum\nolimits_{t=1}^{N-k}(x_t - \bar{x})(x_{t+k} - \bar{x}) \tag{12.1a}$$

$$\gamma_k = \frac{c_k}{c_0}; c_0 = \frac{1}{N}\sum\nolimits_{t=1}^{N}(x_t - \bar{x})^2 \tag{12.1b}$$

The cumulative periodogram $[C(f_k)]$ for time series x_t can be written as follows:

$$I(f_j) = \frac{2}{n}\left|\sum\nolimits_{t=1}^{N} x_t(-2\pi i f_j t)\right| = \frac{2}{N}\left[\left(\sum\nolimits_{t=1}^{N} x_t \cos(2\pi f_j t)\right)^2 + \left(\sum\nolimits_{t=1}^{N} x_t \sin(2\pi f_j t)\right)^2\right]^{\frac{1}{2}} \tag{12.2a}$$

$$C(f_k) = \frac{\sum_{j=1}^{k} I(f_j)}{N\widehat{\sigma_x^2}} \tag{12.2b}$$

In Equations (12.2a) and (12.2b), $I(f_j)$ stands for the periodogram function; $f_j = \frac{j}{N}$, $j = 1, \ldots \lfloor \frac{N}{2} \rfloor$; $\widehat{\sigma_x^2}$ is the estimated variance for the time series.

Applying Equations (12.1) and (12.2), Figure 12.3 plots the sample autocorrelation function and cumulative periodogram for the TPN and DO at station C70. From the sample autocorrelation function plots in Figure 12.1, we clearly see that both DO and TPN have a 12-month cycle. From the cumulative periodogram plot for TPN at C70, we notice a discontinuity at frequency $f = 0.0833 \approx \frac{1}{12}$. The discontinuity of cumulative periodogram indicates the existence of periodicity (or seasonality). From the cumulative periodogram plot for DO at C70, again we see the discontinuity at the same frequency as that of TPN; we see another very small discontinuity at frequency $f = 0.1667 \approx 1/6$, which means six-month period may also exist for the DO sequence. Comparatively speaking, the six-month

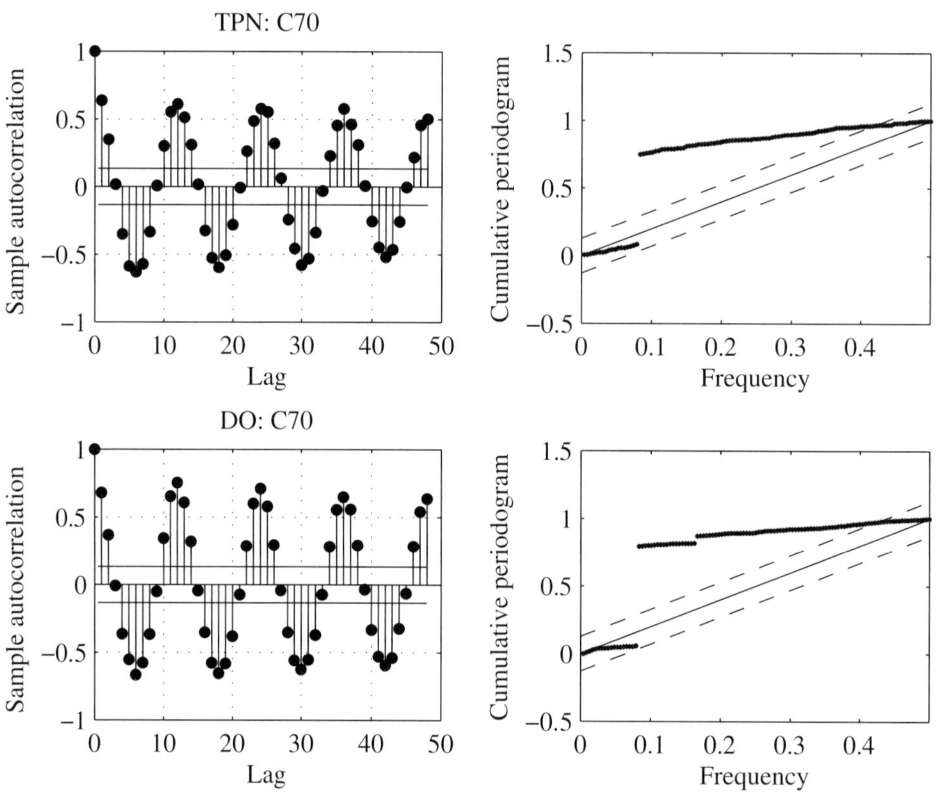

Figure 12.3 Autocorrelation and cumulative periodogram plots for original monthly TPN and DO series.

subcycle is not significant, and we will only deal with the dominating 12-month periodicity for both TPN and DO sequences.

To remove the periodicity, we will introduce a simple but effective method (called the full deseasonalization method). For our monthly water quality study, we will actually remove the monthly average and monthly standard deviation from the water quality time series using the following:

$$x_{r,m}^{de-season} = \frac{x_{r,m} - \hat{\mu}_m}{\hat{\sigma}_m}, m = 1,2, \ldots S \quad (12.3)$$

In this case study, we have $S = 12$ to show that we have monthly period. After applying Equation (12.3), we can then use the deseasonalized sequence to reevaluate whether the periodicity has been successfully removed as shown in Figure 12.4. As seen in Figure 12.4, the periodicity has been successfully removed. Table 12.2 tabulates the monthly sample mean and sample standard deviation for TPN and DO time series, respectively.

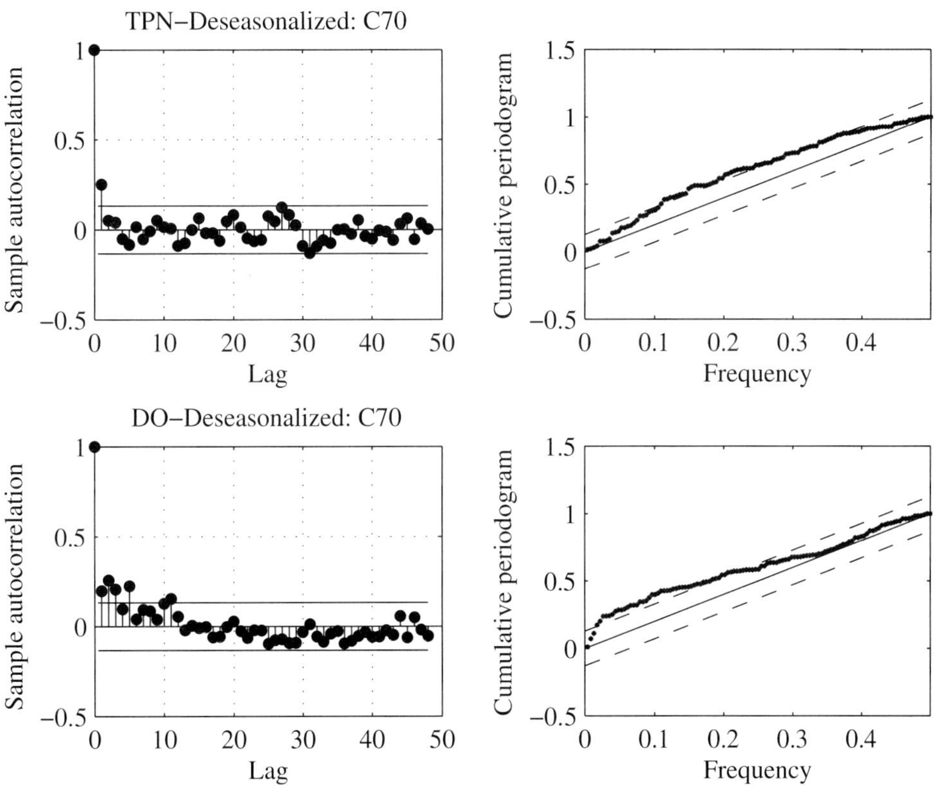

Figure 12.4 Autocorrelation and cumulative periodogram plots for deseasonalized TPN and DO series.

Table 12.2. *Monthly sample mean and standard deviation of TPN and DO series.*

Month	TPN (mg/L) $\hat{\mu}$	TPN (mg/L) $\hat{\sigma}$	DO (mg/L) $\hat{\mu}$	DO (mg/L) $\hat{\sigma}$
January	0.26	0.06	12.94	0.36
February	0.22	0.07	12.87	0.47
March	0.21	0.05	12.67	0.46
April	0.16	0.04	12.31	0.33
May	0.10	0.03	12.10	0.35
June	0.07	0.02	11.50	0.44
July	0.07	0.02	10.65	0.48
August	0.09	0.04	10.09	0.58
September	0.14	0.04	10.48	0.53
October	0.21	0.07	11.28	0.52
November	0.24	0.07	12.21	0.82
December	0.26	0.06	12.71	0.47

With the successful removal of periodicity, we can now proceed to study the temporal dependence using the copula-based Markov process. As stated in Chapter 9, with the application of the copula-based Markov process, the time series does not need to belong or transform to the Gaussian process. In addition, the marginals and serial dependence can be studied separately to avoid possible misidentification. Following the discussion in Sections 9.3–9.5, we will illustrate the application of the copula-based Markov process to the water quality time series. As stated in Chapter 9, the procedure involved for the copula-based Markov process is as follows:

i. Identify the Markov order for the stationary time series.
ii. Investigate the marginal distribution of the Markov process.
iii. Study the serial dependence using copula.
iv. Perform one-step ahead forecasting with the copula-based Markov process.

Identification of the Proper Markov Order for the Deseasonalized TPN and DO Time Series

The Markov order will be identified using the method discussed in Section 9.5.2. The meta-Gaussian copula is applied as the building block for the order identification purpose only. The kernel density method is applied to estimate the marginals nonparametrically. Following the order identification procedure, we obtain that the deseasonalized TPN and DO may be modeled using the first- and second-order Markov process, respectively (as listed in Table 12.3).

With the identified Markov order, we can move on to choose the best-fitted copula functions. For the deseasonalized TPN series, the most common bivariate copulas (i.e.,

Table 12.3. *Markov order identification using the meta-Gaussian copula.*

	F_t, F_{t-1}		$F_{t\|t-1}, F_{t-2\|t-1}$		$F_{t\|t-1,t-2}, F_{t-3\|t-1,t-2}$		
Variable	τ	p-Val	τ	p-Val	τ	p-Val	Order
TPN	0.18	< 0.01	0.02	0.64	—	—	1
DO	0.14	< 0.01	0.16	< 0.01	0.07	0.14	2

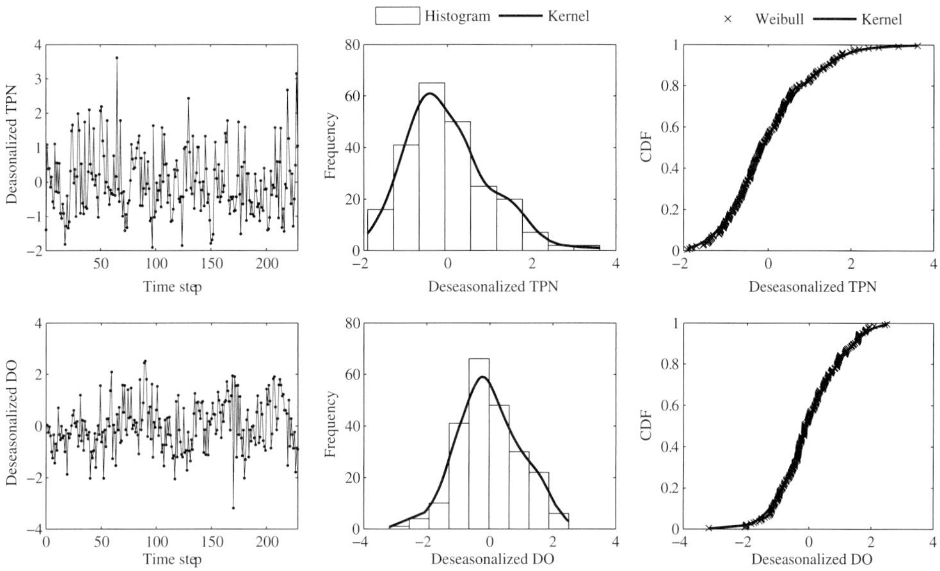

Figure 12.5 Plots of deseasonalized TPN and DO time series, kernel density, as well as the CDF computed from kernel density.

Gumbel, meta-Gaussian, meta-Student *t*, and Frank) will be selected as the candidates. For the deseasonlized DO series, the D-vine copula application to time series discussed in Chapter 9 will be selected. The pseudo-MLE discussed in Section 9.5.3 is applied for parameter estimation with the use of empirical distribution estimated from kernel densities. To illustrate the empirical distribution with the use of kernel density, we selected a simple Gaussian kernel with the bandwidth of 0.3097 and 0.3507 for deseasonalized TPN and DO, respectively. As shown in Figure 12.5, the kernel density fits the histogram very well. The CDF computed from kernel density also fits the empirical CDF computed with the use of Weibull plotting-position formula very well. Figure 12.5 verifies that the kernel density may be applied to model the marginal distribution of time series.

Parameter Estimation for the Deseasonalized TPN and DO Series

Deseasonalized TPN Series Table 12.4 lists the parameter, likelihood, and AIC values estimated using the four previously discussed copula candidates. From Table 12.4, it is seen that the Gaussian copula is the best choice based on the AIC criterion. Results of the SnB goodness-of-fit test (SnB = 0.034, P = 0.23) further confirm that the Gaussian copula may properly model the deseasonalized TPN series.

Deseasonalized DO Series As discussed in Chapter 9, the copula-based second-order Markov process is fully governed by the joint distribution of $(DO_{t-2}, DO_{t-1}, DO_t)$ through the trivariate copula, i.e., three-dimensional D-vine copula shown as Figure 9.10 in Chapter 9. In this structure, (DO_{t-1}, DO_t) and (DO_{t-2}, DO_{t-1}) for the lag-1 dependence possess the same copula. Table 12.5 lists the results for parameter estimation, including the SnB goodness-of-fit statistical test. The results in Table 12.5 show that (1) the Gumbel–Hougaard copula can be applied to model the lag-1 temporal dependence; and (2) the

Table 12.4. *Results from the four copula candidates for first-order deaseasonalized TPN series.*

	Gumbel–Houggard θ	Gaussian ρ	Student t $[\rho, v]$	Frank θ
Parameters	1.23	**0.29**	[0.30, 11.40]	1.97
ML	6.83	**9.34**	9.71	8.73
AIC	−11.67	**−16.68**	−15.43	−15.48

Table 12.5. *Results from the four copula candidates for second-order deseasonalized DO series.*

T1	Gumbel–Hougaard	Gaussian	Student t	Frank
Parameters	**1.18**	0.16	[0.18, 4.01]	1.24
ML	**6.83**	2.85	7.52	3.41
AIC	**−11.66**	−3.71	−11.04	−4.82
SnB = 0.024, P = 0.56 (t,t-1)				
SnB = 0.024, P = 0.59 (t-2,t-1)				
T2	Gumbel–Hougaard	Gaussian	Student t	Frank
Parameters	1.16	**0.22**	[0.23, 3E+06]	1.45
ML	4.77	**5.15**	5.15	4.89
AIC	−7.55	**−8.31**	−6.31	−7.78

Note: SnB = 0.033, P = 0.25 (t|t-1, t-2|t-1)

Gaussian copula can be applied to model the conditional dependence of ($t|t$-1 and t-2$|t$-1). With the selected copula models (i.e., Gaussian for deseasonalized TPN, Gumbel–Gaussian for deseasonalized DO), we will show the simulation and forecast in what follows.

Monthly TPN and DO Simulation and Forecast

Deseasonalized TPN Series The simulation method discussed in Section 9.4.3 is applied to the first-order TPN series. Likewise, Section 9.4.4 is applied for the one-step ahead median and VaR forecasts. Using one simple example, we will show the inversion of simulated variate in the frequency domain back to the real domain.

Suppose that we simulated $U_{June} = 0.8$ from the Guassian copula fitted to the first-order deseasonalized TPN series. Looking up the empirical CDF computed from the kernel density function, we see the simulated $U_{June} = 0.8$ is bounded by [CDF, TPN] in {[0.787, 0.761], [0.804, 0.844]}. Applying the interpolation, we compute the simulated deseasonalized TPN as follows:

$$\mathrm{TPN}_{\mathrm{sim}}^{\mathrm{deseason}} = 0.761 + \frac{0.844 - 0.761}{0.804 - 0.787}(0.8 - 0.787) = 0.8245.$$

Adding back the monthly average and standard deviation for the month of June, we can compute the simulated TPN of June as follows:

$$\mathrm{TPN}_{\mathrm{sim}} = 0.8245(0.0175) + 0.0666 = 0.0811 \text{ mg/L}.$$

Applying the one-step ahead forecast discussed in Example 9.3, we can proceed with the median forecast as well as the 95% and 5% VaR. To compute the VaR, Equation (9.22) can be rewritten as follows:

$$Z_{t+1}^{95\%} = F_n^{-1}\left(C_{F_n(z_{t+1})|F_n(z_t)}(0.95|F_n(z_t);\hat{\boldsymbol{\alpha}}\right) \tag{12.4a}$$

$$Z_{t+1}^{5\%} = F_n^{-1}\left(C_{F_n(z_{t+1})|F_n(z_t)}(0.05|F_n(z_t);\hat{\boldsymbol{\alpha}}\right) \tag{12.4b}$$

Figure 12.6 plots the comparison of simulated monthly TPN with the observed TPN. It also plots the forecasted monthly TPN, its 5% and 95% VaR versus the observed monthly TPN. Figure 12.6 indicates that (a) the simulated deseasonal TPN from the fitted Gaussian copulas well presents the lag-1 temporal dependence compared to the observed deseasonal TPN series; (b) simulated monthly TPN also well presents the dependence of the observed monthly TPN series; (c) the one-step ahead monthly TPN forecast captured the main trend of monthly TPN; and (d) though there is an obvious error for the extreme TPN values, the VaR values may help identify these extreme values. The forecasted and VaR values are listed in Table 12.6.

Deseasonalized DO Series Applying the methods discussed in Sections 9.5.4 and 9.5.5, we can simulate and forecast the DO series, which may be modeled as a second-order Markov process. Substituting the median probability of 0.5 (for forecast purposes) with the

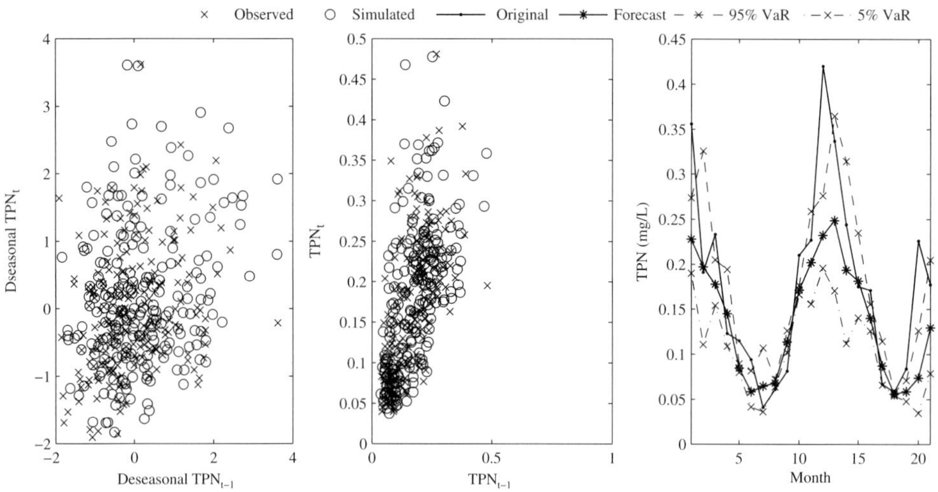

Figure 12.6 Simulations of deseasonal monthly, monthly TPN, and monthly TPN forecast with 95% and 5% VaRs.

conditional probability of 0.05 and 0.95, we will be able to compute 5% and 95% VaRs. For the second-order Markov process, its median forecast, 5% and 95% VaRs can be written as follows:

$$\hat{Z}_t = F_n^{-1}\left(C_{F_n(z_t)|F_n(z_{t-1}),F_n(z_{t-2}}(0.5|F_n(z_{t-1}),F_n(z_{t-2});\hat{\boldsymbol{\alpha}}\right) \tag{12.5a}$$

$$\hat{Z}_t^{5\%} = F_n^{-1}\left(C_{F_n(z_t)|F_n(z_{t-1}),F_n(z_{t-2}}(0.05|F_n(z_{t-1}),F_n(z_{t-2});\hat{\boldsymbol{\alpha}}\right) \tag{12.5b}$$

$$\hat{Z}_t^{95\%} = F_n^{-1}\left(C_{F_n(z_t)|F_n(z_{t-1}),F_n(z_{t-2}}(0.95|F_n(z_{t-1}),F_n(z_{t-2});\hat{\boldsymbol{\alpha}}\right) \tag{12.5c}$$

Figure 12.7 plots the comparison of simulated monthly DO with the observed DO series. It also plots the forecasted monthly DO, its 5% and 95% VaR versus the observed monthly DO. Figure 12.7 indicates that (a) the simulated deseasonal DO from the fitted Gaussian copulas well presents the lag-1 and lag-2 temporal dependence compared to the observed deseasonal DO series; (b) the simulated monthly DO also well presents the dependence of the observed monthly DO series; (c) the fitted second-order copula-based Markov process (i.e., the Gumbel–Gaussian vine copula) well represents the lag-1 and lag-2 dependence that is statistically significant; and (d) for the one-step ahead forecast, the fitted second-order copula-based model performs well. The forecast and VaR values are listed in Table 12.6. Additionally, from Figures 12.6 and 12.7, it is seen that the second-order copula-based DO model yields a better forecast than does TPN. Part of the reason could be that the TPN is more influenced by human activities, etc. (e.g., agriculture), than is DO.

Table 12.6. *Forecast and VaR results computed from the fitted copula-based Markov model.*

	TPN (mg/L)				DO (mg/L)			
Date	Observed	Forecast	5%VaR	95%VaR	Observed	forecast	5%VaR	95%VaR
12-Jan	0.356	0.253	0.176	0.366	13.2	13.29	12.58	13.80
12-Feb	0.191	0.242	0.141	0.382	12.8	13.12	12.37	13.84
12-Mar	0.233	0.195	0.135	0.284	12.8	12.69	12.02	13.47
12-Apr	0.123	0.165	0.115	0.236	12.83	12.29	11.81	12.84
12-May	0.115	0.089	0.058	0.136	12.73	12.24	11.64	12.81
12-Jun	0.094	0.068	0.044	0.101	12.2	11.86	11.04	12.50
12-Jul	0.041	0.080	0.046	0.126	11.2	11.02	10.17	11.70
12-Aug	0.061	0.072	0.021	0.146	10.8	10.40	9.46	11.28
12-Sep	0.081	0.124	0.076	0.195	10.8	10.71	9.84	11.55
12-Oct	0.21	0.175	0.091	0.298	11.6	11.44	10.64	12.28
12-Nov	0.227	0.231	0.144	0.359	12.6	12.35	11.11	13.69
12-Dec	0.42	0.256	0.182	0.366	12	12.78	12.08	13.55
13-Jan	0.337	0.297	0.207	0.420	13.3	12.87	12.34	13.51
13-Feb	0.244	0.237	0.137	0.375	12.5	12.76	12.03	13.58
13-Mar	0.175	0.205	0.142	0.295	12.8	12.67	12.01	13.46
13-Apr	0.171	0.151	0.105	0.221	12.2	12.25	11.77	12.80
13-May	0.067	0.100	0.066	0.148	11.8	12.08	11.57	12.67
13-Jun	0.058	0.058	0.037	0.089	11.4	11.38	10.75	12.14
13-Jul	0.084	0.067	0.037	0.111	9.8	10.50	9.82	11.31
13-Aug	0.226	0.095	0.040	0.175	9.6	9.89	9.05	10.90
13-Sep	0.177	0.163	0.104	0.244	10	10.16	9.38	11.04

12.2.2 Spatial–Temporal Distribution of Water Quality of the Snohomish River Watershed Using Meta-Elliptical Copulas

In this section, we will study the spatial–temporal water quality distribution. As discussed in Chapter 9, we will study the time series and copula separately, that is, the DO time series is treated as a univariate time series and fitted using the classical time series modeling approach first, and then the copula will be applied to the model residual (or also called innovation). This type of model may also be called time series-copula model with the following procedure:

i. Investigate the univariate water quality time series.
ii. Investigate the spatial dependence of the water quality time series through the model residuals.
iii. Perform the simulation and one-step ahead forecast, based on the derived time series-copula model from steps iand ii.

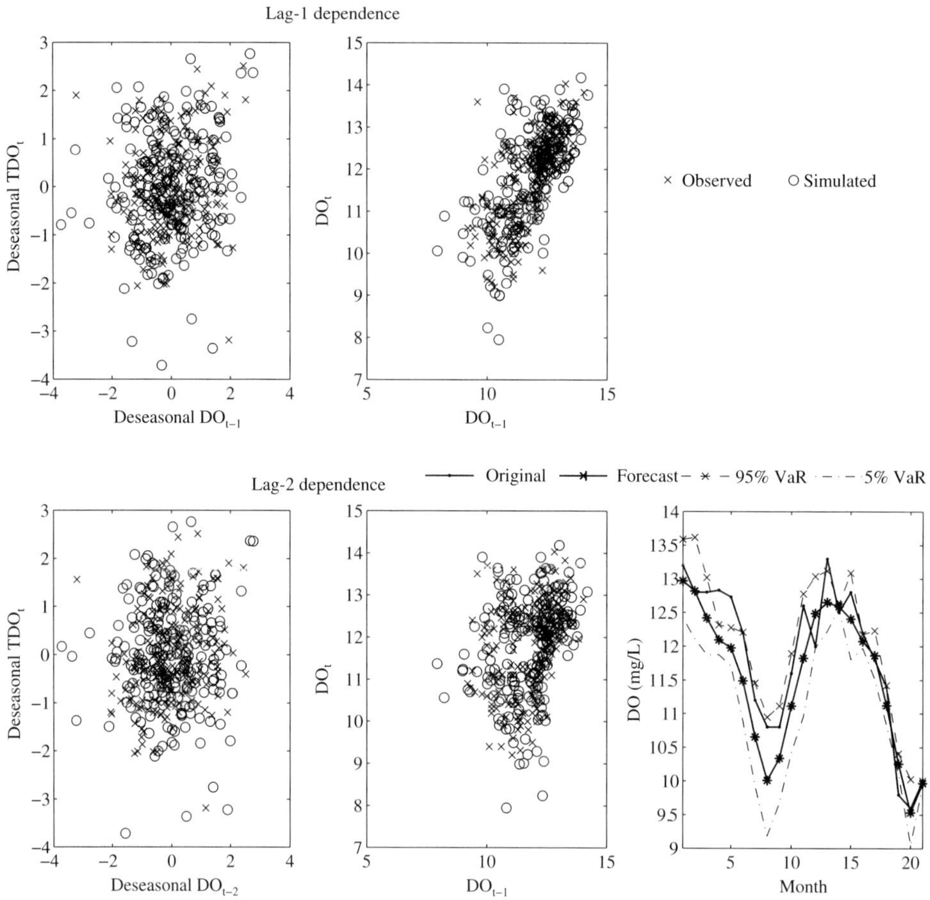

Figure 12.7 Simulations for deseasonal monthly, monthly DO, and monthly DO forecast with 95% and 5% VaRs.

Univariate Time Series Models for the Monthly DO at the Snohomish Watershed

Besides monthly DO at station C70, monthly DO at stations D50, D130, and A90 are also selected for the study. Similar to the monthly DO at C70, we first deseasonalize the monthly DOs using the full-deseasonalization method (Equation 12.3). Table 12.7 lists the monthly average and monthly standard deviation of DO for stations D50, D130, and A90.

After taking the monthly average and monthly standard deviation out of the monthly DO sequence, Table 12.8 lists the sample statistics of deseasonalized DO sequence. Figure 12.8 plots the histograms of the deseasonalized DO sequence. The purpose is to assess whether the deseasonalized time series belongs to the Gaussian process. Results in Table 12.8 and plots in Figure 12.8 show that the deseasonalized monthly DO sequence

Table 12.7. *Monthly average and standard deviation for stations D50, D130, and A90.*

D50		D130		A90	
$\hat{\mu}$(mg/L)	$\hat{\sigma}$(mg/L)	$\hat{\mu}$(mg/L)	$\hat{\sigma}$(mg/L)	$\hat{\mu}$(mg/L)	$\hat{\sigma}$(mg/L)
12.17	0.48	12.47	0.47	12.44	0.41
12.10	0.45	12.47	0.58	12.42	0.47
12.00	0.45	12.48	0.55	12.14	0.47
11.70	0.50	12.12	0.49	11.75	0.53
11.32	0.52	11.95	0.35	11.45	0.47
10.68	0.65	11.20	0.46	10.91	0.55
9.50	0.56	9.99	0.60	9.77	0.76
9.22	0.45	9.45	0.46	9.41	0.51
9.82	0.35	9.93	0.50	9.61	0.59
11.02	0.51	11.15	0.54	10.94	0.50
11.67	0.49	11.95	0.48	11.79	0.52
12.27	0.60	12.38	0.52	12.39	0.56

Table 12.8. *Sample statistics of deseasonalized DO sequences.*

Station	$\hat{\mu}$	$\hat{\sigma}$	Skewness	Kurtosis
C70	−2.49E−16	0.98	0.11	2.87
D50	2.64E−16	0.98	0.18	2.52
D130	−4.08E−16	0.98	0.18	3.23
A90	−8.26E−16	0.98	−0.04	2.72

may be modeled with the time series modeling approach as introduced in Chapter 9 (Box et al., 2007).

Following the proper procedure of model identification, (i) stationarity test, (ii) model order identification, and (iii) test of model residual, Table 12.9 lists the model identification results and Figure 12.9 plots the sample ACF and PACF plots.

Using the model order identified in Table 12.9, the AR(2) model is fitted to the time series at station D50 after differencing. The parameters estimated are listed in Table 12.10. Applying the KS test to stations C70, D130, and A90, the test statistic indicates that the DO series after differencing may be properly modeled with Gaussian distribution ($H = 0$, $P = 0.47$).

Spatial Dependence Study with Meta-Elliptical Copulas

Rather than directly applying the observed time series as the copula-based Markov process discussed in Section 12.2.1, the fitted model residuals computed from the preceding subsection will be applied to study the spatial pattern of DO for the four sampling

Table 12.9. *Model identification results.*

	KPSS test	ADF test	Differencing order	Model order[b]
C70[c]	H = 1[a]	H = 1	0.24	—
D50	H = 1	H = 1	0.43	AR(2)
D130[c]	H = 1	H = 1	0.28	—
A90[c]	H = 1	H = 1	0.32	—

Notes: [a] Reject the null hypothesis; [b] model order for sequence after differencing; [c] the time series after differencing may be considered a random variable.

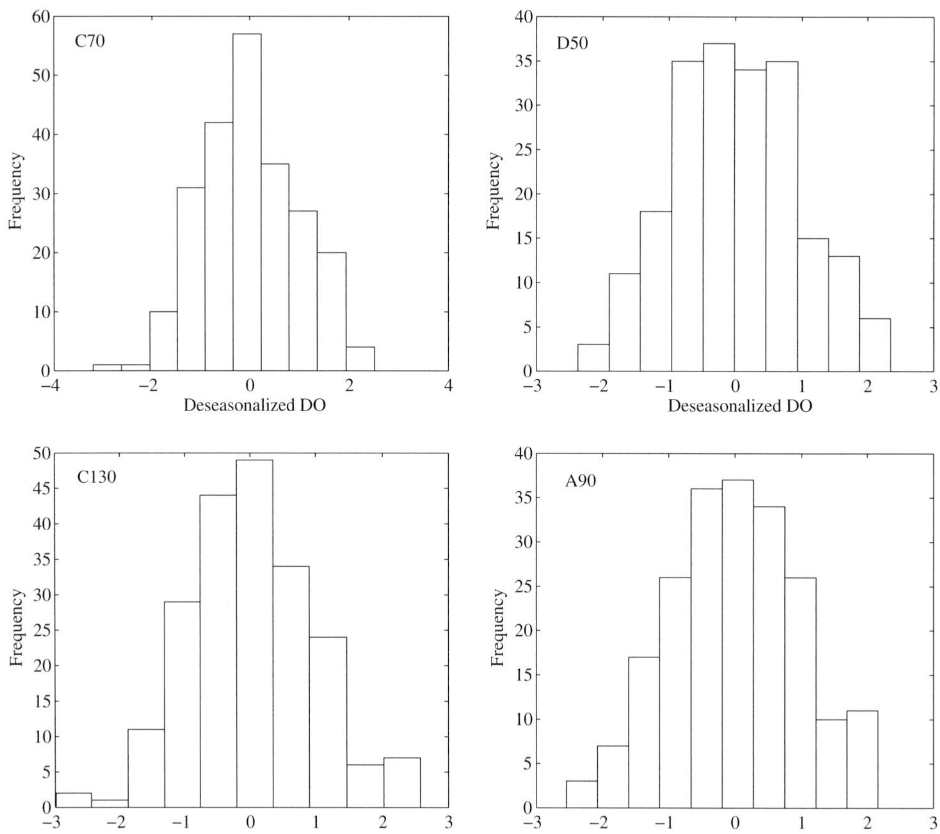

Figure 12.8 Histograms of deseasonalized DO series.

locations. Table 12.11 lists the rank-based Kendall coefficient of correlation matrix. From Table 12.11, it is shown that DO at all locations is positively correlated.

Given the model residuals for C70, D50, and A90 also modeled with Gaussian distribution, the meta-elliptical copula is applied to model the spatial dependence. More

Table 12.10. *Parameter estimated for univariate DO water quality time series.*

Station	Constant	ϕ_1	ϕ_1	σ_e^2	White noise check[a]
C70					$H = 0 \ N(0, 0.94^2)$
D50	–0.0009	–0.26	–0.19	0.88	$H = 0$
D130					$H = 0 \ N(0, 0.95^2)$
A90					$H = 0 \ N(0, 0.94^2)$

Note: [a] Check whether model residual follows $N(0, \sigma_e^2)$, and H = 0 represents the null hypothesis is accepted for KS test.

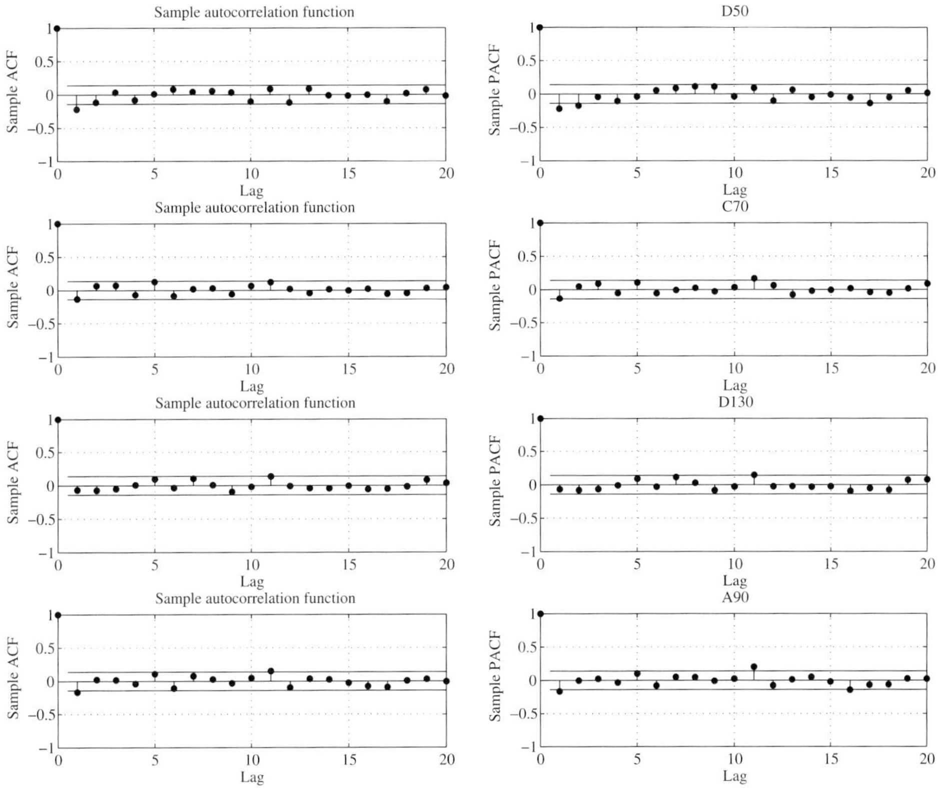

Figure 12.9 ACF and PACF plots.

specifically, meta-Gaussian and meta-Student t copulas are applied for the analysis. The parameters estimated for the meta-elliptical copula candidates are listed in Table 12.12. Figure 12.10 compares the simulated variates with the time series model residuals. It indicates that both meta-Gaussian (SnB = 0.028, $P = 0.29$) and meta-Student t (SnB = 0.019, $P = 0.96$)

Table 12.11. *Rank-based Kendall coefficient of correlation.*

Stations	C70	D50	D130	A90
C70	1	0.41	0.52	0.52
D50	0.41	1	0.48	0.50
D130	0.52	0.48	1	0.52
A90	0.52	0.50	0.52	1

Table 12.12. *Parameters estimated for meta-Gaussian and meta-Student t copulas.*

	Meta-Gaussian				Meta-Student t ($\nu = 5.90$)			
	C70	D50	D130	A90	C70	D50	D130	A90
C70	1	0.58	0.72	0.71	1	0.60	0.73	0.73
D50	0.58	1	0.67	0.69	0.60	1	0.68	0.70
D130	0.72	0.67	1	0.73	0.73	0.68	1	0.74
A90	0.71	0.69	0.73	1	0.73	0.70	0.74	1

copulas may be applied to model the spatial dependence of DO at the Snohomish River watershed.

Figures 12.11 and 12.12 plot the range of monthly DO simulated from the meta-Gaussian and meta-Student *t* copulas. The simulation plots clearly indicate that the fitted meta-Gaussian and meta-Student t copula well preserves the spatial dependence among the DOs at all four stations. At the same time, the range of simulated DO well represents the observed monthly DOs at all four stations.

One-Step Ahead DO Forecast

From the geographical location of four stations, C70, D50, D130, and A90, shown in Figure 12.1, stations D130 and C70 are the two most upstream locations sampled from two different tributaries, and station D50 is at the downstream of station D130 along the same stream. A90 is the most downstream location.

Thus, to perform the forecast, we will assume that we know the DO information at two most upstream locations, i.e., D130 and C70. We will proceed with the one-step ahead forecast for stations D50 and A90 as follows: (i) using D130 as known information to forecast D50, where two stations are along the same stream; and (ii) using D130, C70, and D50 as known information to forecast A90.

Using D130 as Known Information to Forecast D50 To forecast DO at station D50 from the DO at station D130, we need to use the copula function of $C(U_{D130}, U_{D50})$. Previously, we have developed the four-dimensional copula to study the spatial–temporal

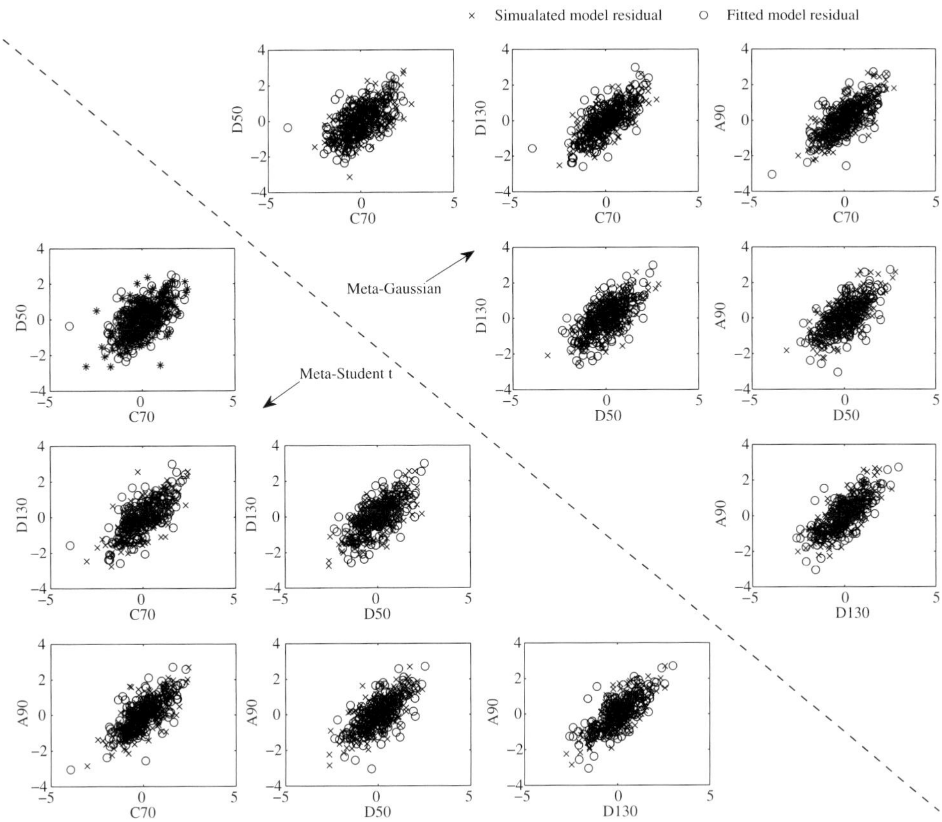

Figure 12.10 Comparison of simulated variates to the time series model residuals.

dependence among the stations of D_{130}, C_{70}, D_{50}, and A_{50}. Let $U_{C70} = U_{A90} = 1$; the four-dimensional copula may be reduced to bivariate copula following the probability theory:

$$\begin{aligned} C(U_{D130}, U_{D50}, 1, 1) &= \int_0^1 \int_0^1 \int_0^{U_{D130}} \int_0^{U_{D50}} c(U_{D130}, U_{D50}, U_{C70}, U_{A50}) dU_{D130} dU_{D50} dU_{C70} dU_{A50} \\ &= \int_0^{U_{D130}} \int_0^{U_{D50}} \int_0^1 c_1(U_{D130}, U_{D50}, U_{C70}) dU_{D130} dU_{D50} dU_{C70} \\ &= \int_0^{U_{D130}} \int_0^{U_{D50}} c_2(U_{D130}, U_{D50}) dU_{D130} dU_{D50} \\ &= C_2(U_{D130}, U_{D50}) \end{aligned} \tag{12.6}$$

In Equation (12.6), $U_{D130}, U_{D50}, U_{C70}, U_{A50}$ are the univariate CDFs for the fitted model residuals of each univariate monthly DO time series at four stations; c, c_1, c_2 are the copula density functions; and C, C_2 are the copula functions. The one-step ahead forecast for D50 is now given as follows:

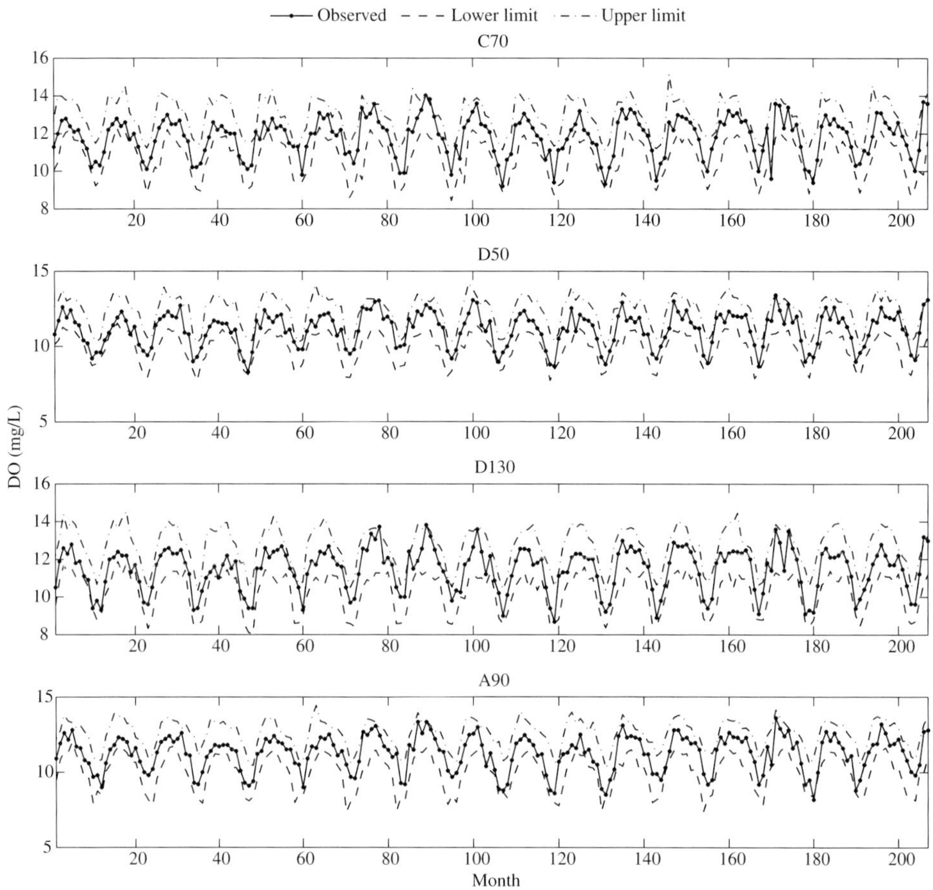

Figure 12.11 Monthly DO simulated from meta-Gaussian copula.

$$\hat{U}_{D50(t+1)} = C^{-1}\left(0.5 \,|\, U_{D130} = \hat{F}\left(DO_{D130(t+1)}\right)\right) \tag{12.7}$$

Using the forecast for January 2012 as an example, we will show how to forecast DO at station D50 in detail. On January 2012, it is assumed that we know DO at the upstream locations of D130 (12.8 mg/L) using meta-Gaussian and meta-Student t copulas.

1. Substituting the DO value at D130 into the corresponding univariate time series model, we compute the fitted model residual as follows: $D130 : r_{jan,2012} = -0.344$.
2. Applying the interpolation to the empirical distribution (or kernel density function), we compute the corresponding probability as follows: $P\left(r_{D130} \leq r_{D130,\, jan,2012} = 0.345\right)$.

   For both meta-Gaussian and meta-Student t copulas, the first two steps are identical. In step 3, we will discuss how to proceed with meta-Gaussian and meta-Student t copulas separately.

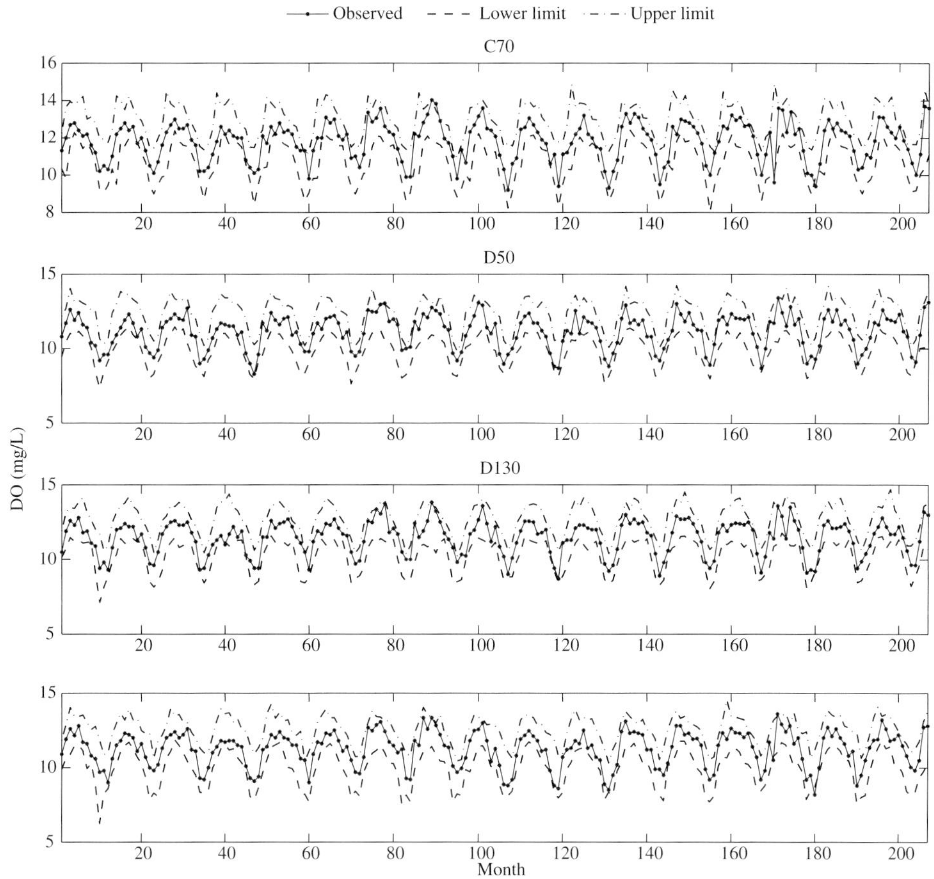

Figure 12.12 Monthly DO simulated from meta-Student T copula.

3a. (Meta-Gaussian copula): Applying the meta-Gaussian copula, we know the conditional copula of $D50 \mid D130$ is a univariate Gaussian distribution that can be estimated from the covariance matrix partition as follows:

$$\mathbf{U}_{DO} = \begin{bmatrix} U_{D50} \\ U_{D130} \end{bmatrix} = \begin{bmatrix} Y_1 \\ 0.345 \end{bmatrix}; \boldsymbol{\mu} = \begin{bmatrix} 0 \\ 0 \end{bmatrix} = \begin{bmatrix} \mu_1 \\ \mu_2 \end{bmatrix}; \boldsymbol{\Sigma} = \begin{bmatrix} 1 & 0.67 \\ 0.67 & 1 \end{bmatrix} = \begin{bmatrix} \Sigma_{11} & \Sigma_{12} \\ \Sigma_{21} & \Sigma_{22} \end{bmatrix} \tag{12.8}$$

In Equation (12.8):

$$\Sigma_{11} = 1; \Sigma_{12} = \Sigma_{21} = 0.67; \Sigma_{22} = 1$$

Similar to Example 7.9, we compute the conditional mean and conditional variance of $D50 \mid D130$ as follows:

$$\mu_{con} = \mu_1 + \Sigma_{12}\Sigma_{22}^{-1}(y_2 - \mu_2) = 0.67\Phi^{-1}(0.345) = -0.267 \tag{12.9a}$$

$$V_{con} = \Sigma_{11} - \Sigma_{12}\Sigma_{22}^{-1}\Sigma_{21} = 1 - 0.67^2 = 0.546 \tag{12.9b}$$

Then, D50|D130 follows the Gaussian distribution with $\mu = -0.267$, $variance = 0.546$.

Setting the conditional probability equal to 0.5, we estimate the model error of D50 in January 2012.

3b. (Meta-Student T copula): Similar to the meta-Gaussian copula, the conditional copula of D50|D13 (obtained from the meta-Student t copula) is the univariate Student t distribution, which can also be computed from the matrix partition, as shown in Section 7.2.2. Following Kotz and Nadarajah (2004), we know that the distribution of $\mathbf{U_{DO}} = [U_{D50}, U_{D130}]^T$ follows the bivariate Student t copula with a degree of $v = 5.9$, which is the same as the degree of freedom estimated for the fitted four-dimensional Student t copula. The pertinent parameters for the conditional distribution of D50|D130 are the following:

$$\mathbf{U_{DO}} = \begin{bmatrix} U_{D50} \\ U_{D130} \end{bmatrix} = \begin{bmatrix} Y_1 \\ 0.345 \end{bmatrix}; \boldsymbol{\mu} = \begin{bmatrix} 0 \\ 0 \end{bmatrix} = \begin{bmatrix} \mu_1 \\ \mu_2 \end{bmatrix}; \boldsymbol{\Sigma} = \begin{bmatrix} 1 & 0.68 \\ 0.68 & 1 \end{bmatrix} = \begin{bmatrix} \Sigma_{11} & \Sigma_{12} \\ \Sigma_{21} & \Sigma_{22} \end{bmatrix} \tag{12.10a}$$

$$v_{D50|D130} = v + 1 = 6.9 \tag{12.10b}$$

In Equation (12.10a), $\Sigma_{11} = 1; \Sigma_{12} = \Sigma_{21} = 0.68; \Sigma_{22} = 1$.

Now following Equation (7.54), the conditional mean and conditional variance of D50|D130 can be given as follows:

$$\mu_{D50|D130} = \left(1 - 0.68^2\right)\left(\frac{0.68}{1 - 0.68^2}\right)T^{-1}(0.345, 5.9) = -0.285 \tag{12.10c}$$

$$\Sigma_{D50|D130} = \frac{5.9 + T^{-1}(0.345, 5.9)^2}{5.9 + 1}\left(1 - 0.68^2\right) = 0.469 = 0.473 \tag{12.10d}$$

Now D50|D130 follows the noncentral Student t distribution with $\mu = -0.285$, $variance = 0.473$. Setting the conditional probability as 0.5, we can compute the estimated median error for station D50 with the known information at station D

4. (Compute the median forecast of DO at station D50): This differs from the forecast of a time series model (Box et al., 2007) with the model error set as 0 for the forecast; the model error estimated from the copula-based model may be different from zero. The median estimation of the model error is computed from the conditional copula with the use of steps 1–3. With the computed model error, the one-step ahead forecast of deseasonalized D50 (DD50) can be given as follows:

$$\widehat{DD50}_{(1)} = \psi(B)e_t = \hat{e}_{(t+1)} + \sum\nolimits_{j=1}^{\infty} \psi_j e_{(t+1-j)} \tag{12.11a}$$

$$\psi(B) = \phi^{-1}(B)(1 - B)^{-d}, d = 0.43 \tag{12.11b}$$

In Equation (12.11b), the parameters for autoregressive component is given in Table 12.10.

5. The final step is to transform the deseasonalized forecast obtained from step 4 back to the seasonal state using the following:

$$\hat{D}_{50(1)} = \widehat{DD50}_{(1)}\sigma_i + \mu_i \tag{12.12}$$

In Equation (12.12), $\{\sigma_i, \mu_i\}$ represents the seasonal deviation and seasonal mean for the forecasted month.

Applying the preceding five steps, Figure 12.13 plots the one-step ahead forecast with 5% and 95% VaRs of station D50 using the known information from D130. Table 12.13 lists the one-step ahead forecast results using meta-Gaussian and meta-Student *t* copulas. The results in Table 12.13 and Figure 12.13 indicate that the forecast follows the observed value well. The DO at the downstream location (D50) may be reasonably forecasted using the DO information at the upstream location (D130). In addition, results show that there is

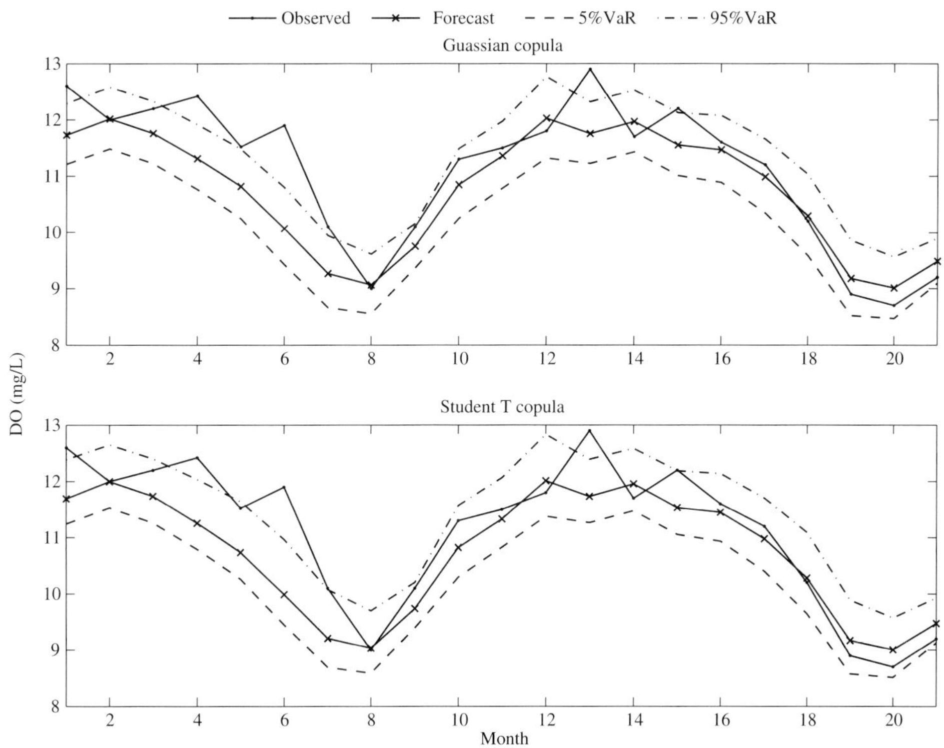

Figure 12.13 One-step ahead DO forecast, 5% and 95% VaRs for station D50 using DO information from D130.

Table 12.13. *One-step ahead DO forecast for station D50 (mg/L).*

Month	Obs.	Gaussian copula Forecast	Gaussian copula 5% VaR	Gaussian copula 95% VaR	Student *t* copula Forecast	Student *t* copula 5% VaR	Student *t* copula 95% VaR
Jan 2012	**12.6**	**11.728**	11.211	12.292	**11.692**	11.243	12.387
Feb 2012	**12**	**12.016**	11.484	12.579	**11.996**	11.525	12.647
Mar 2012	**12.2**	**11.758**	11.220	12.330	**11.736**	11.262	12.402
Apr 2012	**12.42**	**11.309**	10.764	11.915	**11.256**	10.791	12.035
May 2012	**11.52**	**10.816**	10.244	11.477	**10.734**	10.259	11.638
Jun 2012	**11.9**	**10.068**	9.425	10.800	**9.988**	9.448	10.966
Jul 2012	**10.1**	**9.268**	8.662	9.942	**9.209**	8.692	10.076
Aug 2012	**9**	**9.063**	8.554	9.614	**9.031**	8.588	9.701
Sep 2012	**10.1**	**9.752**	9.365	10.158	**9.740**	9.397	10.201
Oct 2012	**11.3**	**10.849**	10.250	11.487	**10.823**	10.295	11.570
Nov 2012	**11.5**	**11.359**	10.785	11.975	**11.328**	10.826	12.065
Dec 2012	**11.8**	**12.027**	11.319	12.768	**12.011**	11.379	12.834
Jan 2013	**12.9**	**11.754**	11.224	12.318	**11.732**	11.264	12.391
Feb 2013	**11.7**	**11.968**	11.429	12.532	**11.954**	11.474	12.583
Mar 2013	**12.2**	**11.551**	11.010	12.122	**11.532**	11.053	12.187
Apr 2013	**11.6**	**11.466**	10.887	12.074	**11.448**	10.934	12.138
May 2013	**11.2**	**10.988**	10.344	11.656	**10.977**	10.399	11.703
Jun 2013	**10.2**	**10.294**	9.585	11.032	**10.280**	9.646	11.090
Jul 2013	**8.9**	**9.178**	8.520	9.856	**9.171**	8.576	9.889
Aug 2013	**8.7**	**9.009**	8.469	9.562	**9.006**	8.513	9.574
Sep 2013	**9.2**	**9.486**	9.097	9.893	**9.477**	9.130	9.930

minimal difference in regard to the performance between the meta-Gaussian copula and the meta-Student *t* copula.

Using D130, C70, and D50 as Known Information to Forecast A90 Previously, we have illustrated the spatial–temporal dependence for the bivariate case (i.e., spatial dependence of D130 at the upstream and D50 at the downstream locations). Here, we will illustrate the multivariate spatial–temporal dependence. As shown in Figure 12.1, station A90 is the most downstream sampling location with stations D130, C70, and D50 as the upstream sampling locations. Here, we will show whether it is possible to perform a one-step ahead DO forecast for station A90 with the use of DOs at all three upstream sampling locations. Similar to the previous case, we will need to proceed as follows:

1. Compute the model error from the fitted univariate time series models for D130, C70, and D50.
2. Compute the probability for the model error obtained from *Step 1*.
3. Derive and compute $P(A90|D130, C70, D50)$ from the fitted meta-Gaussian and meta-Student t copulas (the fitted copula parameters are listed in Table 12.12). As discussed

previously, the conditional density function should follow the univariate Gaussian distribution (the meta-Gaussian copula) and univariate noncentral Student t distribution (meta-Student t copula), respectively.

In what follows, we will show the results of derived conditional distribution functions.

From the Meta-Gaussian Copula

$$\boldsymbol{U_{DO}} = \begin{bmatrix} U_{C70} \\ U_{D50} \\ U_{D130} \\ U^{*}_{A90} \end{bmatrix}; \boldsymbol{E_{DO}} = \begin{bmatrix} \Phi^{-1}(U_{C70}) \\ \Phi^{-1}(U_{D50}) \\ \Phi^{-1}(U_{D130}) \\ E^{*}_{A90} \end{bmatrix} = \begin{bmatrix} \boldsymbol{X}_1 \\ X_2 \end{bmatrix}; \tag{12.13a}$$

$$\boldsymbol{\Sigma} = \begin{bmatrix} 1 & 0.58 & 0.72 & 0.71 \\ 0.58 & 1 & 0.67 & 0.69 \\ 0.72 & 0.67 & 1 & 0.73 \\ 0.71 & 0.69 & 0.73 & 1 \end{bmatrix} = \begin{bmatrix} \boldsymbol{\Sigma}_{11} & \boldsymbol{\Sigma}_{12} \\ \boldsymbol{\Sigma}_{21} & \boldsymbol{\Sigma}_{22} \end{bmatrix} \tag{12.13b}$$

In Equations (12.13), $\boldsymbol{X}_1 = \left[\Phi^{-1}(U_{C70}), \Phi^{-1}(U_{D50}), \Phi^{-1}(D_{130})\right]^{\boldsymbol{T}}$ is the conditioning vector; $\boldsymbol{\Sigma}_{11} = \begin{bmatrix} 1 & 0.58 & 0.72 \\ 0.58 & 1 & 0.67 \\ 0.72 & 0.67 & 1 \end{bmatrix}; \boldsymbol{\Sigma}_{12} = \boldsymbol{\Sigma}_{21}^{\boldsymbol{T}}, \boldsymbol{\Sigma}_{21} = [0.71, 0.69, 0.73]; \boldsymbol{\Sigma}_{22} = 1.$

As discussed in Chapter 7, and after some algebra, we have the following:

$$\mu_{A90|C_{70}, D50, D130} = \boldsymbol{\Sigma}_{21}\boldsymbol{\Sigma}_{11}^{-1}\boldsymbol{X}_1 = 0.33\Phi^{-1}(U_{C70}) + 0.30\Phi^{-1}(U_{D50}) + 0.28\Phi^{-1}(U_{D130}) \tag{12.14a}$$

$$\boldsymbol{\Sigma}_{A90|C70,D50,D130} = \boldsymbol{\Sigma}_{22} - \boldsymbol{\Sigma}_{21}\boldsymbol{\Sigma}_{11}^{-1}\boldsymbol{\Sigma}_{12} = 0.3457 \tag{12.14b}$$

From the Meta-Student t Copula Similar to the meta-Gaussian copula, the maginal CDF vector in Equation (12.13a) will be first transformed to Student t distribution with the degree of freedom v. From Section 7.2.2 and Kotz and Nadarajah (2004), Equations (12.13) and (12.14) can be rewritten as follows:

$$\boldsymbol{X} = \left(\boldsymbol{X}_1^{\boldsymbol{T}}; \boldsymbol{X}_2^{\boldsymbol{T}}\right) = \left(\left[T^{-1}(U_{c70}, v), T^{-1}(U_{D50}, v), T^{-1}(U_{D130}, v)\right]^{T}; T^{-1}\left(U^{*}_{A90}, v\right)\right) \tag{12.15a}$$

$$\mu_{2|1} = \mu_{A90|C_{70},D50,D130} = \boldsymbol{\Sigma}_{21}\boldsymbol{\Sigma}_{11}^{-1}\boldsymbol{X}_1 \tag{12.15b}$$

$$\boldsymbol{\Sigma}_{2|1} = \boldsymbol{\Sigma}_{A90|C70,D50,D130} = \frac{v + \boldsymbol{X}_1^{\boldsymbol{T}}\boldsymbol{\Sigma}_{11}^{-1}\boldsymbol{X}_1}{v+3}\left(\boldsymbol{\Sigma}_{22} - \boldsymbol{\Sigma}_{21}\boldsymbol{\Sigma}_{11}^{-1}\boldsymbol{\Sigma}_{12}\right) \tag{12.15c}$$

$$v_{2|1} = v_{A90|C70,D50,D130} = v + 3 \tag{12.15d}$$

Table 12.14. *One-step ahead DO forecast for station A90 (mg/L).*

Month	Obs.	Gaussian copula Forecast	Gaussian copula 5% VaR	Gaussian copula 95% VaR	Student *t* copula Forecast	Student *t* copula 5% VaR	Student *t* copula 95% VaR
Jan 2012	**12.8**	**12.203**	11.853	12.805	**12.193**	11.881	12.666
Feb 2012	**12.8**	**12.449**	12.110	12.845	**12.446**	12.124	12.880
Mar 2012	**13.2**	**12.059**	11.722	12.458	**12.054**	11.738	12.496
Apr 2012	**12.73**	**11.737**	11.306	12.227	**11.727**	11.329	12.320
May 2012	**12.02**	**11.464**	11.056	11.924	**11.447**	11.009	12.078
Jun 2012	**11.5**	**10.756**	10.198	11.256	**10.737**	10.252	11.357
Jul 2012	**10**	**9.596**	8.901	10.336	**9.575**	8.979	10.443
Aug 2012	**9.6**	**9.288**	8.857	9.803	**9.276**	8.883	9.860
Sep 2012	**10.16**	**9.395**	8.898	9.986	**9.385**	8.978	9.974
Oct 2012	**11.2**	**10.821**	10.404	11.319	**10.811**	10.453	11.337
Nov 2012	**11.7**	**11.705**	11.261	12.228	**11.689**	11.262	12.298
Dec 2012	**12.1**	**12.392**	11.868	12.938	**12.389**	11.852	13.003
Jan 2013	**12.7**	**12.073**	11.732	12.481	**12.060**	11.758	12.509
Feb 2013	**11.3**	**12.471**	12.114	12.853	**12.468**	12.111	12.890
Mar 2013	**11.7**	**11.582**	11.234	11.979	**11.573**	11.249	12.007
Apr 2013	**11.8**	**11.219**	10.773	11.706	**11.208**	10.773	11.753
May 2013	**11.3**	**11.254**	10.817	11.703	**11.247**	10.825	11.727
Jun 2013	**10.3**	**10.721**	10.218	11.231	**10.713**	10.207	11.278
Jul 2013	**9**	**9.522**	8.768	10.244	**9.527**	8.720	10.321
Aug 2013	**8.9**	**9.255**	8.744	9.744	**9.261**	8.732	9.762
Sep 2013	**9.3**	**9.357**	8.770	9.921	**9.369**	8.613	10.117

In Equation (12.15), we have the following:

$$\mathbf{\Sigma}_{11} = \begin{bmatrix} 1 & 0.60 & 0.73 \\ 0.60 & 1 & 0.68 \\ 0.73 & 0.68 & 1 \end{bmatrix}; \mathbf{\Sigma}_{12} = \mathbf{\Sigma}_{21}^{T},\ \mathbf{\Sigma}_{21} = [0.73, 0.70, 0.74]; \mathbf{\Sigma}_{22} = 1; v = 5.9 \tag{12.15e}$$

To this end, the conditional density function can be given as follows:

$$f(X_2|\boldsymbol{X}_1) = \frac{\Gamma\left(\frac{v_{2|1}+1}{2}\right)}{\Gamma\left(\frac{v_{2|1}}{2}\right)\sqrt{v_{2|1}\pi}\left|\mathbf{\Sigma}_{2|1}\right|^{\frac{1}{2}}}\left(1+\frac{1}{v_{2|1}}\left(A90-\mu_{2|1}\right)^{T}\mathbf{\Sigma}_{2|1}^{-1}\left(A_{90}-\mu_{2|1}\right)\right)^{-\frac{v_{2|1}+1}{2}} \tag{12.16}$$

As shown previously, $\boldsymbol{X}_1 = \left[T^{-1}(U_{c70}, v), T^{-1}(U_{D50}, v), T^{-1}(U_{D130}, v)\right]^T; X_2 = T^{-1}\left(U_{A90}^{*}, v\right)$.

From Equations (12.15) and (12.16), it is seen that the conditional variance is scalar. Equation (12.16) may also be called the scaled and shifted univariate Student *t* distribution.

Let $t'_{A90} = \frac{X_2-\mu_{2|1}}{|\Sigma_{2|1}|^{\frac{1}{2}}}$; t'_{A90} will now follow the standard univariate Student *t* distribution: $T\left(t'_{A90}, v_{2|1}\right)$.

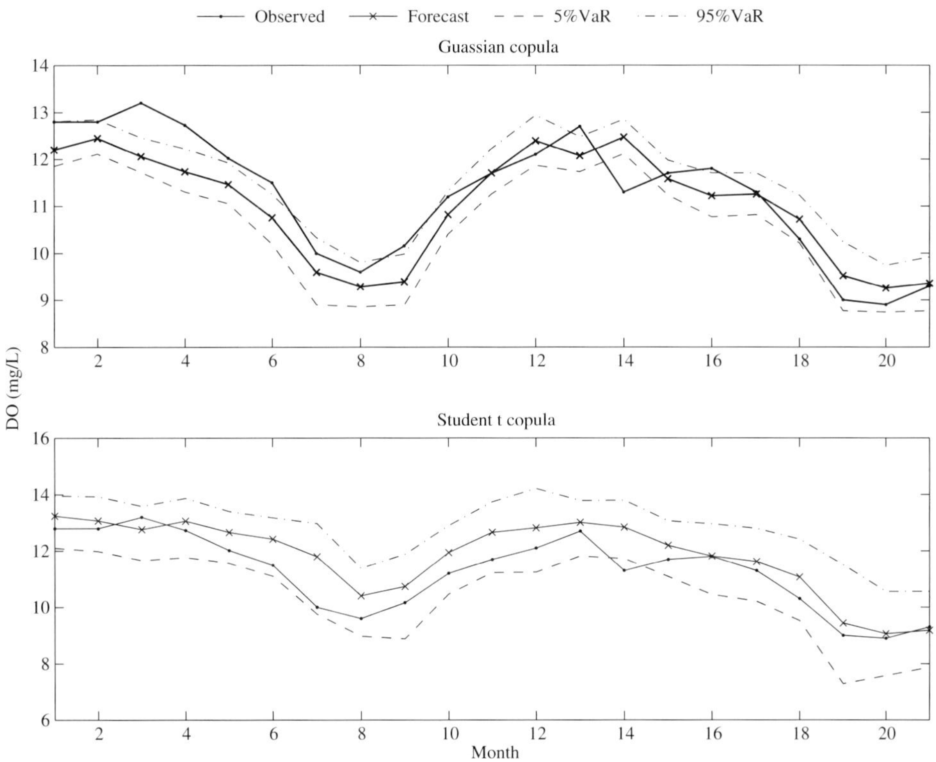

Figure 12.14 Comparison of one-step ahead forecast with the monthly observed DO values at station A90.

Applying the previously discussed approach, Table 12.14 lists the one-step ahead forecast results. Figure 12.14 compares the one-step ahead forecast with the observed monthly DO values. The results again indicate that the DO forecasts for station A90 closely follow the corresponding observed DO values. The monthly DO at station A90 may be reasonably forecasted using the monthly DO at upstream locations (i.e., C70, D50, and D130). In addition, similar to the forecast at station D50, there is minimal difference in regard to the performance between the meta-Gaussian copula and the meta-Student *t* copula. We may safely choose the meta-Gaussian copula as the only candidate in this case.

12.3 Dependence Study for the Chattahoochee River Watershed

According to the availability of the water quality dataset published by USGS, temperature, DO, and pH are selected for the upstream location, i.e., USGS2332017 (Belton Bridge). Besides temperature, DO, and pH, phosphorus is also selected for the downstream location, i.e., USGS2338000 (Whitesburg). For both locations, the period with continuous measurements are selected, that is, September 6–September 12. Similar to the case study for the Snohomish River watershed, we will first study the temporal dependence using the copula-based Markov process followed by the study of spatial dependence. Table 12.15 lists the water quality data selected, in which the water quality measurements of 2012 are used for forecast and calibration purposes.

12.3.1 Temporal Dependence of the Univariate Water Quality Series with the Copula-Based Markov Process

Before we proceed with the study of temporal dependence with copula-based Markov process, we first investigate whether there exists seasonality in water quality parameters. Applying frequency analysis, Figures 12.15 and 12.16 plot the cumulative periodogram for the water quality parameters listed in Table 12.15. The plots show that a 12-month seasonality exists for water temperature and DO, while no obvious seasonality is observed for pH and phosphorus. Table 12.16 lists monthly average and standard deviations for DO and temperature with 12-month seasonality.

To study the temporal dependence, we will choose temperature, pH, and phosphorus at the downstream location (Whitesburg, USGS2338000) as an example. For the downstream temperature dataset, we will first perform the full deseasonalization. Applying the Markov order identification approach discussed in Section 9.5.2, Table 12.17 lists the Markov order identified for the selected downstream water quality time series using the meta-Gaussian copula for identification purposes.

The results in Table 12.17 indicate the following:

1. The deseasonalized temperature and pH may be modeled with a second-order copula-based Markov process.
2. Phosphorus may be considered a random variable; this result is in agreement with the cumulative periodogram plot for phosphorus shown in Figure 12.16.

Now we will only look into the serial dependence for the downstream temperature and pH. For the second-order process, the D-vine copula will again be applied with the Gumbel–Hougaard, Gaussian, Student *t*, Frank and Clayton copulas as candidates. Tables 12.18 and 12.19 list the results for the five copula candidates.

From the results in the tables, we see the following:

- *Deseasonal temperature (downstream):* The Gumbel–Hougaard copula is selected as the best-fitted copula function for both T_1 (lag-1 dependence) and T_2 (conditional dependence of $T|T\text{-}1$ and $T\text{-}2|T\text{-}1$);
- *pH (downstream):* The Clayton copula is selected as the best-fitted copula function for T_1 (lag-1 dependence), while the Frank copula is selected for T_2 (conditional dependence of $T|T1$ and $T\text{-}2|T\text{-}1$).

Using the fitted Gumbel–Gumbel copula model for the second-order deseasonalized temperature series and the Clayton–Frank model for the second-order pH series at the Whitesburg station, Figure 12.17 plots the range of simulated temperature and pH series. Figure 12.18 plots the lag-1 and lag-2 scatter plots to compare the serial dependence of

Table 12.15. *Monthly water quality measurements for the Chattahoochee River watershed.*

Time	USGS2332017 Temperature (°C)	USGS2332017 DO (mg/L)	USGS2332017 pH	USGS2338000 Temperature (°C)	USGS2338000 DO (mg/L)	USGS2338000 pH	USGS2338000 Phosphorus (mg/L)
Sep-06	18.1	8.4	7.1	26.5	5.8	6.6	0.127
Oct-06	9	10.9	7.2	23.4	8.3	7.5	0.056
Nov-06	8	11.1	7.1	18.4	8.2	7	0.082
Dec-06	7.2	11.2	7	13.1	9.2	7.2	0.05
Jan-07	7.1	11.9	7.2	12.6	8.6	6.8	0.163
Feb-07	3	13.3	7.3	11.5	10.8	7.1	0.065
Mar-07	9.1	11.1	7	21.9	8.4	7.3	0.041
Apr-07	11.4	11	7.2	17.4	7.8	7	0.081
May-07	19.5	8.5	6.9	23	9.1	7.7	0.065
Jun-07	21.3	7.1	7.4	25.9	6.3	6.9	0.104
Jul-07	22.3	7.5	7.5	27.2	7	7.2	0.083
Aug-07	26.1	6.5	7.4	29.3	6.3	7.1	0.055
Sep-07	19.3	7.9	7.3	20.6	7.8	7	0.079
Oct-07	17.6	8.4	6.9	20.9	7.9	6.8	0.07
Nov-07	8	10.7	7.7	15.4	8.7	6.9	0.064
Dec-07	12.7	10	7.2	16.7	9	7.2	0.038
Jan-08	5	13.5	7.2	7.5	10.6	6.9	0.072
Feb-08	8.2	10.9	7.1	12.8	9.2	6.7	0.069
Mar-08	10.8	10.4	6.9	15.1	8.5	6.8	0.124
Apr-08	12	9.8	7.2	15.3	8.5	6.6	0.059
May-08	19.3	10.3	7.2	22.6	6.7	7	0.073
Jun-08	26.9	7.4	7.2	26	6.9	7.2	0.067
Jul-08	26.8	7.9	7.4	26.5	6.4	6.7	0.112
Aug-08	23.8	7.3	7.1	27.6	6.5	6.7	0.061
Sep-08	23.6	7.2	7.2	20.3	7.9	7.5	0.079
Oct-08	11.6	10.8	7.2	20.4	7.6	7.3	0.08
Nov-08	9.4	11.9	7.1	9.8	11.1	7.7	0.086
Dec-08	6.5	12.8	7.1	12.5	8.8	6.8	0.256
Jan-09	12.4	9.5	6.8	10.8	10.4	6.9	0.215
Feb-09	4	12.3	7.3	10.3	10.1	7	0.149
Mar-09	6.1	12.3	7.2	7.8	12.3	7.1	0.103
Apr-09	11	10.6	7.2	15.5	7.8	7	0.074
May-09	19.5	8	7	22.4	7.1	7.2	0.092
Jun-09	21.9	8.4	7.1	25.6	6.5	7.2	0.095
Jul-09	23.4	7.2	7.1	24.6	7.2	7.2	0.074
Aug-09	26.3	7.9	7.2	27.4	7.2	7.3	0.096
Sep-09	23.4	8.2	7.3	21.6	6.2	6.4	0.542
Oct-09	15.8	8.7	7.2	15.1	8.4	7	0.103
Nov-09	10.8	10.3	7.1	13.8	8.8	6.6	0.03
Dec-09	9.4	9.8	7.2	11.6	9.2	6.9	0.035

Table 12.15. (*cont.*)

	USGS2332017			USGS2338000			
Time	Temperature (°C)	DO (mg/L)	pH	Temperature (°C)	DO (mg/L)	pH	Phosphorus (mg/L)
Jan-10	3.1	11.6	6.9	8	10.9	7	0.078
Feb-10	5.3	10.9	7.3	6.6	11.8	6.8	0.051
Mar-10	10.5	10	7.1	9.6	10.8	7	0.054
Apr-10	16.9	9.6	7	14.6	9	7.1	0.078
May-10	20.3	8.6	7	16.1	9.3	7	0.068
Jun-10	23.9	7.2	6.9	25	7.3	7.2	0.092
Jul-10	26.4	6.6	6.9	28.3	7.4	7.3	0.065
Aug-10	24	7.3	7.1	25.4	7.6	7.1	0.056
Sep-10	22.3	7.7	7	24.4	7.6	7.2	0.025
Oct-10	15.2	9.8	7.2	18.3	9.2	7.4	0.054
Nov-10	12.2	9.8	7.2	10.7	10.3	7.2	0.036
Dec-10	10.8	9.1	6.9	5.6	12.2	7.2	0.032
Jan-11	5.9	12.2	7.2	7	11.5	7.5	0.034
Feb-11	7.1	11.6	7.2	7.3	11.2	7.3	0.062
Mar-11	11.1	10.6	7	11.6	9.6	7	0.23
Apr-11	13.7	10.7	7.1	14.6	8.8	7	0.08
May-11	20.8	8.1	7.1	22.2	7.9	6.8	0.027
Jun-11	23.2	7.8	7	25.9	6.8	6.6	0.115
Jul-11	28.8	8.4	7.5	27.4	6.3	6.5	0.047
Aug-11	23.6	7.1	7.3	27.3	6.6	6.6	0.063
Sep-11	19.6	8.7	7.3	21.7	7.4	6.6	0.133
Oct-11	16.6	9.6	7.2	17.7	8.6	6.8	0.029
Nov-11	10.7	10.6	6.9	16.6	7	6.8	0.131
Dec-11	9.5	10.5	6.9	10.8	10.3	6.7	0.034
Jan-12	10	9.5	6.4	9.1	10.3	7.3	0.048
Feb-12	10.2	10.9	7.2	11.2	10.1	6.6	0.042
Mar-12	16.1	10.1	7	20.4	7.6	6.6	0.044
Apr-12	18.8	8.8	7.1	18.3	8.4	6.9	0.057
May-12	21.8	8.7	7.4	26.4	6.8	6.9	0.047
Jun-12	26.4	8.4	7.1	25.8	6.9	7	0.122
Jul-12	27.9	7.5	7.2	27.5	6.6	6.8	0.081
Aug-12	23.7	8.2	7.3	25.3	7.1	6.8	0.066
Sep-12	20	8.2	7	21.6	7.9	6.8	0.07

observed versus simulated water quality series. Figure 12.17 shows that the observed water quality series falls into the range of simulation for both the monthly temperature and pH series. Figure 12.18 further indicates the lag-1 and lag-2 serial dependence are well captured by the fitted copula-based second-order Markov process.

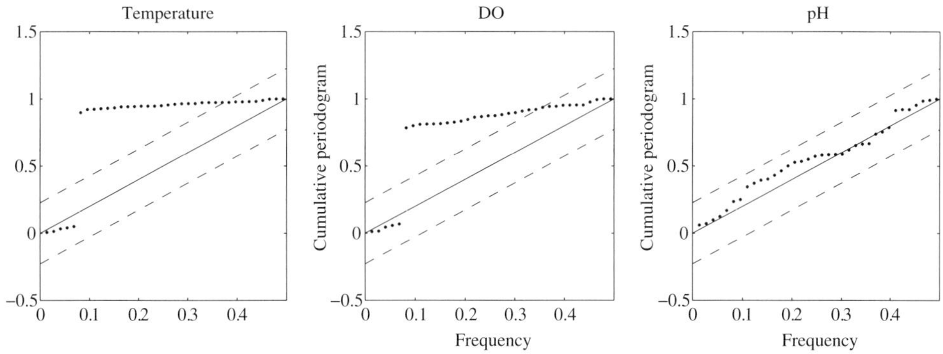

Figure 12.15 Cumulative periodograms of the upstream (Belton Bridge) water quality parameters.

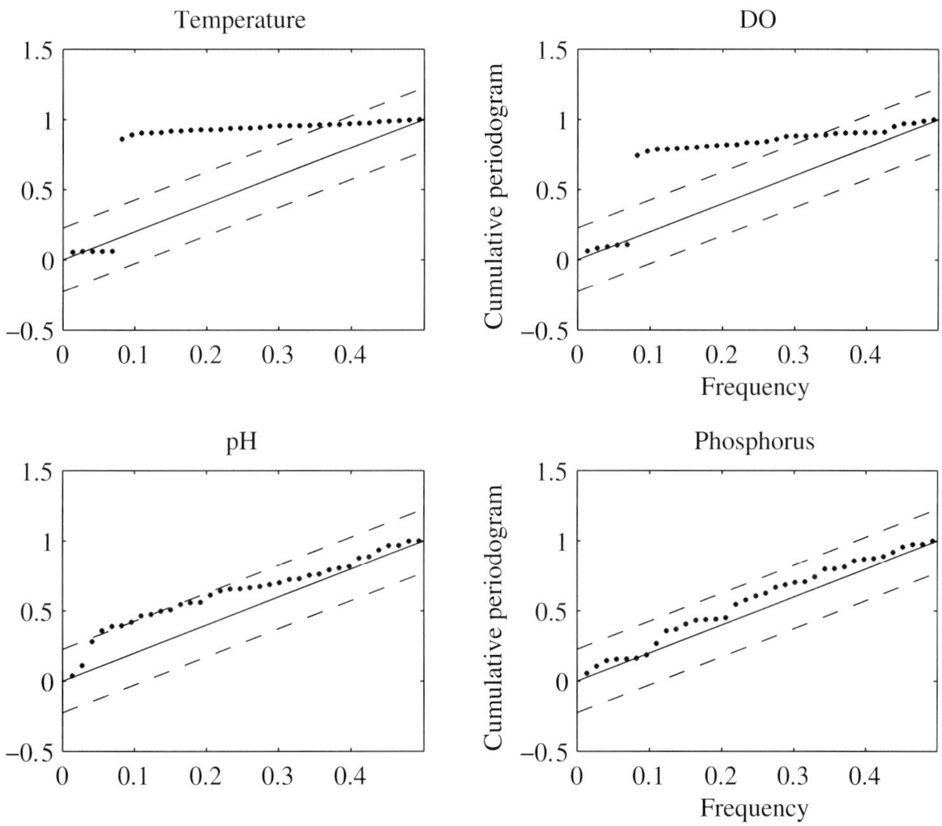

Figure 12.16 Cumulative periodograms of the downstream (Whitesburg) water quality parameters.

Table 12.16. *Monthly average and standard deviation for DO and temperature.*

	Upstream				Downstream			
	DO (mg/L)		Temperature (°C)		DO (mg/L)		Temperature (°C)	
	μ	σ	μ	σ	μ	σ	μ	σ
Jan	11.37	1.58	7.25	3.41	10.38	0.97	9.17	2.16
Feb	11.65	0.98	6.30	2.71	10.53	0.92	9.95	2.47
Mar	10.75	0.85	10.62	3.26	9.53	1.75	14.40	5.78
Apr	10.08	0.83	13.97	3.20	8.38	0.50	15.95	1.54
May	8.70	0.83	20.20	0.97	7.82	1.15	22.12	3.34
Jun	7.72	0.58	23.93	2.30	6.78	0.35	25.70	0.37
Jul	7.52	0.61	25.93	2.56	6.82	0.45	26.92	1.27
Aug	7.38	0.60	24.58	1.26	6.88	0.50	27.05	1.51
Sep	8.04	0.49	20.90	2.18	7.23	0.87	22.39	2.24
Oct	9.70	1.04	14.30	3.31	8.33	0.56	19.30	2.89
Nov	10.73	0.72	9.85	1.68	9.02	1.47	14.12	3.36
Dec	10.57	1.30	9.35	2.28	9.78	1.29	11.72	3.62

Table 12.17. *Markov order identification for the water quality time series.*

	$\boldsymbol{F}_t, \boldsymbol{F}_{t-1}$		$\boldsymbol{F}_{t\|t-1}, \boldsymbol{F}_{t-2\|t-1}$		$\boldsymbol{F}_{t\|t-1,t-2}, \boldsymbol{F}_{t-3\|t-1,t-2}$		
Variable	T	p-Val	τ	p-Val	τ	p-Val	Order
Temperature	0.22	<0.01	0.24	<0.01	0.078	0.34	2
pH	0.25	<0.01	0.19	0.03	0.035	0.67	2
Phosphorus	—		—		—		0

Table 12.18. *Results from five copula candidates for the deseasonalized temperature series.*

*T*1	Gumbel–Hougaard	Gaussian	Student *t*	Frank	Clayton
Parameters	**1.27**	0.23	[0.31, 6.24]	1.67	0.43
ML	**2.10**	1.84	2.27	1.76	2.00
AIC	**–2.20**	–1.68	–0.55	–1.53	–2.01
*T*2	Gumbel–Hougaard	Gaussian	Student *t*	Frank	Clayton
Parameters	**1.42**	0.37	[0.43, 10.36]	2.60	2.28
ML	**6.08**	4.69	4.99	4.64	2.28
AIC	**–10.16**	–7.38	–5.97	–7.28	–2.57

Table 12.19. *Results from five copula candidates for the pH series.*

*T*1	Gumbel–Hougaard	Gaussian	Student *t*	Frank	Clayton
Parameters	1.33	0.33	[0.46, 5.01]	3.09	**0.80**
ML	3.04	3.63	4.72	4.41	**5.06**
AIC	–4.08	–5.26	–5.45	–6.82	**–8.12**
*T*2	Gumbel–Hougaard	Gaussian	Student *t*	Frank	Clayton
Parameters	1.18	0.31	[0.34, 1E+7]	**2.25**	2.37
ML	1.75	3.29	3.32	**3.38**	2.37
AIC	–1.51	–4.57	–2.64	**–4.76**	–2.74

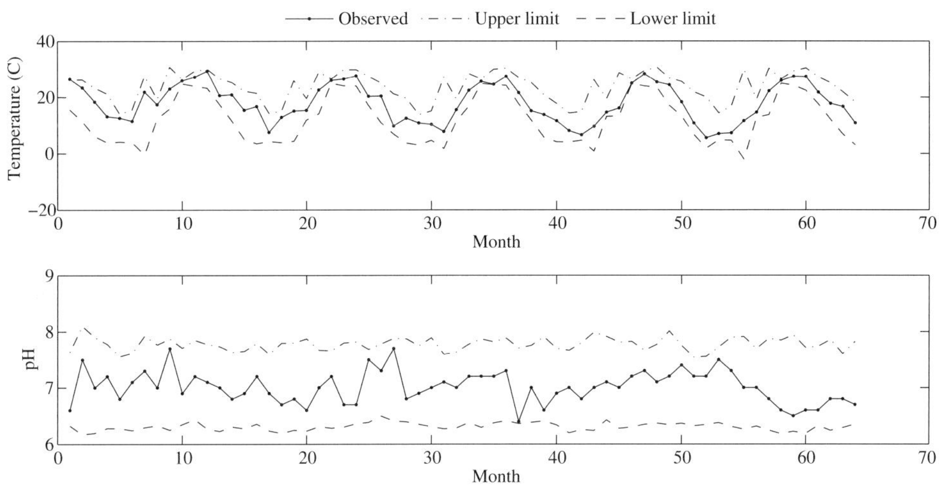

Figure 12.17 Comparison of simulation versus observed measurements at the Whitesburg station.

We have shown that the copula-based second-order Markov process may be applied to model monthly temperature and pH at the Whitesburg station. Now we will evaluate the forecast/prediction capability of the copula-based Markov process through a one-step ahead forecast following the same procedure as discussed in the previous case study for the Snohomish River watershed. Figure 12.19 compares the one-step ahead forecast to the observed monthly temperature and pH for the first nine months of 2012. Figure 12.19 shows that the copula-based second-order Markov process provides reasonable forecasts for temperature and pH at the downstream Whitesburg station. The forecast results, listed in Table 12.20, show that maximum biases are 30% and 6% for temperature and pH, respectively.

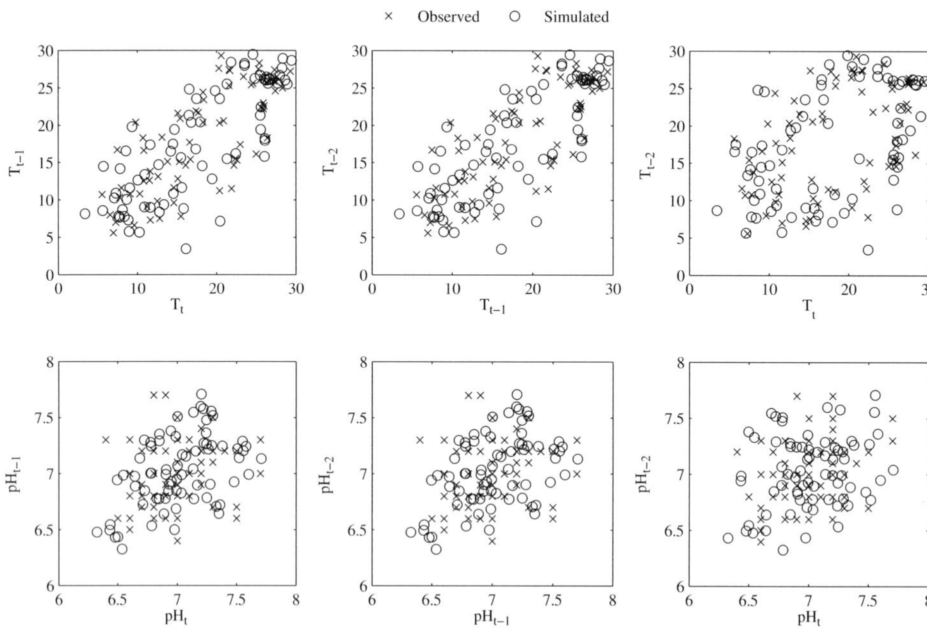

Figure 12.18 Comparison of serial dependence of observed versus simulated monthly water quality time series (i.e., temperature [T] and pH) at the Whitesburg station.

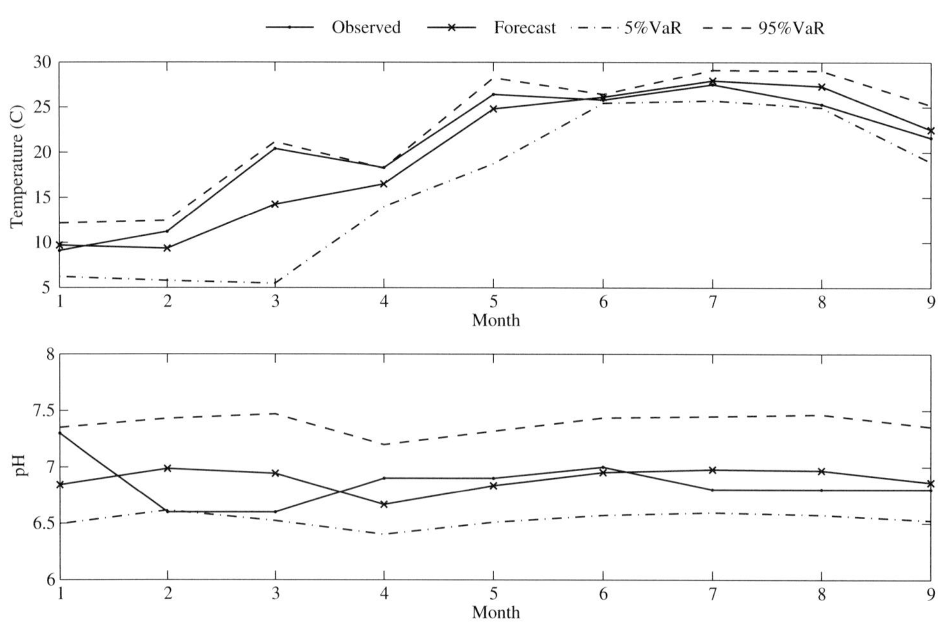

Figure 12.19 Comparison of one-step forecast with the observed monthly water quality series.

Table 12.20. *One-step ahead forecast results for temperature and pH at the Whitesburg station.*

	Temperature (°C)				pH			
Month	Obs.	Forecast	5%VaR	95% VaR	Obs.	Forecast	5%VaR	95%VaR
Jan 2012	**9.1**	**9.7**	6.2	12.2	**7.3**	**6.8**	6.5	7.4
Feb 2012	**11.2**	**9.3**	5.8	12.4	**6.6**	**7.0**	6.6	7.4
Mar 2012	**20.4**	**14.2**	5.5	21.2	**6.6**	**6.9**	6.5	7.5
Apr 2012	**18.3**	**16.5**	14.0	18.2	**6.9**	**6.7**	6.4	7.2
May 2012	**26.4**	**24.8**	18.7	28.2	**6.9**	**6.8**	6.5	7.3
Jun 2012	**25.8**	**26.1**	25.4	26.4	**7**	**7.0**	6.6	7.4
Jul 2012	**27.5**	**27.9**	25.7	29.1	**6.8**	**7.0**	6.6	7.4
Aug 2012	**25.3**	**27.3**	24.9	29.0	**6.8**	**7.0**	6.6	7.5
Sep 2012	**21.6**	**22.5**	18.9	25.2	**6.8**	**6.9**	6.5	7.4

12.3.2 Spatial–Temporal Dependence of the Water Quality Time Series for the Chattahoochee River Watershed

As introduced in Section 12.1, the subwatershed upstream of the Belton Bridge station may be considered a forest watershed. With the major metropolitan area (the city of Atlanta) located between the Belton Bridge station and the Whitesburg station, the LULC is changed from mainly forest to urban developed watershed. To study the spatial dependence, we will use a major water quality parameter DO as an example.

As discussed in Section 12.3.1, there exists 12-month periodicity in the DO time series (Table 12.16, Figures 12.15 and 12.16). We will proceed with the same procedure as follows: (i) perform full deseasonalization on the upstream and downstream monthly DO series; (ii) build a univariate time series model for the deseasonalized DO series; (iii) study the spatial dependence through the fitted model residuals; and (iv) perform the one-step ahead DO forecast for Whitesburg with the use of DO at the upstream Belton Bridge station. Again, the monthly DO series before the year of 2012 is applied to build the time series model, and the monthly DO series after 2012 is used for forecast and validation purposes.

After full deseasonalization, Table 12.21 lists parameters estimated for the selected AR (2) model. Using the fitted-model residual, Table 12.21 lists the rank-based Kendall's tau dependence measure of model residuals. The computed Kendall's tau suggests that the fitted-model residual may be considered independent (with P-value = 0.82). Computing the Kendall's tau for deseasonlized DO series, we have $\tau = -0.08, P = 0.31$, which also indicates independence. The scatter plots shown in Figure 12.20 also indicate the independence pattern.

These results suggest that the change of LULC among the subwatersheds may significantly impact the spatial distribution pattern of water quality parameters that one may also expect. To this end, we will simply apply the product copula (i.e., independent copula:

Table 12.21. *Parameters estimated of the AR(2) model for the DO series.*

Station	Constant	ϕ_1	ϕ_2	σ_e^2	White noise check
Belton Bridge	0.021	0.062	0.123	0.808	H = 0, $N(0, 0.899^2)$
Whitesburg	–0.0002	0.049	0.51	0.681	H = 0, $N(0, 0.825^2)$
$\tau = -0.0198$;	$P = 0.82$				

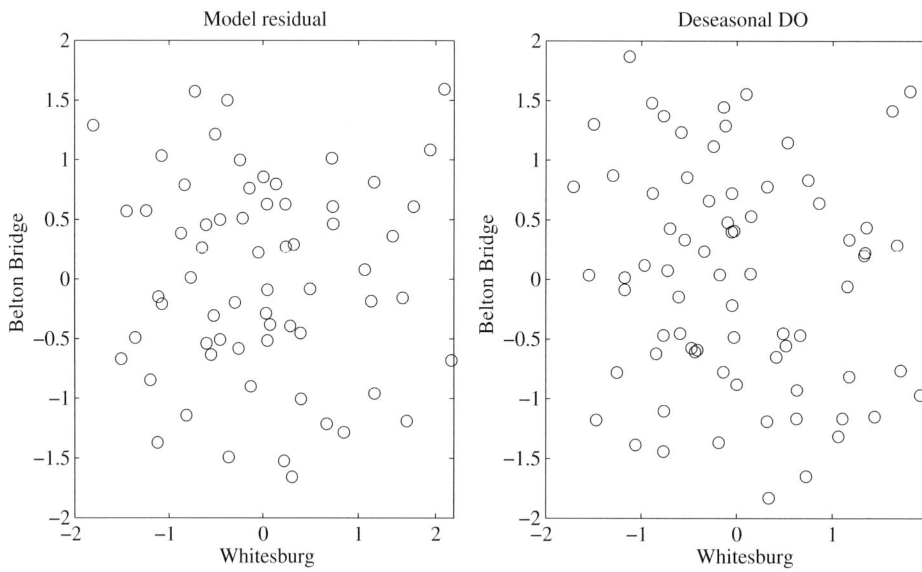

Figure 12.20 Scatter plots for fitted-model residual and deaseasonalized DO series.

$\pi = uv$) and the meta-Gaussian copula for the one-step ahead forecast. In addition, we compare the results from the copula to that computed from the univariate one-step ahead forecast for DO at Whitesburg. Figure 12.21 plots the one-step ahead forecast, and Table 12.22 lists the numerical forecast results. The one-step ahead forecast results show that the Gaussian copula yields similar forecast results as those from the univariate time series. The maximum absolute bias is about 24% (forecast of March 2012 from the Gaussian copula); while the product copula yields the largest root mean square error (RMSE) (0.84 mg/L). Overall, both the Gaussian copula and the product copula provide reasonable forecasts.

Above all, we have shown that even for the subwatershed with significantly different LULC, the copula method may still be applied to investigate the spatial–temporal dependence and provide reasonable forecasting results that may be useful for the watershed engineers to make proper judgment ahead.

Table 12.22. *Comparison of forecast results.*

Month	Observed	1-Step ahead DO forecast at Whitesburg (mg/L) Univariate	Gaussian copula	Product copula
Jan 2012	10.3	9.72	9.75	8.38
Feb 2012	10.1	10.72	10.73	9.78
Mar 2012	7.6	9.42	9.44	7.77
Apr 2012	8.4	8.24	8.25	7.45
May 2012	6.8	7.17	7.16	6.64
Jun 2012	6.9	6.77	6.76	6.75
Jul 2012	6.6	6.62	6.62	6.34
Aug 2012	7.1	6.96	6.94	6.91
Sep 2012	7.9	7.03	7.03	6.65
Maximum bias		*23.9%*	*24.2%*	*18.6%*
RMSE (mg/L)		*0.742*	*0.749*	*0.844*

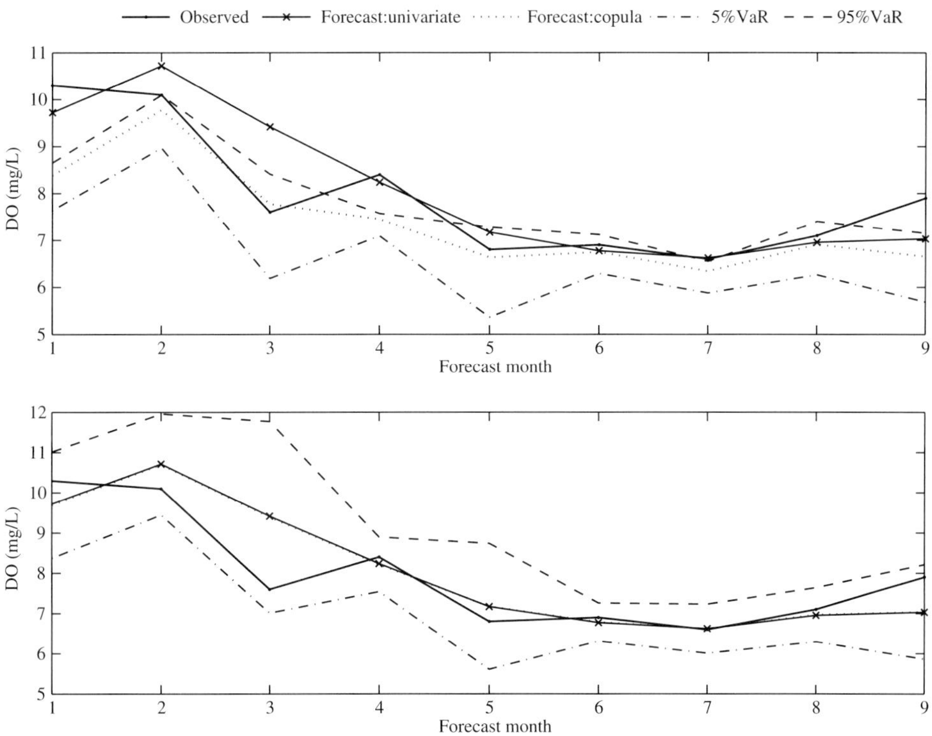

Figure 12.21 Comparison of observed DO with one-step ahead forecast through univariate, product, and Gaussian copulas.

In water quality studies, there exists one more type of dependence that is similar to that in at-site flood, rainfall, or drought frequency analysis. In water quality studies, the at-site dependence study may provide important information. Given the water quality information at Whitesburg, it will be useful if we can responsibly forecast phosphorus with the commonly monitored temperature, DO, and pH. Thus, in the following section, we will focus on the at-site multivariate water quality study.

12.4 At-Site Multivariate Water Quality Dependence Study

For the at-site multivariate water quality dependence study, we will choose the Whitesburg station as a case study. From the previous sections, we have shown temperature and pH may be modeled using the copula-based second-order Markov process, while phosphorus may be considered a random variable (Table 12.17). DO at Whitesburg may be modeled with the AR(2) univariate time series, so we will first evaluate whether temperature and pH may also be modeled with the AR(2) univariate time series. The results listed in Table 12.23 indicate that pH may be modeled using the AR(2) model, while the deseasonalized temperature may be modeled with ARMA(2,1) instead. Table 12.24 lists the Kendall's tau correlation matrix for the residuals computed from the time series model of temperature (T), pH, DO, and observed phosphorus (P), which may be considered a random variable. Table 12.24 indicates the negative correlation of P with both DO and pH, while P is nearly independent of T.

Table 12.23. *Classic time series modeling results for temperature and pH.*

	Model	White noise check
Temperature	ARMA(2,1) $C = -0.04, \phi_1 = 0.86, \phi_2 = 0.07, \theta_1 = -1,\ \sigma_e^2 = 0.50$	$H_0 = 0, P = 0.56$
pH	AR(2) $C = 3.61, \phi_1 = 0.18, \phi_2 = 0.31, \sigma_e^2 = 0.07$	$H_0 = 0, P = 0.85$

Table 12.24. *Rank-based Kendall's tau correlation matrix.*

	DO	pH	T	P
DO	1	0.33	–0.39	–0.19
pH	0.33	1	–0.14	–0.16
T	–0.39	–0.14	1	0.04
P	–0.19	–0.16	0.04	1

Table 12.25. *Parameter estimated for the full model using meta-elliptical copulas.*

	Meta-Gaussian				Meta-Student t			
	DO	pH	T	P	DO	pH	T	P
DO	1	0.49	–0.56	–0.33	1	0.57	–0.63	–0.40
pH	0.49	1	–0.18	–0.33	0.57	1	–0.27	–0.41
T	–0.56	–0.18	1	0.06	–0.63	–0.27	1	0.14
P	–0.33	–0.33	0.06	1	–0.40	–0.41	0.14	1
				$v = 1.12 \times 10^7$				

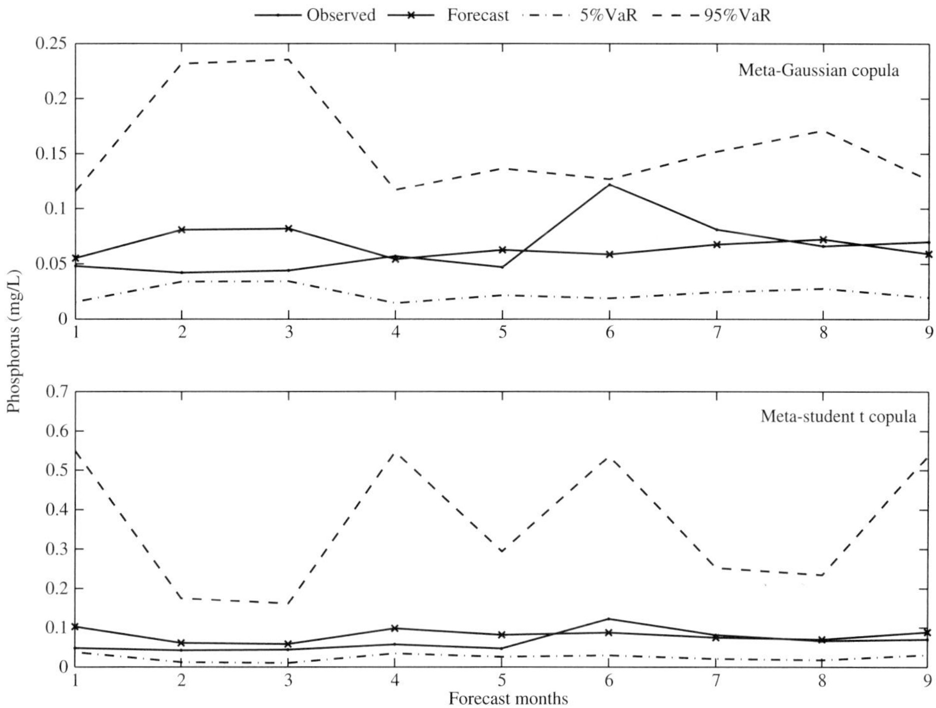

Figure 12.22 Comparison of the one-step ahead forecast with the phosphorus samplings.

With the preceding information, we will build two models, i.e., (i) a full model of $f(DO, pH, T, P)$; and (ii) a reduced model of $f(DO, pH, P)$.

Full model. For the full model, we will simply apply the meta-Gaussian and meta-Student t copula to illustrate the analysis. Table 12.25 lists the parameters estimated for the full model.

Using DO, PH, and T as conditioning variables, Figure 12.22 compares the phosphorus forecast with its observations. The comparison shows that the phosphorus observations are

Table 12.26. *Rank-based Kendall's tau correlation matrix for reduced model.*

	DO	pH	Phosphorus
DO	1.00	0.33	–0.19
pH	0.33	1.00	–0.16
Phosphorus	–0.19	–0.16	1.00

Table 12.27. *Parameters estimated from meta-Gaussian and meta-Student t copulas.*

	Meta-Gaussian			Meta-Student t		
	DO	pH	Phosphorus	DO	pH	Phosphorus
DO	1.00	0.49	–0.33	1.00	0.55	–0.38
Ph	0.49	1.00	–0.33	0.55	1.00	–0.38
Phosphrous	–0.33	–0.33	1.00	–0.38	–0.38	1.00
				$v = 10.56$		

within the range of 5% and 95% VaRs. The obvious differences are seen between the forecasts and observations for the very high and very low sampled phosphorus values.

Reduced model. In the case of reduced model, Table 12.26 tabulates the rank-based Kendall correlation matrix. As seen in Table 12.26, the negative relation is found between phosphorus and DO as well as phosphorus and pH. Meta-elliptical and vine copulas will be applied to evaluate the reduced model.

Applying the meta-elliptic copulas, Table 12.27 tabulates the estimated parameters for meta-Gaussian and meta-Student t copulas with the one-step ahead forecast plotted in Figure 12.23. Comparing Figure 12.23 to Figure 12.22, we may reach the following conclusions:

i. For the meta-Gaussian model, the median forecast and 5% (95% VaRs) yield similar results for both full and reduced models.
ii. For the meta-Student t model, the median forecast and 5% VaRs yield similar results for both the full and reduce models. However, there exist noticeable differences in regard to the 95% VaRs estimated from the full and reduced models.
iii. The observations fall into the region bounded by 5% and 95% VaRs.

Applying the vine copulas, we choose either DO or pH as the center. Results indicate that phosphorus may be studied only using DO or pH rather than both of the variables as follows:

pH as center: The Clayton copula is selected to study DO and pH, while the Gaussian copula is selected to study DO and phosphorus for the first level T_1. The rank-based

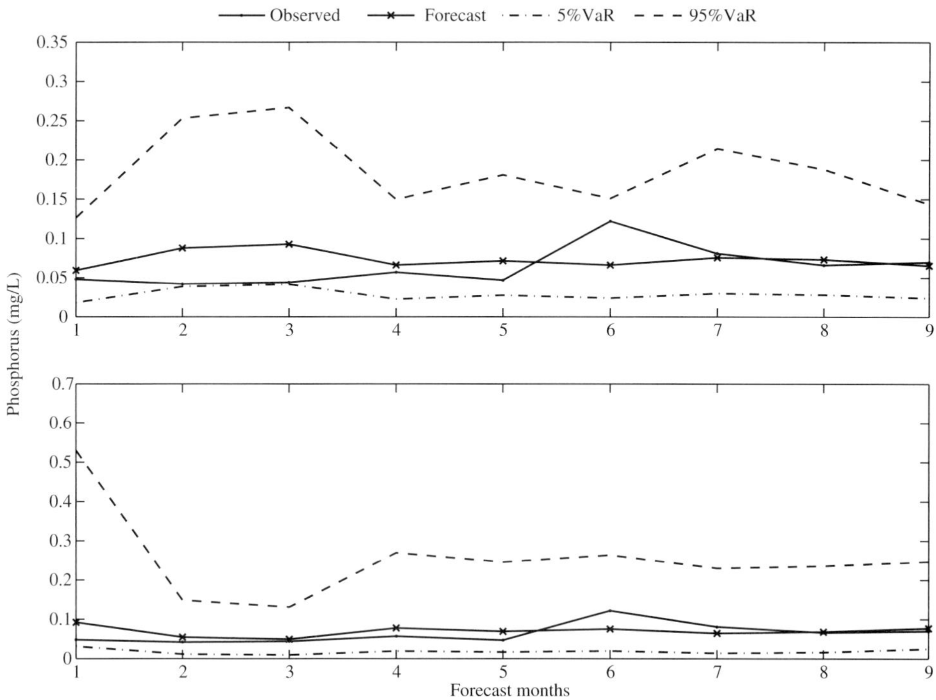

Figure 12.23 Comparison of the one-step ahead forecast with the phosphorus samplings for the reduced model using meta-elliptical copulas.

correlation for the second-level T_2 is computed as –0.03 with a *P*-value of 0.75 for pH|DO and P|DO.

DO as center: With the Clayton copula selected to study DO and pH, the Gaussian copula is again found as the proper copula to study pH and phosphorus for the first-level T_1. The rank-based correlation of T_2 is computed as –0.08 with a *P*-value of 0.35 for DO|pH and P|pH.

Thus, the model may be further reduced to the bivariate model with the use of the meta-Gaussian copula for (P, DO with parameter $\rho = -0.326$) or (P, pH with parameter $\rho = -0.332$). Figure 12.24 plots the one-step ahead forecast of phosphorus using pH and DO, respectively.

Figure 12.24 again shows the similar results comparing the full model and reduced model with the use of meta-elliptical copulas. To further compare all three models, Table 12.28 lists the results of comparison, which indicate that (i) the largest error occurs for June forecast from both full and reduced models; (ii) the full model results in the smallest RMSE compared to the reduced models; (iii) comparing the further reduced bivariate model using either pH or DO with the reduced model using pH, DO, and temperature through meta-elliptic copulas, there exist minimal differences; and

Table 12.28. *Prediction results from full and reduced model.*

	Full model		Reduced model			
Observations	Meta-Gaussian	Meta-Student *t*	Meta-Gaussian	Meta-Student *t*	Through pH	Though DO
Jan, 0.048	0.055	0.052	0.059	0.058	0.059	0.067
Feb, 0.042	0.081	0.083	0.088	0.090	0.087	0.080
Mar, 0.044	0.082	0.085	0.093	0.097	0.090	0.084
Apr, 0.057	0.054	0.052	0.066	0.066	0.067	0.068
May, 0.047	0.063	0.062	0.072	0.072	0.069	0.074
Jun, 0.122	0.059	0.057	0.066	0.066	0.068	0.066
Jul, 0.081	0.068	0.067	0.076	0.076	0.077	0.071
Aug, 0.066	0.072	0.072	0.073	0.073	0.077	0.068
Sep, 0.07	0.059	0.057	0.066	0.065	0.074	0.060
RMSE	0.005	0.003	0.027	0.028	0.031	0.020
Max absolute error	0.063	0.065	0.056	0.056	0.054	0.056
Min absolute error	0.003	0.004	0.004	0.005	0.004	0.002

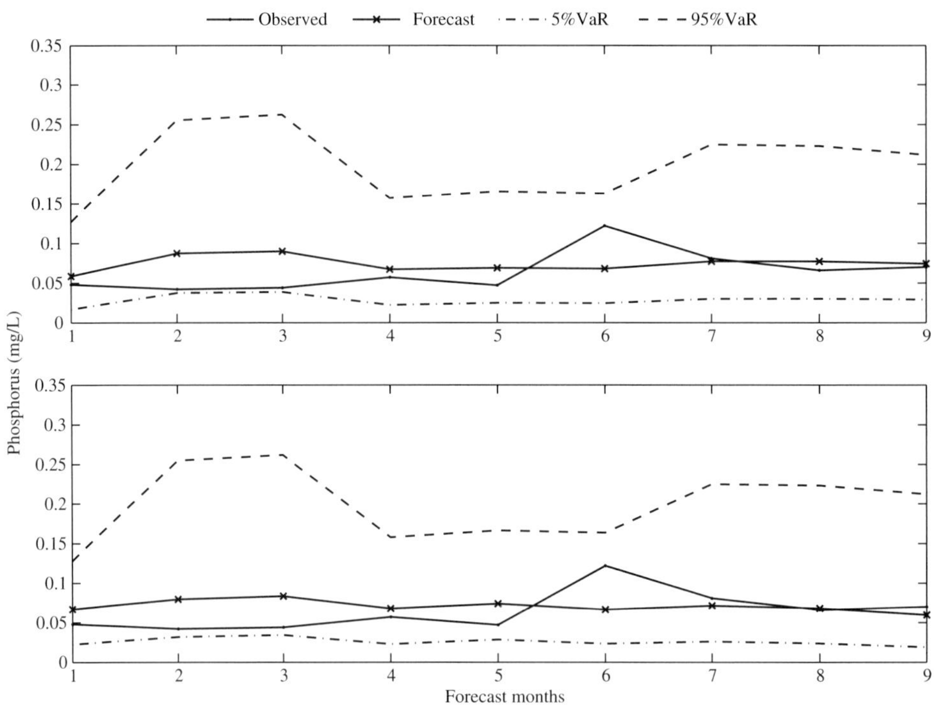

Figure 12.24 Comparison of the one-step ahead forecast with the phosphorus sampling for reduced model through (a) pH and (b) DO.

(iv) the negative relations between pH and phosphorous, as well as DO and phosphorous, are in line with natural phenomena.

12.5 Summary

In this chapter, we have introduced the copula application in water quality analysis. Two types of watersheds have been considered: (1) natural watershed and (2) urban watershed. The two case studies indicate the following:

I. The copula-based Markov process (CMP) is more robust compared to the classic time series with Gaussian innovations. The serial dependence structure may be well preserved through D-vine copula. The simulation study shows the ability to apply CMP for water quality management. The forecast study indicates the good forecast ability of CMP. In addition, compared to the urban watershed (the watershed in Georgia), the forecast accuracy of CMP is higher for the natural watershed (the watershed in Washington).

II. In the case of the spatial analysis for the natural watershed in Washington, the spatial distribution pattern of DO is well preserved. Given the characteristic of the water quality data information (i.e., monthly data), the meta-elliptic copulas perform very well to capture the spatial dependence as well as the one-step ahead water quality forecast.

III. In the case of at-site multivariate water quality frequency analysis for the watershed in Georgia, the forecast ability is acceptable but not as good as that for the natural watershed in Washington. The largest forecast error (phosphorus) occurs in June for both full and reduced models. This may be due to the following: (1) the human activities are usually more active during the late spring/early summer (i.e., agricultural practice); (2) runoff into the system from rainfall events or irrigation may bring more nutrients into the system, which result in high values; and (3) with more information of human activity, the forecast ability of the model may be improved.

IV. For the watershed in Georgia, there is no obvious dependence structure between the upstream (Belton Bridge) and downstream (Whitesburg) subwatersheds due to the significant change of LULC (i.e., natural [or forest] for the upstream subwatershed, while urban for the downstream [the city of Atlanta]). In addition, the behavior of phosphorus at the downstream (i.e., random) may be considered another indicator of disturbance due to the human activities within the watershed.

V. Overall, the copula-based approach provides a rule-of-thumb for watershed engineers in regard to water quality parameters that need to be monitored and to make the protocol for management purposes.

VI. Due to the limitations of the water quality data, the monthly water quality data have been applied for both of the case-study watersheds, and the monthly model can be easily converted to the model for water quality with higher sampling frequencies (i.e., weekly, daily, etc.) to provide more immediate evaluation for decision making (e.g., algal control protocol through water quality measurement).

References

Box, G. E. P., Jenkis, G. M., and Reinsel, G. C. (2007) *Time Series Analysis: Forecasting and Control*, 5th edition. John Wiley and Sons, Inc., Hoboken.

Kotz, S., and Nadarajah, S. (2004) *Multivariate t Distributions and Their Applications*. Cambridge University Press, Cambridge.

Additional Reading

http://ecy.wa.gov/programs/WQ/tmdl/SnohomishTribs/index.html
http://nwis.waterdata.usgs.gov/usa/nwis/qwdata
http://mlrc.gov
http://nhd.usgs.gov

13

Drought Analysis

ABSTRACT

In this chapter, we focus on the copula applications to at-site bivariate/trivariate drought analysis. In a case study, drought variables are separated from long-term daily streamflow series, i.e., drought severity, drought duration, drought interarrival time, and maximum drought intensity. Drought severity and duration are applied for bivariate drought frequency analysis. Drought severity, duration, and maximum intensity are applied for trivariate drought frequency analysis. The Archimedean, meta-elliptical, and vine copulas are adopted for the bivariate/trivariate analyses. The case study shows that the copula approach may be properly applied for drought analysis.

13.1 Introduction

Droughts may be identified with the following five types: (1) agricultural drought, (2) meteorological drought, (3) hydrological drought, (4) groundwater drought, and (5) social-economic drought. The commonly applied drought indices include Palmer drought severity index (PDSI; Palmer, 1965), crop moisture index (CMI; Palmer, 1968), standard precipitation index (SPI; McKee et al., 1993), and standard runoff index (SRI; Shukla and Wood, 2008). There are many other indices discussed and compared in an extensive review paper by Mishra and Singh (2010). According to the drought index applied, drought events may then be determined using the run theory proposed by Yevjevich (1967). For each drought event, there are three characteristics: drought severity (S: total deficit), drought duration (D), and drought intensity (the average intensity is usually considered: $I = S/D$). There is one more variable for two consecutive independent drought events: interarrival time (IT). The IT represents the time span from the onset of the first drought event to the onset of the second drought event (i.e., dry period + wet period). Figure 13.1 depicts the drought characteristics.

13.2 Copula Applications in Drought Studies

Conventionally, drought has been studied with the use of univariate frequency analysis (Van Rooy, 1965; Palmer, 1968; Santos, 1983; Rao and Padmanabhan, 1984; Voss et al.

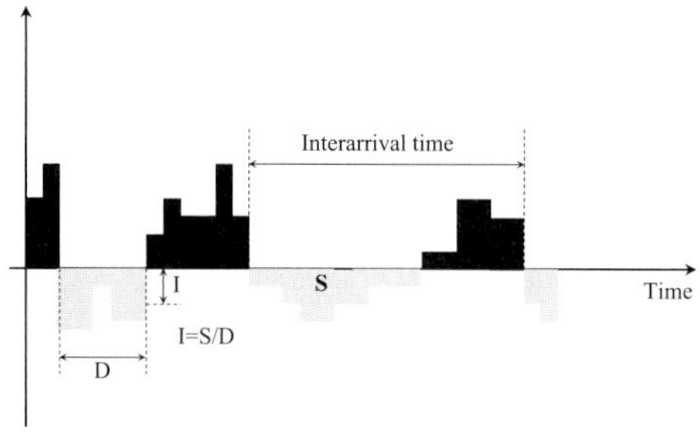

Figure 13.1 Schematic of drought characteristics.

2002; among others). With the increasing popularity of copula application in hydrology and water resources engineering, copulas have been applied to model bivariate and trivariate drought frequency analyses (Chen et al., 2013; Yoo et al., 2013; Hao and AghaKouchak, 2014; Janga et al., 2014; AghaKouchak, 2015; Salvadori and De Michele, 2015; Zhang et al., 2015; Kwak et al., 2016; Hao et al., 2016; Tu et al., 2016; among others). Here we first review some recent studies, followed by examples applying copulas to drought analysis.

Kao and Govindaraju (2010) proposed a joint deficit index (JDI) for drought analysis. In their study, monthly precipitation and streamflow data, computed from daily values, were applied for meteorological and hydrological drought analysis with the temporal window ranging from one month to 12 months. As a result, a 12-dimensional empirical copula was constructed to compute the Kendall distribution of $K_C(t) = P(C(u_1, u_2, \ldots, u_{12}) \leq t)$. The JDI was defined using the standardized normal distribution transformation as follows: $JDI = \Phi^{-1}[K_C(t)]$. In their method, there was no need to separate the drought events, based on a separation criterion (e.g., the threshold for flow computed from the flow-duration curve). Using the trivariate Plackett copula with the genetic algorithm for parameter estimation, Song and Singh (2010a) investigated the dependence among three drought characteristics (i.e., severity, duration, and interarrival time) using the trivariate Plackett copula. In their study, the Weibull distribution was applied to model drought duration and interarrival time, while the gamma distribution was applied to model drought severity. Song and Singh (2010b) investigated the drought frequency with the meta-elliptical copula.

Madadgar and Moradkhani (2013) investigated drought under climate change by readjusting the SDI (streamflow drought index) with different moving windows. In their study, the impact of climate change on drought was studied through the future climate scenarios generated from the General Circulation Models (GCMs). The Student t and Gumbel–Hougaard copulas were applied to study the dependence structure.

Chen et al. (2013) studied four drought characteristics, using the SPI index. The Archimedean and meta-elliptical copulas were chosen as the candidates to model the association of drought characteristics, i.e., drought severity, drought duration, interval time, and minimum SPI.

Rather than using the well-known drought characteristics of drought duration, drought severity, and interarrival time, Xu et al. (2015) applied the affected area, drought duration, and drought severity as the drought indicators for bivariate and trivariate drought frequency analyses to capture the spatial–temporal variability. Similar to other studies, drought variables were considered as random variables.

13.3 Hydrological Drought with the Use of Daily Streamflow: A Case Study

In this case study, we illustrate the copula application using daily streamflow (from December 1, 1942, to February 7, 2017) from the Nueces River near Tilden. Located in Texas, the Nueces River is about 315 miles in length and 16,800 square miles in drainage area (average annual runoff of about 620,000 acre-feet). The Nueces River flows through the central and southern parts of Texas and empties into the Gulf of Mexico. The unregulated USGS gauging station near Tilden (28°18'31"N, 98°33'25"W, i.e., USGS08194500) is located upstream (i.e., west) of the first major reservoir (i.e., the Chock Canyon Reservoir). In addition, as stated in the *Handbook of Texas*, the Nueces River watershed is predominantly a rural area, with the only metropolis of Corpus Christi located at the mouth (Texas State Historical Association, n.d.).

Daily (or monthly) streamflow statistics were also readily available from the USGS website that can be applied to determine the threshold of drought severity. Among all the available daily steamflow data from December 1, 1942, to February 2, 2017, daily streamflow data from August 19, 2009, to September 30, 2009, were not available.

13.3.1 Determination of Drought Severity, Duration, and Interarrival Time

The theory of runs, proposed by Yevjevich (1967), was used to determine the drought severity (i.e., total flow deficit in the case of hydrological drought). Following Yevjevich (1967), the threshold equation can be written as follows:

$$\overline{X_{0i}} = \mu_i + \alpha\sigma_i \tag{13.1}$$

where μ_i, σ_i represent the estimated long-term mean and standard deviation (daily or monthly), depending on the streamflow records.

Given the high fluctuation of daily streamflow, we will use the long-term daily average streamflow as a threshold. Additionally, following Zelenhasic and Salvai (1987), (a) minor drought events were ignored under the condition of $S_i \leq 0.005 \max(S), i = 1, 2, \ldots, N$, where N represents the total number of drought events identified and S represents the drought severity; and (b) the two consecutive drought events were pooled into a single drought event if the interevent wet period was relatively short and the ratio of surplus of pervious drought severity was small, the pooled event can be given as follows:

$$S = S_i + S_{i+1}, D = D_i + D_{i+1}; \Delta = SP_{i,i+1}/S_i \tag{13.2}$$

where S_i, S_{i+1} represent drought severity of two consecutive drought events; D_i, D_{i+1} represent the drought duration of two consecutive drought events; and $SP_{i,i+1}$ represents the total amount of streamflow above the threshold in the wet period between the two consecutive droughts.

The rules to pool the events are as follows:

i. The consecutive drought events are pooled into one drought event, if the interevent wet period is less than seven days.
ii. To pool the consecutive drought events if the ratio $\Delta \leq 0.05$ in Equation (13.2) such that the total surplus in the interevent period cannot relieve the dry condition.

To further illustrate the process, we show a simple example using daily streamflow from June 20, 1943, to May 2, 1944. Table 13.1 lists observed daily streamflow and its difference from the long-term daily average. The surplus of daily streamflow is in bold Italic. Figure 13.2 graphs the streamflow time series and the differences from daily thresholds. It is seen from Figure 13.2 that there existed flow deficit for most of the time in the period from June 20, 1943, to May 2, 1944.

Without either ignoring or pooling drought events, Table 13.2 lists the drought events by adding the continuous flow deficit (negative flow differences). In addition, before pooling the drought events together, we compute the maximum drought deficit, which is 2.36E+05 cfs.day from all the available daily streamflow data investigated in the case study. With the maximum drought deficit, the deficit less than $0.005 \max(\textit{deficit}) = 0.005(2.36 \times 10^5) = 1181.2$ cfs.day. With this criterion, the minor drought event from March 23, 1944, to March 27, 1944, is ignored, which is in bold Italic (Table 13.2). After ignoring minor droughts, Table 13.3 lists the remaining drought events. These remaining drought events are then further pooled using the rules of pooling discussed earlier. In addition, the last two droughts also need to be pooled, since $\Delta = 0.03 \leq 0.05$. Finally, all the individual droughts in Table 13.3 need to be pooled into one drought as follows:

Severity = 89257.53 cfs.day; Duration = 293 days; Inter-arrival time=334 days; Starting time = 06/20/1943; Ending time = 05/02/1944.

Using the identification procedure for drought events explained previously, we identified a total of 115 drought events. Table 13.4 lists the statistics of these identified drought events. As shown in Table 13.4, the fluctuations of drought variables are significant with a heavy tail. The drought events, such identified, are assumed as independent random variables. To further ensure the assumption, we will apply correlation by lag, as shown in Figure 13.3. The autocorrelation function plot with lag indicates that there is no serial dependence within each individual drought variable. As a result, drought variables can be considered as random variables for analysis. More specifically, drought variables are assumed to be continuous random variables throughout the analysis.

Table 13.1. *Sample daily streamflow and the difference from the long-term daily average.*

Time	(1)	(2)	Time	(1)	(2)	Time	(1)	(2)
6/20/1943	260	–475.14	10/4/1943	527	–17.39	1/18/1944	0.3	–77.15
6/21/1943	134	–569.28	10/5/1943	231	–238.80	1/19/1944	0.2	–67.45
6/22/1943	249	–415.00	10/6/1943	192	–267.71	1/20/1944	0.1	–71.63
6/23/1943	214	–331.42	10/7/1943	368	–189.20	1/21/1944	0.1	–73.92
6/24/1943	118	–346.54	10/8/1943	476	–492.46	1/22/1944	0.1	–58.30
6/25/1943	91	–298.27	10/9/1943	512	–699.65	1/23/1944	0.1	–58.64
6/26/1943	58	–299.96	10/10/1943	440	–575.82	1/24/1944	0.1	–71.94
6/27/1943	35	–329.05	10/11/1943	272	–1,264.61	1/25/1944	0.1	–74.61
6/28/1943	21	–368.70	10/12/1943	169	–969.71	1/26/1944	0.1	–78.00
6/29/1943	13	–519.24	10/13/1943	118	–924.32	1/27/1944	0.1	–87.55
6/30/1943	8.6	–606.69	10/14/1943	84	–917.10	1/28/1944	0.1	–95.74
7/1/1943	5.7	–651.45	10/15/1943	72	–1,014.33	1/29/1944	0.1	–98.22
7/2/1943	4	–731.88	10/16/1943	58	–1,224.35	1/30/1944	0.1	–80.38
7/3/1943	2.7	–860.08	10/17/1943	36	–1,291.34	1/31/1944	0.1	–66.76
7/4/1943	1.8	–880.55	10/18/1943	22	–1,212.70	2/1/1944	0.2	–60.67
7/5/1943	1.1	–1,032.43	10/19/1943	15	–1,352.55	2/2/1944	0.1	–57.44
7/6/1943	0.7	–1,106.26	10/20/1943	10	–1,029.42	2/3/1944	0.1	–51.09
7/7/1943	0.5	–1,059.12	10/21/1943	6.8	–881.09	2/4/1944	0.1	–61.35
7/8/1943	0.4	–1,035.25	10/22/1943	5	–769.01	2/5/1944	0.1	–76.43
7/9/1943	0.4	–993.28	10/23/1943	4	–699.60	2/6/1944	0.1	–86.45
7/10/1943	0.7	–877.82	10/24/1943	2.7	–668.94	2/7/1944	0.1	–78.81
7/11/1943	0.6	–737.09	10/25/1943	1.6	–744.00	2/8/1944	0.1	–71.32
7/12/1943	0.3	–662.59	10/26/1943	1	–703.39	2/9/1944	0.1	–62.20
7/13/1943	314	–284.57	10/27/1943	0.7	–650.11	2/10/1944	0.1	–51.69
7/14/1943	***676***	***200.28***	10/28/1943	0.5	–633.03	2/11/1944	0	–48.69
7/15/1943	304	–165.87	10/29/1943	0.3	–597.11	2/12/1944	0	–52.46
7/16/1943	424	–176.39	10/30/1943	0.2	–559.78	2/13/1944	0	–52.09
7/17/1943	***608***	***121.57***	10/31/1943	0.1	–540.33	2/14/1944	0	–51.05
7/18/1943	495	–43.63	11/1/1943	0.1	–612.66	2/15/1944	0	–49.88
7/19/1943	285	–258.77	11/2/1943	0.1	–587.60	2/16/1944	0	–46.82
7/20/1943	285	–158.28	11/3/1943	0.1	–536.40	2/17/1944	0	–43.75
7/21/1943	315	–32.26	11/4/1943	0.1	–479.11	2/18/1944	0	–41.93
7/22/1943	268	–32.48	11/5/1943	119	–301.67	2/19/1944	0	–40.49
7/23/1943	149	–134.54	11/6/1943	150	–221.10	2/20/1944	0	–41.64
7/24/1943	72	–174.14	11/7/1943	53	–272.51	2/21/1944	0	–57.31
7/25/1943	36	–176.77	11/8/1943	22	–268.47	2/22/1944	0	–90.26
7/26/1943	20	–198.86	11/9/1943	10	–256.47	2/23/1944	0	–320.83
7/27/1943	13	–231.18	11/10/1943	5.7	–257.31	2/24/1944	0	–628.09
7/28/1943	8.8	–229.18	11/11/1943	3.5	–261.08	2/25/1944	0	–421.89
7/29/1943	6.8	–231.50	11/12/1943	75	–184.71	2/26/1944	0	–292.50
7/30/1943	5.2	–239.47	11/13/1943	121	–99.92	2/27/1944	0	–195.38

Table 13.1. (*cont.*)

Time	(1)	(2)	Time	(1)	(2)	Time	(1)	(2)
7/31/1943	3.8	–255.87	11/14/1943	87	–99.26	2/28/1944	0	–126.72
8/1/1943	2.6	–246.38	11/15/1943	61	–76.85	2/29/1944	0	–72.65
8/2/1943	1.7	–219.91	11/16/1943	49	–61.05	3/1/1944	0	–101.54
8/3/1943	1.1	–207.64	11/17/1943	34	–69.37	3/2/1944	0	–92.08
8/4/1943	0.8	–198.74	11/18/1943	133	–8.79	3/3/1944	0	–80.73
8/5/1943	0.6	–211.95	***11/19/1943***	***268***	***96.80***	3/4/1944	0	–79.80
8/6/1943	0.5	–219.54	11/20/1943	102	–95.73	3/5/1944	0	–80.14
8/7/1943	0.4	–214.45	11/21/1943	36	–220.37	3/6/1944	0	–97.36
8/8/1943	0.4	–208.52	11/22/1943	17	–208.21	3/7/1944	0	–235.27
8/9/1943	0.4	–236.16	11/23/1943	10	–181.14	3/8/1944	0	–237.32
8/10/1943	0.3	–320.24	11/24/1943	7.2	–172.71	3/9/1944	0	–192.98
8/11/1943	0.2	–315.58	11/25/1943	5	–179.02	3/10/1944	0.1	–154.09
8/12/1943	0.2	–294.99	11/26/1943	3.6	–174.14	3/11/1944	0	–116.15
8/13/1943	0.2	–355.40	11/27/1943	3.5	–159.81	3/12/1944	0	–100.34
8/14/1943	0.1	–366.51	11/28/1943	4.6	–130.02	3/13/1944	0	–101.19
8/15/1943	0.1	–441.86	11/29/1943	4	–116.98	3/14/1944	0	–99.40
8/16/1943	0.1	–377.74	11/30/1943	4.6	–123.94	3/15/1944	0	–92.68
8/17/1943	0.1	–279.99	12/1/1943	4.8	–128.70	3/16/1944	0	–93.21
8/18/1943	0.1	–236.28	12/2/1943	3.1	–139.21	3/17/1944	0	–91.94
8/19/1943	0	–228.42	12/3/1943	3.1	–129.13	3/18/1944	0	–89.59
8/20/1943	0	–343.15	12/4/1943	12	–115.58	3/19/1944	0.1	–90.63
8/21/1943	0	–457.47	12/5/1943	80	–40.66	3/20/1944	40	–51.56
8/22/1943	0	–445.06	***12/6/1943***	***180***	***72.08***	***3/21/1944***	***216***	***111.83***
8/23/1943	0	–395.05	***12/7/1943***	***260***	***162.21***	***3/22/1944***	***144***	***52.34***
8/24/1943	0	–354.06	***12/8/1943***	***146***	***59.89***	3/23/1944	62	–14.97
8/25/1943	0	–321.86	***12/9/1943***	***90***	***12.07***	3/24/1944	27	–43.74
8/26/1943	0	–299.22	12/10/1943	45	–28.66	3/25/1944	9.4	–58.92
8/27/1943	0	–279.80	12/11/1943	30	–43.74	3/26/1944	4.8	–65.01
8/28/1943	0	–273.29	12/12/1943	31	–40.46	3/27/1944	37	–23.18
8/29/1943	0	–263.71	12/13/1943	25	–42.85	***3/28/1944***	***140***	***85.44***
8/30/1943	0	–254.53	12/14/1943	16	–56.60	***3/29/1944***	***245***	***182.52***
8/31/1943	0	–237.63	12/15/1943	9.5	–70.99	***3/30/1944***	***196***	***120.86***
9/1/1943	0	–268.43	12/16/1943	7.7	–70.07	***3/31/1944***	***102***	***32.65***
9/2/1943	0	–289.05	12/17/1943	5.7	–71.02	4/1/1944	60	–9.06
9/3/1943	159	–462.74	12/18/1943	4	–73.18	4/2/1944	39	–50.95
9/4/1943	505	–218.31	12/19/1943	2.9	–81.02	4/3/1944	29	–62.65
9/5/1943	***749***	***2.21***	12/20/1943	2.1	–91.09	4/4/1944	19	–68.82
9/6/1943	***1130***	***382.68***	12/21/1943	1.8	–94.72	4/5/1944	15	–68.02
9/7/1943	***1760***	***1056.62***	12/22/1943	1.6	–116.84	4/6/1944	11	–74.92
9/8/1943	***1690***	***953.12***	12/23/1943	1.4	–116.90	4/7/1944	6.8	–82.04
9/9/1943	352	–414.57	12/24/1943	1.1	–101.90	4/8/1944	5	–90.34

Table 13.1. (*cont.*)

Time	(1)	(2)	Time	(1)	(2)	Time	(1)	(2)
9/10/1943	186	–626.36	12/25/1943	0.9	–92.09	4/9/1944	3.5	–97.91
9/11/1943	348	–229.78	12/26/1943	0.9	–89.79	4/10/1944	2	–107.93
9/12/1943	305	–238.43	12/27/1943	0.7	–82.95	4/11/1944	1	–132.69
9/13/1943	146	–725.07	12/28/1943	0.6	–74.80	4/12/1944	0.7	–139.90
9/14/1943	60	–952.46	12/29/1943	0.5	–90.51	4/13/1944	0.4	–116.94
9/15/1943	31	–931.12	12/30/1943	0.4	–91.35	4/14/1944	0.3	–90.52
9/16/1943	25	–997.26	12/31/1943	0.4	–89.75	4/15/1944	0.2	–86.47
9/17/1943	22	–921.83	1/1/1944	0.5	–93.26	4/16/1944	0.1	–81.40
9/18/1943	14	–766.21	1/2/1944	0.5	–86.10	4/17/1944	0.1	–89.15
9/19/1943	10	–578.15	1/3/1944	0.4	–85.01	4/18/1944	0.1	–105.27
9/20/1943	7.7	–440.76	1/4/1944	0.4	–78.94	4/19/1944	0.1	–105.92
9/21/1943	5.4	–681.06	1/5/1944	0.4	–94.52	4/20/1944	0.1	–91.85
9/22/1943	3.8	–1,140.30	1/6/1944	0.3	–106.63	4/21/1944	0.1	–126.40
9/23/1943	2.4	–1,072.43	1/7/1944	0.3	–116.32	4/22/1944	0.1	–179.66
9/24/1943	87	–1,349.54	1/8/1944	0.3	–139.38	4/23/1944	0	–218.81
9/25/1943	93	–1,060.46	1/9/1944	0.4	–183.61	4/24/1944	0	–210.94
9/26/1943	62	–714.50	1/10/1944	0.4	–168.02	4/25/1944	0	–244.29
9/27/1943	46	–563.73	1/11/1944	0.4	-131.69	4/26/1944	0	–365.98
9/28/1943	24	–530.45	1/12/1944	0.4	-113.08	4/27/1944	0	–432.56
9/29/1943	116	–394.19	1/13/1944	0.5	-112.36	4/28/1944	0	–394.68
9/30/1943	75	–507.08	1/14/1944	0.6	-103.62	4/29/1944	0	–414.73
10/1/1943	101	–566.52	1/15/1944	0.5	-95.25	4/30/1944	0	–661.52
10/2/1943	451	–202.83	1/16/1944	0.4	-89.38	5/1/1944	0	–733.76
10/3/1943	***629***	***20.89***	1/17/1944	0.3	-88.86	5/2/1944	245	–339.09

Note: (1): Observed daily streamflow in cfs; (2) difference from the long-term daily average using observed-daily average in cfs in which the negative values represent the flow deficit.

13.3.2 Univariate Drought Frequency Analysis

Before we proceed to the bivariate and trivariate drought analyses, we will first perform univariate drought frequency analysis. The fitted parametric univariate distribution will be applied for risk analysis. To study the univariate drought frequency analysis, drought variables are conventionally assumed as continuous random variables. It is also known that there may be ties that commonly exist in drought duration and drought interarrival time variables. To avoid the impact of ties in the univariate analysis, the parametric univariate distribution is fitted to the unique values (e.g., if there is more than one 30-day duration drought, only one 30-day duration will be used for fitting the univariate distribution). We obtain that (i) if there is no tie in drought severity, all 115 drought severity values are applied for univariate analysis; (ii) there are 15 values that are repeated

Table 13.2. *Drought identified before ignoring minor droughts and pooling droughts.*

Start	End	Severity (cfs.day)	Duration (day)	Interarrival (day)	Surplus (cfs.day)	Interevent time (days)
06/20/1943	07/13/1943	15,471.67	24	28	200.28	4
07/18/1943	09/04/1943	12,740.57	49	53	2,394.63	4
09/09/1943	10/02/1943	16,605.10	24	25	20.89	1
10/04/1943	11/18/1943	25,782.17	46	47	96.80	1
11/20/1943	12/05/1943	2,315.34	16	20	306.24	1
12/10/1943	03/20/1944	10,267.52	102	113	164.16	11
03/23/1944	***03/27/1944***	***205.82***	***5***	***9***	***421.46***	
04/01/1944	05/02/1944	6,075.16	32	41	5,709.02	9

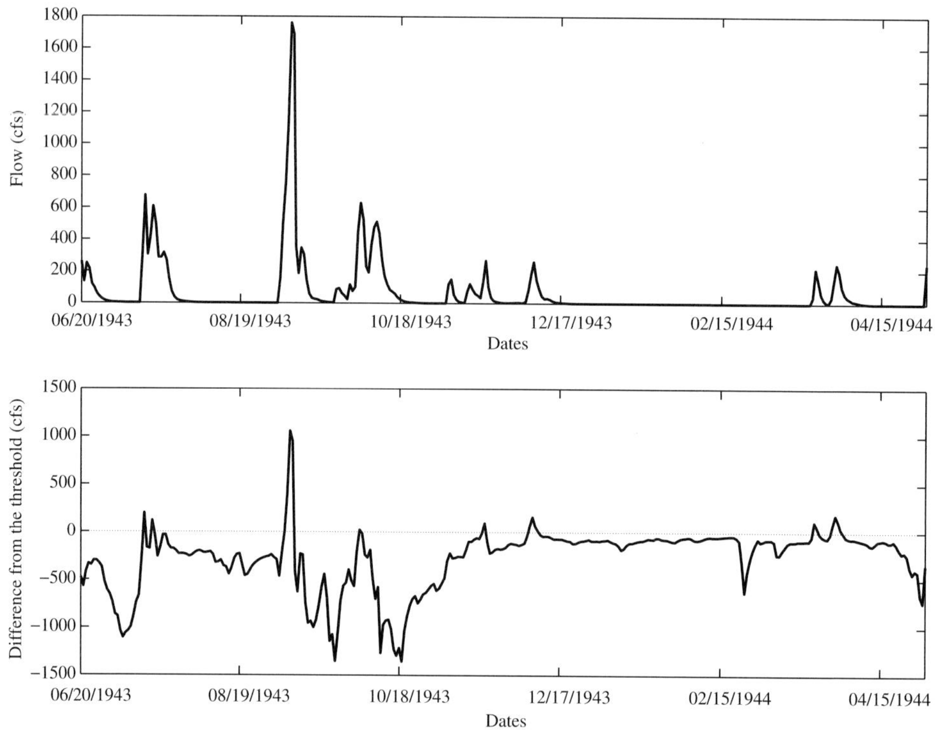

Figure 13.2 Daily streamflow and its difference from the long-term daily threshold from June 20, 1943, to May 2, 1944.

more than once for the drought duration, and 93 unique duration values are applied for univariate analysis; (iii) there are 13 values that are repeated more than once for drought interarrival times, and 98 unique interarrival time values are applied for the univariate analysis.

Table 13.3. *Drought events by ignoring minor droughts.*

Start	End	Severity (cfs.day)	Duration (day)	Surplus (cfs.day)	Interevent (day)	*Eq.*(12.2) : Δ
06/20/1943	07/13/1943	15,471.67	24	200.28	4	0.02
07/18/1943	09/04/1943	12,740.57	49	2,394.63	4	0.14
09/09/1943	10/02/1943	16,605.10	24	20.89	1	0.00
10/04/1943	11/18/1943	25,782.17	46	96.80	1	0.04
11/20/1943	12/05/1943	2,315.34	16	306.24	1	0.03
12/10/1943	03/20/1944	10,267.52	102	164.16	11	0.03
04/01/1944	05/02/1944	6,075.16	32	5,709.02	9	0.85

Table 13.4. *Statistics of identified drought events.*

	Mean	Std.	Skewness	Kurtosis
Severity (S: cfs.day)	63189.05	83782.42	2.42	9.73
Duration (D: days)	183.24	229.57	2.20	8.60
Interarrival time (IT: days)	263.90	286.64	2.29	9.16

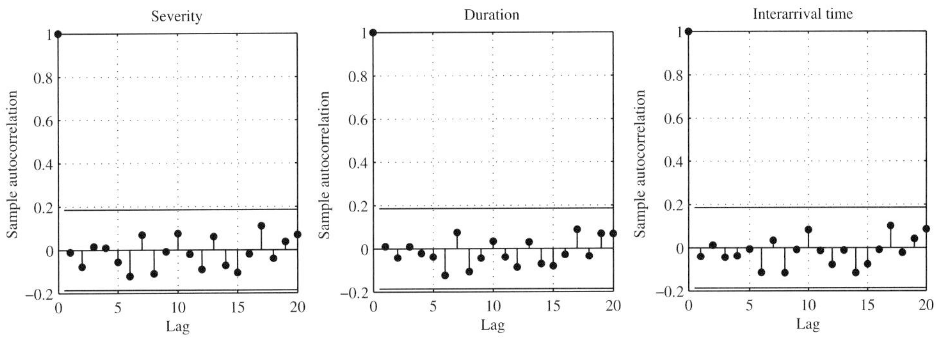

Figure 13.3 Check for independence of drought variables.

Based on the previous studies, the following parametric univariate distributions are considered as candidates: gamma, exponential, and Weibull distributions. In addition, the log-normal distribution has been commonly applied to model drought severity (i.e., streamflow deficit), while the Weibull distribution has been commonly applied to model drought duration and drought interarrival time. Table 13.5 lists the fitted univariate distributions as well as the formal goodness-of-fit (GoF) statistics using the Kolmogorov–Smirnov (KS) test. Figure 13.4 compares the parametric marginal distributions with the empirical distributions. From Table 13.5, it is seen that the log-normal (for

Table 13.5. *Fitted univariate distribution and GoF test results.*

	Severity (cfs.day)		Duration (days)		Interarrival time (days)	
	Parameters	GoF [S, P][a]	Parameters	GoF [S, P]	Parameters	GoF [S, P]
Log-normal	**[10.20, 1.46]**	**[0.042, 0.68]**	**[4.69, 1.32]**	**[0.04, 0.93]**	*[5.17, 1.03]*	*[0.059, 0.33]*
Exponential	63189	[0.17, 0]	216.18	[0.12, 0.01]	**285.6**	**[0.046, 0.66]**
Weibull	*[54413, 0.78]*	*[0.074, 0.067]*	*[203.8, 0.89]*	*[0.079, 0.077]*	*[291.6, 1.05]*	*[0.064, 0.24]*

Note: [a] S: KS test statistic, P: *P*-value computed using the parametric bootstrap method.

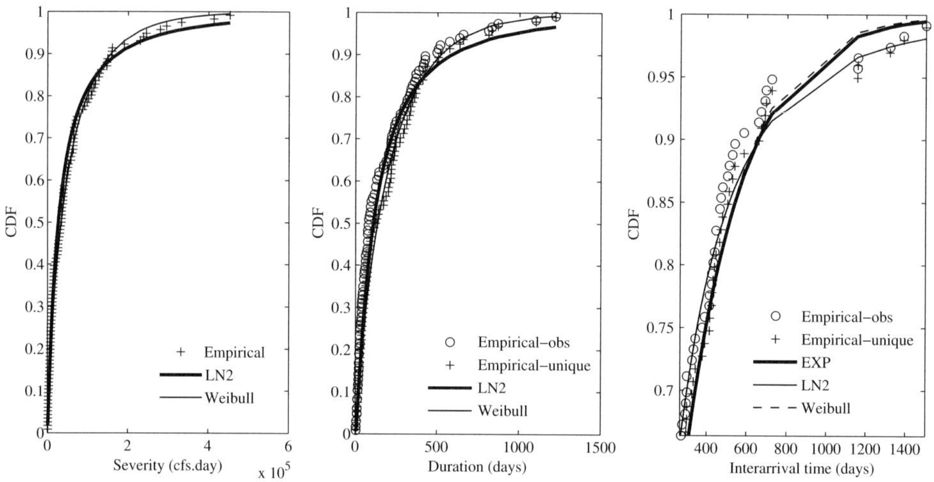

Figure 13.4 Fitted parametric marginal distributions versus empirical distribution.

severity and duration) and exponential (for interarrival time) distributions yield the smallest KS test statistics (i.e., the smallest distance between the parametric and empirical distributions). However, comparisons in Figure 13.4 show that (i) in the case of drought severity and drought duration, the Weibull distribution fits the upper tail better than the lognormal distribution; and (ii) in the case of drought interarrival time, there is minimal difference in fitting the upper tail for log-normal and Weibull distributions. To comply with the conventional univariate drought analysis, we will use the conventional marginal distributions for illustration, i.e., log-normal distribution for drought severity and Weibull distribution for drought duration and drought interarrival time. One other reason of applying the conventional distributions is that both log-normal and Weibull distributions pass the formal GoF KS test.

13.3.3 Bivariate Drought Frequency Analysis

Fitting Copula Functions to Bivariate Drought Variables

Using the identified drought events, Figure 13.5 plots show the scatter plots of the paired random variables. Figure 13.5 clearly shows the positive dependence between drought severity and drought duration; between drought severity and drought interarrival time; as well as between drought duration and drought interarrival time.

As introduced in Chapter 3, we will investigate the dependence separately from the marginals. The empirical marginals (i.e., using the Weibull plotting position formula) are applied to study the dependence among drought variables. The parametric marginal distributions, fitted in the previous section, will be applied for risk analysis through joint (and/or conditional) return periods. In addition, we will choose the Archimedean and meta-elliptical copulas for the bivariate drought analysis, while only meta-elliptical copulas will be chosen for trivariate drought analysis. Given the positive dependence structure between drought variables, the selected Archimedean copulas are the Gumbel–Hougaard (which belongs to the extreme value family), Clayton, and Frank copulas. One may also choose other Archimedean copulas as candidates. The Gaussian and Student t copulas are selected as candidates from the meta-elliptical family.

Applying the pseudo-MLE to the bivariate drought variables, Table 13.6 lists the estimated copula parameters as well as the GoF test results through the improved Rosenblatt transform following the procedures discussed in Section 3.8.3 (i.e., the SnB test; Genest, et al., 2007) with the following steps briefly outlined:

i. For d-dimensional random variables $\boldsymbol{u} = [u_1, u_2, \ldots, u_d]$; setting $\boldsymbol{Z} = [Z_1, Z_2, \ldots, Z_d]$, where

$$Z_1 = u_1 = F_1(x_1); Z_2 = \frac{\partial C(u_1, u_2)}{\partial u_1}; \ldots, Z_d = \frac{\partial C^{d-1}(u_1, \ldots, u_d)}{\partial u_1 \partial u_2 \ldots \partial u_{d-1}} \Big/ \frac{\partial C^{d-1}(u_1, \ldots u_{d-1})}{\partial u_1 \partial u_2 \ldots \partial u_{d-1}}$$

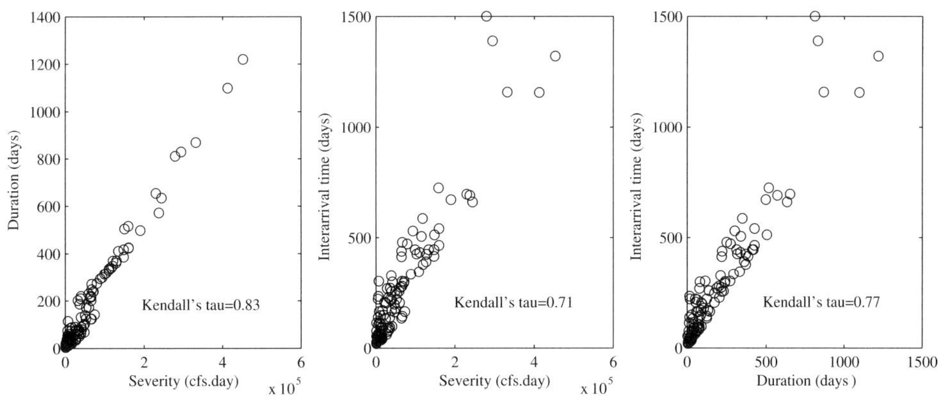

Figure 13.5 Scatter plots of paired drought variables.

Table 13.6. *Parameters estimated and GoF test for copula candidates.*

	S and D		S and INT		D and INT	
	(1)	(2)	(1)	(2)	(1)	(2)
GH[a]	[6.2, 157.31]	[0.17, 0.24]	[3.30, 90.00]	[0.06, 0.81]	[3.97, 109.02]	[0.27, 0.08]
Clayton	[4.11, 92.71]	[0.26, 0.08]	[2.10, 49.23]	[0.21, 0.13]	[2.80, 64.50]	[0.42, 0.02]
Frank	[18.88, 126.33]	[0.17, 0.23]	[11.21, 80.73]	[0.10, 0.56]	[14.76, 80.73]	[0.32, 0.04]
Gaussian	[0.95, 137.11]	[0.33, 0.15]	[0.87, 80.77]	[0.08, 0.71]	[0.91, 99.78]	[0.26, 0.10]
Student t	(0.95, 1.47)[b] 144.68[c]	[0.24, 0.10]	(0.88, 5.79) 83.59	[0.10, 0.54]	(0.92, 4.99) 104.73	[0.31, 0.04]

Notes: (1) estimated parameter and CL; (2) SnB test statistics and P-value;
[a] GH represents the Gumbel–Hougaard copula, [b]correlation and degree of freedom, [c] CL for Student t copula.

The null hypothesis is that $\mathbf{u} = [u_1, u_2, \ldots, u_d] \sim C_\theta \Rightarrow \mathbf{Z} = [Z_1, Z_2, \ldots, Z_d]$ are independent.

ii. Compute the test statistic using

$$S_n^{(B)} = \frac{n}{3^d} - \frac{1}{2^{d-1}} \sum_{i=1}^{n} \sum_{k=1}^{d} \left(1 - E_{ik}^2\right) + \frac{1}{n} \sum_{i=1}^{n} \sum_{j=1}^{n} \prod_{k=1}^{d} \left(1 - \max\left(Z_{ik}, Z_{jk}\right)\right) \tag{13.3}$$

iii. Generate random variables from the fitted copula function with the same sample size as the sample data.

iv. Reestimate the copula parameters from the tested copula function and recompute the test statistics.

v. Repeat steps ii–iv for a large number of times (N) and approximate the P-value using $P_{value} = \frac{1}{N} \sum_{k=1}^{N} 1\left(S_{n,k}^{(B)*} > S_n^{(B)}\right)$, where $S_{n,k}^{(B)*}$ represents the test statistic from step iv.

From Table 13.6 we obtain that (i) all the copula candidates from Archimedean and meta-elliptical families may be applied to model drought severity (S) and drought duration (D) as well as drought severity (S) and drought interarrival time (INT); (ii) the Gumbel–Hougaard and Guassian copulas are the only two copula functions that may be applied to model drought duration and drought interarrival time.

Joint and Conditional Return Period for Bivariate Drought Analysis

In this section, we will apply the Gumbel–Hougaard copula (which belongs to the extreme value family) and the Gaussian copula to investigate risk through joint and conditional return periods. Here, we will only focus on drought severity and drought duration.

Joint Return Period of Bivariate Drought Analysis As discussed in Section 3.10.2, the bivariate joint return period may be represented with either the "AND" case or "OR" case. Here we will focus on the "AND" case only. Equation (3.139) in Chapter 3 can be revised as follows:

$$T_{AND}(s,d) = \frac{E(INT)}{1 - F_S(s) - F_D(d) + C_{12}(F_S(s), F_D(d))} \tag{13.4}$$

where E(INT) represents the expected drought interarrival time in years, E(INT) = 0.723 year.

Using Equation (3.139), Table 13.7 lists the "AND" case joint return periods with the fitted parametric marginal distributions. To further illustrate the computation, we will show how to compute the return period for $T(S > 8000\ cfs.day, D > 30\ days)$:

- $F_S(S < 8000) = 0.2035$ from the fitted log-normal distribution $S \sim LN2(10.1992, 1.4614)$.
- $F_D(D < 30) = 0.1661$ from the fitted Weibull distribution $D \sim Weibull(203.80, 0.8903)$.
- $F_{S,D}(S \leq 8000, D \leq 30) = C(0.2035, 0.1661; \theta = 6.2015) = 0.1479$ from the fitted Gumbel–Hougaard copula for drought severity and drought duration.
- The exceedance probability:

$$\begin{aligned} F(S > 8000, D > 30) &= 1 - F_S(8000) - F_D(30) + C(F_S, F_D) \\ &= 1 - 0.2035 - 0.1661 + 0.1479 = 0.7783 \end{aligned}$$

- The "AND" case joint return period $T(S > 8000, D > 30) = \dfrac{0.723}{0.7783} \approx 0.93\ yr.$

Figure 13. 6 shows the Joint return period of the "AND" case for drought severity and drought duration.

Conditional Return Period of Bivariate Drought Variables There are two commonly applied approaches to study the conditional return period:

$T(X_1 > x_1|X_2 > x_2)$ and $T(X_1 > x_1|X_2 = x_2)$. Here, we will investigate both conditional return periods for drought severity and drought duration, with the use of drought duration as the conditioning variable.

$T(S > s|D > d)$In this case, the exceedance conditional probability of S given D exceeding a given duration (d) can be written through the copula as follows:

$$P(S > s|D > d) = \frac{P(S > s, D > d)}{P(D > d)} = \frac{1 - F_S(s) - F_D(d) + C_{12}(F_S(s), F_D(d))}{1 - F_D(d)} \tag{13.5}$$

The conditional return period can then be written as follows:

$$T(S > s|D > d) = \frac{E(INT)}{(1 - F_D(d))(1 - F_S(s) - F_D(d) + C_{12}(F_S(s), F_D(d)))} \tag{13.6}$$

Equation (13.5) may also tell whether there exists the right tail increasing (RTI) property. The RTI property exists if the conditional exceedance probability is a nondecreasing function of drought duration for all drought severity values.

Table 13.7. *Joint return period of drought severity and duration ("AND" case).*

		D (days) Gumbel–Hougaard copula						*D* (days) Gaussian copula					
		30	120	365	520	700	1120	30	120	365	520	700	1120
	8000	0.93	1.35	3.88	7.23	14.52	69.01	0.93	1.35	3.88	7.23	14.52	69.01
	28000	1.48	1.55	3.88	7.23	14.52	69.01	1.48	1.57	3.88	7.23	14.52	69.01
S	92000	3.62	3.62	4.20	7.25	14.52	69.01	3.62	3.62	4.54	7.41	14.54	69.01
(cfs.day)	180000	7.48	7.48	7.51	8.29	14.58	69.01	7.48	7.48	7.71	9.38	15.34	69.03
	300000	14.64	14.64	14.64	14.68	16.47	69.01	14.64	14.64	14.67	15.36	19.48	69.65
	800000	71.44	71.44	71.44	71.44	71.44	79.63	71.44	71.44	71.44	71.45	72.02	103.25

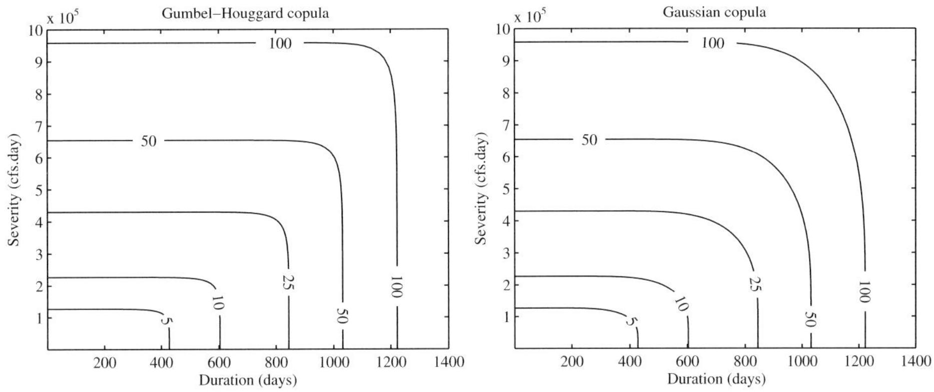

Figure 13.6 Joint return period of the "AND" case for drought severity and drought duration.

Using Equations (13.5) and (13.6) and the Gumbel–Hougaard copula as an illustrative example, Table 13.8 lists the conditional exceedance probability and conditional return period. Figure 13.7 plots the conditional exceedance probability and conditional return period. From Table 13.8 and Figure 13.7, it is seen that the exceedance probability is a nondecreasing function of duration, i.e., with the increase of drought duration, the exceedance probability of $S>s|D>d$ is nondecreasing. The RTI property indicates that it is more likely for the drought severity exceeding a given threshold conditioned on a higher drought duration than that conditioned on a lower drought duration. Using $S > 8000$ cfs.day in Table 13.8 as an example, we have the following:

$$P(S > 8000|D > 30) < P(S > 8000|D > 120) = P(S > 8000|D > 365)$$
$$= P(S > 8000|D > 520) = P(S > 8000\,|D > 700 = P(S > 8000|D > 1120)$$

From Table 13.8 and Figure 13.7, it is also seen that for a given drought duration, the exceedance probability decreases with the increase of drought severity. To illustrate the computation, we will show the procedure to compute $P(S > 8000|D > 30)$ and $T(S > 8000|D > 30)$:

- Previously we have computed $P(S > 8000, D > 30) = 0.7783$ for the "AND" case.
- The exceedance conditional probability is as follows: $P(S > 8000|D > 30) = \dfrac{P(S > 8000, D > 30)}{P(D > 30)} = \dfrac{0.7783}{1 - 0.1661} = 0.933$
- The conditional return period is as follows:

$$T(S>8000|D>30) = \frac{E(INT)}{(1-F_D(30))P(S>8000.D>30)} = \frac{0.723}{(1-0.1661)0.7783} \approx 1.11\,yr$$

Table 13.8. *Conditional exceedance probability and conditional return period using drought duration as the conditioning variable.*

		D (days) $P(S > s \mid D > d)$						D (days) $T(S > s \mid D > d)$					
		30	120	365	520	700	1,120	30	120	365	520	700	1,120
	8,000	0.93	1.00	1.00	1.00	1.00	1.00	1.11	2.52	20.82	72.27	291.60	6586.07
	28,000	0.59	0.87	1.00	1.00	1.00	1.00	1.77	2.88	20.82	72.27	291.60	6586.07
S	92,000	0.24	0.37	0.92	1.00	1.00	1.00	4.34	6.75	22.54	72.46	291.61	6586.07
(cfs.day)	180,000	0.12	0.18	0.52	0.87	1.00	1.00	8.98	13.97	40.30	82.84	292.81	6586.08
	300,000	0.06	0.09	0.27	0.49	0.88	1.00	17.55	27.32	78.54	146.80	330.78	6586.36
	800,000	0.01	0.02	0.05	0.10	0.20	0.87	85.66	133.33	383.31	714.17	1434.63	7600.17

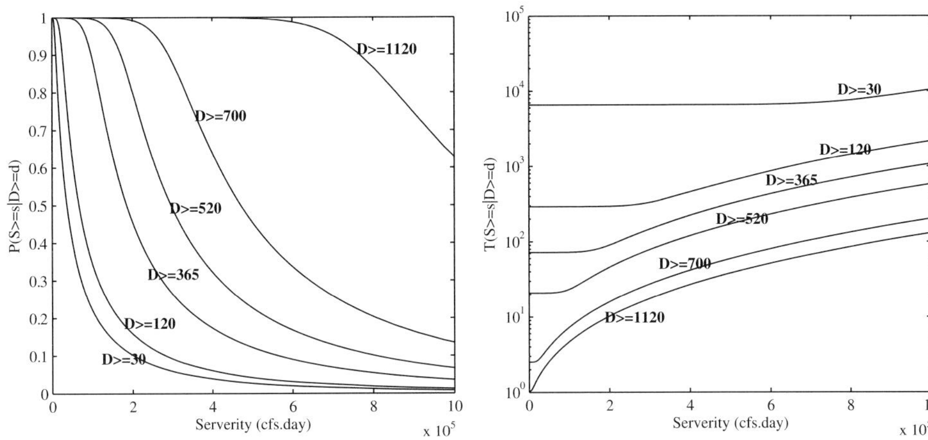

Figure 13.7 Conditional exceedance probability and conditional return period of $\boldsymbol{S > s \mid D > d}$.

$T(S > s|D = d)$. In this case, the drought duration is the fixed conditioning variable. The exceedance conditional probability may be written as follows:

$$P(S > s|D = d) = 1 - P(S \leq s|D = d)$$
$$= 1 - C(F_S|F_D = F_D(d)) = 1 - \left.\frac{\partial C_{12}(F_S, F_D)}{\partial F_D(d)}\right|_{F_D(d)} \tag{13.7}$$

The conditional return period can then be written as follows:

$$T(S > s|D = d) = \frac{E(INT)}{P(S > s|D = d)} = \frac{E(INT)}{1 - \left.\frac{\partial C_{12}(F_S,F_D)}{\partial F_D(d)}\right|_{F_D(d)}} \tag{13.8}$$

According to Nelson (2006), the stochastic increasing (SI) property exists if $P(S > s|D = d)$ is a nondecreasing function of drought duration for all drought severity values.

Using Equations (13.7) and (13.8) and the Gumbel–Hougaard copula as an illustrative example, Table 13.9 lists the exceedance conditional probability and conditional return period. Figure 13.8 plots the exceedance probability and conditional return period. Figure 13.8 clearly shows that for $D = d$, the exceedance probability $P(S > s|D = d)$ is a nondecreasing function of drought duration, i.e., $P(S > s|D = d_1) \leq P(S > s|D = d_2), d_1 < d_2$.

Dynamic Return Period for a Given Drought Episode

In the previous sections, we have investigated the joint and conditional return periods of bivariate drought variables, namely drought severity and drought duration. Following De Michele et al. (2013), we will investigate the evolution of drought within a given drought event (or simply called drought episode). It is worth mentioning that the copula function

Table 13.9. *Exceedance conditional probability and conditional return period of* $S > s|D = d$.

		D (days) $P(S > s\|D = d)$					D (days) $T(S > s\|D = d)$				
		30	120	365	520	700	30	120	365	520	700
	8000	0.36	0.99	1.00	1.00	1.00	2.02	0.73	0.72	0.72	0.72
	28000	0.00	0.29	1.00	1.00	1.00	286.86	2.45	0.72	0.72	0.72
S	92000	2.73E-06	0.00	0.57	0.98	1.00	2.65E+05	1.60E+03	1.27	0.74	0.72
(cfs.day)	180000	2.08E-08	3.45E-06	0.01	0.39	0.97	3.48E+07	2.10E+05	67.70	1.83	0.74
	300000	2.78E-10	4.61E-08	1.44E-04	0.01	0.43	2.60E+09	1.57E+07	5.01E+03	80.12	1.67
	800000	1.33E-14	2.19E-12	6.85E-09	4.32E-07	3.82E-05	5.43E+13	3.31E+11	1.06E+08	1.67E+06	1.89E+04

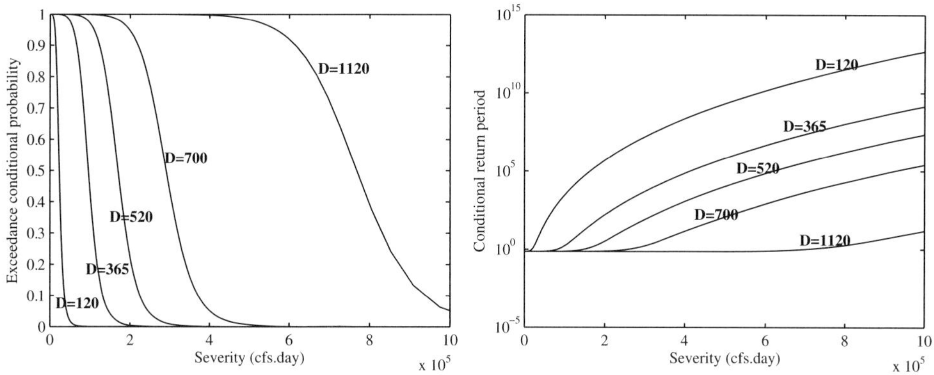

Figure 13.8 Exceedance conditional probability and conditional return period plot.

fitted to the drought severity and drought duration will not be applicable here. The empirical copula will be applied to study the dynamic return period for the given drought episode.

As discussed in De Michele et al. (2013), the dynamic return period is estimated through the Survival Kendall Distribution (also called DSKRP). As introduced in Section 4.5.1, the Kendall distribution may be considered as univariate realization of the copula function. In the case of bivariate analysis, the Kendall distribution may be simply written as follows:

$$K_C(t) = P(C(F_{X_1}(x_1), F_{X_2}(x_2)) \leq t) \tag{13.9}$$

and the survival Kendall distribution $[\bar{K}_C(t)]$ may be written as follows:

$$\bar{K}_C(t) = P(\bar{C}(F_{X_1}(x_1), F_{X_2}(x_2)) \geq t) \tag{13.10}$$

In Equation (13.10), $\bar{C}$ represents the survival copula, and $\overline{F_{X_i}}(x_i) = 1 - F_{X_i}(x_i), i = 1, 2.$

The DSKRP can then be written as follows:

$$T_{DSKRP} = \frac{\mu}{1 - \bar{K}_C(t)} \tag{13.11}$$

In Equation (13.11), μ represents the average interarrival time of the drought event $(\mu = 0.723\ yr)$.

Furthermore, to investigate the DSKRP for a given drought episode, the average running drought intensity (I) will be applied. The average running drought intensity is computed as the average drought deficit starting from the initiation of a drought episode until day k into the drought. With this in mind, the new bivariate drought variable is given by pair as $(I_k, k), k = 1, 2, \ldots, m,$ where m represents the total number of days of the drought episode. To illustrate the DSKRP method, we will use the recent 21-day drought episode identified (i.e., September 7–27, 2016) as an example. Table 13.10 lists the daily streamflow and streamflow deficit during this dry period.

Table 13.10. *Daily streamflow and flow deficit from September 7 to September 27, 2016.*

Date	Flow (cfs)	Deficit (cfs)	I_k
7-Sep-2016	444	259.38	259.38
8-Sep-2016	154	582.88	421.13
9-Sep-2016	102	664.57	502.28
10-Sep-2016	73	739.36	561.55
11-Sep-2016	56	521.78	553.60
12-Sep-2016	46	497.43	544.24
13-Sep-2016	36	835.07	585.78
14-Sep-2016	29	983.46	635.49
15-Sep-2016	24	938.12	669.12
16-Sep-2016	20	1,002.26	702.43
17-Sep-2016	17	926.83	722.83
18-Sep-2016	15	765.21	726.36
19-Sep-2016	13	575.15	714.73
20-Sep-2016	11	437.46	694.93
21-Sep-2016	8.8	677.66	693.78
22-Sep-2016	7.4	1,136.70	721.46
23-Sep-2016	6.5	1,068.33	741.86
24-Sep-2016	5.5	1,431.04	780.15
25-Sep-2016	6.4	1,147.06	799.46
26-Sep-2016	74	702.50	794.61
27-Sep-2016	366	243.73	768.38

As introduced earlier, the running average I_k can be computed as follows:

$$I_k = \frac{\sum_{i=1}^{k} deficit(i)}{k}, k = 1, 2, \ldots, 21.$$

For example, for the drought period of day 3, we have the following: $I_3 = \frac{259.38 + 582.88 + 664.57}{3} = 502.28\ cfs.$

Figure 13.9 plots the flow deficit running average, as well as DSKRP and return period computed from the univariate flow deficit (RPFD) for the recent dry period. The plot on the left shows that the running average fluctuates within the drought episode. This fluctuation may reflect the severity of the state of drought on a given day. The plot on the right shows the DSKRP and the RPFD. It shows that within the drought episode, DSKRP and RPFD for the state of drought share a similar pattern and reflect the fluctuation of flow deficit. Table 13.11 lists the computed survival Kendall distribution and the corresponding DSKRP. As an illustration, here we will show how to compute DSKRP for $(I_1, 1) = (259.38, 1)$:

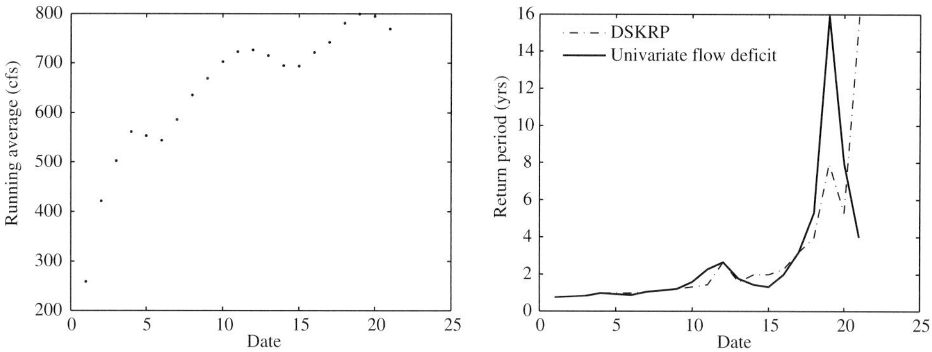

Figure 13.9 Flow deficit running average and DSKRP for 2016 event from September 7–27.

1. To compute the exceedance empirical marginal probability, use Weibull plotting position formula as follows:

$$\overline{F_{I_k}}(I_1) = 1 - F_{I_k}(I_k \le I_1) = 1 - \frac{1}{21+1} \approx 0.9545; \bar{F}_k(1) = 1 - \frac{1}{21+1} \approx 0.9545.$$

2. To compute the empirical survival copula, i.e., $\bar{C}$, apply the following formula:

$$\begin{aligned}\bar{C}(F_{I_k}(I_1), F_k(1)) &= \overline{F_{I_k}}(I_1) + \bar{F}_k(1) - 1 + C(F_{I_k}(I_1), F_k(1)) \\ &= 0.9545 + 0.9545 - 1 + C\left(\frac{1}{22}, \frac{1}{22}\right) \\ &= 0.9545 + 0.9545 - 1 + 0.0476 \approx 0.9567\end{aligned}$$

   Here, $C(F_{I_k}(I_1), F_k(1))$ is estimated using the empirical copula formula given as Equation (2.59). The first pair $(I_1, 1) = (259.38, 1)$ is the smallest pair among all $(I_k, k), k = 1, 2, \ldots$.. and we have $C\left(\frac{1}{22}, \frac{1}{22}\right) = \frac{1}{21} \approx 0.0476$.

3. To compute the empirical survival Kendall distribution $\bar{K}_C$, Equation (13.10) is applied to compute the empirical survival Kendall distribution. Again using $\bar{C}(F_{I_k}(I_1), F_k(1))$ as an example, we have $\bar{C}(F_{I_k}(I_1), F_k(1)) \approx 0.957$, which is the largest one among all the pairs, then, $\bar{K}_C(0.957) = \frac{1}{21+1} \approx 0.045$.
4. To compute DSKRP and RPFD, DSKRP = E(INT)/ $[1 - \bar{K}_C(t)]$; RPFD=E(INT)/ [1-F_{I_k}].

13.3.4 Trivariate Hydrological Drought Frequency Analysis

Marginal Distribution of Maximum Drought Intensity

In the previous section, we studied bivariate hydrological drought frequency analysis with the use of drought severity and drought duration as an example. In this section, we will

Table 13.11. *Results for estimating DSKRP and RPFD.*

Date	I_k (cfs)	F_{I_k}	F_k	Rank (I_k, k)	$C(F_{I_k}, F_k)$	$\hat{C}(F_{I_k}, F_k)$	$\bar{K}_C(t)$	DSKRP (yrs)	RPFD (yrs)
7-Sep-16	259.38	0.045	0.045	1	0.048	0.957	0.045	0.757	0.757
8-Sep-16	421.13	0.091	0.091	2	0.095	0.913	0.091	0.795	0.795
9-Sep-16	502.28	0.136	0.136	3	0.143	0.870	0.136	0.837	0.837
10-Sep-16	561.55	0.273	0.182	4	0.190	0.736	0.273	0.994	0.994
11-Sep-16	553.60	0.227	0.227	4	0.190	0.736	0.273	0.994	0.936
12-Sep-16	544.24	0.182	0.273	4	0.190	0.736	0.273	0.994	0.884
13-Sep-16	585.78	0.318	0.318	7	0.333	0.697	0.318	1.060	1.060
14-Sep-16	635.49	0.364	0.364	8	0.381	0.654	0.364	1.136	1.136
15-Sep-16	669.12	0.409	0.409	9	0.429	0.610	0.409	1.224	1.224
16-Sep-16	702.43	0.545	0.455	10	0.476	0.476	0.455	1.326	1.591
17-Sep-16	722.83	0.682	0.500	11	0.524	0.342	0.500	1.446	2.272
18-Sep-16	726.36	0.727	0.545	12	0.571	0.299	0.727	2.651	2.651
19-Sep-16	714.73	0.591	0.591	11	0.524	0.342	0.545	1.591	1.767
20-Sep-16	694.93	0.500	0.636	10	0.476	0.340	0.636	1.988	1.446
21-Sep-16	693.78	0.455	0.682	10	0.476	0.340	0.636	1.988	1.326
22-Sep-16	721.46	0.636	0.727	14	0.667	0.303	0.682	2.272	1.988
23-Sep-16	741.86	0.773	0.773	17	0.810	0.264	0.773	3.181	3.181
24-Sep-16	780.15	0.864	0.818	18	0.857	0.175	0.818	3.977	5.302
25-Sep-16	799.46	0.955	0.864	19	0.905	0.087	0.909	7.953	15.907
26-Sep-16	794.61	0.909	0.909	19	0.905	0.087	0.864	5.302	7.953
27-Sep-16	768.38	0.818	0.955	18	0.857	0.084	0.955	15.907	3.977

Table 13.12. *Kendall's correlation coefficient for drought severity, duration, and MDI.*

Variables	Severity	Duration	MDI
Severity	1.00	0.83	0.71
Duration	0.83	1	0.58
MDI	0.71	0.58	1.00

Table 13.13. *Results of copula candidates for drought severity and MDI.*

	GH	Clayton	Frank	Gaussian	Student T[a]
Parameter	2.77	2.70	**11.47**	0.87	[0.88, 3.07E + 06]
MLE	70.05	63.53	**83.54**	79.99	80.17
SnB test statistics	0.10	0.08	**0.03**	0.05	0.04
P-value of SnB	0.50	0.68	**> 0.99**	0.95	0.98

Note: [a] With high degree of freedom estimated, the Student t copula converges to the Gaussian copula.

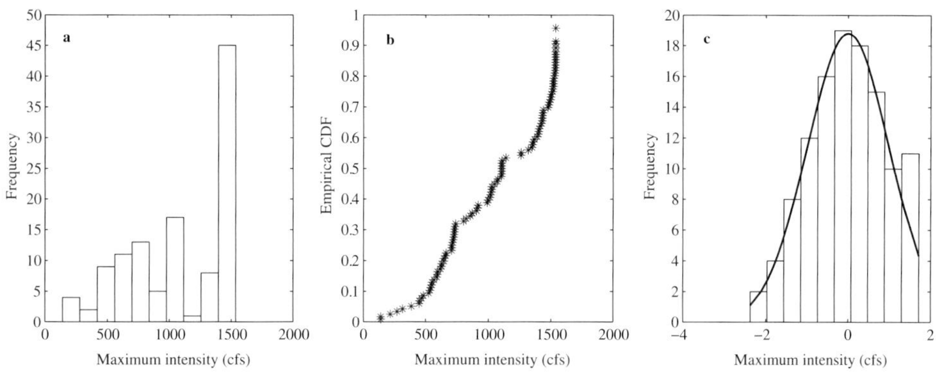

Figure 13.10 Plots to study the maximum drought intensity (MDI): (a) histogram of MDI; (b) empirical distribution of DMI; (c) histogram and fitted N(0,1) for transformed MDI.

study the trivariate drought frequency analysis by applying both vine and meta-elliptical copulas. The variables considered are drought severity (S), drought duration (D), and maximum drought intensity (MDI, i.e., the maximum flow deficit of a drought episode). With S and D fitted by log-normal and Weibull distributions, here we only need to investigate MDI. The histogram of MDI in Figure 13.10(a) clearly shows that its density function is skewed to the left with a long left tail. Thus, to reduce the complexity of fitting univariate distribution, the meta-Gaussian transformation is applied, which is the same as the preparation of marginals for the meta-elliptic copula approach as follows: Variable (X)→empirical distribution

(F_n, e.g., Weibull plotting-positon formula)→$X^T \sim \Phi^{-1}(F_n)$. The empirical distribution and density after transformation are shown in Figure 13.10(b)–(c).

Vine-Copula Approach to Model Trivariate Drought Variables

Table 13.12 lists Kendall's correlation coefficient for S, D, and MDI. As expected, all three drought variables are positively dependent. According to the degree of dependence, we will use severity as the center variable to construct the three-dimensional vine copula using the following structure: $T_1 : D - S - MDI$; $T_2 : D|S - MDI|S$. As introduced in Chapter 5, there are three bivariate copula functions involved in this analysis: (S and D); (S and MDI); and (D|S, MDI|S). All three bivariate copula functions are estimated separately and allowed to be fitted with different copula functions. In the bivariate analysis, we have investigated the Gumbel–Hougaard and Gaussian copulas to S and D. Applying the same copula candidates as the bivariate analysis, the best-fitted copula function will be selected to model the positively dependent S and MDI. Results for the copula candidates are listed in Table 13.13.

As shown in Table 13.13, all copula candidates may be applied to model S and MDI. Among all the copula candidates, the Frank, Gaussian, and Student t copulas yield very similar performance and outperform the Gumbel–Hougaard and Clayton copulas. In addition, the degree of freedom v is estimated as 3.07E + 06; this high degree of freedom suggests the convergence to the Gaussian copula. Based on the MLE and SnB test statistics, the Frank copula (with the largest MLE and smallest SnB statistics) is chosen to model S and MDI. Figure 13.11 compares the pseudo-observations (empirical CDF) with those simulated from the fitted Frank copula. In Figure 13.11, we also provide a plot of comparison with observed variables in the real domain. Comparisons show the appropriateness to apply the fitted copula functions visually.

With Gumbel–Hougaard and Frank copulas chosen for T_1, the copula candidate for T_2 is selected, based on the conditional copulas (i.e., $C^{GH}_{S,D}(F_D|F_S = F_S(s)); C^{Frank}_{MDI,S}(F_{MDI}|F_S = F_S(s))$). From the computed conditional copulas, we compute the Kendall correlation coefficient for T_2 as follows: $\tau_n\left(C^{GH}_{S,D}(F_D|F_S = F_S(s)), C_{MDI,S}(F_{MDI}|F_S = F_S(s))\right) \approx -0.27$. With the negative Kendall's tau computed, only Frank, Gaussian, and Student t copulas are used to evaluate for T_2, with the results listed in Table 13.14. The table reveals the following:

i. All three copulas may properly model the variables for T_2.
ii. The Student t copula again converges to the Gaussian copula.
iii. SnB test statistics increase significantly, compared to those for T_1.
iv. With the increase of SnB statistics, the corresponding *P*-value decreases.
v. Based on the *P*-value as well as the computed MLE, the Gaussian copula is the fitted copula for T_2.

To this end, we have completed the construction of the vine copula for the drought variables as follows:

$$\mathrm{T}_1\colon \{(\mathrm{D,S}) : \mathrm{GH}(6.2); (\mathrm{S,MDI})\colon \mathrm{Frank}(11.47)\} \text{ and } \mathrm{T}_2\colon \{(\mathrm{D|S,MDI|S})\colon \mathrm{Gaussian}\colon \rho = -0.418\}.$$

Table 13.14. *Results of copula candidates for T_2.*

	Frank	Gaussian	Student t
Parameter	–2.389	**–0.418**	[–0.427, 2.67E+06]
MLE	8.427	**10.970**	10.978
SnB test statistics	0.195	**0.188**	0.193
P-value of SnB	0.148	**0.168**	0.167

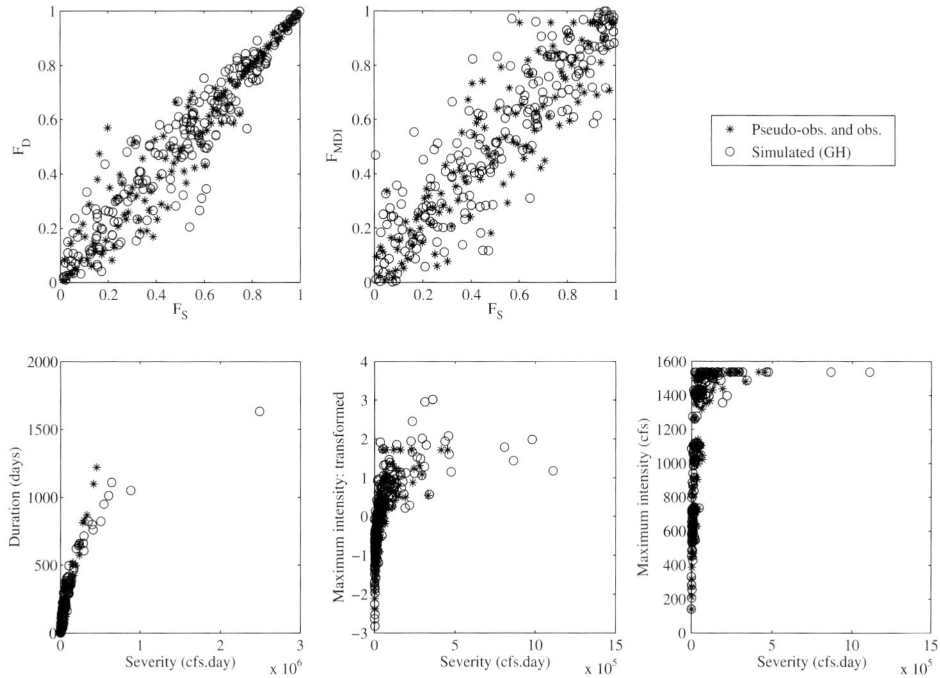

Figure 13.11 Comparison of pseudo-observations and real observations with those simulated from the Gumbel–Hougaard (S and D) and Frank (S and MDI) copulas for T1.

The joint distribution may then be computed using Equation (5.60). Figure 13.12 compares pseudo-observations (through parametric conditional copula) with simulations from the Gaussian copula of T_2. Figure 13.12 also compares the empirical copula and the parametric joint distribution from the vine copula. It is shown that the vine copula fits the trivariate drought variable reasonably well.

Simulation from the Fitted Vine Copula Following the simulation algorithms (Aas et al., 2009), Figure 13.13 shows the comparison of observations with drought variables

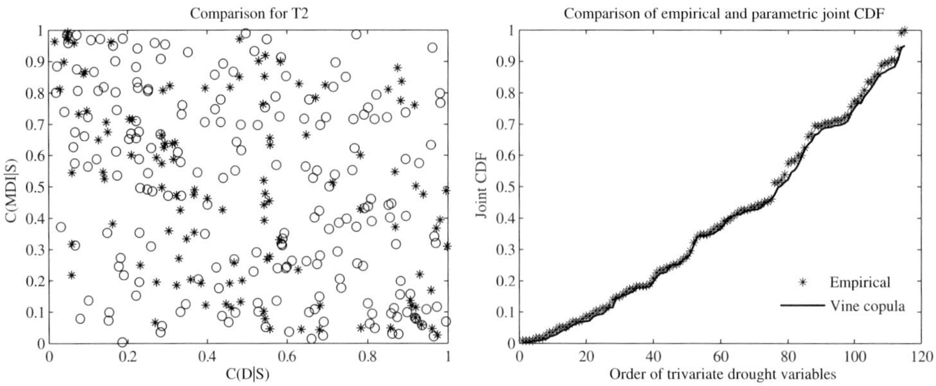

Figure 13.12 Comparison plots for T_2 and joint CDF.

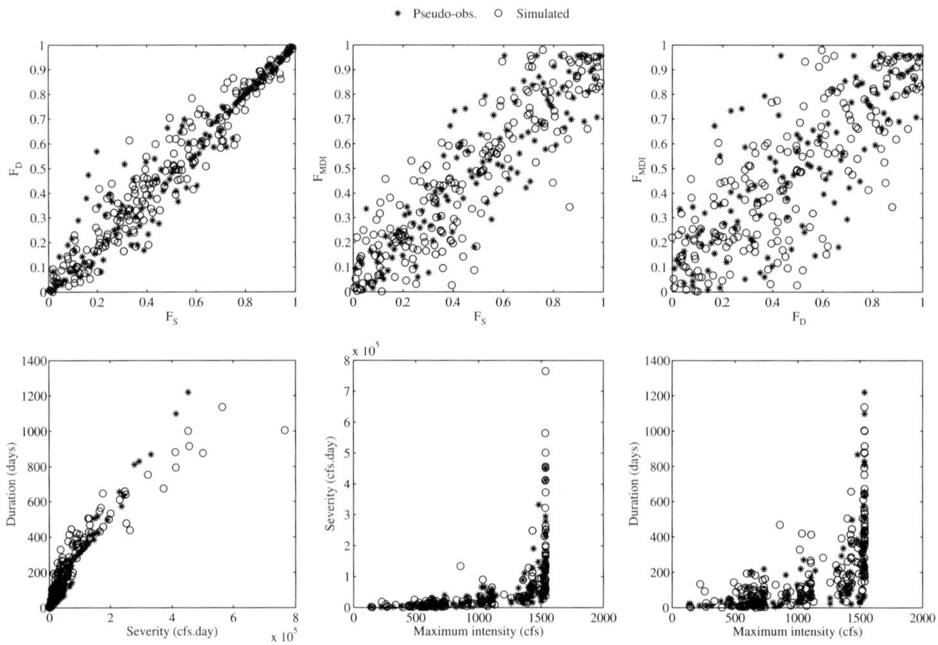

Figure 13.13 Comparison of observed drought variables with simulations from the fitted vine copula.

simulated from the fitted vine copula. Here we will again illustrate how to simulate the random variable from the vine copula with the fitted GH–Frank–Gaussian copulas.

1. Generate three independent, uniformly distributed random variables: $w = [0.7372, 0.7869, 0.6537]$, where

$$w(1) = U(1); w(2) = C_{12}(U(2)|U(1)); w(3) = C_{3|12}(U(3)|U(1), U(2)).$$

In this example, $U(1) = F_D(d); U(2) = F_S(s); U(3) = F_{MDI}(mdi)$.

2. Compute $U(2)$, i.e., $F_D(d)$:

   From step 1, we have $U(1) = F_D(d) = w(1) = 0.7372$ and $C_{12}(U(2)|U(1)) = C_{SD}^{GH}(F_S|F_D(d) = 0.7372) = w(2) = 0.7869$.

   With the fitted Gumbel–Hougaard copula ($\theta = 6.2$) for drought severity and duration, we will need to compute U(2) (i.e., $F_S(s)$) using the following:

$$C_{SD}^{GH}(F_S|F_D(d) = 0.7372) = \left.\frac{\partial C_{SD}^{GH}(F_D, F_S; 6.2)}{\partial F_D}\right|_{F_D=0.7372}$$

   To be consistent with the discussion in Chapter 5, we assign the conditional copula as the h function: $C_{SD}^{GH}(F_S|F_D(d) = 0.7372) = h_{GH}(F_S, F_D; 6.2) = 0.7869$. The conditional copula for the Gumbel–Hougaard copula is listed as No. 4 in Table 4.2. As seen in Table 4.2, U(2) needs to be solved for numerically using the root-finding (e.g., bisection method) technique. Using the bisection method, we have the following:

$$U(2) = F_S(s) = h^{-1}(0.7869, 0.7372; 6.2) = 0.777.$$

3. Compute $U(3)$, i.e., F_{MDI}:

   From the vine structure, we have the following:

$$w_3 = C_{3|12}(U(3), |U(1), U(2)) = \frac{\partial C_{13|2}(C_{23}(U(3)|U(2)), C_{12}\ (U(1)|U(2)))}{\partial C_{12}(U(1)|U(2))}$$

$$= \frac{\partial_{D,MDI|S}^{Gaussian}\left(C_{MDI,S}^{Frank}(F_{MDI}|F_S(s); 11.47), C_{S,D}^{GH}(F_D|F_S(s); 6.2); -0.418\right)}{\partial C_{S,D}^{GH}(F_D|F_S(s); 6.2)}$$

   and $U(3)$ can then be computed with the following three steps:

   a. Compute the conditional copula $C_{S,D}^{GH}(F_D|F_S)$. From the first two steps, we have $F_D = 0.7372, F_S = 0.777$; $C_{S,D}^{GH}(F_S|F_D; 6.2)$ can then be computed by substituting $F_D = 0.7372, F_S = 0.777, \theta = 6.2$ into No. 4 conditional copula in Table 4.2. We obtain the following:

$$C_{S,D}^{GH}(F_D|F_S; 6.2) = C(0.7372|0.777; 6.2) = 0.2791$$

   b. Compute the conditional copula of $C_{MDI,S}^{Frank}(F_{MDI}|F_S(s); 11.47)$ with the use of the meta-Gaussian copula fitted to T_2 by setting the h function as follows:

$$w(3) = h_{gaussian}\left(C_{MDI,S}^{Frank}(F_{MDI}|F_S; 11.47), C_{S,D}^{GH}(F_D|F_S; 6.2); -0.418\right) = 0.6537$$

   and we have the following: $C_{MDI,S}^{Frank}(F_{MDI}|F_S; 11.47) = h_{gaussian}^{-1}(0.6537, 0.2791; -0.418)$

   For the Gaussian copula, its conditional copula is the univariate normal distribution. The derivation of the conditional copula is given as Equation (7.42). In this particular problem, Equation (7.42a) can be rewritten as follows:

$$h_{gaussian}\left(C_{MDI,S}^{Frank}(F_{MDI}|F_S; 11.47), C_{S,D}^{GH}(F_D|F_S; 6.2); -0.418\right)$$
$$\sim \Phi\left(\frac{\Phi^{-1}\left(C_{MDI,S}^{Frank}(F_{MDI}|F_S; 11.47)\right) - \rho\Phi^{-1}\left(C_{S,D}^{GH}(F_D|F_S; 6.2)\right)}{(1-\rho^2)^{0.5}}\right)$$

Let $h_{MDI,S} = C^{Frank}_{MDI,S}(F_{MDI}|F_S; 11.47)$ we have:

$$\Phi^{-1}(h_{MDI,S}) = \Phi^{-1}(w(3))\left(1-\rho^2\right)^{0.5} + \rho\Phi^{-1}\left(C^{GH}_{S,D}(F_D|F_S(s); 6.2)\right.$$
$$= \left(1-(-0.418)^2\right)^{0.5}\Phi^{-1}(0.6537) - 0.418\Phi^{-1}(0.2791) = 0.571$$

and $h_{MDI,S} = C^{Frank}_{MDI,S}(F_{MDI}|F_S; 11.47) = \Phi(0.571) = 0.716$.

c. Compute F_{MDI} from $C^{Frank}_{MDI,S}(F_{MDI}|F_S; 11.47)$.

In steps b, we have computed $C^{Frank}_{MDI,S}(F_{MDI}|F_S; 11.47) = 0.716$. Using the conditional Frank copula in Table 4.2, we have the following:

$$U(3) = F_{MDI}(mdi) = h^{-1}(0.716, 0.777; 11.47) = 0.8421.$$

To this end, we have finished one simulation:

$$[F_D, F_S, F_{MDI}] = [0.7372, 0.777, 0.8421].$$

To repeat the preceding procedure, we can simulate the random variables of size *N*.

By applying the fitted marginal distributions we obtain the simulation in the real domain as follows:

$$S_{simu} = lognormal^{-1}\left(0.777, 10.1992, 1.4614^2\right) = 8.19 \times 10^4 cfs.day$$

$$D_{simu} = weibull^{-1}(0.7372; 203.8, 0.89) = 282.26\ day$$

$$MDI^{tranformed}_{simu} = \Phi^{-1}(0.8421) = 1.0031$$

Applying the linear interpolation of F_{MDI} to [MDI, empirical distribution of MDI], we have the following:

$MDI_{simu} = 1534.9\ cfs$. Finally, we have the corresponding simulation in the real domain as follows:

$$[D_{simu}, S_{simu}, MDI_{simu}] = \left[282.26\ day, 8.19 \times 10^4 cfs.day, 1534.9 cfs\right].$$

Figure 13.14 compares the sample Kendall's tau of the observed trivariate drought variables with those computed using the simulated trivariate variables from the vine

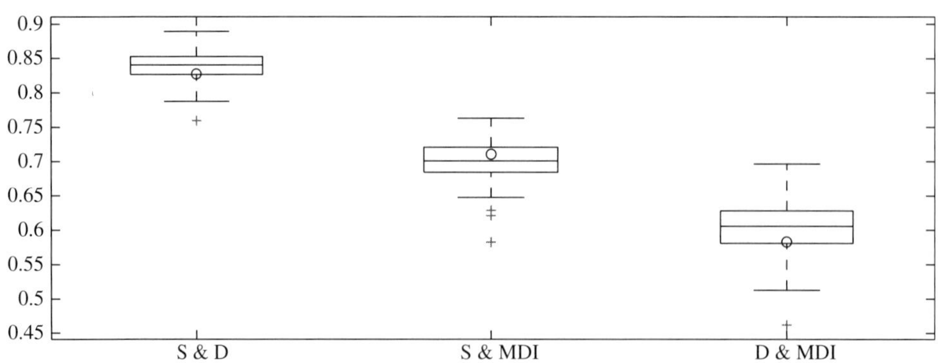

Figure 13.14 Comparison of sample Kendall's tau and those simulated from the fitted vine copula.

copula. Comparison shows that the dependence structure of the drought variables is well preserved. The fitted vine copula may be applied further for risk analysis.

Joint and Conditional Return Period through Vine Copula In this section, we will proceed with risk analysis through the joint and conditional return period. In the case of joint return period, we will only investigate the "AND" case.

Joint Return Period "AND" Case Similar to the bivariate case, the trivariate return period of the "AND" case can be given as follows:

$$T(S > s \cap D > d \cap MDI > mdi) = \frac{E(INT)}{P(S > s \cap D > d \cap MDI > mdi)} = \frac{E(INT)}{C(F_S > F_S(s) \cap F_D > F_D(d) \cap F_{MDI} > F_{MDI}(mdi))} \tag{13.12}$$

where

$$\begin{aligned} &C(F_S > F_S(s) \cap F_D > F_D(d) \cap F_{MDI} > F_{MDI}(mdi)) \\ &= 1 - F_S(s) - F_D(d) - F_{MDI}(mdi) + C^{GH}_{S,D}(F_S(s), F_D(d)) + C^{Frank}_{S,MDI}(F_S(s), F_{MDI}(mdi) \\ &\quad + C_{D,MDI}(F_D(d), F_{MDI}(mdi)) - C^{GH-Frank-Gaussian}_{S,D,MDI}(F_S(s), F_D(d), F_{MDI}(mdi)) \end{aligned} \tag{13.12a}$$

To compute using Equation (13.12a), we will need to first evaluate $C_{D,MDI}(F_D, F_{MDI})$. $C_{D,MDI}(F_D, F_{MDI})$ is the 2-margin of the trivariate copula as follows:

$$C_{D,MDI}(F_D, F_{MDI}) = C_{S,D,MDI}(1, F_D, F_{MDI}) = \int_0^1 C_{S,D,MDI}(F_D, F_{MDI}|t)f(t)dt \tag{13.13}$$

where $f(t) = 1$ for the uniformly distributed random variable on [0, 1].

Figure 13.15 compares the empirical copula and that computed using Equation (13.13). In Figure 13.15, we also plot the joint CDF of D and MDI estimated separately using the Gumbel–Hougaard copula ($\theta = 2.22$). The goodness-of-fit study ensures the appropriateness with $SnB = 0.05; P_{value} = 0.91$. From Figure 13.15, we see that there is minimal difference between the computed joint CDF (i.e., for D and DMI) using Equation (13.13) and that computed directly from the Gumbel–Hougaard copula (the maximum absolute difference is 0.028). To reduce the complexity of computation, we will directly apply the Gumbel–Hougaard copula to D and MDI. Setting $MDI \geq 1526.1$ *cfs* (i.e., equivalent to $F_{MDI}(mdi \leq 1526.1) = 0.8$), Figures 13.16 and 13.17 plot the joint exceedance probability (i.e., equivalent to survival copula) and the corresponding joint return period. Table 13.15 lists the sample results for computing the exceedance joint probability and its joint return period. As shown in Figures 13.16 and 13.17, the exceedance probability shows the concave shape and the joint return period shows the convex shape with $MDI \geq 1.53E+03$ *cfs* as the example graphed. It shows the low exceedance probability and high return period for a long duration with low drought severity, as well as for the short duration with high drought severity. The moderate drought duration and drought severity yield a shorter

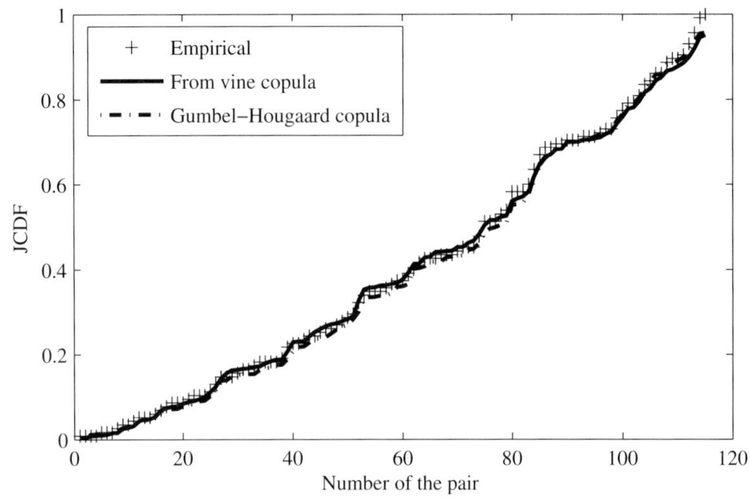

Figure 13.15 Comparison of joint CDF computed for D and MDI.

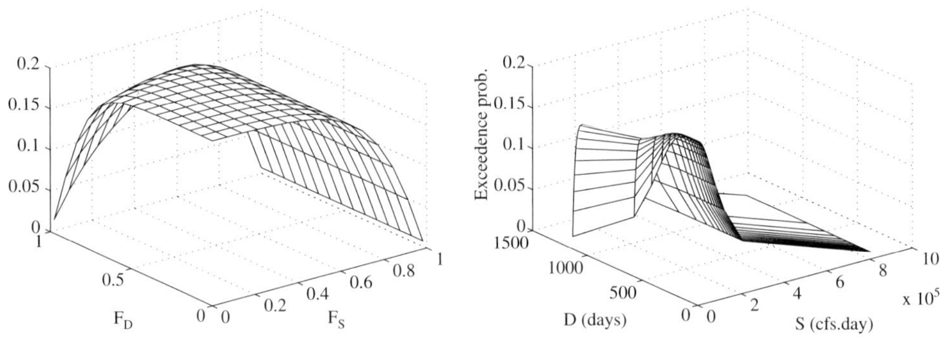

Figure 13.16 Exceedance probability for $\boldsymbol{C(D \geq d, S \geq s, MDI \geq 1.53E + 03cfs)}$.

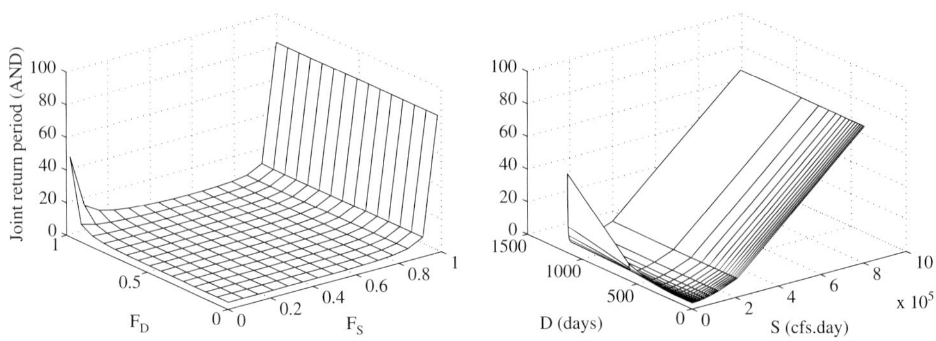

Figure 13.17 Joint return period "AND" case for $\boldsymbol{D \geq d, S \geq s, MDI \geq 1.53E + 03}$.

Table 13.15. *Exceedance joint CDF and joint return period ("AND" case) with* $MDI \geq 1.53E+03$ *cfs.*

			F_D				
Frequency domain			0.2	0.5	0.9	0.95	0.99
Exceed. joint CDF	F_S	0.2	0.199	0.192	0.078	0.042	0.009
		0.5	0.199	0.192	0.078	0.042	0.009
		0.9	0.136	0.170	0.078	0.042	0.009
		0.95	0.109	0.156	0.077	0.041	0.009
		0.99	0.084	0.142	0.074	0.040	0.009
Real domain			D (days)[a]				
			38	135	520	699	1133
Joint return period (yrs)	S	7860	3.626	3.774	9.281	17.402	83.468
	(cfs.day)	26880	3.633	3.774	9.281	17.402	83.468
		174920	5.298	4.265	9.292	17.402	83.468
		297460	6.620	4.634	9.424	17.424	83.468
		805320	8.595	5.078	9.727	17.875	83.546

[a] The duration is rounded to the nearest integer number.

return period than does high drought severity with a short duration. We should note that with the change of plotted examples for *MDI* (or *D* or *S*), the shape may change accordingly.

Conditional Return Period with the Constructed Vine Copula Here we will investigate the following cases for the conditional return period as examples:

(i) $D > d \cap MDI > mdi \,|\, S \leq s$
(ii) $D > d \cap MDI > mdi \,|\, S = s$
(iii) $D > d \cup MDI > mdi \,|\, S \leq s$
(iv) $D > d \cup MDI > mdi \,|\, S = s$
(iv) $D > d \,|\, MDI \leq mdi \cap S \leq s$
(vi) $D > d \,|\, MDI = mdi \cap S = s$

Cases (i) and (ii): $\boldsymbol{D > d \cap MDI > mdi \,|\, S \leq S}$ **and** $\boldsymbol{D > d \cap MDI > mdi \,|\, S = s}$ Cases (i) and (ii) investigate the impact of drought severity on drought duration and MDI during the drought episode under the condition of both D and MDI exceeding the corresponding critical level.

The conditional exceedance probability $P(D > d \cap MDI > mdi|S \leq s)$ for case (i) may be written as follows:

$$\begin{aligned} &P(D > d \cap MDI > mdi|S \leq s) \\ &\quad = 1 - P(D \leq d|S \leq s) - P(MDI \leq mdi|S \leq s) + P(D \leq d, MDI \leq mdi|S \leq s) \\ &\quad = 1 - \frac{C_{DS}(F_D, F_S)}{F_S} - \frac{C_{MDI,S}(F_{MDI}, F_S)}{F_S} + \frac{C_{D,MDI,S}(F_D, F_{MDI}, F_S)}{F_S} \end{aligned} \tag{13.14}$$

and the corresponding conditional return period can simply be given as follows:

$$T(D > d \cap MDI > mdi|S \leq s) = \frac{E(INT)}{P(D > d \cap MDI > mdi|S \leq s)} \tag{13.14a}$$

The conditional exceedance probability $P(D > d \cap MDI > mdi|S = s)$ may be written as follows:

$$\begin{aligned} &P(D > d \cap MDI > mdi|S = s) \\ &\quad = 1 - P(D \leq d|S = s) - P(MDI \leq mdi|S = s) + P(D \leq d, MDI \leq mdi|S = s) \end{aligned} \tag{13.15}$$

where

$$P(D \leq d|S = s) = \left.\frac{\partial C_{DS}(F_D, F_S)}{\partial F_S}\right|_{F_S = F_S(s)}; \; P(MDI \leq mdi|S = s) = \left.\frac{\partial C_{MDI,S}(F_{MDI}, F_S)}{\partial F_S}\right|_{F_S = F_S(s)} \tag{13.15a}$$

$$P(D \leq d, MDI \leq mdi|S = s) = C(F_D, F_{MDI}|F_S = F_S(s)) = \left.\frac{\partial C(F_D, F_{MDI}, F_S)}{\partial F_S}\right|_{F_S = F_S(s)} \tag{13.15b}$$

Applying $F_S = [0.2, 0.5, 0.9, 0.95, 0.99]$ for the conditioning severity, we have the drought severity estimated as $S \approx [7860, 26880, 174920, 297460, 805320]$cfs.day.

Table 13.16 lists and Figures 13.18 and 13.19 plot the conditional return period for cases (i) and (ii) using S = 174920 cfs.day as an example.

Cases (iii) and (iv): $\boldsymbol{D > d \cup MDI > mdi\,|S \leq s; D > d \cup MDI > mdi\,|S = S}$ Cases (iii) and (iv) again investigate the impact of drought severity on drought duration and MDI but under different conditions, i.e., at least one drought variable (D or MDI) exceeding the corresponding critical level.

The conditional exceedance probability $P(D > d \cup MDI > mdi|S > s)$ of case (iii) may be rewritten with the following set of equations as follows:

$$\begin{aligned} &P(D > d \cup MDI > mdi|S \leq s) \\ &\quad = 1 - P(D \leq d \cap MDI \leq mdi|S \leq s) = 1 - \frac{C(F_D(d), F_{MDI}(mdi), F_S(s))}{F_S(s)} \end{aligned} \tag{13.16}$$

Table 13.16. *Conditional exceedance probability and conditional return period: cases (i) and (ii).*

			Case (i)			Case (ii)		
			$MDI > mdi$ (cfs)			$MDI > mdi$ (cfs)		
			646	1106	1537	646	1106	1537
Conditional exceedance prob.	$D >$ d (day)	38	0.573	0.295	0.005	1.00	1.00	0.45
		135	0.295	0.131	0.001	0.99	0.99	0.44
		520	0.023	0.007	1.84E−05	0.41	0.41	0.12
		699	0.008	0.002	4.27E−06	0.20	0.20	0.043
		1133	0.001	2.38E−04	2.46E−07	0.037	0.037	0.005
Conditional return period	$D >$ d (day)	38	1.26	2.45	160.04	0.72	0.72	1.62
		135	2.45	5.52	636.34	0.73	0.73	1.65
		520	31.24	103.24	3.94E+04	1.78	1.78	6.22
		699	85.11	316.69	1.69E+05	3.66	3.66	16.67
		1133	658.89	3038.20	2.93E+06	19.45	19.46	151.40

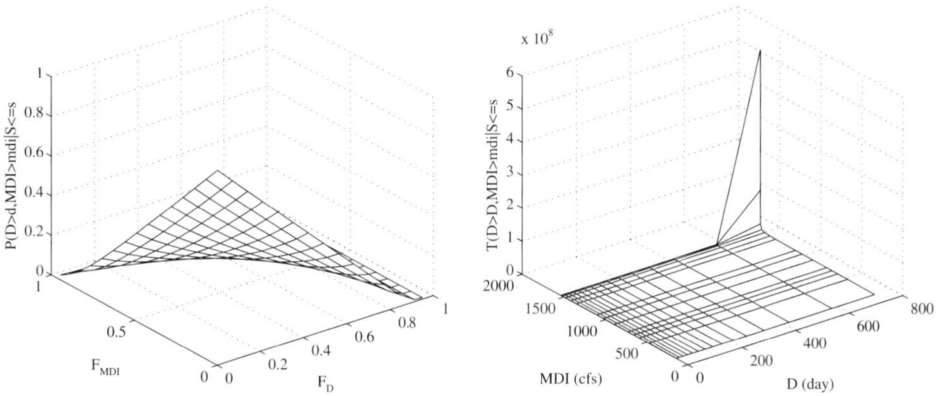

Figure 13.18 Conditional exceedance probability and conditional return period: case (i).

The corresponding joint return period is then written as follows:

$$T(D > d \cup MDI > mdi|S \leq s) = \frac{E(INT)}{1 - P(D > d \cup MDI > mdi|S \leq s)} \tag{13.16a}$$

The conditional exceedance probability $P(D > d \cup MDI > mdi|S = s)$ of case (iv) can be written as follows:

$$\begin{aligned} &P(D > d \cup MDI > mdi|S = s) \\ &\quad = 1 - P(D \leq d \cap MDI \leq mdi|S = s) = 1 - C(F_D(d), F_{MDI}(mdi)|F_S(s)) \end{aligned} \tag{13.17}$$

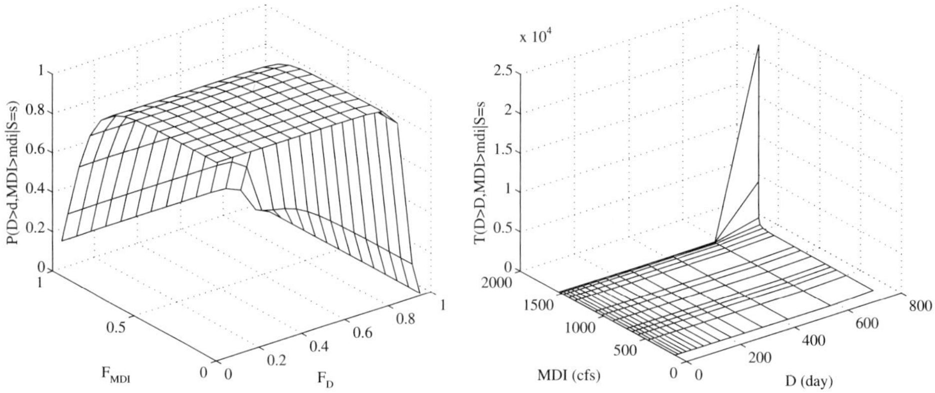

Figure 13.19 Conditional exceedance probability and conditional return period: case (ii).

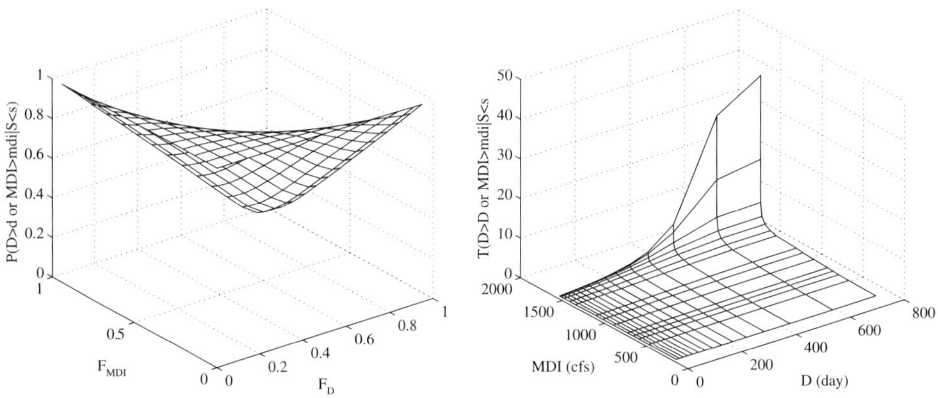

Figure 13.20 Conditional exceedance probability and the corresponding conditional joint return period for case (iii).

Here it is obvious that $C_{D,MDI|S}$ in Equation (13.17) is nothing but the copula function at T_2 of the vine structure. The corresponding conditional return period can be written as follows:

$$T(D > d \cup MDI > mdi|S = s) = \frac{E(INT)}{1 - C(F_D(d), F_{MDI}(mdi)|F_S(s))} \tag{13.17a}$$

Figures 13.20 and 13.21 plot the conditional exceedance probability and the conditional return period for cases (iii) and (iv) using S = 174920 cfs.day as an illustrative sample. Table 13.17 lists the sample results.

Cases (v) and (vi): $\boldsymbol{D > d|MDI \leq mdi \cap S \leq s}$ **and** $\boldsymbol{D > d|MDI = mdi \cap S = s}$ Cases (v) and (vi) investigate the combined impact of maximum drought intensity and severity on drought duration.

Table 13.17. *Conditional exceedance probability and conditional return period: cases (iii) and (iv).*

			Case (iii)			Case (iv)		
			$MDI > mdi$ (cfs)			$MDI > mdi$ (cfs)		
			646	1106	1537	646	1106	1537
Conditional exceedance prob.	$D > d$ (day)	38	0.98	0.93	0.79	1.00	1.00	1.00
		135	0.93	0.76	0.46	1.00	1.00	1.00
		520	0.81	0.49	0.06	1.00	1.00	0.74
		699	0.79	0.46	0.03	1.00	1.00	0.60
		1,133	0.78	0.45	0.02	1.00	1.00	0.48
Conditional return period	$D > d$ (day)	38	0.74	0.78	0.92	0.72	0.72	0.72
		135	0.78	0.95	1.58	0.72	0.72	0.72
		520	0.90	1.48	11.52	0.72	0.72	0.98
		699	0.91	1.56	21.46	0.72	0.72	1.20
		1,133	0.93	1.61	45.04	0.72	0.72	1.51

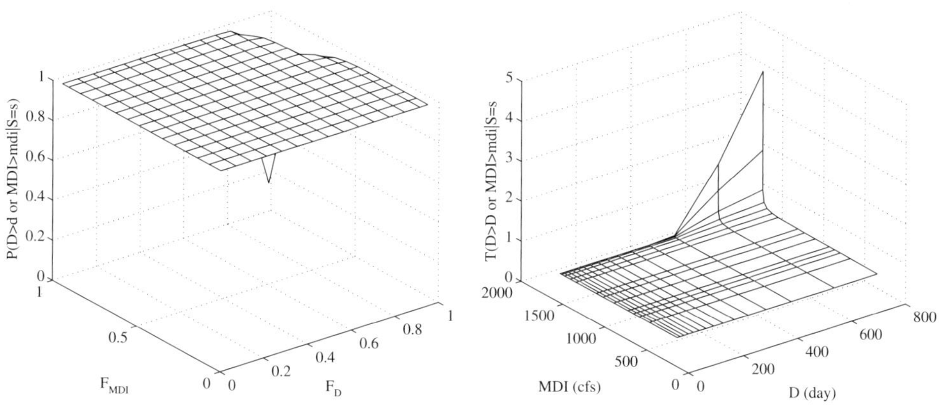

Figure 13.21 Conditional exceedance probability and corresponding conditional joint return period for case (iv).

For case (v), its conditional exceedance probability may be written as follows:

$$P(D > d|MDI \leq mdi \cap S \leq s) = \frac{P(MDI \leq mdi, S \leq s) - P(D \leq d, MDI \leq mdi, S \leq s)}{P(MDI \leq mdi, S \leq s)}$$

$$= \frac{C^{FRANK}_{MDI,S}(F_{MDI}, F_S) - C(F_D, F_S, F_{MDI})}{C^{FRANK}_{MDI,S}(F_{MDI}, F_S)} \tag{13.18}$$

and the corresponding conditional joint return period can be written as follows:

$$T(D > d|MDI \leq mdi \cap S \leq s) = \frac{EINT}{P(D > d|MDI \leq mdi \cap S \leq s)} \tag{13.18a}$$

For case (VI), its conditional exceedance probability can be written as follows:

$$\begin{aligned} P(D > d|MDI = mdi \cap S = s) &= 1 - P(D \leq d|MDI = mdi \cap S = s) \\ &= 1 - \frac{\partial C_{MDI,D|S}(C_{MDI,S}(F_{MDI}|F_S(s)), C_{D,S}(F_D(d)|F_S(s)))}{\partial(C_{MDI,S}(F_{MDI}(mdi)|F_S(s))} \end{aligned} \tag{13.19}$$

The corresponding conditional return period may be estimated through the vine copula as follows:

$$T(D > d|MDI = mdi \cap S = s) = \frac{EINT}{P(D > d|MDI = mdi \cap S = s)} \tag{13.19a}$$

The copula functions in Equation (13.19) directly reflect the constructed vine copula, i.e., $C_{MDI,D|S}$ is Gaussian copula of T_2, and $C_{D,S}$ and $C_{MDI,S}$ are the Gumbel–Hougaard and Frank copula of T_1.

Table 13.18 lists the sample results using [S = 174920 cfs.day, MDI = 1526 cfs], [S = 25000 cfs.day, MDI = 800 cfs], and [S = 10000 cfs.day, MDI = 1000 cfs] as the illustration samples. Figures 13.22 and 13.23 plot the conditional exceedance probability and the corresponding conditional return period for cases (v) and (vi).

Elliptical-Copula Approach to Model Trivariate Drought Variables

In this section, we apply the elliptical copula to model the trivariate drought variables as well as evaluate the corresponding risk through return period. Not to complicate the process, the Gaussian and Student t copulas are adopted for analysis. Applying MLE, Table 13.19 lists the parameters estimated from both the copulas with the use of empirical marginals. From this point on, we will use the Student t copula as an example to illustrate the application. Figures 13.24 and 13.25 compare the simulated variables with the observed variables in both frequency and real domains. Figure 13.24 indicates a good comparison between pseudo-observations and simulations. In Figure 13.25 (real domain comparison), the comparison does not look as good as that in Figure 13.24 due to the ties in drought variables: D and DMI.

Joint and Conditional Return Period from the Student T Copula With the chosen Student t copula, we will again evaluate the joint return period ("AND") and all six cases of the conditional return period that are applied to the vine copula.

Table 13.18. *Sample results for cases (v) and (vi).*

			Case (v)			Case (vi)		
			$S \leq s$ (*cfs.day*); $MDI \leq mdi$ (cfs)			$S = s; MDI = mdi$		
			S = 174920; MDI = 1526	*S = 25000; MDI = 800*	*S = 10000; MDI = 1000*	*S = 174920; MDI = 1526*	*S = 25000; MDI = 800*	*S = 10000; MDI = 1000*
Conditional exceedance prob.	D > d (day)	38	0.81	0.68	0.28	1.00	1.00	0.60
		135	0.48	0.09	3.16E-03	1.00	0.58	0.002
		520	0.01	1.07E-06	2.68E-08	0.56	6.99E-06	3.80E-10
		699	2.24E-04	1.24E-08	3.09E-10	0.01	5.14E-08	1.18E-12
		1133	9.16E-09	5.04E-13	1.26E-14	1.40E-07	6.09E-13	0
Conditional return period	D > d (day)	38	0.89	1.06	2.59	0.72	0.72	1.21
		135	1.50	8.19	229.10	0.72	1.24	369.24
		520	51.20	6.75E+05	2.69E+07	1.29	1.03E+05	1.90E+09
		699	3234.94	5.85E+07	2.34E+09	63.30	1.41E+07	6.12E+11
		1133	7.89E+07	1.43E+12	5.74E+13	5.16E+06	1.19E+12	Inf

Table 13.19. *Parameters estimated for elliptical copulas.*

	Gaussian			*T*		
	S	*D*	*DMI*	*S*	*D*	*DMI*
S	1	0.95	0.87	1	0.96	0.87
D	0.95	1	0.76	0.96	1	0.78
DMI	0.87	0.76	1	0.87	0.78	1
				$v = 19.14$		

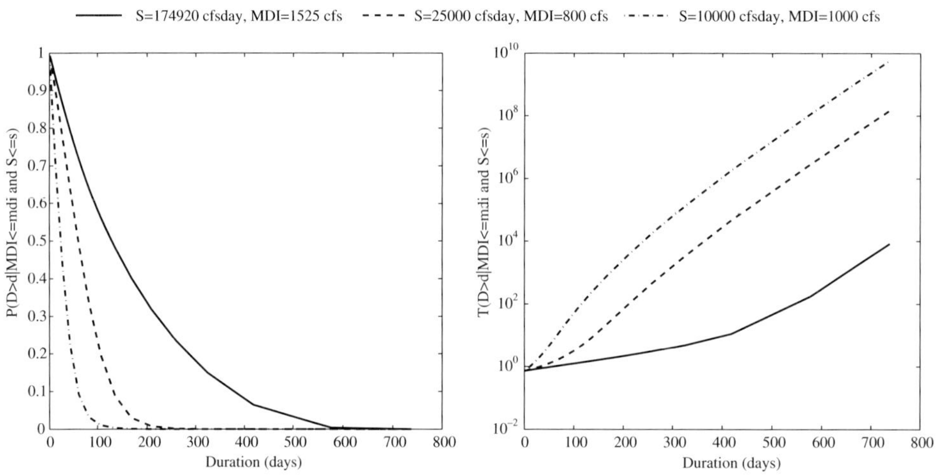

Figure 13.22 Conditional exceedance probability and conditional return period: case (v).

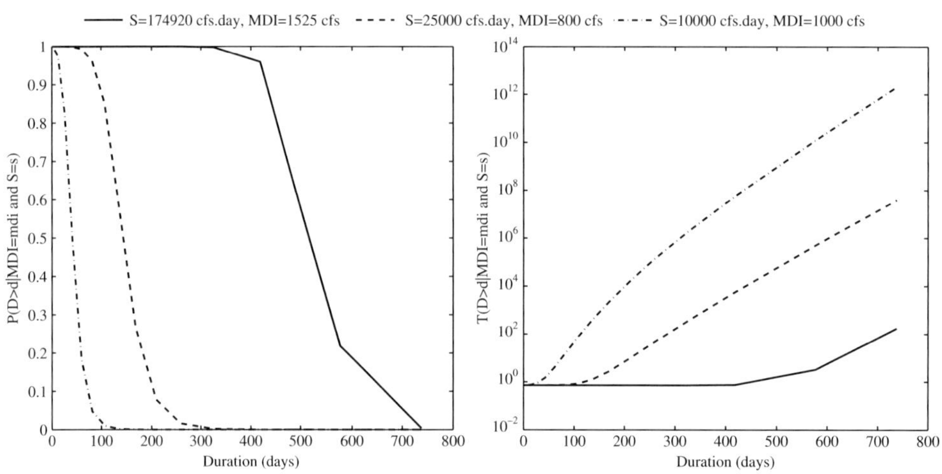

Figure 13.23 Conditional exceedance probability and conditional return period: case (vi).

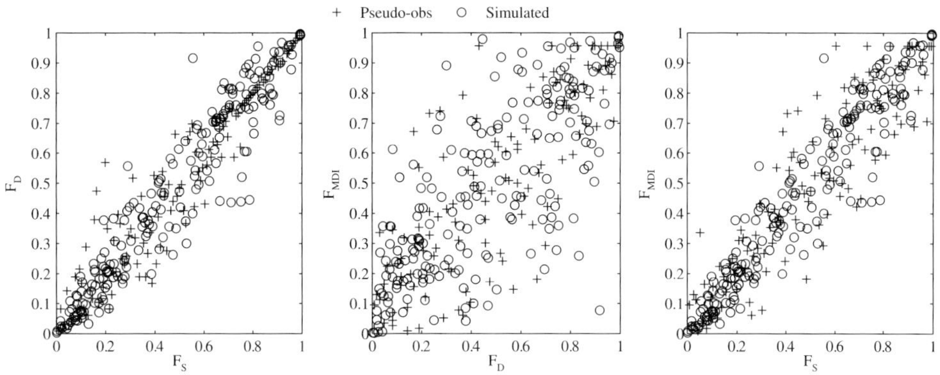

Figure 13.24 Comparison of simulated random variables with the pseudo-observations.

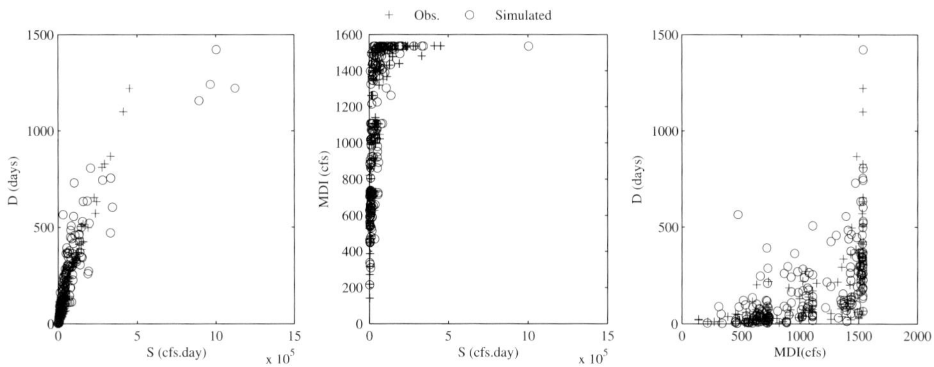

Figure 13.25 Comparison of simulated drought variables with observed drought variables.

Joint Return Period ("AND") Case As shown previously, we will need to apply Equation (13.13), in which the bivariate copula margins are as follows: $C(F_S, F_D) = C(F_S, F_D, 1)$; $C(F_S, F_{MDI}) = C(F_S, 1, F_{MDI})$; $C(F_D, F_{MDI}) = C(1, F_S, F_{MDI})$. In the case of Student t copula, the two margins are also bivariate copulas with the degree of freedom being the same as that of multivariate student t copula. As an example, $C(F_S, F_D)$is a bivariate strudent copula with $\Sigma = \begin{bmatrix} 1 & 0.96 \\ 0.96 & 1 \end{bmatrix}, \nu = 19.14$. Applying Equation (13.13), Figure 13.26 plots the joint exceedance probability and the corresponding joint return period. As before, we set $F_{MDI} = 0.8$ as an illustrative example. Table 13.20 lists the sample results.

Conditional Return Period Estimated Using the Student t Copula Rather than considering all six cases as those for the Vine copula approach, here we will only investigate the following three cases:

1. Case: $D > d \cap MDI > mdi | S \leq s$

Table 13.20. *Sample results of the joint return period computed from the Student t copula.*

			F_D				
Frequency domain			0.2	0.5	0.9	0.95	0.99
Exceed. joint CDF	F_S	0.2	0.199	0.187	0.078	0.043	0.010
		0.5	0.196	0.187	0.078	0.043	0.010
		0.9	0.088	0.088	0.070	0.043	0.010
		0.95	0.048	0.048	0.046	0.037	0.010
		0.99	0.010	0.010	0.010	0.010	0.007
Real domain			D (days)[1]				
			38	135	520	699	1133
Joint return period (yrs)	S (cfs.day)	7860	3.63	3.87	9.33	16.65	75.10
		26880	3.69	3.87	9.33	16.65	75.10
		174920	8.20	8.20	10.30	16.85	75.10
		297460	15.14	15.14	15.78	19.77	75.25
		805320	72.66	72.66	72.68	73.09	101.74

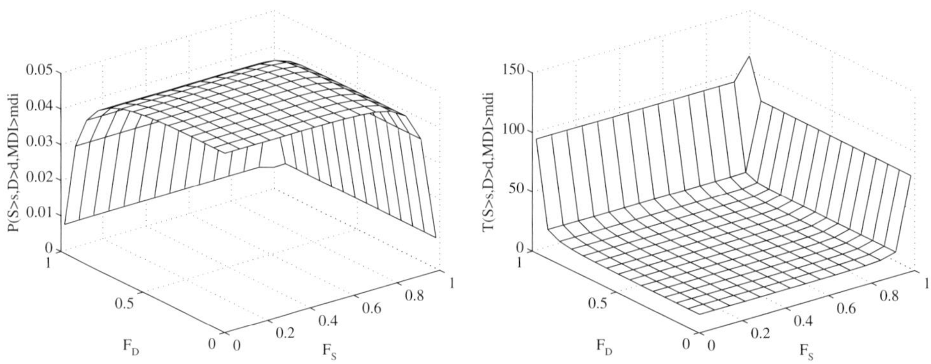

Figure 13.26 Plot of joint exceedance probability and the corresponding return period.

Identical to the vine copula approach discussed earlier, the survival copula and the two margins need to be assessed, so as to evaluate the conditional return period. As shown in the joint return period "AND" case, the two margins of the Student t copula can be easily computed. Applying Equation (13.13) and using S = 174920 cfs.day as an example, Table 13.21 lists the sample results. Figure 13.27 provides the sample plots for conditional exceedance probability and conditional return period.

2. Case: $D > d \cup MDI > mdi | S \leq s$

   Equation (13.15) is applied to compute the conditional exceedance probability and conditional return period. Using $S = 174290$ cfs.day as an example, Table 13.22 lists the sample results. Figure 13.28 provides sample plots.

Table 13.21. *Conditional exceedance probability and conditional return periods:* $D > d \cap MDI > mdi | S \leq s$.

			$MDI > mdi$ (cfs)		
			646	1106	1537
Conditional exceedance prob.	$D > d$ (day)	38	0.696	0.430	0.022
		135	0.430	0.325	0.019
		520	0.038	0.035	0.002
		699	0.010	0.009	0.001
		1,133	3.01E-04	2.69E-04	9.92E-06
Conditional return period	$D > d$ (day)	38	1.04	1.68	32.93
		135	1.68	2.22	37.10
		520	19.12	20.75	301.30
		699	74.15	79.86	1,387.011
		1,133	2.41E+03	2.69E+03	7.29E+04

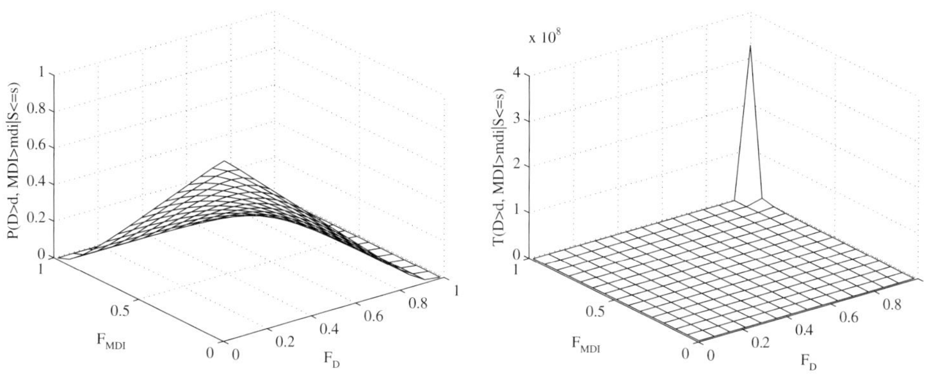

Figure 13.27 Conditional exceedance probability and conditional return period: $\boldsymbol{D > d \cap MDI > mdi | S \leq s}$.

3. Case: $MDI > mdi | D \leq d \cap S \leq s$

   Equation (13.17) (i.e., the general form) is applied to compute the conditional exceedance probability and conditional return period for this case. Table 13.23 lists sample results. Figure 13.29 provides sample plots.

13.3.5 Comparison of Vine Copula and Student T Copula for Trivariate Drought Analysis

In this section, we will further compare the differences yielded from the fitted vine copula and simple trivariate Student t copula by (1) overall performance through the joint CDF; (2) joint return period of the "AND" case; and (iii) six cases of the conditional return period.

Table 13.22. *Conditional exceedance probability and conditional return periods:* $D > d \cup MDI > mdi|S \leq s$.

			$MDI > mdi$ (cfs)		
			646	1106	1537
Conditional exceedance prob.	$D > d$ (day)	38	0.860	0.792	0.778
		135	0.792	0.564	0.448
		520	0.778	0.448	0.058
		699	0.778	0.445	0.031
		1,133	0.778	0.444	0.022
Conditional return period	$D > d$ (day)	38	0.84	0.91	0.93
		135	0.91	1.28	1.62
		520	0.93	1.62	12.52
		699	0.93	1.62	23.02
		1,133	0.93	1.63	32.24

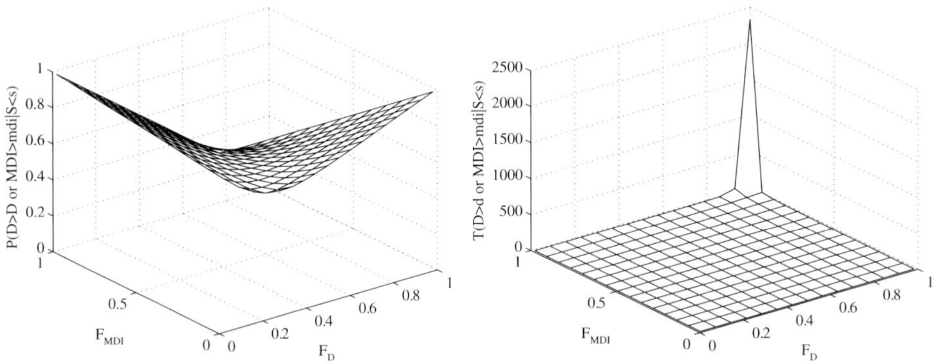

Figure 13.28 Conditional exceedance probability and conditional return period: $\boldsymbol{D > d \cup MDI > mdi|S \leq s}$.

Overall Performance through Joint CDF

To compare the overall difference, Figure 13.30 compares the empirical CDF with the joint CDF computed from the fitted GH–Frank–Gaussian vine and Student t copulas. As shown in Figure 13.30, the overall performance is very similar to the fitted Student t copula and vine copula. Figure 13.30 indicates the possible underestimation of the joint probability for higher drought severity, duration, and MDI from both fitted vine and Student t copulas. The RMSE is computed as 0.019 and 0.017 for the fitted vine and Student t copulas, respectively, which further explains their similar overall performance.

Table 13.23. *Conditional exceedance probability and conditional return periods:* $D > d|MDI \leq mdi \cap S \leq s$.

			Case (v)		
			$S \leq s$ (*cfs.day*); $MDI \leq mdi$ **(cfs)**		
			S = 174920; MDI = 1526	*S = 25000; MDI = 800*	*S = 10000; MDI = 1000*
Conditional exceedance prob.	$D > d$ (day)	38	0.75	0.46	0.26
		135	0.38	0.06	2.56E-03
		520	0.02	6.26E-06	8.10E-08
		699	1.83E-03	2.86E-07	5.50E-09
		1,133	4.33E-06	1.24E-09	5.13E-11
Conditional return period	$D > d$ (day)	38	0.97	1.56	2.80
		135	1.89	1.16E+01	2.83E+02
		520	45.14	1.15E+05	8.92E+06
		699	3.95E+02	2.52E+06	1.32E+08
		1,133	1.67E+05	5.85E+08	1.41E+10

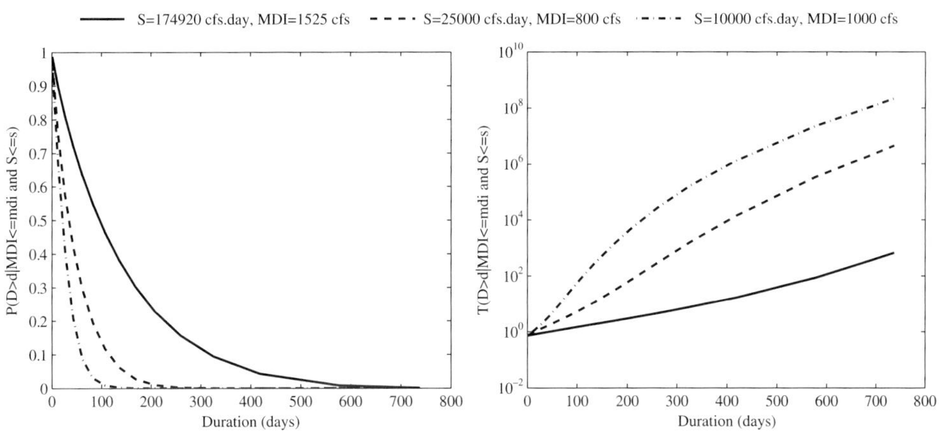

Figure 13.29 Conditional exceedance probability and conditional return period: $\boldsymbol{D > d|MDI \leq mdi \cap S \leq s}$.

Joint Return Period "AND" Case

Tables 13.15 and 13.20 list the sample results for the joint return period "AND" case computed from the fitted vine copula and Student t copula, respectively. Figures 13.16(a), 13.17(a), and 13.26 provide the sample plot for the joint return period "AND" case. $MDI = 1.53E03$ cfs (i.e., $F_{MDI}(MDI \leq 1.53E03\ cfs) = 0.8$) is applied for the table results

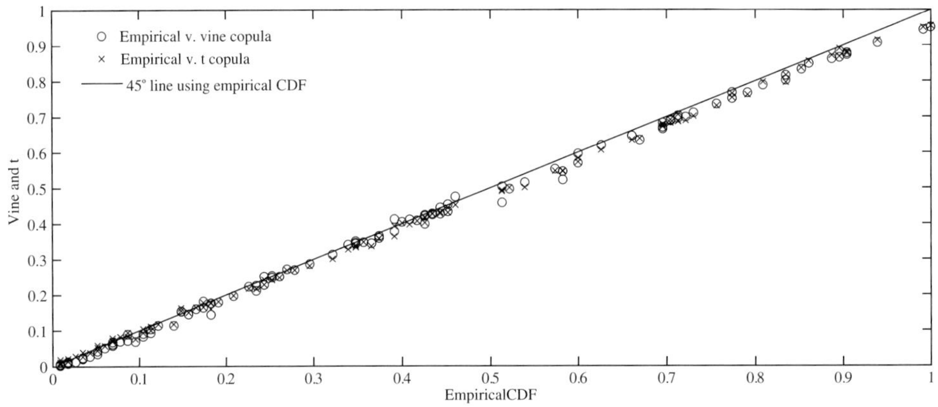

Figure 13.30 Comparison of JCDF computed from T and vine copulas to the empirical copula.

and sample plots. Results show similar joint return periods (i.e., the risk of D, S, and MDI all exceeding the critical threshold). Let us consider the univariate D and S with non-exceedance probability of 0.99 as an example. As discussed earlier in Section 13.3.2, we have $D_{F(d)=0.99} = 1133\ day, S_{F(s)=0.99} = 805320\ cfs.day$ using the Weibull and log-normal distributions fitted to D and S. From the fitted vine copula and Student t copula, we obtain the following:

Vine copula: $T(D \geq 1133\ day \cap S > 805320\ cfs.day \cap MDI > 1530\ cfs) \approx 83.5$yrs.
Student t copula: $T(D \geq 1133\ day \cap S > 805320\ cfs.day \cap MDI > 1530\ cfs) \approx 102$ yrs.

It is seen that the vine copula yields a smaller return period (i.e., higher risk) for all three drought variables exceeding the threshold values compared to the Student t copula. It is partly due to the negative correlation of T_2 ($\rho \approx -0.42$) for the fitted Gaussian copula at T_2, while the positive variance-covariance structure is shown for Student t copula (Table 13.19). Both vine and Student t copulas show that it is more realistic to study the dependence than assuming that the variables are independent. With the assumption of the independence, we will have $T_{and} = EINT/\ (1 - F_S)(1 - F_D)(1 - F_{MDI})$, and substituting $EINT = 0.73, F_S = 0.99, F_D = 0.99, F_{MDI} = 0.8$, we get $T_{and} \approx 36500\ yr$.

In one aspect, considering the fitted GH–Frank–Gaussian vine copula for drought variables *D*, *S*, and *MDI*, we have Gumbel–Hougaard, Frank, and Gaussian copulas applied to model $\{D, S\}$, $\{S, MDI\}$, and $\{D|S, MDI|S\}$, respectively. This is done, purely based on the dependence (i.e., degree of association) among drought variables. Compared to the sample rank-based Kendall correlations among all three drought variables, the drought severity has higher dependence on drought duration (0.83) and *MDI* (0.71). As the result, S is set as the center variable as shown in the section "Vine-Copula Approach to Model Trivariate Drought Variables." In addition, to estimate the joint exceedance probability and the corresponding joint return period "AND" case, we will need to estimate the copula (i.e., JCDF) for $\{D, MDI\}$. The copula of $\{D, MDI\}$ is also called 2-margins of the trivariate copula. Since $\{D, MDI\}$ is not directly linked with the fitted vine copula, the numerical integration is involved (i.e., Equation (13.13)). The numerical integration may further accumulate the computational error (or may also be called computational uncertainty).

In the other aspect, belonging to the meta-elliptical copula family, the Student t copula is constructed upon the correlation matrix directly. As the result, it is not needed to rearrange the variables, while rearranging variables is a common case for the vine copula. In addition, the two margins of the multivariate Student t copula are the bivariate Student t copula with the same degree of freedom as that of multivariate Student t copula. In the case of computing the joint exceedance probability (and joint return period "AND" case), its computation is simpler than that for the fitted vine copula.

Conditional Return Period of Cases (i), (iii), and (v)

- Cases (i): $D > d \cap MDI > mdi | S \leq s$

  As discussed in the previous sections, case (i) investigates the risk of both D and MDI exceeding the given thresholds with S smaller than the given threshold. For this case, the conditional return periods obtained from the vine copula are generally higher than those obtained from the Student t copula, which means the risk of exceedance is lower from the vine copula than that from the Student t copula.
- Cases (iii): $D > d \quad \cup \quad MDI > mdi | S \leq s$

  Case (iii) investigates the risk of at least one of D or MDI exceeding the given threshold with S smaller than its given threshold. As expected, the risk is higher for at least one of D or MDI exceeding the given threshold from cases (iii) than that from cases (i). For example, with $D > D_{F(d)=0.99} \quad \cup \quad MDI > MDI_{F(mdi)=0.1} \mid S \leq S_{F(s)=0.8}$, we compute the return period as about 45 and 32 years from the vine and Student t copulas for case (iii). The corresponding risk obtained from the Student t copula is higher than that from the vine copula.
- Cases (v): $D > d | MDI \leq mdi \cap S \leq s$

  Case (v) investigates the risk of D exceedance under the condition of both MDI and S smaller than their threshold. Different from cases (i) and (iii), the return period reduces to the univariate case under the given condition for cases (v). The sample results in the table indicate a high risk for D to exceed a lower threshold value. With the increase of MDI and S, the risk is significantly reduced for D to exceed a higher threshold value.

13.4 Summary

In this chapter, we apply the copula theory to drought frequency analysis, including bivariate and trivariate cases. For the bivariate drought frequency analysis (drought duration and drought severity), the Archimedean and meta-elliptical (Gaussian and Student t) copulas are applied. For trivariate drought frequency analysis (drought duration, drought severity, and MDI of the drought event), the vine and meta-elliptical copulas are applied. The bivariate Archimedean and meta-elliptical copulas are applied as the candidates to construct the vine copula. Throughout this case study, we reach the following conclusions:

1. Similar to many other investigations, the log-normal and Weibull distributions are fitted to drought severity and drought duration, respectively. Due to the difficulty to fit a proper distribution directly to the MDI, the nonlinear meta-Gaussian transformation is applied to model the MDI such that standard Gaussian distribution may be applied to model the transformed variable
2. The Gumbel–Hougaard copula is most proper to model drought severity and drought duration. Conceptually, it is understandable for the applicability of this particular copula: (i) the GH copula belongs to the extreme value family, which may better represent the extremes in the nature of drought events; and (ii) the upper-tail dependence of the GH copula may better evaluate the risk of $S>s|D>d$ (or $S>s|D=d$) and vice versa.
3. The dynamic return period may be assessed for the evolution of a certain drought episode. The case example shows that as the drought episode evolves, the dynamic return period goes up and down as well.
4. Both vine and Student t copulas are applied to model the trivariate drought variables. Compared to the vine copula, the Student t copula may be easier to apply with less computational burden to study risk. In addition, the design based on the risk computed from the Student t copula could be more conservative, since for a given condition, the risk from the Student t copula (lower return period) is generally higher than that from the vine copula (higher return period).
5. Similar to other investigations, the case study presented here assumes all drought variables as continuous random variables. However, for daily (or monthly) values, the duration is actually discrete and may be with many ties (i.e., one duration may be associated with at least two different drought severities). Compared to the commonly applied drought analysis with the use of monthly values, the analysis with daily values significantly cuts down the ties existing within the dataset. It may be worth the effort to actually try to model the duration as discrete variables.

References

Aas, K., Czado, C., Frigessi, A., and Bakken, H. (2009). Pair-copula constructions of multiple dependence. *Insurance: Mathematics and Economics*, 44, 182–198, doi:10.1016/j.insmatheco.2007.02.001.

AghaKouchak, A. (2015). A multivariate approach for persistence-based drought prediction: application to the 2010–2011 East Africa drought. *Journal of Hydrology*, 526, 127–135. doi:10.1016/j.jhydrol.2014.09.063.

Chen, L., Singh, V. P., Guo, S., Mishra, A. K., and Guo, J. (2013) Drought analysis using copulas. *Journal of Hydrologic Engineering*, 18(7), 797–808. doi:10.1061/(ASCE)HE.1943-5584.0000697.

Chen, Y. D., Zhang, Q., Xiao, M., and Singh, V. P. (2013). Evaluation of risk of hydrological droughts by the trivariate Plackett copula in the East River basin (China). *Natural Hazards*, 68, 529–547.

De Michele, C., Salvadori, G., Vezzoli, R., and Pecora, S. (2013). Multivariate assessment of droughts: frequency analysis and dynamic return period. *Water Resources Research*, 49, 6985–6994. doi:10.1002/wrcr.20551.

Genest, C., Rémillard, B., and Beaudoin, D. (2007). Goodness-of-fit tests for copulas: a review and a power study. *Insurance: Mathematics and Economics*. doi:10.1016/j.insmatheco.2007.10.1005.

Hao, Z. and AghaKouchak, A. (2014). A nonparametric multivariate multi-index drought monitoring framework. *Journal of Hydrometeorology*, 15, 89–101. doi:10.1175/JHM-D-12-0160.1.

Hao, Z., Hao, F., Singh, V. P., Sun, A. Y. and Xia, Y. (2016). Probabilistic prediction of hydrologic drought using a conditional probability approach based on the meta-Gaussian model. *Journal of Hydrology*, 542, 772–780. doi:10.1016/j.jhydrol.2016.09.048.

Kao, S-C. and Govindaraju, R. S. (2010). A copula-based joint deficit index for droughts. *Journal of Hydrology*, 380, 121–134. doi:10.1016/j.jhydrol.2009.10.029.

Janga Reddy, M. and Singh, V. P. (2014). Multivariate modeling of droughts using copulas and meta-heuristic methods. *Stochastic Environmental Research an Risk Assessment*, 28, 475–489.

Kwak, J., Kim, S., Kim, G., Singh, V. P., Park, J., and Kim, H. S. (2016). Bivariate drought analysis using tree ring streamflow reconstruction in the Sacramento Basin, California, USA: a case study. *Water*, 8(122), 1–16. doi:10.3390/w8040122.

Madadgar, S. and Moradkhani, H. (2013). Drought analysis under climate change using copula. *Journal of Hydrologic Engineering*, 18(7), 746–759. doi:10.1061/(ASCE)HE.1943-5584.0000532.

McKee, T. B., Doesken, N. J., and Kleist, J. (1993). The relationship of drought frequency and duration to time scales. *8th Conference on Applied Climatology*, American Meteorological Society, Anaheim. www.droughtmanagement.info/literature/AMS_Relationship_Drought_Frequency_Duration_Time_Scales_1993.pdf.

Mishra, A. K. and Singh, V. P. (2010). A review of drought concepts. *Journal of Hydrology*, 391, 202–216. doi:10.1016/j.jhydrol.2010.07.012.

Palmer, W. C. (1965). Meteorologic drought. US Department of Commerce, Weather Bureau, Research paper No. 45.

Palmer, W.C. (1968). Keeping track of crop moisture conditions, nationwide: the new crop moisture index. *Weatherwise*, 21, 156–161.

Rao, A. R. and Padamanabhan, G. (1984). Analysis and modeling of Palmer's drought index series. *Journal of Hydrology*, 68, 211–229.

Salvadori, G. and De Michele, C. (2015). Multivariate real-time assessment of droughts via copula-based multi-site hazard trajectories and fans. *Journal of Hydrology*, 526, 101–115. doi:10.1016/j.jhydrol.2014.11.056.

Salvadori, G., Durante, F., and De Michele, C. (2013). Multivariate return period calculation via survival functions. *Water Resources research*, 49, 2308–2311. doi:10.1002/wrcr.20204.

Santos, M. A. (1983). Regional droughts: a stochastic characterization. *Journal of Hydrology*, 66, 183–211.

Shukla, S. and Wood, A. W. (2008). Use of a standardized runoff index for characterizing hydrologic drought. *Geophysical Research Letters*, 35, L02405. doi:10.1029/2007GL032487.

Song, S. and Singh, V. P. (2010a). Frequency analysis of droughts using the Plackett copula and parameter estimation by genetic algorithm. *Stochastic Environmental Research and Risk Assessment*, 24(5), 783–805. doi:10.1007/s00477-010-0364-5.

Song, S. and Singh, V. P. (2010b). Meta-elliptical copulas for drought frequency analysis of periodic hydrologic data. *Stochastic Environmental Research and Risk Assessment*, 24(3), 425–444.

Texas State Historical Association. (n.d.). Nueces River, *Handbook of Texas,* www.tshaonline.org/handbook/online/articles/rnn15.

Tu, X, Singh, V. P., Chen, X., Ma, M., Zhang, Q., and Zhao, Y. (2016) Uncertainty and variability in bivariate modeling of hydrological droughts. *Stochastic Environmental Research and Risk Assessment*, 30, 1317–1334.

Van Rooy, M. P. (1965). A rainfall anomaly index independent of time and space. *Notos*, 14, 43–48.

Voss, R., May, W., and Roeckner, E. (2002). Enhanced resoluation modeling study on anthropogenic climate change: changes in extremes of the hydrological cycle. *International Journal of Climatology*, 22, 755–777.

Xu, K., Yang, D., Xu, X., and Lei, H. (2015). Copula based drought frequency analysis considering the spatio-temporal variability in Southwest China. *Journal of Hydrology*, 527, 630–640. doi:10.1016/j.jhydrol.2015.05.030.

Yevjevich, V. (1967). An objective approach to definitions and investigations of continental hydrologic droughts. Hydrology Papers, Colorado State University, Fort Collins.

Yoo, J. Y., Shin, J. Y., Kim, D. K., and Kim, T.-W. (2013). Drought risk analysis using stochastic rainfall generation model and copula functions. *Journal of Korea Water Resource Association*, 46(4), 425–437.

Zelenhasic, E. and Salvai, A. (1987). A method of streamflow drought analysis. *Water Resources Research*, 23(1), 156–168.

Zhang, Q., Xiao, M., and Singh, V. P. (2015) Uncertainty evaluation of copula analysis of hydrological droughts in the East River Basin, China. *Global and Planetary Change*, 129, 1–9.

14
Compound Extremes

ABSTRACT

In this chapter, the copula modeling is applied to analyze compound extremes. The number of warm days (NWDs) and monthly precipitation are applied for the case study. The time-varying generalized extreme value (GEV) distribution with a linear trend in the location parameter is applied to model the NWDs after the change. The time-varying copula is applied to model the compound risk of hot and dry, as well as wet and cold days.

14.1 Introduction

Extreme events (e.g., peak flow, heat wave, etc.) have been conventionally analyzed as univariate variables with the use of such distributions as generalized extreme value (GEV) distribution. These events have also been analyzed in bivariate (multivariate) frameworks considering their intrinsic characteristics (e.g., peak discharge, flood volume and flood duration in flood frequency analysis; drought severity, duration, and interarrival time in drought frequency analysis). This multivariate framework applies the intrinsic properties to better represent the risk induced by the events. However, there may be other variables (factors) that may either increase or decrease the risk of occurrence of extreme events. For example, heat wave (or high temperature) in general increases drought severity, stresses plant growth, increases evapotranspiration, impacts bacterial or viral activity, etc. When more variables (or extremes of different types) than one are analyzed, analysis of extremes is called compound (or concurrent) analysis. In what follows, we will first briefly review recent studies.

Using the hypothesis of flood and sea surge being more likely to occur concurrently on the east coast of Britain than the north coast, Svensson and Jones (2002) proposed the χ empirical dependence measure to evaluate the flood, surge, and precipitation for the spatial dependence of flood, surge, or precipitation of different stations, as well as for the cross variable, with the assumption of flood, surge, and precipitation being independent identically distributed (i.i.d.) random variables. The proposed χ dependence measure may be applied to investigate the concurrence of extremes, i.e., the probability of one variable being extreme provided the other one is extreme.

Hao et al. (2013) evaluated the occurrence of the compounding monthly precipitation and temperature extremes using the data from the Climate Research unit, University of Delaware, and the simulations from CMIP5 models. Pertaining to precipitation and temperature, four combinations were considered for evaluation: wet/warm (P75/T75); dry/warm (P25/T75); wet/cold (P75/T25); and dry/cold (P25/T25). Their investigation concluded the increasing occurrences of wet/warm and dry/warm for some regions in the world with the decreasing occurrences of wet/cold and dry/cold for a majority of the world.

Wahl et al. (2015) studied the compound flooding risk from storm surge and heavy rainfall for major coastal cities in the United States. Using rank-based correlation, their study revealed that the compounding flood risk was higher at the Atlantic/Gulf coast than at the Pacific coast. Additionally, the number of events increased due to the long-term sea level rise in the past century (Wahl et al., 2015). Using the copula theory, Miao et al. (2016) studied the stochastic relation of precipitation and temperature in the Loess Plateau in China.

Sedlmeier et al. (2016) investigated compound extremes under climate change. In their study, heavy precipitation and low temperature in winter, and high temperature and dry days in summer, were applied for compound extreme analysis using the Markov Chain method. Through the study, they were able to identify three regions that may be more likely to be impacted due to the future change in terms of heavy precipitation and low temperature in the winter. They also identified one region likely to be impacted by the future change of dry and hot summer. In this chapter, we will focus on applying the copula theory to analyze compound extremes.

14.2 Dataset

To illustrate the analysis, maximum daily temperature and daily precipitation were collected from NOAA at USC00411720 (Choke Canyon Dam, Texas). The range of data was from water year 1983 (October 1, 1983–April 7, 2017). In the data collected from NOAA, there were five months of missing data as listed in Table 14.1.

To obtain the complete time series, the nearby station, i.e., USC00411337 (Calliham, Texas), close to USC00411720, is chosen to fill the missing precipitation and temperature. By replacing the missing precipitation and temperature with those at USC00411337, we see that the missing precipitation is successfully replaced. However, the missing temperature cannot be successfully filled for the months listed in Table 14.1 except for October 2003. Thus, to keep the continuity of daily precipitation and temperature, daily information starting from the calendar year of 1990 is applied for analysis.

Besides the missing values listed in Table 14.1, Table 14.2 lists the days with missing precipitation (and/or temperature) as well as the replaced values. These missing values are filled, with the rules as follows:

Table 14.1. *The entire month of missing precipitation and temperature data.*

Jan. 1985	Oct. 1986	Aug. 1988	Dec. 1989	Oct. 2003

Table 14.2. *Days of missing daily precipitation and temperature after 1990.*

Precipitation[a] (mm/day)						Temperature[b] (°C)	
01/13/1997	0	02/13/2012	0.5	03/29/2012	39.1	02/04/2011	2.2
09/18/2011	0	03/09/2012	5.1	07/11/2012	5.1	05/05/2011	27.8
12/11/2011	14.5	03/10/2012	5.1	09/14/2012	44.5	05/25/2014	30.6
01/25/2012	9.1	03/11/2012	17.8	09/29/2012	81.3	04/05/2015	20

Note: [a] Applied rule (i); [b] applied rule (ii).

i. Replacing the missing precipitation (and/or temperature) with the available observation at USC00411337 on the same day;
ii, Otherwise, replacing the missing precipitation (and/or temperature) with the average values of one day before and one day after of both two stations. Using February 4, 2011, as an example, the missing temperature of that day is filled using the temperatures of February 3, 2011, and February 5, 2011, at both stations USC00411720 and USC00411337.

With missing daily precipitation and maximum temperature data filled, we may compute monthly precipitation and the number of warm days (NWD) for each month. The NWD is computed as follows:

$$NWD_{i,j} = \sum_{k=1}^{n_j} \mathbf{1}\left(T_{i,j,k} > \overline{T_j}\right) \tag{14.1}$$

in which: *i, j* represent the year and month of observation, n_j represents the number of days for month *j*, and $\overline{T_j}$ represents the sample average monthly maximum temperature computed from the entire dataset.

Figure 14.1 plots the individual time series and the scatter plot. The scatter plot indicates the negative relation between monthly precipitation and NWDs. The negative relation is supported by the rank-based sample Kendall's tau coefficient of correlation, and we get $\tau_N \approx -0.38$. To assess the stationarity for the time series, the Kwiatkowski–Phillips–Schmidt–Shin (KPSS) and Mann–Kendall tests are performed.

The null hypothesis of the KPSS test is that the time series is trend stationary (or level stationary, i.e., no trend). The alternative hypothesis of KPSS test is that the time series is a unit-root process. To perform the KPSS test, the time series $\{X_t : t = 1, 2, \ldots, n\}$ is expressed as a sum of three components, deterministic trend, random walk, and stationary residual, as follows:

$$X_t = \alpha t + r_t + e_{1t} \tag{14.1a}$$

$$r_t = r_{t-1} + e_{2t} \tag{14.1b}$$

In Equation (14.1), α represents the deterministic trend with $\alpha = 0$ for the test of level stationary; r_t represents the random walk; e_{1t} represents the stationary process; and $e_{2t} \sim i.i.d.(0, \sigma^2)$. With Equation (14.1), the null hypothesis may be rewritten as follows:

$$H_0 : \alpha \neq 0, \sigma^2 = 0, \text{ for trend stationary}; \alpha = 0,\ \sigma^2 = 0 \text{ for level stationary} \tag{14.1c}$$

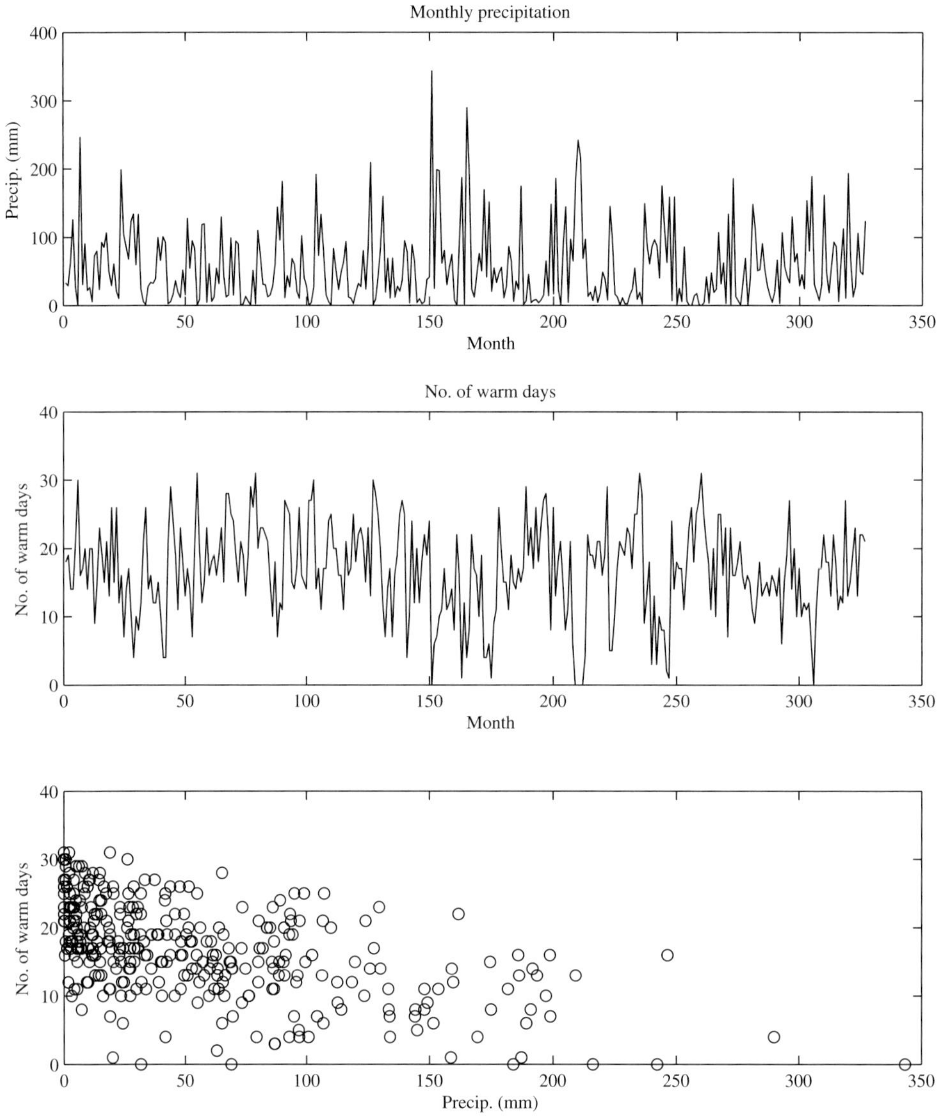

Figure 14.1 Time series of monthly precipitation and NWD.

To assess the stationarity of the univariate time series, we can directly apply KPSS test function in MATLAB using the following: [*h*, *P*-Value, Statistics, Critical Value] =KPSS test(X, 'lags', a, 'trend' true/false, 'alpha', alpha), where *X* is the time series tested; *a* is the number of lag considered; 'trend', true represents the trend stationary (default) and false represents the level stationary; and 'alpha' represents the significance level (default = 0.05)].

Originally proposed by Mann (1945) and Kendall (1970), the nonparametric Mann–Kendall test evaluates whether there exists a monotonic trend in the dataset. The null hypothesis is that the data are i.i.d. random variables with the alternative hypothesis of monotonic trend existing in the dataset. The Mann–Kendall test statistics is computed using the S-score as follows:

$$S = \sum_{k=1}^{n-1} \sum_{j=k+1}^{n} sign(X_j - X_i), \begin{cases} sign(\ ) = 1; & X_j - X_i > 0 \\ sign(\ ) = 0; & X_j - X_i = 0 \\ sign(\ ) = -1; & X_j - X_i < 0 \end{cases} \tag{14.2a}$$

The S-score in Equation (14.2a) has the following statistics:

$$E(S) = 0; \sigma_S^2 = \left(n(n-1)(2n+5) - \sum_{j=1}^{p} t_j(t_j - 1)(2t_j + 5)\right)/18 \tag{14.2b}$$

In Equation (14.2b), p represents the number of tied groups in the dataset; and t_j represents the number of data in the jth tied group. Furthermore, the test statistics S may be transformed to Z-score (i.e., following the standard normal distribution) as follows:

$$Z^* = \begin{cases} \dfrac{S-1}{\sigma_S}, \ if\ S > 0 \\ 0, \ if\ S = 0 \\ \dfrac{S+1}{\sigma_S}, \ if\ S < 0 \end{cases} \tag{14.2c}$$

The P-value can then be computed by computing the exceedance probability as follows:

$$P_{value} = 1 - \Phi^{-1}(Z^*) \tag{14.2d}$$

Based on the sample autocorrelation and partial autocorrelation plots shown in Figure 14.2, the KPSS test is performed up to a two-month lag for the monthly time series using the matlab function (kpsstest). Table 14.3 lists the results of KPSS test with the null hypothesis of level stationary, and Mann–Kendall test with the null hypothesis of observed data being i.i.d. random variables. Results listed in Table 14.3 show that (1) monthly precipitation may be viewed as a stationary time series (i.e., level stationary at all lags and monotonic trend is not detected by Mann–Kendall test); and (2) there exists a trend in the NWD per month. Applying a linear regression of NWD with respect to sequential month, we have $NWD = b_1 + b_2 x; x = 1, \ldots, 327; b_1 = 17.774, b_2 = -0.007, P_{value} = 0.076$. The P-value computed is slightly higher than 0.05, which means the null hypothesis may not be rejected with significance level of $\alpha = 0.05$; however, the rejection by the Mann–Kendall test suggests that there may be a monotonic trend or a sudden change existing in the NWDs.

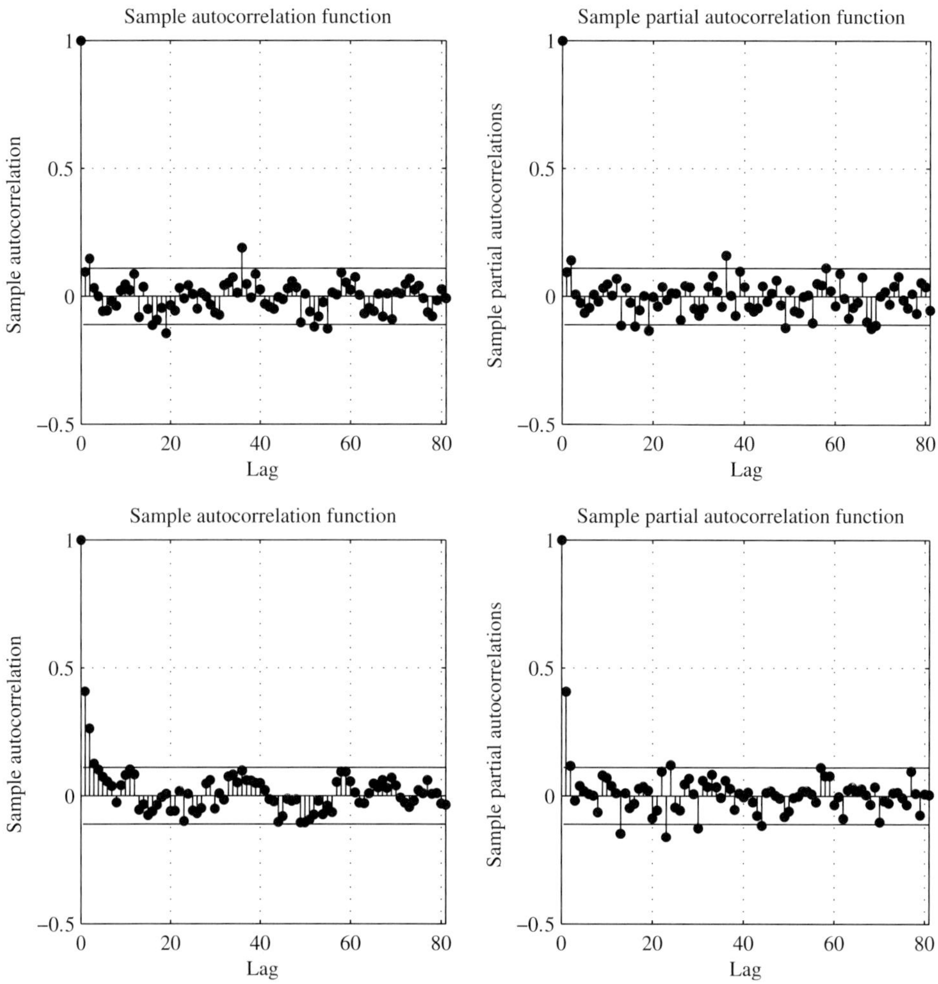

Figure 14.2 Sample autocorrelation and partial autocorrelation plots for monthly precipitation and number of the warm days.

In this case study, the Pettitt test (Pettitt, 1979) is applied to detect the change point of NWDs. The Pettitt test is a version of Mann–Whitney's U-test. The null hypothesis of the Pettitt test is that there is no change point detected. Similar to Mann–Kendall test, the U-score of the Pettitt test is given as follows:

$$U_{t,N} = U_{t-1,N} + \sum_{j=1}^{N} sign\left(X_t - X_j\right), t = 2, \ldots, N \tag{14.3a}$$

The test statistic is then given as follows:

$$k(t) = \max_{1 \le t \le N} \mid U_{t,N} \mid \tag{14.3b}$$

Table 14.3. *Results of KPSS and Mann–Kendall tests.*

Variables		KPSS			Mann–Kendall	
		H	Stat.	Cri.	S_score	P-value
Precipitation	Lag = 0	0	0.059	0.463	–0.6	0.5456
	Lag = 1	0	0.054	0.463		
	Lag = 2	0	0.048	0.463		
NWDs per month	Lag = 0	1	0.014	0.463	–1.99	0.047
	Lag = 1	1	0.043	0.463		
	Lag = 2	0	0.074	0.463		

and the P-value is approximated as follows:

$$p \cong e^{-\left(\frac{6k(t)^2}{N^3+N^2}\right)} \tag{14.3c}$$

In Equation (14.3), N is the sample size, and X is the observed series.

Applying the Pettitt test, we detect the change point at month 150 (i.e., June 2002). Now, with the initial analysis, we can proceed to further analyze the monthly precipitation and NWDs.

14.3 Univariate Analysis of Monthly Precipitation and NWDs

In the previous section, we have shown that monthly precipitation belongs to stationary signal, while there exists a changing point at month 150 (June 2002) for NWDs. To this end, the exponential distribution is fitted to model monthly precipitation and the time-varying GEV distribution is applied to model the NWDs. In the case of GEV distribution applied, we only consider a linear change in the location parameter. Table 14.4 lists the fitted parameters and GoF statistics for the fitted univariate distributions, and Figure 14.3 plots the histogram and fitted probability density functions as well as the change of the location parameter for the NWDs after month 150 (June 2002).

14.4 Bivariate Analysis of Monthly Precipitation and NWDs

The bivariate analysis of monthly precipitation and NWDs is investigated with the use of copula theory. Unlike stationary copula models applied in the previous chapters, the time-varying copula is applied to model monthly precipitation and NWDs. The time-varying copula may be written using $C(u, v; \theta_t)$, where the stationary copula is applied before June 2002 (month 150 before the change) and the time-varying copula with a moving average window size 1 applied after the change. Figure 14.4 plots the sample Kendall's tau coefficients for the monthly precipitation and NWDs before the change for the entire dataset, assuming the NWDs as stationary, and those with the moving average window size 1 after the change point. Figure 14.4 shows a decreasing trend after June 2002, i.e.,

Table 14.4. *Results of univariate analysis.*

Variables		Distribution	Parameters	GoF Test stat.	GoF *P*-value
Monthly precipitation		Exponential[a]	$\mu = 64.65\ mm$	0.05	0.46
NWDs	Before June 2002	GEV[b]	$k = -0.32,\ s = 8.21,\ \mu = 15.70$	0.21	0.19
NWDs	After June 2002	Trend	$\mu_t = 16.91 - 0.014t,\ t = 151 : 327$		

Notes: [a] KS test for GoF evaluation; [b] generalized extreme value distribution.

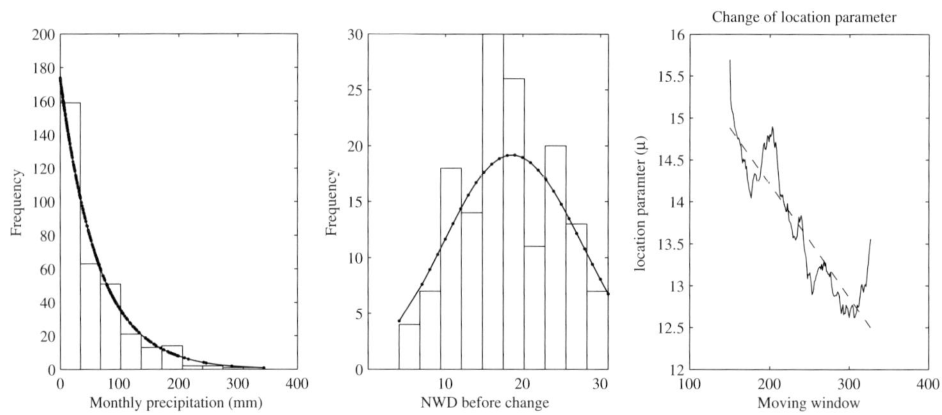

Figure 14.3 Fitted distributions for monthly precipitation, NWDs, as well as the change of location parameter of GEV distribution for NWDs after month 150 with moving window size 1.

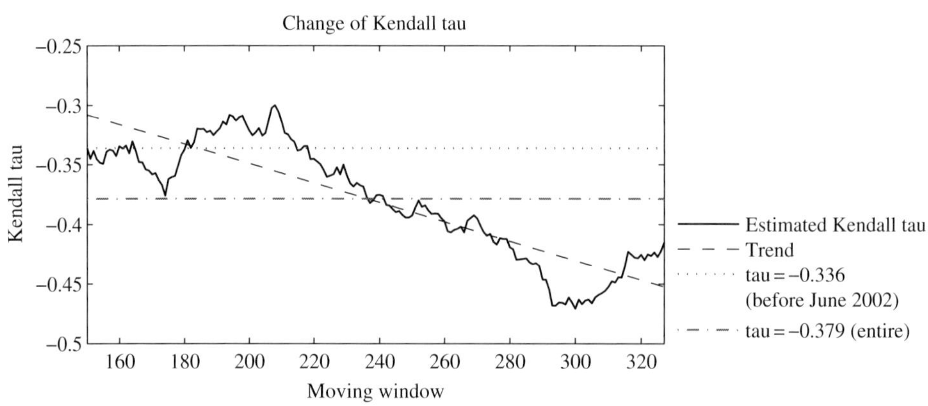

Figure 14.4 Sample Kendall correlation coefficients computed.

monthly precipitation and NWDs get more negatively correlated, or equivalently longer (severer) drought may be expected with less precipitation.

With the negative Kendall correlation coefficient estimated, the Frank copula (Archimedean family) and meta-Student *t* and meta-Gaussian copulas (the meta-elliptic family) are applied to model the monthly precipitation and NWDs. The stationary copula is applied for the bivariate data before June 2002, while the time-varying copula is applied for the bivariate data after June 2002.

Applying the pseudo-MLE to the monthly precipitation and NWDs before June 2002, Table 14.5 lists the parameter and log-likelihood estimated for each copula candidate. It is seen from Table 14.5 that the meta-Student t copula converges to the meta-Gaussian copula. From comparison of log-likelihood values obtained from all three candidates, the meta-Gaussian copula is applied to model the monthly precipitation and NWDs before June 2002 ($S_n^B = 0.028, P = 0.623$). Figure 14.5 compares simulated variables with observed variables before June 2002. Comparison shows that the Gaussian copula properly models monthly precipitation and NWDs before the change point.

With the moving window size 1, the time-varying Gaussian copula is applied to monthly precipitation and NWDs after the changing point with the estimated parameters plotted in Figure 14.6. Figure 14.6 shows the overall decreasing trend as that of the Kendall correlation coefficient.

Table 14.5. *Estimated parameters and corresponding LogLs.*

	Frank	Gaussian	Student *t*
Parameters	–3.282	–0.513	$[-0.532, 4.67 \times 10^6]$
Log*L*	19.25	22.86	22.91

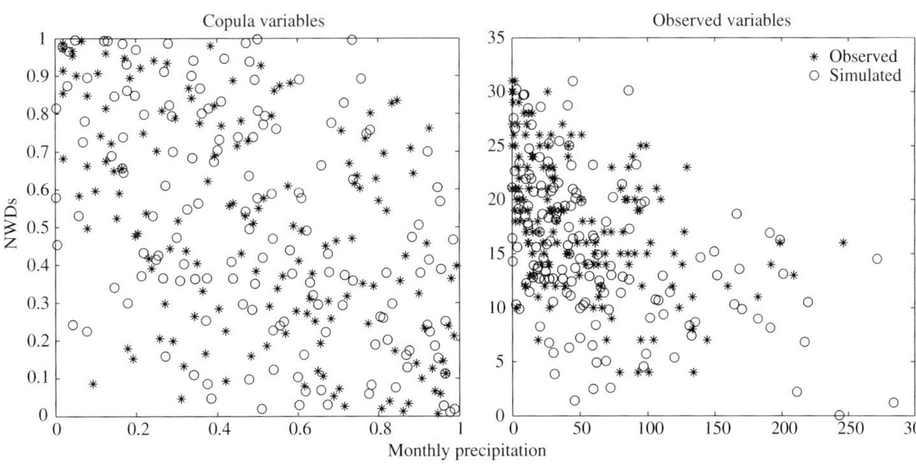

Figure 14.5 Comparison of simulated variables with observed variables before June 2002.

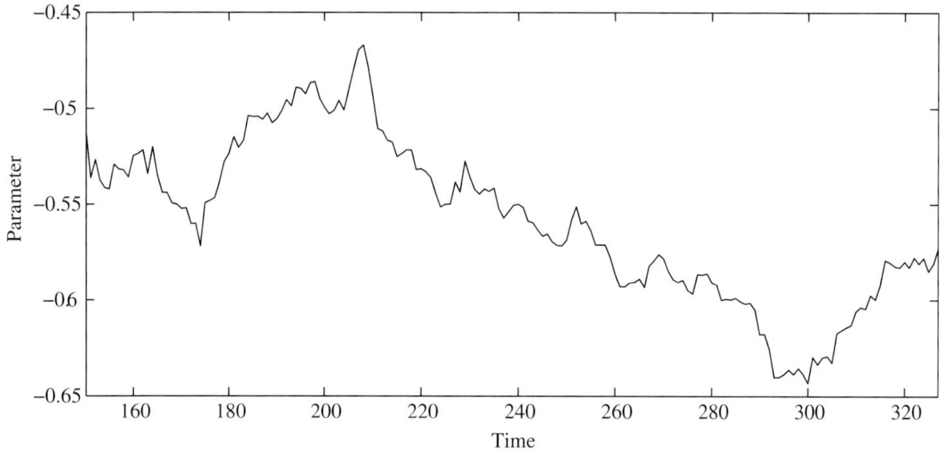

Figure 14.6 Parameters estimated after the change point with moving window size 1 (the meta-Gaussian copula).

14.5 Risk Analysis with Meta-Gaussian Copula

To assess the compound risk, 25 and 75 percentiles of monthly precipitation and NWDs are computed from the original dataset as follows:

$$Precip_{25} = 12.2\ mm, Precip_{75} = 86.25\ mm, NWDs_{25} = 12\ days, NWDs_{75} = 21\ days.$$

To assess the duration and severity of drought, one may look at two types of compound risks, i.e.,

$$Prob(NWDs \geq NWDs_{75} \cap Precip \leq Precip_{25}) \tag{14.4a}$$

$$Prob(NWDs \geq NWDs_{75} | Precip \leq Precip_{25}) \tag{14.4b}$$

Equations (14.4a)-(14.4b) indicates the risk (probability) of occurrence of dry condition and warm days as well as the warm days conditioned on the dry condition. According to return periods discussed in Chapter 3, Equations (14.4a)-(14.4b) may be rewritten as follows:

$$\begin{aligned} &Prob(NWDs \geq NWDs_{75} \cap Precip \leq Precip_{25}) \\ &= P(Precip \leq Precip_{25}) - P(NWDs \leq NWDs_{75} \cap Precip \leq Precip_{25}) \\ &= P(Precip \leq Precip_{25}) - C(F(NWDs \leq NWDs_{75}), F(Precip \leq Precip_{25}); \theta_t) \end{aligned} \tag{14.5a}$$

$$\begin{aligned} &Prob(NWDs \geq NWDs_{75} | Precip \leq Precip_{25}) \\ &= \frac{P(NWDs \geq NWDs_{75} \cap Precip \leq Precip_{25})}{P(Precip \leq Precip_{25})} \\ &= \frac{P(Precip \leq Precip_{25}) - C(F(NWDs \leq NWDs_{75}), F(Precip \leq Precip_{25}); \theta_t)}{P(Precip \leq Precip_{25})} \end{aligned} \tag{14.5b}$$

In Equations (14.5a)-(14.5b), the fitted time-varying Gaussian copula is applied in which the copula parameter is constant before June 2002 and changes with moving window size 1 after June 2002. Before June 2002, the joint probability is computed as $P = 0.112$ using Equation (14.5a), and the conditional probability is computed as $P = 0.649$ using Equation (14.5b), with the use of the stationary meta-Gaussian copula with parameter $\theta = -0.513$. After June 2002, the joint and conditional probabilities are computed for each moving window and plotted in Figure 14.7. Figure 14.7 shows that the joint probabilities of concurrence of NWDs and dry conditions are within the range of [0.092, 0.111] with the average of 0.099, and the conditional probabilities (i.e., NWDs provided dry weather conditions) are within the range of [0.532, 0.642] with the average of 0.578. Comparing to the conditional probabilities computed, the joint probabilities are more stable. The risk of having more abnormal warmer days in a month is higher providing the dry weather conditions (i.e., monthly precipitation is at lower 25 percentile).

Wet and cold are another compound risk in which one may be interested, especially in the case of a long cold and wet winter, using the following:

$$Prob(NWDs \leq NWDs_{25} \cap Precip \geq Precip_{75}) \tag{14.6a}$$

$$Prob(NWDs \leq NWDs_{25} | Precip \geq Precip_{75}) \tag{14.6b}$$

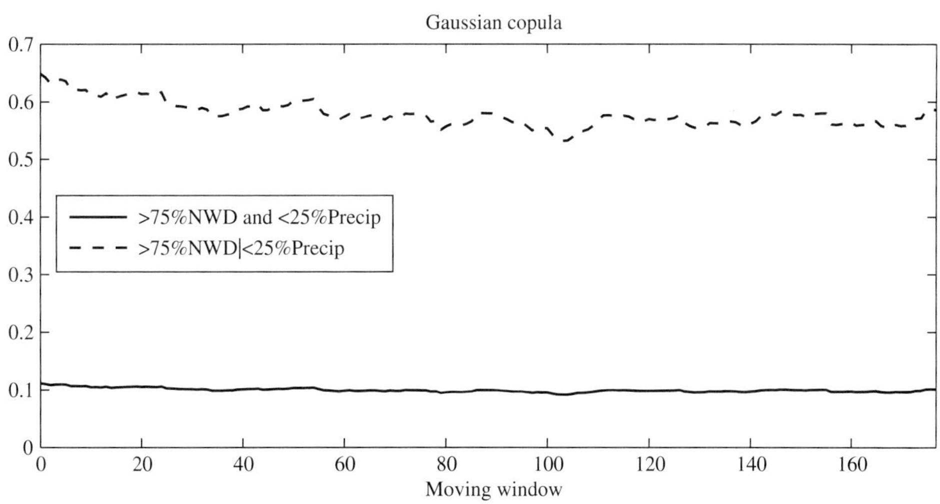

Figure 14.7 Time-varying joint and conditional probability assessed with the time-varying Gaussian copula.

Similar to the hot and dry conditions, Equations (14.6) may be rewritten as follows:

$$
\begin{aligned}
&Prob(NWDs \leq NWDs_{25} \cap Precip \geq Precip_{75}) \\
&= P(NWDs \leq NWDs_{25}) - P(NWDs \leq NWDs_{25} \cap Precip \leq Precip_{75}) \\
&= P(NWDs \leq NWDs_{25}) - C(F(NWDs \leq NWDs_{25}), F(Precip \leq Precip_{75}); \theta_t)
\end{aligned}
\tag{14.7a}
$$

$$
\begin{aligned}
&Prob(NWDs \leq NWDs_{25} | Precip \geq Precip_{75}) \\
&= \frac{P(NWDs \leq NWDs_{25}, Precip \geq Precip_{75})}{P(precip \geq precip_{75})} \\
&= \frac{P(NWDs \leq NWDs_{25}) - C(F(NWDs \leq NWDs_{25}), F(Precip \leq Precip_{75}); \theta_t)}{1 - P(precip \leq precip_{75})}
\end{aligned}
\tag{14.7b}
$$

Similar to the risk analysis for the dry and hot condition, the joint probability using Equation (14.7a) is computed as $P = 0.121$, and the conditional probability using Equation (14.7b) is computed as $P = 0.158$, with the use of the stationary meta-Gaussian copula ($\theta = -0.514$). After June 2002, the joint and conditional probabilities are computed for each moving window using the time-varying meta-Gaussian copula and plotted in Figure 14.8.

In the case of the compound risk of wet and cold, Figure 14.8 shows that the joint probabilities are within the range of [0.129, 0.184] with the average of 0.156, and the

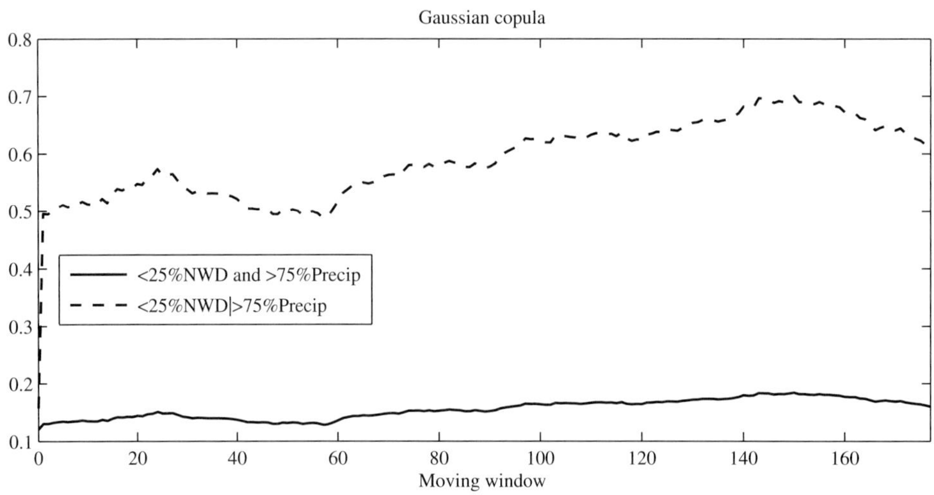

Figure 14.8 Probability of compound risk for wet and cold.

conditional probabilities are within the range of [0.489, 0.700] with the average of 0.592. Again, the joint probabilities computed are more stable than are the conditional probabilities. The risk of having fewer warm days in a month is higher under the condition of monthly precipitation higher than its 75 percentile than the concurrent joint probability.

14.6 Summary

In this chapter, we have applied copula theory to compound risk analysis. Throughout the study, the NWDs and monthly precipitation are applied to assess the following compound risk:

1. Hot and dry conditions, which may be considered as a compounding factor for severe draught
2. Cold and wet conditions, which may be considered as a compounding factor for cold winter

The application shows that the joint probabilities for both wet/cold and dry/warm conditions are smaller than the marginal exceedance probabilities; while the conditional probabilities for warm condition given dry conditions and cold conditions given wet conditions fall in between marginal exceedance probabilities. In addition, the study of compound risk may better investigate extreme events such as drought and winter storms.

References

Hao, Z., AghaKouchak, A., and Phillips T. J. (2013). Changes in concurrent monthly precipitation and temperature extremes. *Environmental Research Letters*, 8, 034014.

Kendall, M. G. (1970). *Rank Correlation Methods*, 2nd edition. Hafner, New York.

Mann, H. B. (1945). Nonparametric tests against trend. *Econometrica*, 13, 245–259.

Miao, C., Sun, Q., Duan, Q., and Wang, Y. (2016). Joint analysis of changes in temperature and precipitation on the Loess Plateau during the period 1961–2011. *Climate Dynamics*. doi:10.1007/s00382–016–3022-x.

Pettitt, A. N. (1979). A non-parametric approach to the change-point problem. *Journal of Applied Statistics*, 18, 126–135.

Sedlmeier, K., Mieruch, S., Shädler, G., and Kottmeier, C. (2016). Compound extremes in a changing climate – a Markov chain approach. *Nonlinear Processes in Geophysics*, 23, 375–390.

Svensson, C. and Jones, D. A. (2002). Dependence between extreme sea surge, river flow and precipitation in eastern Britain. *International Journal of Climatology*, 22, 1149–1168.

Wahl, T., Jain, S., Bender, J., Meyers, S. D., and Luther, M. E. (2015) Increasing risk of compound flooding from storm surge and rainfall or major US cities. *Nature Climate Change*, 5, 1093–1097. doi:10.1038/NCLIMATE2736.

15

Network Design

ABSTRACT

In this chapter, we apply copulas to network evaluation and design. The network is considered to be comprised of rain gauges that are located in the southwest (seven gauges) and east central (three gauges) parts of Louisiana. To select proper rain gauges for network design, the kernel density is applied to model the marginal rainfall variables as that studied for rainfall analysis in Chapter 10. For the simplicity of illustrating the copula-based network design, meta-elliptical copulas (i.e., meta-Gaussian and meta-Student t) are applied to model the spatial dependence among rain gauges. The network design case study shows the appropriateness of the copula-based network design.

15.1 Introduction

A majority of studies on network design and evaluation have applied the multivariate normal distribution. Krstanovic and Singh (1992a, b) applied the entropy theory to evaluate the rainfall network in Louisiana. They studied both spatial and temporal rainfall network design. For spatial investigation, they imposed the assumption of no temporal dependence for the univariate rainfall record of a given rain gauge station and vice versa. The multivariate Gaussian distribution was applied in the evaluation procedure.

Chow and Liu (1968) evaluated the dependence tree with discrete probability distributions and mutual information between any given two (or pair of) random variables. They proposed an optimization of n-dimensional probability distribution with the product of 1 univariate distribution and n-2 bivariate conditional distributions. Applying the gamma distribution to rainfall variables at each gauging station and bivariate normal distribution to model rainfall variables at the paired stations, Al-Zahrani and Husain (1998) studied rainfall network reduction and expansion. Using extreme flow data in southern Manitoba, Yang and Burn (1994) proposed directional information transfer (DIT) to study the information transmitted between the paired gauging stations. They used DIT to group streamflow gauges.

Dong et al. (2005) studied the impact of the density of rain gauges on the streamflow simulation accuracy based on the cross-correlation coefficient (with lag k) between areal rainfall and discharge at Yuxiakuo of the Qingjiang River basin, located in the south of Three Gorges area of the Yangtze River, China. They found that with the increase of number of rain gauges, the variance of areal rainfall decreased hyperbolically. Inversely, with the increase of number of rain gauges, the cross-correlation increased hyperbolically between areal rainfall and discharge.

Yeh et al. (2006) studied the optimization of the groundwater quality monitoring network with factorial kriging and genetic algorithms with a case study of Pingtung Plain in Taiwan. They found that Gaussian models (with a range of 28.5 km) and spherical model (with a range of 40 km) may be applied for the modeling of short and long spatial variations. Mishra and Coulibaly (2009) reviewed and discussed hydrometric network designs. Xu et al. (2015) applied entropy theory to rain gauge network analysis, using the XiangJiang River (a tributary of the Yangze River) as a case study. Among 184 rain gauges in the basin, combinations of 8 gauges were investigated. Three measures (i.e., information of the bicombinations, bias, and Nash–Sutcliffe coefficient) were applied to identify the best network combination. Based on the good and best subnetwork obtained from different combinations of the rainfall networks and using Xinanjiang and Soil and Water Assessment Tool (SWAT) models, the authors compared streamflow hydrographs generated from the subnetwork of the rain gauges and all 184 rain gauges. Li et al. (2012) proposed entropy criterion: maximum information minimum redundancy (MIMR) for hydrometric network design, which maximized the joint entropy within the optimal set, as well as the transinformation between stations within and outside of the optimal set. Additionally, the optimal set should possess the minimal duplication of information.

Using the Pijnacker region in the Netherlands as a case-study example, Alfonso et al. (2010) proposed a water level monitor network design using information theory of discrete case. They also applied the mutual information and DIT for water level monitoring. Additionally, they estimated the total correlation of the network using the following: $TC(X_1, X_2, \ldots, X_N) = \sum_{i=1}^{N} H(X_i) - H(X_1, X_2, \ldots X_N)$

As stated in Markus et al. (2003), the difficulties in conventional DIT are (1) the joint distribution must be constructed to compute the mutual information I and (2) for the multivariate case, several simplifications are made, by analyzing mutual information of pairs of stations and analyzing the resulting two-dimensional transinformation matrices or by assuming a normal distribution to calculate the multivariate joint entropy. These difficulties lead to the limitations of the previous studies: (1) an inappropriate distribution function may be selected, as a result of limited sample size available to characterize the multivariate distribution; (2) involvement of comparing different probability distribution functions represents another subjective aspect of the problem; and (3) a high level of skill and experience is needed to deal with the conventional multivariate distribution functions.

To overcome these difficulties and limitations, Xu et al. (2017) investigated the gauge network design using a two-phase copula entropy-based model. In this chapter, the copula-based network design is presented using the rainfall network from Southwest and East Central Louisiana as a case study to answer the following questions:

1. How much information is retained by a random variable (station)?
2. What is the information conveyed by several variables (stations) together?

3. How much information of the random variable (station) can be inferred from the knowledge of other stations through transinformation (i.e., mutual information) with the use of copula theory?

15.2 Dataset

Based on the study by Krstanovic and Singh (1992a, b), daily precipitation data from East Central and Southwest Louisiana are selected for the case study. Table 15.1 lists the names of the rain gauges and the lengths of records. To simplify computation, the common annual rainfall record from 1980–2015 are computed from the daily record and applied for rainfall network analysis. Figure 15.1 maps the 10 rain gauges selected. As stated in Krstanovic and Singh (1992a, b), rain gauge numbers 2, 8, and 10 are located in the East Central region, and the rest of the stations are located in the Southwest region. Table 15.2 lists the

Table 15.1. *List of the rain gauges.*

No.	Stations	Record range	No.	Stations	Record length
1	Abbeville	1923–2016	6	Lake Charles	1973–2016
2	Baton Rouge	1930–2016	7	Leland Bowman	1951–2016
3	Crowley	1927–2016	8	Livington	1980–2016
4	De Ridder	1915–2015	9	Rockfeller	1965–2016
5	Jennings	1917–2016	10	Slidell	1974–2016

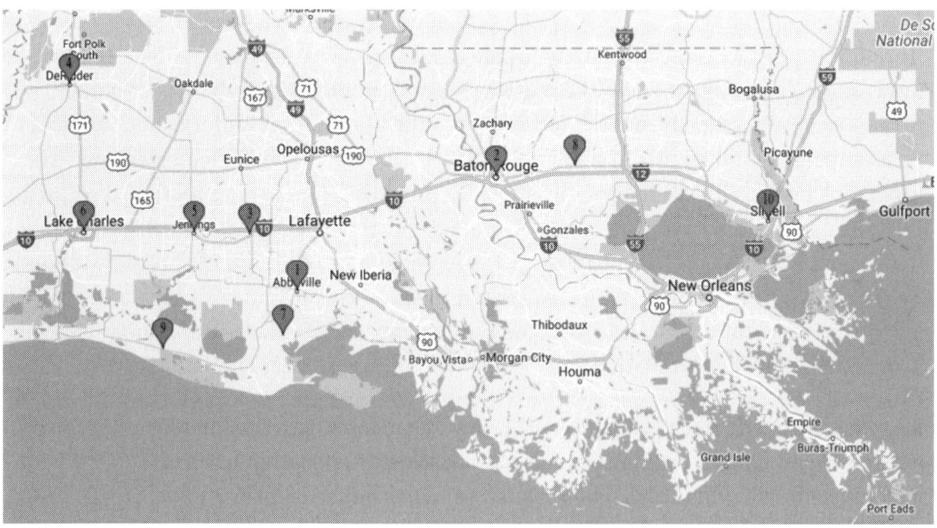

Figure 15.1 Locations of selected rain gauges (retrieved from http://maps.google.com).

Table 15.2. *Sample statistics of annual rainfall record.*

Stations	Mean (mm)	Standard deviation (mm)	Skewness	Kurtosis
Abbeville	1561.86	297.62	–0.02	2.55
Baton Rouge	1562.76	288.44	0.12	2.63
Crowley	1540.08	262.58	–0.23	2.17
De Ridder	1520.89	363.61	–0.38	3.45
Jennings	1545.28	258.95	0.11	2.25
Lake Charles	1513.91	307.29	–0.13	2.45
Leland Bowman	1559.30	318.48	–0.53	3.07
Livington	1615.64	311.20	–0.35	2.23
Rockfeller	1482.09	315.78	–0.07	2.88
Slidell	1611.07	344.99	0.48	3.36

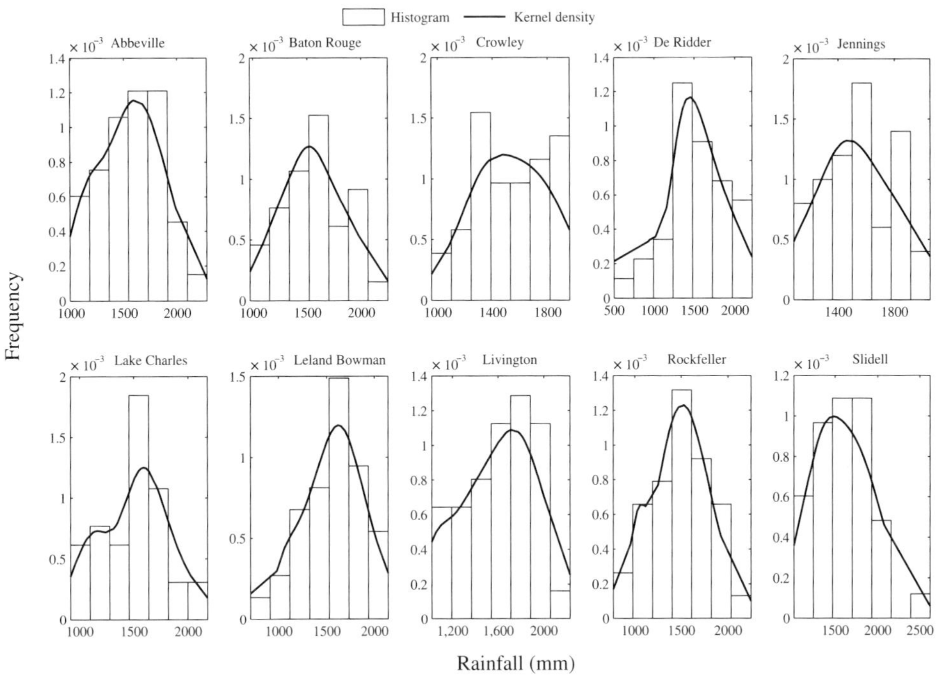

Figure 15.2 Histogram of annual rainfall at each rain gauge.

sample statistics of each rain gauge. It is seen that the annual rainfall variable (except at stations Baton Rouge, Jennings, and Slidell) is slightly skewed to the left. The histograms in Figure 15.2 show that the univariate Gaussian distribution may not be the appropriate candidate to model the marginal rainfall variables. As a result, the kernel density function is applied to model the marginal rainfall variables, which is also shown in Figure 15.2.

15.3 Methodology for Rainfall Network Design

15.3.1 Assumptions and Evaluation Procedures

Following Krstanovic and Singh (1992a, b) and Alfonso et al. (2010), the methodology for network design is based on the assumptions that (i) the stations are as independent as possible, (ii) a station should yield high marginal entropy, and (iii) the mutual information should be minimized (or in other words, maximize the nontransferred information).

The design procedure can be outlined as follows:

1. Compute the marginal entropy (H_{X_i}) and choose the station yielding the maximum marginal entropy as the center station (X_{m_1}) that needs to be added to the network.
2. Determine the second station (X_{m_2}) by minimizing the transinformation (i.e., mutual information) or maximize the nontransferred information between station m_1 and the remaining stations using the following:

$$X_{m_2} \in \min\left(I(X_{m_1};X_i)\right) i \in 1, \ldots, M, \quad i \neq m_1 \tag{15.1a}$$

or

$$X_{m_2} \in \max\left(1 - \frac{I(X_{m_1};X_i)}{H(X_{m_1})}\right) = t_1 \tag{15.1b}$$

where the mutual information I is symmetric.

$$I(X_{m_1};X_i) = H(X_{m_1}) - H(X_{m_1}|X_i) \tag{15.1c}$$

3. Determine the third station (X_{m_3}) conditioning on X_{m_1} and X_{m_2} by minimizing the transinformation (mutual information in a multivariate case) or maximize the coefficient of nontransformed information:

$$X_{m_3} \in \min\left(H(X_{m_1}, X_{m_2}) - H(X_{m_1}, X_{m_2}|X_i)\right), i = 1, \ldots, M, i \notin (m_1, m_2) \tag{15.2a}$$

Comparing with Equation (15.1c), Equation (15.2a) is equivalent to

$$X_{m_3} \in \min\left(I((X_{m_1}, X_{m_2}); X_{m_3})\right), i = 1, \ldots, M, i \notin (m_1, m_2) \tag{15.2b}$$

or

$$X_{m_3} \in \max\left(1 - \frac{H(X_{m_1}, X_{m_2}) - H(X_{m_1}, X_{m_2}|X_i)}{H(X_{m_1}, X_{m_2})}\right) = t_2, i = 1, \ldots, M, i \notin (m_1, m_2) \tag{15.2c}$$

4. Similarly, one can determine X_{m_i} using

$$\begin{aligned} X_{m_i} &\in \min\left(H(X_{m_1}, \ldots, X_{m_{i-1}}) - H(X_{m_1}, \ldots, X\_(m_{i-1})|X_i)\right) \\ &\in \min\left(I((X_{m_1}, \ldots, X_{m_{i-1}}); X_i)\right), i = 1, \ldots, M, i \notin (m_1, \ldots, m_{i-1}) \end{aligned} \tag{15.3a}$$

or

$$X_{m_i} \in \max\left(1 - \frac{H(X_{m_1}, \ldots, X_{m_{i-1}}) - H(X_{m_1}, \ldots, X_{m_{i-1}}|X_{m_i})}{H(X_{m_1}, \ldots, X_{m_{i-1}})}\right) = t_{i-1} \tag{15.3b}$$

The coefficient of nontransformed information should fulfill the following condition:

$$t_i < t_{i-1} < \cdots < t_1 \leq 1 \tag{15.4}$$

No more station is needed when $t_i \leq t_{i+1}$, i.e., the repetitive information exists at station $X_{m_{i+1}}$ such that only first $X_{m_1}, \ldots, X_{m_i}$ stations are necessary for the network with initial M stations. In what follows, we will describe the procedure of rainfall network design using the procedures discussed in this section.

15.3.2 Estimation of Marginal Entropy

As stated in Section 15.3.1, the marginal entropy needs to be first estimated, and the station that yields the largest entropy will be chosen as the center station. As stated earlier, the empirical kernel density is applied to model the marginal rainfall variables in order to avoid the possible probability distribution misidentification. Furthermore, with the characteristic of rainfall records, the kernel density with the positive support is applied for analysis.

Following Beirlant et al. (2001), the marginal entropy is written as follows:

$$H_{m_i} = -\frac{1}{n}\sum_{j=1}^{n} \ln\left(f_{m_i}^{ker}(x_{m_i}(j))\right) = -E\left(\ln f_{m_i}^{ker}\right) \tag{15.5}$$

where H_{m_i} represents the marginal entropy of rain gauge m_i; n represents the length of rainfall record; and $f_{m_i}^{ker}$represents the kernel density function with positive supports.

15.3.3 Estimation of Mutual Information and Coefficient of Nontransferrable Information

In Chapter 8, we have discussed that the mutual information between two correlated random variables (X, Y) may be expressed through the copula entropy as follows:

$$I(X;Y) = -H_C(u,v) = -E[\ln c(u,v)]; u = F_X(x), V = F_Y(y) \tag{15.6}$$

and Equation (15.1b) may be rewritten through the copula entropy as follows:

$$t_1 = \max\left(1 + \frac{H_C(u,v)}{H(X_{m_1})}\right), u = F_n(X_{m_1}); v = F_n(X_i), i = 1, \ldots, M, i \neq m_1 \tag{15.7}$$

As stated in Yang and Burn (1994) and Alfonso et al. (2010), $-H_C(u,v)/H(X_{m_1})$ represents the information inferred by first station m_1 for another station $X_i, i \neq m_1$ (on in other words, the information of m_1 maintained in $X_i, i \neq m_1$).

In a similar vine, the general equation (i.e., Equation (15.3a)) may be written as follows:

$$\begin{aligned} &H(X_{m_1}, \ldots, X_{m_{i-1}}) - H(X_{m_1}, \ldots, X\_(m_{i-1})|X_i) \\ &= H(X_{m_1}, \ldots, X_{m_{i-1}}) - [H(X_{m_1}, \ldots, X_{m_{i-1}}, X_{m_i}) - H(X_{m_i})] \end{aligned} \tag{15.8a}$$

Applying the copula theory to Equation (15.8a), we have the following:

$$H(X_{m_1}, \ldots, X_{m_{i-1}}) = \sum_j H(X_{m_j}) + H_C(U_1, \ldots, U_j, \ldots, U_{i-1}) \tag{15.8b}$$

where $U_j = F^n_{m_j}(X_{m_j}), j = 1, \ldots, i-1$ is estimated from the kernel density with the positive support.

$$\begin{aligned} H(X_{m_1}, \ldots, X\_(m_{i-1})|X_i) &= H(X_{m_1}, \ldots, X_{m_{i-1}}, X_{m_i}) - H(X_{m_i}) \\ &= \sum_j H(X_{m_j}) + H_C(U_1, \ldots, U_j, \ldots, U_i) - H(X_{m_i}), j = 1, \ldots, i \end{aligned} \tag{15.8c}$$

Equation (15.8a) may be written using the copula entropy as follows:

$$\begin{aligned} & H(X_{m_1}, \ldots, X_{m_{i-1}}) - H(X_{m_1}, \ldots, X_{m_{i-1}}|X_i) \\ &= H_C(U_1, \ldots, U_{i-1}) - H_C(U_1, \ldots, U_i) = I(X_{m_1}, \ldots, X_{m_{i-1}}|X_i) \end{aligned} \tag{15.9}$$

and Equation (15.3b) may be rewritten as follows:

$$t_{i-1} = \frac{\sum_j H(X_{m_j}) + H_C(U_1, \ldots, U_i)}{\sum_j H(X_{m_j}) + H_C(U_1, \ldots, U_j, \ldots U_{i-1})}, j = 1, \ldots, i-1 \tag{15.10}$$

It is seen from Equations (15.8)–(15.10) that the copula theory has a unique advantage of separating the marginal distribution from its joint distribution such that one may easily compute the joint and conditional entropies through the summation of marginal entropy and copula entropy.

15.4 Evaluation of Rainfall Network

To evaluate the rainfall network using the rainfall stations in Table 15.1 and Figure 15.1, the meta-elliptic (meta-Gaussian and meta-Student *t*) copulas are applied for illustrative purposes. It is worth mentioning that one may apply other copulas, including empirical copulas, to evaluate the network design.

15.4.1 Evaluation of the Rainfall Network with All Rainfall Stations

Applying Equation (15.5) for the marginal entropy estimation with the use of kernel density, Table 15.3 lists the computed marginal entropy for all 10 stations located in Louisiana. Table 15.3 shows that station Slidell (located in East Central Louisiana) yields the largest marginal entropy. As a result, Slidell is chosen as the center (the first) rain gauge station in the network.

Applying Equation (15.6), Tables 15.4 and 15.5 list the mutual information of *Slidell* with respect to the rest of the stations with the fitted meta-Gaussian and meta-Student *t* copulas, respectively. Using Slidell versus Abbeville as an illustrative example, the mutual

Table 15.3. *Estimated marginal entropy for the annual rainfall variable.*

Stations	Marginal H	Stations	Marginal H
Abbeville	7.0920	Lake Charles	7.0935
Baton Rouge	7.0760	Leland Bowman	7.1398
Crowley	6.9883	Livington	7.1209
De Ridder	7.2280	Rockfeller	7.1328
Jennings	6.9674	**Slidell**	**7.2361**

Table 15.4. *Mutual information as well as parameter estimated with respect to rain gauge Slidell (meta-Gaussian copula).*

Stations	Copula parameter	I	Stations	Copula parameter	I
Abbeville	0.542	0.17	Lake Charles	0.510	0.15
Baton Rouge	0.544	0.18	Leland Bowman	0.429	0.10
Crowley	0.685	0.32	Livington	0.716	0.36
De Ridder	0.233	0.03	Rockfeller	0.652	0.28
Jennings	0.696	0.33			

Table 15.5. *Mutual information as well as the parameter estimated with respect to rain gauge Slidell (meta-Student* t *copula).*

Stations	Copula parameter	I	Stations	Copula parameter	I
Abbeville	[0.639, 4.67E06]	0.186	Lake Charles	[0.605, 1.47E07]	0.160
Baton Rouge	[0.673, 3.18]	0.243	Leland Bowman	[0.529, 4.67E06]	0.109
Crowley	[0.764, 4.67E06]	0.334	Livington	[0.800, 7.80]	0.389
De Ridder	[0.299, 25.90]	0.030	Rockfeller	[0.726, 9.09]	0.295
Jennings	[0.772, 1.14E07]	0.349			

information between Slidell and Abbeville may be estimated using bivariate meta-Gaussian copula ($\theta = 0.542$) by taking the expectation for the copula density (i.e., bivariate Gaussian density) in the logarithm domain ($H_C = -0.17$), which results in the mutual information $I = -H_C = 0.17$. Applying the meta-Gaussian and meta-Student t copulas, Tables 15.4 and 15.5 yield similar results with De Ridder (located in Southwest Louisiana) identified as the second station needed in the network.

With Slidell and De Ridder identified as the first two stations, the third station may be identified using Equation (15.9) by setting $i = 3$ and minimizing $I(X_1, X_2|X_3) = H_C(U_1, U_2) - H_C(U_1, U_2, U_3)$, which is estimated similarly as the bivariate case. Using the stations Slidell, De Ridder, and Abbeville as an illustrative example, we compute the copula entropy of Slidell (U_1), De Ridder (U_2), and Abbeville (U_3) with the fitted bivariate and trivariate meta-Gaussian copula as follows:

Bivariate (Slidell and De Ridder):

$$\theta = 0.2329, H_C(U_1, U_2) = -0.0279;$$

Trivariate (Slidell, De Ridder, and Abbeville):

$$\boldsymbol{\theta} = \begin{bmatrix} 1 & 0.2329 & 0.5424 \\ 0.2329 & 1 & 0.5907 \\ 0.5424 & 0.5907 & 1 \end{bmatrix}, H_C(U_1, U_2, U_3) = -E[\ln c(u_1, u_2, u_3)] = -0.3970;$$

Conditional mutual information (Slidell, De Ridder|Abbeville):

$$I(X_1, X_2|X_3) = H_C(U_1, U_2) - H_C(U_1, U_2, U_3) = -0.0279 + 0.3970 = 0.3691.$$

Tables 15.6 and 15.7 list all the computed conditional mutual information using the fitted meta-Gaussian and meta-Student *t* copulas. As seen in Tables 15.6 and 15.7, the meta-Gaussian and meta-Student *t* copulas again are in agreement that Baton Rouge is the third station needed.

Proceeding with the same procedure, we will add more stations to the network until the criterion described in Equation (15.4) is no longer valid. The final results are listed in Table 15.8 using the fitted meta-Gaussian copula as an example. The same stations are

Table 15.6. *Mutual information computed with respect to rain gauges Slidell (X_1) and De Ridder (X_2) (meta-Gaussian copula).*

$X_1, X_2 \mid X_3$	$-H_C(X_1, X_2) = I(X_1, X_2) = 0.0279$		
$I(X_1, X_2 \mid Abbeville)$	0.369	$I(X_1, X_2 \mid Lake\ Charles)$	0.388
$I(X_1, X_2 \mid Baton\ Rouge)$	0.177	$I(X_1, X_2 \mid Leland\ Bowman)$	0.286
$I(X_1, X_2 \mid Crowley)$	0.414	$I(X_1, X_2 \mid Livgington)$	0.360
$I(X_1, X_2 \mid Jennings)$	0.508	$I(X_1, X_2 \mid Rockfeller)$	0.364

Table 15.7. *Mutual information computed with respect to rain gauges Slidell (X_1) and De Ridder (X_2) (meta-Student* t *copula).*

$X_1, X_2 \mid X_3$	$-H_C(X_1, X_2) = I(X_1, X_2) = 0.0279$		
$I(X_1, X_2 \mid Abbeville)$	0.369	$I(X_1, X_2 \mid Lake\ Charles)$	0.388
$I(X_1, X_2 \mid Baton\ Rouge)$	0.177	$I(X_1, X_2 \mid Leland\ Bowman)$	0.286
$I(X_1, X_2 \mid Crowley)$	0.414	$I(X_1, X_2 \mid Livgington)$	0.360
$I(X_1, X_2 \mid Jennings)$	0.508	$I(X_1, X_2 \mid Rockfeller)$	0.364

Table 15.8. *Final results for the rainfall network design (meta-Gaussian copula).*

Stations already identified	Station added	$H(X_1, .., X_i)$	$H(X_1, \ldots, X_i \mid X_{i+1})$	$I((X_1, \ldots, X_i); X_{i+1})$	t
—	Slidell	7.236	—	—	1
Slidell	De Ridder	7.236	7.208	0.0279	0.996
Slidell, De Ridder	Baton Rouge	14.436	14.259	0.177	0.987
Slidell, De Ridder Baton Rouge	Leland Bowman	21.336	20.969	0.367	0.983
Slidell, De Ridder Baton Rouge, Leland Bowman	Livington	28.108	27.519	0.589	0.979

Figure 15.3 Rain gauges needed for the network (retrieved from http://maps.google.com).

obtained with the use of the Student t copula. Figure 15.3 plots the identified rain gauges on the map. As shown in Figure 15.3, all three rain gauges located in East Central Louisiana are needed for rainfall network design, while only two of seven rain gauges are needed for those located in Southwest Louisiana. This information may indicate more uncertainty within East Central Louisiana than that within Southwest Louisiana.

15.4.2 Evaluation of Rain Gauges Located in Southwest Louisiana Only

From Figure 15.1, there are seven stations located in the Southwest region. Applying the same procedure as that for all the rain gauges, we can identify the rain gauges needed for the Southwest region only. Table 15.9 lists the final four identified rain gauges with the use

Table 15.9. *Final results for rainfall network design (Southwest region) (meta-Gaussian copula).*

Stations already identified	Station added	$H(X_1, .., X_i)$	$H(X_1, ..., X_i \vert X_{i+1})$	$I((X_1, ..., X_i); X_{i+1})$	t
—	De Ridder	7.228	—	—	1
De Ridder	Rockfeller	7.228	7.117	0.112	0.985
De Ridder, Rockfeller	Crowley	14.249	13.881	0.368	0.974
De Ridder, Rockfeller Crowley	Abbeville	20.87	20.275	0.595	0.972

Figure 15.4 Final identification of rain gauges needed for Southwest Louisiana (retrieved from http://maps.google.com).

of the meta-Gaussian copula as an example. Figure 15.4 maps the stations identified for the Southwest region. Comparing to the final result of the Southwest region with that for combined Southwest and East Central regions, station De Ridder is identified in both cases. In addition, there is only about 19-mile distance between Leland Bowman (selected for Southwest and East Central only) and Abbeville (Southwest only).

15.5 Summary

In this case study, rain gauges located in the East Central and Southwest regions of Louisiana are applied for the rainfall network design. Considering the East Central and Southwest regions together, the needed rain gauges reduce from 10 to 5. All three rain gauges in the East Central region are needed, while only De Ridder and Leland Bowman (about 19 miles southwest of Abbeville) are needed for the Southwest region.

Considering Southwest Louisiana only, four out of seven stations are needed. Of the four stations needed, station De Ridder is the common station identified for both cases. Besides the De Ridder station, the fourth added station (Abbeville) is geographically close to Leland Bowman station.

The spatial distribution of rain gauges, for the East Central and Southwest regions, and the Southwest region only, well covers the region studied respectively. Investigation of the network results in the reduction of the number of rain gauges.

Application of the empirical marginal distributions (kernel density) for the marginal rainfall may avoid the misidentification of the marginal distributions. Application of the copula theory eases the complexity of estimating the joint and conditional entropies; in higher dimensions, the estimation may be made by separately assessing the marginal entropy and the copula entropy.

The network design with the copula theory may be applied not only in the rainfall network, it may also be easily applied to other network design problems (streamflow gauges, sewer monitoring program, etc.). In addition, it may be applied to add an additional point if the current monitor program may not properly represent the system.

References

Al-Zahrani, M. and Husain, T. (1998). An algorithm for designing a precipitation network in south-western region of Saudi Arabia. *Journal of Hydrology*, 205, 205–216.

Alfonso, L., Lobbrecht, A., and Rice, R. (2010). Information theory-based approach for location of monitoring water level gauges in polders. *Water Resources Research*, 46, W03528, doi:10.1029/2009WR008101.

Beirlant, J., Dudewicz, E. J., Gyorfi, L., and Van deMeulen, E. C. (2001). Nonparametric entropy estimation: an overview. http://jimbeck.caltech.edu/summerlectures/references/Entropy%20estimation.pdf.

Chow, C. K. and Liu, C. N. (1968). Approximating discrete probability distributions with dependence tree. *IEEE Transactions on Information Theory*, IT-14(3), 462–467.

Dong, X., Dohmen-Janssen, C. M., and Booij, M. J. (2005). Approximate spatial sampling of rainfall for flow simulation. *Hydrological Sciences Journal*, 50(2), 279–298.

Krstanovic, R. F. and Singh, V. P. (1992a). Evaluation of rainfall networks using entropy: I. Theoretical development. *Water Resources Management,* 6, 279–293.

Krstanovic, R.F. and Singh, V.P. (1992b). Evaluation of rainfall networks using entropy: II. Application. *Water Resources Management,* 6, 295–314.

Li, C., Singh, V. P., and Mishra, A. K. (2012). Entropy theory-based criterion for hydrometric network evaluation and design: maximum information minimum redundancy. *Water Resources Research*, 48, W05521, doi:10.1029/2011WR011251.

Markus, M., Knapp, H. V., and Tasker, G. D. (2003). Entropy and generalized least square methods in assessment of the regional value of streamgages. *Journal of Hydrology*, 283, 107–121, doi: 10.1016/S0022–1694(03)00224–0.

Mishra, A. K. and Coulibaly, P. (2009). Developments in hydrometric network design: a review, *Reviews of Geophysics*, 47, RG2001, doi:10.1029/2007RG000243.

Xu, H., Xu, C.-Y., Sælthun, N. R., Xu, Y., Zhou, B., and Chen, H. (2015). Entropy theory based multi-criteria resampling of rain gauge networks for hydrological modelling: a case study of humid area in southern China, *Journal of Hydrology*, 525, 138–151.

Xu, P. C., Wang, D., Singh, V. P., et al. (2017). A two-phase copula entropy-based multiobjective optimization approach to hydrometeorological gauge network design. *Journal of Hydrology*, 555, 328–341.

Yang, Y. and Burn, D. H. (1994). An entropy approach to data collection network design, *Journal of Hydrology*, 157, 307–324.

Yeh, M. S., Lin, Y. P., and Chang, L. C. (2006) Designing an optimal multivariate geostatistical groundwater quality monitoring network using factorial kriging and genetic algorithms, *Environmental Geology*, 50, 101–121, doi:10.1007/s00254–006–0190–8.

16

Suspended Sediment Yield Analysis

ABSTRACT

In the previous chapters, we have briefly introduced applications of copulas to analyses of rainfall, streamflow, drought, water quality, and compound extremes, as well as network design. In this chapter, we will introduce suspended sediment transport. Two case studies will be discussed to (i) apply copulas to construct the discharge-sediment rating curve using the Yellow River dataset; and (ii) investigate the dependence among precipitation, discharge, and sediment yield using the event-based dataset retrieved from the flume #3 at Santa Rita experimental watershed.

16.1 Discharge-Sediment Rating Curve Construction

In this section, application of copulas will be illustrated for modeling suspended sediment transport in regard to discharge-sediment rating curve by:

i. Identifying the study region and collecting discharge and suspended sediment data
ii. Applying the mixed copula that may reasonably capture the upper, lower, and overall dependence to model the discharge-suspended sediment data
iii. Comparing the performance of the copula approach to the classic USGS discharge-suspended sediment rating equations
iv. Interpreting the results according to the underlying surface and channel characteristics

The Yellow River, the most famous sediment-rich river system in the world, is used here as a case study for discharge-sediment rating curves. More specifically, sediment transport is studied for the middle reach of the Yellow River basin. The middle reach is the major sediment source area supplying 90% of the total sediment yield and 38% of the total water resources of the middle Yellow River (MYR) basin. Geologywise, the underlying surface materials change dramatically from north to south, varying from soft rock (i.e., Pisha Rock by locals), sand, loess, to rock mountains (Li and Li, 1994; Wang et al., 2007; Ni et al., 2008) with the median particle size decreasing from north (0.088 mm) to south (0.018 mm). To evaluate the study, we choose four representative stations for each geologically identified unique underlying surface: (1) typical loess – SuiDe, (2) soft-rock sand – Wangdaohengta, (3) rock mountain – Liujiahe, and (4) sand – Gaojiabao. Throughout the case study, we will apply both the classic USGS rating equation and copulas and also compare their performance. To prepare the dataset, the discharge values less than average discharge will be dropped out of the dataset, since we are more concerned with the large amount of suspended sediment transported during runoff events.

The discharge-sediment rating curve has been commonly applied to forecast suspended sediment yield or concentration. The classic USGS sediment rating curve (i.e., Equation (16.1)) through either power function or log-linear function has been commonly applied to achieve this end:

$$SSL = aQ^b \quad or \quad \ln(SSL) = a^* + b\ln(Q) : a^* = \ln(a) \tag{16.1}$$

In the previous chapters, we have discussed different types of copulas as well as their applications. In this section, we will apply mixed copula. Let U and V represent the marginals of discharge and suspended sediment, respectively; then we can write the mixed copula as follows:

$$C(u, v; \boldsymbol{\theta}) = w_1 C_1(u, v; \boldsymbol{\theta}_1) + w_2 C_2(u, v; \boldsymbol{\theta}_2) + w_3 C_3(u, v; \boldsymbol{\theta}_3); \exists w_1 + w_2 + w_3 = 1 \tag{16.2}$$

Commonly, copulas chosen are the copula with $\lambda_L \neq 0$, the meta-Gaussian copula, and the copula with $\lambda_U \neq 0$. Here we will choose the survival Gumbel–Hougaard copula, Gumbel–Hougaard copula, and Gaussian copula as the mixture. The survival Gumbel–Hougaard copula mirrors the Gumbel–Hougaard copula. As a result, the survival Gumbel–Hougaard copula has the lower-tail dependence. With the preceding mixture, Equation (16.2) may be rewritten as follows:

$$C_{mix}(u, v; \boldsymbol{\theta}) = w_1 C^{GH}(u, v; \theta_1) + w_2 C^{Gaussian}(u, v; \theta_2) + w_3 C_S^{GH}(u, v; \theta_3); \theta_1, \theta_3 \geq 1 \tag{16.3}$$

In Equation (16.3), the survival Gumbel–Hougaard copula (C_S^{GH}) can be expressed as follows:

$$C_S^{GH}(u, v; \theta_3) = u + v - 1 + C^{GH}(1 - u, 1 - v; \theta_3) \tag{16.3a}$$

Again, denoting c as the copula density, the copula density of C_S^{GH} can be given as follows:

$$c_S^{GH}(u, v; \theta_3) = c^{GH}(1 - u, 1 - v; \theta_3) \tag{16.3b}$$

The lower- and upper-tail dependence coefficient is given for the survival Gumbel–Hougaard and Gumbel–Hougaard copula as follows:

$$SGH : \lambda_L = 2 - 2^{\frac{1}{\theta_3}}, \quad GH : \lambda_U = 2 - 2^{\frac{1}{\theta_1}} \tag{16.3c}$$

Substituting Equation (16.3a) into Equation (16.3), the density function for the mixture copula can be given as follows:

$$c_{mix}(u, v; \boldsymbol{\theta}) = w_1 c^{GH}(u, v; \theta_1) + w_2 c^{Gaussian}(u, v; \theta_2) + w_3 c^{GH}(1 - u, 1 - v) \tag{16.4}$$

Similar to the discussions in other application chapters, to predict the suspended sediment, one needs to estimate the expected suspended discharge with $E(SSL|Q = q)$ from

$$P(SSL \leq ssl|Q = q) = 0.5 \tag{16.5}$$

Equation (16.5) may be also called as the median forecast. Applying the copula theory, Equation (16.5) may be rewritten as follows:

$$P(SSL \leq ssl|Q = q) = P\left(F_{SSL} \leq F_{SSL}(ssl)\middle|F_Q = F_Q(q)\right) = 0.5 \tag{16.5a}$$

In Equation (16.5a), $P\left(F_{SSL} \leq F_{SSL}(ssl)\middle|F_Q = F_Q(q)\right)$ can be written through copula as follows:

$$P\left(F_{SSL} \leq F_{SSL}(ssl)\middle|F_Q = F_Q(q)\right) = \left.\frac{\partial C(F_{SSL}, F_Q; \boldsymbol{\theta})}{\partial F_Q}\right|_{F_Q = F_Q(q)} \tag{16.5b}$$

Let $h(F_{SSL}, F_Q; \boldsymbol{\theta}) = \left.\frac{\partial C(F_{SSL}, F_Q; \boldsymbol{\theta})}{\partial F_Q}\right|_{F_Q = F_Q(q)}$; we will have the following:

$$F_{SSL} = h^{-1}(0.5, F_Q(q), \boldsymbol{\theta}) \tag{16.6}$$

The 90% bound may be written through VaR(5%) (i.e., lower bound) and VaR(95%) (i.e., upper bound) as follows:

$$F^{5\%}{}_{SSL} = h^{-1}(0.05, F_Q(q), \boldsymbol{\theta}); F_{SSL}^{95\%} = h^{-1}(0.95, F_Q(q), \boldsymbol{\theta}) \tag{16.7}$$

With the computed $F_{SSL}, F_{SSL}^{5\%}, F_{SSL}^{95\%}$, we usually compute the corresponding value in the real domain with the fitted parametric marginal distributions. In this case, we will use the kernel density function (with a normal kernel) with the positive support to estimate the projected sediment yield. Application of the kernel density function for prediction avoids the possible ill-identification of the marginal distributions due to the very high skewness. From the sample statistics listed in Table 16.1, it is seen that sediment yield is heavily skewed and tailed.

Parameters of the mixture copula are estimated with the pseudo-MLE. The Weibull plotting-position formula is applied to compute the empirical marginals. In the case of model parsimony, the copula will be dropped out of the mixture if its corresponding weight

Table 16.1. *Sample statistics of sediment yield for four selected stations.*

Station	Mean (kg/s)	Standard deviation (kg/s)	Skewness	Kurtosis
Suide	11.262.25	32.037.33	7.54	86.83
Gaojiabao	1.273.16	7.481.30	12.24	185.30
Wangdaoheng	6.260.81	29.372.95	7.46	62.33
LiujiaHe	11.698.75	46.025.54	9.28	115.36

Table 16.2. *Parameters estimated for the discharge-sediment rating curve from the copula.*

Underlying Surface	Geographic characteristics Station	Area (km^2)	Slope	Copula Gaussian	GH	SGH	USGS $[a^*, b]$
Loess	Suide	3893	22.99[a]	–	3.071	–	[1.388, 2.254]
Sand	Gaojiabao	2095	0.74	–	1.863	–	[–2.373, 2.822]
Soft-rock	Wangdaohengta	3390	0.29	–	2.955	–	[–0.023, 2.134]
Rock mountain	Liujiahe	2361	44.25[a]	–	2.544	–	[–0.629, 2.595]

Note: [a] The percentage of the slope steeper than 1.5%.

is less than 0.1. For example, if the weights for both the Gaussian (w_2) and survival Gumbel–Hougaard copula (w_3) are less than 0.1, the mixture copula will be reduced to the Gumbel–Hougaard copula. With this procedure and the maximum likelihood applied to pseudo-observations (i.e., the empirical marginal of discharge and suspended sediment), Table 16.2 lists the estimated parameters of four stations from both copula approach as well as the USGS log-linear regression equations. The results of copula approach in Table 16.2 indicate that the Gumbel–Hougaard copula is the only copula needed based on the model selection procedure (i.e., only consider the copulas with the weight higher than 10% in the mixture). This is quite understandable due to the procedure of data processing: (1) omit the [discharge, sediment] pair when the discharge is lower than the average discharge; and (2) omit the [discharge, sediment] when the sediment yield is less than 0.5% of the average sediment yield.

To visually compare the copula approach with the USGS equation, Figure 16.1 compares the fitted copula function and USGS equation (in Figure 16.1A) as well as their forecast power (in Figure 16.1B). The forecast results (Figure 16.1B) are listed in Table 16.3.

From Figure 16.1 and Table 16.3, we can reach the following conclusions.

16.1.1 Stations Suide and Liujiahe

These two stations yield similar results as follows:

a. The relation between discharge and suspended sediment yield is obviously nonlinear for the discharge lower than 100 cms and the relation tends to be linear in logarithm domain for the discharge higher than 100 cms. The fitted copula function properly follows this observed trend.
b. The linear USGS regression equation in the logarithm domain (or the power function in real domain) tends to overestimate the suspended sediment yield for discharge higher than 100 cms at both locations.
c. From the forecast results, it again shows that the copula-based approach provides a very reasonable forecast for sediment yield, especially in the case of high sediment yield due

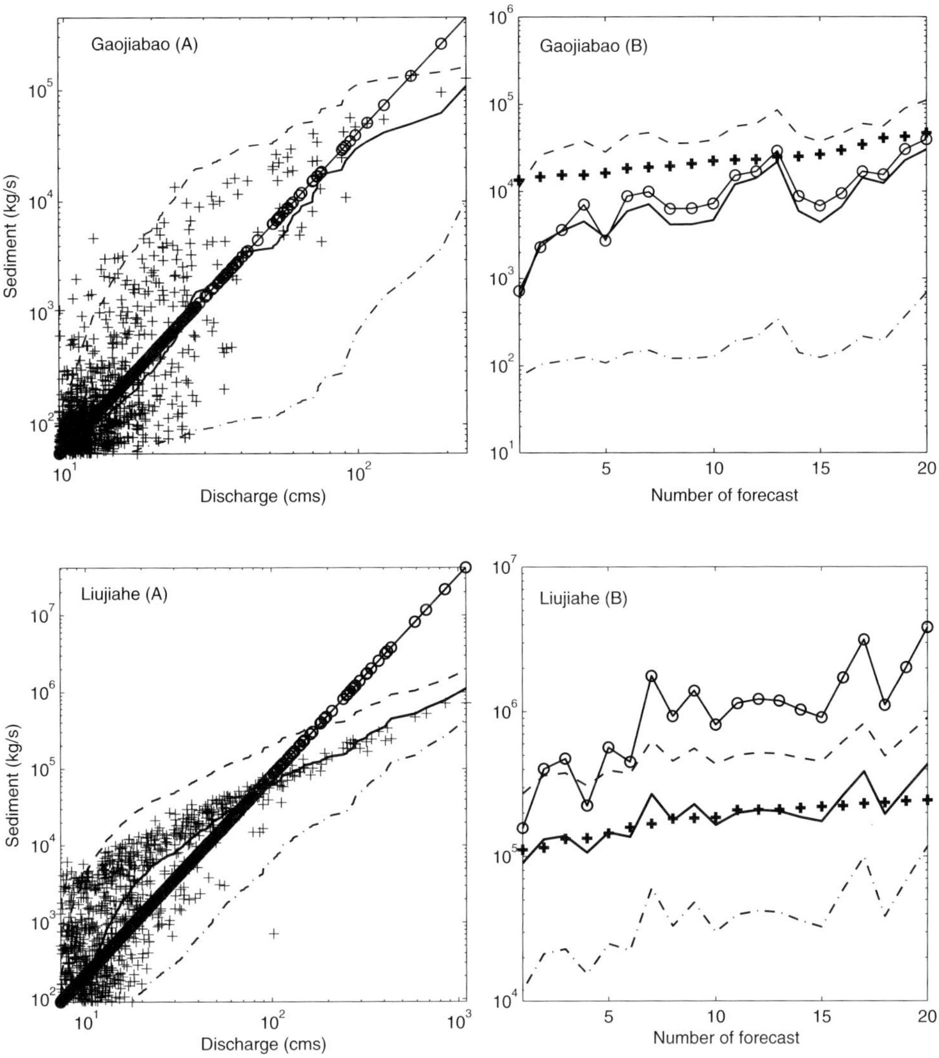

Figure 16.1 *Plots of discharge and suspended sediment yield for all four stations: (A) fitted copula function and USGS equation; (B) comparison of the forecast power between the copula function and USGS equation).*

to high runoff events. During the flood season, one may directly use the copula approach to project the possible suspended sediment rushing downhill.

d. For the same dataset for testing the forecast power, the USGS equation significantly overestimates sediment yield. The results obtained directly from the USGS equation are not reliable.

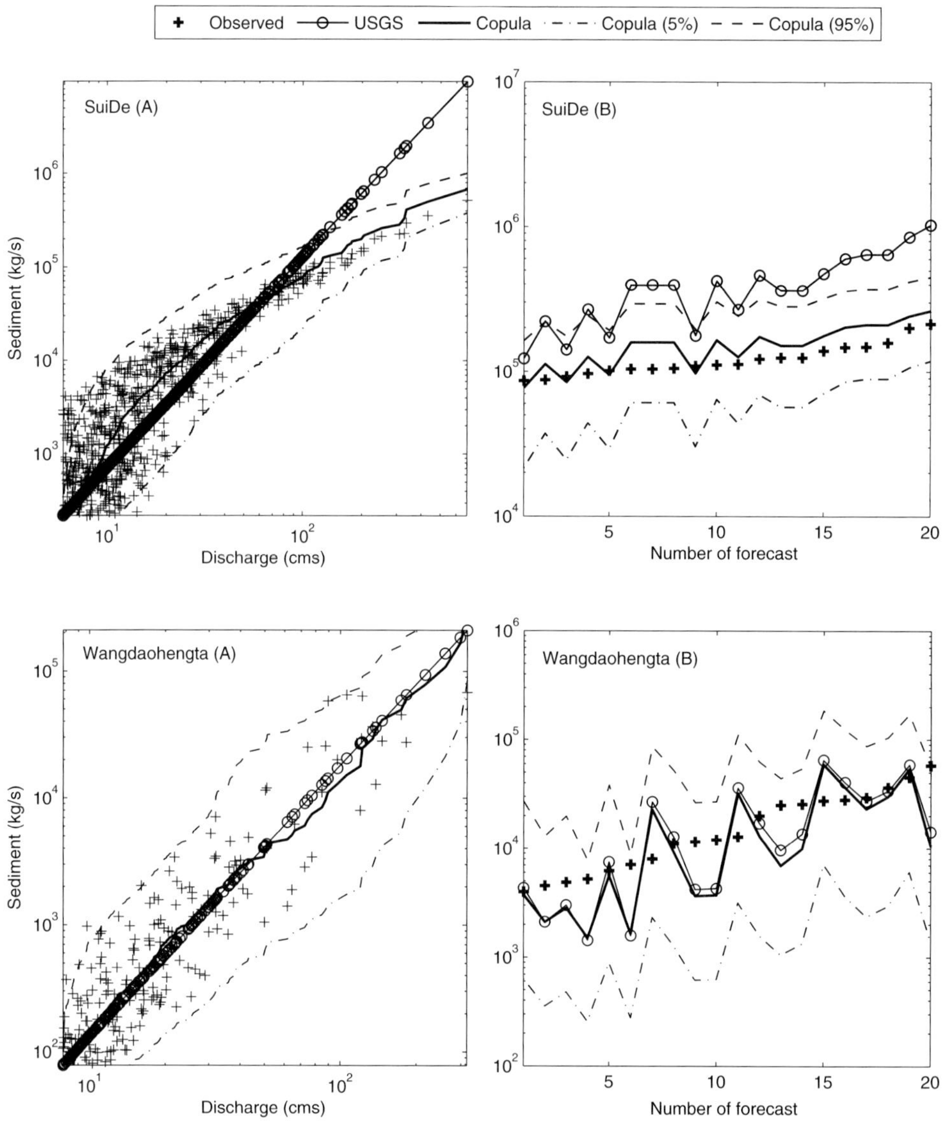

Figure 16.1 (*cont.*)

16.1.2 Stations Gaojiabao and Wangdaohengta

Like the stations at Suide and Liujiahe, similar results are obtained for Gaojiabao and Wangdaohengta stations as discussed in what follows:

a. Unlike the Suide and Liujiahe stations, the scatterplots tend to be more linear at Gaojiabao and Wangdaohengta stations. As a result, both the copula and USGS equation approaches follow the trend reasonably well.

Table 16.3. *Comparison of forecast power of copula and USGS approaches.*

Q (cms)[a]	SSY(kg/s)[a,b]	Copula[c]	USGS[c]	Q (cms)	SSY(kg/s)	Copula	USGS
Suide station				Liujiahe station			
97.7	86,073.7	76,957.20	122,621.87	128	110,976	89,958.95	156,852.25
127	87,249	112,057.81	221,483.23	184	114,632	131,095.07	402,279.65
104	92,040	84,049.35	141,170.48	196	13,1908	138,174.95	473,956.12
138	96,462	125,758.45	267,092.44	147	133,182	106,072.00	224,639.69
113	101,022	94,727.58	170,214.45	210	144,060	145,989.60	566,893.19
164	103,812	158,278.90	394,136.30	192	158,976	135,863.41	449,260.01
164	103,812	158,278.90	394,136.30	326	167,890	269,808.11	1774,991.57
164	104,960	158,278.90	394,136.30	254	182,626	176,613.58	928,772.23
115	109,020	971,64.83	177,081.01	297	184,140	229,288.33	1,393,760.25
169	111,033	164,714.59	421,742.65	241	185,811	165,330.82	810,388.83
138	112,056	125,758.45	267,092.44	275	209,000	200,007.38	1,141,415.85
176	121,968	173,833.17	462,146.64	282	210,090	208,958.40	1,218,358.53
158	124,346	150,666.33	362,375.14	279	210,924	205,069.38	1,185,005.20
158	124,346	150,666.33	362,375.14	264	215,952	187,018.04	1,026,674.31
178	139,018	176,451.53	474,069.33	252	221,004	174,722.80	909,911.49
198	146,916	202,558.00	602,680.75	322	224,112	264,266.82	1,719,020.60
204	147,900	210,140.26	644,633.46	407	231,990	385,953.85	3,157,341.74
204	157,896	210,140.26	644,633.46	272	236,096	196,331.14	1,109,380.45
231	200,046	241,611.05	853,096.43	343	242,158	293,295.70	2,025,308.37
251	213,099	262,330.06	1,028,698.61	438	244,842	433,235.20	3,819,957.26
309	224,952	328,976.59	1,643,641.81	375	304,125	338,053.86	2,552,848.84
Gaojiabao Station				Wangdaohengta Station			
23.9	13,503.5	6,01.10	723.16	51.1	3,980.69	3,729.80	4,313.90
36.1	14,692.7	2,573.53	2315.61	36.5	4,526	2,102.76	2,104.27
42.3	15,397.2	3,527.72	3621.66	43.1	4,870.3	2,825.67	2,999.91

Table 16.3. (*cont.*)

Q (cms)[a]	SSY(kg/s)[a,b]	Copula[c]	USGS[c]	Q (cms)	SSY(kg/s)	Copula	USGS
53.5	15,408	4,535.20	7027.18	30.5	5,215.5	1,503.63	1,434.50
38.3	16,200.9	2,973.66	2736.31	66	6,217.2	5,577.68	7,446.53
57.9	18,296.4	5,916.65	8783.01	31.9	7,081.8	1,636.34	1,578.65
60.3	18,873.9	7,124.40	9849.59	120	7,992	22,696.57	2,6662.30
51.5	19,415.5	4,171.17	6310.86	84.8	11,108.8	9,176.62	12,711.38
51.6	20,691.6	4,185.74	6345.50	50.4	11,491.2	3,650.05	4,188.80
54.1	22,289.2	4,674.29	7251.86	51	11,985	3,718.39	4,295.91
70.3	22,917.8	11,992.07	15186.77	138	12,806.4	32,135.92	35,925.11
72.9	23,182.2	14,130.16	16825.71	97.4	19,967	13,019.05	17,082.53
88.4	24,663.6	21,679.14	28989.32	74.4	25,072.8	6,926.16	9,615.22
57.9	25,012.8	5,916.65	8783.01	87.6	25,579.2	9,919.53	13,623.63
52.9	26,502.9	4,409.92	6807.06	182	27,482	59,230.34	64,838.33
59.4	29,581.2	6,633.95	9440.36	146	28,032	36,673.52	40,514.79
73	34,310	14,219.50	16890.92	121	29,403	23,187.85	27,138.58
70.8	40,780.8	12,360.18	15493.56	134	36,448	29,939.41	33,739.90
89.7	423,38.4	22,507.14	30208.52	174	45,240	53,886.25	58,908.88
98.5	46,886	30,084.53	39338.98	89.6	57,881.6	10,489.26	14,295.85
108	50,220	37,758.66	51011.10	122	63,440	23,683.93	27,619.34

Notes: [a] Observed; [b]: SSY stands for suspended sediment yield; [c] forecast from discharge (Q) in column 1.

b. The forecast performance using both approaches reach very similar results. The suspended sediment at Wangdaohengta is better forecasted than that at Gaojiabao (which exhibits as more of a nonlinear relation than that at Wangdaohengta).

Besides the obtaining the preceding results by grouping the stations with similar results together, we may gain more information from geomorphological and geological aspects:

i. The slope of the river channel in different subreaches may be the main deciding factor for the relation between discharge and suspended sediment. The higher the percentage of slope (i.e., steeper than 1.5%), the more nonlinear the relation becomes. As the flattest station, Wangdaohengta reaches the most linear relation compared to all other stations (Figure 16.1; Wangdaohengta (A)). With the increase of slope, the nonlinearity becomes more and more obvious from low to high as Gaojiabao, Suide, and Liujiahe.
ii. For the sections with steep river channels, the sediment from the upper subreach may be easier to be transported as suspended sediment. The local sediment may also be easier to be picked up and transported as suspended sediment. Suide station, located in the typical Loess region, may be a good example to carry both sediment from the upper subreach and local sediment as suspended sediment during flood events. Liujiahe station, located in the rock mountain region, may be a good example to carry sediment from upper subreach as suspended sediment.
iii. For the sections with flat river channels, for the same amount of runoff, the flow velocity and the corresponding shear stress will be significantly reduced. As a result, the suspended sediment from the upper subreach may be deposited as a bed load rather than suspended sediment. Wangdaohengta station in the soft-rock region may be a good example to illustrate sediment deposition in a flat river channel and linear relation in the logarithm domain. Gaojiabao station in the sand region, which is slightly more nonlinear than Wangdaohengta station, may be a good example to explain the deposition with more suspended sediment transported than Wangdaohengta: (a) the overall slope at Gaojiabao is slightly steeper than that at Wangdaohengta and (b) the particle size of the underlying sand surface is generally smaller than that of soft-rock.

Above all, this case study suggests to apply the copula approach to construct the discharge-sediment rating curve for steeper channels, while the USGS regression equation in the logarithm domain may be safely applied for flatter channels.

Applying the sediment-rich middle reach of Yellow River with four major underlying surfaces, the case study indicates the following:

- The USGS regression equation may work as well as the copula-based method when the channel is flat. In this case, sediment may be harder to be transported or it may actually move as bed load rather than suspended sediment due to low flow velocity and low shear force. For the flat channel, the type of underlying surface does not seem to be the dominating factor for the suspended sediment transport.
- The copula-based method works much better than the USGS regression equation when the channel is steep. For the steep channel, the relation of discharge and suspended

sediment in the logarithm domain is no longer linear. The nonlinearity for the discharge-suspended sediment rating curve is dependent on the underlying surface when the discharge is lower than a certain threshold. However, for the discharge higher than the certain threshold, the nonlinear relation seems to be replaced by the linear relation similar to the linear relation for the flatter channel.

16.2 Dependence Study of Precipitation, Discharge, and Sediment Yield

16.2.1 Event-Based Sediment Dataset

Precipitation, discharge, and sediment yield are related. In this section, we are going to use the events collected from Flume #3 at Santa Rita experimental watershed (1975–2018) as the case study to further evaluate this interrelation. In the available dataset, sediment data were missing from 1990–2000. Figure 16.2 shows the watershed map retrieved from tucson.ars.ag.gov (USDA, n.d.). As shown in Figure 16.2, Flume 3 covers a small area of 6.81 acres. Table 16.4 lists the events selected from the available dataset.

16.2.2 Empirical Analysis of Sediment Dataset

Before we proceed with the copula application, we first perform an empirical analysis using the sediment dataset. The sample statistics listed in Table 16.5 show that the target variables are skewed to the right and tailed. As shown in Figure 16.3, the histogram and kernel density frequency yield the same results. It should be noted that the support of the kernel density needs to be positive based on the nature of dataset. Table 16.6 lists the rank-based Kendall and Spearman correlation coefficients of the pair variables. Results indicate the positive dependence for all pair variables. The scatter plots and chi- and K-plots further confirm the positive dependence structure for the pair variables, as shown in Figures 16.4 and 16.5.

Now, with the dependence structure identified, we may further study the dependence with the use of copula theory.

16.2.3 Dependence Study of Runoff Volume and Sediment Yield with Copula Theory

We have shown that the pair variables are positively dependent. Additionally, there exists a higher degree of association between sediment yield versus runoff volume and sediment yield versus peak runoff than that between sediment yield versus rainfall depth. To illustrate the copula application in sediment analysis, we will first do bivariate analysis using sediment yield and runoff volume followed by the multivariate sediment analysis.

Before we apply the copula theory to the sediment yield and runoff volume, we will first compute the tail dependence coefficient. Applying all three empirical tail dependence coefficients LOG, SEC, and CFG, we compute the following: $\widehat{\lambda}_U^{SEC} = 0.696$,

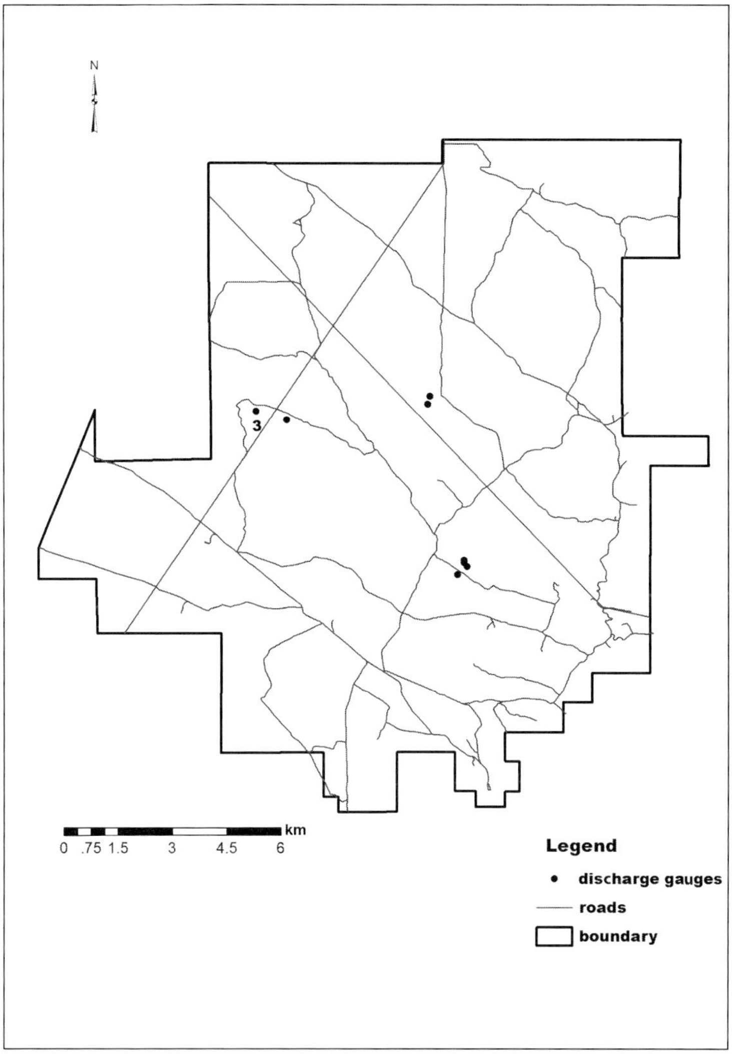

Figure 16.2 Santa Rita experimental watershed (retrieved from tucson.ars.ag.gov).

$\widehat{\lambda}_U^{LOG} = 0.683$, $\widehat{\lambda}_U^{CFG} = 0.686$. From the computed empirical upper-tail dependence coefficient, there is minimal difference in the results with the use of three approaches. Thus, we may apply the copula belonging to the extreme value family to do bivariate analysis. As discussed in Chapter 4, we will apply the Gumbel–Hougaard copula.

The empirical probabilities computed from the kernel density with positive support (Table 16.7) are applied to estimate the Gumbel–Hougaard copula parameter. Additionally, Figure 16.6 plots the comparison of kernel density–based CDF with that computed using the Weibull plotting-position formula. The scatter plots of CDFs (i.e., Figure 16.6(c))

Table 16.4. *Selected rainfall, discharge, and sediment data (flume 3, Santa Rita watershed).*

Dates	Sediment yield (lb)	Rainfall depth (inch)	Runoff volume (ft^3)	Peak runoff (cfs)
9/1/75	5,512	1.33	4,759	7.506
8/22/76	1,329	0.35	2,002	4.909
9/5/77	1,426	0.41	3,116	4.526
9/11/77	2,279	0.45	4,027	4.157
10/6/77	9,974	2.32	16,180	14.7
12/28/77	5,833	0.36	4,948	4.464
8/10/83	54,092	1.32	17,350	12.91
10/19/83	14,826	0.93	11,800	8.337
10/3/84	1,990	0.51	3,604	4.328
7/10/88	12,696	1.22	9,240	8.171
8/7/89	4,481	2.1	17,060	6.309
8/11/89	20,256	1.43	14,260	10.06
6/22/00	2,015	0.69	4,073	5.798
8/17/00	1,924	0.68	5,733	5.26
7/22/02	4,698	1.15	8,385	7.565
8/28/02	5,515	0.455	5,453	6.96
9/6/02	14,820	2.765	20,360	21.51
7/22/03	17,146	0.86	10,820	14.84
4/1/04	4,869	0.565	4,483	5.981
7/13/04	14,654	1.185	11,420	13.22
7/23/04	69,546	2.2	35,770	23.22
8/5/04	10,980	0.605	6,282	7.565
7/31/05	4,256	0.5	4,426	5.616
8/2/05	13,187	0.64	6,508	11.67
8/14/05	14,623	0.84	5,868	5.26
8/23/05	14,358	1.31	9,362	4.395
9/8/05	4,748	0.355	2,730	4.912
7/4/06	11,577	1.04	7,295	10.69
7/29/06	11,087	0.93	7,734	7.979
9/8/06	5,134	0.305	3,175	5.437
9/12/06	10,304	0.305	3,416	4.737
7/12/07	12,520	0.87	5,979	6.96
7/31/07	12,011	0.765	5,056	5.616
8/1/07	26,797	1.19	17,140	14.29
9/16/07	9,385	0.635	5,486	6.96
7/19/08	54,921	3.425	41,500	29.04
7/22/08	34,092	1.17	13,740	11.18
8/1/08	36,473	1.335	21,320	16.55
8/31/08	23,429	1.365	16,120	13.22
9/10/08	26,849	1.45	16,320	13.75

Table 16.4. (*cont.*)

Dates	Sediment yield (lb)	Rainfall depth (inch)	Runoff volume (ft^3)	Peak runoff (cfs)
9/11/08	13,059	1.02	8,666	5.616
7/3/09	5,730	2.135	23,870	15.97
7/3/09	38,128	2.135	23,870	15.97
7/8/09	29,706	1.505	16,130	10.45
9/7/10	8,498	0.61	5,520	5.26
7/20/11	21,671	1.06	11,750	9.977
8/18/11	14,785	0.84	7,442	10.69
9/9/11	11,197	1.42	6,409	7.771
9/9/11	4,460	1.42	6,409	7.771
9/13/11	19,549	0.88	13,820	15.12
9/13/11	22,942	0.88	13,820	15.12
7/14/12	12,690	0.69	6,930	8.407
9/4/12	6,072	0.4	3,527	3.902
3/1/14	14,610	1.49	9,354	4.912
7/9/14	27,849	1.325	14,660	15.69
7/11/14	10,553	0.525	4,029	4.395
7/13/14	39,104	1.66	19,180	18.03
9/6/14	14,009	0.55	5,492	10.45
10/8/14	7,612	0.415	3,315	4.737
6/30/15	3,737	0.36	2,053	4.229
6/30/15	2,538	0.36	2,053	4.229
9/3/15	10,678	0.97	4,356	4.065
12/12/15	14,549	1.095	7,623	7.565
6/10/16	10,971	0.53	4,786	9.067
6/30/16	10,124	0.44	5,451	9.291
7/1/16	9,442	0.61	7,146	8.407
7/26/16	12,959	0.685	3,850	5.616
7/27/16	13,071	0.645	5,682	11.42
7/30/16	10,930	0.755	6,323	7.565

Table 16.5. *Sample statistics for rainfall, runoff, and sediment events.*

Variables	Mean	Standard deviation	Skewness	Kurtosis
Sediment yield (lb)	14,896.16	13,192.28	2.01	7.54
Rainfall depth (inch)	1.01	0.62	1.48	5.61
Runoff volume (ft^3)	9,678.49	7,562.99	1.94	7.58
Peak runoff (cfs)	9.31	5.11	1.43	5.31

Table 16.6. *Sample rank-based Kendall and Spearman correlation coefficients.*

	Sediment yield	Rainfall depth	Runoff volume	Peak runoff
Sediment yield	1	0.48	0.627	0.544
Rainfall depth	*0.626*	1	0.724	0.494
Runoff volume	*0.767*	*0.881*	1	0.655
Peak runoff	*0.726*	*0.667*	*0.828*	1

Note: Values in italic denotes the Spearman correlation coefficients.

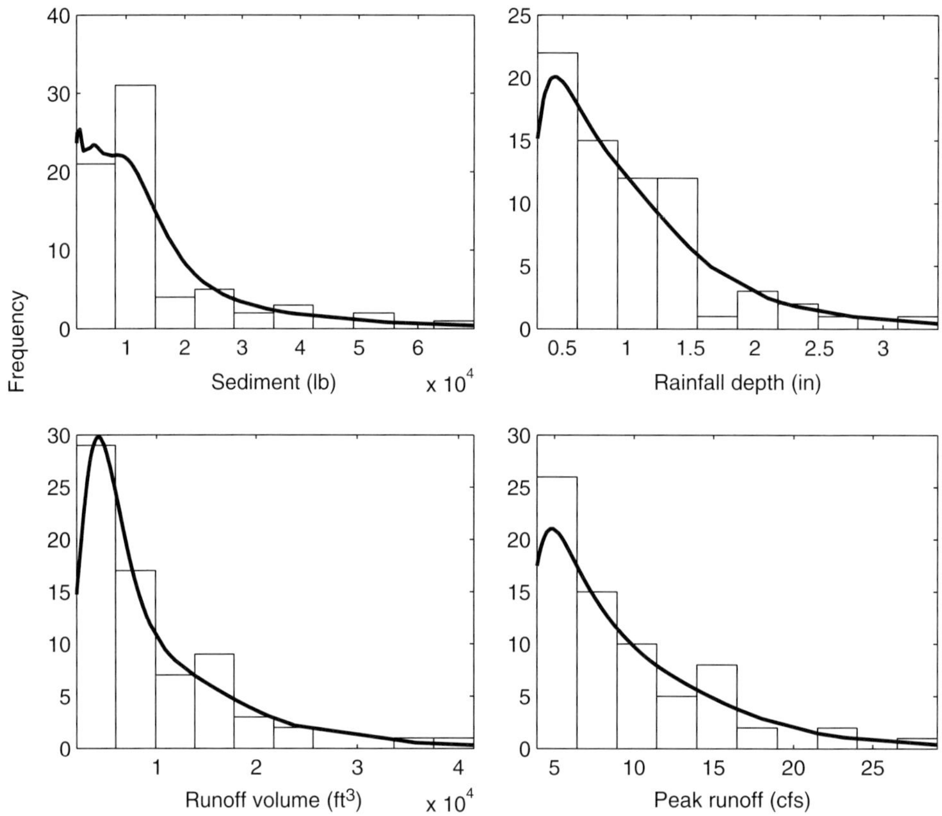

Figure 16.3 Histogram and kernel density frequencies for the univariate variables.

visually indicates the upper-tail dependence between sediment yield and runoff volume. Applying the pseudo-MLE, the parameter is estimated as $\theta = 2.822$. The corresponding theoretical upper-tail dependence coefficient is computed as $\lambda_U^{GH} = 2 - 2^{\frac{1}{\theta}} = 2 - 2^{\frac{1}{2.822}} = 0.722$, which is slightly higher than its empirical estimation. Applying the Rosenblatt goodness-of-fit test discussed in Chapters 3 and 4, we compute $P = 0.31, S_n^B = 0.15$. The

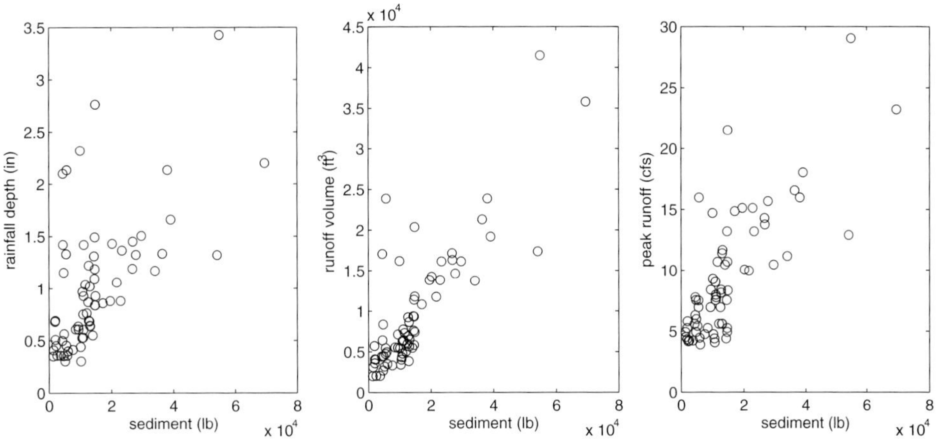

Figure 16.4 Scatter plots of sediment yields versus rainfall depth, runoff volume, and peak runoff.

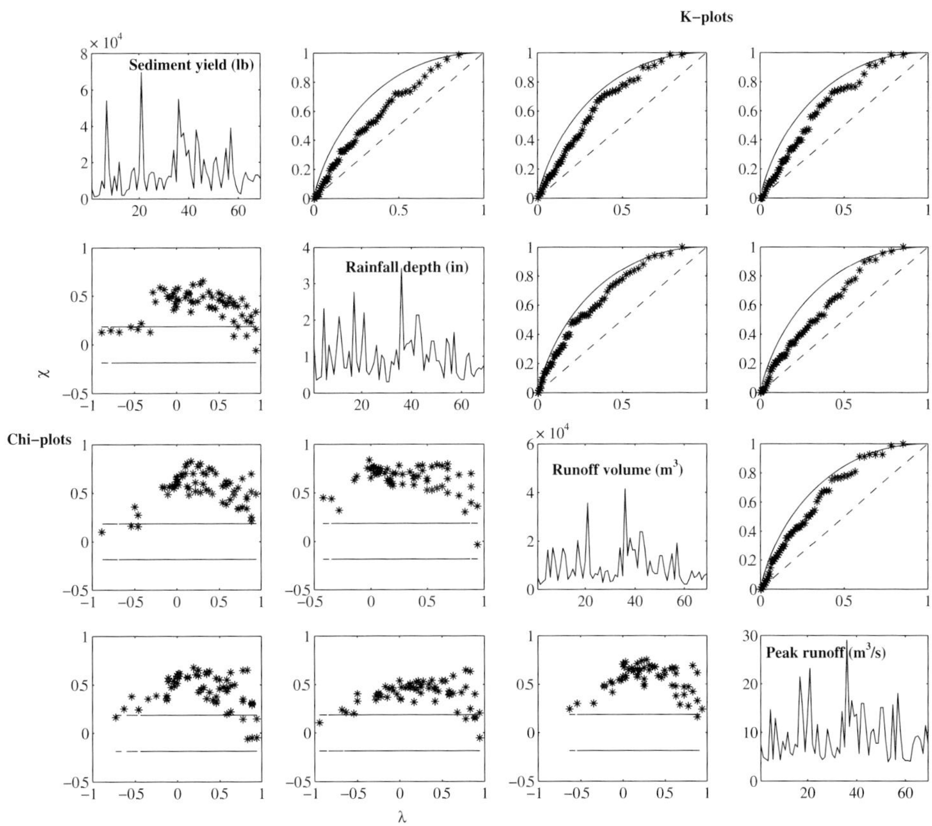

Figure 16.5 K- and chi-plots for the pair variables.

Table 16.7. *Empirical probability computed using kernel density.*

Pair no.	Sediment yield	Runoff volume	Pair no.	Sediment yield	Runoff volume
1	0.230	0.288	36	0.964	0.986
2	0.022	0.030	37	0.898	0.766
3	0.027	0.114	38	0.910	0.903
4	0.072	0.208	39	0.810	0.822
5	0.440	0.823	40	0.847	0.826
6	0.245	0.308	41	0.572	0.591
7	0.962	0.845	42	0.240	0.927
8	0.634	0.712	43	0.917	0.927
9	0.057	0.163	44	0.870	0.822
10	0.558	0.618	45	0.371	0.366
11	0.179	0.840	46	0.787	0.710
12	0.764	0.780	47	0.633	0.522
13	0.059	0.213	48	0.495	0.446
14	0.054	0.386	49	0.178	0.446
15	0.190	0.577	50	0.751	0.769
16	0.230	0.360	51	0.804	0.769
17	0.634	0.892	52	0.557	0.486
18	0.699	0.680	53	0.257	0.155
19	0.198	0.258	54	0.627	0.623
20	0.628	0.700	55	0.856	0.789
21	0.982	0.977	56	0.466	0.209
22	0.485	0.435	57	0.921	0.875
23	0.168	0.252	58	0.606	0.363
24	0.577	0.454	59	0.329	0.134
25	0.627	0.399	60	0.142	0.033
26	0.618	0.623	61	0.085	0.033
27	0.192	0.080	62	0.472	0.244
28	0.511	0.512	63	0.625	0.533
29	0.490	0.540	64	0.485	0.291
30	0.212	0.120	65	0.447	0.359
31	0.455	0.144	66	0.415	0.502
32	0.551	0.409	67	0.568	0.189
33	0.530	0.319	68	0.572	0.382
34	0.846	0.842	69	0.483	0.439
35	0.412	0.363			

results of the goodness-of-fit test further confirm that the Gumbel–Hougaard copula may be properly applied to do bivariate sediment analysis.

The plots in Figure 16.7 yield the same results in regard to the comparison of (i) empirical Kendall distribution versus parametric Kendall distribution for the fitted

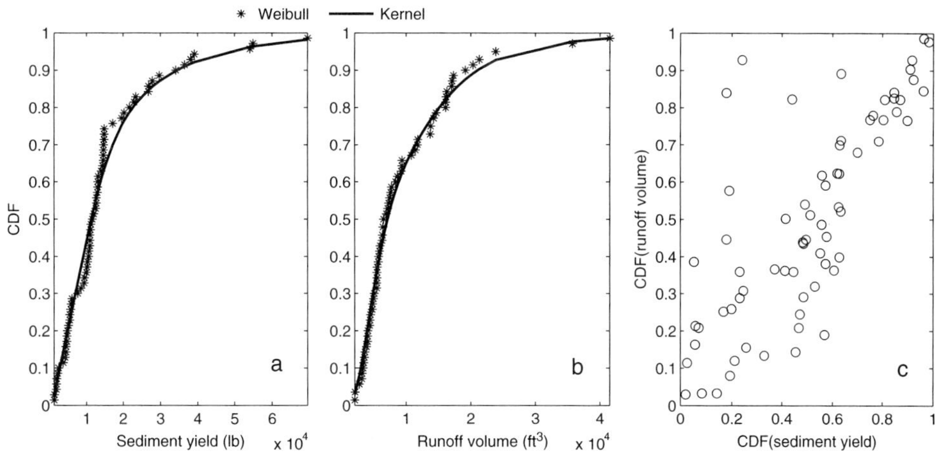

Figure 16.6 Comparison of kernel density with the Weibull plotting-position formula and scatter plots of empirical CDFs.

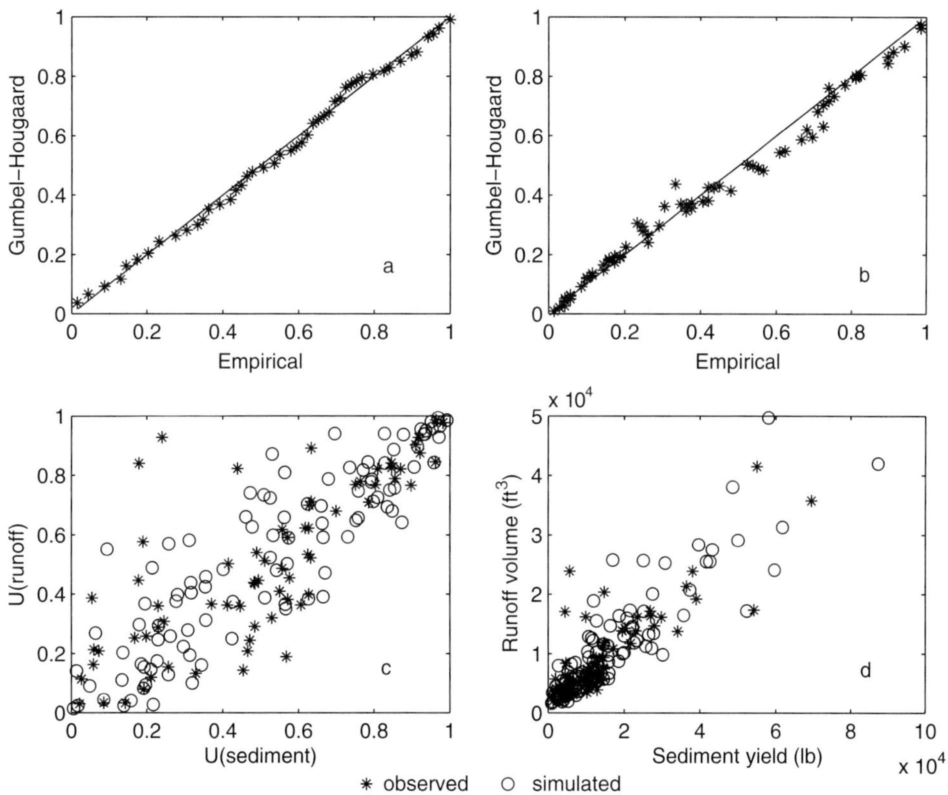

Figure 16.7 Comparison of the fitted Gumbel–Hougaard copula with observations: (a) Kendall distribution, (b) copula, (c): variables in the frequency domain, (d) variables in real domain.

Gumbel–Hougaard copula, (ii) the empirical copula versus the fitted Gumbel–Hougaard copula, (iii) the empirical CDF (kernel density approach) versus simulated variates from the fitted Gumbel–Hougaard copula (frequency domain), and (iv) observations versus simulated variates (real domain).

To this end, we have shown that a copula can successfully model the dependence of runoff volume and sediment yield. To better understand the interrelation among rainfall, runoff, and sediment yield, we will study multivariate dependence using rainfall depth, runoff volume, and sediment yield in the following section.

16.2.4 Multivariate Dependence Study of Rainfall Depth, Runoff Volume, and Sediment Yield

To study the multivariate dependence among rainfall depth, runoff volume, and sediment yield, we will apply two approaches: (i) meta-Student t copula and (ii) the vine copula with runoff volume as the center variable. To investigate the ability of using rainfall depth and runoff volume to predict sediment yield, we use the first 59 observations (Table 16.7) to build the dependence model and the last 10 observations to study the predictability of rainfall depth and runoff volume.

Dependence Model with the Meta-Student t Copula

Applying the meta-Student t copula discussed in Section 7.2.2, Table 16.8 lists the estimated parameter (i.e., the correlation matrix and degree of freedom) with pseudo-MLE in which the empirical CDF is again computed with the kernel density function approach. The application of kernel density function allows us to find the empirical CDF of "future" values continuously across $\mathbb{R}^+$ (listed in Table 16.8 for rainfall depth and runoff volume).

We have shown in other chapters that the meta-elliptical copula (i.e., meta-Student t copula here) may be successfully applied to model the dependence structure, since the meta-elliptical copula is built on the transformed data (i.e., univariate meta-Student t transformation) as discussed in Chapter 7. Here, we will proceed to the forecast study for the sediment yield directly using the parameters listed in Table 16.8 and variates in Table 16.9.

Table 16.8. *Estimated parameter for the meta-Student* t *copula with pseudo-MLE.*

	Correlation matrix			
	Sediment yield	Rainfall depth	Runoff volume	Degree of freedom
Sediment yield	1	0.709	0.857	
Rainfall depth	0.709	1	0.915	$\nu = 5.817$
Runoff volume	0.857	0.915	1	

Table 16.9. *Univariate CDF computed for the last 10 observations using the kernel density function built with the first 59 observations.*

Rainfall depth		Runoff volume	
Inch	CDF	ft^3	CDF
0.36	0.086	2,053	0.032
0.36	0.086	2,053	0.032
0.97	0.525	4,356	0.226
1.095	0.595	7,623	0.477
0.53	0.225	4,786	0.266
0.44	0.152	5,451	0.324
0.61	0.287	7,146	0.448
0.685	0.342	3,850	0.179
0.645	0.313	5,682	0.343
0.755	0.390	6,323	0.392

With rainfall depth and runoff volume known, the conditional distribution of sediment yield for the given rainfall depth and runoff volume may be written using Equation (7.56) as follows:

$$\sum_{11} = \begin{bmatrix} 1 & 0.915 \\ 0.915 & 1 \end{bmatrix}, \sum_{12} = \sum_{21}^{T} = [0.709, 0.857], \sum_{22} = 1, v_{2|1} = 7.817.$$

Applying the median forecast (i.e., P(sediment yield<=S|Rainfall depth = *d*, Runoff volume = *v*) = 0.5), we have $T\left(\frac{S-\mu_{s|d,v}}{\left|\Sigma_{s|d,v}\right|^{0.5}} v_{S|d,v}\right) = 0.5$, where S is the sediment yield forecasted. Using the same approach as that in Example 7.11, the forecasted sediment yields are plotted in Figure 16.8. Figure 16.8 also plots the forecasted sediment yields from bivariate sediment analysis. It is seen that the meta-Student *t* copula and GH copula (using the same parameter estimated for the entire dataset in Section 16.2.3) yield similar forecast results. Even though the observations fall into the 95% confidence interval constructed using the Student *t* copula, there exist visible differences among the observations and forecasts.

Dependence Model with Vine Copula

As in the preceding section, the first 59 observations are used to build the vine copula (shown in Figure 16.9), and the last 10 observations are applied for forecast purposes. The Gumbel–Hougaard copula is applied to model the runoff volume and sediment yield as well as rainfall depth and runoff volume of T1, and the Frank copula is applied for T2. The parameters estimated are also shown in Figure 16.9. It is noted that S, V, and D represent sediment yield, runoff volume, and rainfall depth, respectively. Comparisons shown in Figures 16.10 and 16.11 indicate the appropriateness of applying the fitted vine copula to

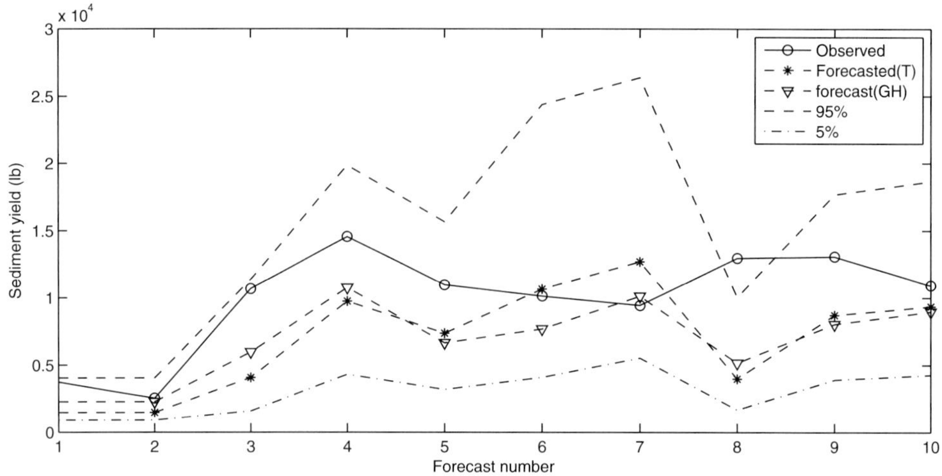

Figure 16.8 The forecasted sediment yields from trivariate and bivariate sediment analysis.

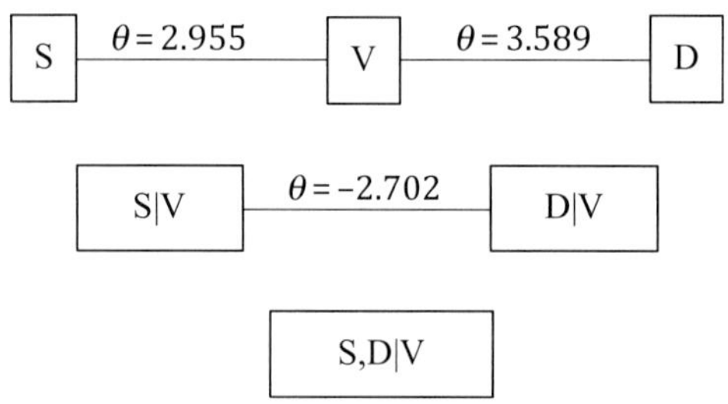

Figure 16.9 Vine structure and the parameter estimated.

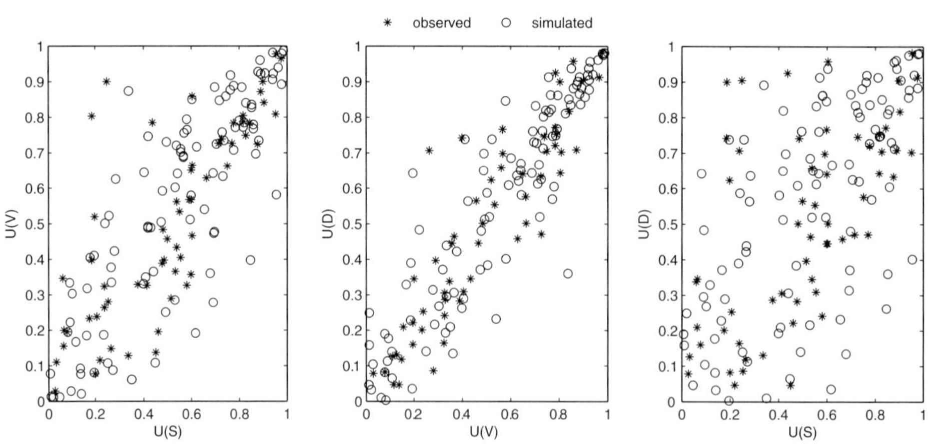

Figure 16.10 Comparison of observations to the simulated variates in frequency domain.

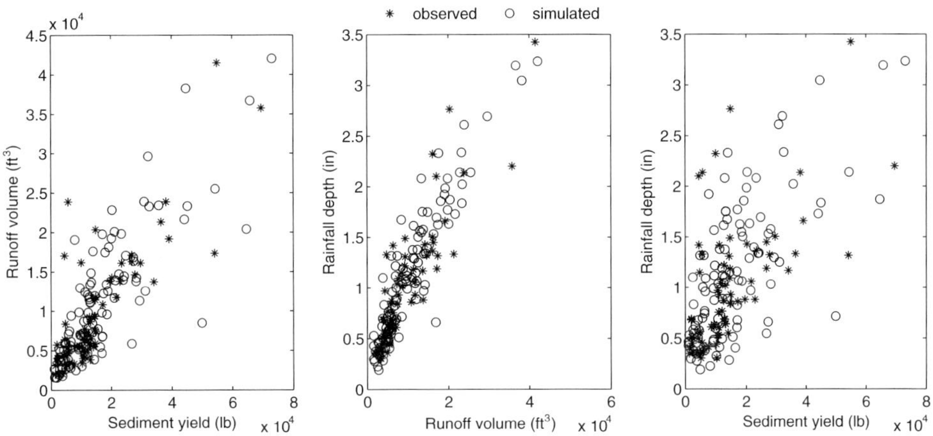

Figure 16.11 Comparison of observations and simulated variates in real domain.

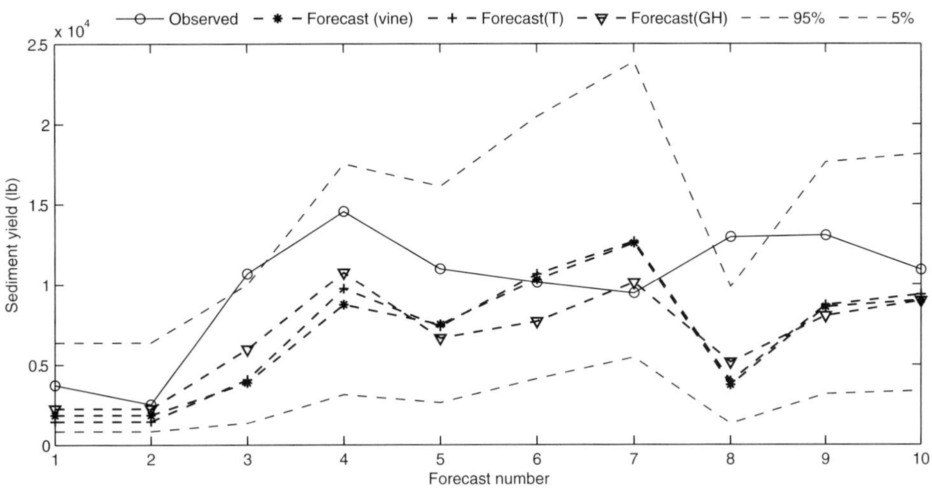

Figure 16.12 Comparison of the forecasted sediment yield from vine, Student t, and Gumbel–Hougaard copulas with the observed sediment yield.

do the trivariate sediment analysis. The Rosenblatt goodness-of-fit test also confirms that the fitted vine copula may properly model the dependence ($S_n^B = 0.059, P = 0.635$).

With the fitted vine copula, the median forecast of the sediment yield may be computed using the similar procedure for the second-order copula-based Markov process. In detail, the forecast of sediment yield is performed with the following steps:

1. Compute the empirical CDF for the last 10 observations of rainfall depth (D) and runoff volumes (V), using the kernel density function fitted to the first 59 observed rainfall

depth (listed in Table 16.9). Here we apply the assumption that variables are random and the first 59 observations may represent the population statistics.

2. Compute conditional probability $p_1 = P(D \leq d|V = v) = \frac{\partial C_{GH}(F(d),F(v);3.589)}{\partial F(v)}$.
3. Compute the conditional probability $p_2 = P(S \leq s|V = v)$ from $P(S \leq s|D = d, V = v) = 0.5$ as: $p_2 = P(S \leq s|V = v) = P^{-1}(p_1, 0.5) = C^{-1}_{Frank-condtional}(p_1, 0.5; -2.702)$. The conditional Frank copula is listed in Chapter 4.
4. Compute the probability of the forecasted sediment yield by setting $p_2 = P(S \leq s|V = v)$ and $F(s) = C^{-1}_{GH-conditional}(F(v), p_2; 2.955)$.
5. Interpolate the forecasted sediment yield in the real domain with the use of the kernel density fitted to the first 59 observations.

Comparisons in Figure 16.12 indicate a very similar performance between the meta-Student t and vine copula.

16.3 Summary

In this chapter, we apply copulas to (1) suspended sediment analysis by constructing the copula-based discharge-sediment rating curve and (2) bivariate and trivariate sediment analysis with the use of meta-Student t and vine copulas.

Applying the sediment-rich middle reach of Yellow River with four major underlying surfaces, the case study of the discharge-sediment rating curve indicates the following:

- The USGS regression equation may work as well as the copula-based method when the channel is flat. In this case, the sediment may be harder to transport or it may actually move as bed load rather than suspended sediment due to low flow velocity and low shear force. For the flat channel, the type of underlying surface does not seem to be the dominating factor for the suspended sediment transport.
- The copula-based method works much better than the USGS regression equation when the channel is steep. For the steep channel, the relation of discharge and suspended sediment in the logarithm domain is no longer linear. The nonlinearity for the discharge-suspended sediment rating curve is dependent on the underlying surface when the discharge is lower than a certain threshold. However, for the discharge higher than the certain threshold, the nonlinear relation seems to be replaced by the linear relation similar to the linear relation for the flatter channel.

Using the Flume #3 at Santa Rita experimental watershed as a case study example, the GH copula is applied to model the dependence of sediment yield and runoff volume. It is shown that the GH copula may properly capture both the upper-tail and overall dependence of the sediment yield and runoff volume. According to the nature of the sediment yield, runoff volume, and rainfall depth, the meta-Student t and GH–GH–Frank vine copula are applied to model the trivariate dependence. As shown in the case study, the runoff volume is considered the center variable, and the GH copula may also properly model the bivariate dependence of runoff volume and rainfall depth. The goodness-of-fit studies prove that the meta-Student t and GH–GH–Frank copulas may properly model the dependence. It is also confirmed visually from the simulation study.

Applying the median forecast of sediment yield with known runoff volume and rainfall depth, the forecast study shows the meta-Student t and vine copulas yield very similar forecast results. The forecast study shows there exist significant differences between forecasted and observed sediment yields for some particular events.

References

Li, B. Y. and Li, J. Z. (1994). *Geomorphological Maps of China (1:4,000,000)*. Beijing Science Press, Beijing.

Ni, H. B., Zhang, L. P., Wu, X. Y. and Fu, X. T. (2008). Weathering of the Pisha-Sandstones in the wind-water erosion crisscross region on the Loess Plateau. *Journal of Mountain Science*, 5, 340–349.

USDA, United States Department of Agriculture, Agricultural Research Service. www.tucson.ars.ag.gov/dap/.

Wang, Y. C., Wu, Y. H., Kou, Q., Min, D. A., Chang, Y. Z., and Zhang, R. J. (2007). Definition of arsenic rock zone borderline and its classification. *Science of Soil and Water Conservation*, 5(1): 14–18 (in Chinese).

17

Interbasin Transfer

ABSTRACT

In this chapter, we will introduce the last application of the book, i.e., interbasin transfer. In this process, there are two main components: donor and receiver basins. The purpose of interbasin transfer to redistribute water from a water-rich region to the region with water shortage. The interbasin transfer may help reducing the impact of dry conditions in the region with water shortage.

17.1 Case-Study Site and Dataset

In this chapter, we will provide a synthetic analysis for interbasin transfer using the river systems in Texas, the United States, as an example. Based on the description of Texas, the climate in Texas varies from arid in the west to humid in the east (as shown in Figure 17.1). The major river systems as well as major cities are shown in Figure 17.1. Among all the major river systems, the Brazos, Sabine, and Trinity Rivers carry the largest annual runoff of 6,074,000; 5,864,000; and 5,127,000 acre feet, respectively. Climatewise, the eastern coastal region is in the tropic humid climate region with abundant precipitation throughout the year. However, the central and western parts of Texas are within the arid/semi-arid climate region and may not receive enough precipitation. Thus, it is viable to transport the abundant water from the eastern part of Texas to central and western parts of Texas under the conditions of no or minimum negative impact on the highly developed eastern coastal area of Texas. In this case study, we will choose Lake Houston (USGS 08072000) as a donor reservoir, and E. V. Spence Reservoir (USGS 08123950) as a receiver reservoir to evaluate the possibility of interbasin transfer.

Lake Houston was constructed in 1953 and is currently serving as the primary source of water supply for the city of Houston. The E. V. Spence Reservoir is located west of Robert Lee, Texas. In the normal years, the E. V. Spence Reservoir may be sufficient to provide the water supply for Robert Lee and surrounding communities in Coke County. However, during the recent drought, the reservoir storage decreased to less than 0.76% of its capacity. As of June 2016, the lake was back up to 10.4% of its capacity (Wikipedia.org). To illustrate the process, monthly storage is applied. The full capacity storage of Lake Houston is 134,313 acre feet. The full capacity storage of the E. V. Spence Reservoir is 135,704 acre feet. Based on the availability of the dataset (USGS), the data from water year

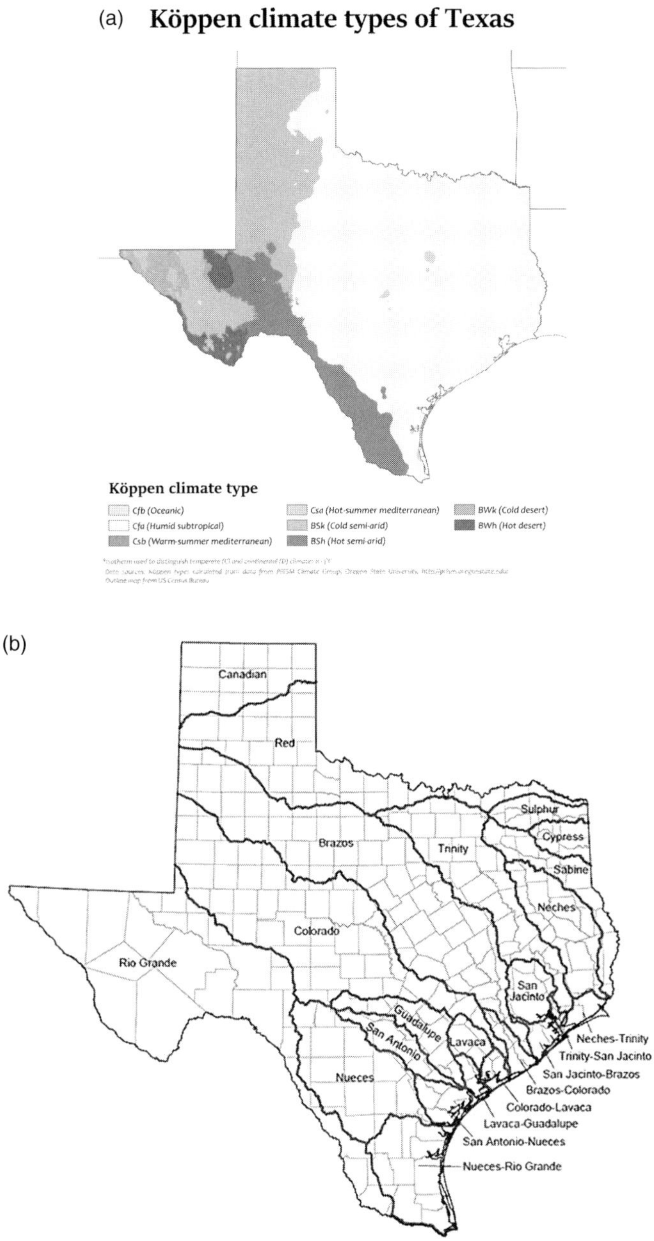

Figure 17.1 (a) Köppen climate types of Texas (retrieved from https://commons.wikimedia.org/wiki/File:Texas_K%C3%B6ppen.svg).
(b) Major rivers and cities in Texas (retrieved from www.twdb.texas.gov/surfacewater/rivers/index.asp, courtesy of Texas Water Development Board). A black and white version of this figure will appear in some formats. For the color version, please refer to the plate section.

of 2000 to 2016 are applied for analysis. In addition, there is one data value missing for Lake Houston (May 2015) and one for E. V. Spence Reservoir (May 2004). The missing value at Lake Houston and E. V. Spence Reservoir is filled based on the recent drought. The missing value at Lake Houston is filled with the average flow of May, while the missing value at E.V. Spence Reservoir is filled with the average flow of May before water year 2010 (i.e., before the 2010–2013 drought in the southern United States and Mexico). The entire dataset is listed in Table 17.1.

With the collected reservoir storage dataset listed in Table 17.1, the procedure to assess the interbasin transfer is outlined as follows:

i. Investigate the univariate time series.
ii. Apply the time series-copula approach to study the bivariate analysis.
iii. Set the rule for interbasin transfer and assess the interbasin transfer probability using the time series-copula developed in step ii.

17.2 Investigation of Univariate Storage Time Series

Given that monthly reservoir storage may not be considered a random variable, the autocorrelation and partial autocorrelation are plotted in Figure 17.2 first to evaluate the stochastic behavior of the monthly storage at USGS08072000 (Lake Houston) and at USGS08123950 (E. V. Spence Reservoir). In Figure 17.2, we see the following:

1. There is no obvious seasonality at both locations.
2. The storage at USGS08072000 (Lake Houston) may be considered a stationary signal.
3. The storage at USGS08123950 (E. V. Spence Reservoir) is clearly nonstationary.

Table 17.2 lists the sample statistics of the storage series. The sample statistics show that the storage series is clearly skewed with a heavy tail. In addition to the sample statistics listed in Table 17.2, the histogram plotted in Figure 17.3 also suggests that the storage series at both locations do not follow the Gaussian process.

As a result, we are applying the meta-Gaussian transformation to the storage series before we proceed. To avoid the future interpolation of the order series discussed in Ayra and Zhang (2014), the kernel density approach is adopted here to assess the univariate distribution rather than the empirical distribution with plotting position. The univariate storage time series is then transformed as follows:

1. Apply the kernel density with positive support to estimate the empirical CDF of S_i as $F_n(S_i)$.
2. Apply the meta-Gaussian transformation to obtain the transformed storage variable S_i^T as

$$S_i^T = \Phi^{-1}(F_n(S_i); 0, 1) \tag{17.1}$$

With the application of meta-Gaussian transformation, Figure 17.4 plots the histogram of storage time series after transformation. Figure 17.5 plots the sample autocorrelation and

Table 17.1. *Storage at Lake Houston and E. V. Spence Reservoir (acre feet).*

Year	Month	USGS08072000	USGS08123950	Year	Month	USGS0807200	USGS08123950
2000	10	100,800	83,700	2008	10	141,900	59,510
2000	11	136,800	88,260	2008	11	143,700	56,880
2000	12	144,200	86,270	2008	12	142,600	54,680
2001	1	147,200	84,840	2009	1	140,600	52,590
2001	2	145,400	83,870	2009	2	137,700	51,010
2001	3	144,600	82,740	2009	3	141,600	49,320
2001	4	144,100	80,930	2009	4	145,100	46,400
2001	5	141,600	78,060	2009	5	143,100	43,710
2001	6	147,400	74,020	2009	6	134,300	40,920
2001	7	143,500	69,470	2009	7	134,900	37,290
2001	8	140,200	64,540	2009	8	137,200	34,180
2001	9	143,500	61,800	2009	9	138,600	31,080
2001	10	140,600	58,570	2009	10	144,700	28,530
2001	11	134,000	58,300	2009	11	144,400	26,170
2001	12	148,300	61,510	2009	12	145,700	24,840
2002	1	145,000	59,730	2010	1	143,300	23,950
2002	2	144,300	57,840	2010	2	143,700	24,410
2002	3	141,800	55,190	2010	3	142,600	23,900
2002	4	143,000	53,860	2010	4	140,300	22,990
2002	5	136,900	52,880	2010	5	141,600	25,400
2002	6	137,300	55,000	2010	6	142,700	25,100
2002	7	142,900	54,760	2010	7	143,500	24,530
2002	8	140,900	51,230	2010	8	137,100	22,680
2002	9	141,200	47,390	2010	9	140,600	19,950
2002	10	146,500	45,600	2010	10	132,300	20,810
2002	11	150,300	44,740	2010	11	134,800	18,550
2002	12	147,500	42,940	2010	12	133,100	16,360
2003	1	145,500	41,410	2011	1	142,100	14,600
2003	2	148,900	40,000	2011	2	138,300	13,650
2003	3	144,900	38,420	2011	3	136,300	12,060
2003	4	143,000	35,540	2011	4	130,300	10,130
2003	5	139,900	32,770	2011	5	118,600	8,026
2003	6	141,500	51,950	2011	6	106,600	6,104
2003	7	144,000	57,600	2011	7	99,040	4,027
2003	8	140,900	53,110	2011	8	87,870	2,847
2003	9	143,900	53,040	2011	9	84,700	2,500
2003	10	144,700	51,590	2011	10	96,990	2,362
2003	11	144,600	49,320	2011	11	114,400	2,231
2003	12	142,000	46,790	2011	12	136,500	2,194
2004	1	146,500	44,630	2012	1	141,700	2,249
2004	2	145,700	42,870	2012	2	144,800	2,323
2004	3	141,000	46,050	2012	3	143,800	2,320
2004	4	141,200	49,040	2012	4	139,000	2,215
2004	5	145,300	63,819	2012	5	134,900	2,100

Table 17.1. (*cont.*)

Year	Month	USGS08072000	USGS08123950	Year	Month	USGS0807200	USGS08123950
2004	6	143,900	44,640	2012	6	137,300	1,839
2004	7	140,200	42,840	2012	7	143,700	1,463
2004	8	138,500	43,160	2012	8	139,100	1,164
2004	9	138,100	43,590	2012	9	131,800	1,111
2004	10	123,300	39,490	2012	10	135,700	28,440
2004	11	146,400	53,810	2012	11	131,900	28,840
2004	12	143,700	79,410	2012	12	128,600	28,480
2005	1	144,500	78,630	2013	1	138,300	27,800
2005	2	150,000	78,460	2013	2	139,800	27,210
2005	3	143,500	78,490	2013	3	135,400	26,000
2005	4	141,200	76,190	2013	4	139,100	24,790
2005	5	140,300	73,600	2013	5	138,900	23,370
2005	6	139,600	72,450	2013	6	139,800	26,680
2005	7	141,600	68,290	2013	7	134,000	28,510
2005	8	141,900	85,160	2013	8	132,800	27,570
2005	9	133,800	101,900	2013	9	132,300	25,770
2005	10	138,300	99,360	2013	10	141,700	24,170
2005	11	138,700	97,300	2013	11	140,900	22,800
2005	12	141,700	94,870	2013	12	140,200	20,990
2006	1	140,600	92,960	2014	1	139,300	19,030
2006	2	142,800	91,210	2014	2	139,900	17,400
2006	3	140,700	89,790	2014	3	143,300	15,550
2006	4	140,100	88,650	2014	4	138,100	13,410
2006	5	142,300	87,120	2014	5	140,600	11,330
2006	6	143,200	82,880	2014	6	143,500	11,790
2006	7	142,900	77,900	2014	7	141,400	10,450
2006	8	140,900	72,880	2014	8	136,500	8,627
2006	9	138,700	76,020	2014	9	138,300	8,025
2006	10	146,700	73,890	2014	10	140,100	14,670
2006	11	136,700	71,120	2014	11	137,400	13,040
2006	12	135,400	69,300	2014	12	140,700	11,760
2007	1	140,300	68,460	2015	1	143,300	10,730
2007	2	134,300	67,650	2015	2	139,400	10,460
2007	3	140,400	67,200	2015	3	143,800	96,50
2007	4	141,900	70,340	2015	4	143,200	11,410
2007	5	144,100	74,250	2015	5	140,330	18,260
2007	6	143,300	74,490	2015	6	145,500	28,370
2007	7	138,600	72,600	2015	7	141,100	38,570
2007	8	136,500	77,170	2015	8	137,800	39,670
2007	9	134,300	84,430	2015	9	140,100	36,760
2007	10	134,200	80,920	2015	10	134,300	37,830
2007	11	138,100	77,460	2015	11	144,600	46,240
2007	12	140,400	75,980	2015	12	145,400	50,110
2008	1	140,200	74,570	2016	1	144,300	50,840

Table 17.1. (*cont.*)

Year	Month	USGS08072000	USGS08123950	Year	Month	USGS0807200	USGS08123950
2008	2	143,700	72,700	2016	2	141,800	49,390
2008	3	144,100	71,540	2016	3	147,400	48,020
2008	4	141,500	70,370	2016	4	151,100	48,130
2008	5	142,400	68,160	2016	5	154,400	51,810
2008	6	138,200	66,990	2016	6	150,100	53,790
2008	7	135,400	65,340	2016	7	142,100	53,430
2008	8	141,300	63,400	2016	8	144,200	50,120
2008	9	144,000	62,440	2016	9	143,100	49,630

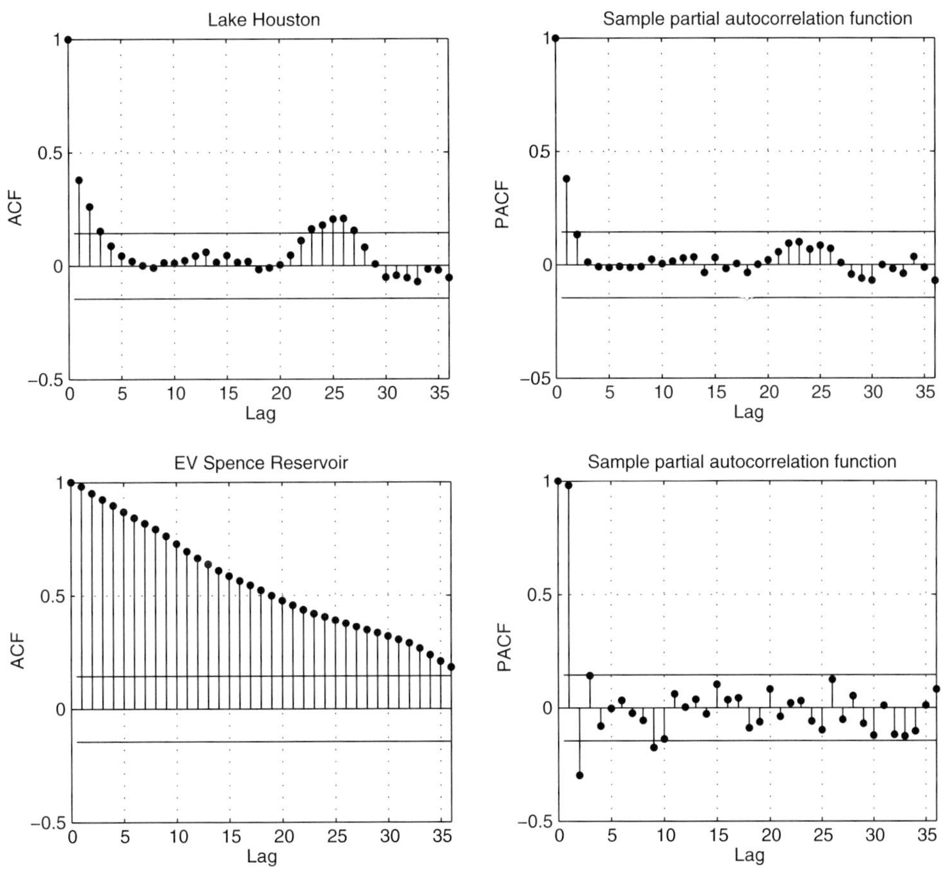

Figure 17.2 Sample autocorrelation and partial autocorrelation plots for the stations at Lake Houston and E. V. Spence Reservoir.

Table 17.2. *Sample statistics of observed the storage series at USGS08072000, US08123950, as well as USGS08123950 after the first-order difference.*

Station	Mean	Standard deviation	Skewness	Kurtosis
USGS08072000	139311.98	9335.10	–3.52	17.98
USGS08123950	45319.61	26711.91	0.04	1.95

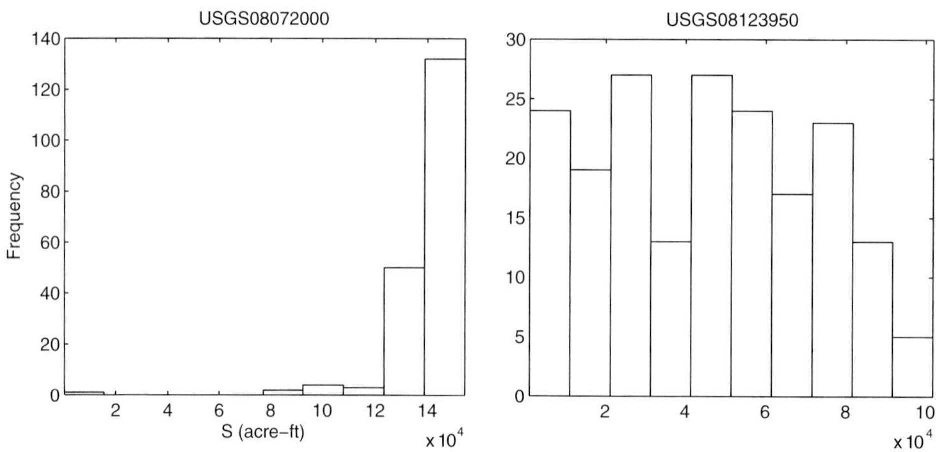

Figure 17.3 Histogram of observed storage series.

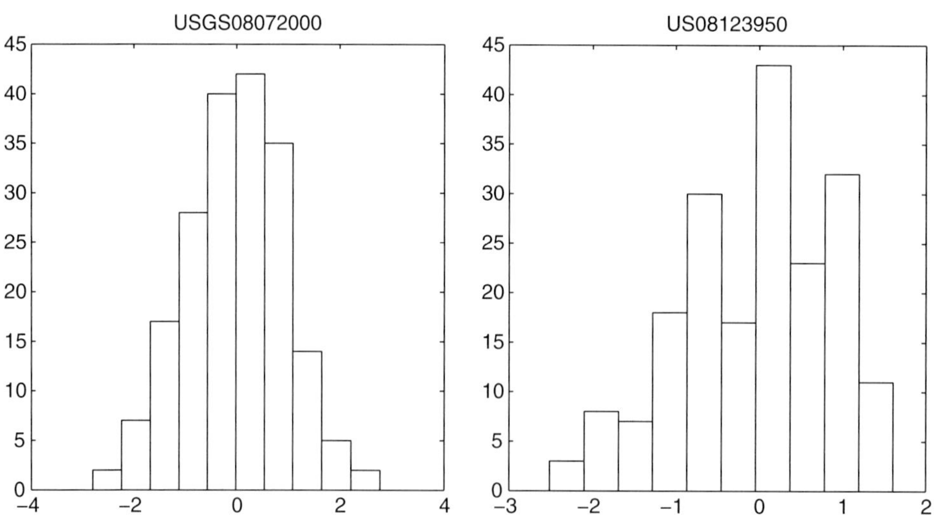

Figure 17.4 Histogram of storage series after meta-Gaussian transformation.

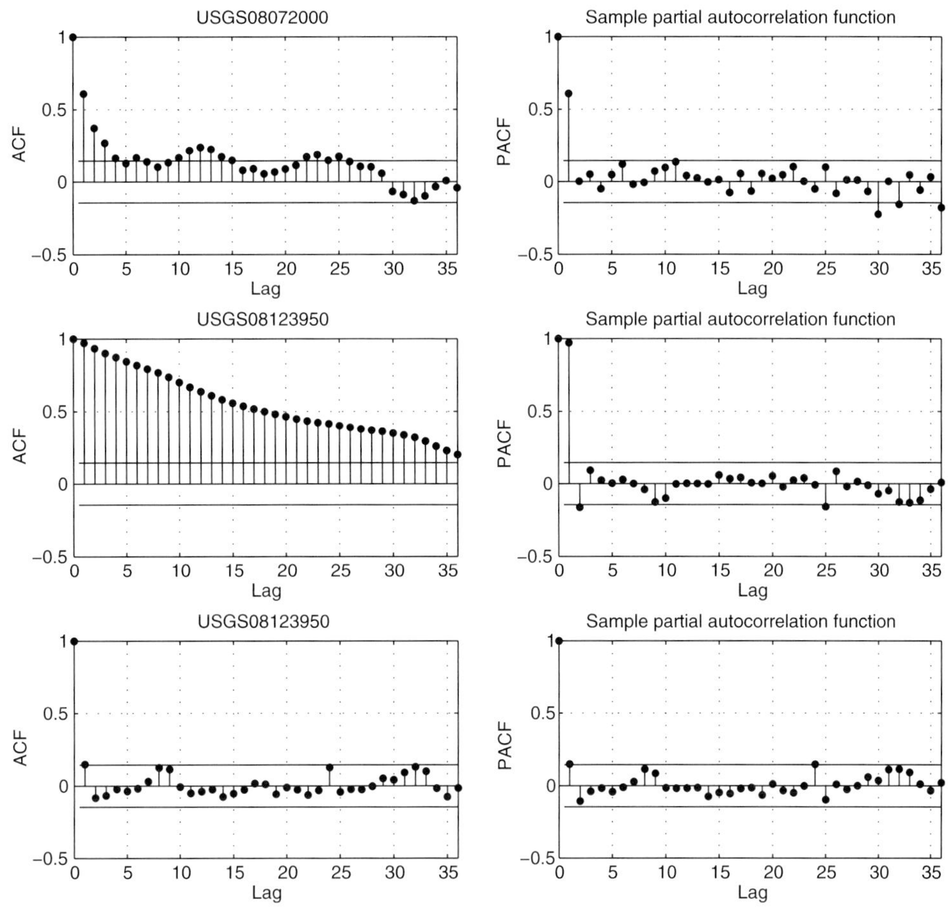

Figure 17.5 Sample autocorrelation and partial autocorrelation of the transformed series.

partial autocorrelation of the transformed storage series at USGS08072000, USGS08123950, and USGS08123950 after the first-order differencing. Figure 17.4 shows that the transformed series is now closer to the Gaussian process. Figure 17.5 shows that the overall structure of the time series is not changed after transformation. Thus, we can safely study the transformed series as a representation for the observed storage series directly. As shown in Figure 17.3 and Figure 17.5, the AR(1) model may be applied to both the transformed storage at USGS08072000 and the first-order differenced transformed storage at USGS08123950 with parameters estimated as listed in Table 17.3. The diagnostics are given in Table 17.4 for the assessment of linear independence (Ljung-Box Q-test), second-order independence (ARCH test), and white Gaussian noise test for the model residuals. The results in Table 17.4 show that the normality test for USGS08123950 fails even after the application of meta-Gaussian transformation; however, the parameters listed in Table 17.3 are still valid estimates for USGS08123950 if the

Table 17.3. *Parameter estimated for univariate storage time series.*

Station	Model	c	ϕ	Variance (σ_e^2)
USGS08072000	AR(1)	–0.0018	0.637	0.55
USGS08123950	ARIMA(1,1,0)	–0.004	0.149	0.04

Table 17.4. *Diagnostic test for model residuals.*

Diagnostic	LBQ (H, P)	ARCH (H, P)	$N(0, \sigma_e^2)$
USGS08072000	[0, 0.87]	[0, 0.33]	[0, 0.80]
USGS08123950	[0, 0.96]	[0, 0.84]	[1, 0]

model residual may be fitted using the stable distribution (DuMouchel, 1973). Applying the stable distribution, the parameters estimated for the model residual at USGS08123950 are as follows:

$$\alpha = 1.159,\ \beta = 0.441,\ \gamma = 0.032,\ \delta = -0.04.$$

After performing the KS goodness-of-fit study, we have $D = 0.042$, $P = 0.8799$. To this end, we have successfully constructed the univariate time series model for the storage time series at USGS08072000 and USGS08123950 as follows:

1. The AR(1) model may properly model the storage series at USGS08072000 after the meta-Gaussian transformation.
2. ARIMA(1,1,0) with stable distributed residues may properly model the storage series at USGS08123950 after the meta-Gaussian transformation.

17.3 Investigation of Storage at USGS08072000 and USGS08123950 with Bivariate Analysis

With the univariate time series constructed, we will now be able to study their joint dependence structure and assess the possible water transfer from USGS08072000 (Lake Houston) to USGS08123950 (E. V. Spence Reservoir) by applying copulas to the fitted model residuals. The Kendall correlation coefficient computed with the use of the fitted model residuals at both locations is computed as $\tau_n = 0.087$. From the computed Kendall correlation, it is seen that USGS08072000 and USGS08123950 are close to being independent. This may be understood by the different climate regions of USGS08072000 (Lake Houston: humid) and USGS08123950 (E. V. Spence Reservoir, semi-arid).

Applying the Frank, Clayton, and Gaussian copula, Figure 17.6 plots the comparison of simulated random variables from the copula candidates to the fitted model residuals. As seen from Figure 17.6, there are no visually significant differences among three copula functions.

Table 17.5. *Parameter and GoF results for the fitted copula functions.*

Copula	Frank	Clayton	Gaussian
Parameter (θ)	1.076	0.274	0.156
Gof(S_n^B, P_{value})	(0.085, 0.241)	(0.049, 0.299)	(0.072, 0.253)

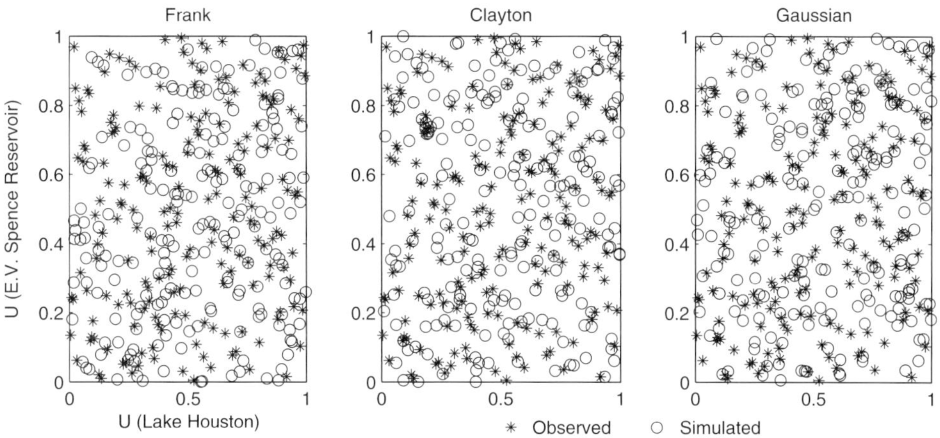

Figure 17.6 Comparison of marginals from the fitted model residuals to the random variables simulated from the fitted copula candidates.

Table 17.5 lists the estimated parameters as well as the GoF results with the Rosenblatt transform (Genest et al., 2007). Based on the GoF results, all three copula candidates pass the test, with the Clayton copula yielding the smallest test statistics. Thus, the Clayton copula is chosen for further assessment.

17.4 Assessment of Interbasin Transfer

The interbasin transfer is evaluated using the following rules:

1. No interbasin transfer will be allowed if the storage at donor (USGS08072000: Lake Houston) is less than 70% of its capacity.
2. Interbasin transfer will be allowed if the storage at donor is greater than 70% of its capacity and the storage at the receiver (USGS08123950) is less than 30%.
3. Interbasin transfer is not necessary if the storage at receiver is higher than 60% of its capacity.

With the preceding rules, we know the univariate analysis may be applied to evaluate rules 1 and 3, and the bivariate study will be needed for rule 2. Now to evaluate rule 2, we may compute the joint probability of $P\left(S_r^t \leq 0.4S_r^{Full} \cap S_d^t \geq 0.7S_d^{Full}\right)$ and the conditional probability of $P\left(S_d^t \geq 0.7S_d^{Full} | S_r^t \leq 0.4S_r^{Full}\right)$ as follows:

$$R_1 : P\left(S_r^t \leq 0.4S_r^{Full} \cap S_d^t \geq 0.7S_d^{Full}\right) = P\left(S_r^t\right) - C\left(P\left(S_r^t\right), P\left(S_d^t\right)\right) \tag{17.2}$$

$$\begin{aligned} R_2 : P\left(S_d^t \geq 0.7S_d^{Full} | S_r^t \leq 0.3S_r^{Full}\right) &= \frac{P\left(S_r^t \leq 0.3S_r^{Full} \cap S_d^t \geq 0.7S_d^{Full}\right)}{P\left(S_r^t\right)} \\ &= \frac{P\left(S_r^t\right) - C\left(P\left(S_r^t\right), P\left(S_d^t\right)\right)}{P\left(S_r^t\right)} \end{aligned} \tag{17.3}$$

In Equations (17.2) and (17.3), the marginals are evaluated from the univariate time series model through the fitted model residuals as follows:

$$\text{USGS08072000} : e_d(t) = S_d^T(t) + 0.0018 - 0.637S_d^T(t-1)$$

$$\text{USGS08123950} : e_r(t) = S_r^T(t) + 0.004 - 1.149S_r^T(t-1) - 0.149S_r^T(t-2)$$

Additionally, from the raw data, we see 190 out of 192 months with the storage higher than 70% of the capacity (except September and October of 2011) at USGS08072000 (Lake Houston). However, 121 out of 192 months were found with storage less than 40% of the capacity at USGS 08123950 (E. V. Spence Reservoir). To this end, we conclude that it is viable to transfer the water from Lake Houston to E. V. Spence Reservoir without imposing negative impacts on the communities served by Lake Houston.

Let $R = 1$ (no transfer is available) if the storage at USGS0807000 is less than 70% for rule 1; and $R = 0$ (no transfer is necessary) if the storage at USGS08123950 is greater than 60% for rule 3. In the case of rule 2, the joint probability and conditional probability are computed using Equations (17.2) and (17.3). Figure 17.7 plots the probability of rule 2 in conjunction of rules 1 and 3. Figure 17.7 indicates the following:

1. The receiver reservoir (USGS08123950) may not receive any water from the donor (USGS08072000) for September and October of 2011 regardless of the situation of the receiver reservoir, due to insufficient water storage in the donor reservoir.
2. The receiver reservoir has enough water, and no interbasin transfer is necessary for the periods of October 2000–July 2002, July 2003, April 2005, and December 2004–December 2008.
3. The receiver reservoir is in need of water from the donor Lake Houston. It is seen for most cases that the receiver may receive water from Lake Houston except for September and October of 2011. This coincides with the southern and Mexico drought of 2010–2013, and Lake Houston itself was experiencing the decrease of the storage due to drought.

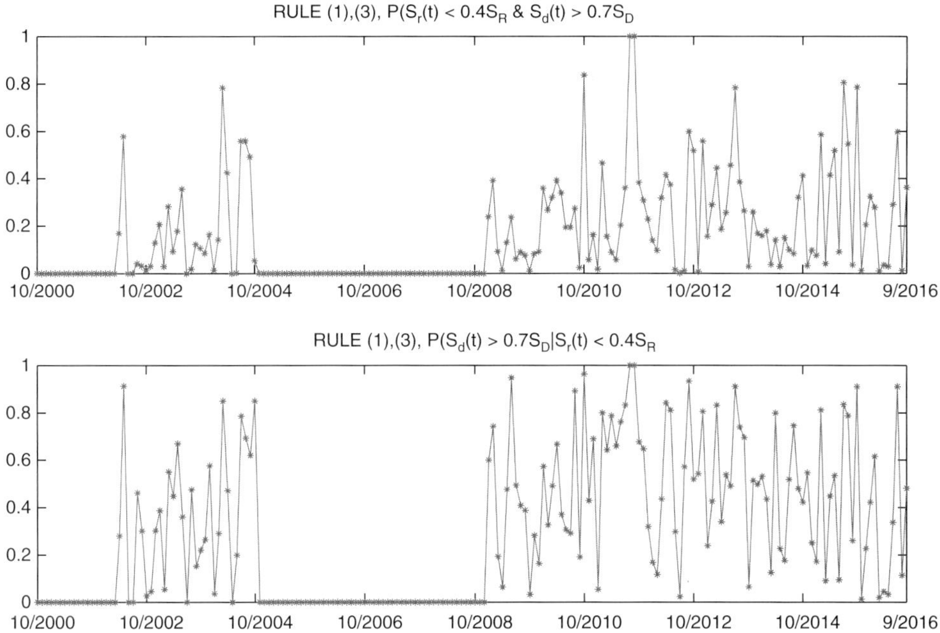

Figure 17.7 Probability of rule 2 and in conjunction with rules 1 and 3.

17.5 Forecast of Interbasin Transfer

In this section, we will provide a simple example to illustrate the procedure of interbasin transfer forecast.

1. One-month ahead storage forecast with the use of the fitted univariate time series model for the time series with meta-Gaussian transformation (i.e., S_D^T/S_R^T):
   USGS08072000:
   The forecast equation may be written as follows:

$$\left[S_D^T(t+1)\right] = c_D + \phi_D S_D^T(t) \tag{17.4}$$

   Substituting $c = -0.0018$, $\phi = 0.637$ $S_D^T(192) = 0.4835$ into Equation (17.4), we have the following:

$$Oct.\ 2016: S_D^T(t+1) = S_D^T(193) = -0.0018 + 0.637(0.4835) = 0.3062$$

   With the results obtained from the meta-Gaussian transformation, we may reestimate the storage of USGS08072000 through its inverse:

$$P = \Phi(0.3062, 0, 1) = 0.6203$$

With the probability computed in the preceding, we may finally estimate the storage for October 2016 through the kernel density function as follows:

$$\begin{aligned} S_D(Oct.\ 2016) &= 142410\ acre.\ ft \\ &= 1.06\ full\ capacity\ of\ Lake\ Houston\ (134313\ acre.\ ft) \end{aligned}$$

USGS08123950:

Similar to that for USGS8072000, the forecast equation for USGS081239500 may be written as follows:

$$\left[S_R^T(t+1)\right] = c_R + (1+\phi_R)S_R^T(t) - \phi_R S_R^T(t-1) \tag{17.5}$$

Substituting $c_R = -0.004$, $\phi_R = 0.149$, $S_R^T(192) = 0.1646; S_R^T(191) = 0.1791$ into Equation (17.5), we have the following:

$$S_R^T(Oct.\ 2016) = 0.1583; P = \Phi(0.1583) = 0.5629;$$

Finally, we have

$$\begin{aligned} S_R(Oct.\ 2016) &= 50877\ acre.\ ft \\ &= 0.37\ full\ capacity\ of\ E.V.Spence\ Reservoir\ (135704\ acre.\ ft). \end{aligned}$$

2. Probability of interbasin transfer for the coming month.

Previously we have estimated the storage for October 2016 as 142,410 acre feet and 5,0877 acre feet for Lake Houston (USGS08072000) and E. V. Spence Reservoir (USGS08123950), respectively. As compared to the full capacity of Lake Houston and E. V. Spence Reservoir, the storage condition falls into rule 2, that is, water is needed from Lake Houston to replenish E. V. Spence Reservoir. Based on rule 2, we can further compute the corresponding joint and conditional probability. It is known that when we proceed for the forecast, we assume $e_t = 0$ for median forecast.

As discussed earlier, the USGS08072000 may be fitted by the classic AR(1) model with Gaussian white noise and we have $P_D(e_t) = N(0, 0, 0.545) = 0.5$. For the stable distribution-driven ARIMA(1,1,0) model for USGS08123950, we can compute the probability numerically as $P_R(e_t) = 0.7349$.

Finally we have the joint probability and conditional probability as *R1* = 0.348 and *R2* = 0.473. The probability obtained for rule 2 tells us the following:

i. The probability of the receiver having less storage (i.e., the storage being less than 40%) and the storage at the donor being higher than the estimated storage above the 70% cutoff limit is about 34.8% (i.e., *R1*).
ii. The probability of donor with storage higher than 70% given the receiver basin with less than 30% (full storage) is about 47.3% (i.e., *R2*).
iii. The probability computed suggests the preparation for basin transfer.

17.6 Summary

In this chapter, we introduced the applications of copula to interbasin transfer study. Applying USGS08072000 (Lake Houston) and USGS08123950 (E. V. Spence Reservoir) as an example, the near real-time interbasin transfer is explained. Lake Houston is located in southeastern Texas within the humid climate region, while E. V. Spence Reservoir is located in central western Texas within the semi-arid region. In this case study, the monthly storage is applied for analysis. The seasonality is not found within the storage series. The analysis shows the following:

- With the highly skewed and heavy tailed structure of the time series, the meta-Gaussian transformation is first applied with the empirical frequency assessed by the kernel density function with positive support.
- The storage at USGS08072000 is stationary, while the storage at USGS08123950 is nonstationary. This may be understood, as for the humid region in Texas, the overall weather pattern throughout the year is more consistent than in central western Texas in the semi-arid region.
- With the meta-Gaussian transformation, the AR(1) model with white Gaussian noise may be applied to model the storage series at USGS08072000, and ARIMA(1,1,0) with stable distributed noise may be applied to model the storage series at USGS08123950.
- With the storage series being time series rather than the random variable, the copula is applied to the model residuals, which are random.
- Application of copula to the model residuals shows that the fitted model residuals at two locations is about 0.087, which is close to being independent. This is understandable due to the geographical distance as well as different climate regions.
- With the time series copula approach, it is possible to forecast the probability of interbasin transfer of the following month with the use of one-month ahead forecast.

References

Arya, F. K. and Zhang, L. (2004). Time series analysis of water quality parameters at Stillaguamish River using order series method. *Stochastic Environmental Research and Risk Assessment*. doi:10.1007/s00477–014–0907–2. Climate of Texas, https://commons.wikimedia.org/wiki/File:Texas_K%C3%B6ppen.svg.

DuMouchel, W. H. (1973). On the asymptotic normality of the maximum-likelihood estimate when sampling from a stable distribution. *Annals of Statistics,* 1(5), 948–957.

Genest, C., Remillard, B., and Beaudoin, C. (2007). Goodness-of-fit tests for copulas: a review and a power study. *Insurance: Mathematics and Economics*. doi:10.1016/j\.insmatheco.2007.10.1005.

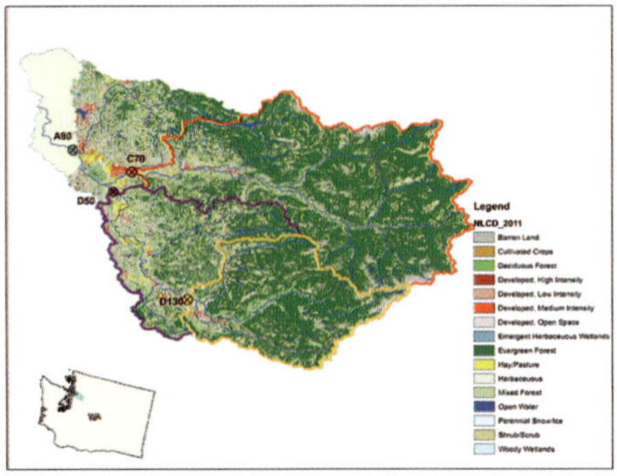

Figure 12.1 Snohomish watershed map and its LULC in 2011(retrieved from USGS and NLCD). A black and white version of this figure will appear in some formats.

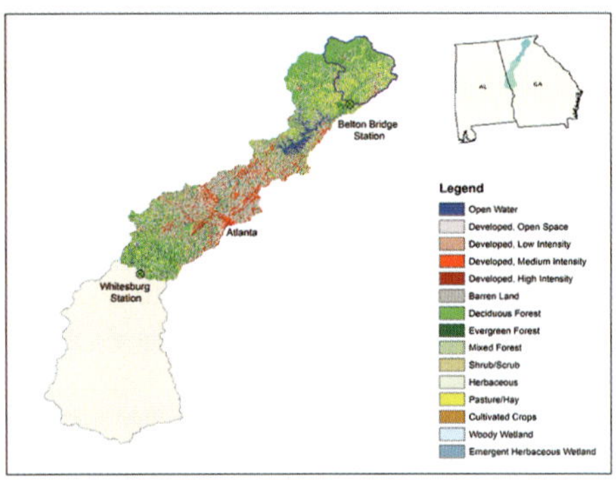

Figure 12.2 Chattahoochee River watershed upstream of the Whitesburg station and its LULC in 2011 (retrieved from USGS and NLCD). A black and white version of this figure will appear in some formats.

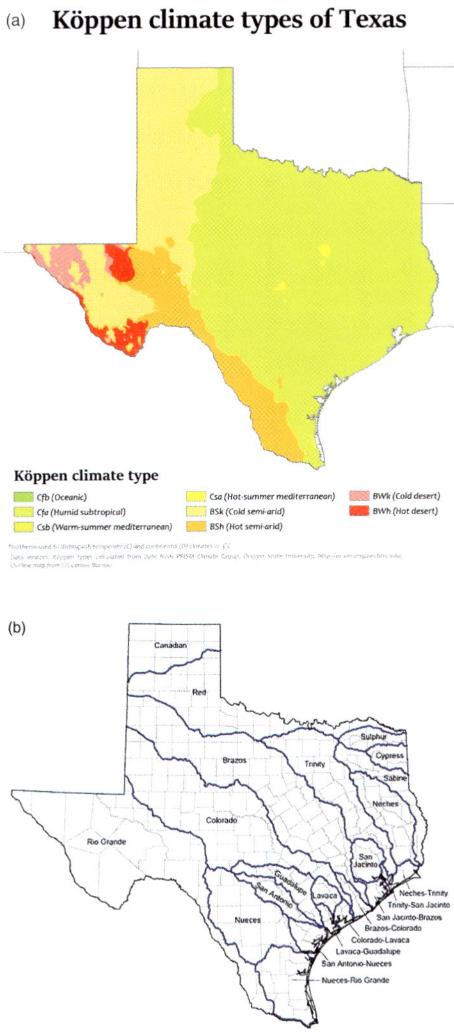

Figure 17.1 (a) Köppen climate types of Texas (retrieved from https://commons.wikimedia.org/wiki/File:Texas_K%C3%B6ppen.svg).
(b) Major rivers and cities in Texas (retrieved from www.twdb.texas.gov/surfacewater/rivers/index.asp, courtesy of Texas Water Development Board). A black and white version of this figure will appear in some formats.

Index